The series Lecture Notes in Computer Science (LNCS), including its subseries Lecture Notes in Artificial Intelligence (LNAI) and Lecture Notes in Bioinformatics (LNBI), has established itself as a medium for the publication of new developments in computer science and information technology research, teaching, and education.

LNCS enjoys close cooperation with the computer science R & D community, the series counts many renowned academics among its volume editors and paper authors, and collaborates with prestigious societies. Its mission is to serve this international community by providing an invaluable service, mainly focused on the publication of conference and workshop proceedings and postproceedings. LNCS commenced publication in 1973.

Vasco Brattka · Henning Fernau ·
Lorenzo Galeotti
Editors

Timeless Machines: Computability Across Eras

22nd Conference on Computability in Europe, CiE 2026
Trier, Germany, July 27–31, 2026
Proceedings

Springer

Editors
Vasco Brattka
Universität der Bundeswehr München
Neubiberg, Germany

Henning Fernau
Universität Trier
Trier, Germany

Lorenzo Galeotti
University of Amsterdam
Amsterdam, The Netherlands

ISSN 0302-9743 ISSN 1611-3349 (electronic)
Lecture Notes in Computer Science
ISBN 978-3-032-31347-8 ISBN 978-3-032-31348-5 (eBook)
https://doi.org/10.1007/978-3-032-31348-5

This Springer imprint is published by the registered company Springer Nature Switzerland AG
The registered company address is: Gewerbestrasse 11, 6330 Cham, Switzerland

Preface

Computability in Europe 2026: Timeless Machines – Computability Across Eras
Trier, Germany
27–31 July 2026

The conference *Computability in Europe* (CiE) is organized yearly under the auspices of the Association CiE, a European association of mathematicians, logicians, computer scientists, philosophers, physicists, biologists, historians, and others interested in new developments in computability and their underlying significance for the real world. CiE promotes the development of computability-related science, ranging over mathematics, computer science, and applications in various natural and engineering sciences, such as physics and biology, as well as related fields, such as philosophy and history of computing. CiE 2026 had as its motto *Timeless Machines – Computability Across Eras*, which alludes to the fact that in 2026 we are celebrating 90 years of Turing's seminal 1936 paper in which he introduced his now famous machine model. It is also a testament to Trier's eventful history, which dates back to Roman times.

CiE 2026 was the 22nd conference in the series, and the third to take place in Germany — the first took place in Heidelberg in 2009 and the second one in Kiel in 2018. The previous CiE conferences have been held in Amsterdam (The Netherlands) in 2005, Swansea (Wales) in 2006, Siena (Italy) in 2007, Athens (Greece) in 2008, Heidelberg (Germany) in 2009, Ponta Delgada (Portugal) in 2010, Sofia (Bulgaria) in 2011, Cambridge (UK) in 2012, Milan (Italy) in 2013, Budapest (Hungary) in 2014, Bucharest (Romania) in 2015, Paris (France) in 2016, Turku (Finland) in 2017, Kiel (Germany) in 2018, Durham (UK) in 2019, Salerno (Italy – virtually) in 2020, Ghent (Belgium – virtually) in 2021, Swansea (UK) in 2022, Batumi (Georgia) in 2023, Amsterdam (The Netherlands) in 2024, and Lisbon (Portugal) in 2025. Currently, the annual CiE conference is the largest international meeting focused on computability-theoretic issues. The proceedings containing the best submitted papers, as well as extended abstracts of some invited, tutorial, and special session speakers, for all these meetings are published in the Springer series *Lecture Notes in Computer Science*.

Structure and Program of the Conference

The PC invited seven speakers to give plenary lectures at CiE 2026, the first two of whom presented tutorials: Olivier Bournez (Paris, France), Georg Zetzsche (Kaiserslautern, Germany), Albert Atserias (Barcelona, Spain), Johanna Franklin (Hempstead, NY, USA), Mathieu Hoyrup (Nancy, France), Luca San Mauro (Bari, Italy), and Francesca Zaffora Blando (Pittsburgh, USA).

In addition, the conference had six special sessions:

Learning Theory Meets Computability Theory
Organizers: Cameron Freer and Sandra Zilles
Speakers: Dana Fisman, Kevin Kelly, Alexander Kozachinskiy, and Ruth Urner.

Groups and Computability
Organizers: Laura Ciobanu and André Nies
Speakers: Arman Darbinyan, Turbo Ho, Emmanuel Rauzy, and Alex Thumm.

At the Borderline of Universality
Organizers: Erzsébet Csuhaj-Varjú and Serghei Verlan
Speakers: Artiom Alhazov, Matthew Cook, Gemma De les Coves, and Victor Mitrana.

Natural Computation and Bioinformatics
Organizers: Karel Brinda and Giuditta Franco
Speakers: Vincenzo Bonnici, Ragnar Groot Koerkamp, David Orellana Martín, and Joanna Sułkowska.

Quantum Computing and Information
Organizer: Mika Hirvensalo
Speakers: Jaroslaw Miszczak, Giuseppe Santoro, and Abuzer Yakaryilmaz.

HaPoC (History and Philosophy of Computing): New French Philosophy of Computing
Organizer: Hajo Greif
Speakers: Thierry Coquand, Paula Quinon, Thomas Seiller, and Loïc Colson.

Speakers in these special sessions were selected by the respective special session organizers and were invited to contribute papers to this volume.

The members of the Program Committee (PC) of CiE 2026 selected 30 of the 57 non-invited submitted papers for publication in this volume and for presentation at the conference. Each paper received at least three reviews by the PC and their subreviewers. The reviewing process was single-blind. In addition to the accepted contributed papers, this volume contains six (longer) invited papers and further abstracts of invited speakers. The production of the volume would have been impossible without the diligent work of our expert referees, both PC members and subreviewers. We would like to thank all of them for their excellent work.

We thank Springer for publishing the proceedings of CiE 2026 in the ARCoSS subline of their LNCS series and for their financial support towards a Best Student Paper Award. The winner(s) of this award was revealed during the conference.

All authors who contributed to this conference were encouraged to submit significantly extended versions of their papers, with additional unpublished research content, to *Computability: The Journal of the Association CiE.*

The Steering Committee of the conference series CiE is concerned about the representation of women researchers in the field of computability. In order to increase women's participation, the series started the *Women in Computability* (WiC) program in 2007.

In 2016, after the new constitution of the Association CiE allowed for the possibility of creating special interest groups, a special interest group named *Women in Computability* was established. The WiC program includes a workshop, the annual WiC dinner, the mentorship program, and a grant program for young women researchers. The Women in Computability workshop continued in 2026, coordinated by Johanna Franklin.

The whole CiE 2026 week also saw three more conferences and workshops happening at the same place and time, namely, MCU 2026: Machines, Computability, Universality, CCA 2026: Computability and Complexity in Analysis, and GSW 2026: Grammar Systems Workshop.

The organizers of CiE 2026 would like to acknowledge and thank the following entities for their financial support (in alphabetical order): the *Association Computability in Europe* (ACiE), the *Deutsche Forschungsgemeinschaft* (DFG), FE 560/13-1, the *Association for Symbolic Logic* (ASL), the *Women in Computability Program*, and *Trier University*.

We extend our gratitude to all members of the CiE 2026 Organizing Committee for their work towards making the conference a successful event.

May 2026

Vasco Brattka
Henning Fernau
Lorenzo Galeotti

Organization

Program Committee Chairs

Vasco Brattka	Universität der Bundeswehr Munich, Germany
Henning Fernau	Trier University, Germany

Steering Committee Chair

Lorenzo Galeotti	University of Amsterdam, The Netherlands

Organizing Committee Chair

Henning Fernau	Trier University, Germany

Steering Committee Members

Verónica Becher	University of Buenos Aires, Argentina
Gianluca Della Vedova	University of Milano-Bicocca, Italy
Lorenzo Galeotti	University of Amsterdam, the Netherlands
Benedikt Lowe	University of Hamburg, Germany & Cambridge University, UK
Florin Manea	University of Göttingen, Germany
Klaus Meer	Brandenburg University of Technology Cottbus-Senftenberg, Germany
Russell Miller	City University of New York, USA
Isabel Oitavem	NOVA University Lisbon, Portugal
Ludovic Levy Patey	Mathematics Institute of Jussieu-Paris Rive Gauche, France
Mariya Soskova	University of Wisconsin-Madison, USA

Program Committee Members

Verónica Becher	Universidad de Buenos Aires, Argentina
Paola Bonizzoni	Università di Milano-Bicocca, Italy

Vasco Brattka	Universität der Bundeswehr Munich, Germany
Rod Downey	Victoria University of Wellington, New Zealand
Damir D. Dzhafarov	University of Connecticut, USA
Henning Fernau	Trier University, Germany
Gilda Ferreira	Universidade Aberta, Portugal
Lorenzo Galeotti	University of Amsterdam, The Netherlands
Daniel Graça	University of Algarve, Portugal
Gabriel Istrate	University of Bucharest, Romania
Jarkko Kari	University of Turku, Finland
Lila Kari	University of Waterloo, Canada
Akitoshi Kawamura	Kyoto University, Japan
Takayuki Kihara	Nagoya University, Japan
Alberto Marcone	Università di Udine, Italy
Elvira Mayordomo	Universidad de Zaragoza, Spain
Wolfgang Merkle	Universität Heidelberg, Germany
Cécilia Pradic	Swansea University, UK
Giuseppe Primero	University of Milan, Italy
Paul Shafer	University of Leeds, UK
Tomasz Steifer	Polish Academy of Sciences, Poland
Frank Stephan	National University of Singapore, Singapore
Manlio Valenti	Swansea University, UK
Serghei Verlan	University Paris-Est Créteil, LACL, France
Andreas Weiermann	Ghent University, Belgium
Tomoyuki Yamakami	Fukui University, Japan

Short Abstracts of Invited Talks

In the following, we list the invited abstracts in alphabetical order of the Invited Speakers, including a short report on the HaPoC session written by the organizer himself. Longer abstracts will follow below in a separate chapter of these proceedings. All these Invited Speakers could be either Invited Speakers of the CiE conference itself, or Invited Speakers of some Special Session.

Busy Beaver Problem for Small Deterministic Register Machines

Artiom Alhazov

The busy beaver problem for deterministic Turing machines consists of finding, for a given the number of states (and either 2 or a given number of symbols) the maximal number of non-blank symbols produced by Turing machines starting on a blank tape when they halt. The difficulty of the problem consists in very fast growth of the number of machines of a given size, as well as that the halting is undecidable. We consider the busy beaver problem for deterministic register machines, for a given number of instructions (and either 2 or an unrestricted number of registers). The goal is to produce, starting from empty registers, a value as large as possible when a register machine halts. We propose a number of champions for small values of instructions.

Universality and Unreachability in Physics, Mathematics, Computer Science and Slightly Beyond

Gemma De les Coves

ICREA, Institució Catalana d'Estudis i Recerca Avançats, Passeig Lluís Companys 23, 08010, Barcelona
Departament d'Enginyeria, Universitat Pompeu Fabra, Carrer Tànger 122, 08018, Barcelona
gemma.delescoves@upf.edu

Certain statistical physics models are universal (compare the technical report https://arxiv.org/abs/2407.13428) in a similar way to universal Turing machines and the universal approximation theorem for machine learning models. I will share our work to relate these forms of universality by direct comparison (see https://arxiv.org/abs/2006.03529, https://arxiv.org/abs/2507.07669) and via a categorical framework (cf. https://arxiv.org/abs/2307.06851). I will also talk about their relation to forms of unreachability such as uncomputability or undecidability, which are instances of Lawvere's fixed-point theorem.

Highness for Scott Rank

Johanna N. Y. Franklin

Hofstra University, Hempstead, NY, 11549 USA
johanna.n.franklin@hofstra.edu
http://johannafranklin.net/

A Turing degree is said to be *high* in a particular relativizable context if it behaves as a maximal oracle in that context. In this talk, I will discuss some ways in which a Turing degree can be high in the context of computable structure theory.

A Turing degree is said to be high for isomorphism if it can compute an isomorphism between any pair of isomorphic computably presented structures. I will focus on a notion derived from this based on Scott rank. A degree is said to be high for Scott rank α if, for every computable structure $\mathcal{A}$ of Scott rank α and all computable $\mathcal{B}$ isomorphic to $\mathcal{A}$, it computes an isomorphism from $\mathcal{B}$ to $\mathcal{A}$. We define highness for computably defined Scott rank α similarly, and we characterize the degrees with these properties for $0 < \alpha < \omega_1^{ck}$. I will also briefly discuss highness for Scott ranks ω_1^{ck} and $\omega_1^{ck} + 1$.

This work is joint with Dino Rossegger and Dan Turetsky.

HaPoC Special Session: New French Philosophy of Computing

Hajo Greif

Warsaw University of Technology, Poland
hans-joachim.greif@pw.edu.pl

Abstract. This Special Session is supported by the Commission for the History and Philosophy of Computing (HaPoC) and includes contributions from

- Loïc Colson (Université de Lorraine)
- Thierry Coquand (Chalmers University of Technology and University of Gothenburg)
- Paula Quinon (Warsaw University of Technology and University of Gothenburg)
- Thomas Seiller (CNRS)

Our session will mostly proceed in discursive fashion, including an interview and a panel discussion.

Keywords: French philosophy of computing · constrained computation · primitive recursion · Curry–Howard correspondence · implicit computational complexity · Geometry of Interaction · intensionality

1 Topic

Our Special Session is devoted to ideas of a distinctive cluster of French work on constrained computation: primitive recursive functions, sub-recursive hierarchies, strict termination guarantees, and systems designed to ensure that every definition is total by construction. This work shares a common attitude: where the Church–Turing tradition asks "What is *computable*?" (which functions belong to the class?), this approach asks "What is *computation*?" (what structure must a function have to count as genuinely computed?). The label "school" is justified not because these researchers all work together, but because their work exhibits a common sensibility: a preference for transparency over power, and for understanding computation through its internal structure.

2 Session Content

Paula Quinon will introduce the main tenets in recent French philosophy of computing and defend the notion that the common characteristics of the approaches warrant being jointly referred to as a New French School of Philosophy of Computing.

Loïc Colson will offer an "Introduction to Functional Systems". He will discuss the philosophical implications of intensional incompleteness and the "ultimate obstination" theorem that he introduced in earlier work on functional systems [1, 2], and that was later given a constructive formulation by Coquand [3]. His analysis will help to distinguish and explicate various forms of Platonism versus Intuitionism in mathematics.

Thomas Seiller will present a paper on "Mathematical Informatics: Algorithms", in which he argues that programs and computations, rather than functions, constitute the primary objects of inquiries into computability. He introduces a formal notion of an algorithm as a finite directed graph whose edges are labelled by partial maps over an abstract data structure, with the aim of offering a precise notion of implementation and situating algorithms as abstract partial specifications of computational behaviour.

Besides the short presentations from Quinon, Colson, and Seiller, the session will comprise an interview conducted by Paula Quinon with Thierry Coquand, followed by a panel discussion between all participants.

Acknowledgments. This Special Session is supported by the Commission for the History and Philosophy of Computing (HaPoC), https://hapoc.org/.

Disclosure of Interests. The authors have no competing interests to declare that are relevant to the content of this article.

References

1. Colson, L.: About primitive recursive algorithms. In: Ausiello, G., Dezani-Ciancaglini, M., Della Rocca, S.R. (eds) Automata, Languages and Programming. ICALP 1989, LNCS, vol. 372, pp. 194–206. Springer, Heidelberg (1989). https://doi.org/10.1007/BFb0035761. Journal Version in Theoret. Comput. Sci. **83**, 57–69 (1991)
2. Colson, L.: Représentation intentionnelle d'algorithmes dans les systèmes fonctionnels: une étude de cas. Ph.D. thesis, Université Paris 7 (1991)
3. Coquand, T.: Une preuve directe du théorème d'ultime obstination. Comptes Rendus de l'Académie des Sciences, Paris, Série I **314**, 489–492 (1992)

Word Problems and the Structure of Computably Enumerable Equivalence Relations

Uri Andrews[1], Meng-Che "Turbo" Ho[2], and Luca San Mauro[3]

[1] University of Wisconsin, Madison, WI, 53706-1388 USA
[2] New College of Florida, Sarasota, FL, 34243 USA
[3] University of Bari, 70121 Bari, BA, Italy

In 1911, Dehn [7] proposed three decision problems for finitely presented groups: the word problem, the conjugacy problem, and the isomorphism problem. These problems have been central to both group theory and computability theory, and were each proven to be undecidable in the 1950s [1, 4, 12, 13]. Classically, Turing degrees — the complexity classes induced by Turing reduction — are used to study the complexity of word problems. Every word problem of a finitely presented group has computably enumerable Turing degree. Conversely, results by Fridman [10], Clapham [6], Boone [5], and Bokut [3], all independently, state that every computably enumerable Turing degree contains the word problem of some finitely presented group.

In this paper, we refine this picture by studying group word problems as computably enumerable equivalence relations (ceers), ordered by computable reducibility. Ceers provide a finer measurement than Turing degrees and can distinguish word problems that are Turing equivalent. For instance, in contrast with the previous theorem, not all ceer degrees contain a word problem [8].

Formally, a ceer is a computably enumerable equivalence relation on ω, and we say a ceer E reduces to another ceer F, denoted $E \leq_c F$, if there is a computable function $g : \omega \to \omega$ such that $iEj \Leftarrow g(i)Fg(j)$ for every $i, j \in \omega$ [9, 11]. *Id* is the ceer defined as the equality relation on ω, and a ceer is *dark* if *Id* does not reduce to it. A ceer is *dark minimal* if no ceer (other than itself) with infinitely many classes reduces to it. A ceer is *universal* if all ceers reduce to it [2].

Given a ceer E, we define several natural group word problems generated by E. For $n \geq 2$, let $\mathbb{Z}_n E$ be the free $\mathbb{Z}_n$-module (considered as an abelian group) over the set of E-classes, and let $\mathbb{Z}_n^0 E$ be the quotient in which the class of 0 is identified with the identity. We note that $E \leq_c \mathbb{Z}_n^0 E \leq_c \mathbb{Z}_n E$.

Although these constructions are algebraically simple and have the same Turing degree as E, their ceer degrees can behave very differently. In particular, the behavior of $\mathbb{Z}_n E$ depends not only on E, but also n.

Theorem 1. *For every prime p, there is a dark minimal ceer E such that $\mathbb{Z}_p^0 E$ is universal, while $\mathbb{Z}_n E$ is dark whenever $p \nmid n$.*

More generally, for any Σ_3^0 subset X of the primes, one can construct a dark minimal ceer E such that $\mathbb{Z}_p^0 E$ is universal precisely for the primes in X, and $\mathbb{Z}_n E$ is dark if n is not divisible by any prime in X.

We also study minimality of group word problems above a fixed ceer. For suitable E, the group word problems $\mathbb{Z}_p^0 E$ are minimal among group word problems, in a strong structural sense.

Theorem 2. *There exists a ceer E such that, for every prime p and every group word problem H, if* $E \leq_c H \leq_c \mathbb{Z}_p^0 E$, *then H is computably isomorphic to* $\mathbb{Z}_p^0 E$.

In particular, $\mathbb{Z}_p^0 E$ in the previous theorem is the only group word problem (up to computable isomorphism) in its ceer degree.

For this same E, the family $\{\mathbb{Z}_n^0 E : n \geq 2\}$ encodes the divisibility order.

Theorem 3. *For the ceer E above,* $\mathbb{Z}_n^0 E \leq_c \mathbb{Z}_m^0 E$ *if and only if* $n|m$.

Finally, we analyze upper cones. In sharp contrast with the Turing-degree setting, there are ceer upper cones containing no non-universal group word problems at all.

Theorem 4. *There is a dark minimal ceer E such that every group word problem G with* $E \leq_c G$ *is universal.*

These results demonstrate that the ceer degrees provide a refined setting in which group word problems display structure invisible to Turing degrees, opening a program for classifying group word problems within the ceer degrees.

Disclosure of Interests. The authors have no competing interests to declare that are relevant to the content of this article.

References

1. Adyan, S.I.: Algorithmic unsolvability of problems of recognition of certain properties of groups. Dokl. Akad. Nauk SSSR (N.S.) **103**, 533–535 (1955)
2. Andrews, U., Sorbi, A.: Joins and meets in the structure of ceers. Computability **8**(3-4), 193–241 (2019)
3. Bokut, L.A.: On a property of the Boone groups. Algebra i Logika Sem. **5**(5), 5–23 (1966)
4. Boone, W.W.: Certain simple, unsolvable problems of group theory. V, VI, Indag. Math. **19**, 22–27, 227–232 (1957)
5. Boone, W.W.: Word problems and recursively enumerable degrees of unsolvability. A sequel on finitely presented groups. Ann. Math. **84**(2), 49–84 (1966)
6. Clapham, C.R.J.: Finitely presented groups with word problems of arbitrary degrees of insolubility. Proc. London Math. Soc. **14**(3), 633–676 (1964)
7. Dehn, M.: Über unendliche diskontinuierliche Gruppen. Math. Ann. **71**(1), 116–144 (1911)
8. Delle Rose, V., San Mauro, L., Sorbi, A.: Classifying word problems of finitely generated algebras via computable reducibility. Internat. J. Algebra Comput. **33**(4), 751–768 (2023)
9. Ershov, Y.L.: Positive equivalences, Algebra i Logika **10**, 620–650 (1971)
10. Fridman, A.A.: Degrees of unsolvability of the word problem for finitely presented groups. Dokl. Akad. Nauk SSSR **147**, 805–808 (1962)
11. Gao, S., Gerdes, P.M.: Computably enumerable equivalence relations. Studia Logica **67**(1), 27–59 (2001)
12. Novikov, P.S.: On the algorithmic unsolvability of the word problem in group theory (Russian) Akad. Nauk SSSR, Moscow (1955)
13. Rabin, M.O.: Recursive unsolvability of group theoretic problems. Ann. Math. **67**(2), 172–194 (1958)

Presentations of Topological Spaces

Mathieu Hoyrup

Université de Lorraine, CNRS, Inria, LORIA, 54000 Nancy, France

Abstract. Computable analysis is the study of the computable aspects of mathematical objects. In order to manipulate such objects algorithmically, one needs to choose a symbolic representation of them. That choice has a deep impact on the abilities of a computer program to extract information about the objects, which leads to investigating the computability-theoretic content of representations.

In this talk, the mathematical objects to be represented are topological spaces, notably compact Polish spaces. There are many ways of presenting them, and comparing the computable contents of these presentations has been an active research topic in the recent years. I will give an overview of some of the results obtained by several research groups.

Topological Foundations for Inductive Learnability

Kevin T. Kelly

Carnegie Mellon University, Pittsburgh, PA, 15213 USA
kk3n@andrew.cmu.edu

Abstract. It is proposed that topology, or rather the more fundamental theory of Moore interior and closure, provides a very general, elegant, and fundamental perspective on empirical verifiability, refutability, and inductive learnability.

Keywords: Learnability · Topology · Simplicity

When it comes to theories of scientific verifiability and refutability, logic and probability theory naturally come to mind as the relevant background. I suggest that, while logic and probability theory are both relevant, the best fits to the subject are topology and computability. One usually thinks of topology as rubber geometry, but that is incidental: it is really the pure math of verifiability and refutability. It applies to analysis because analysis is built on the idea of approximation of the infinite with finite precision, which is exactly what is at stake both in inductive reasoning and in the theory of computability. Topological elegance comes at a price: it weakens "soon" to "eventually". But that can be a feature: quick results call for strong assumptions and compromised ambitions. Conceding that small signals can have big consequences allows one to re-focus on the real aims: learning the underlying structures that make skills transfer across contexts.

Assume a set W of possible worlds whose subsets are propositions. Information about the world is obtained from physical signals. Let Σ be the set of possible signals. An observation event involves receipt of a finite set $O \subseteq \Sigma$ of signals. Each world affords some set $s(w)$ of possible observations. Observation O rules out all worlds that cannot generate it, and the set of all worlds remaining is the corresponding propositional information state:

$$\begin{aligned} &\textit{information state}: && E_O := \{w \in W \mid O \in s(w)\}.\\ &\textit{information states in } w: && \varepsilon_w := \{E_O \mid O \in s(w)\}.\\ &\textit{information states}: && \varepsilon := \bigcup_{w \in W} \varepsilon_w. \end{aligned}$$

It follows immediately that an information state is true in w iff it can also possibly arise in w :[1]

$$\textit{informationprinciple}: \varepsilon_w = \{E \in \varepsilon \mid w \in E\}.$$

[1] Because $w \in E_O$ iff $O \in s(w)$.

That covers a range of examples, such as the function and language learning settings of Gold [10], and the more realistic example of finitely many open coordinate rectangles incident to an underlying function on the real numbers. It also applies to very abstract learning models, such as lower bounds on poset points, e.g., in a Scott domain [1]. Most of the examples satisfy the standard axioms of a topological basis.

$$\textit{covering} : W = \bigcup \varepsilon.$$

$$\textit{cumulativity}: E, F \in \varepsilon_w \Rightarrow (\exists G \in \varepsilon_w) G \subseteq E \cap F.$$

But both can fail (e.g., lower bounds in posets), so it makes sense to bracket them. Covering tacitly assumes that the learner exists to receive signals, and more importantly, cumulativity assumes that one will not observe one's way into an informational dead-end, which is more of a norm or ideal than a literal semantic truth.

By the information principle, verifiability in world w is:

$$\textit{verifiability in}\, w : w \in \Delta A \Leftrightarrow (\exists E \in \varepsilon)\, w \in E \subseteq A.$$

It follows that Δ is a Moore-interior operator [15]: monotone, idempotent, and intensive. Its dual Moore-closure operator ∇ expresses that A is not refutable in w. Given covering and cumulativity, they are topological interior and closure operators, satisfying the Kuratowski axioms, which add the usual principles $\Delta \mathrm{T} = \mathrm{T}$ and $\Delta \mathrm{T} = \mathrm{T}$ and $\Delta(A \cap B) = \Delta A \cap \Delta B$. So topology is not as deeply rooted in epistemology as in Moore theory, which may be called ur-topology.

The Kuratowski interior axioms are equivalent to the modal system S4. In addition to the standard designations for modal principles, let C be distribution of interior over binary conjunction, and let M be monotonicity. Then although verifiability semantics and standard Kripke semantics meet in the middle at the topological system $S4$, their basic principles are almost entirely disjoint (Table 1).

Table 1. Verifiability Semantics vs. Standard Semantics

semantics	system			
verifiability	*minimal*	T4	M	
both	S4	T 4	M	N K C
standard	*minimal*		M	N K C

The connection between empirical verifiability and topology extends naturally to inductive learning. Verifiable/refutable are open/closed. Scientific models (e.g., polynomial degree $D = n$) are locally closed. They provide a semantics for default reasoning:

when trigger O is verified, conclude $O \backslash O'$ until defeater O' is verified. The inferential pattern is no-yes-no or 010, whereas the pattern for verifiable (open) is just 01. So one may say that open is 1-open and locally closed is 2-open. That generalizes to n-open and n-closed, which gives rise to the Hausdorff finite difference hierarchy.

Scientific paradigms (e.g., polynomial vs. trigonometric polynomial) are countable disjunctions of 2-open propositions, so they may be said to be ω-open [6]. That collapses to Σ_2^0 Borel for Hausdorff spaces, but the poset and curve-fitting examples are weaker, calling for a more general descriptive set theory [17]. Paradigms, i.e., ω-open propositions, characterize the outer reaches of inductive learnability: Angluin's [2] characterization of language learnability extends to arbitrary propositions of difference complexity ω.[2] The positive argument depends on covering and cumulativity, but the negative side does not.

Topology sheds fresh light on empirical simplicity and its role in inductive learning. The proposition that A is simpler than B is [6]:

$$\textit{empirical simplicity}: A \lhd B := A \cap \partial B \backslash B, \tag{1}$$

which is a one-way trap-door from A to B through which nature can drag an arbitrary inductive learner. Ockham's razor is mandated for the simple (appropriately enough) reason that concluding B before $A \lhd B$ is refuted opens one to inductive risk that nature is obligated to eliminate.[3] So don't.

It goes without saying that the underlying signaling model is ideally set up for an effective refinement. A larger hurdle is chance. Science, for better or worse, is now formulated almost exclusively in terms of probability. And notoriously, sample information literally rules out no possible worlds: the connection is entirely in chance. But that objection has been bridged: Genin [5, 7] has shown in detail that the collection of statistical propositions verifiable in chance is itself a topology (the weak topology on measures). Furthermore, that topology has a countable basis.[4] It follows that all of the above learning applications (e.g., the characterization of inductive learnability and the argument for Ockham's razor) apply literally to statistical inference, providing an elegant and transparent understanding of what is essential. That opens the door to serious statistical learning applications, such as statistical retractions of causal orientation [8]. As the limitations of purely correlation-based robots and chat-bots become increasingly clear, topological inductive learning of counterfactual structure may play a larger role in the theory of machine learning.

Disclosure of Interests. The author declares no competing interests.

---

[2] There is a long progression of learnability characterizations leading up to that result [18, 19, 20, 22, 25, 27, 30, 32].

[3] As a consequence, nature can force you through extra retractions, can force you to retract the truth, and can force you into a cycle of opinions [28, 29].

[4] Genin assumes IID sampling, which is a very strong statistical version of cumulativity. It remains to consider the consequences of weakening it.

References

1. Abramsky, S., Jung, A.: Domain theory. In: Abramsky, S., Gabbay, D.M., Maibaum, T.S.E. (eds.) Handbook of Logic in Computer Science. vol. 3, pp. 1–168. Oxford University Press, Oxford (1994)
2. Angluin, D.: Inductive inference of formal languages from positive data. Inf. Control **45**, 117–135 (1980).
3. Baltag, A., Gierasimczuk, N., Smets, S.: On the solvability of inductive problems: a study in epistemic topology. Electron. Proc. Theor. Comput. Sci. **215**, 81–98 (2016). https://doi.org/10.4204/EPTCS.215.7
4. de Brecht, M., Yamamoto, A.: Interpreting learners as realizers for Σ^0_2-measurable functions. Spec. Interest Group Fundam. Prob. Artif. Intell. (SIG-FPAI) **74**, 39–44 (2009). https://www.iip.ist.i.kyoto-u.ac.jp/member/matthew/Sigma2_full_version.pdf
5. Genin, K.: The Topology of Statistical Inquiry. PhD thesis, Carnegie Mellon University (2018)
6. Genin, K., Kelly, K.T.: Theory choice, theory change, and inductive truth-conduciveness. In: Ramanujam, R. (ed.) Proceedings of the Fifteenth Conference on Theoretical Aspects of Rationality and Knowledge (TARK), pp. 111–121 (2015). https://www.imsc.res.in/tark/TARK2015-proceedings.pdf
7. Genin, K., Kelly, K.T.: The topology of statistical verifiability. In: Lang, J. (ed.) Proceedings of the Sixteenth Conference on Theoretical Aspects of Rationality and Knowledge (TARK), pp. 236–250 (2017). https://doi.org/10.4204/EPTCS.251.17
8. Kelly, K.T., Mayo-Wilson, C.: Causal conclusions that flip repeatedly and their justification. In: Proceedings of the Twenty-Sixth Conference on Uncertainty in Artificial Intelligence (UAI 2010), pp. 277–286. AUAI Press (2010)
9. Gierasimczuk, N.: Knowing One's Limits: Logical Analysis of Inductive Inference. PhD thesis, University of Amsterdam (2010)
10. Gold, E.M.: Language identification in the limit. Inf. Control **10**, 447–474 (1967)
11. Kelly, K.T.: The Logic of Reliable Inquiry. Oxford University Press, Oxford (1996)
12. Kelly, K.T.: Justification as truth-finding efficiency: how Ockham's razor works. Minds Mach. **14**(4), 485–505 (2004). https://doi.org/10.1023/B:MIND.0000045993.31233.63
13. Kelly, K.T., Glymour, C.: Why probability does not capture the logic of scientific justification. In: Hitchcock, C. (ed.) Contemporary Debates in the Philosophy of Science, pp. 94–114. Blackwell (2004)
14. Kugel, P.: Induction: Pure and simple. Inf. Control **35**(4), 276–336 (1977)
15. Moore, E.H.: Introduction to a Form of General Analysis. The New Haven Mathematical Colloquium, pp. 1–150. Yale University Press (1910)
16. Putnam, H.: Trial and error predicates. J. Symb. Logic **30**, 49–57 (1965)
17. Selivanov, V.L.: Towards a descriptive set theory for domain-like structures. Theor. Comput. Sci. **365**, 258–282 (2006)

Efficient Dimensions for Efficient Learning and Generation

Alexander Kozachinskiy

CENIA, Chile
alexander.kozachinskyi@cenia.cl

Abstract. In this talk, we survey some recent works on studying computable analogs of classical machine learning models. We cover PAC learning, Online Learning, and a recently introduce Language Generation framework, with the emphasis on the effective versions of classical combinatorial dimensions like VC and Littlestone dimensions.

Keywords: Computable PAC Learning · Computable Online Learning · Language Generation.

Recently, many classical machine learning models have been revisited from the computability viewpoint. The idea of this research line is to examine how computability requirements affect a given model – does the set of learnable concept classes become smaller, do classical results about the model admit their effective reformulations, etc. By "computability requirements" one usually means that the Learner in the computable case, has to be implementable by a Turing machine.

A lot of attention within this line of research initially has been given to the PAC learning model [1, 2, 12]. This framework starts with a concept class C of functions $f : \mathbb{N} \to \{0, 1\}$. The class C is known to the Learner. There is also a function $f \in C$, not known to the Learner. The Learner can sample pairs of the form $(x, f(x))$, where x is taken at random according to some distribution D over $\mathbb{N}$ (also not known to the Learner). Additionally, the Learner receives a numerical parameter $\varepsilon > 0$, and its goal is to output (with high probability) some function $\widehat{f} : \mathbb{N} \to \{0, 1\}$ such that $\Pr_{x\sim D}[f(x) \neq \widehat{f}(x)] \leq \varepsilon$. A Learner achieving this is called a PAC-learner for C.

A classical result of machine learning states that classes admitting a PAC learner are exactly classes whose *VC dimension* is finite [5], where the latter means that for some $k \in \mathbb{N}$ and for all $x_1, \ldots, x_k \in \mathbb{N}$ we have:

$$\left|\{f(x_1)\ldots f(x_k) \in \{0, 1\}^k | f \in C\}\right| < 2^k.$$

Surprisingly, this result has a very clean "computable" counterpart. Imagine that we additionally want a PAC learner to be implementable by a Turing machine, assuming that it outputs the function $\widehat{f}$ in the form of a program. It turns out that this corresponds to the following "effectivization" of VC dimension: a class C has finite *effective* VC dimension if for some $k \in \mathbb{N}$ there is an algorithm that, given $x_1, \ldots, x_k \in \mathbb{N}$, finds some string $w \in \{0, 1\}^k$ such that

$$f(x_1)\ldots f(x_k) \neq w$$

for all $f \in C$.

Theorem 1 ([6]). *A concept class C admits a computable PAC learner if and only if its effective VC dimension is finite.*

Here, the PAC learner has to be computable by a total Turing machine, which, in particular, has to output something even if it gets input-output pairs not realizable by any function $f \in C$. Additionally, the output function $\widehat{f}$ may not belong to C. One can show that these specifications are important, and without them, the result no longer holds. This makes the computable setting less robust than the classical one.

Afterwards, there was an attempt [7, 9] to obtain a similar characterization of computable *online learning*, another classical machine learning model known to be characterized by the so-called Littlestone dimension [11]. Littlestone dimension can be effectivized in a similar manner as VC dimension, but it leads to a notion weaker than any reasonable version of computable online learning. Nevertheless, some interesting results are still possible in this framework; for example, even this weak requirement of having finite effective Littlestone dimension implies that all functions from the concept class are computable.

Recently, Kleinberg and Mullainathan [10] have come up with the *language generation model*, a less restrictive refinement of the classical language identification framework of Gold [8] (completely characterized, both in the classical and the computable regimes, by Angluin [3]). In this setting, the Learner gets an enumeration of the preimage of 1 for some unknown function $f \in C$ (all functions from the class are assumed to have an infinite preimage of 1). The goal of the Learner is to correctly indicate, starting from some moment, an infinite subset of this preimage (as opposed to the Gold-Angluin framework where one has to exactly identify f). As Kleinberg and Mullainathan show, this is possible for any countable C, although it might require to see an unfeasible number of elements of the preimage to start getting the problem right, even for simple finite concept classes [4]. When C is uniformly computable (given an index of f in C and $x \in \mathbb{N}$, one can compute $f(x)$), this result admits a computable counterpart, although the exact characterization of classes that admit a computable learner in this framework remains open. Even under a uniform computability assumption, it is not clear how to obtain a computable learner that outputs a subset of $f^{-1}(1)$ of positive density, while uncomputably this is possible.

Acknowledgments. The author is funded by the National Center for Artificial Intelligence CENIA FB210017, Basal ANID, and ANID Fondecyt Iniciación grant 11250060.

Disclosure of Interests. The author has no competing interests to declare that are relevant to the content of this article.

References

1. Agarwal, S., Ananthakrishnan, N., Ben-David, S., Lechner, T., Urner, R.: On learnability wih computable learners. In: Algorithmic Learning Theory, pp. 48–60. PMLR (2020)
2. Agarwal, S., Ananthakrishnan, N., Ben-David, S., Lechner, T., Urner, R.: Open problem: Are all VC-classes CPAC learnable? In: Conference on Learning Theory, pp. 4636–4641. PMLR (2021)
3. Angluin, D.: Inductive inference of formal languages from positive data. Inf. Control **45**(2), 117–135 (1980)
4. Arenas, M., Barceló, P., Cofré, L., Kozachinskiy, A.: Language generation: complexity barriers and implications for learning. arXiv preprint arXiv:2511.05759 (2025)
5. Blumer, A., Ehrenfeucht, A., Haussler, D., Warmuth, M.K.: Learnability and the Vapnik-Chervonenkis dimension. J. ACM (JACM) **36**(4), 929–965 (1989)
6. Delle Rose, V., Kozachinskiy, A., Rojas, C., Steifer, T.: Find a witness or shatter: the landscape of computable PAC learning. In: The Thirty Sixth Annual Conference on Learning Theory, pp. 511–524. PMLR (2023)
7. Delle Rose, V., Kozachinskiy, A., Steifer, T.: Effective Littlestone dimension. In: Algorithmic Learning Theor,. pp. 405–417. PMLR (2025)
8. Gold, E.M.: Language identification in the limit. Inf. Control **10**(5), 447–474 (1967)
9. Hasrati, N., Ben-David, S.: On computable online learning. In: International Conference on Algorithmic Learning Theory, pp. 707–725. PMLR (2023)
10. Kleinberg, J., Mullainathan, S.: Language generation in the limit. In: Advances in Neural Information Processing Systems. vol. 37, pp. 66058–66079 (2024)
11. Littlestone, N.: Learning quickly when irrelevant attributes abound: a new linear-threshold algorithm. Mach. Learn. **2**(4), 285–318 (1988)
12. Sterkenburg, T.F.: On characterizations of learnability with computable learners. In: Conference on Learning Theory, pp. 3365–3379. PMLR (2022)

Quantum Random Numbers and Where to Find Them (Extended Abstract)

Jaroslaw Miszczak

Institute of Theoretical and Applied Informatics, Polish Academy of Sciences, Baltycka 5, 44-100 Gliwice, Poland
jmiszczak@iitis.pl, jarek@miszczak.eu

Abstract. The ability to access sources of randomness is crucial across almost every scientific and technical field. Among the proposed sources of *good* random numbers, Quantum Random Number Generators (QRNGs) are one of the most promising, especially considering their intrinsic randomness based on the laws of quantum physics. In this talk we review the state of the art regarding QRNGs, recent developments regarding their availability, and the quality of the obtained data. We focus on the possibility of employing QNRG sources for cryptographic applications. To provide a baseline for the quality of QRNGs, we use the cryptographically secure randomness source available in the Java platform.

Keywords: post-quantum cryptography · random number generators

1 Introduction

The use of randomness might seem counterintuitive in a world built on logic and precision. However, the ability to use random numbers is of fundamental importance for almost every scientific and technical field. One of the most important motivations is the ability to harness Monte Carlo simulations to predict the behavior of complex systems [14].

Moreover, nearly every cryptographic protocol – key generation, nonces, salts, initialization vectors – depends on unpredictability. For this reason, secure random number generation is a foundational problem in cryptography [12]. The utilization of *good* random numbers is crucial even in quantum communication protocols like BB84 [8], where some part of the procedure relies on the random selection of the measurement. Hence, the ability to use unbiased and predictable data can lead to serious security risks, even if the security of the underlying protocols is ensured by the laws of physics.

Finally, an area of application gaining considerable attention during recent years is machine learning and artificial intelligence. The training of AI systems requires initializing the artificial neural networks with random weights. This prevents the system from getting stuck and helps it learn more effectively from the data.

The goal of this talk is to provide a description of the current state of the art in the field of quantum random number generators, especially focusing on practical considerations, availability, and utilization in cybersecurity.

This talk is based on the software developed for the purpose of integrating QRNG with post-quantum cryptographic methods in the scope of the Q-Fence project [6]. We describe use cases based on QRNGs provided by Fraunhofer IPMS [5] and Quantis devices from ID Quantique [4]. We also use the functionality delivered by the Java Cryptography Architecture dedicated to providing a cryptographically secure source of random numbers in Java.

The following part of this talk is organized as follows. In Section 2 we discuss the motivation for using QRNGs, details regarding the used hardware back-ends, and the issues related to the availability. In Section 3 we describe sources of random numbers and some implementation details for accessing selected hardware back-ends. The provided implementations are used to describe the results of the statistical tests which are presented in Section 4.

2 Motivation

Randomness sources available in typical computer systems are based on pseudo-random number generators (PRNGs) – computational algorithms designed to produce sequences of numbers that appear random but are entirely determined by an initial value [11]. Because PRNGs are deterministic, if the seed or internal state is known or guessed, outputs can be reproduced, and an attacker who discovers the seed or the algorithm can perfectly predict every subsequent number. This has historically led to catastrophic failures in cryptographic systems [13].

To mitigate these risks in sensitive applications, developers often turn to Cryptographically Secure PRNGs (CSPRNGs), which add complexity to ensure that past outputs cannot be used to determine future ones. For high-stakes environments like lottery systems or quantum cryptography, the ultimate alternative is a True Random Number Generator (TRNG), which derives randomness from unpredictable physical phenomena like atmospheric noise or radioactive decay.

Finally, as the basic laws of quantum mechanics guarantee that the quantum measurement process is probabilistic, they guarantee a truly random source of data. Hence, the utilization of real quantum hardware is very attractive regarding the quality aspects of the obtained random numbers, especially as some research organizations provide online access to their hardware.

3 Sources of Random Numbers

Modern systems, including Java, do not rely purely on mathematical PRGs. They combine true entropy sources, which are slow but unpredictable, with cryptographic PRNGs, which are fast but deterministic. The entropy source can be provided by the underlying operating system (e.g., via /dev/urandom in Linux), hardware noise (timing, interrupts), or specialized hardware.

Considering the hardware availability and costs, one of the best methods for utilizing truly random numbers generated using quantum mechanics is by using on-line services [15].

Currently, qrandom.io [3] and ANU QRNG [1, 17] are popular online platforms providing free access to true random numbers generated via quantum mechanics. qrandom.io allows users to generate random numbers, strings, and shuffled cards up to 100 Mb per day, with results signed with post-quantum cryptography. ANU QRNG offers real-time numbers derived from vacuum fluctuations, available via a web interface or an API hosted on AWS.

Other accessible options include HotBits, which has provided genuine random numbers since 1996 by timing radioactive decays, and Randonautica, which uses quantum randomness to generate coordinates for exploration. For developers seeking API integration, QRANGE offers commercial solutions for cheaper, faster, and more secure devices, while Quantum Base has developed practical generators embeddable in electronic products.

In this talk we describe resultants based on QRNGs provided by Fraunhofer IPMS and Quantis devices from ID Quantique. The devices are based on different technologies, offer different functionality, and differ regarding efficiency. However, both ensure the speed and availability required for the purpose of testing the quality of obtained numbers.

4 Quality and Functionality Tests

Ensuring high-quality entropy is not just a technical formality, but a fundamental requirement for digital trust and experimental accuracy. Test suites like Dieharder or NIST tests are employed to ensure that an RNG's output is indistinguishable from genuine noise [10]. Regarding QRNGs, in-depth tests of the hardware and firmware of the commercial Quantis quantum random number generator ID were also performed [16].

NIST SP 800-22 [2, 7] focuses specifically on cryptographic applications and includes 15 standardized tests with clear pass/fail criteria based on p-values. It is widely used for evaluating random number generators in security contexts.

Dieharder [9] is a widely recognized open-source statistical test suite designed to evaluate the randomness of both hardware- and software-based random number generators. It includes an expanded and more rigorous test suite compared to NIST SP 800-22, incorporating all the original Diehard tests, additional tests from other sources, including some from NIST SP 800-22, as well as newly developed ones. It is designed for more stringent statistical validation and supports automated test repetition and p-value distribution analysis.

Even though Dieharder offers greater comprehensiveness and sensitivity, NIST SP 800-22 remains the benchmark for compliance in cryptographic standards. However, as Dieharder is suitable for detecting subtle flaws that NIST tests might miss, we deploy it for the purpose of the presented work.

Acknowledgments. This work was partially supported by the EU grant Q-FENCE *Securing Tomorrow's Digital Infrastructure with Quantum-Resistant Cryptography* (project number 101225708). The author would like to acknowledge support in obtaining data required for this work received from Mateusz Żarski, Stanisław Miszczak, Benjamin Mustin and Christoph Posenau.

References

1. ANU QRNG – quantum random numbers. https://qrng.anu.edu.au/. Accessed 30 Mar 2026
2. The NIST statistical test suite. https://github.com/terrillmoore/NIST-Statistical-Test-Suite
3. Quantum Random Number Generator, true random number and data generator based on quantum physics - qrandom.io. https://qrandom.io/. Accessed 30 Mar 2026
4. Quantis QRNG: Standard certified USB quantum random number generation (QRNG) module (2001). https://www.idquantique.com/random-number-generation/products/quantis-random-number-generator/
5. Photonic quantum chip for fast and reliable random number generation (2023). https://www.ipms.fraunhofer.de/en/press-media/press/2023/Photonic-quantum-chip.html
6. Q-fence: Securing Tomorrow's Digital Infrastructure with Quantum-Resistant Cryptography (2025). https://qfence-project.eu/
7. Bassham, L., et al.: A statistical test suite for random and pseudorandom number generators for cryptographic applications. Technical report, National Institute of Standards and Technology (2008)
8. Bennett, C.H., Brassard, G.: Quantum cryptography: public key distribution and coin tossing. Theor. Comput. Sci. **560**, 7–11 (2014). https://doi.org/10.1016/j.tcs.2014.05.025
9. Brown, R.G.: Dieharder: A random number test suite (2004). https://webhome.phy.duke.edu/rgb/General/dieharder.php
10. Crocetti, L., Nannipieri, P., Di Matteo, S., Fanucci, L., Saponara, S.: Review of methodologies and metrics for assessing the quality of random number generators. Electronics **12**(3), 723 (2023). https://doi.org/10.3390/electronics12030723
11. Goldreich, O.: Pseudorandom Generators: A Primer, University Lecture Series, vol. 55. AMS (2010). https://www.wisdom.weizmann.ac.il/oded/prg-primer.html
12. Kietzmann, P., Schmidt, T.C., Wählisch, M.: A guideline on pseudorandom number generation (PRNG) in the IoT. ACM Comput. Surv. **54**(6), 1–38 (2021). https://doi.org/10.1145/3453159
13. Markowsky, G.: The sad history of random bits. J. Cyber Secur. Mobility **3**(1), 1–24 (2014). https://doi.org/10.13052/jcsm2245-1439.311
14. Metropolis, N., Ulam, S.: The Monte Carlo method. J. Am. Stat. Assoc. **44**(247), 335–341 (1949). https://doi.org/10.1080/01621459.1949.10483310
15. Miszczak, J.A.: Employing online quantum random number generators for generating truly random quantum states in Mathematica. Comput. Phys. Commun. **184**(1), 257–258 (2013). https://doi.org/10.1016/j.cpc.2012.08.012

16. Petrov, M., Radchenko, I., Steiger, D., Renner, R., Troyer, M., Makarov, V.: Independent quality assessment of a commercial quantum random number generator. EPJ Quant. Technol. **9**(1) (2022). https://doi.org/10.1140/epjqt/s40507-022-00136-z
17. Symul, T., Assad, S.M., Lam, P.K.: Real time demonstration of high bitrate quantum random number generation with coherent laser light. Appl. Phys. Lett. **98**(23) (2011). https://doi.org/10.1063/1.3597793

Residual Properties of Groups in the Continuous Weihrauch Lattice

Emmanuel Rauzy

LACL, Université Paris-Est Créteil
emmanuel-rauzy@u-pec.fr

1 Motivation

The present work is motivated by the study of the computability of quotients of finitely generated groups. Let $\mathcal{C}$ be a set of recursively presented groups, and G a group with a finite generating set S. We say that (G, S) has computable quotients in $\mathcal{C}$ if there is an algorithm that, given a recursive presentation of a group H in $\mathcal{C}$ and a function f from S to the generators of H, stops if and only if the function f can be extended to a group homomorphism. This property is in fact independent of the choice of a generating set S.

The following was established in [5] for finite quotients and in [1] for nilpotent quotients:

Theorem 1 ([5,1]). *Let $\mathcal{C}$ be either the set of finitely generated nilpotent groups or the set of finite groups. There is a finitely generated group G, residually $\mathcal{C}$, with solvable word problem, but which does not have computable quotients in $\mathcal{C}$.*

The above result is the only known source of recursively presented groups which do not admit presentations that are EDT0L [1]. On the other hand:

Proposition 1 ([1]). *Let $\mathcal{C}$ be either the set of metabelian groups, the set of k-nilpotent groups for $k \in \mathbb{N}$, or the set of free groups. Then every finitely generated group has computable quotients in $\mathcal{C}$.*

Let $\mathcal{C}$ be a set of groups. Every finitely generated group G has a greatest residually $\mathcal{C}$ quotient, namely the quotient of G by the intersection of all kernels of group homomorphisms going from G to groups in $\mathcal{C}$. We denote by $\text{Res}_{\mathcal{C}}$ the function that maps a group to its greatest residually $\mathcal{C}$ quotient. This function is our main topic of interest.

Upon inspecting its proof, one realizes that Theorem 1 relies on the fact that the function $\text{Res}_{\mathcal{C}}$ is discontinuous in the topology of the space of marked groups (when C is either the set of nilpotent groups or that of finite groups). Yet, the function $\text{Res}_{\mathcal{C}}$ is also discontinuous when C is one of the sets of free, metabelian, or k-nilpotent groups, for a fixed $k \in \mathbb{N}$, and Proposition 1 contrasts with Theorem 1. In some sense, for free or metabelian or k-nilpotent groups, the function $\text{Res}_{\mathcal{C}}$ is "not discontinuous enough" to obtain a result akin to Theorem 1.

Following these remarks, it is very natural to try to study the map $\text{Res}_{\mathcal{C}}$ via Weihrauch reducibility. Indeed, Weihrauch reducibility is a tool that enables one to quantify "how discontinuous" a function is, and to relate this discontinuity with computability-theoretic phenomena [3]. Our goal is thus to answer the following problem:

Problem 1. What are the properties of $\mathcal{C}$ which determine the Weihrauch degree of $\mathrm{Res}_{\mathcal{C}}$?

2 Weihrauch Degrees

For an introduction to Weihrauch degrees, see [3]. Our results mostly involve well-known degrees:

1. The degree *Limit on Baire space*, denoted by Lim, which is equivalent to closed choice on Baire space [3]. It is also Σ_2^0-measurable-complete [2].
2. The jump of Limit on Baire space, denoted by Lim', is the composition of two limit operators, Lim $\circ$ Lim. It is Σ_3^0-measurable-complete [2].
3. Limit on the natural numbers, $\mathrm{Lim}_{\mathbb{N}}$, is the problem of mapping a converging sequence in $\mathbb{N}$ to its limit. A problem P with $P\leq_W \mathrm{Lim}_{\mathbb{N}}$ is *finite-mind-changes computable* [3].

We also need some less-studied degrees, related to the *level of discontinuity* of problems, as introduced by Hertling in [4]. We now define them.

Let α be a countable ordinal. Let $f : \mathbb{N} \to \alpha$ be a surjection. We consider the following problem:

$$\mathrm{Lim}_{\searrow(\alpha,f)} :\subseteq \mathbb{N}^{\mathbb{N}} \rightrightarrows \mathbb{N},$$

$$\mathrm{dom}(\mathrm{Lim}_{\searrow(\alpha,f)}) = \{(u_n)_{n\in\mathbb{N}} \in \mathbb{N}^{\mathbb{N}} | \forall n \in \mathbb{N}, f(u_{n+1}) \leq f(u_n)\},$$

$$\forall (u_n)_{n\in\mathbb{N}} \in \mathrm{dom}(\mathrm{Lim}_{\searrow(\alpha,f)}), \mathrm{Lim}_{\searrow(\alpha,f)}((u_n)_{n\in\mathbb{N}}) = \{t \in \mathbb{N} | f(t) = \inf_{n\in\mathbb{N}}(f(u_n))\}.$$

Thus, the input to $\mathrm{Lim}_{\searrow(\alpha,f)}$ is a weakly descending sequence in α, and the output is its limit. It is easy to see that the continuous Weihrauch degree of the above problem does not depend on the choice of f. Thus we can define $\mathrm{Lim}_{\searrow\alpha}$ as the degree of $\mathrm{Lim}_{\searrow(\alpha,f)}$ for any surjection f.

With these notions, we can state our results.

3 Results

For a set $\mathcal{C}$ of groups, we denote by $\mathcal{RC}$ the class of residually $\mathcal{C}$ groups.

Theorem 2. *We have* $\mathrm{Res}_{\mathcal{C}} \leq_W$ Lim *if and only if* $\mathcal{RC}$ *is a quasivariety.*

Theorem 3. *For each of the following quasivarieties,* $\mathrm{Res}_{\mathcal{C}} \equiv_W \mathrm{Lim}$:

the set of finitely generated k-solvable groups, for $k > 2$,
- the set of finitely generated LEF groups,
- the set of finitely generated torsion-free groups,
- the set of finitely generated left-orderable groups.

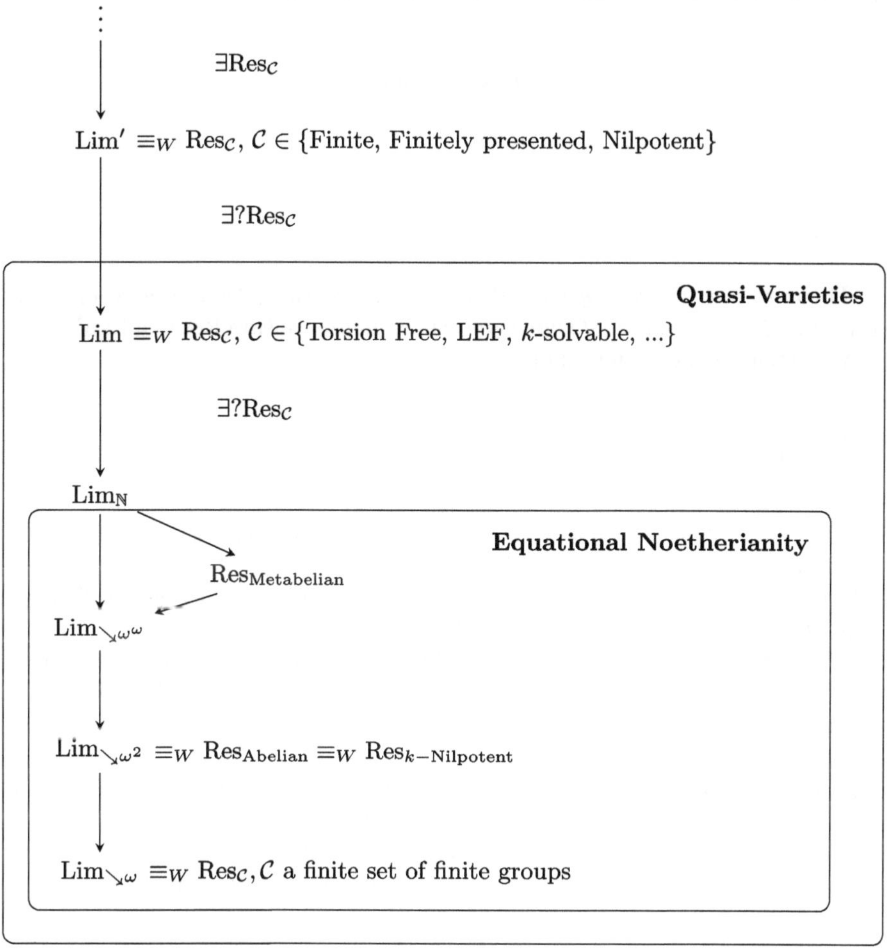

Fig. 1. Summary

Theorem 4. *The Weihrauch reduction*

$$\mathrm{Res}_{\mathcal{C}} \geq_W \mathrm{Lim}_{\mathbb{N}}$$

holds if and only if $\mathcal{C}$ is not equationally Noetherian.

Theorem 5. *We get the following Weihrauch reductions:*

- *For $\mathcal{C}$ equal to the set of metabelian groups:*

$$\mathrm{Res}_{\mathcal{C}} \geq_W \mathrm{Lim}_{\searrow \omega^\omega}.$$

- *For $\mathcal{C}$ equal to the set of p-nilpotent groups, for any $p \geq 1$,*

$$\mathrm{Res}_{\mathcal{C}} \equiv_W \mathrm{Lim}_{\searrow \omega^2}.$$

- *For $\mathcal{C}$ equal to a finite set of finite groups,*

$$\mathrm{Res}_{\mathcal{C}} \equiv_W \mathrm{Lim}_{\searrow \omega}.$$

Theorem 6. *For $\mathcal{C}$ any of the following sets: the sets of finite groups, of finitely generated nilpotent groups, of finitely presented groups, we have* $\mathrm{Res}_{\mathcal{C}} \equiv_W \mathrm{Lim}'$.

We summarize our results in Fig. 1.

References

1. Bartholdi, L., Pernak, L., Rauzy, E.: Groups with presentations in EDT0L. In: Groups, Geometry, and Dynamics (2025). https://doi.org/10.4171/ggd/911
2. Brattka, V.: Effective Borel measurability and reducibility of functions. Math. Logic Q. **51**(1), 19–44 (2004). https://doi.org/10.1002/malq.200310125
3. Brattka, V., Gherardi, G., Pauly, A.: Weihrauch complexity in computable analysis. In: Brattka, V., Hertling, P. (eds.) Handbook of Computability and Complexity in Analysis, pp. 367–417. Springer, Cham (2021). https://doi.org/10.1007/978-3-030-59234-9_11
4. Hertling, P.: Topological complexity with continuous operations. J. Complex. **12**(4), 315–338 (1996). https://doi.org/10.1006/jcom.1996.0021
5. Rauzy, E.: Computability of finite quotients of finitely generated groups. J. Group Theory **25**(2), 217–246 (2022). https://doi.org/10.1515/jgth-2020-0029

Left and Right Effective Symmetries on Baire and Cantor Space

Luca San Mauro

University of Bari
luca.sanmauro@gmail.com

Keywords: countable Borel equivalence relations, computable reducibility, recursive isomorphism, fully primitive recursive permutations

The symmetric group $\mathrm{Sym}(\omega)$ acts on Baire space ω^ω and Cantor space 2^ω in two canonical ways: on the left, by relabeling values, and on the right, by reindexing positions. Restricting to effective subgroups—most notably the computable permutations and the fully primitive recursive ones—yields natural countable Borel equivalence relations. Writing E_L^c and E_R^c for the orbit equivalence relations of the computable left and right actions, $E_R^c(2^\omega)$ is exactly recursive isomorphism, equivalently 1-equivalence, on subsets of ω. Classical results show that $E_R^c(\omega^\omega)$ is a universal countable Borel equivalence relation [2]; universality has been progressively pushed down to $E_R^c(k^\omega)$ for $k \geq 5$ [1] and then $k > 3$ [4]; the case $k = 2$ remains a long-standing open problem [4]. By contrast, the corresponding left action—equally natural from a group-theoretic standpoint—has, to our knowledge, not been studied systematically.

In this talk, we report on a recent project that begins to fill this gap, by comparing the orbit equivalence relations arising from the left and right actions of the computable and fully primitive recursive symmetric groups on 2^ω and ω^ω under three main notions of reducibility: computable, continuous, and Baire class 1. The resulting picture is unexpectedly rich: these notions induce qualitatively different landscapes, with collapses appearing as the reducibility class is enlarged. At the computable level, the structure is rigid. While $E_L^c(2^\omega)$ is easily seen to be smooth, on Baire space the two actions are computably incomparable, with E_0 reducing to $E_L^c(\omega^\omega)$ but not to $E_R^c(\omega^\omega)$; moreover, neither relation is computably reducible to any effectively Π_3^0 equivalence relation. Weakening the notion of reduction substantially alters the picture: on Baire space, the computable left action becomes Baire class 1 reducible to the computable right action.

The proofs draw on tools from both computability theory and descriptive set theory. The work thus fits into a broader recent interest in finer notions of effective reducibility for equivalence relations [5, 3], and isolates a natural family in which the two perspectives genuinely interact.

This is joint work with Jun Le Goh, Heer Tern Koh, and Keng Meng Ng.

References

1. Andretta, A., Camerlo, R., Hjorth, G.: The relation of recursive isomorphism for countable structures. J. Symb. Logic **67**(2), 879–895 (2002)
2. Dougherty, R., Kechris, A.S.: How many Turing degrees are there? In: Computability Theory and Its Applications. Contemporary Mathematics. vol. 257, pp. 83–94. American Mathematical Society (2000)
3. Ho, M.-C., Jackson, S.C., Lempp, S., Miller, R.G., Schweber, N.D.: Countable and finitary reductions on equivalence relations. Fund. Math. **270**(3), 201–221 (2025)
4. Marks, A.S.: Uniformity, universality, and computability theory. J. Math. Log. **17**(1), 1750003 (2017)
5. Miller, R., Ng, K.M.: Finitary reducibility on equivalence relations. J. Symb. Logic **81**(4), 1225–1254 (2016)

Novel Knotted Solenoid Fold with Order-Shifted Coil Arrangement

Joanna Sułkowska

Centre of New Technologies, University of Warsaw, Banacha 2c, 02-097 Warsaw, Poland
j.sulkowska@cent.uw.edu.pl

This presentation will address the topic of the novel Knotted Solenoid fold. I will present crystal structures of proteins adopting a fold not identified among known solenoids in current structural databases, a Knotted Solenoid. These proteins exhibit a characteristic solenoidal architecture, closely resembling solenoid proteins. However, a unique "skip-and-backtrack" shift is observed: One coil skips a rotation while the next realigns, distinguishing this fold from previously described solenoids and allowing the formation of a 3_1 (trefoil) knot. Moreover, the proteins form homodimers, and conservation of interface residues suggests a shared oligomerization state across all Knotted Solenoids. This fold is exclusive to a specific group of bacteria and remains structurally conserved despite high sequence variability (pairwise identities down to 6%). Conserved residues are observed at the beginning of each coil and within the knot core, suggesting functional or structural significance, and an independent evolutionary path to unknotted solenoids. In vitro chemical and thermal stability studies showed fully reversible unfolding in urea, and no additional transitions even in high concentrations of guanidinium chloride. The far-ultraviolet circular dichroism unfolding kinetics showed relatively rapid unfolding. Explicit-solvent molecular dynamics simulations and a generative deep learning model show that topological constraints stabilize "skip-and-backtrack" shift in the Knotted Solenoid. A monomeric unit can self-tie into the native 3_1 knotted state through a slipknot intermediate. It then interacts with another chain via its hydrophobic surface, promoting second chain folding and dimerization. The identification of novel knotted proteins within a previously considered unknotted fold provides an opportunity to investigate the evolutionary pressures and functional implications of knotting in shaping protein architecture.

Reference

1. Sikora, M., et al.: Novel Knotted Solenoid fold with order-shifted coil arrangement leads to nontrivial 3_1 topology. Proc. Natl. Acad. Sci. U.S.A. **123**(17), e2525920123. https://doi.org/10.1073/pnas.2525920123 (2026)

CPAC: A Brief Tale about a New Paradigm of Learnability

Ruth Urner

York University and Vector Institute, Toronto, Canada
uruth@yorku.ca

Abstract. We trace the development of the CPAC framework for learnability with computable learners from its inception [1] to recent results.

Keywords: Probably Approximately Correct (PAC) Learning · Computability · CPAC Learning.

1 Computable PAC (CPAC) Learning

Statistical learning is concerned with deriving insights about unknown processes from finite observational data. Formally, the unknown process is modeled as a probability distribution P over some domain $\mathcal{Z}$ and the observations are modeled as iid samples $S = (z_1, z_2, \ldots, z_n)$ from this distribution. The type of insight we are seeking to gain is modeled by means of a class $\mathcal{H}$ of hypotheses.

PAC learnability The Probably Approximately Correct (PAC) framework [15] provides well-established criteria for evaluating whether a type of insight is *learnable*. It requires sample sizes that are sufficient for learning success (approximate up to an error parameter ϵ) with high probability over the drawn data (up to a confidence parameter δ), to be uniformly bounded over all data generating distributions (as a function of ϵ, δ and the hypothesis class $\mathcal{H}$ only). Thus, a success guarantee is provided with sample sizes depending on parameters that are in control of the user (ϵ, δ and $\mathcal{H}$) rather than allowing dependence on the unknown component, the data-generating distribution P.

For example, in the case of classification tasks, where $\mathcal{Z} = \mathcal{X} \times \mathcal{Y}$ is the product of a *feature space* $\mathcal{X}$ and a *label space* $\mathcal{Y}$, $\mathcal{H} \subseteq \mathcal{Y}^{\mathcal{X}}$ is a class of classifiers $h : \mathcal{X} \to \mathcal{Y}$, and success is measured in terms of the *expected loss* $\mathcal{L}_P(h) = \mathbb{P}_{(x,y)\sim P}[h(x) \neq y]$, PAC learnability is the following requirement:

Definition 1 (PAC Learnability). *A hypothesis class* $\mathcal{H} \subseteq \mathcal{Y}^{\mathcal{X}}$ *is* PAC learnable in the agnostic case *if there exists a learner* $\mathcal{A} : (\mathcal{X} \times \mathcal{Y})^* \to \mathcal{Y}^{\mathcal{X}}$ *and a function* $m : (0, 1)^2 \to \mathbb{N}$ *such that for all* $\epsilon, \delta \in (0, 1)$, *for all distributions* P *on* $\mathcal{X} \times \mathcal{Y}$ *and for all* $m \geq m(\epsilon, \delta)$ *we have*

$$\mathbb{P}_{S\sim P^m}\big[\mathcal{L}(\mathcal{A}(S)) \geq \text{opt}(\mathcal{H}) + \epsilon\big] \geq 1 - \delta,$$

where $\text{opt}_P(\mathcal{H}) = \inf_{h\in\mathcal{H}}\mathcal{L}_P(h)$ *is the smallest loss achievable with a hypothesis from the class* $\mathcal{H}$. *The class* $\mathcal{H}$ *is* PAC learnable in the realizable case *if the above holds in case* $\text{opt}_P(\mathcal{H}) = 0$. *The class* $\mathcal{H}$ *is called* properly *PAC learnable if the above is satisfied for a learner* $\mathcal{A} : (\mathcal{X} \times \mathcal{Y})^* \to \mathcal{H}$ *that always outputs a hypothesis from* $\mathcal{H}$.

It has long been established, and is sometimes referred to as the *fundamental theorem of statistical learning* [13], that for binary classification, i.e. $\mathcal{Y} = \{0, 1\}$, a hypothesis class $\mathcal{H}$ is PAC learnable (in the realizable and agnostic case) if and only if it has finite VC-dimension [13, 16]. Moreover, in this case every learner $\mathcal{A}$ that is an *empirical risk minimizer (ERM)*, i.e. $\mathcal{A}(S) \in \inf_{h\in\mathcal{H}} \mathcal{L}_S(h)$ for all $S = ((x_1, y_1), (x_2, y_2), \ldots (x_n, y_n))$, where $\mathcal{L}_S(h) = \sum_{i=1}^{|S|} 1\left[h(x_i) \neq y_i\right]$, is a succesful PAC learner for a hypothesis class of finite VC-dimension [13, 16].

Establishing similar clean and general characterizations of learnability by a dimension of the hypothesis class for other statistical learning problems has proven to be challenging and in some cases impossible. Even for multi-class classification (where $|\mathcal{Y}| > 2$ and possibly $|\mathcal{Y}| = \infty$), there are hypothesis classes that are learnable, but not by an ERM learner (and some classes are learnable by some, but not all ERM learners) [5]. A characterizing dimension for multi-class learnability has only recently been found [4. For certain distribution learning problems (where the hypothesis class consists of distributions and success is measured in terms of the total variation distance between the data-generating distribution and a learner's output) it has been proven that there cannot exist a dimension that characterizes learnability [8, 9]. For PAC learnability under an adversarial loss, similar discrepancies between general learnability and proper learnability as for multiclass-classification have been established [10] and the only currently known characterization is in terms of a property of the infinite one-inclusion graph [11] (which does not satisfy requirements that have been stipulated for a property to qualify as a dimension [8, 9]).

Perhaps most strikingly, for some proper learning problems, namely the general *expectation maximization* (EMX) problem, it has been shown that there exist hypothesis classes for which the question of learnability is equivalent to the continuum hypothesis and thus independent of the axioms of set theory [2, 3]. This independence result triggered an inquiry into the definition of statistical (PAC) learning [1]. The proof of independence of learnability from set theory [2, 3] hinges on the notion of a learner in Definition. Namely, a learner is modeled there as a function mapping finite samples to hypotheses. However, if the formal settings of statistical learning, and the PAC framework in particular, are to serve as formal foundations for machine learning, then clearly both learners and hypotheses need to be *computable* objects. Would such an (arguably minimal) restriction on learners and hypotheses change what is learnable?

*CPAC learnability*We addressed the above question by introducing the framework of *Computable Probably Approximately Correct* (CPAC) learnability [1], requiring both learners and hypotheses to be computable functions. We furthermore introduced the following computability requirements for hypothesis classes:

Definition 2 (Computable Representation [1]). *A hypothesis class* $\mathcal{H} \subseteq \mathcal{Y}^{\mathcal{X}}$ *is called* decidably representable (DR) *if there exists a decidable set of programs* $\mathcal{P}$ *such that the set of all functions computed by programs in* $\mathcal{P}$ *equals* $\mathcal{H}$. *The class* $\mathcal{H}$ *is called* recursively enumerably representable (RER) *if there exists such a set of programs that is recursively enumerable.*

Learners in the CPAC framework are required to be computable functions that halt on all inputs and output total computable hypotheses:

Definition 3 (CPAC Learnability [1]). *A class* $\mathcal{H} \subseteq \mathcal{Y}^{\mathcal{X}}$ *is (agnostic)* CPAC learnable *if there exists a computable (agnostic) PAC learner for* $\mathcal{H}$ *that outputs total computable hypotheses.*

Focusing on binary classification, the initial work on CPAC learnability showed that the added computability requirement indeed revealed a different landscape of learnability: For DR and RER classes PAC and CPAC learnability are equivalent in the realizable case, but not in the agnostic case (and without the basic restriction of a DR or RER encoding, the equivalence even fails in the realizable case) [1]. Follow-up studies established a more complete picture of CPAC learnability for binary classification [12, 14]. These works introduced a computable version of the VC dimension, the *effective VC dimension* $\mathrm{evc}(\mathcal{H})$ of a hypothesis class $\mathcal{H}$ [12], as well as a notion of *strong CPAC* (SCPAC) learnability where a computable bound on the sample complexity is added as a requirement [14], yielding the following equivalences [12]:

$$\mathrm{evc}(\mathcal{H}) < \infty \Leftrightarrow \mathcal{H}\text{ CPAC learnable} \Leftrightarrow \mathcal{H}\text{ SCPAC learnable}$$

$$\Leftrightarrow \mathcal{H} \subseteq \widehat{\mathcal{H}}\text{ properly SCPAC learnable}$$

$$\Leftrightarrow \mathcal{H} \subseteq \widehat{\mathcal{H}}\text{ properly CPAC learnable}$$

For proper SCAPC learnability the following equivalence was established [14]:

$$\mathcal{H}\text{ proper SCPAC} \Leftrightarrow \mathrm{vc}(\mathcal{H}) < \infty \text{ and } \mathcal{H}\text{ has a computable ERM}$$

and for proper CPAC learnability [12]:

$$\mathcal{H}\text{ proper CPAC} \Leftrightarrow \mathrm{vc}(\mathcal{H}) < \infty \text{ and } \mathcal{H}\text{ has a computable asymptotic ERM}$$

These results can be extended to multiclass learnability in case the label class $\mathcal{Y}$ is finite [7]. While the above landscape might seem at least reminiscent of the fundamental theorem for statistical learning, incorporating computability requirements into more complex settings of learnability exhibits some perhaps surprising phenomena. When requiring adversarial robustness, there are classes that are both CPAC and robustly PAC learnable but are not robustly CPAC learnable, and for robust CPAC learnability it is not necessary that the robust loss is computably evaluable [6]. These invite further study into where, in contemporary statistical learning settings, computability requirements matter.

Acknowledgments. This presentation is based on joint works [1, 6, 7] with Sushant Agarwal, Nivasini Ananthakrishnan, Shai Ben-David, Pascale Gourdeau and Tosca Lechner. The author's research program is supported by an NSERC discovery grant.

References

1 Agarwal, S., Ananthakrishnan, N., Ben-David, S., Lechner, T., Urner, R.: On learnability wih computable learners. In: Algorithmic Learning Theory, ALT 2020, pp. 48–60. Proceedings of Machine Learning Research, PMLR (2020)
2. Ben-David, S., Hrubes, P., Moran, S., Shpilka, A., Yehudayoff, A.: A learning problem that is independent of the set theory ZFC axioms. CoRR **abs/1711.05195** (2017)
3. Ben-David, S., Hrubes, P., Moran, S., Shpilka, A., Yehudayoff, A.: Learnability can be undecidable. Nat. Mach. Intell. **1**(1), 44–48 (2019)
4. Brukhim, N., Carmon, D., Dinur, I., Moran, S., Yehudayoff, A.: A characterization of multiclass learnability. In: 63rd IEEE Annual Symposium on Foundations of Computer Science, FOCS 2022, Denver, CO, USA, October 31 — November 3, 2022, pp. 943–955. IEEE (2022)
5. Daniely, A., Sabato, S., Ben-David, S., Shalev-Shwartz, S.: Multiclass learnability and the ERM principle. J. Mach. Learn. Res. **16**, 2377–2404 (2015)
6. Gourdeau, P., Lechner, T., Urner, R.: On the computability of robust PAC learning. In: The Thirty Seventh Annual Conference on Learning Theory, COLT 2024, pp. 2092–2121. Proceedings of Machine Learning Research, PMLR (2024)
7. Gourdeau, P., Lechner, T., Urner, R.: On the computability of multiclass PAC learning. CoRR **abs/2502.06089** (2025)
8. Lechner, T., Ben-David, S.: Inherent limitations of dimensions for characterizing learnability of distribution classes. In: The Thirty Seventh Annual Conference on Learning Theory COLT 2024, pp. 3353–3374. Proceedings of Machine Learning Research, PMLR (2024)
9. Lechner, T., Ben-David, S.: No scale sensitive dimension for distribution learning. In: 37th International Conference on Algorithmic Learning Theory (2026)
10. Montasser, O., Hanneke, S., Srebro, N.: VC classes are adversarially robustly learnable, but only improperly. In: Conference on Learning Theory, COLT 2019, pp. 2512–2530. Proceedings of Machine Learning Research, PMLR (2019)
11. Montasser, O., Hanneke, S., Srebro, N.: Adversarially robust learning: a generic minimax optimal learner and characterization. In: Advances in Neural Information Processing Systems 35: Annual Conference on Neural Information Processing Systems NeurIPS 2022 (2022)
12. Rose, V.D., Kozachinskiy, A., Rojas, C., Steifer, T.: Find a witness or shatter: the landscape of computable PAC learning. In: The Thirty Sixth Annual Conference on Learning Theory, COLT 2023, pp. 511–524. Proceedings of Machine Learning Research, PMLR (2023)
13. Shalev-Shwartz, S., Ben-David, S.: Understanding Machine Learning - From Theory to Algorithms. Cambridge University Press (2014)
14. Sterkenburg, T.F.: On characterizations of learnability with computable learners. In: Conference on Learning Theory, COLT 2022, pp. 3365–3379. Proceedings of Machine Learning Research, PMLR (2022)
15. Valiant, L.G.: A theory of the learnable. Commun. ACM **27**(11), 1134–1142 (1984)
16. Vapnik, V., Chervonenkis, A.: On the uniform convergence of relative frequencies of events to their probabilities. In: Theory of Probability and Its Applications (1971)

$QIP \subseteq AM(2QCFA)$

Abuzer Yakaryılmaz

Center for Quantum Computer Science, University of Latvia, Latvia
abuzer.yakaryilmaz@lu.lv

Proof systems play a central role in classical and quantum complexity theory [3, 8]. In polynomial-time computation, interactive proof systems (IPSs) [5, 6], Arthur-Merlin (AM) proof systems [1], and quantum interactive proof systems [7] have the same verification power, i.e., they can verify all and only languages in PSPACE.

In constant-space, IPSs with two-way probabilistic finite-state automaton (2PFA) verifiers can verify any language in ASPACE(n) (i.e., alternating linear space) [4]. On the other hand, AM systems with 2PFA verifiers are much weaker: any language verified by them is in P, and they cannot verify some languages in L such as PAL $= \{w \in \{a, b\}^* | w = w^{reverse}\}$ [2, 4].

More than a decade ago, Yakaryılmaz introduced and investigated AM systems where the verifier is a two-way automaton with quantum and classical states (2QCFA) [9, 10]. We denote $AM_1(2QCFA)$ as the class of the languages verified by these systems with perfect-completeness (i.e., member strings are accepted by the verifier with probability 1). Yakaryılmaz showed that $AM_1(2QCFA)$ contains NP, ASPACE(n), and an NEXP-complete problem. Here we show that $AM_1(2QCFA)$ also contains PSPACE. We use modified techniques given in [9, 10].

References

1. Babai, L.: Trading group theory for randomness. In: STOC'85: Proceedings of the 17th Annual ACM Symposium on Theory of Computing, pp. 421–429 (1985). https://doi.org/10.1145/22145.22192
2. Condon, A.: Space-bounded probabilistic game automata. J. ACM **38**(2), 472–494 (1991). https://doi.org/10.1145/103516.128681
3. Condon, A.: Complexity Theory: current research, chap. The complexity of space bounded interactive proof systems, pp. 147–190. Cambridge University Press, USA (1993). https://dl.acm.org/doi/10.5555/183589.183728
4. Dwork, C., Stockmeyer, L.: Finite state verifiers I: The power of interaction. J. ACM **39**(4), 800–828 (1992). https://doi.org/10.1145/
5. Goldwasser, S., Micali, S., Rackoff, C.: The knowledge complexity of interactive proof-systems (extended abstract). In: STOC'85: Proceedings of the 17th Annual ACM Symposium on Theory of Computing, pp. 291–304 (1985). https://doi.org/10.1145/22145.22178

6. Goldwasser, S., Micali, S., Rackoff, C.: The knowledge complexity of interactive proof systems. SIAM J. Comput. **18**(1), 186–208 (1989). https://doi.org/10.1137/0218012
7. Watrous, J.: PSPACE has constant-round quantum interactive proof systems. In: FOCS'99: Proceedings of the 40th Annual Symposium on Foundations of Computer Science, pp. 112–119 (1999). https://doi.org/10.1109/SFFCS.1999.814583
8. Watrous, J.: Quantum computational complexity. In: Meyers, R.A. (ed.) Encyclopedia of Complexity and Systems Science, pp. 7174–7201. Springer, New York (2009). https://doi.org/10.1007/978-0-387-30440-3_428
9. Yakaryılmaz, A.: Public-qubits versus private-coins. Electron. Colloquium Comput. Complex. **TR12-130** (2012). https://eccc.weizmann.ac.il/report/2012/130
10. Yakaryılmaz, A.: Public qubits versus private coins. In: The Proceedings of Workshop on Quantum and Classical Complexity, pp. 45–60. Univeristy of Latvia Press (2013). https://users.utu.fi/mikhirve/workshop/Proceedings.pdf

Algorithmic Randomness and Weak Merging of Opinions

Francesca Zaffora Blando

Carnegie Mellon University, Pittsburgh, PA, USA
fzaffora@andrew.cmu.edu

Abstract. In this talk, I will present a general framework for studying notions of algorithmic randomness defined in terms of the weak merging of computable probability measures.

Keywords: Algorithmic Randomness · Merging of Opinions · Computable Probability Theory.

The goal of this talk is to present a general framework for studying "merging randomness": namely, notions of algorithmic randomness defined via suitably effectivized merging-of-opinions theorems. The results I will discuss are joint work with Simon Huttegger (UC Irvine) and Sean Walsh (UCLA) and they appear in the article "Algorithmic randomness and the weak merging of computable probability measures" [2].

Arguably, the most well-known merging-of-opinions theorem is the Blackwell-Dubins Theorem [1], which establishes that, given two prior probability measures μ and ν, if μ is absolutely continuous with respect to ν, then, with μ-probability one, the posteriors induced by ν are guaranteed to become arbitrarily close to the posteriors induced by μ as the shared evidence accumulates. Crucially, to quantify the divergence between these posteriors, the Blackwell-Dubins Theorem relies on the total variation distance, a statistical distance defined in terms of the probabilities assigned to all measurable events. In this talk, I will instead focus on a kind of merging between probability measures called *weak merging*, which was studied extensively by Kalai and Lehrer [3] and which was originally defined via a variant of the total variation distance restricted to one-step-ahead predictions (equivalently, to probability assignments to finite-horizon events). In addition to this type of weak merging, I will also introduce notions of weak merging defined in terms of the Hellinger distance and the Kullback-Leibler divergence, appropriately restricted to one-step-ahead predictions, and I will discuss the notions of merging randomness that all these different variants of weak merging yield. The main results I will present are novel characterizations of Martin-Löf randomness [4] and Schnorr randomness [5] in terms of weak merging of opinions and the Kullback-Leibler divergence.

Let $\mathfrak{F}_n$ denote the sub-σ-algebra of the Borel σ-algebra on $2^{\mathbb{N}}$ generated by the cylinder sets associated to strings of length n. Given a strictly positive probability measure μ on $2^{\mathbb{N}}$ and $X \in 2^{\mathbb{N}}$, let $\mu_{X,n}$ be an abbreviation for the conditional probability $\mu(\cdot \mid [\![X \upharpoonright n]\!])$ obtained by conditioning μ on the cylinder set $[\![X \upharpoonright n]\!]$, restricted to the events in $\mathfrak{F}_{n+1}$. For two strictly positive probability measures μ, ν on $2^{\mathbb{N}}$ and $X \in 2^{\mathbb{N}}$, the Kullback-Leibler divergence between $\mu_{X,n}$ and $\nu_{X,n}$ is then given by $D(\mu_{X,n}|\nu_{X,n}) = \mathbb{E}_{\mu_{X,n}}[\ln \frac{d\mu_{X,n}}{d\nu_{X,n}}]$, while the summable Kullback-Leibler divergence between $\mu_{X,n}$ and $\nu_{X,n}$ is given by $\sum_{n\in\mathbb{N}} D(\mu_{X,n}|\nu_{X,n})$. Our characterizations of Martin-Löf randomness and Schnorr randomness in terms of the Kullback-Leibler divergence are then as follows:

Theorem 1 ([2]). *Let μ be a computable strictly positive probability measure on $2^{\mathbb{N}}$. Then, the following are equivalent for $X \in 2^{\mathbb{N}}$:*

1. *X is Martin-Löf random relative to μ;*
2. *for all computable strictly positive probability measures ν on $2^{\mathbb{N}}$ such that $\sup_n \mathbb{E}_\mu \ln \frac{\mu([\![\cdot\restriction n]\!])}{\nu([\![\cdot\restriction n]\!])}$ is finite,[5] one has that $\sum_{n\in\mathbb{N}} D(\mu_{X,n}|\nu_{X,n})$ is finite.*

Theorem 2 ([2]). *Let μ be a computable strictly positive probability measure on $2^{\mathbb{N}}$. Then, the following are equivalent for $X \in 2^{\mathbb{N}}$:*

1. *X is Schnorr random relative to μ;*
2. *for all computable strictly positive probability measures ν on $2^{\mathbb{N}}$ such that $\sup_n \mathbb{E}_\mu \ln \frac{\mu([\![\cdot\restriction n]\!])}{\nu([\![\cdot\restriction n]\!])}$ is finite and computable, one has that $\sum_{n\in\mathbb{N}} D(\mu_{X,n}|\nu_{X,n})$ is finite.*

As we will see, this work reveals a rich landscape of weak-merging randomness notions whose relationship with the traditional algorithmic randomness hierarchy is still largely unexplored and whose import extends to the philosophical study of probabilistic (Bayesian) learning and inter-subjective agreement (see, for instance, [6]).

References

1. Blackwell, D., Dubins, L.: Merging of opinions with increasing information. Ann. Math. Stat. **33**(3), 882–886 (1962)
2. Huttegger, S. M., Walsh, S., Zaffora Blando, F.: Algorithmic randomness and the weak merging of computable probability measures. Ann. Pure Appl. Logic **177**(8), 103737, 1–27 (2026)
3. Kalai, E., Lehrer, E.: Weak and strong merging of opinions. J. Math. Econ. **23**, 73–86 (1994)
4. Martin-Löf, P.: The definition of random sequences. Inf. Control **9**(6), 602–619 (1966)
5. Schnorr, C.-P.: A unified approach to the definition of a random sequence. Math. Syst. Theory **5**, 246–258 (1971)
6. Zaffora Blando, F.: Algorithmic randomness and computable Bayesianism. In: Huttegger, S. M., Romeijn, J.-W. (eds.), Bayesianism, Philosophical Movements Revisited. Cambridge, Cambridge University Press (forthcoming)

[5] Note that this is a strengthening of absolute continuity: that is, if $\sup_n \mathbb{E}_\mu \ln \frac{\mu(\cdot\restriction n)}{\nu(\cdot\restriction n)}$ is finite, then μ is absolutely continuous with respect to ν.

General Decidability Results on Unboundedness Problems

Georg Zetzsche

Max Planck Institute for Software Systems (MPI-SWS), Kaiserslautern, Germany
georg@mpi-sws.org

We will discuss decidability of unboundedness problems for formal languages. Informally, unboundedness problems ask about the existence of infinitely many words (satisfying certain properties) in a formal language. For example: Is a given language infinite? Or: Does a given language have super-polynomial growth? These came into focus in recent years because of their connections to downward closure computation and separability problems. It turns out that there are techniques that are at the same time conceptually very simple, but also apply to a surprisingly wide variety of language classes or concrete unboundedness problems. The talk will survey recent results (and techniques) concerning unboundedness problems, and will concentrate on two types of general decidability results: (i) decidability of an unboundedness problem for a variety of language classes and (ii) decidability of a variety of unboundedness problems for specific prominent language classes.

Contents

Invited Papers

Ordinary Differential Equations as a Universal Language for Computability and Complexity: From Polynomial Time to the Hyperarithmetical Hierarchy ... 3
Olivier Bournez

Learning Omega-Regular Languages: A Tour of Learning Results and Canonical Representations ... 23
Dana Fisman, Elina Sudit, and Oded Zimerman

Multisets in Reaction Systems ... 48
Victor Mitrana

Computing with Spikes and Membranes: Computability and Complexity in Spiking Neural P Systems ... 60
David Orellana-Martín

The New French School of Philosophy of Computing ... 78
Paula Quinon

Mathematical Informatics: Algorithms ... 86
Thomas Seiller

Contributed Papers

Computable Cofinal Fraïssé Limits ... 109
Nathanael Ackerman, Cameron Freer, and Mostafa Mirabi

Bishop's (Up)Crossing Inequality and Lower Semicomputable Random Reals Revisited ... 124
Mikhail Andreev and Alexander Shen

What is a Polynomial-Time Computable Square-Integrable Function? ... 135
Aras Bacho, Svetlana Selivanova, and Martin Ziegler

On Computability of Ideal Lattices ... 150
Nikolay Bazhenov, Manat Mustafa, and Stanislav Yun

Glivenko's Theorem Underneath Structure 162
Riccardo Borsetto, Giulio Fellin, Tarmo Uustalu, and Cheng-Syuan Wan

Complexities of Effective Reductions with Ordinal Turing Machines 178
Merlin Carl

Weihrauch Reducibility Between Ramsey-Type Theorems and Well-Ordering Principles at the Level of Σ_2^0-Induction 193
Lorenzo Carlucci and Giordano Celli

Free Sets, Thin Sets and Rainbows for Barriers 208
Lorenzo Carlucci and Oriola Gjetaj

Adaptive Multi-Head Finite-State Gamblers 224
Julianne Cruz, Sho Glashausser, Xiaoyuan Li, and Neil Lutz

A Note on Computable Étale Spaces 240
Matthew de Brecht

Placing Green Bridges Optimally for Robust Habitat Reconnection 256
Gero Ellmies and Till Fluschnik

Nuclear Shifts for Conservation 272
Giulio Fellin, Sara Negri, and Peter Schuster

On the Σ-Hierarchy of the Logical Principles Over Intuitionistic Predicate Logic 289
Makoto Fujiwara

Towards Settling the Complexity of the Lettericity Problem 304
Mario Grobler, Nils Morawietz, and Silas Cato Sacher

Closure Properties of Left NP Real Numbers and NP Real Functions 320
Peter Hertling and Leonard Schulte-Michels

Counting Random Oracles for the Polynomial-Time Hierarchy and Quantum Complexity Classes 335
John M. Hitchcock, Adewale Sekoni, and Hadi Shafei

Tadaki Numbers and Speedability 350
Philip Janicki

Adversarial Synchronization and Reset Length 358
Anton E. Lipin and Mikhail V. Volkov

Lines in Every Direction with No ee-Random Points 374
Neil Lutz, Spencer Park Martin, and Rain White

Self-assembly of Strings and Languages Revisited: Efficient Membership Algorithms 389
Katalin Anna Lázár, Florin Manea, Stefan Siemer, and Timo Specht

Efficiently Finding All Shortest Absent Subsequences in a String 405
Florin Manea, Tina Ringleb, Stefan Siemer, and Maximilian Winkler

Comparing the Effective Content of Subshifts 421
A. Nakid Cordero and I. Scott

Termination of Real Linear Loops 437
Eike Neumann and Margret Tembo

Safety and Reachability in *k*-Control Games on Integer Vector Addition Systems with States 453
Reino Niskanen, Igor Potapov, and James Topley

Mergeable Represented Spaces 469
Arno Pauly

On Normality and Equidistribution for Separator Enumerators 482
Subin Pulari

On the Reverse Mathematics of Darboux's Supremum Principle 498
Sam Sanders

On the Computational Properties of Ambivalent Sets and Functions 514
Dag Normann and Sam Sanders

A Polynomial-Time Algorithm for the Automatic Baire Property 528
Ludwig Staiger

The More Gödels, the Better 537
Patrick Uftring

Author Index 553

Invited Papers

Ordinary Differential Equations as a Universal Language for Computability and Complexity: From Polynomial Time to the Hyperarithmetical Hierarchy

Olivier Bournez(✉)

LIX, CNRS, École polytechnique, Institut Polytechnique de Paris, Palaiseau, France
olivier.bournez@lix.polytechnique.fr

Abstract. We argue that ordinary differential equations form a universal language in which the geography of computation, from polynomial time at the bottom up to the hyperarithmetical hierarchy at the top, finds a natural and uniform expression. The control parameters of this language are the regularity of the right-hand side, the length of the solution, the precision admitted on the input, and the continuous or discrete nature of the iteration. Models historically treated as separate, including the General Purpose Analog Computer, recurrent neural networks under the deep-narrow regime, and chemical reaction networks under mass-action kinetics, all appear as slices of one mathematical object, the polynomial ODE. The technical heart of the picture is a four-way equivalence at the polynomial time level: Turing time, ODE solution length, fixed-width neural network depth, and Boolean circuit size are polynomially related as resources for computing the same functions, with each substrate admitting a universal device that simulates the others at polynomial cost. The upward extension to the hyperarithmetical hierarchy is obtained by relaxing analytic regularity to a class of solvable discontinuous fields. The tutorial surveys the recent results that establish this picture and identifies some of the directions that remain open.

1 Introduction

Ordinary differential equations are a universal language for computability and complexity.

The parameters of this language, regularity of the right-hand side, length of the solution, precision admitted, and continuous or discrete nature of the iteration, function as control knobs that together span computation from FPTIME to the hyperarithmetical hierarchy. Models historically discussed in isolation, mechanical differential analysers, recurrent neural networks, chemical reaction networks, hybrid automata, piecewise-affine systems, are slices of one mathematical object. The aim of this tutorial is to draw the resulting map.

V. Brattka et al. (Eds.): CiE 2026, LNCS 16674, pp. 3–22, 2027.
https://doi.org/10.1007/978-3-032-31348-5_1

Two Classical Questions

The initial motivation comes from a question that has organised the analog-computation literature for nearly a century: *which real functions are physically realisable as solutions of an analog device?* This was Shannon's question when he formalised Bush's mechanical differential analysers [27] and identified the class of functions they generate [73]. The comparison with Turing computability split it into two: *can analog computation reach above the Turing barrier?*, and *when it does not, at what cost can a Turing machine simulate it?* The modern answer depends sharply on the regularity of the right-hand side, which calibrates the entire vertical axis. Polynomial right-hand sides yield computable analysis; discontinuous but solvable ones reach the hyperarithmetical hierarchy. Within the polynomial regime, solution length and admitted precision dial finely between FPTIME and FPSPACE. A recent survey of entry points into this picture is [49].

A Four-Way Equivalence in the Polynomial Regime

A central claim of this tutorial, established in Sect. 6, is that at the polynomial-time level the picture is genuinely tight: Turing running time, polynomial-ODE solution length, fixed-width neural network depth, and Boolean circuit size are all polynomially related as resources for computing the same functions, with each substrate admitting a universal device that simulates the others at polynomial cost. The four substrates therefore offer four equally fundamental descriptions of the same complexity class.

Why Now

The tutorial is written at a moment when the analog substrate is being rebuilt as engineering. The German company Anabrid markets the THAT and LUCIDAC reconfigurable analog hardware platforms [62], and broader surveys of analog computing for domain-specific accelerators [9] document parallel activity across other research groups. On the theoretical side, undecidability results in mathematical physics have proliferated since the early 2020s [67], and fluid-mechanical realisations of Turing-complete dynamics form an active research line [34,35]. The geographic reading proposed here is, in this context, both a theoretical synthesis and a practical map.

Roadmap

Sections 2 and 3 develop the GPAC and the polynomial-ODE foundations, with the time-equals-length and space-equals-precision correspondences that calibrate the polynomial regime. Section 4 reads the same algebra with a discrete derivative and recovers implicit complexity classes from FAC^0 upward. Section 5 extends the picture above Turing, into the hyperarithmetical hierarchy via discontinuous solvable ODEs. Section 6 closes the polynomial regime through the four-way equivalence with neural networks. Section 7 traces the chemical incarnation, and Sect. 8 gathers open questions and ongoing developments.

2 GPAC and Polynomial ODEs

The General Purpose Analog Computer (GPAC) is the formal idealisation of the mechanical differential analysers built by Bush [27]. It was introduced by Shannon [73] as the class of real-valued functions generated by a finite circuit of adders, multipliers, integrators and constants. Despite its elementary syntax, the GPAC turns out to be the natural setting for the entire computability and complexity picture this tutorial draws. The reason is that GPAC-generable functions admit a clean modern characterisation as components of solutions of *polynomial initial value problems.*

2.1 From Mechanical Analysers to Polynomial IVPs

A polynomial initial value problem (pODE) is a system of the form

$$\mathbf{y}'(t) = \mathbf{p}(\mathbf{y}(t)), \qquad \mathbf{y}(0) = \mathbf{y}_0, \tag{1}$$

where $\mathbf{p}$ is a vector of polynomials and $\mathbf{y}_0$ a vector of real initial values. A function $f\colon I \to \mathbb{R}$ is *GPAC-generable* if it appears as a component of $\mathbf{y}$ in some such system on I. The equivalence between Shannon-style GPAC circuits and pODEs in the form (1) was established through a series of works in the early 2000s, notably by Graça and Costa [50] and Graça [52]; the two formulations are now used interchangeably. Polynomiality is essential: the celebrated result of Pour-El [70] on which functions a more permissive analyser generates required additional regularity assumptions, subsequently revisited by Lipshitz and Rubel [66].

The class of GPAC-generable functions, however, is not the right object to compare with computable analysis. It contains every solution of a polynomial IVP with computable initial data, including sin, exp, and the elementary functions of analysis, but it misses obviously computable functions such as the gamma function or Riemann's zeta. The mismatch is structural: GPAC-generable functions are produced *directly* as values of trajectories, whereas computable analysis produces a value $f(x)$ as the *limit* of an approximation process. The correct notion was introduced by Graça [52].

Definition 1 ([16,52]). *A function $f\colon I \to \mathbb{R}$ is* GPAC-computable *if there exist a pODE* (1) *whose initial data depends on the input $x \in I$ and two distinguished components y_1, y_2 of its solution such that, for every $x \in I$ and every $t \geq 0$,*

$$|y_1(t) - f(x)| \leq y_2(t), \qquad y_2(t) \to 0 \ \text{as}\ t \to \infty.$$

The shift from generable to computable, replacing direct production by limiting approximation, is exactly the move that aligns the analog model with the standard notion of computability over the reals.

2.2 GPAC-Computability Equals Computable Analysis

The first crisp result identifies one direction of the equivalence.

Theorem 1 ([16]). *Every GPAC-computable function on a compact interval is computable in the sense of computable analysis.*

The converse direction, completed by Bournez, Graça and Pouly [16], requires a careful treatment of unbounded domains and of uniformity in the input. Together the two directions yield the following equivalence, which is the foundational statement of the tutorial.

Theorem 2 ([16]). *A function $f\colon \mathbb{R} \to \mathbb{R}$ is GPAC-computable if and only if it is computable in the sense of computable analysis.*

The classical analog model of Shannon and Bush, once equipped with the right output convention, computes exactly what one normally wishes to call computable on the reals, no more and no less. Several refinements of this picture have appeared since.

Poças and Zucker [69] have related GPAC-generable functions to the tracking computability of metric algebras through the L-GPAC, an extension of Shannon's model that incorporates limit operations and applies beyond real-valued data streams. Collins [42] has given an expository treatment of computable analysis for dynamical systems and control, with a representation-theoretic foundation in the Weihrauch tradition. The handbook chapter [26] collects the modern view in a single place, and the recent course notes [49] provide a pedagogical entry point into the surrounding literature.

2.3 A Universal Ordinary Differential Equation

A result established by Bournez and Pouly, exhibits an explicit form of universality at the level of pODEs themselves.

Theorem 3 ([24,25]). *There exists a fixed polynomial ODE in finitely many variables such that, for every function $f\colon \mathbb{R} \to \mathbb{R}$ and every $\epsilon > 0$, an initial condition can be chosen so that the corresponding solution approximates f everywhere to within ϵ.*

A fixed list of polynomials, of fixed degree, thus generates every real function as the initial condition is varied. The result is the analog counterpart of Turing's universality theorem, but it is not a simulation argument: no universal machine is encoded inside the system. The pODE plays the role of a universal computational substrate by its dynamics alone.

2.4 Compilation, Quadratisation and Present-Day Uses

The pODE substrate has become a target language for compilation in its own right. Hemery, Fages and Soliman [55] have analysed the complexity of quadratising arbitrary polynomial ODE systems, that is, transforming them into equivalent systems of degree at most two, and Cai and Pogudin [30] have extended this to dissipative quadratisations that preserve energy-decay properties of the original system. Huang and Migunov [57] have built an automatic compiler from GPAC descriptions to population protocols, suggesting that the GPAC is becoming a useful intermediate representation across substrates.

3 Complexity in Continuous Time

The equivalence of Theorem 2 settles the computability question: GPAC-generable functions are exactly the computable real functions. The complexity question is more delicate. Time on a Turing machine is a discrete primitive; on a pODE the independent variable t has no operational meaning, since rescaling $t \mapsto \lambda t$ trivially shrinks any computation. The first task of a complexity theory of analog computation is therefore to identify the right resources. Two of them, the *length of the solution curve* and the *numerical precision required to track it*, turn out to recover Turing time and space exactly.

3.1 Time Equals Length

The natural measure of computational time on a pODE is not the parameter t but the arc length of the solution trajectory. An input precision 2^{-p} is reached when the projection of the solution lies within 2^{-p} of the target output, and the cost is the length of the curve travelled to get there.

Theorem 4 ([22]). *A function $f\colon \mathbb{R} \to \mathbb{R}$ is computable in polynomial time on a Turing machine if and only if it is generable by a polynomial IVP whose solution curve, on input x and precision 2^{-p}, has length polynomial in $|x|$ and p.*

The Turing-to-pODE direction transcribes any polynomial-time computation as a polynomial vector field whose trajectory tracks the tape contents in real time, with length consumption matching time. The pODE-to-Turing direction adapts a numerical solver to maintain precision proportional to the curve length. Together, they identify FPTIME on the reals as an intrinsic property of the dynamics rather than a feature of any particular discrete model.

3.2 Space Equals Precision

The space resource on a Turing machine measures the number of bits held simultaneously. Its analog counterpart turns out to be the *precision* required to track the solution: how many digits of $\mathbf{y}(t)$ must be maintained at any instant for the computation to remain correct. Without further constraints, however, a pODE may exhibit pathological sensitivity to perturbations of its field or initial data, breaking any naive correspondence with a discrete space resource. The right setting is therefore that of *robust* pODEs, which behave continuously with respect to small perturbations.

Theorem 5 ([12,13]). *A function $f: \mathbb{R} \to \mathbb{R}$ is computable in polynomial space on a Turing machine if and only if it is generable by a robust polynomial IVP that can be tracked, on input x and precision 2^{-p}, with a working precision polynomial in $|x|$ and p.*

The robustness assumption is essential: it rules out pathological fields whose dynamics depend non-polynomially on perturbations. Once it is in place, the polynomial-resource picture is closed: time is length, space is precision, and the two analog resources behave with respect to each other exactly as Turing time behaves with respect to Turing space. A complementary continuous characterisation of PSPACE via pODEs of polynomial length and bounded state-space, due to Bournez, Gozzi, Graça and Pouly [21], fills out the picture from a different angle.

3.3 The Wider Ecosystem

The two characterisations above have triggered several extensions and refinements. Graça and Zhong [51] have identified the minimum dimension at which analytic dynamical systems robustly simulate Turing machines: one for analytic maps and two for analytic ODEs, with a two-dimensional C^∞ realisation on the sphere S^2 as a corollary. Biswas [10] has established that real-function computability by oracle pointer machines implies a real-time simulation property, recovering a Turing-style characterisation in a non-Turing model of computation. From a more deductive angle, Platzer and Qian [68] have proved completeness of an axiomatisation, in differential dynamic logic, for open safety, open liveness and existence properties of compact polynomial IVPs, unifying symbolic deduction and rigorous numerical analysis on the same dynamical objects.

The picture has been enriched on adjacent fronts. Selivanova, Steinberg, Thies and Ziegler [72] have given exact-real-computation procedures, with bit-complexity guarantees, for solution operators of linear analytic systems of PDEs, extending the resource-conscious approach of computable analysis beyond ODEs. Bychkov, Issan, Pogudin and Krämer [28] have developed an exact and optimal quadratisation algorithm that transforms arbitrary smooth finite-dimensional ODEs into equivalent systems of degree at most two; this provides a practical preprocessing step that brings general nonlinear dynamics into the polynomial form on which the theorems above are stated.

4 Discrete ODEs and Implicit Complexity

The polynomial-time and polynomial-space classes of classical complexity admit several machine-independent characterisations from the implicit complexity tradition: Cobham's algebra of bounded recursion on notation [41], Bellantoni and

Cook's safe recursion [8], and the various ramified or tiered schemes that followed [64,65]. Each of these uses a recursion schema as the core primitive and adjusts the closure conditions to recover a target complexity class; see [40] for a panorama. The thread of the present section is that *taking a discrete ODE as the recursion primitive* produces an algebra that is equally natural and slightly different in flavour, and that this single algebra captures classes ranging from FAC^0 to $\mathsf{FPSPACE}$ once the structural conditions are tuned.

4.1 The Discrete Derivative and Length-ODE Recursion

The key primitive is the discrete derivative $\delta f(x) = f(x+1) - f(x)$, defined on functions of an integer or dyadic argument. A *length-ODE* is a recurrence of the form

$$\delta_\ell\, y(x,\ell) = h(y(x,\ell),\, x,\, \ell), \qquad y(x,0) = g(x),$$

where δ_ℓ denotes the discrete derivative with respect to the iteration variable ℓ and h belongs to some background algebra. The terminology *length* reflects that the output is read at $\ell = \text{poly}(|x|)$, with the iteration count thus scaling polynomially with the bit-length of the input. A length-ODE schema is in this sense the discrete counterpart of a continuous ODE driven for a time polynomial in $|x|$: the two formulations characterise the same class FPTIME, continuously by the polynomial length of the solution curve (Sect. 3) and discretely by closure under the length-ODE schema (Sect. 4.2). The discrete and continuous frameworks are therefore not parallel theories but two presentations of one and the same recursion mechanism.

4.2 Polynomial Time and Polynomial Space

The first sharp characterisation along this line is the following.

Theorem 6 ([18]). *The class of functions $\mathbb{N}^k \to \mathbb{N}$ in FPTIME coincides with the algebra generated from constants $0, 1$, projections, the binary-length function ℓ, basic arithmetic and the sign function under composition and the linear length-ODE schema.*

FPTIME already admits many characterisations; the contribution here is the primitive used to obtain it. The schema is structurally a *linear* length-ODE, with the right-hand side linear. This single algebraic constraint replaces the explicit bounded-recursion conditions of Cobham, the safe/normal separation of Bellantoni and Cook, and the ramification or tiering of Leivant: no syntactic stratification is needed, only linearity in the iterated variable. The same line of thought extends to functions over the reals: the conference paper [11] characterises both FPTIME and $\mathsf{FPSPACE}$ over $\mathbb{R}$ by adding to the discrete-ODE algebra an effective limit operator and allowing analytic discrete ODEs in the schema. The space characterisation is the first one of $\mathsf{FPSPACE}$ via discrete ODEs and avoids the explicit bounding of classical Cobham-style schemes by relying instead on numerically stable schemes, which are natural in the ODE context.

The bridge with Sect. 3 is then complete: FPSPACE admits two characterisations through the same dynamical primitive, continuously through the length-precision correspondence of [12] and discretely through the analytic discrete-ODE algebra above.

4.3 Refining the Picture Toward Parallel Classes

A parallel programme by Antonelli, Durand and Kontinen has pushed the discrete-ODE language down into small parallel circuit classes. The starting point [5] characterises FAC^0 and FTC^0 through shifting schemas that combine derivation along the logarithmic length function ℓ with a refined notion of polynomial degree on the right-hand side. Derivation along ℓ keeps the number of induction steps logarithmic, which is the parallel-complexity counterpart of the length-ODE linearity used in [18]. The follow-up [6] systematises the picture across the chain

$$\mathsf{FAC}^0 \subseteq \mathsf{FACC}[2] \subseteq \mathsf{FTC}^0 \subseteq \mathsf{FNC}^1 \subseteq \mathsf{FAC}^1 \subseteq \mathsf{FPTIME} \subseteq \mathsf{FPSPACE},$$

providing in particular the first ODE-based characterisations of FACC[2] and FNC^1. A striking observation of the same paper is that the linear length-ODE schema $\partial f/\partial\lambda = A \cdot f + B$ characterises FPTIME when $\lambda = \ell$ and FNC^1 when $\lambda = \ell \circ \ell$, suggesting a way to read parallelism as a slowdown of the underlying iteration. The whole chain is thus captured within a single algebraic language, with each class distinguished by the shape of the recurrence schema, the polynomial degree admitted, and the function controlling the speed of iteration. The polynomial-time level of this chain is exactly the level at which Sect. 6

5 Above the Turing Barrier: Discontinuous ODEs and the Hyperarithmetical Hierarchy

The continuous-time picture developed in Sects. 2–3 developed the continuous-time picture under polynomial regularity. The dynamics was real-analytic and the resulting classes were classical Turing ones. Relaxing that regularity extends the picture upward in a controlled way: a sharply identified class of *discontinuous* IVPs with unique solutions corresponds exactly to computation over the hyperarithmetical hierarchy. The same ODE language, with regularity as the knob, traces out a single hierarchy from FPTIME at the bottom up to Π^1_1 at the top.

5.1 Why Discontinuity, Why Uniqueness

Above Turing, computable right-hand sides are not enough. Sections 2–3 sit entirely inside computable analysis: polynomial right-hand sides generate analytic dynamics, whose solutions are computable by Theorem 2. The same conclusion holds, more generally, for any computable right-hand side under suitable regularity. Reaching beyond this level requires admitting fields that are themselves not computable.

A natural way to do so is to relax continuity, since a discontinuous right-hand side is no longer constrained to lie inside computable analysis. But unrestricted discontinuity is too much: a discontinuous IVP typically loses uniqueness of solutions, and a non-unique solution cannot count as the output of a programmable device. The interesting regime sits between: the discontinuous right-hand sides for which the IVP keeps a unique solution. The task is to identify a structural condition that admits genuine discontinuity while still allowing an analytic construction of the unique solution.

5.2 Solvable Right-Hand Sides

A structural condition that captures the regime is the following.

Definition 2 ([19,20]). *Let $E \subseteq \mathbb{R}^r$ be a compact domain and $f\colon E \to \mathbb{R}^m$ a function. The function f is* solvable *if it is of Baire class one and, for every closed $K \subseteq E$, the set of discontinuity points of the restriction $f|_K$ is closed in K.*

Iterating the discontinuity operator on f produces a transfinite decreasing sequence $E = E_0 \supseteq E_1 \supseteq \cdots$ of f-removed sets, where each $E_{\alpha+1}$ is the discontinuity set of f restricted to E_α. The solvability condition is exactly the structural assumption that makes this sequence reach the empty set at some countable ordinal $\alpha < \omega_1$ [20].

The condition admits two compatible readings. From the descriptive set theory side, it places f at a low level of the Baire hierarchy with a regularity-of-singularities constraint. From the dynamical side, it is precisely what allows the unique solution of the IVP to be reconstructed analytically by transfinite recursion in countably many steps, through a transfinite generalisation of the ten-thousand-monkeys algorithm of [43]. The method is inspired by Denjoy's totalisation procedure for recovering a function from its derivative beyond the reach of Lebesgue integration. Concrete examples include cases whose solution at an integer time encodes the halting set of Turing machines, with right-hand sides discontinuous on a single line [20].

5.3 Reaching the Hyperarithmetical Hierarchy

The main result of [20] is a constructive solution method for solvable IVPs.

Theorem 7 ([20]). *The unique solution of a solvable IVP can be obtained analytically by transfinite recursion of countable length, through a transfinite generalisation of the Collins-Graça algorithm.*

This constructive guarantee is the foundation of a higher-recursion reading of solvable systems. A non-trivial example exhibited in the same paper is a solvable IVP, with right-hand side discontinuous on a single line and computable initial data, whose solution at an integer time encodes the halting set of Turing machines. By a rescaling-and-concatenation construction analogous to the one used to climb the levels E_α of the discontinuity hierarchy, this single example serves as a building block for encoding any Turing jump along a recursive ordinal, and conjecturally all of the hyperarithmetical reals. The full correspondence between solvable IVPs and the hyperarithmetical hierarchy is the natural analog counterpart of the polynomial-IVP/computable-analysis correspondence of Sect. 2, with regularity replaced by solvability and Turing computation replaced by transfinite recursion.

A finer descriptive complexity reading, based on a coanalytic rank on solvable functions, is developed in [48] and will be discussed in Sect. 8.

6 Recurrent Neural Networks as Dynamical Systems

A feedforward neural network with L layers $\mathbf{z}_{i+1} = \rho(W_i\mathbf{z}_i + \mathbf{b}_i)$ defines a discrete-time dynamical system on the state vector $\mathbf{z}_i \in \mathbb{R}^w$. When w is held fixed and L grows, the network maintains a constant-size real state updated layer by layer, exactly the pattern of a register machine: depth plays the role of time and width the role of space. This *deep-narrow* reading is developed in detail in [15]. The continuous-depth limit is the Neural ODE of [37]: the bridge with the residual update $\mathbf{z}_{i+1} = \mathbf{z}_i + g_i(\mathbf{z}_i)$, identified independently in [82], makes the transition a forward Euler step [60].

6.1 The Siegelmann-Sontag Lineage and Uniformity

Two questions guide the modern reading. The first goes back to Siegelmann and Sontag, who established that recurrent networks with sigmoidal activations and rational weights are Turing-equivalent [77], and that real weights push the model up to P/poly [76,78]. These results are existence statements: for each computable function or each language, a network exists. The classical universal approximation theorems [44] are likewise density statements with no constructive content. Both leave open the *uniformity* question: does there exist a single fixed architecture that, given a description of the target function and a desired precision, performs the corresponding computation by simply iterating? The second question is the cost: how does the iteration depth required to reach precision ϵ depend on the complexity of the target?

6.2 A Universal Uniform Approximation Theorem

Both questions are answered in [17]. There exists a fixed RELU network of constant width with the following property: given oracle access to a code $\langle f \rangle$ for any computable function $f\colon [0,1]^d \to [0,1]^d$ and to a precision $\epsilon = 2^{-p}$, there is an iteration count $T \in \mathbb{N}$ such that the T-fold self-composition of the network approximates f uniformly on $[0,1]^d$ to within ϵ. The construction is uniform in f and ϵ, lives entirely inside computable analysis, and the iteration count T is polynomial in p and in any running-time bound for f. The technical heart of the construction is a RELU-net implementation of 1,3-quadric encodings that bypass the discontinuity of binary digit extraction. A consequence is that fixed-width RELU networks form a clean substrate for register-machine simulation.

6.3 A Four-Way Equivalence

Combining [17] with the GPAC results of Sect. 3 closes the unifying picture proposed in this tutorial. For computable $f\colon [0,1]^d \to [0,1]^d$, the following resources are polynomially related: the running time of a Turing machine computing a 2^{-p}-approximation; the iteration depth T of the universal deep-narrow network of [17]; the length of the solution of a polynomial ODE in the sense of [22]; and the size of a Boolean circuit family for the same function [81]. Each of the first three has a universal device, a universal Turing machine, a universal ODE, a universal network, that simulates the others at polynomial cost. Via [2,3], all of these are moreover polynomially related to suitable variants of Kolmogorov complexity (KT unrelativized, C relative to the halting set). The four substrates are thus genuinely unified at this level: the same complexity class admits four equally fundamental descriptions.

6.4 Recent Ecosystem

A recent connected line of work studies primitive recursion as a composition of dynamical systems [14], providing a continuous-dynamics counterpart to classical recursion-theoretic hierarchies that complements the discrete-ODE picture of Sect. 4. Cabessa and Strozecki [29] characterise the super-Turing power of analog, evolving and stochastic recurrent networks via the Kolmogorov complexity of their real weights, producing an infinite hierarchy of complexity classes located between P and P/poly and extending the older Siegelmann-Sontag analog hierarchy into a quantitative theory. Fan, Zhang, Gao and Dochain [46] push the deep-narrow viewpoint into the chemical substrate, designing a programmable biochemical reaction network that realises a fully connected neural network with both forward propagation and backpropagation, with explicit convergence and error bounds. This connects directly to Sect. 7 and illustrates how the same dynamical primitives traverse the neural and the molecular charts. Sidl et al. [75] take the opposite vantage and read biological computation through computational complexity, arguing that evolution confines living systems to subsets of instances solvable by efficient algorithms hardcoded in the system, with RNA secondary structure prediction as a tractable surrogate. Two questions stand out and motivate Sect. 8: how hard is it to verify properties of fixed-width RELU networks, and what is the precise complexity content of the analog hierarchy under fixed computable precision.

7 Chemical and Molecular Computation

The pODE machinery has a natural physical incarnation: the mass-action dynamics of a chemical reaction network (CRN). A program is a finite list of reactions over a finite set of molecular species, and the differential semantics produces a polynomial ODE on the positive orthant. The connection with the rest of this tutorial is mechanical: under suitable encodings, results about pODE computability and complexity transfer to mass-action chemistry, and CRN constructions read back as concrete pODE programs.

7.1 Turing-Completeness and Compilation

The first sharp result is due to Fages, Le Guludec, Bournez and Pouly [45]: continuous mass-action CRN over a finite set of molecular species are *Turing-complete in the strong sense of uniform computability.* The proof is constructive and yields a compiler from mathematical functions to elementary CRN. A subsequent line by Hemery, Fages and Soliman has refined this compiler into a programmatic pipeline: a polynomialization step turns systems of elementary differential equations into pODE [56], and a quadratization step then restricts the resulting pODE to elementary reactions with at most two reactants. The same group characterises the functions that an online mass-action CRN can stabilise as exactly the real algebraic functions [53], and gives an explicit CRN module for the analog operation of differentiation [54].

7.2 Variations and the Wider Ecosystem

Several distinct programmes refine and reorganise the picture. Chen, Doty, Reeves and Soloveichik [36] characterise the functions computed by continuous CRN *independently of the specific rate constants*: under the dual-rail representation these are exactly the continuous piecewise linear functions with rational coefficients, and a separate but related characterisation holds for the direct representation. Calabrese and Doty [31] have since shown that allowing inhibitory reactions in this rate-independent setting recovers full Turing-universality, separating cleanly the expressive role of inhibition from that of fine-tuned kinetics. The surface CRN model, in which molecules are placed on a lattice and react only with their immediate neighbours, was introduced by Qian and Winfree [71]. Clamons, Qian and Winfree [39] explore this model in depth, showing that surface CRNs can emulate deterministic cellular automata, implement continuously active Boolean circuits, generate spatial patterns from simple initial conditions, and realise swarm-robotic behaviours through purely local rules. More recent contributions to the universality side include a minimal ten-species, ten-reaction surface CRN, restricted to bimolecular reactions, that achieves universality through an embedding of delay-insensitive circuits [83], and a systematic comparison of surface CRN with other models of distributed computation [63].

A parallel programming-languages effort designs surface languages that compile down to CRN: CRN++ [80] is an early imperative entry, while *Reactamole* [61], by Klinge, Lathrop, Osera and Rogers, is a recent functional reactive variant published in *Natural Computing.* On the verification side, Fletcher, Klinge, Lathrop and collaborators have studied robust real-time computing in deterministic

CRN [47]. Throughout, the discrete-ODE and pODE language of earlier sections can be seen as a connecting tissue.

7.3 The Neural-Chemical Bridge and 2026 Perspectives

The compilation of neural networks into chemical reaction networks makes the connection between Sects. 6 and 7 concrete. Vasić, Chalk, Khurshid and Soloveichik [79] established a tight correspondence between binary-weight ReLU networks (BinaryConnect, trained in silico) and rate-independent CRN, yielding a compilation procedure from such networks to robust chemical implementations. Fan, Zhang, Gao and Dochain [46] take the question further on the chemical side, designing a programmable bimolecular CRN that realises both the forward propagation and the backpropagation of a fully connected neural network, with explicit convergence guarantees and error bounds.

The pace of recent work suggests that the chemical substrate is becoming an active site for computation research. A *Nature Reviews Chemistry* comment by Ji, Chen, Yu and Nijhuis [58] surveys non-steady-state chemical dynamics as a substrate for neuromorphic and reservoir computing. The first volume of *Unconventional Computing* contains a programmatic study by Johnson, Bohm Agostini, Cannon and collaborators [59] on energy-efficient scientific computing using chemical reservoirs, motivated by the post-exascale energy challenge. On the control side, Shi and Gao [74] use biochemical relaxation oscillators to generate clock signals that sequence reaction modules in molecular computation. Compilation infrastructure is catching up: the ChemComp framework [1] provides an end-to-end pipeline for compiling programs into CRN. Beyond chemistry, the *physical-oracle* programme of Beggs, Costa and Tucker [4,7] reads computation as the interaction between a Turing machine and a continuous physical experiment, providing a complementary external-oracle perspective on the internal substrate view of this section.

8 Perspectives

The unifying picture drawn in this tutorial is technically closed at one specific point: the polynomial-time level, where the four-way equivalence between Turing time, ODE solution length, fixed-width network depth and Boolean circuit size, established in Sect. 6, identifies the same complexity class on four substrates. Around this anchor, several directions are open or in active development.

Refining the Polynomial Regime

Within the polynomial-time level, the standard finer hierarchies remain to be calibrated: the polynomial hierarchy and EXPTIME do not yet have clean pODE characterisations, and finding a knob analogous to length or precision that produces them is a natural question. Verification of fixed-width RELU networks appears to be an early site of hardness in this fine-grained landscape; ongoing work on the complexity of reachability and equivalence for networks of width as small as four suggests that classical worst-case lower bounds appear surprisingly low in the deep-narrow regime, and that the boundary between tractability and hardness may sit at depth-four or three rather than at the larger widths suggested by classical analyses. Quantitative robustness, in the sense studied in [13], is another knob whose interaction with the basic resources is not yet understood.

Effective Content of the Transfinite

Above Turing, Theorem 7 identifies the class of solvable IVPs with the hyperarithmetical hierarchy, but the descriptive set theory of solvable functions inside this class is only beginning to be mapped. A coanalytic-rank refinement is being developed in [48] that should give a graded structure on the class, with each rank corresponding to a level in a ranked transfinite computation model. Connections with classical results of Kechris and Woodin and with the rank theories of Denjoy and Zalcwasser are being investigated, and an extended journal treatment of [20] that develops these threads is in preparation [19]. Independently, the discrete-ODE side of the picture admits a primitive-recursion reading through three dynamical perspectives that an arXiv preprint formalises [14], and the *reverse mathematics* stratification of solvable IVPs has just been worked out in the ISSAC 2026 contribution [23], which calibrates the strength of the correspondence in subsystems of second-order arithmetic.

Lateral and Physical Extensions

Several substrates extend the picture without aiming above Turing. On the quantum side, recent uses of pODE-style complexity inside quantum simulation [38] suggest that the same unification carries across to analogue quantum devices, where "analogue" refers to a physical system targeted at reproducing the dynamics of a quantum model, not to the analog/digital distinction of this tutorial. Biologically motivated lines, from chemical-reservoir computing to the computational reading of evolution [75], push the picture into life sciences. The emerging analog hardware programme, surveyed in [9], makes the engineering side a serious counterpart to the theoretical map drawn here.

A separate strand realises Turing-complete dynamics inside steady solutions of the Euler equations of incompressible fluid mechanics, using techniques from contact topology and Beltrami flows. Cardona, Miranda, Peralta-Salas and Presas [32,35] construct a stationary Euler flow on the Riemannian three-sphere whose particle paths simulate a universal Turing machine. The follow-up [33] produces Beltrami fields in Euclidean space and on the flat three-torus that simulate universal and tape-bounded Turing machines respectively, with a companion programme towards a physically realisable fluid computer [34]. A recent review by Perales-Eceiza, Cubitt, Gu, Pérez-García and Wolf [67] surveys the broader landscape of undecidability results in physics, of which the Euler-flow constructions are one strand.

The Unifying Picture, Revisited

The opening of this tutorial proposed reading polynomial ODEs as a universal language for computability and complexity. The bulk of the tutorial has unpacked four readings of this language: the classical pODE substrate of Sects. 2 and 3, the discrete-ODE substrate of Sect. 4, the deep-narrow neural substrate of Sect. 6, and the chemical substrate of Sect. 7, plus the upward extension of Sect. 5. The polynomial-time four-way equivalence is the technical demonstration that these are not just analogies but genuine identifications of the same complexity class. Whether comparable identifications hold at other levels of the hierarchy, from FAC^0 to $\mathsf{FPSPACE}$ and beyond into the hyperarithmetical, is the natural next question.

References

1. Agostini, N.B., Johnson, C., Cannon, W.R., Tumeo, A.: Chemcomp: a compilation framework for computing with chemical reaction networks (2025). https://doi.org/10.1145/3658617.3703315
2. Allender, E.: Circuit complexity, kolmogorov complexity, and prospects for lower bounds. In: Câmpeanu, C., Pighizzini, G. (eds.) 10th International Workshop on Descriptional Complexity of Formal Systems, DCFS 2008, pp. 7–13. University of Prince Edward Island, Citeseer (2008)
3. Allender, E., Buhrman, H., Koucký, M., van Melkebeek, D., Ronneburger, D.: Power from random strings. SIAM J. Comput. **35**(6), 1467–1493 (2006). https://doi.org/10.1137/050628994
4. Ambaram, T., Beggs, E., Costa, J.F., Poças, D., Tucker, J.V.: An analogue-digital model of computation: turing machines with physical oracles. In: Advances in Unconventional Computing, pp. 73–115. Springer (2017)
5. Antonelli, M., Durand, A., Kontinen, J.: A new characterization of FAC^0 via discrete ordinary differential equations. In: Královic, R., Kucera, A. (eds.) 49th International Symposium on Mathematical Foundations of Computer Science, MFCS 2024, Bratislava, Slovakia, 26–30 August 2024. LIPIcs, vol. 306, pp. 10:1–10:18. Schloss Dagstuhl - Leibniz-Zentrum für Informatik (2024). https://doi.org/10.4230/LIPICS.MFCS.2024.10
6. Antonelli, M., Durand, A., Kontinen, J.: Characterizing small circuit classes from FAC^0 to FAC^1 via discrete ordinary differential equations. In: Gawrychowski, P., Mazowiecki, F., Skrzypczak, M. (eds.) 50th International Symposium on Mathematical Foundations of Computer Science, MFCS 2025. LIPIcs, vol. 345, pp. 10:1–10:18 (2025). https://doi.org/10.4230/LIPICS.MFCS.2025.10
7. Beggs, E., Costa, J.F., Loff, B., Tucker, J.V.: Computational complexity with experiments as oracles. Proc. Roy. Soc. London-A **464**(2098), 2777 (2008). https://doi.org/10.1098/rspa.2008.0085
8. Bellantoni, S., Cook, S.: A new recursion-theoretic characterization of the polytime functions. Comput. Complex. **2**(2), 97–110 (1992). https://doi.org/10.1007/bf01201998
9. Belostotski, L., Uddin, A., Madanayake, A., Mandal, S.: A survey of analog computing for domain-specific accelerators. Electronics (2025). https://doi.org/10.3390/electronics14163159

10. Biswas, S.: Computability of real functions with oracle pointer machines implies real-time simulation of chemical reaction networks. Lecture Notes in Computer Science (2025). https://doi.org/10.1007/978-3-031-95908-0_13
11. Blanc, M., Bournez, O.: A characterisation of functions computable in polynomial time and space over the reals with discrete ordinary differential equations: simulation of turing machines with analytic discrete ODEs. In: Leroux, J., Lombardy, S., Peleg, D. (eds.) 48th International Symposium on Mathematical Foundations of Computer Science, MFCS 2023. LIPIcs, vol. 272, pp. 21:1–21:15 (2023). https://doi.org/10.4230/LIPIcs.MFCS.2023.21
12. Blanc, M., Bournez, O.: The complexity of computing in continuous time: space complexity is precision. In: Bringmann, K., Grohe, M., Puppis, G., Svensson, O. (eds.) 51st International Colloquium on Automata, Languages, and Programming, ICALP 2024. LIPIcs, vol. 297, pp. 129:1–129:22 (2024). https://doi.org/10.4230/LIPIcs.ICALP.2024.129
13. Blanc, M., Bournez, O.: Quantifiying the robustness of dynamical systems. Relating time and space to length and precision. In: Murano, A., Silva, A. (eds.) 32nd EACSL Annual Conference on Computer Science Logic, CSL 2024. LIPIcs, vol. 288, pp. 17:1–17:20. Dagstuhl, Germany (2024). https://doi.org/10.4230/LIPIcs.CSL.2024.17
14. Bournez, O.: Primitive recursion without composition: dynamical characterizations, from neural networks to polynomial odes. Technical report (2026). https://arxiv.org/abs/2604.24356
15. Bournez, O.: Uniform Computation and Complexity in Deep–Narrow Neural Networks (2026)
16. Bournez, O., Campagnolo, M.L., Graça, D.S., Hainry, E.: Polynomial differential equations compute all real computable functions on computable compact intervals. J. Complex. **23**(3), 317–335 (2007). https://doi.org/10.1016/j.jco.2006.12.005
17. Bournez, O., Cohen, J., Wurm, A.: A universal uniform approximation theorem for neural networks. In: Gawrychowski, P., Mazowiecki, F., Skrzypczak, M. (eds.) 50th International Symposium on Mathematical Foundations of Computer Science, MFCS 2025. LIPIcs, vol. 345, pp. 29:1–29:20. Dagstuhl, Germany (2025). https://doi.org/10.4230/LIPIcs.MFCS.2025.29
18. Bournez, O., Durand, A.: A characterization of functions over the integers computable in polynomial time using discrete ordinary differential equations. Comput. Complex. **32**(2), 7 (2023). https://doi.org/10.1007/s00037-023-00240-1
19. Bournez, O., Gozzi, R.: Solvable initial value problems ruled by discontinuous ordinary differential equations. Technical report (2024). https://doi.org/10.48550/arXiv.2405.00165
20. Bournez, O., Gozzi, R.: Solving discontinuous initial value problems with unique solutions is equivalent to computing over the transfinite. In: Symposium on Theoretical Aspects of Computer Science (STACS), Clermont-Ferrand, France (2024). https://doi.org/10.4230/LIPIcs.STACS.2024.20
21. Bournez, O., Gozzi, R., Graça, D.S., Pouly, A.: A continuous characterization of PSPACE using polynomial ordinary differential equations. J. Complex. **77**, 101755 (2023). https://doi.org/10.1016/j.jco.2023.101755
22. Bournez, O., Graça, D.S., Pouly, A.: Polynomial time corresponds to solutions of polynomial ordinary differential equations of polynomial length. J. ACM **64**(6), 38:1–38:76 (2017). https://doi.org/10.1145/3127496
23. Bournez, O., Núñez, A.: Relating the computational and logical difficulty of solving odes: from polynomial to discontinuous right-hand sides. In: Proceedings of

International Symposium on Symbolic and Algebraic Computation (ISSAC 2026) (2026)

24. Bournez, O., Pouly, A.: A universal ordinary differential equation. In: Chatzigiannakis, I., Indyk, P., Kuhn, F., Muscholl, A. (eds.) 44th International Colloquium on Automata, Languages, and Programming, ICALP 2017. LIPIcs, vol. 80, pp. 116:1–116:14 (2017). https://doi.org/10.4230/LIPIcs.ICALP.2017.116
25. Bournez, O., Pouly, A.: A universal ordinary differential equation. Log. Methods Comput. Sci. **16**(1) (2020). https://doi.org/10.23638/LMCS-16(1:28)2020
26. Bournez, O., Pouly, A.: A survey on analog models of computation. In: Brattka, V., Hertling, P. (eds.) Handbook of Computability and Complexity in Analysis. Springer (2021). https://doi.org/10.1007/978-3-030-59234-9_6
27. Bush, V.: The differential analyser. J. Franklin Inst. **212**(4), 447–488 (1931)
28. Bychkov, A., Issan, O., Pogudin, G., Krämer, B.: Exact and optimal quadratization of nonlinear finite-dimensional nonautonomous dynamical systems. SIAM J. Appl. Dyn. Syst. (2024). https://doi.org/10.1137/23m1561129
29. Cabessa, J., Strozecki, Y.: Refined kolmogorov complexity of analog, evolving and stochastic recurrent neural networks. Inf. Sci. (2025). https://doi.org/10.1016/j.ins.2025.122104
30. Cai, Y., Pogudin, G.: Dissipative quadratizations of polynomial ode systems. Lecture Notes in Computer Science (2024). https://doi.org/10.1007/978-3-031-57249-4_16
31. Calabrese, K., Doty, D.: Rate-independent continuous inhibitory chemical reaction networks are turing-universal. Lecture Notes in Computer Science (2024). https://doi.org/10.1007/978-3-031-63742-1_8
32. Cardona, R., Miranda, E., Peralta-Salas, D.: Turing universality of the incompressible euler equations and a conjecture of moore. Int. Math. Res. Not. (2021). https://doi.org/10.1093/imrn/rnab233
33. Cardona, R., Miranda, E., Peralta-Salas, D.: Computability and beltrami fields in euclidean space. Journal de Mathématiques Pures et Appliquées (2022). https://doi.org/10.1016/j.matpur.2022.11.007
34. Cardona, R., Miranda, E., Peralta-Salas, D.: Towards a fluid computer. Found. Comput. Math. (2025). https://doi.org/10.1007/s10208-025-09699-6
35. Cardona, R., Miranda, E., Peralta-Salas, D., Presas, F.: Constructing turing complete euler flows in dimension 3. Proc. Natl. Acad. Sci. (2021). https://doi.org/10.1073/pnas.2026818118
36. Chen, H.L., Doty, D., Reeves, W., Soloveichik, D.: Rate-independent computation in continuous chemical reaction networks. J. ACM (2023). https://doi.org/10.1145/3590776
37. Chen, T.Q., Rubanova, Y., Bettencourt, J., Duvenaud, D.: Neural ordinary differential equations. In: Bengio, S., Wallach, H.M., Larochelle, H., Grauman, K., Cesa-Bianchi, N., Garnett, R. (eds.) Advances in Neural Information Processing Systems 31: Annual Conference on Neural Information Processing Systems 2018, NeurIPS 2018. pp. 6572–6583 (2018). https://proceedings.neurips.cc/paper/2018/hash/69386f6bb1dfed68692a24c8686939b9-Abstract.html
38. Chu, M.T.: Preparing hamiltonians for quantum simulation: a computational framework for cartan decomposition via lax dynamics. Math. Comput. (2024). https://doi.org/10.1090/mcom/4056
39. Clamons, S., Qian, L., Winfree, E.: Programming and simulating chemical reaction networks on a surface. J. R. Soc. Interface (2020). https://doi.org/10.1098/rsif.2019.0790

40. Clote, P., Kranakis, E.: Boolean Functions and Computation Models. Texts in Theoretical Computer Science. An EATCS Series. Springer (2002). https://doi.org/10.1007/978-3-662-04943-3
41. Cobham, A.: The intrinsic computational difficulty of functions. In: Bar-Hillel, Y. (ed.) Proceedings of the International Conference on Logic, Methodology, and Philosophy of Science, pp. 24–30. North-Holland, Amsterdam (1962)
42. Collins, P.: Computable analysis with applications to dynamic systems. Math. Struct. Comput. Sci. (2020). https://doi.org/10.1017/s096012952000002x
43. Collins, P., Graça, D.S.: Effective computability of solutions of differential inclusions the ten thousand monkeys approach. J. Univ. Comput. Sci. **15**(6), 1162–1185 (2009). https://doi.org/10.3217/jucs-015-06-1162
44. Cybenko, G.: Approximation by superpositions of a sigmoidal function. Math. Control. Signals Syst. **2**(4), 303–314 (1989). https://doi.org/10.1007/BF02551274
45. Fages, F., Le Guludec, G., Bournez, O., Pouly, A.: Strong turing completeness of continuous chemical reaction networks and compilation of mixed analog-digital programs. In: Feret, J., Koeppl, H. (eds.) CMSB 2017. LNCS, vol. 10545, pp. 108–127. Springer, Cham (2017). https://doi.org/10.1007/978-3-319-67471-1_7
46. Fan, Y., Zhang, X., Gao, C., Dochain, D.: Automatic implementation of neural networks through reaction networks–part i: circuit design and convergence analysis. IEEE Trans. Autom. Control (2025). https://doi.org/10.1109/tac.2025.3554428
47. Fletcher, W., Klinge, T.H., Lathrop, J.I., Nye, D.A., Rayman, M.: Real-time computing and robust memory with deterministic chemical reaction networks. Nat. Comput. (2024). https://doi.org/10.1007/s11047-024-09994-1
48. Gozzi, R., Bournez, O.: Set descriptive complexity of solvable functions. Technical report (2024). https://doi.org/10.48550/arXiv.2405.19304
49. Graça, D.S.: Computation with real numbers and continuous-time dynamical systems. In: Crossroads of Computability and Logic: Insights, Inspirations, and Innovations: 21st Conference on Computability in Europe, CiE 2025 (2025). https://doi.org/10.1007/978-3-031-95908-0_5
50. Graça, D.S., Costa, J.F.: Analog computers and recursive functions over the reals. J. Complex. **19**(5), 644–664 (2003). https://doi.org/10.1016/s0885-064x(03)00034-7
51. Graça, D.S., Zhong, N.: Analytic one-dimensional maps and two-dimensional ordinary differential equations can robustly simulate turing machines. Computability (2022). https://doi.org/10.3233/com-210381
52. Graça, D.S.: Some recent developments on shannon's general purpose analog computer. Math. Logic Q. **50**(4-5), 473–485 (2004). https://doi.org/10.1002/malq.200310113
53. Hemery, M., Fages, F.: Algebraic biochemistry: A framework for analog online computation in cells. Lecture Notes in Computer Science (2022). https://doi.org/10.1007/978-3-031-15034-0_1
54. Hemery, M., Fages, F.: On estimating derivatives of input signals in biochemistry. Lecture Notes in Computer Science (2023). https://doi.org/10.1007/978-3-031-42697-1_6
55. Hemery, M., Fages, F., Soliman, S.: On the complexity of quadratization for polynomial differential equations. In: Abate, A., Petrov, T., Wolf, V. (eds.) CMSB 2020. LNCS, vol. 12314, pp. 120–140. Springer, Cham (2020). https://doi.org/10.1007/978-3-030-60327-4_7
56. Hemery, M., Fages, F., Soliman, S.: Compiling elementary mathematical functions into finite chemical reaction networks via a polynomialization algorithm for ODEs.

In: Cinquemani, E., Paulevé, L. (eds.) CMSB 2021. LNCS, vol. 12881, pp. 74–90. Springer, Cham (2021). https://doi.org/10.1007/978-3-030-85633-5_5
57. Huang, X., Migunov, A.N.: A general purpose analog computer to population protocol compiler (2024). https://doi.org/10.1145/3637543.3654758
58. Ji, X., Chen, Y., Yu, X., Nijhuis, C.A.: Making chemistry compute with non-steady-state chemical dynamics. Nat. Rev. Chem. (2026). https://doi.org/10.1038/s41570-026-00796-w
59. Johnson, C., Agostini, N.B., Cannon, W.R., Tumeo, A.: Energy-efficient scientific computing using chemical reservoirs. NPJ Unconventional Computing (2026). https://doi.org/10.1038/s44335-026-00053-9
60. Kidger, P.: On neural differential equations. CoRR abs/2202.02435 (2022). https://doi.org/10.48550/arXiv.2202.02435. https://arxiv.org/abs/2202.02435
61. Klinge, T.H., Lathrop, J.I., Osera, P.M., Rogers, A.: Reactamole: functional reactive molecular programming. Nat. Comput. (2024). https://doi.org/10.1007/s11047-024-09982-5
62. Köppel, S., Ulmann, B., Heimann, L., Killat, D.: Using analog computers in today's largest computational challenges. Adv. Radio Sci. (2021). https://doi.org/10.5194/ars-19-105-2021
63. Lee, Y.X., Chen, H.L.: On the simulation power of surface chemical reaction networks. Lecture Notes in Computer Science (2024). https://doi.org/10.1007/978-3-031-63742-1_11
64. Leivant, D., Marion, J.: A characterization of alternating log time by ramified recurrence. Theor. Comput. Sci. **236**(1-2), 193–208 (2000). https://doi.org/10.1016/S0304-3975(99)00209-1
65. Leivant, D., Marion, J.Y.: Math. Struct. Comput. Sci. (2020). https://doi.org/10.1017/s0960129519000112
66. Lipshitz, L., Rubel, L.A.: A differentially algebraic replacement theorem, and analog computability. Proc. Am. Math. Soc. **99**(2), 367–372 (1987). https://doi.org/10.1090/s0002-9939-1987-0870803-1
67. Perales-Eceiza, Á., Cubitt, T., Gu, M., Pérez-García, D., Wolf, M.M.: Undecidability in physics: a review. Phys. Rep. (2025). https://doi.org/10.1016/j.physrep.2025.06.004
68. Platzer, A., Qian, L.: Axiomatization of compact initial value problems: open properties. J. ACM (2025). https://doi.org/10.1145/3763228
69. Poças, D., Zucker, J.I.: Tracking computability of GPAC-generable functions. J. Log. Comput. (2020). https://doi.org/10.1093/logcom/cxaa081
70. Pour-El, M.B.: Abstract computability and its relation to the general purpose analog computer (some connections between logic, differential equations and analog computers). Trans. Am. Math. Soc. **199**, 1–28 (1974). https://doi.org/10.2307/1996870
71. Qian, L., Winfree, E.: Parallel and scalable computation and spatial dynamics with DNA-based chemical reaction networks on a surface. In: Murata, S., Kobayashi, S. (eds.) DNA 2014. LNCS, vol. 8727, pp. 114–131. Springer, Cham (2014). https://doi.org/10.1007/978-3-319-11295-4_8
72. Selivanova, S., Steinberg, F., Thies, H., Ziegler, M.: Exact real computation of solution operators for linear analytic systems of partial differential equations. In: Boulier, F., England, M., Sadykov, T.M., Vorozhtsov, E.V. (eds.) CASC 2021. LNCS, vol. 12865, pp. 370–390. Springer, Cham (2021). https://doi.org/10.1007/978-3-030-85165-1_21
73. Shannon, C.E.: Mathematical theory of the differential analyser. J. Math. Phys. MIT **20**, 337–354 (1941). https://doi.org/10.1002/sapm1941201337

74. Shi, X., Gao, C., Dochain, D., Fan, Y.: Biochemical relaxation oscillator designed to control molecular computation. Syst. Control Lett. (2026). https://doi.org/10.1016/j.sysconle.2026.106359
75. Sidl, L., et al.: Computational complexity, algorithmic scope, and evolution. J. Phys. Complex. (2025). https://doi.org/10.1088/2632-072x/adb928
76. Siegelmann, H.T.: Neural networks and analog computation - beyond the Turing limit. Progress in theoretical computer science, Birkhäuser (1999)
77. Siegelmann, H.T., Sontag, E.D.: Turing computability with neural nets. Appl. Math. Lett. **4**(6), 77–80 (1991). https://doi.org/10.1016/0893-9659(91)90080-f
78. Siegelmann, H.T., Sontag, E.D.: Analog computation via neural networks. Theor. Comput. Sci. **131**(2), 331–360 (1994). https://doi.org/10.1016/0304-3975(94)90178-3
79. Vasić, M., Chalk, C., Khurshid, S., Soloveichik, D.: Deep molecular programming: a natural implementation of binary-weight relu neural networks. arXiv (Cornell University) (2020). https://doi.org/10.48550/arxiv.2003.13720
80. Vasić, M., Soloveichik, D., Khurshid, S.: *CRN++*: molecular programming language. Nat. Comput. **19**(2), 391–407 (2020). https://doi.org/10.1007/s11047-019-09775-1
81. Vollmer, H.: Introduction to Circuit Complexity: A Uniform Approach. Springer (1999)
82. E, W.: A proposal on machine learning via dynamical systems. Commun. Math. Stat. **5**(1), 1–11 (2017). https://doi.org/10.1007/s40304-017-0103-z
83. Yu, S., Lee, J., Isokawa, T.: Universality of a surface chemical reaction network using only bi-molecular reactions. J. Membr. Comput. (2024). https://doi.org/10.1007/s41965-024-00149-w

Learning Omega-Regular Languages: A Tour of Learning Results and Canonical Representations

Dana Fisman(✉), Elina Sudit, and Oded Zimerman

Institute for the Theory of Computing, Stein Faculty of Computer and Information Science, Ben-Gurion University of the Negev, Beer-Sheva, Israel
dana@bgu.ac.il

Abstract. Learning regular languages of finite words is guided by the Myhill–Nerode congruence and the minimal DFA. For ω-regular languages, the picture is more fragmented: different learning results rely on different representations, each with its own algorithmic and succinctness properties. This survey presents the main paradigms and results for learning ω-regular languages, including passive learning, active query learning, polynomial predictability, characteristic samples, and efficient teachability. We cover negative results for nondeterministic ω-automata, positive results for informative and weak classes, learning through FDFAs, SUBAs and M2MAs, and recent canonical models based on history determinism and natural colors. Throughout, we emphasize that learnability guarantees must be understood together with the succinctness of the target representation.

1 Introduction

Learning automata from examples and queries is a classical topic at the interface of automata theory, formal languages, and machine learning. For regular languages of finite words, the theory is by now well understood. The Myhill–Nerode congruence provides a canonical representation of a language by its minimal DFA, and this correspondence underlies many of the central learning results: passive learning algorithms such as RPNI, active learning algorithms such as $\mathbf{L}^*$, and information-theoretic notions such as characteristic samples.

The situation is substantially more intricate for ω-regular languages. These languages, which describe sets of infinite words, and model reactive systems and temporal logic are central in formal verification and synthesis. They can be represented by several types of ω-automata, including Büchi, coBüchi, parity, Muller, Rabin, and Streett automata, as well as by more recent models such as families of DFAs, strongly unambiguous Büchi automata and chains of coBüchi automata. Unlike in the finitary case, however, there is no single Myhill–Nerode-style object that simultaneously provides a canonical, succinct, and learnable representation for the full class.

V. Brattka et al. (Eds.): CiE 2026, LNCS 16674, pp. 23–47, 2027.
https://doi.org/10.1007/978-3-032-31348-5_2

This survey presents the main approaches to learning ω-regular languages, with an emphasis on how learning results depend on the chosen representation. We discuss passive learning, active query learning, polynomial predictability, and information-theoretic notions such as characteristic samples and efficient teachability. We also highlight the role of ultimately periodic words and their different decompositions, which provide finite representations of infinite words and suffice to distinguish ω-regular languages.

Learnability results cannot be fully evaluated without considering the succinctness of the target representation. Since the efficiency of a learning algorithm, or the size of a characteristic sample, is measured with respect to the size of this representation, different automata models may lead to very different guarantees. We therefore discuss each result together with the model it learns or assumes, and the known succinctness relations between models.

The survey is organized as follows. We first recall ω-regular languages, the basic learning paradigms and the foundations of learning regular languages. We then describe the main challenges that arise when moving from finitary to infinitary languages, and survey negative results for nondeterministic ω-automata. We next discuss positive results for informative classes, weak ω-regular languages, FDFAs, SUBAs and M2MAs, and finally recent canonical and prismatic models based on history determinism and natural colors. We conclude with a discussion of the current landscape and several directions for future work.

2 Regular ω-Languages

We use Σ^* for the set of finite words over an alphabet Σ, and Σ^ω for the set of infinite words, also called ω-words, over Σ. A language is a set of words. A language L is termed *finitary* (resp. *infinitary*) if it is a subset of Σ^* (resp. Σ^ω).

An *automaton* is a tuple $\mathcal{A} = (\Sigma, Q, q_\iota, \delta, \alpha)$ consisting of a finite alphabet Σ of symbols, a finite set Q of states, an initial state $q_\iota \in Q$, a transition function $\delta : Q \times \Sigma \to 2^Q$, and an accepting condition α. A run of an automaton is defined in the usual manner. For $u \in \Sigma^*$, we use $\mathcal{A}(u)$ for the state that $\mathcal{A}$ reaches on reading u. For regular languages the acceptance conditions is a set $F \subseteq Q$ and a run is accepting if it ends in a state in F. We will shortly elaborate on the acceptance conditions used for ω-regular languages. An automaton accepts a word if at least one of the runs on that word is accepting. We use $[\![\mathcal{A}]\!]$ for the set of words accepted by $\mathcal{A}$.

$\mathcal{A}$ is *deterministic* if $|\delta(q,\sigma)| = 1$ for every $q \in Q$ and $\sigma \in \Sigma$. For deterministic automata we abbreviate $\delta(q,\sigma) = \{q'\}$ as $\delta(q,\sigma) = q'$. We refer to the first four components of an automaton as an *automaton structure*. Two automata structures $\mathcal{A}$ and $\mathcal{A}'$ over the same alphabet Σ are *isomorphic* if there exists a bijection f from the states Q of $\mathcal{A}$ to the states Q' of $\mathcal{A}'$ such that $f(q_\iota) = q'_\iota$ and for every $q \in Q$ and $\sigma \in \Sigma$, $\{f(r) \mid r \in \delta(q,\sigma)\} = \delta'(f(q),\sigma)$.

Acceptance conditions for ω-automata typically depend on the set of states visited infinitely often along a run ρ, which we denote by $\inf(\rho)$. Sometimes the acceptance condition is defined instead with respect to the set of transitions visited infinitely often; in this case, we call the automaton *transition-based.* Note that, since the automaton has finitely many states, every run eventually stabilizes in some strongly connected component (SCC). Here, a singleton state is an SCC iff it has a self-loop.

A *parity* acceptance condition is a map $\kappa : Q \to \mathbb{N}$ assigning to each state a natural number termed a rank. A run of a parity automaton is accepting if the **minimum** rank visited infinitely often is **even**. When the image of κ is $\{0, 1\}$ the automaton is termed a *Büchi* automaton, and a run is accepting iff the rank 0 is visited infinitely often. When the image of κ is $\{1, 2\}$ the automaton is termed *coBüchi* and a run is accepting iff the rank 1 is visited finitely often. The *Muller* acceptance condition is a set of sets of states $\alpha = \{F_1, \ldots, F_k\}$ and a run ρ is accepting iff $\inf(\rho) \in \alpha$. That is the set of states visited infinitely often is exactly one of the sets listed in α.

A *weak parity* acceptance condition is also a map $\kappa : Q \to \mathbb{N}$ but a run of weak parity automaton is accepting if the minimum color visited (rather than visited infinitely often) is even. *Weak Büchi* and *weak coBüchi* automata are defined in the same manner.

There are other common types of ω-automata, in particular *Rabin* and *Streett.* Their definition is not required for the rest of the paper. We use three or four letter acronyms for referring to ω-automata. The first letter is in $\{D, N\}$ and represents the transition type – deterministic or non-deterministic. The last letter is always A. The middle letter(s) is in $\{P, B, C, M, R, S, WP, WB, WC\}$ and refers to the acceptance type. For instance NBA stands for non-deterministic Büchi automaton, and DPA for deterministic parity automaton. We use blackboard bold font for the respective classes of languages. E.g. $\mathbb{DCA}$ stands for the class of languages recognized by deterministic coBüchi automata.

The class of languages recognized by deterministic parity automata is called the class of *ω-regular languages.* We use the terms ω-regular languages and *infinitary regular languages* interchangeably.

$\mathbb{DPA}$ is as expressive as $\mathbb{DMA}$, $\mathbb{DRA}$ and $\mathbb{DSA}$. $\mathbb{DBA}$ and $\mathbb{DCA}$ are less expressive than $\mathbb{DPA}$. Considering the non-deterministic version of Büchi, we have that $\mathbb{DPA}$ are as expressive as $\mathbb{NBA}$. A language L is in $\mathbb{DBA}$ iff its complement, namely $\Sigma^\omega \setminus L$ is in $\mathbb{DCA}$. For coBüchi non-determinism does not add expressive power, that is $\mathbb{DCA}$ and $\mathbb{NCA}$ are equally expressive. Last, $\mathbb{DWPA}$ is strictly less expressive than both $\mathbb{DCA}$ and $\mathbb{DBA}$ and more expressive than both $\mathbb{DWCA}$ and $\mathbb{DWBA}$. These relations are captured by the Manna-Pnueli hierarchy of temporal properties depicted in Fig. 1. This hierarchy is further refined by the infinite Wagner hierarchy, which is based on inclusion patterns among accepting and rejecting SCCs and is tightly related to the minimal number of colors required by a parity automaton for the language [52, 20].

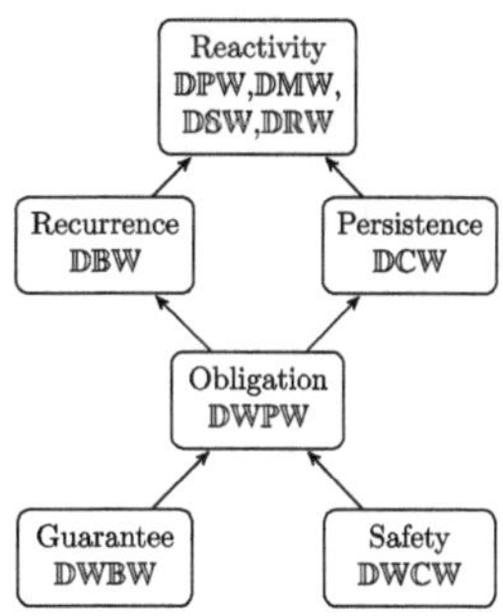

Fig. 1. The expressiveness hierarchy of ω-automata [43].

An *ultimately periodic word*, also called a *lasso word*, is a word of the form $u(v)^\omega$, where $u \in \Sigma^*$ and $v \in \Sigma^+$. By McNaughton's theorem, two ω-regular languages are equivalent iff they agree on the set of ultimately periodic words. Thus, lasso words play an important role in learning: they provide finite representations of infinite words, while still sufficing to distinguish ω-regular languages. A *decomposition* of a lasso word uv^ω is a pair (x, y) such that $xy^\omega = uv^\omega$. Note that a lasso word may have several distinct decompositions. For example, $ba(bab)^\omega$ can be decomposed as (ba, bab), as well as (bab, abb), $(baba, bbabba)$, and so on. Henceforth we use Σ^{UP} to abbreviate $\Sigma^* \times \Sigma^+$, that is Σ^{UP} consists of all decompositions of ultimately periodic words.

Comparing deterministic ω-automata with DFAs, we note that for DFAs, complementation is constant-time, and union and intersection are quadratic. For deterministic ω-automata, the picture is more subtle. Only deterministic parity automata admit constant-time complementation. Furthermore, none of the standard models simultaneously has polynomial constructions for both union and intersection. Indeed, parity, Muller, Rabin, and Streett automata incur an exponential blowup for at least one of these operations [15].

3 Learning Paradigms

Several learning paradigms have been studied in the realm of formal languages. We focus here on a few representative paradigms. We first note that all paradigms assume a class of languages $\mathbb{L}$ and a class of representations $\mathbb{R}$ for languages in $\mathbb{L}$. For instance, when $\mathbb{L}$ is the class of finitary regular languages, one may take DFAs as the representation class, or instead take regular expressions as the representation class. For a representation $\mathcal{R} \in \mathbb{R}$, we use $[\![\mathcal{R}]\!]$ for the corresponding language, which is a member of $\mathbb{L}$.

Learning results may differ substantially for different representations, since some representations may be easier to infer than others, and since the requirements of the learning paradigm are stated with respect to the size of the representation, and different representations may differ in succinctness. For instance, regular languages can be learned in polynomial time using membership and equivalence queries when represented by DFAs, but not when represented by NFAs. This is a particular instance of a general phenomena, where usually the more succinct a representation is, the harder it tends to be to learn.

The paradigm of *passive learning* concerns the problem of inferring a representation that is consistent with a given sample. A *sample* is a finite set S of words labeled according to their membership in the language. Usually, one views $S = (S^+, S^-)$ as consisting of two sets of positively and negatively labeled words. The aim is not only to learn a representation that is consistent with the sample, but also to do so efficiently. That is, one prefers learners that use less time and infer more concise representations.

Active learning paradigms assume that the learner can interact with an *oracle* that answers certain types of queries. The most commonly considered queries are *membership queries* (MQ) and *equivalence queries* (EQ). We refer to this paradigm as (MQ, EQ)-query learning.

Another active learning paradigm is *polynomial predictability* with MQ. Roughly speaking, in this paradigm the learner may ask, in addition to MQs, *draw queries*, which return a randomly drawn word together with its membership label. At some point, the learner asks for a test word to predict, and learning succeeds if it classifies it correctly with probability at least $1-\varepsilon$. The (MQ, EQ)-query-learning paradigm implies polynomial predictability.

For regular languages, deciding, given a sample $S=(S^+,S^-)$ and an integer k, whether there exists a DFA of size at most k consistent with S is NP-hard [31]. In contrast, there is an active learning algorithm, $\mathbf{L}^*$, that learns a minimal DFA for a regular language using MQ and EQ, and runs in polynomial time [1].

Information-theoretic paradigms are more concerned with the size of a set of examples from which a language can be correctly inferred. Their definitions usually build on the notions of a teacher and a learner. A *teacher* for $\mathbb{R}$ is a function $\mathbf{T}$ from $\mathbb{R}$ to samples such that $\mathbf{T}(\mathcal{R})$ is consistent with $[\![\mathcal{R}]\!]$. A *learner* for $\mathbb{R}$ is a function $\mathbf{L}$ from samples to $\mathbb{R}$ such that every sample S is consistent with the representation $\mathbf{L}(S)$.

A sample T is a *characteristic sample* for $\mathcal{R}$ and a learner $\mathbf{L}$ if T is consistent with $[\![\mathcal{R}]\!]$ and, for every sample $T' \supseteq T$ consistent with $[\![\mathcal{R}]\!]$, we have $[\![\mathbf{L}(T')]\!] = [\![\mathcal{R}]\!]$. That is, the learner correctly infers a representation for the target language from the sample generated by the teacher, as well as from any sample that extends it, as long as the extended sample remains consistent with $[\![\mathcal{R}]\!]$. The requirement for considering all subsuming T' is meant to cope with the problem of collusion, e.g. that the teacher places in the sample the encoding of the representation $\mathcal{R}$ itself. See the seminal paper of Goldman and Mathias [32] for the fascinating challenges involved in defining meaningful collusion-free information-theoretic notions.

Naturally, one is interested in producing characteristic samples of small size. It is shown by Barzdin and Freivalds [11] that the *halving algorithm* can be used in the context of predictions to bound the number of mistakes of prediction linearly in the size of the target concept (cf. surveys [30,55]). This implies that the cardinality of the sample can always be made polynomial (cf. [8, Thm. 3.5]). However, this may come at the expense of having words of non-polynomial length in the sample. Thus, to obtain meaningful results regarding sample size, the *size of a sample* is measured as the sum of the lengths of its words.

We note that the definition of a characteristic sample does not require the teacher and learner to be computable. In fact, it is shown in [27] that there exists a class $\mathbb{R}$ that has characteristics samples but there is no corresponding computable teacher $\mathbf{T}$. The characteristic samples in this class have cardinality 2 but there is no computable bound on their length. This motivates the definition of the following learning paradigms.

A class $\mathbb{R}$ is *identifiable in the limit using polynomial time and data* if there exist a polynomial $p(n)$, a teacher $\mathbf{T}$, and a polynomial time learner $\mathbf{L}$ such that for every $\mathcal{R} \in \mathbb{R}$, the sample $\mathbf{T}(\mathcal{R})$ is a characteristic sample for $\mathcal{R}$ and $\mathbf{L}$, and $|\mathbf{T}(\mathcal{R})| \leq p(\mathit{size}(\mathcal{R}))$. That is, this paradigm makes requirement for a polynomial time learner, and a polynomial sized characteristic sample, but it makes no requirements on the teacher.

In contrast, a class of representations $\mathbb{R}$ is said to be *efficiently teachable* if there exists a polynomial time teacher $\mathbf{T}$ and a learner $\mathbf{L}$ such that $\mathbf{T}(\mathcal{R})$ is a characteristic sample for $\mathcal{R}$ and $\mathbf{L}$ for every $\mathcal{R} \in \mathbb{R}$. Note that the bound on running time of the teacher implies that the characteristic sample itself is of polynomial size. This paradigm makes no requirement on the learner.

A class is termed *efficiently teachable/learnable* if both the learner and teacher are required to run in polynomial time (which implies also a polynomial bound on the characteristic sample). It is shown in [8] that if a class $\mathbb{R}$ has a polynomial time algorithm to decide membership, and a default acceptor that agrees with a sample can be generated in polynomial time, then efficient teachability implies efficient teachable/learnable. For a comprehensive discussion of learning paradigms for formal languages and the relationships between them, see [8, §3].

4 Foundations of Learning Regular Languages

We introduce basic notation and survey fundamental results on regular languages that underlie many learning algorithms, including those for ω-regular languages.

Inferring a DFA Consistent with a Sample. Let S be a sample of positively and negatively labeled examples. We first note that returning a DFA that agrees with S can be done by enumerating all DFAs from small to large, and returning the first one, according to the enumeration order, that is consistent with the sample. The issue with this method, of course, is that it is not polynomial.[1]

There is, however, a much simpler construction: arrange the words in the sample in a trie data structure, so as to exploit shared prefixes. Then transform the trie into a DFA by classifying states as accepting or rejecting according to the label of the word that reaches them, and direct all missing transitions to an added sink state. This automaton is termed the *prefix-tree automaton.* The prefix-tree automaton clearly agrees with the sample and can be constructed in polynomial time. However, it does not necessarily generalize well; more formally, it is typically not the minimal DFA for the target language.

The RPNI passive learning algorithm [51, 45] starts by constructing the prefix-tree automaton, and then merges states to obtain a smaller consistent DFA. To this end, it enumerates the states using BFS, so smaller indices correspond to states closer to the root. It then iterates over pairs $1 \leq i \leq n$ and $1 \leq j < i$, trying to merge q_i and q_j. If the resulting automaton is consistent with the sample, the merge succeeds; otherwise, it fails. This method is not guaranteed to return the minimal DFA, but it does so when the sample subsumes a characteristic sample. Moreover, it works well in practice and is widely used.

[1] The number of DFAs of size n over an alphabet of size k is $n2^{n-1}n^{(k-1)n}$ [22].

Right Congruences and Automata. With a right congruence relation $\sim$ one can associate an automaton-structure $\mathcal{A}[\sim]$ as follows: $\mathcal{A}[\sim] = (\Sigma, Q, q_0, \delta)$ where Q are the equivalence classes of $\sim$, q_0 is the equivalence class of $[\varepsilon]$, and $\delta([x], \sigma) = [x\sigma]$. Similarly, an automaton structure $\mathcal{A} = (\Sigma, Q, q_0, \delta)$ defines a relation $\sim_{\mathcal{A}}$ as follows: $x \sim_{\mathcal{A}} y$ if $\mathcal{A}(x) = \mathcal{A}(y)$.

A language $L \subseteq \Sigma^*$ induces the right congruence equivalence relation $\sim_L$ defined as follows. Let $x, y \in \Sigma^*$. Then $x \sim_L y$ iff for all $z \in \Sigma^*$ we have $xz \in L \iff yz \in L$. When $x \not\sim_L y$ then there exists $z \in \Sigma^*$ such that $xz \in L \iff yz \notin L$, the word z is referred to as a *distinguishing suffix*. The number of equivalence classes in $\sim_L$ is termed the *index* of L.

The famous Myhill–Nerode theorem states that $\mathcal{A}[\sim_L]$ is the minimal DFA for L. Hence the number of states in a minimal DFA for L is the index of L and there is a one-to-one correspondence between its states and the equivalence relation $\sim_L$.

For an automaton $\mathcal{A} = (\Sigma, Q, q_\iota, \delta, \alpha)$ and a state $q \in Q$, we use $\mathcal{A}_{[q}$ for the automaton obtained from $\mathcal{A}$ by making q initial, namely $\mathcal{A}_{[q} = (\Sigma, Q, q, \delta, \alpha)$. Similarly, $\mathcal{A}_{q]} = (\Sigma, Q, q_\iota, \delta, \{q\})$ is obtained by making q accepting; this applies when $\alpha \subseteq Q$, as for regular languages. Words in $[\![\mathcal{A}_{q]}]\!]$ are often termed *access words* to state q. The following claim states that in a minimal DFA, all access words to a given state are equivalent, and they are inequivalent to all others.

Claim 4.1. *If $\mathcal{A}$ is a minimal DFA then $u \sim_L v$ iff u and v are access words to the same state of $\mathcal{A}$.*

Words are often ordered by the *shortlex order*: first by length, and then lexicographically among words of the same length. As the following claim observes, the set of shortlex access words is prefix closed.

Claim 4.2. *Let $\mathcal{A}$ be a minimal DFA and let $w_1, \ldots, w_n$ be the shortlex access words to its states. Then the set $\{w_1, \ldots, w_n\}$ is prefix closed.*

The next claim asserts that distinguishing suffixes can be found in polynomial time. Indeed, for access words to states p and q, construct $\mathcal{A}_{[q} \times \mathcal{A}_{[p}$ and search for a state accepting in one component and rejecting in the other. If $\mathcal{A}$ has n states, the corresponding word has length at most n^2.

Claim 4.3 (Finding Distinguishing Suffixes). *Given a DFA $\mathcal{A}$ and two states q, p a distinguishing suffix between access words to p and q can be found in quadratic time.*

From these claims we can establish the construction of characteristic samples for the class of DFAs.

Claim 4.4 (Characteristic Sample). *Let $\mathcal{A}$ be minimal DFA. A characteristic sample for $\mathcal{A}$ can be constructed by letting S be the set of shortlex access words to states of $\mathcal{A}$, letting D be a set consisting of a distinguishing suffix for every pair of strings in S, defining $W = SD \cup S\Sigma D$ and letting the characteristic sample consists of all words in W and their correct membership label.*

Observation Table, and DFA Reconstruction. From the characteristic sample a learner can infer the DFA as follows. Let B be the prefixes of W, and E the suffixes of W. Let T be a table with rows labeled by B and columns by E. For $b \in B$ and $e \in E$ the entry $T(b, e)$ is filled with ? if be is not in the sample. Otherwise $T(b, e)$ is filled with + (resp. −) if be is labeled positively (resp. negatively). Two rows b, b' are *inequivalent* if there exists $e \in E$ such that $T(b, e) \neq T(b', e)$ and both are different from ?. If b and b' are different then $b \not\sim_L b'$ since e is a distinguishing suffix. By the construction of W the learner can observe a set S of shortlex access words to all the states of $\mathcal{A}$ and determine they are inequivalent. Furthermore, their one letters extension can be observed and the equivalent state in the basis can be identified. Consider the DFA $\mathcal{A} = (\Sigma, S, \varepsilon, \delta, F)$ where $\delta(s, \sigma) = s'$ if s' is the state of S that $s\sigma$ is indistinguishable from, and F is the subset of S corresponding to words in L (namely, the rows s such that the entry $T(s, \varepsilon) = +$). Then $\mathcal{A}$ recognizes L.

The idea behind L*. The idea behind the $\mathbf{L}^*$ (MQ,EQ)-learning algorithm [1] is to use membership queries to gradually fill a table as above (aka an *observation table*) in order to both differentiate strings and collect information to reconstruct the DFA. To this aim a *basis* B for the table is a prefix closed set of inequivalent rows. A table is termed *closed* if for every $b \in B$ the string $b\sigma$ labels some row. The learner uses MQs to fill the table until it is closed, in which case he can construct a DFA consistent with the table, and ask an equivalence query. If the answer is "yes" the learner returns this DFA, otherwise it uses the provided counterexample to extract new information. In more details, the learner adds the counterexample and all its suffixes to the columns, and fills the new entries. It can be shown that this process reveals at least one new inequivalent row (cf. [26, p.32]) and thus the process converges and returns the correct minimal DFA.

5 Challenges in Learning ω-Regular Languages

Before surveying the results on learning ω-regular languages, let us pause to ask why this problem is challenging. Indeed, while the answers for finitary regular languages largely culminated in the 1980s, the quest for analogous results for infinitary regular languages is still ongoing.

Recall that most learning algorithms for regular languages are funded on the relation $\sim_L$ which relates two words x and y if they do not have a distinguishing suffix z. In the case of finitary languages the words x, y, z are finite. In the case of infinitary languages we still want to relate two finite words x and y, since they correlate to words reaching certain states in the automaton, but the word distinguishes them, z, should be infinite, that is for infinitary words we define: $x \sim_L y$ iff for all $z \in \Sigma^\omega$ we have $xz \in L \iff yz \in L$.

One challenge in going from finitary to infinitary languages is that $\sim_L$ does not have a one-to-one correspondence with the states of any common ω-automata model. To see this, consider the finitary regular language $L = \Sigma^* b^+$ over $\Sigma = \{a, b\}$. The relation $\sim_L$ has two equivalence classes: words ending in b, and words ending in a. These indeed correspond to the states of the minimal DFA for the language. If we consider the very similar infinitary language $L = \Sigma^* b^\omega$, then a deterministic parity automaton over the same structure can be defined. However, $\sim_L$ has only one equivalence class. Indeed, for any $x, y \in \Sigma^*$ and $z \in \Sigma^\omega$, we have $xz \in L \iff yz \in L$, since membership in L is determined solely by whether z has a suffix of the form b^ω. In particular, it is independent of x and y. Yet there is no one-state DPA for this language, and the same holds for all other deterministic models capable of expressing it. The absence of such a correspondence appears to be a major challenge for learning ω-regular languages.

Another challenge is illustrated by the following closely related languages over $\Sigma_n = \{a_1, \ldots, a_n\}$. The language L_n consists of all words in which all letters in Σ_n occur, while L'_n consists of all words in which all letters occur infinitely often. The minimal DFA for L_n has 2^n states, since it must record the subset of letters seen so far. In contrast, for L'_n there exists a DPA $\mathcal{P}_n$ with n states. While $\mathcal{P}_n$ checks that every letter occurs infinitely often, it need not remember all subsets seen from some point onward. Instead, it fixes an arbitrary order of the letters and repeatedly looks for the next letter.

6 Negative Results

When learning infinitary languages, a first question is how to handle infinite words. Samples and membership queries require finite representations of words. One possibility is to consider prefixes of infinite words. It was shown in [21] that the class of deterministic weak coBüchi automata is identifiable in the limit using polynomial time and data with prefixes, and that furthermore, when using prefixes, this result cannot be extended beyond this class.

To obtain more positive learnability results, one instead uses *ultimately periodic words*. As mentioned above, two ω-regular languages are equivalent iff they agree on the set of ultimately periodic words; restricting attention to such words should therefore not limit the learnability results obtained.

With regard to the learned class, as mentioned above, different classes have different expressive power and different succinctness properties.

Before surveying the positive results, we present the following negative results regarding the learnability of nondeterministic ω-automata.

The first result relies on plausible cryptographic assumptions and follows from an analogous result for NFAs [5]. The proof is for the class $\mathbb{NBA}$s but extends to the other non-deterministic classes, since NBAs are a special case of NPA, NMA, NRA, and NSA.

Theorem 6.1 ([6,8]). *Under plausible cryptographic assumptions,* $\mathbb{NBA}$, $\mathbb{NPA}$, $\mathbb{NMA}$, $\mathbb{NRA}$, *and* $\mathbb{NSA}$ *are not polynomially predictable with* MQs, *and hence are not polynomial-time* (MQ, EQ)*-learnable.*[2]

[2] The assumptions are the computational intractability of any of the following three problems: (1) testing quadratic residues modulo a composite, (2) inverting RSA encryption, or (3) factoring Blum integers.

The next negative result does not rely on cryptographic assumptions. It shows that the classes $\mathbb{NBA}$ and $\mathbb{NCA}$ do not have concise characteristic sets; that is, there need not exist even a noncomputable teacher that can generate a polynomial-size characteristic sample for some learner. The proof constructs a family of languages $\{L_n\}_{n\in\mathbb{N}}$ with acceptors of size quadratic in n, for which every characteristic sample for L_n must include at least one word of exponential length in n. This result, too, extends to the other classes that generalize $\mathbb{NBA}$.

Theorem 6.2 ([9, 8]). *$\mathbb{NBA}$, $\mathbb{NPA}$, $\mathbb{NMA}$, $\mathbb{NCA}$, $\mathbb{NRA}$, and $\mathbb{NSA}$ are neither identifiable in the limit using polynomial time and data nor efficiently teachable.*

7 The Fully Informative Classes

As shown in Sect. 5 the right congruence relation $\sim_L$ associated with an ω-regular language L is in general not informative enough to deduce the structure of a minimal automaton for the language. More precisely, there are ω-regular languages for which the minimal ω-automaton requires more states than there are equivalence classes in $\sim_L$. Put otherwise, no ω-automaton can be defined on the automaton structure $\mathcal{A}[\sim_L]$. In this section we present learnability results for classes in which each language admits an automaton on the structure $\mathcal{A}[\sim_L]$.

7.1 Deterministic Weak Parity Automata

Recall the weak parity acceptance condition and the corresponding class $\mathbb{DWPA}$. The languages in this class are known as the *weak ω-regular languages*. Interestingly, they are exactly the languages that can be recognized both by deterministic Büchi automata and by deterministic co-Büchi automata; that is, $\mathbb{DWPA} = \mathbb{DBA} \cap \mathbb{DCA}$. Staiger showed that, for every weak ω-regular language, there is a one-to-one correspondence between the states of a minimal DWPA and the equivalence classes of $\sim_L$ [49].

Maler and Pnueli showed that the $\mathbf{L}^*$ algorithm can be extended to an algorithm, which we term $\mathbf{L}^{\mathbf{w}\omega}$, that learns the class $\mathbb{DWPA}$ from membership queries (MQ) and equivalence queries (EQ), giving rise to the following theorem.

Theorem 7.1 ([41]). *The class $\mathbb{DWPA}$ is polynomially learnable with membership and equivalence queries.*

The idea is to learn the automaton structure in the same way that $\mathbf{L}^*$ learns the structure of a DFA, using an observation table. In $\mathbf{L}^*$, both row and column labels are words in Σ^*. In $\mathbf{L}^{\mathbf{w}\omega}$, the column labels are ultimately periodic words $x(y)^\omega$, represented as pairs (x, y). Apart from this difference, the idea is similar: the learner gradually fills the table using MQs, every new distinct row is added to the basis, and its one-letter extensions are added to the rows. This continues until the table is closed, at which point an automaton structure can be extracted.

In $\mathbf{L}^*$, once the automaton structure has been extracted, turning it into an automaton only requires classifying states as accepting or rejecting. This is done according to the membership of the state representative in L, namely according to the value of the row in the ε column.

Recall that the acceptance condition for $\mathbb{DWPA}$ is given by a map from states to ranks, where a run is accepting iff the minimal rank visited along the run is even. Since the run of an infinite word eventually stabilizes in some SCC, the ranks can be assigned by decomposing the automaton structure into SCCs and assigning ranks that are monotonically non-increasing along paths in the SCC DAG. States in the same SCC receive the same rank, and a run is accepting iff it eventually gets trapped in an SCC with an even rank.

Thus, when $\mathbf{L}^{\mathbf{w}\omega}$ classifies states as accepting or rejecting, it must verify that there is no state s and no words u and v looping on s for which it is known that $su^\omega \in L \iff sv^\omega \notin L$. If such words exist, a *conflict* is detected, signifying that the automaton structure cannot define a DWPA for the language.

When a conflict is detected, $\mathbf{L}^{\mathbf{w}\omega}$ uses MQs to extract new information that will resolve it. In the observation table, the rows corresponding to s, su, and sv are indistinguishable, since the current structure assumes that both u and v loop on s. This cannot occur in a correct automaton for the language. The conflict-resolution MQs therefore try to distinguish s from su or from sv. A clever argument, using properties of weak DPAs, shows that a word of the form xu^ω or xv^ω, where x is a prefix of $(u^nv^n)^n$ or of $(v^nu^n)^n$ and n is an upper bound on the number of states in the minimal weak DPA for L, must provide the desired distinction. The learner therefore asks membership queries for such words, gradually increasing n.

This class has also been shown to be efficiently teachable, as a corollary of the result we present in the next subsection.

Theorem 7.2 ([8]). *The class* $\mathbb{DWPA}$ *is efficiently teachable.*

7.2 Efficient Teachability of the Fully Informative Classes

Following the discussion in Sect. 7.1, a natural question is whether all languages whose minimal automata have a one-to-one correspondence with the equivalence classes of $\sim_L$ belong to $\mathbb{DWPA}$.

An ω-regular language is termed *fully informative* if a deterministic ω-automaton, of one of the types mentioned in Sect. 2, can be defined on the automaton structure $\mathcal{A}[\sim_L]$. For $X \in \{B, C, P, M, R, S\}$, we say that a language is in $\mathbb{IXA}$ if it is in $\mathbb{DXA}$ and the minimal DXA for L has a one-to-one correspondence with the equivalence classes of $\sim_L$. We refer to these classes as the *fully informative* classes. As a simple illustration of what this condition captures, $\Sigma^*(ab)^\omega$ is not fully informative, whereas the closely related languages $(ab)^\omega$ and $aa^*(ab)^\omega$ are.

At first glance, these classes may seem quite restricted in expressive power. However, it is shown in [9] that every level of the infinite Wagner hierarchy contains fully informative languages.

It was shown in [9, 8] that these classes are learnable in the limit using polynomial time and data. This means that (a) there exists a polynomial-time learner that, for any given sample, returns an automaton of the appropriate type that is consistent with the sample, and (b) for every language L in the class whose minimal automaton has n states, there is a characteristic sample S_L of size polynomial in n. In other words, once the given sample S subsumes S_L, the learner must return an automaton that not only agrees with S but recognizes L exactly.

The construction of an automaton from a sample has two steps: first defining the automaton structure, and then defining the acceptance condition. Words are placed in the sample to help the learner with both tasks. The extraction of the automaton structure, and the set of words required for it, are the same for all informative classes and follow the idea presented in Sect. 4, except that column labels are ultimately periodic words. The words needed for defining the acceptance type and reconstructing it depend, of course, on the acceptance type.

For Muller automata, for every SCC S in the automaton structure, the teacher places in the sample a word w such that $\inf(\rho_w) = S$, where ρ_w is the run on w. Once the automaton structure is reconstructed, the learner can go over each SCC and look for the words in the sample that traverse it. If some words corresponding to the same SCC disagree, then no Muller condition can be defined on this structure. This means that the given sample does not subsume a characteristic sample, and the learner resorts to returning a kind of prefix-tree automaton for ω-words. Otherwise, the Muller condition α consists of all SCCs for which all traversing words are positively labeled.

For parity automata, the task is more challenging. Its development goes through the definition of *the canonical forest of a DPA*, a forest of SCCs of a given DPA that has useful properties and induces a minimal ranking of the states of the parity automaton. For Rabin automata, the development of the words required for defining acceptance goes through a definition of a singleton normal form. See [8] for details. The fact that the constructed samples are of polynomial size relies on polynomial-time algorithms for inclusion and equivalence for the corresponding automata types [8, §§11–14].

Following these results, [14] developed an RPNI-based approach for learning these classes. The algorithm uses transition-based ω-automata and procedures that test whether an acceptance condition of the desired type can be imposed on the inferred structure while remaining consistent with the sample.

Theorem 7.3 ([9, 14, 8]). *The classes* $\mathbb{IBA}, \mathbb{ICA}, \mathbb{IPA}, \mathbb{IMA}, \mathbb{IRA}, \mathbb{ISA}$ *are identifiable in the limit using polynomial time and data.*

This alone does not entail that the corresponding classes are efficiently teachable. The reason is that the class $\mathbb{IBA}$, for instance, also includes DBAs that are not themselves IBAs but have an equivalent IBA. A missing piece of the puzzle is therefore a polynomial-time algorithm that determines whether a DBA has an equivalent IBA; similarly for the other classes [8, §§15–16].

Theorem 7.4 ([8]). *The classes* $\mathbb{IBA}, \mathbb{ICA}, \mathbb{IPA}, \mathbb{IMA}, \mathbb{IRA}, \mathbb{ISA}$ *are efficiently teachable.*

To date, no polynomial-time (MQ, EQ)-learning algorithm is known for these classes. Moreover, it was shown in [14] that, in the setting of (MQ, EQ)-learning, restricting attention to the fully informative classes does not make the problem easier.

Theorem 7.5 ([14]). *For $\mathbb{X} \in \{\mathbb{B}, \mathbb{P}, \mathbb{R}\}$, there is a polynomial-time* (MQ, EQ)*-learning algorithm for the class $\mathbb{IXA}$ iff there is a polynomial-time* (MQ, EQ)*-learning algorithm for the class $\mathbb{DXA}$.*

8 Families of DFAs

As discussed in Sect. 4, many learning algorithms for regular languages rely on the correspondence between $\sim_L$ and the minimal automaton for L. A long line of work has sought analogous correspondences for ω-regular languages, independently of learning motivations [10, 39, 50, 33, 42].

One issue with $\sim_L$ is that it does not capture the constraints imposed on the periodic parts of words in L. One may instead define the relation $x \dot{\sim}_L y$ iff, for every $u, v \in \Sigma^*$, we have $u(xv)^\omega \in L \iff u(yv)^\omega \in L$. For the language $\Sigma^*(ba)^\omega$, this relation distinguishes $(ba)^*$ from $(ba)^*b$, unlike $\sim_L$, which has only one equivalence class. However, for the language $L' = (a + aac)(b)^\omega$, it does not distinguish a from aa, or from aaa, although an automaton for the language clearly needs to differentiate between these words.

One may then try to combine this relation with $\sim_L$, defining $x \ddot{\sim}_L y$ iff both $x \sim_L y$ and $x \dot{\sim}_L y$. This works well for the language L' above, but not for $L'' = a^*b(aab)^\omega$: while we would like to distinguish a from aa in the loop, we do not want to distinguish them in the transient part.

These issues motivated Maler and Staiger to introduce the notion of a *family of right congruences* (FORC) [42]. Loosely speaking, a FORC consists of a leading right congruence $\sim$ and, for each equivalence class u of $\sim$, a dedicated right congruence $\approx_u$. This allows one to separate the treatment of the transient and periodic parts of words. They showed that an ω-language is regular iff it is recognized by a finite FORC, and that a particular FORC, the *syntactic FORC*, is the minimal FORC for the language.

We next present a closely related model, *families of DFAs* (FDFAs) [4], for which learning results have been obtained. Although FDFAs were introduced with learning motivations in mind, they were later shown to have good properties as acceptors for ω-regular languages. In terms of succinctness, they offer a middle ground between deterministic and nondeterministic ω-automata. We elaborate on these in the sequel.

FDFAs. An FDFA is a pair $\mathcal{F} = (\mathcal{Q}, \{\mathcal{P}_q\}_{q \in Q})$ consisting of a *leading* automaton structure $\mathcal{Q}$ and, for each state q of $\mathcal{Q}$, a *progress* DFA $\mathcal{P}_q$.[3]

An FDFA receives an ultimately periodic word represented as a pair $(u, v) \in \Sigma^{\mathrm{UP}}$. Intuitively, it first performs some normalization, yielding a decomposition

[3] The FDFA model can be viewed as an automata-based variant of FORCs. The correspondence is not exact: FORCs require each progress right congruence to refine the leading right congruence, whereas the FDFA model considered here imposes no analogous requirement. We note that the term FDFA is used in [35] for a DFA-based model that does impose this additional requirement.

(x, y) such that $uv^\omega = xy^\omega$. It then reads x in the leading automaton $\mathcal{Q}$. If $\mathcal{Q}(x) = \tilde{x}$, namely if $\tilde{x}$ is the state reached by $\mathcal{Q}$ after reading x, then acceptance is determined by whether y is accepted by the progress DFA $\mathcal{P}_{\tilde{x}}$. Henceforth, we write $\mathcal{P}_x$ for $\mathcal{P}_{\tilde{x}}$.

Several acceptance notions are defined for FDFAs. They differ in the normalization they perform, or equivalently, in the decomposition they use to determine acceptance. Recall that $\Sigma^{\text{UP}} = \Sigma^* \times \Sigma^+$. Formally, we say that $(u, v) \in \Sigma^{\text{UP}}$ is accepted by an FDFA under α-normalization if there exists an α-decomposition (x, y) of (u, v) such that $\mathcal{P}_x$ accepts y. Three decomposition types are considered in the literature: exact decompositions due to [18], normalized decompositions due to [42], and duo-normalized decompositions due to [28].

Definition 8.1 (FDFA decomposition types). *Let $\mathcal{F} = (\mathcal{Q}, \{\mathcal{P}_q\}_{q \in Q})$ be an FDFA. Let $(u, v) \in \Sigma^{\text{UP}}$ and let $w = uv^\omega$.*

- *Any decomposition $(x, y) \in \Sigma^{\text{UP}}$ of w is termed* exact.
- *A decomposition (x, y) is* normalized *if $\mathcal{Q}(x) = \mathcal{Q}(xy)$.*
- *A normalized decomposition is* duo-normalized *if $\mathcal{P}_x(y) = \mathcal{P}_x(yy)$.*

For each $\alpha \in \{\text{exact, normalized, duo-normalized}\}$, there exist i, j quadratic in the size of $\mathcal{F}$ such that the decomposition (uv^i, v^j) is an α-decomposition.

An FDFA $\mathcal{F}$ is said to be α-*saturated* if, for every ultimately periodic word w, all α-decompositions of w agree on whether they are accepted by $\mathcal{F}$. When saturation holds, α-acceptance can equivalently be defined by first computing a particular α-normalized decomposition (x, y) and then testing whether $y \in [\![\mathcal{P}_x]\!]$. Henceforth, all FDFAs are assumed to be saturated.

It is immediate that if $\mathcal{F}$ is saturated under exact acceptance and recognizes L, then it is also saturated and recognizes L under normalized acceptance. Similarly, saturation and recognition are preserved when moving from normalized acceptance to duo-normalized acceptance.

Several canonical representations exist for FDFAs: the *periodic FDFA* [18], the *syntactic FDFA* [42], the *recurrent FDFA* [3], its dual, the *limit FDFA* [37], and the colorful FDFA [28]. The periodic FDFA uses exact acceptance; the syntactic, recurrent, and limit FDFAs use normalized acceptance; and the colorful FDFA uses duo-normalized acceptance.

For brevity, we refer to them by their initials, P,S,R,L,C. All these canonical representations use $\sim_L$ for the leading automaton. They differ in the progress DFAs: for each $\text{X} \in \{\text{P,S,R,L,C}\}$, the progress DFA $\mathcal{P}_u^{\text{X}}$ is defined via a relation $\approx_u^{\text{X}}$, namely $\mathcal{P}_u^{\text{X}} = \mathcal{A}[\approx_u^{\text{X}}]$ is the induced DFA. We now define the relations $\approx_u^{\text{X}}$. In all these definitions, the language L is implicit.

Definition 8.2 ($\approx_u^{\text{X}}$). *Let $L \subseteq \Sigma^\omega$, and $x, y, u \in \Sigma^*$.*

- $x \approx_u^{\text{P}} y$ *if* $\forall v {\in} \Sigma^*.\ u(xv)^\omega \in L \iff u(yv)^\omega \in L$
- $x \approx_u^{\text{S}} y$ *if* $ux \sim_L uy$ *and* $\forall v {\in} \Sigma^*$ *if* $uxv \sim_L u$ *then* $u(xv)^\omega \in L \iff u(yv)^\omega \in L$
- $x \approx_u^{\text{R}} y$ *if* $\forall v {\in} \Sigma^*.\ (uxv \sim_L u) \wedge u(xv)^\omega \in L \iff (uyv \sim_L u) \wedge u(yv)^\omega \in L$

– $x \approx^{\mathrm{L}}_u y$ *if* $\forall v{\in}\Sigma^*.\ (uxv \sim_L u) {\Rightarrow} u(xv)^\omega \in L \iff (uyv \sim_L u) {\Rightarrow} u(yv)^\omega \in L$
– $x \approx^{\mathrm{C}}_u y$ *if* $\forall v{\in}\Sigma^*.\ \Omega^{\text{infix}}_L(u, xv) = \Omega^{\text{infix}}_L(u, yv)$
where $\Omega^{\text{infix}}_L(u, z)$ *is the natural infix color of* z *wrt* u *see Sect. 10.*

For $\mathrm{X} \in \{\mathrm{P,S,R,L,C}\}$ *the* X-*FDFA is* $(\mathcal{A}[\sim_L], \{\mathcal{A}[\approx^{\mathrm{X}}_u]\}_{u \in \sim_L})$.

The relative succinctness of the canonical FDFAs was studied in [4, 37, 28] and is summarized in Fig. 2. A double-line (resp. single-line) arrow indicates an exponential (resp. quadratic) succinctness gap in the direction of the arrow. As the figure shows, the exponential gaps arise from the different acceptance types; among canonical representations that use the same acceptance type, the gaps are only quadratic.

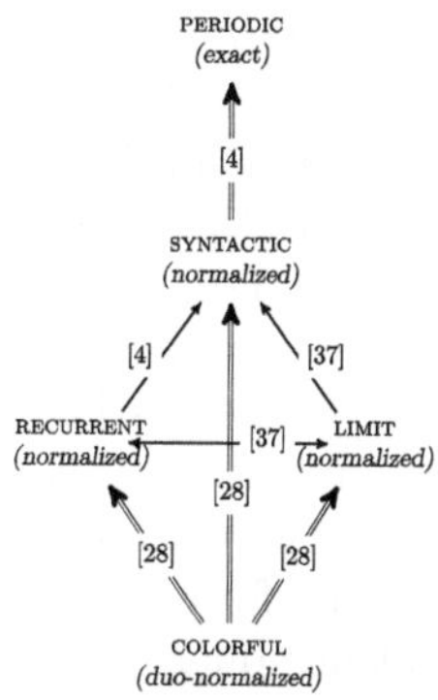

Fig. 2. Succinctness results for canonical FDFAs.

To see why different normalizations may yield more succinct representations, observe that, with respect to a given language, some decompositions may be *easier* to spot than others. Consider, for instance, the language $L = (aba + bab)^\omega$. It is immediate that $(\varepsilon, aba \cdot bab)$ is accepted, but it is less immediate that (ε, ab) is accepted as well.

Consider now the language L_n over $\Sigma_n = \{0, 1, \ldots, n\}$ given by $(\bigcup_{i \in \Sigma \setminus \{0\}} i \cdot \{0, 1, \ldots, n\}^* \cdot i)^\omega$. Some accepted periods are easy to recognize, for example $(11)^\omega$, $(33)^\omega$, $(12032001)^\omega$, and $(1203200133)^\omega$. Other accepted periods, such as $(2023)^\omega$, are harder to identify. To see why $(2023)^\omega$ is a good period, note that it is equal to $(202 \cdot 32023)^\omega$. This family of languages was used in [4] to show the exponential gap between recurrent and periodic FDFAs.

Algorithmic Properties and Succinctness. Before going into learing of FDFA we mention some of their good qualities.

FDFAs have been shown to have good algorithmic properties as acceptors for ω-regular languages: complementation can be computed in constant space; intersection, union, and membership can be computed in logarithmic space; and emptiness, universality, containment, and equivalence can be computed in non-deterministic logarithmic space [2, 28].

Their succinctness relative to ω-automata was studied in [2]. It is shown that FDFAs can be exponentially more succinct than deterministic parity automata, whereas nondeterministic Büchi automata can be exponentially more succinct than FDFAs; see Fig. 3. A double-line (resp. single-line) arrow indicates an exponential (resp. polynomial) succinctness gap in the direction of the arrow. A dashed line indicates a constant time transformation.

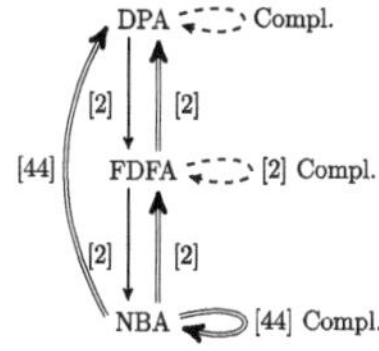

Fig. 3. Succinctness of FDFAs

In some sense, FDFAs are semi-deterministic: although both the leading and progress automata are deterministic, there remains a nondeterministic choice in selecting a decomposition.

Learning FDFAs. Calbrix, Nivat, and Podelski [18] proposed a reduction from regular ω-languages to finitary regular languages. Given $L \subseteq \Sigma^\omega$, they define $(L)_\$ = \{u\$v \mid u(v)^\omega \in L\}$. The periodic FDFA can be viewed as an FDFA representation of $(L)_\$$. Building on this reduction, Farzan et al. [25] proposed an algorithm for learning regular ω-languages that uses $\mathbf{L}^*$ at its core. The drawback of this approach is succinctness: if L has an NBA with n states, then the corresponding DFA for $(L)_\$$ may require $2^n + 3n^2$ states [36].

Query-learning algorithms for the periodic, syntactic, and recurrent FDFAs were developed in [3,38], we term them $\mathbf{L}^\omega$. The idea is to exploit the tight correspondence between the underlying relations $\sim_L$ and $\approx_u^{\text{X}}$ and the states of the FDFAs. Note, however, that this does not reduce the problem to simply running $\mathbf{L}^*$ for each of these relations, since the oracle answers membership and equivalence queries with respect to the unknown language L, rather than with respect to the languages of the corresponding automata.

The approach of $\mathbf{L}^\omega$ is to maintain two kinds of tables. The first is a leading table, with rows in Σ^* and columns in Σ^{UP}, as in $\mathbf{L}^{\mathbf{w}\omega}$, whose goal is to discover $\sim_L$. In addition, for each discovered equivalence class u of $\sim_L$, the algorithm maintains a progress table T_u whose goal is to discover $\approx_u^{\text{X}}$. The rows in T_u, are as usual, words in Σ^*; the columns are words in Σ^+. An entry (x, v) of the table T_u is filled with a pair $(b_1, b_2) \in \{\text{T}, \text{F}\} \times \{+, -\}$. The bit b_1 is set to T if $uxv \sim_L u$, and the bit b_2 records the membership of $u(xv)^\omega$ in L. Then a definition of when two rows are different follows the respective definition of $\approx_u^{\text{X}}$.

It is shown in [4] that, for the three representations, the algorithm terminates in time polynomial in the size of the periodic FDFA. The reason is that, before the full relation $\sim_L$ is discovered, a progress table may work harder than necessary, trying to identify periods whose acceptance becomes apparent only after choosing a suitable decomposition. If the relation $\sim_L$ is known in advance, then the running time is polynomial in the size of the respective FDFA. Experiments reported in [4] indicate that, on randomly generated DMAs, the phenomenon of having an intermediate representation larger than the final one rarely occurs.

Theorem 8.3 ([4]). *For every* $\text{X} \in \{\text{P},\text{S},\text{R},\text{L}\}$, *there is an* (MQ, EQ)*-learning algorithm for* X*-FDFAs running in time polynomial in* $|\approx^{\text{P}}|$. *If* $\sim_L$ *is known in advance, the running time is polynomial in* $|\approx^{\text{X}}|$. [4]

The tables used by the active learner can also be used to construct characteristic samples for X-FDFAs. These samples have size polynomial in the size of the X-FDFA itself, and allow a learner to reconstruct the corresponding X-FDFA. This yields the following corollary.

[4] The proof in [4] is for $\text{X} \in \{\text{P},\text{S},\text{R}\}$. The L-FDFA was defined later, in [37], and the corresponding result follows from its duality with the R-FDFA [37].

Corollary 8.4 *For every* X $\in$ {P,S,R,L}, X-*FDFAs are learnable in the limit using polynomial time and data, and are efficiently teachable.*

9 Strong Unambiguity and Multiplicity Automata

In Sect. 6, we presented hardness results for learning nondeterministic ω-automata. In many other contexts, well-behaved forms of nondeterministic automata have been studied, with the aim of combining the succinctness gained from nondeterminism with the good algorithmic properties of deterministic models. In light of the hardness results above, it is natural to consider such restricted nondeterministic models also from a learning perspective.

A common restriction on nondeterminism is *unambiguity.* An NFA is unambiguous, denoted UFA, if every accepted word has a unique accepting run. This definition extends naturally to Büchi automata, yielding unambiguous Büchi automata (UBAs).

Carton and Michel [19] introduced the notion of a complete strongly unambiguous Büchi automaton (CUBA). Their definition requires uniqueness of the accepting run independently of the initial state: in a CUBA for L, every word $w \in L$ is accepted from exactly one state. This entails that the transition relation is reverse deterministic, making CUBAs a promising candidate for learning.[5] The definition also requires the transition relation to be complete.

Carton and Michel proved that every regular ω-language can be accepted by a CUBA. Bousquet and Löding [17] later introduced the relaxed notion of a strongly unambiguous Büchi automaton (SUBA), which drops the requirement that the transition relation be complete. Since a CUBA is also a SUBA, every ω-regular language has a SUBA. They showed that containment and equivalence for SUBAs are solvable in polynomial time. Their proof reduces these problems for SUBAs to containment and equivalence for UFAs, which had been shown to be solvable in polynomial time by Stearns and Hunt [23].

We note that, although SUBAs are not commonly discussed in the literature, the standard translation of a linear temporal logic (LTL) formula into a generalized Büchi automaton (GNBA) produces a generalized SUBA (GSUBA). Thus, despite receiving relatively little explicit attention, this model plays a central role in formal verification. See the papers of Wilke [53,54] for more on the relationship between LTL and strongly unambiguous Büchi automata.

We also note that SUBAs are not canonical: there may be two non-isomorphic minimal SUBAs for the same language. Moreover, a DBA need not be a SUBA, simply because DBAs are not, in general, reverse deterministic.

It is shown in [6] that SUBAs are polynomially predictable with MQ. The approach is to give an (MQ, EQ)-query-learning algorithm that uses non-proper equivalence queries; that is, the hypotheses submitted in equivalence queries are represented by a model other than SUBAs.

[5] A transition relation is *reverse-deterministic* if reversing all transitions yields a deterministic transition relation.

More concretely, the approach builds on the fact that SUBAs can be reduced to UFAs, which in turn can be viewed as a special type of mod-2 multiplicity automata. Multiplicity automata admit an (MQ, EQ)-query-learning algorithm [12].

Multiplicity automata (MA) are an algebraic form of automata. In general, they define functions from words in Σ^* to values in some field $\mathcal{K}$. *Mod-2 multiplicity automata* (M2MA) are multiplicity automata over the Galois field $\mathcal{K} = GF(2)$.

Formally, a *multiplicity automaton* is a tuple $\mathcal{M} = (\Sigma, d, v_I, \{\mu_\sigma\}_{\sigma\in\Sigma}, v_F)$, where Σ is the input alphabet, d is the dimension, $v_I \in \mathcal{K}^d$ is the initial vector, for each letter $\sigma \in \Sigma$ the matrix $\mu_\sigma \in \mathcal{K}^{d\times d}$ is the transition map on σ, and $v_F \in \mathcal{K}^d$ is the final vector. Intuitively, the value of a word is obtained by multiplying the matrices corresponding to its letters, with the initial and final vectors applied at the two ends.

More formally, the transition matrices define a function $\mu : \Sigma^* \to \mathcal{K}^{d\times d}$ as follows. For the empty word ε, $\mu(\varepsilon)$ is the $d \times d$ identity matrix. For a word $w = \sigma_1\sigma_2\cdots\sigma_n$, we define $\mu(w) = \mu_{\sigma_1}\mu_{\sigma_2}\cdots\mu_{\sigma_n}$. The value computed by $\mathcal{M}$ on w is then $f(w) = v_I^\top \mu(w) v_F$, where v_I and v_F are viewed as column vectors.

We note that, although M2MAs are deterministic, they are not canonical either: a vector space over $GF(2)$ may have many different bases, and therefore the same language may admit two non-isomorphic minimal M2MAs.

The discussion above yields the following learning result.

Theorem 9.1 ([6,8]). *The class* $\mathbb{SUBA}$ *can be learned in polynomial time using* MQ *and non-proper* EQ. *It is also polynomially predictable using* MQs*, and is efficiently teachable.*

This result marks an important milestone: via the SUBA/M2MA representation, it gives the first learning algorithm for a representation of the full class of ω-regular languages. Thus, it goes beyond the previously learnable subclasses. The active-learning result is due to [6]; the efficient-teachability statement follows from [8], using the fact that a consistent SUBA can be constructed from any given sample.

The succinctness of SUBAs relative to DBAs and DMAs was studied in [17], and their succinctness relative to other learnable models was studied in [6]. These results are summarized on the right. As usual a double-line arrow indicates an exponential succinctness gap in the direction of the arrow.

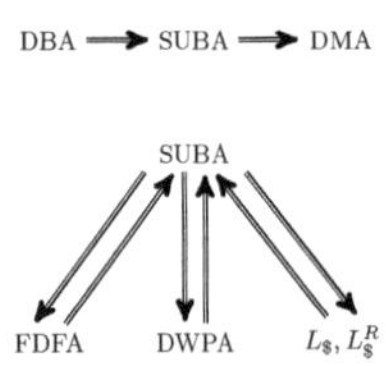

Motivated by the reduction from SUBAs to M2MAs, Angluin et al. [7] studied M2MAs as acceptors for ω-regular languages. In addition to being deterministic, M2MAs admit polynomial-time algorithms for equivalence, union, intersection, complementation, and reversal. They can also be minimized in cubic time [47,48]. Their succinctness relative to standard representations of regular languages, namely DFAs, UFAs, and NFAs, was studied in [34,7].

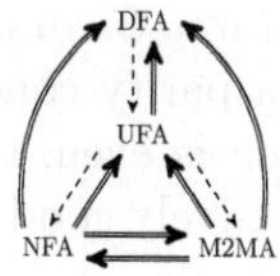

These results show that M2MAs can be more succinct than both DFAs and UFAs. In comparison with NFAs, the succinctness relation is incomparable in the worst case: there are exponential gaps in both directions (as depicted on the right).

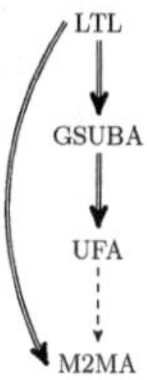

M2MAs can be used to represent ω-regular languages by considering the associated regular language $(L)_\$$. Chaining the translations from LTL to GSUBA and from GSUBA to UFA, which can then be viewed as an M2MA, yields a doubly exponential upper bound. Angluin et al. [7] provide a more direct translation from LTL to M2MAs with only a single exponential blowup: a formula of size n is translated into an M2MA of size $O(2^{2n})$. Thus, in a sense, M2MAs are as succinct as NBAs for representing LTL formulas, making them an attractive representation of ω-regular languages in the context of formal verification.

10 Prismatic Canonical Models

In this section, we present several recently introduced canonical models. Since these models are closely related to the notion of *natural colors of infinite words*, we refer to them as *prismatic models*.

History-Deterministic coBüchi (HDCA). Continuing the discussion of restricted nondeterministic models with good algorithmic properties, another line of work studies *history-deterministic* automata, also known as *good-for-games* automata; see the survey [16]. In the setting of ω-regular languages, Abu Radi and Kupferman showed that *transition-based history-deterministic coBüchi automata*, henceforth HDCA, form a canonical model [46]. Recall from Sect. 2 that nondeterministic coBüchi automata (NCA) are strictly less expressive than the full class of ω-regular languages, and have the same expressive power as their deterministic counterpart, DCA. HDCAs recognize exactly this same class.

Chains of HDCA (COCOA). Ehlers and Schewe [24] provided a canonical representation for all ω-regular languages using *chains of coBüchi automata*, henceforth COCOA. Very roughly, the idea is to classify words according to the rank they require in a DPA representation, and to use the canonical HDCA model for each class in this classification. To make this precise, they define the rank required by a word with respect to a given ω-regular language, a notion they call the *natural color* of a word. Using this notion they show that an ω-regular language L can be represented by a chain of coBüchi automata $\mathcal{C}_0, \mathcal{C}_1, \ldots, \mathcal{C}_d$, where d is the maximal natural color of L. Letting $L_i = [\![\mathcal{C}_i]\!]$, the chain satisfies $L_0 \supset L_1 \supset \cdots \supset L_d$, where L_i consists of all words whose natural color is at least i. It is further shown that given a DPA for L with n states, it is possible to extract corresponding HDCA $\mathcal{C}_i$ of size at most n.

Natural Colors. We briefly give some intuition for natural colors. Recall that, in a parity condition, a run is accepting iff the minimal color visited infinitely often is even. Consider a loop in a DPA. If the loop contains an event that is decisively good, making the word accepted regardless of what else happens in the loop, then the loop should receive a minimal even color. Dually, if it contains an event that is decisively bad, making the word rejected regardless of what else happens, then it should receive a minimal odd color. If a good event has occurred, but can still be spoiled later in the loop by a bad event, then the color should be even but not minimal. Similarly, if a bad event has occurred but can still be repaired later in the loop by a good event, then the color should be odd but not minimal.

For example, consider the language $\infty aa \wedge \neg\infty bb$, requiring aa to occur infinitely often and bb to occur only finitely often. The word $(b)^\omega$ is rejected, and an occurrence of bb in a loop is unrecoverable; hence its natural color is 1. The word $(a)^\omega$ is accepted, and an occurrence of aa in a loop is a good event. However, this can still be spoiled by a later occurrence of bb within the same loop, and therefore the natural color of $(a)^\omega$ is 2. Finally, the natural color of $(ab)^\omega$ is 3: the word is not in the language, but can be repaired by a later occurrence of aa in the loop, and the resulting loop can again be spoiled by a later occurrence of bb. See [24] for the formal definition.

Related notions were subsequently developed in other contexts. Bohn and Löding [13] introduced a notion of natural color for an *infix* of a loop, and Fisman and Sudit [29] introduced a related notion for the natural color of a *letter* in an infinite word.

Prismatic Canonical Models. The natural infix color was used in [13] to define the *precise DPA*, in the context of learning DPAs, and in [28] to define the *colorful FDFA* (from Sect. 8) in the context of analyzing the position of an FDFA in the Wagner hierarchy. The natural letter color was used in [29] to define the *robustness DPA*, in the context of giving a quantitative interpretation to ω-regular languages: words, whether satisfying or violating, are assigned values from an infinite domain, with higher values corresponding to better and more robust satisfaction of the language.

Precise DPAs and Passive Learning. The precise DPA was used to obtain a passive learning algorithm for DPAs [13]. This result also yields a notion of characteristic sample, but the size of the sample is measured with respect to the precise DPA, which may be exponentially larger than a minimal DPA. It is noted in [13] that if both the number of ranks of a DPA, denoted d, and the number of pairwise language-equivalent states, denoted k, are fixed, then the characteristic sample has size polynomial in the size of the given DPA. Using (k, d)-$\mathbb{DPA}$ to denote this class, we obtain the following.

Theorem 10.1 ([13]). *The class (k, d)-$\mathbb{DPA}$ is learnable in the limit using polynomial time and data.*

Learning HDCA. Most recently, Löding and Walukiewicz [40] characterized the right congruence relations of the canonical minimal HDCA. Based on this characterization, they developed a passive learning algorithm for $\mathbb{HDCA}$ and showed that these automata have polynomial-sized characteristic samples. This is a remarkable result: the relevant right congruences, denoted $\approx^{\mathrm{H}}$, form only a logarithmic-size subset of the right congruences $\approx^{\mathrm{S}}$ and $\approx^{\mathrm{C}}$ used in their development, which in general may be exponential in the size of a minimal DCA.

Theorem 10.2 ([40]). *The class* $\mathbb{HDCA}$ *is learnable in the limit using polynomial time and data.*

This raises the question of whether the class $\mathbb{COCOA}$ also admits polynomial-sized characteristic samples. A natural approach to learning the full class $\mathbb{COCOA}$ would be to enrich the sample with words that allow the learner to identify the natural color of a word. Once these colors are identified, the learner could classify words accordingly and, for each color and the corresponding HDCA, use a characteristic sample for that HDCA to learn the language L_i of words whose color is at least i. The main question is therefore whether, and how, the natural color of a word can be taught.

11 Discussion

The results surveyed above show that the learnability landscape for ω-regular languages is richer than for finitary regular languages. For finitary regular languages, the Myhill–Nerode congruence provides a canonical target representation, and many learning results can be understood relative to it. For ω-regular languages, no single representation plays this role uniformly.

For the full class of ω-regular languages, several positive active-learning results are known. The algorithm $\mathbf{L}^{\omega}$ provides (MQ, EQ)-query learning algorithms for periodic, syntactic, recurrent, and limit FDFAs. However, in the absence of prior knowledge of $\sim_L$, the worst-case running time is polynomial in the size of the periodic FDFA, which may be exponentially larger than the other canonical FDFAs. A different route is provided by $\mathbb{SUBA}$ and $\mathbb{M2MA}$: via M2MAs, one obtains polynomial predictability for the full class of ω-regular languages represented as SUBA. On the passive-learning side, the strongest general positive result is the extension of RPNI to construct DPAs, which is complete for the class of ω-regular languages.

From an information-theoretic perspective, characteristic samples of polynomial size are known with respect to several representations: periodic, syntactic, recurrent, and limit FDFAs; SUBAs; M2MAs for $(L)_\$$; and the precise DPA. This makes their relative succinctness particularly important. Existing succinctness results compare most of these models, except for the precise DPA, and show bidirectional exponential gaps between many pairs. For the precise DPA, it is known that it may be exponentially larger than a minimal DPA, and also exponentially larger than the syntactic FDFA unless the number of colors is fixed. A direct comparison with SUBAs would therefore be interesting.

From the perspective of canonical representations, the COCOA model, based on chains of history-deterministic coBüchi automata (HDCA), is especially intriguing. HDCAs are canonical and compact, and their underlying right congruences have recently been characterized. This characterization led to an algorithm that learns HDCAs in the limit using polynomial time and data, and in particular to polynomial-size characteristic samples. This naturally raises the question of whether $\mathbb{COCOA}$ also admits polynomial-sized characteristic samples.

Overall, the picture that emerges is that learning ω-regular languages is not governed by a single canonical automaton model. Rather, it depends on a delicate interaction between representation, succinctness, algorithmic properties, and the type of information available to the learner. A central direction for future work is therefore to identify representations that simultaneously support strong learning guarantees, compact characteristic samples, and a semantic explanation of the learned structure.

References

1. Angluin, D.: Learning regular sets from queries and counterexamples. Inf. Comput. **75**(2), 87–106 (1987)
2. Angluin, D., Boker, U., Fisman, D.: Families of DFAs as acceptors of ω-regular languages. In: 41st International Symposium on Mathematical Foundations of Computer Science, MFCS 2016, 22–26 August 2016 - Kraków, Poland, pp. 11:1–11:14 (2016)
3. Angluin, D., Fisman, D.: Learning regular omega languages. In: Auer, P., Clark, A., Zeugmann, T., Zilles, S. (eds.) ALT 2014. LNCS (LNAI), vol. 8776, pp. 125–139. Springer, Cham (2014). https://doi.org/10.1007/978-3-319-11662-4_10
4. Angluin, D., Fisman, D.: Learning regular omega languages. Theor. Comput. Sci. **650**, 57–72 (2016)
5. Angluin, D., Kharitonov, M.: When won't membership queries help? J. Comput. Syst. Sci. **50**(2), 336–355 (1995)
6. Angluin, D., Antonopoulos, T., Fisman, D.: Strongly unambiguous büchi automata are polynomially predictable with membership queries. In: Fernández, M., Muscholl, A. (eds.) 28th EACSL Annual Conference on Computer Science Logic, CSL 2020, Barcelona, Spain, 13–16 January 2020, LIPIcs, pp. 8:1–8:17. Schloss Dagstuhl - Leibniz-Zentrum für Informatik (2020)
7. Angluin, D., Antonopoulos, T., Fisman, D., George, N.: Representing regular languages of infinite words using mod 2 multiplicity automata. In: FoSSaCS 2022. LNCS, vol. 13242, pp. 1–20. Springer, Cham (2022). https://doi.org/10.1007/978-3-030-99253-8_1
8. Angluin, D., Fisman, D.: Constructing concise characteristic samples for acceptors of omega regular languages. Log. Methods Comput. Sci. **20**(4) (2024)
9. Angluin, D., Fisman, D., Shoval, Y.: Polynomial identification of ω-automata. In: TACAS 2020. LNCS, vol. 12079, pp. 325–343. Springer, Cham (2020). https://doi.org/10.1007/978-3-030-45237-7_20
10. Arnold, A.: A syntactic congruence for rational ω-languages. Theor. Comput. Sci. **39**, 333–335 (1985)

11. Barzdin, J.M., Freivalds, R.V.: On the prediction of total recursive functions. Dokl. Akad. Nauk SSSR **206**(3), 521–524 (1972). (in Russian)
12. Beimel, A., Bergadano, F., Bshouty, N.H., Kushilevitz, E., Varricchio, S.: Learning functions represented as multiplicity automata. J. ACM **47**(3), 506–530 (2000)
13. Bohn, L., Löding, C.: Constructing deterministic parity automata from positive and negative examples. TheoretiCS **3** (2024)
14. Bohn, L., Löding, C.: Constructing deterministic ω-automata from examples by an extension of the RPNI algorithm. In: Bonchi, F., Puglisi, S.J. (eds.) 46th International Symposium on Mathematical Foundations of Computer Science, MFCS 2021, Tallinn, Estonia, 23–27 August 2021, LIPIcs, pp. 20:1–20:18. Schloss Dagstuhl - Leibniz-Zentrum für Informatik (2021)
15. Boker, U.: Why these automata types? In: Barthe, G., Sutcliffe, G., Veanes, M. (eds.) LPAR-22. 22nd International Conference on Logic for Programming, Artificial Intelligence and Reasoning, Awassa, Ethiopia, 16–21 November 2018, EPiC Series in Computing, pp. 143–163. EasyChair (2018)
16. Boker, U., Lehtinen, K.: When a little nondeterminism goes a long way: an introduction to history-determinism. ACM SIGLOG News **10**(1), 24–51 (2023)
17. Bousquet, N., Löding, C.: Equivalence and inclusion problem for strongly unambiguous Büchi automata. In: Dediu, A.-H., Fernau, H., Martín-Vide, C. (eds.) LATA 2010. LNCS, vol. 6031, pp. 118–129. Springer, Heidelberg (2010). https://doi.org/10.1007/978-3-642-13089-2_10
18. Calbrix, H., Nivat, M., Podelski, A.: Ultimately periodic words of rational ω-languages. In: Brookes, S., Main, M., Melton, A., Mislove, M., Schmidt, D. (eds.) MFPS 1993. LNCS, vol. 802, pp. 554–566. Springer, Heidelberg (1994). https://doi.org/10.1007/3-540-58027-1_27
19. Carton, O., Michel, M.: Unambiguous büchi automata. Theor. Comput. Sci. **297**(1–3), 37–81 (2003)
20. Pin, J.-E., Perrin, D.: Infinite Words: Automata, Semigroups, Logic and Games. Springer (2004)
21. de la Higuera, C., Janodet, J.-C.: Inference of ω-languages from prefixes. Theor. Comput. Sci. **313**(2), 295–312 (2004)
22. Domaratzki, M., Kisman, D., Shallit, J.: On the number of distinct languages accepted by finite automata with n states. In: Third International Workshop on Descriptional Complexity of Automata, Grammars and Related Structures - DCAGRS 2001, Vienna, Austria, 20–22 July 2001. Preproceedings, pp. 67–78 (2001)
23. Stearns, E.E., Hunt, B.: On the equivalence and containment problems for unambiguous regular expressions, regular grammars and finite automata. SIAM J. Comput. **14**, 598–611 (1985)
24. Ehlers, R., Schewe, S.: Natural colors of infinite words. In: Dawar, A., Guruswami, V. (eds.) 42nd IARCS Annual Conference on Foundations of Software Technology and Theoretical Computer Science, FSTTCS 2022, IIT Madras, Chennai, India, 18–20 December 2022, LIPIcs, pp. 36:1–36:17. Schloss Dagstuhl - Leibniz-Zentrum für Informatik (2022)
25. Farzan, A., Chen, Y.-F., Clarke, E.M., Tsay, Y.-K., Wang, B.-Y.: Extending automated compositional verification to the full class of ω-regular languages. In: Ramakrishnan, C.R., Rehof, J. (eds.) Tools and Algorithms for the Construction and Analysis of Systems. Lecture Notes in Computer Science, vol. 4963, pp. 2–17. Springer, Heidelberg (2008)
26. Fisman, D.: Inferring regular languages and ω-languages. J. Log. Algebr. Meth. Program. **98**, 27–49 (2018)

27. Fisman, D., Frenkel, H., Zilles, S.: Inferring symbolic automata. Log. Methods Comput. Sci. **19**(2) (2023)
28. Fisman, D., Goldberg, E., Zimerman, O.: A robust measure on FDFAs following duo-normalized acceptance. In: Královic, R., Kucera, A. (eds.) 49th International Symposium on Mathematical Foundations of Computer Science, MFCS 2024, 26–30 August 2024, Bratislava, Slovakia. LIPIcs, , vol. 306, pp. 53:1–53:17. Schloss Dagstuhl - Leibniz-Zentrum für Informatik (2024)
29. Fisman, D., Sudit, E.: Omega-regular robustness. CoRR, abs/2503.12631 (2025)
30. Freivalds, R., Bārzdiņš, J., Podnieks, K.: Inductive inference of recursive functions: complexity bounds. In: Bārzdinš, J., Bjørner, D. (eds.) Baltic Computer Science. LNCS, vol. 502, pp. 111–155. Springer, Heidelberg (1991). https://doi.org/10.1007/BFb0019358
31. Gold, E.M.: Complexity of automaton identification from given data. Inf. Control **37**(3), 302–320 (1978)
32. Goldman, S.A., Mathias, H.D.: Teaching a smarter learner. J. Comput. Syst. Sci. **52**(2), 255–267 (1996)
33. Jürgensen, H., Thierrin, G.: On ω-languages whose syntactic monoid is trivial. Int. J. Parallel Prog. **12**(5), 359–365 (1983)
34. Kaznatcheev, A., Panangaden, P.: Weighted automata are compact and actively learnable. Inf. Process. Lett. **171**, 106133 (2021)
35. Klarlund, N.: A homomorphism concepts for omega-regularity. In: Computer Science Logic, 8th International Workshop, CSL 1994, Kazimierz, Poland, 25–30 September 1994, Selected Papers, pp. 471–485 (1994)
36. Kuperberg, D., Pinault, L., Pous, D.: Coinductive algorithms for Büchi automata. Fundam. Informaticae **180**(4), 351–373 (2021)
37. Li, Y., Schewe, S., Tang, Q.: A novel family of finite automata for recognizing and learning ømega-regular languages. In: Automated Technology for Verification and Analysis - 21st International Symposium, ATVA, pp. 53–73 (2023)
38. Li, Y., Sun, X., Turrini, A., Chen, Y.-F., Xu, J.: ROLL 1.0: ω-regular language learning library. In: Vojnar, T., Zhang, L. (eds.) TACAS 2019. LNCS, vol. 11427, pp. 365–371. Springer, Cham (2019). https://doi.org/10.1007/978-3-030-17462-0_23
39. Lindner, R., Staiger, L.: Eine Bemerkung über nichtkonstantenfreie sequentielle operatoren. Elektronische Informationsverarbeitung und Kybernetik **10**(4), 195–202 (1974)
40. Löding, C., Walukiewicz, I.: Minimal history-deterministic co-büchi automata: congruences and passive learning. In: 40th Annual ACM/IEEE Symposium on Logic in Computer Science, LICS 2025, Singapore, 23–26 June 2025, pp. 431–443. IEEE (2025)
41. Maler, O., Pnueli, A.: On the learnability of infinitary regular sets. Inf. Comput. **118**(2), 316–326 (1995)
42. Maler, O., Staiger, L.: On syntactic congruences for omega-languages. Theor. Comput. Sci. **183**(1), 93–112 (1997)
43. Manna, Z., Pnueli, A.: A hierarchy of temporal properties. In: Proceedings of the Ninth Annual ACM Symposium on Principles of Distributed Computing, Quebec City, Quebec, Canada, 22–24 August 1990, pp. 377–410 (1990)
44. Michel, M.: Complementation is much more difficult with automata on infinite words. In: Manuscript, CNET (1988)
45. Oncina, J., Garcia, P.: Identifying regular languages in polynomial time. In: Advances in Structural and Syntactic Pattern Recognition. Series in Machine Perception and Artificial Intelligence, vol. 5, pp. 99–108. World Scientific (1992)

46. Radi, B.A., Kupferman, O.: Minimization and canonization of GFG transition-based automata. Log. Methods Comput. Sci. **18**(3) (2022)
47. Sakarovitch, J.: Elements of Automata Theory. Cambridge University Press (2009)
48. Sakarovitch, J.: Automata and rational expressions. In: Pin, J.-É. (ed.) Handbook of Automata Theory, pp. 39–78. European Mathematical Society Publishing House, Zürich, Switzerland (2021)
49. Staiger, L.: Finite-state omega-languages. J. Comput. Syst. Sci. **27**(3), 434–448 (1983)
50. Trakhtenbrot, B.: Finite automata and monadic second order logic. Siberian Math J. 103–131 (1962)
51. Trakhtenbrot, B.A., Barzdin, J.M.: Finite Automata: Behavior and Synthesis. Fundamental Studies in Computer Science, vol. 1. North-Holland Publishing Company, Amsterdam (1973). Translated from Russian by D. Louvish
52. Wagner, K.: A hierarchy of regular sequence sets. In: Bečvář, J. (ed.) MFCS 1975. LNCS, vol. 32, pp. 445–449. Springer, Heidelberg (1975). https://doi.org/10.1007/3-540-07389-2_231
53. Wilke, T.: Past, present, and infinite future. In: 43rd International Colloquium on Automata, Languages, and Programming, ICALP 2016, 11–15 July 2016, Rome, Italy, pp. 95:1–95:14 (2016)
54. Wilke, T.: Backward deterministic Büchi automata on infinite words. In: FSTTCS (2017)
55. Zeugmann, T., Zilles, S.: Learning recursive functions: a survey. Theor. Comput. Sci. **397**(1), 4–56 (2008). Forty Years of Inductive Inference: Dedicated to the 60th Birthday of Rolf Wiehagen

Multisets in Reaction Systems

Victor Mitrana(✉)

National Institute of Research and Development for Biological Sciences,
296 Independentei Bd. District 6, 060031 Bucharest, Romania
victor.mitrana@incdsb.ro

Abstract. This work is a survey of several variants of reaction systems employing multisets instead of classic sets. The effect of replacing sets by multisets in reaction systems, which is a quantifying transformation, is both theoretically and practically important. There are a few theoretical developments allowed by the new variants surveyed in this paper. First, there are many ways of enabling a multiset reaction or a multiset of reactions, each one defining a specific evolution. Second, the new variants may be used as multiset computing devices. Third, the new variants are more adequate and efficient than the classic ones for modeling, numerical and statistical analysis. We consider here four variants of multiset reaction systems: basic, quantitative, valuated, and weighted. We survey the main results regarding their simulation capacity and computational power.

Keywords: Multiset · Multiset reaction system · Quantitative multiset reaction system · Valuated multiset reaction system · Weighted multiset reaction system · Simulation · Computational power

1 Introduction

An abstract computational model inspired by the chemical reaction metaphor have been introduced in [12]. The computation is defined as a succession of chemical reactions that interact with the environment where they are located. Each such reaction consumes various substances from the environment and produces new ones. The substances of each reaction are divided into *reactants*, that promote the reaction, and *inhibitors*, that block the reaction. The result of enabling a reaction is a set of *products* that are placed back in the environment. The environment may enable or block a reaction depending on the presence or absence of reactants and the absence of inhibitors. In a classic chemical reaction, the substances in the left side of a chemical equation are typically reactants while the products appear in the right side of the reaction. The model introduced in [12,13] is actually a qualitative model which means that if a substance is present in the

This work was supported by the National Authority of Research through the Core Program of the National Research, Development and Innovation Plan 2022-2027, project no. PN 23020101, Contract No. 7N/2023.

V. Brattka et al. (Eds.): CiE 2026, LNCS 16674, pp. 48–59, 2027.
https://doi.org/10.1007/978-3-032-31348-5_3

environment, it is available in unlimited amounts, hence the environment can be defined as a set. This principle is known as the *non-threshold principle.* Another principle of the model is the *non-permanency principle* [10] which states that if a compound is not explicitly produced by any reaction, it will disappear from the environment. An early overview can be found in [11]. Although this model is simple and elegant, it does not capture some important aspects of chemical and biochemical reaction systems: due to the boolean presence of reactants it can't capture the concentration or rate of substances, the non-threshold principle forces artificial "self-production" rules just to maintain elements, it also does not allow any competition on resources between reactions, the non-permanency principle may be unrealistic for many reaction systems where entities persist naturally.

One of the simplest feature that could be introduced in the definition of a reaction system is to consider *multiset.* The first step is to define the environment as a multiset. The idea of considering this type of environment has appeared in several works: Γ model [1,2], chemical abstract machine [4], quantitative reaction systems [14,15], multiset reaction systems [6], valuating reaction systems [7], etc. Furthermore, this setting offers the possibility to renounce the aforementioned two principles. Even more, the sets in the definition of a reaction may be replaced by multisets. The introduction of multisets has several consequences: reactions have now access to counted (hence, limited) resources and will compete among themselves on the resources, such that the resources may not be enough for all enabled reactions to be executed. Additionally, we also adopt the principle that resources are removed from the system only as a result of being consumed/used by an enabled reaction that got to be executed; otherwise, they are left to persist in the successor state. Using multisets in bioinspired computing models is not a new idea. In the literature, there are many contribution in this respect to various areas: computer science (programmed parallel machines [1], networks of bio-inspired processors [5], multiset automata [9]), artificial life (chemical abstract machine [4], artificial proto-cells [3], symbolic chemical systems [21]), numerous variants of membrane system [20], etc.

In this work we survey several variants of multiset reaction systems and recall the main results regarding their computational and simulation power. The survey is based on the papers [6,7,15–17]. These papers try to take into account various quantitative aspects. To this aim, we give here a short and informal description of these models. A model that aborts both the non-permanency and non-threshold principles was introduced in [6] under the name of multiset reaction system. The main novelty is that both the environment as well as all the parameters of a reaction are described by multisets. Starting from the existence of different types of inhibitors: competitive (they compete with the reactant substrate for the active site of the enzyme), non-competitive (they react with the enzyme-substrate complex slowing down the rate of reaction), and uncompetitive (they bind to the enzyme and enhances the binding of substrate, having an overall effect similar to the non-competitive inhibitors), multiset reaction systems taking into account this aspect have been considered in [15] and [16]. Each reaction contains a set of pairs such that a reactant is paired with a specific inhibitor. Assigning valences (numerical values) to the molecules makes that competition graded instead of absolute. In this way, both reactants and inhibitors do their job with some relative strength. A related model computing sets of strings was proposed in [19]. In [7], it was introduced and investigated a new model of multiset reaction system called valuated multiset reaction system. The function that assigns weights to the compounds in [7] was called valence. Actually, the valence function used in [7] should be seen as a function that assigns a capacity to promote or oppose causation of each reactant or inhibitor, respectively. Furthermore, it makes sense to discuss about the weight of a reaction which might be previously defined or it might be dynamically computed by the valence of reactants and inhibitors of that reaction that are present in the environment. Such approach has been considered in [18].

After presenting the main concepts and notations, accompanied with some examples, we recall the main results regarding the capacity of each variant to simulate different evolutions. Afterwards, these models are considered as multiset computational models and investigate their computational power. Finally, we discuss possible directions for further investigation.

2 Formal Preliminaries

2.1 Multisets

For a finite set X, $card(X)$ indicates its cardinality. We now recall the definition of a multiset. A multiset is an extension of the concept of a set, in which each element is accompanied by its number of copies. Formally, given a finite set U, we define a *multiset* X over U via a *membership function* $\sigma_X : U \longrightarrow \mathbf{N}$, where $\sigma_X(x)$ gives the number of copies of element $x \in U$ in X. For the sake of simplicity, we represent a multiset over U as a string over U considered as an alphabet. Clearly, this string is a representative of the equivalence class with respect to permutation. With the string notation, for a multiset X, we have $\sigma_X(x) = |X|_x$, the number of occurrences of x in X. Furthermore, the support of the multiset X is set $alph(X) = \{x \in U \mid \sigma_X(x) > 0\}$, and the weight of X is actually the length of string representation, that is $|X| = \sum_{x \in U} |X|_x$.

The *empty multiset* is denoted by ε, that is, $|\varepsilon|_a = 0$ for all $a \in U$. The set of all multisets on U is denoted by U° while the set of all nonempty multisets is denoted by $U^\#$. A subset of U° is called *macroset*.

We recall the standard operations on multisets defined as operations on strings. For two multisets X, Y over a set U, we define:

- the *inclusion* relation $X \sqsubseteq Y$, where $|X|_x \leq |Y|_x$, for all $x \in U$;
- the *addition* of multisets X and Y is just the concatenation XY of strings X and Y. Note that $|XY|_x = |X|_x + |Y|_x$, for all $x \in U$, hence the order of concatenation does not matter;
- the *difference* of multisets X and Y is denoted by X/Y, where $|X/Y|_x = |X|_x - |Y|_x$ for each $x \in U$, provided that $Y \sqsubseteq X$;

- the *scalar multiplication* of multiset X by the positive integer k is X^k, where $|X^k|_x = k \cdot |X|_x$, for all $x \in U$.

2.2 Multiset Reaction Systems

We assume that the reader is familiar with the reaction system model as introduced in [12,13]. Briefly, for a background set S, a *reaction over* S is a triple of subsets of S, $a = (R_a, I_a, P_a)$, such that $R_a \cap I_a = \emptyset$. The elements of R are called reactants, those of I are called inhibitors, while P is a multiset of products. A *reaction system* is then a pair (S, A) for some set of reactions A over S.

The main difference between a reaction system and a *multiset reaction system* is that each component of a multiset reaction is a multiset, and not a standard set. Furthermore, the states of a multiset reaction system are multisets as well, they contain the quantities of each reactant present in the environment. In a classic reaction system, an entity can participate in arbitrarily many reactions simultaneously, the resource competition as well as the capacity of the entities is not captured. Furthermore, presence is boolean: an entity is either there or not. Consequently, following [6], a *multiset reaction* (MR for short) over a set S is actually a triple $a = (R, I, P)$, where R, I and P are multisets over a background set S and $alph(R) \cap alph(I) = \emptyset$. Similarly to reaction systems, a multiset reaction system (MRS for short) is a pair $\mathcal{A} = (S, A)$, where S is the background set and A is a finite set of multiset reactions over S.

Given a multiset $T \in S^\#$, which we refer to as environment or current state, we say that T enables a, denoted the boolean function $\mathsf{men}_a(T)$, if $|T|_x \geq |R|_x$, for all $x \in alph(R)$, and $|T|_y = 0$, for all $y \in I$. The result of enabling a is $\mathsf{mres}_a(T) = P(T/R)$.

2.3 Quantitative Multiset Reaction Systems

A quantitative multiset reaction system is an extension of a multiset reaction system with a set of relationships reactant-inhibitor that intends to capture the phenomenon of selective inhibitors. A *quantitative multiset reaction (QMR)* α over a background set S has four parameters, $\alpha = (R, I, X, P)$, where (R, I, P) is a MR over S, and $X \subseteq alph(R) \times alph(I)$ is the set of selective interactions. Each pair $(x, y) \in X$ means that the inhibitor y acts selectively on the reactant x. We now continue with the necessary conditions for a multiset $T \in S^\#$ to enable a quantitative multiset reaction $\alpha = (R, I, X, P)$, following [7,15]. The first step is to compute the multiset $T(\alpha)$ by the next algorithm.

Algorithm 1

```
begin
for each (x, y) ∈ X
   while ((|T|_x ≥ 1)&(|T|_y ≥ 1))
      T := T/xy;
   endwhile
```

endfor;
$T(\alpha) := T$;
end

After computing $T(\alpha)$ by Algorithm 1, we say that T enables α, and write $\mathsf{qmen}(\alpha, T)$, if $R \sqsubseteq T(\alpha)$, as well as $|T(\alpha)|_y = 0$, for all $y \in alph(I)$. The result of enabling α by T is $\mathsf{qmres}(\alpha, T) = P(T(\alpha)/R)$.

Example 1. *Let $S = \{a, b, c, d\}$ be a background set. Now, let $\alpha = (a^2c^3, bd, \{(a,b)(a,d)\}, d^2)$ be a quantitative multiset reaction over S. With $T = a^5b^2c^7$ we get $T(\alpha) = a^3c^7$ and $\mathsf{qmres}(\alpha, T) = ac^4d^2$.*

A quantitative multiset reaction system is a pair (S, A), where A is a set of quantitative multiset reactions over S. It is obvious that a quantitative multiset reaction system is an extension of a multiset reaction system because if the sets of selective interactions of every reaction is empty we actually have a multiset reaction system.

2.4 Valuated Multiset Reaction Systems

To define the notion of valence, we consider that each multiset reaction, having a multiset of inhibitors, defines proper levels of expression for its entities as a function that assigns natural numbers to the entities.

Given $\alpha = (R, I, P)$ a MR over a background set S, a *valence function* associated with α is a function $\vartheta_\alpha : alph(RI) \to \mathbb{N}$. The pair $vmr(\alpha) = (\alpha, \vartheta_\alpha)$ is said to be a *valuated multiset reaction* (VMR).

The next definition extends the notion of enabling for valuated multiset reactions. With $T \in S^\#$, $\alpha = (R, I, P)$ a multiset reaction, and $vmr(\alpha) = (\alpha, \vartheta_\alpha)$ a VMR over S let

$$\nu_R(\alpha, T) = \sum_{x \in alph(R)} \vartheta_\alpha(x) \cdot \min(|R|_x, |T|_x),$$
$$\nu_I(\alpha, T) = \sum_{y \in alph(I)} \vartheta_\alpha(y) \cdot \min(|I|_y, |T|_y).$$

Then, $\Upsilon(vmr(\alpha), T) = (\nu_R(\alpha, T), \nu_I(\alpha, T))$ is the *valuation* of $vmr(\alpha)$ *in* T.

We say that $vmr(\alpha)$ is enabled by T, and write $\mathsf{vmen}(vmr(\alpha), T)$, if $R \sqsubseteq T$ and $\nu_R(\alpha, T) > \nu_I(\alpha, T)$. Moreover, the effect of *applying* $vmr(\alpha)$ on T is $\mathsf{vmres}(vmr(\alpha), T) = (T/R)P$.

Example 2. *We consider the same background set S and the MR in the previous example with the valence function defined by $\vartheta_\alpha(a) = 1$, $\vartheta_\alpha(b) = 2$, $\vartheta_\alpha(c) = 1$, $\vartheta_\alpha(d) = 1$. For $T = a^5b^3c^7$ we have*

$$\nu_R(\alpha, T) = \sum_{x \in \{a,c\}} \vartheta_\alpha(x) \cdot \min(|R|_x, |T|_x) = 2 + 3 = 5.$$
$$\nu_I(\alpha, T) = \sum_{y \in \{b,d\}} \vartheta_\alpha(y) \cdot \min(|I|_y, |T|_y) = 2.$$

As $R \sqsubseteq T$ *and* $\nu_R(\alpha, T) > \nu_I(\alpha, T)$*, it follows that* $vmr(\alpha)$ *is enabled by* T *and* $\mathsf{vmres}((vmr(\alpha), T) = a^3b^3c^4d^2$. □

As one can easily see, the difference regarding the enabling between multiset reactions and valuated multiset reactions is the replacement of the condition that requires the absence of inhibitors by a condition which requires that the valuation of reactants overcomes that of inhibitors. This seems to mirror some real regulatory systems: weak inhibitors may be overridden by strong activation, multiple activators can compensate for partial inhibition, thresholds matter more than absolute presence. A few explanation why to consider that each reaction has its own valence function is more realistic that a global valence function. Obviously, a global valence function implicitly asserts that an entity contributes the same activating (or inhibiting) power in every reaction. That is rarely realistic because a protein can strongly activate one pathway and weakly activate another, the same molecule can be a strong inhibitor in one reaction and a mild one in another, binding affinity is interaction-specific. So reaction-local weights are strictly more expressive.

As usual, a valuated multiset reaction system (VMRS) over a background set S is a pair (S, A), where A is a set of valuated multiset reactions.

2.5 Weighted Multiset Reaction Systems

A weighted multiset reaction (WMR) over a background set S is a pair $(vmr(\alpha), \mu)$, where $vmr(\alpha)$ is a VMR over S and μ is a non-negative integer which is the *weight* of $vmr(\alpha)$. As one can see, a WMR is a slight extension of a VMR by adding a weight.

We say that $T \in S^\#$ enables the WMR $(vmr(\alpha), \mu)$ with the threshold $k > 1$, and write $\mathsf{vmen}_k(vmr(\alpha), \mu)$, if T enables $vmr(\alpha)$ and $\frac{(\nu_R(\alpha,T) - \nu_I(\alpha,T))}{\mu} \geq k$. The result of this activation is $\mathsf{vmres}((vmr(\alpha), \mu), T) = (T - R) + P$.

Example 3. *In Example 2,* T *enables* $(vmr(\alpha), 1)$ *with any threshold* $1 < k \leq 3$*; it enables* $(vmr(\alpha), 2)$ *with any threshold* $1 < k \leq 3/2$*, but it can't enable* $(vmr(\alpha), \mu)$ *with any threshold for any* $\mu \geq 3$. □

As usual, a weighted multiset reaction system (WMRS) over a background set S is a pair (S, A), where A is a set of weighted multiset reactions.

2.6 Evolutions

We recall the definition of an arbitrary event for each variant of multiset reaction systems defined above. Let S be a background set and $\mathcal{A} = (S, A)$ be a MRS/QMRS/VMRS/WMRS, with $A = \{\alpha_1, \alpha_2, \ldots, \alpha_n\}$, for some $n \geq 1$, and $\alpha_i = (R_i, I_i, P_i)$, $1 \leq i \leq n$. Furthermore, let $\Delta \in A^\#$ be a multiset of reactions. In the following we will present the definitions for evolutions in a unified manner. For a better reading, the multiset reaction system for which is given the definition appears in a box.

MRS We define the resource and inhibitor multisets of Δ by $R_\Delta = \prod_{1\le i\le n} R_i^{|\Delta|_{\alpha_i}}$, and $I_\Delta = \bigcup_{|\Delta|_{\alpha_j}\ge 1} alph(I_j)$. We say that $T \in S^\#$ *enables simultaneously* the reactions of Δ, denoted $\mathsf{MEN}(\Delta, T)$, if $R_\Delta \sqsubseteq T$ and $|T|_x = 0$ for all $x \in I_\Delta$.

In this case, the result of *applying the multiset of activated reactions* Δ to the state T is given by $\mathsf{MRES}(\Delta, T) = \prod_{1\le i\le n} P_i^{|\Delta|_{\alpha_i}}(T/R_\Delta)$. Please note that, though the brackets does not belong to the string, we use them for a better reading.

QMRS We now define the same event for QMRS. To this aim, we extend the definitions of the reactions in A by $\alpha_i = (R_i, I_i, X_i, P_i)$, $1 \le i \le n$. For $T \subseteq S^\#$ and $\Delta = \alpha_{i_1}\alpha_{i_2}\dots\alpha_{i_k}$, for some $k \ge 1$, we compute the set T_Δ as follows:

Algorithm 2.

```
begin
for j = 1 to k
   T := T(α_{i_j});
endfor
T_Δ := T;
end
```

We now say that T enables simultaneously the reactions of Δ, denoted by $\mathsf{QMEN}(\Delta, T)$ if $R_\Delta \sqsubseteq T_\Delta$ and $|T_\Delta|_x = 0$ for all $x \in I_\Delta$. Note that the order of elements in Δ does not matter. If T enables simultaneously the reactions of Δ, the result of enabling Δ is $\mathsf{QMRES}(\Delta, T) = \prod_{1\le i\le n} P_i^{|\Delta|_{\alpha_i}}(T_\Delta/R_\Delta)$.

VMRS The same event is now defined for a valuated multiset reaction system. Let $\Gamma = (S, A)$ be a VMRS with $A = \{vmr(\alpha_1), vmr(\alpha_2), \dots, vmr(\alpha_n)\}$. For any multiset $\Delta = vmr(\beta_1)vmr(\beta_2)\dots vmr(\beta_k) \in A^\#$, $vmr(\beta_i) = (\beta_i, \vartheta_{\beta_i})$, $\beta_i = (R_{\beta_i}, I_{\beta_i}, P_{\beta_i})$, $1 \le i \le k$, and a multiset $T \in S^\#$, we define the boolean function ν_Δ as follows:

$$\nu_\Delta(T) = \prod_{j=1}^{k}(\nu_{R_{\beta_j}}(\beta_j, T) > \nu_{I_{\beta_j}}(\beta_j, T)).$$

Given a multiset Δ as above and $T \in S^\#$, we say that T enables Δ, and write $\mathsf{VMEN}(\Delta, T)$, if $R_\Delta \sqsubseteq T$ and $\nu_\Delta(T)$ is true. Note that the order of indexes in Δ has no importance. The results of such an activation is

$$\mathsf{VMRES}(\Delta, T) = (T/R_\Delta)\prod_{q=1}^{k} P_{\beta_q}.$$

Example 4. *For a better understanding, we give an example. Let* $S = \{a, b, c, d\}$ *and* α_1, α_2 *be two VMR over* S *defined as follows.*
$\alpha_1 = (a^2c^3, bd, d^2)$, $\quad \vartheta_{\alpha_1}(a) = 1, \quad \vartheta_{\alpha_1}(b) = 2, \quad \vartheta_{\alpha_1}(c) = 1, \quad \vartheta_{\alpha_1}(d) = 1.$
$\alpha_2 = (a^2bc, dd, abd)$, $\quad \vartheta_{\alpha_2}(a) = 2, \quad \vartheta_{\alpha_2}(b) = 1, \quad \vartheta_{\alpha_2}(c) = 2, \quad \vartheta_{\alpha_2}(d) = 3.$
For $T = a^7b^3c^7d$ *and we have*

$$\nu_{R_{\alpha_1}}(\alpha_1, T) = \sum_{x \in \{a,c\}} \vartheta_{\alpha_1}(x) \cdot \min(|R_{\alpha_1}|_x, |T|_x) = 2 + 3 = 5.$$
$$\nu_{I_{\alpha_1}}(\alpha_1, T) = \sum_{y \in \{b,d\}} \vartheta_{\alpha_1}(y) \cdot \min(|I_{\alpha_1}|_y, |T|_y) = 1 + 3 = 4.$$

Analogously,

$$\nu_{R_{\alpha_2}}(\alpha_2, T) = \sum_{x \in \{a,b,c\}} \vartheta_{\alpha_2}(x) \cdot \min(|R_{\alpha_2}|_x, |T|_x) = 4 + 1 + 2 = 7.$$
$$\nu_{I_{\alpha_2}}(\alpha_2, T) = 3.$$

We consider now $\Delta = vmr(\alpha_1)vmr(\alpha_2)vmr(\alpha_1)$. *It follows that* $\nu_\Delta(T)$ *is true because* $5 > 4$ *and* $7 > 3$. *As* $R_\Delta = a^6bc^7$ *it follows that* T *enables* Δ *and* $\mathsf{VMRES}(\Delta, T) = a^2b^3d^6$. □

WMRS Let $\mathcal{A} = (S, A)$ be a WMRS with $A = \{(vmr(\alpha_1), \mu_1), (vmr(\alpha_2), \mu_2), \ldots, (vmr(\alpha_n), \mu_n)\}$. For a multiset $\Delta = (vmr(\beta_1), q_1)(vmr(\beta_2), q_2) \ldots (vmr(\beta_k), q_k) \in A^\#$, and $T \in S^\#$, we define

$$p(\Delta, T) = min\left\{ \frac{\nu_{R_{\beta_j}}(\beta_j, T) - \nu_{I_{\beta_j}}(\beta_j, T))}{q_j} \mid 1 \leq j \leq k \right\}.$$

Now, we say that T enables Δ with threshold t if $R_\Delta \sqsubseteq T$ and $p(\Delta, T) \geq t$.

Example 5. *If we add the weights* $\mu_1 = \mu_2 = 2$ *to the valuated multiset reactions in Example 4,* T *cannot enable* $\Delta' = (vmr(\alpha_1), \mu_1)(vmr(\alpha_2), \mu_2)(vmr(\alpha_1), \mu_1)$ *with any threshold as* $p(\Delta', T) = 1$.
However, T *can enable* $\Gamma = (vmr(\alpha_2), \mu_2)(vmr(\alpha_2), \mu_2)$ *with any threshold* $1 < t \leq 2$. *The result of this activation is* $\mathsf{WMRES}(\Gamma, T) = a^5b^3c^5d^3$. □

We define three types of *events* generated by $T \in S^\#$ for MRS. These events are defined analogously for the other three variants of multiset reaction systems. Let $\mathcal{A} = (S, A)$ be a MRS over S and T be a multiset over S.

- The multiset T over S generates an *arbitrary event* in $\mathcal{A}$ if there exists the multiset $\Delta \in A^\#$ such that $\mathsf{MEN}(\Delta, T)$ is true. The result of this event is $\mathsf{MRES}(\Delta, T)$.
- The multiset T over S generates a *sequential event* in $\mathcal{A}$ if there exists a MR $\alpha \in A$ such that $\mathsf{men}(\alpha, T)$ is true. The result of this event is $\mathsf{mres}(\alpha, T)$.
- The multiset T over S generates a *maximal event* in $\mathcal{A}$ if there exists $\Delta \in A^\#$ such that $\mathsf{MEN}(\Delta, T)$ is true, and there is no $\alpha \in A$ such that $\mathsf{MEN}(\Delta\alpha, T)$ is true. The result of this event is $\mathsf{MRES}(\Delta, T)$. A multiset Δ as above is called a *maximally enabled* multiset.

Note that Δ in Example 4 is a maximally enabled multiset. It is worth mentioning that for defining these events in a WMRS one considers a universal threshold.

Given a MRS $\mathcal{A} = (S, A)$, an arbitrary *evolution* in $\mathcal{A}$, generated by $T \in S^\#$ is the sequence $D_0, D_1, \ldots$ of multisets over S, defined as follows:

1. $D_0 = T$,
2. for $i \geq 0$, $D_{i+1} = \mathsf{MRES}(\Delta_i, D_i)$, for some arbitrary multiset $\Delta_i \in A^\#$. That is D_{i+1} is generated from D_i as the result of an arbitrary event.

The sequential evolution as well as the maximal evolution is defined in the same way as above by imposing that $\Delta_i \in A$ and Δ_i to be a maximally enabled multiset, respectively. Furthermore, the same evolutions have been defined for QMRS, VMRS, and WMRS.

3 Simulations Between Evolutions

We say that arbitrary evolutions can be simulated by sequential evolutions in MRS if for any MRS $\mathcal{A} = (S, A)$ there exists a MRS $\mathcal{A}' = (S', A')$ such that for any $T \in S^\#$ and any arbitrary evolution $D_0, D_1, \ldots$ generated by T in $\mathcal{A}$, there is a sequential evolution $C_0, C_1, \ldots$ generated by T in $\mathcal{A}'$, and a sequence of integers $0 = i_0 < i_1 < \ldots$ such that $D_j = C_{i_j}$, for all $j \geq 0$. This definition can be extended for any pair of evolutions in MRS but also in other variants of multiset reaction systems.

Theorem 1. [6,7,15–18] *Arbitrary as well as maximal evolutions can be simulated by sequential evolutions in any variant of multiset reaction system.*

We now extend the above definition for simulations in different multiset reactions systems. We give the definition for the sequential evolutions, the definition for the other evolutions is defined analogously. Sequential evolutions in a multiset reaction system MRS/QMRS/VMRS/WMRS can be simulated by sequential evolutions in another multiset reaction system MRS/QMRS/VMRS/WMRS if for any $\mathcal{A} = (S, A)$ in $\{MRS, QMRS, VMRS, WMRS\}$ there exists $\mathcal{A}' = (S', A')$ in $\{MRS, QMRS, VMRS, WMRS\}$, $\mathcal{A}$ and $\mathcal{A}'$ are of different type, such that for any $T \in S^\#$ and any sequential evolution $D_0, D_1, \ldots$ generated by T in $\mathcal{A}$, there is a sequential evolution $C_0, C_1, \ldots$ generated by T in $\mathcal{A}'$, and a sequence of integers $0 = i_0 < i_1 < \ldots$ such that $D_j = C_{i_j}$, for all $j \geq 0$.

Theorem 2. [6,7,15–18] *Sequential evolutions in any variant of multiset reaction system can be simulated by sequential evolutions in any other variant of multiset reaction system.*

4 Multiset Reaction Systems as Models of Computation

We now define the computation of macrosets according to different evolutions. To this aim, we introduce to the background set S a special "omnipotent inhibitor", denoted by †, and assume it to be present in the inhibitor set of each reaction. This is similar to the idea of 'dummy' inhibitors, that can be used to switch easily between a setup where inhibitor set of all reactions must be non-empty, and one without this constraint. Such an inhibitor is sufficient for multiset reaction systems and quantitative multiset reaction systems. For valuated multiset reaction systems, each reaction associates a huge valence to this inhibitor such that if it appears in the environment, no event can take place anymore. Obviously, given a valuated multiset reaction system, the valence of this inhibitor can be easily computed as follows. Let $\mathcal{A} = (S, A)$ be a valuated multiset reaction system with $A = \{vmr(\alpha_i, \vartheta_{\alpha_i}) \mid 1 \leq i \leq n\}$; we set

$$\vartheta_{\alpha_i}(\dagger) = max\left\{ \sum_{x \in alph(R_{\alpha_i} I_{\alpha_i})} \vartheta_{\alpha_i}(x) \cdot |R_{\alpha_i} I_{\alpha_i}|_x \mid 1 \leq i \leq n \right\} + 1. \quad (1)$$

Then a MRS/QMRS/VMRS/WMRS with this property is called a computing MRS/QMRS/VMRS/WMRS, respectively. They are denoted by cMRS/cQMRS/cVMRS/cWMRS, respectively. Given a cMRS $\mathcal{A}$, we say that $T \in S^{\#}$ is computed by arbitrary/maximal/sequential evolutions in $\mathcal{A}$ if there exists an arbitrary/maximal/sequential evolution $C_0 = T, C_1, C_2, \ldots, C_m$, for some $m \geq 1$, such that $\dagger \in supp(C_m)$, so that no further event is possible. The macroset of all multisets computed by arbitrary/maximal/sequential evolutions in $\mathcal{A}$ is called the *macroset computed by* $\mathcal{A}$, and is denoted by $MRS_{arb}(\mathcal{A})$, $MRS_{max}(\mathcal{A})$, and $MRS_{seq}(\mathcal{A})$, respectively. Analogously, one defines and denotes the macrosets computed by a cQMRS/cVMRS/cWMRS.

The following results have been proved in [6,7,15–18]:

Theorem 3.
1. The classes of macrosets computed by arbitrary, maximal, and sequential evolutions in cMRS coincide.
2. The classes of macrosets computed by arbitrary, maximal, and sequential evolutions in cQMRS coincide.
3. The classes of macrosets computed by arbitrary, maximal, and sequential evolutions in cVMRS coincide.
4. The classes of macrosets computed by arbitrary, maximal, and sequential evolutions in cWMRS coincide.

Rather surprisingly, we also have:

Theorem 4. *The classes of macrosets computed by sequential evolutions in cMRS, cQMRS, cVMRS, and cWMRS coincide.*

Actually, all these models are equivalent to the multiset Turing machine with detection defined in [9], hence register machine as well.

5 Conclusion

We have surveyed four variants of multiset reaction systems and recall the capabilities of them to simulate various kind of evolutions as well as to simulate each other. In many proofs, the evolutions are simulated by large numbers of intermediate states. We do not know whether these numbers can be reduced to constants. Along the same lines, many proofs are obtained via sequential evolutions. It would be interesting to find direct simulations and their characteristics.

References

1. Banâtre, J.P., Coutant, A., Le Metayer, D.: A parallel machine for multiset transformation and its programming style. Futur. Gener. Comput. Syst. **4**, 133–144 (1988)
2. Banâtre, J.P., Le Metayer, D.: Programming by multiset transformation. Commun. ACM **36**, 98–110 (1993)
3. Bedau, M.A., McCaskill, J.S., Packard, N.H., Rasmussen, S.: Chemical evolution among artificial proto-cells. In: Artificial Life VII: Proceedings of the Seventh International Conference on Artificial Life, pp. 54–63. MIT Press (2000)
4. Berry, G., Boudol, G.: The chemical abstract machine. Theor. Comput. Sci. **96**, 217–248 (1992)
5. Bottoni, P., Labella, A., Mitrana, V.: Networks of polarized multiset processors. J. Comput. Syst. Sci. **85**, 93–103 (2017)
6. Bottoni, P., Mitrana, V., Petre, I.: Multiset reaction systems. In: Jiménez- López, M.D., Vaszil, G. (eds.) Languages of Cooperation and Communication. LNCS, vol. 15840, pp. 179–193. Springer (2025)
7. Bottoni, P., Labella, A., Mitrana, V., Păun, M., Petre, I.: Multiset reaction systems with valences. J. Autom. Lang. Comb. (in press)
8. Bottoni, P., Labella, A., Mitrana, V., Petre, I.: Immersing reaction system specifications in evaluating environments (submitted)
9. Csuhaj-Varjú, E., Martín-Vide, C., Mitrana, V.: Multiset automata. In: Calude, C., Pŭn, G., Rozenberg, G., Salomaa, A. (eds.) Multiset Processing, Mathematical, Computer Science, and Molecular Computing Points of View, WMP 2000. Lecture Notes in Computer Science, vol. 2235, pp. 69–84. Springer, Heidelberg (2001)
10. Ehrenfeucht, A., Main, M.G., Rozenberg, G.: Functions defined by reaction systems. Int. J. Found. Comput. Sci. **22**(1), 167–178 (2011)
11. Ehrenfeucht, A., Petre, I., Rozenberg, G.: Reaction systems: a model of computation inspired by the functioning of the living cell. In: Konstantinidis, S., Moreira, N., Reis, R., Shallit, J. (eds.) The Role of Theory in Computer Science, pp. 1–32. World Scientific (2017)
12. Ehrenfeucht, A., Rozenberg, G.: Basic notions of reaction systems. In: Calude, C.S., Calude, E., Dinneen, M.J. (eds.) DLT 2004. LNCS, vol. 3340, pp. 27–29. Springer, Heidelberg (2004). https://doi.org/10.1007/978-3-540-30550-7_3
13. Ehrenfeucht, A., Rozenberg, G.: Reaction systems. Fundam. Informaticae **75**(1–4), 263–280 (2007)
14. Meski, A., Koutny, M., Penczek, W.: Towards quantitative verification of reaction systems. In: Amos, M., Condon, A. (eds.) UCNC 2016. LNCS, vol. 9726, pp. 142–154. Springer, Cham (2016). https://doi.org/10.1007/978-3-319-41312-9_12
15. Mitrana, V., Păun, M., Petre, I., Prelipcean, A.M.: Quantitative reaction systems. In: 5th International Conference on Innovative Research in Applied Science, Engineering and Technology (IRASET), pp. 1–6. IEEE Press (2025)
16. Mitrana, V., Pŭn, M., Petre, I.: Two new types of evolution in quantitative reaction systems. In: 23rd Jubilee International Symposium on Intelligent Systems and Informatics (SISY), pp. 25–30. IEEE Press (2025)

17. Mitrana, V., Păun, A., Păun, M.: Valuated multiset reaction systems with maximal evolution. In: International Conference on Soft Computing Models in Industrial and Environmental Applications (2026, accepted)
18. Mitrana, V., Păun, A., Păun, M.: Weighted multiset reaction systems. In: International Conference on the Interplay between Natural and. Artificial Computation (2026, accepted)
19. Okubo, F., Kobayashi, S., Yokomori, T.: Reaction automata. Theor. Comput. Sci. **429**, 247–257 (2012)
20. Păun, G., Rozenberg, G., Salomaa, A.: The Oxford Handbook of Membrane Computing. Oxford University Press, London (2009)
21. Suzuki, Y., Tanaka, H.: Symbolic chemical system based on abstract rewriting system and its behavior pattern. Artif. Life Robot. **1**, 211–219 (1997)

Computing with Spikes and Membranes: Computability and Complexity in Spiking Neural P Systems

David Orellana-Martín[1,2(✉)]

[1] Research Group on Natural Computing, Department of Computer Science and Artificial Intelligence, Universidad de Sevilla, Avda. Reina Mercedes s/n, 41012 Sevilla, Spain
dorellana@us.es

[2] SCORE Lab, I3US, Universidad de Sevilla, Avda. Reina Mercedes s/n, 41012 Sevilla, Spain

Abstract. Spiking neural P systems are formal models of computation inspired by the structure and connections of neural networks and how information is transmitted through energy pulses or spikes. This contribution presents a theoretical overview of spiking neural P systems and some interesting properties of this framework which lies in the area of membrane computing. We first recall the basic ingredients of the model, including spikes, rules, synapses, configurations, and derivation modes. We also examine the usual ideas behind the proofs of completeness and universality, mostly by simulation of universal models. Apart from computability issues, some computational complexity aspects, especially concerning time, will be introduced. We then discuss how complexity-theoretic questions can be formulated in this setting, with special attention to time, precomputed resources, uniformity, and the time–space trade-off. Since SNP systems are related to the third generation of neural network models, special attention is paid to the representation of information, including input/output encodings based on spike multiplicities and temporal patterns. Finally, some representative variants and extensions, as well as open problems related to both computability and complexity-theoretic characterizations, will be discussed. In general, spiking neural P systems are presented as a formal laboratory for studying computation through the interaction of signals, time, and structure.

Keywords: Spiking neural P systems · Membrane computing · Natural computing · Computability theory · Computational complexity theory · Universality · Formal models of computation

1 Introduction: Natural Computation and Formal Languages

Natural computing encompasses different perspectives connecting nature and computation; in particular, four different directions are usually taken: (i) Nature

V. Brattka et al. (Eds.): CiE 2026, LNCS 16674, pp. 60–77, 2027.
https://doi.org/10.1007/978-3-032-31348-5_4

can be seen as an information-processing mechanism, interpreting biological processes as computational processes, such as the ones occurring in living cells, taking the genomic regulatory system as a computational system [14]; (ii) Nature can be used as a hardware platform, overcoming the limitations of classical silicon chips, such as in molecular computing [1] or quantum computing [10]; (iii) Nature can be synthesized by means of computing, computation being a way to understand how nature works, characterizing real-life systems with models such as Lindenmayer systems [33]; and (iv) Nature can be used as an inspiration for new models of computation, whose computational properties can be studied and used for modelling or experimental purposes, such as cellular automata [43], evolutionary computation [7], neural computing [9] and membrane computing [29].

Indeed, the last two models are especially relevant throughout this work. On the one hand, neural computing, started in [21] by Warren McCulloch and Walter Pitts, takes inspiration from how neurons are connected and send their signals through the synapses. This foundational model is considered as the first generation of neural networks, which usually return a binary answer. The second generation, which underlies many widely used models nowadays, improves the previous model with new non-linear activation functions, transforming the output into a real value [34]. The third generation corresponds to models closer to real biological neurons, where neurons are activated only after reaching an internal charge potential and where information is represented through temporally distributed spikes rather than only through binary or real-valued activations [19]. Models of computation within this framework are generically called *neural networks*. On the other hand, membrane computing, started in [25,26] by Gheorghe Păun, takes inspiration from the compartmentalized structure of living cells and how chemical components react within the different compartments of a cell [27], or the exchange of chemical components between different cells in living tissues [20]. The models of computation characterizing this framework are called *P systems* or *membrane systems*. A related neural-like model, arising from the interaction between neural computing and membrane computing, is given by the so-called *spiking neural P systems* [12]. These models of computation can be viewed as a membrane computing counterpart of the third generation of neural networks within the framework of membrane computing [30,46]. These models have been studied both from a theoretical [17] and from a practical [45] point of view. In this study, some remarks concerning computability theory and computational complexity theory will be recalled.

In this paper, spiking neural P systems (or SNP systems) are regarded as a formal laboratory for studying parallel and synchronized models of computation based on discrete spikes. The rest of the work is organized as follows: Sect. 2 introduces the basic model of SNP systems. Section 3 recalls some completeness and universality results, while Sect. 4 discusses some solutions to **NP**-complete problems and some limitations of the systems. Section 5 focuses on features that are specific to spike-based membrane computation, such as encoding, synchronization, timing, topology, and rules. In Sect. 6, representative variants and extensions of the basic model are discussed. Section 7 discusses some relations with spiking neural networks and other biological processes. Finally, Sect. 8 outlines open problems and Sect. 9 concludes the paper.

2 Spiking Neural P Systems as Formal Computing Devices

First, we recall the definition of SNP systems given in [12,29].

Definition 1. *A* SNP system *of degree m is the following tuple:*

$$\Pi = (O, \sigma_1, \ldots, \sigma_m, syn, in, out),$$

where

- $O = \{a\}$ *is the singleton alphabet (a is called* spike*);*
- σ_i, $1 \leq i \leq m$ *are* neurons *of the form* (n_i, R_i), *where* $n_i \geq 0$ *is the* initial number of spikes *contained in* σ_i *and* R_i *is a finite set of rules of the following two forms:*
  - *(*spiking rules*)* $E/a^c \to a; d$, *where E is a regular expression over O (we also call it a* context condition*),* $c \geq 1$ *and* $d \geq 0$. *If* $E = a^c$ *or* $d = 0$, *the corresponding parts are omitted from the writing, e.g.,* $a^c \to a$*;*
  - *(*forgetting rules*)* $a^s \to \lambda$, $s \geq 1$, *with the restriction that for any rule of type* $E/a^c \to a; d$ *from* R_i *we have* $a^s \notin L(E)$*;*
- $syn \subseteq \{1, \ldots, m\} \times \{1, \ldots, m\}$ *with* $(i, i) \notin syn$, *for all* $1 \leq i \leq m$, *is the set of* synapses *between neurons. The tuple* $(\{1, \ldots, m\}, syn)$ *forms a directed graph;*
- $in, out \in \{1, \ldots, m\}$ *indicate the* input *and the* output *neurons of* Π. *If there are no input or output neurons, then we denote the corresponding values by* λ.

There is a graphical notation for SNP systems, where each neuron is depicted as a rectangle/node. Inside this node, the initial number of spikes is indicated by the exponent of the letter a, along with the rules governing that neuron. The synapses syn are represented by directed edges between the neurons. The input neuron is marked with an arrow pointing towards the neuron and the output neuron is doubly circled, see Fig. 1. When context makes it clear, a neuron σ_i *will be referred to simply as i.*

A firing rule $E/a^c \to a; d \in R_i$ *is applicable in neuron* σ_i *if it contains* $k \geq c$ *spikes and* $a^k \in L(E)$. *The application of such a rule removes c spikes from* σ_i *and delivers a spike to each connected (via a synapse) neuron after d computational steps (in the same step if* $d = 0$*). In the period after the spikes were removed and before they were emitted (at steps 1 to* $d - 1$*), the neuron remains* closed. *Any spike arriving at a closed neuron is lost, and also no rule can be applied, or even selected, during this period. A forgetting rule just removes the corresponding number of spikes.*

During each step of the computation, all applicable rules are selected for application, with the condition of selecting at most one rule per neuron. If there are several possibilities, then one of them is non-deterministically chosen.

The configuration of the system is the vector $(r_1/t_1, \ldots, r_m/t_m)$*, where* r_i *gives the number of spikes in neuron* σ_i *and* $t_i \geq 0$ *is the number of steps until the neuron* σ_i *will be open* ($t_i = 0$ *if there is no delay associated with the rules). If all rules of the system have no delay, then corresponding* t_i *parts are omitted from the configuration. We will also call the number of spikes in a neuron the* contents *or the* value *of the neuron.*

The computation starts with the initial configuration $(n_1/0, \ldots, n_m/0)$ *and continues by applying at each step rules as described above. The input neuron may receive spikes from an external entity at each time step.*

Several input/output conventions are possible, most of them based on the notion of a *spike train*. For a neuron σ, the spike train is defined as the sequence of the number of spikes in σ at each step, including the initial configuration (at step 0). A *spike train input* is a spike train introduced in the input neuron by an external entity and having the property of being a binary sequence with exactly two occurrences of the symbol 1. The value of the input is the time interval between two consecutive 1s. Since it is not possible to represent zero with this convention, it is possible to consider only the number of zeros between the two 1 markers. If such convention is used, we will explicitly indicate it. For the output there is a similar notion of *spike train output* that is the time difference between two consecutive symbols 1 in the binary spike train of the output neuron. A more detailed discussion of this will be given in Sect. 5.

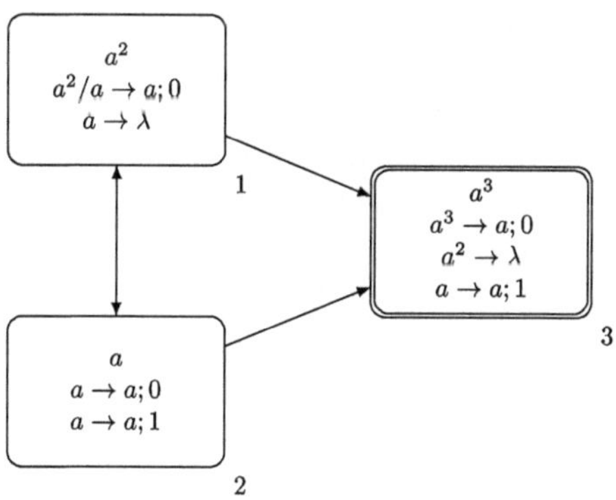

Fig. 1. A basic spiking neural P system generating numbers from $\mathbb{N} \setminus \{0, 1\}$.

3 Computability: Universality, Completeness, and Robustness

As with any formal model of computation, a first theoretical question concerns what SNP systems compute and how the result of a computation is represented. Unlike many artificial neural network models, which are usually considered in connection with classification, inference, or learning tasks, SNP systems have also been studied as formal devices for number generation, language acceptance, and function computation. In this sense, they have been considered as number generators, language acceptors and function-computing devices. The result of a computation can be encoded in different ways by the number of spikes emitted to the environment, the time elapsed between distinguished output spikes, or the contents of a designated neuron.

A common way to show universality of these systems consists in simulating register machines. In such constructions, the multiplicity of spikes stored in a neuron or in a group of neurons represents the value of a register, while additional neurons encode the finite control of the simulated machine. Firing and forgetting rules implement arithmetic updates and conditional branching, and synapses determine the flow of information between instruction modules.

An SNP system Π works as a number generator in the following way. With each computation (halting or not) in an SNP system, a spike train can be associated, describing the sequence of steps when the output neuron fires, and thus several numbers can be associated with such a spike train. A commonly used method for interpreting the outgoing spike train is to consider the number of steps elapsed between the first two spikes emitted by the output neuron; the set of all such numbers associated with all computations of the system Π is denoted by $N_2(\Pi)$. For instance, for the SNP system depicted in Fig. 1, $N_2(\Pi) = \mathbb{N} \setminus \{0, 1\}$. Then, $Spik_2P_m(rule_k; cons_p; forg_q; dely_r; outd_v)$ denotes the family of all sets of natural numbers $N_2(\Pi)$, generated by SNP systems Π with at most $m \geq 1$ neurons, using at most $k \geq 1$ rules in each neuron, with all spiking rules $E/a^c \to a; d$ having $c \leq p$ and $d \leq r$ and all forgetting rules $a^s \to \lambda$ having $s \leq q$, with all systems having the outdegree of the synapses at most v. If one of the parameters, m, k, p, q, r, and v, in the above notations is not bounded, then it is replaced with $*$.

In particular, an interesting result is the following:

$$Spik_{\underline{2}}P_*(rule_2; cons_3; forg_3; dely_1; outd_4) = NRE,$$

where NRE is the family of recursively enumerable sets of numbers. Taking into account that this is exactly the set of numbers that Turing machines can generate, the equivalence between both models is established. For this purpose, the usual idea is to simulate register machines through SNP systems, creating the modules for the three types of instructions of this model: `ADD`, `SUB` and `HALT`. The idea is that for each register, there exists a neuron whose number of spikes represents the value stored in the corresponding register, and for each instruction, some neurons will be connected in such a way as to replicate the behavior of the corresponding instruction following the flow of the original register machine computation.

Besides establishing computational universality, it is also relevant to identify where the "frontiers" of computability are; that is, when a model ceases to be as powerful as a Turing machine. In [12], two interesting results are proved in this sense:

$$Spik_2P_1(rule_*; cons_1; forg_0) = Spik_2P_2(rule_*; cons_*; forg_*) = NFIN,$$

where $NFIN$ is the family of finite sets of numbers. In the first result, taking into account that only one neuron is allowed, then the number of spikes of the system cannot increase throughout the computation. In the second one, since one of the neurons must be the output neuron, the number of spikes in the other neuron can be increased by at most two, and such a neuron cannot fire more rules than the number of initial spikes in such a neuron plus 2, at most. Therefore, the length of all the computations of the system is bounded, making the set of numbers computed by the systems finite.

In this definition, some ingredients such as the number of rules per neuron or the number of spikes in forgetting rules are used, but what happens when other ingredients are taken into account? For instance, one could think about a maximum charge of the neuron or a topological limitation. Let us denote by $Spik_2P_m(rule_k; cons_p; forg_q; bound_\gamma)$ the family of sets of natural numbers generated by SNP systems with at most $m \geq 1$ neurons, using at most $k \geq 1$ rules in any neuron, consuming at most p and forgetting at most q spikes in each rule, and having at most γ spikes present at any time in any neuron (if a computation reaches a configuration where a neuron accumulates more than γ spikes, then it aborts, and such a computation does not provide any result). As usual, if the parameters are not bounded, then they are replaced with $*$, where $bound_*$ denotes that the SNP systems have a bound on the number of spikes present in any neuron, but this bound is not specified. When $bound_\gamma$ is simply removed from the notation, this will mean that the number of spikes in neurons can grow arbitrarily. An interesting result is

$$Spik_2P_*(rule_3; cons_3; forg_3; bound_3) = SL,$$

where SL is the family of semilinear sets of numbers. The idea is, on the one hand, that the number of configurations reached by an SNP system with a bounded number of spikes in each neuron is finite; and on the other hand, that any semilinear set of numbers is the union of a finite set with a finite number of arithmetic progressions. All these results are extracted from [12]. Chapter 2 of [46] and Chap. 3 of [30] contain more insights about computability theory results by means of spiking neural P systems.

In theoretical computer science, the search for frontiers of computability in terms of the capability of devices to solve problems is very interesting. However, another interesting branch looking for more "realistic" frontiers is also investigated with SNP systems, a practical computability theory: *computational complexity theory.*

4 Complexity-Theoretic Issues: Resources, Efficiency, and Uniformity

While P systems have been widely studied as computing models for computability purposes, a very fruitful area in membrane computing is the intersection between this framework and computational complexity theory [35]. In particular, the main idea of this area is to find new tools to tackle the **P** vs **NP** problem. Usually, this approach is developed in the framework of decision problems. Given a problem X, instances are encoded as inputs for a family of P systems, and the computation is interpreted as an accepting or rejecting process. In this way, one can study whether the family decides X within a prescribed time bound, typically polynomial time, and under suitable restrictions on the resources used by the systems. To solve these problems, *recognizer P systems* were introduced [15] as accepting devices capable of recognizing languages/solving decision problems.

Definition 2. *A recognizer P system [15] can be defined as a 3-tuple* (Π, Σ, i_{in}), *where:*

1. $\Pi = (\Gamma, \mu, \mathcal{M}_1, \ldots, \mathcal{M}_q, \mathcal{R}, i_{out})$ *is a membrane system of degree* $q \geq 1$ *such that:*
   - Γ *has two distinguished elements* `yes` *and* `no`*;*
   - $\mathcal{M}_1, \ldots, \mathcal{M}_q \in \Gamma \setminus \Sigma$*;*
   - i_{out} *is the label of the environment (usually* 0 *or* env*).*
2. Σ *is a (finite) input alphabet contained in* Γ*;*
3. i_{in} *is the label of the input region, which must be a compartment of* Π*.*
4. *All the computations of* Π *are halting computations;*
5. *If* $\mathcal{C}$ *is a computation of* Π *then, either the symbol* `yes` *or the symbol* `no` *(but not both) is sent to the environment and only in the last step of the computation.*

If (Π, Σ, i_{in}) *is a recognizer P system, then for each multiset* m *over the input alphabet* Σ*, we consider the P system* Π *with the input multiset* m*. Such a system is denoted by* $\Pi + m$*, and it is characterized by the multisets associated to its initial configuration as follows:* $\mathcal{M}_1, \ldots, \mathcal{M}_{i_{in}} + m, \ldots, \mathcal{M}_q$*; that is, such a configuration is obtained from the initial configuration* $(\mathcal{M}_1, \ldots, \mathcal{M}_{i_{in}}, \ldots, \mathcal{M}_q)$ *by adding the multiset* m *to the initial multiset* $\mathcal{M}_{i_{in}}$*.*

We define the function Output whose domain is the set of computations of Π*. If* $\mathcal{C}$ *is a computation of the P system* Π*, then* $Output(\mathcal{C}) =$ `yes` *(respectively,* $Output(\mathcal{C}) =$ `no`*) if the object* `yes` *(resp.,* `no`*) is sent to the environment associated with the halting configuration of* $\mathcal{C}$ *but does not appear in the environment in any other previous configuration from* $\mathcal{C}$*. We say that a computation* $\mathcal{C}$ *of* Π *is an* accepting computation *(respectively, a* rejecting computation*) if* $Output(\mathcal{C}) =$ `yes` *(resp.,* $Output(\mathcal{C}) =$ `no`*).*

Moreover, recognizer P systems have a restriction. Given a recognizer P system with input $\Pi + m$*, either all the computations are accepting or all the computations are rejecting. Then, it is said that the recognizer P system* $\Pi + m$ *is* confluent. *In this sense, recognizer P systems capture the true concept of algorithm, even when they work in a non-deterministic way, since for a single input m, Π will always return the same answer, no matter which computation has been chosen.*

From this concept, the term *solution of a decision problem* refers to a *family of recognizer P systems solving that problem.* In particular, we are interested not in general solutions, but in *efficient* solutions; that is, solutions that take an feasible number of computing resources.

In the case of recognizer spiking neural P systems, while different interpretations of *accepting* an instance differ, these are usually compatible through simple transformations from one to another [13,18,40]. Besides, efficient solutions need to be constructed in an efficient way, thus the size of such systems should be polynomial. In [16], the *size* of an SNP system is defined as the number of necessary bits to describe it. More precisely, if $\Pi = (O, \sigma_1, \sigma_2, \ldots, \sigma_m, syn, in, out)$, where each $\sigma_i = (n_i, R_i)$, the number of bits necessary to describe Π is calculated as follows:

1. Since O is fixed, no bits are necessary to describe it;
2. Every neuron needs both a natural number n_i, which can be encoded in $\lfloor \log_2 n_i \rfloor + 1$ bits, and each rule from R_i has a bit for indicating the type (firing or forgetting), a regular expression and three natural numbers (for a rule of the form $E/a^c \to a^p; d$).
3. For syn, at most m^2 bits are necessary;
4. Both in and out use at most $\log m$ bits.

In this framework, frontiers of tractability are studied by identifying which ingredients of the systems are responsible for moving from non-efficient to presumably efficient computations. First, we analyze non-efficient classes of SNP systems. More precisely, if $\mathbf{SN}_{\mathcal{R}}(f)$ is the set of problems solvable by means of uniform families of SNP systems of the class $\mathcal{R}$ bounded in time by f, then $\mathcal{R} \in \mathcal{P}(\{-reg, -del\})$, $-reg$ for systems with regular expressions of the form $a^n, n \geq 1$ and $-del$ for systems without delays. Concerning efficiency, one studies computations whose time is bounded by a polynomial function. Then,

$$\mathbf{PSN}_{\mathcal{R}} = \bigcup_{f \text{ polynomial}} \mathbf{SN}_{\mathcal{R}}(f)$$

is the class of problems such that there exists a family of recognizer SNP systems of the class $\mathcal{R}$ that provides a uniform solution to the problem.

In [16], deterministic SNP systems are simulated by deterministic Turing machines in an efficient way; that is, in polynomial time with respect to the number of steps taken and their description size. For this purpose, an algorithm simulating a single computational step is described. The idea is that such an algorithm simulating a computational step of a deterministic SNP system of degree m can be computed by a deterministic Turing machine with $m+1$ tapes. From this result, $\mathbf{PSN}_{-reg,-del} = \mathbf{P}$ can be extracted.

The next step is to find solutions to presumably hard problems (i.e. **NP**-complete problems). If $\mathbf{SN}_{\mathcal{R}}(f)$ is the set of problems solvable by means of uniform families of SNP systems of the class $\mathcal{R}$ bounded in time by f, then $\mathcal{R} \subseteq \{pre\}$, pre for systems with pre-computed resources (usually, an exponential number of resources):

In [6], a uniform solution to SAT by means of a family of SNP systems using pre-computed resources is provided. Let φ be a propositional Boolean formula with n variables and m clauses. The system works as follows[1]:

1. First, the variables are encoded in the corresponding neuron as the number of initial spikes. In this sense, there are nm neurons of the type $\bigcirc_{c_j x_i 1}$ (respectively, $\bigcirc_{c_j x_i 0}$). If x_i (resp., $\neg x_i$) is present in the clause C_j, then 2 spikes (resp., 3 spikes) will be introduced in $\bigcirc_{c_j x_i 1}$ (resp., $\bigcirc_{c_j x_i 0}$), and 1 spike (resp., 4 spikes) will be introduced in $\bigcirc_{c_j x_i 0}$ (resp., $\bigcirc_{c_j x_i 1}$), and they will be sent to the next stage.
2. In the next stage, neurons $\circledcirc_{c_j, bin}$ simulate the OR operation of each clause, being $bin \in \{0,1\}^n$ the corresponding truth assignment encoded in binary. All the clauses are analyzed in parallel for all the possible truth assignments, thus $2^n m$ spikes can be activated. These neurons will spike if the clause C_j is satisfied by the corresponding truth assignment.
3. Following the previous spikes, the 2^n neurons $\square_{bin}$ will fire if and only if m spikes have arrived at them; that is, if all the m clauses have been satisfied by the corresponding truth value.
4. Finally, neuron $\rhd_4$ will fire if it has received at least one spike, which means that the input formula has at least one truth assignment that makes it true.

In this case, the number of initial neurons is $2^n(m+1)+2nm+1$. Although the family is constructed using exponential precomputed resources, by abusing notation, the following holds:

$$\mathbf{NP} \cup \mathbf{coNP} \subseteq \mathbf{PSN}_{pre}$$

Since the construction takes exponential time, the class $\mathbf{PSN}_{pre}$ should not be interpreted as polynomially uniform: the solution is based on an **EXP**-uniform family of SNP systems. For the sake of simplicity and comparability, we will continue with this notation.

In [13], solutions to the QSAT problem and the Q3SAT problem are presented. In this case, while keeping the idea of having pre-computed resources, the input is introduced as a spike train instead of being introduced directly in different neurons. The provided solution follows a similar scheme from the previous one (including an extra quantifier checker stage). Thus, the following holds

$$\mathbf{PSPACE} \subseteq \mathbf{PSN}_{pre}.$$

As it can be noticed, these solutions use an exponential number of neurons precomputed in the initial configuration, making the construction not efficient. A

[1] The notation is the same used in the original paper [6].

good idea would be to construct such exponential workspace during the computation (and, ideally, in an efficient way). More insights about the terms introduced in this section, as well as others, can be found in Chap. 2 of [46].

5 Spike-Based Membrane Computation: Timing, Synchronization, and Structure

In different models of computation, "information" is encoded by different symbols. In the basic model, spikes are the only indistinguishable elements of the *singleton* alphabet of the system. This makes spiking neural P systems symbolically simple, but dynamically rich. In this model, information is not represented by a large alphabet of symbols, but by the multiplicity of spikes, their distribution among neurons and their temporal evolution.

In previous sections of this work, a single encoding of the input information and two different encodings of the output have been presented. The term *spike train* provides the basic idea of encoding in such a framework.

If the input of the system is a natural number, then it can be provided to the system through the time difference between two spikes; that is, if the input is the number 4, then the spike train would be 100001. This means that, in the first step, a spike is sent to the input neuron, from the second to the fifth step, no spikes are sent to the input neuron and finally, in the sixth step, another spike is sent to the input neuron. From that point, the system continues working as it is intended. Another way to encode by means of a spike train is to assign different patterns to different inputs. In [13], the authors follow this procedure: Given a quantified Boolean formula in CNF with n variables and m clauses, let $\alpha_{i,j}$ be the assignment of spikes for the variable x_j in the clause C_i. If the literal x_j appears in C_i, then $\alpha_{i,j} = a$. If the literal $\neg x_j$ appears in C_i, then $\alpha_{i,j} = a^2$. If neither x_j nor $\neg x_j$ appear in C_i, then $\alpha_{i,j} = \lambda$. Then, the formula is encoded as the following spike train[2]:

$$\alpha_{1,1}\alpha_{1,2}\ldots\alpha_{1,n}\alpha_{2,1}\alpha_{2,2}\ldots\alpha_{2,n}\ldots\alpha_{m,1}\alpha_{m,2}\ldots\alpha_{m,n}.$$

Of course, the input is not only encoded by means of spike trains. For example, in [6] the authors encode the input Boolean formula in CNF in differentiated input neurons. If the literal x_i (resp., $\neg x_i$) appears in the clause C_j, then the neuron $\bigcirc_{c_j x_i 1}$ (respectively, $\bigcirc_{c_j x_i 0}$) will have 2 spikes and the neuron $\bigcirc_{c_j x_i 0}$ (respectively, $\bigcirc_{c_j x_i 1}$) will have 1 spike. If neither x_i nor $\neg x_i$ appear in the clause C_j, both $\bigcirc_{c_j x_i 1}$ and $\bigcirc_{c_j x_i 0}$ will not have any spikes.

In the same way, different strategies to encode the output can be found. A single output neuron can be used as a "reservoir" where the number of spikes in the *halting configuration* encodes the output of the system. Usually, instead of this perspective, the output is encoded as a spike train. In Fig. 1, the non-deterministic behavior of the system determines which natural number is generated in each computation. The set of numbers generated by this SNP system

[2] In the paper, the number of variables is $2n$, for the sake of simplicity, we use here only n.

is defined as the set of numbers that are generated in the (infinite number of) computations of the system. This set of numbers is usually written as $N_2(\Pi)$. When one wants k numbers to be generated in a single computation, then the time between $k+1$ spikes is measured, and it is represented by $N_{k+1}(\Pi)$. If all the (possibly infinite) numbers can be generated in a single computation, then it is denoted by $N_\omega(\Pi)$. Apart from that, as explained before, accepting/rejecting computations can be defined as the output of the system.

In general, the local state of each neuron is represented quantitatively, while the global state arises from the distribution of these quantities over the network. As stated in Sect. 2, there exists a global clock that governs the behavior of the systems. The usual dynamics ensure that one rule per neuron (while applicable) is applied, thus introducing a distributed behavior. However, different derivation modes can be found in the literature. In particular, in the *asynchronous mode*, only one rule can be applied in the whole system. Another possible semantics includes another layer of parallelism, making it possible to apply more than one rule per neuron. Besides, different sets of rules can be defined in a specific SNP system, and a set of rules is defined to be applicable if and only if all the rules of the set are applicable at the same time. More details about these and other derivation modes can be found in Chap. 4 of [46].

Another relevant feature of these systems is related to the topology of the network formed by the neurons. While the topology is usually not restricted, some results could be obtained by restricting how neurons are connected, for instance, not allowing the system to have cycles. Besides, while the network is static in the basic model, one may consider including the ability to change the structure of the network. This is introduced in [5] through plasticity rules of the form $E/a^c \rightarrow \alpha k(i, N_j)$, where $c \geq 1, \alpha \in \{+, -, \pm, \mp\}, k \geq 1, 1 \leq j \leq | R_i |, N_j \subseteq \{1, \dots, m\}$. These rules are devoted to creating new synapses between neurons or to removing them, making it possible to change the architecture of the device.

Taking this idea into account, which types of changes can we include in a system? What are the consequences of including new ingredients in terms of computational power? Can they reduce the number of neurons or rules needed to build a universal device? Can we solve **NP**-complete problems in an efficient way with SNP systems that are built in polynomial time? Some of these issues are discussed in the next section.

6 Variants and Extensions: What Changes Computationally?

While some early variants can be obtained by imposing restrictions on the basic model, for instance on regular expressions, on the number of spikes in the left-hand side of rules, or on delays, later variants introduce genuinely new ingredients. Of course, such restrictions usually come with an increase in the number of other ingredients (such as neurons, spikes...) to achieve universality. A natural first extension of the system is the so-called extended SNP systems [3], where the right-hand side of the rule is not limited to a single spike anymore, but the rules have a more general form $E/a^c \to a^k; d$, with $c \geq k, c \geq 1, k \geq 0$, abstracting both firing rules and forgetting rules. Another extension from the perspective of spikes in rules are white hole rules [2], where rules $E/all \to a^k; d$ indicate that all the spikes from the source neuron are removed.

Apart from that, different new ingredients can appear from biology. For instance, synaptic transmission of neurons is usually regulated by means of a specific type of glial cells called *astrocytes* [24]. These are introduced as another type of cell within the spiking neural P system, leaving the device divided into two graphs: the graph of neurons and the graph of astrocytes, which are connected to the *synapses* of the former graph to control when the communication is effective (for instance, allowing the spikes to pass through the synapses only if the amount is greater than a certain threshold).

Considering the amount of information transmitted through synapses, an interesting idea is to see what happens if more than one synapse between two neurons is allowed [31]. Besides, channels can take a leading role if the rules are placed on channels instead of in neurons [36].

There are other types of variants that completely change the perspective of how rules work. Instead of sending spikes through the firing of rules, neurons can request spikes from neurons connected to them. That is the case of spiking neural P systems with communication on request [44]. Such rules are described as E/Qw, where E serves (as usual) as a regular expression over the number of spikes within a specific neuron, and w is a sequence of *requests* for the neurons connected to the neuron, being the query (a^q, j) satisfied if the number of spikes in neuron σ_j is greater than or equal than q. A very close model to this one, in terms of inspiration, is the model of dendrite P systems [22], whose neurons also request spikes from other neurons in some sense, but now the regular expressions of the rules, whose notation is $(E_{i_1}, \ldots, E_{i_{s_i}})/a^p \leftarrow (a^{c_1}, \ldots, a^{c_{s_i}})$ refer to the number of spikes in the preceding neurons and the elements $c_1, \ldots, c_{s_i}$ correspond to the number of spikes removed from such neurons.

New types of spikes can be included to increase the expressivity of the model without including other structural or technical ingredients. Of course, this increases the complexity of representing the system. Taking inspiration for their name from Petri nets, *spiking neural P systems with colored spikes* [37] are introduced to study how different symbols can influence the computational power of the model. Besides, a specific type of spike, called anti-spike, can be introduced as a way to remove spikes "without consuming computation time" [38], represented by the rules $a\bar{a} \to \lambda$.

Other changes to the nature of spikes have been considered, such as introducing fuzzy logic into the behavior of the systems, changing the characterization of spikes from natural numbers to *truth values*, introducing *fuzzy reasoning spiking neural P systems* [32]. Another well-known variant of spiking neural P systems, where the domain of spikes changes from natural numbers to real numbers, is given by *nonlinear spiking neural P systems*, where the rules $T/a^{g(x_i)} \to a^{f(x_i)}$ consume $g(x_i)$ spikes from the source neuron and send $f(x_i)$ spikes to the neighboring neurons, being $g(x_i)$ and $f(x_i)$ nonlinear functions. These two variants, among others, are the basis for many applications of these models, such as fault diagnosis [42] and machine learning [39].

These variants may reduce some parameters, such as the number of neurons or rules, simplify universality constructions, improve descriptional succinctness, or provide more biologically motivated models. However, such advantages usually come at the price of adding information to the description or increasing the semantic complexity of the system. While these ingredients clearly enrich SNP systems, they are not enough to solve presumably hard problems efficiently.

In this sense, as discussed in Sect. 4, a different approach is needed for obtaining efficient solutions to hard problems: the system should be able to generate a large amount of computational workspace during the computation itself (unless $\mathbf{P} = \mathbf{NP}$). Ingredients such as structural plasticity, neuron division, and neuron budding are therefore especially relevant, since they modify the topology of the device during the computation. The first of these was introduced in the previous section, but since it only changes the number of synapses but not the number of neurons, it does not seem sufficient to achieve presumed efficiency. For complexity tasks, division rules and budding rules [23] are investigated.

Division rules, $[\,E\,]_i \rightarrow [\;]_j \parallel [\;]_k$, introduced as their counterparts in other frameworks of membrane computing [27,28], are devoted to generating two neurons from the original one, both sharing all the connections of the original neuron. Thus, if a neuron i is divided, the original neuron i is removed and two new neurons j and k are created. For every $e \in syn$ such that $e = (i, t)$, two new synapses (j, t) and (k, t) are generated. Analogously, for every $e = (t, i)$, two new synapses (t, j) and (t, k) are generated.

Budding rules, $[\,E\,]_i \rightarrow [\;]_i/[\;]_j$, are introduced as an abstraction of *budding* in real-life cells, where a part of a cell splits off and gives rise to a new cell. In this case, if a neuron i is split, then a new neuron j is created. Now, for every $(i, t) \in syn$, it is removed and a new synapse (j, t) is created.

In [23], a solution to the SAT problem is provided by means of a uniform family of *spiking neural P systems with division rules and budding rules*, thus obtaining

$$\mathbf{NP} \cup \mathbf{coNP} \subseteq \mathbf{PSN}_{div,bud},$$

where div, bud denote the use of division rules and budding rules. In [41], the result is even tighter: the need for budding rules is removed, and a solution to SAT by means of *spiking neural P systems with division rules* is provided, resulting in

$$\mathbf{NP} \cup \mathbf{coNP} \subseteq \mathbf{PSN}_{div}.$$

Thus, these variants should not be seen only as independent models. They can be understood as systematic modifications or *extensions* of a common framework, enriching the model and making it possible to compare different variants with respect to the ingredients they use and in terms of computational power and efficiency.

7 Connections with Neural Computation and Bioinformatics

SNP systems, like spiking neural networks, use discrete events as a way to transmit information, but from a different perspective. While SNP systems use discrete spikes in the form of a symbol from a singleton alphabet, spiking neural networks usually involve continuous internal dynamics. Moreover, the idea of spiking neural networks is more related to learning, neural dynamics and neuromorphic computing [11]. While some variants of SNP systems that are more oriented towards practical applications take this perspective, in general they are focused on formal languages, computability and complexity, placing them at a higher level of abstraction with respect to biological neural models. This point is interesting for keeping the balance between biological inspiration and mathematical precision.

Another meaningful connection is the relation between SNP systems and bioinformatics. While it is not immediate, these devices can be seen as formal models of biological information processing, thus allowing us to design models that abstract or simulate the behavior of complex systems such as regulatory networks or signaling pathways. These applications have been found interesting, in general, in the framework of membrane computing [8]; hence, modelling specific processes from the human brain within this framework seems a natural next step.

Several directions can be pursued from here. As stated above, one of the applications could be the use of these models to abstract signal propagation in human brain models. Another is the study of hybrid models combining neurons with rules and learning mechanisms, obtaining algorithms capable of improving the efficiency of designing new models[3]. A final direction concerns the formal analysis of information encoded not in quantity, but in time, making it both the clock of the system and a carrier of information.

These directions suggest that the relevance of spiking neural P systems to bioinformatics and neural computation lies in the formal vocabulary they provide for describing distributed, temporal, and signal-based computation in addition to their possible applications.

8 Open Problems and Future Directions

While the framework of SNP systems has been widely studied [30,46], many open research lines can be found in the literature [4] and many theoretical questions remain open.

While a more settled complexity theory exists for other models within membrane computing [35], in the framework of SNP systems the notion of solving

[3] Some of these ideas are elaborated in the field of *automatic design of spiking neural P systems*. Some information about this research can be found in Chap. 5 of [46].

a problem is still treated in different ways across the literature. A robust computational complexity theory could clarify the role of resources, establish lower bounds, and identify the role of uniformity.

Notions such as uniformity and semi-uniformity offer a natural way to relate families of SNP systems to the framework of circuit complexity. Developing this connection could strengthen the links between the two areas and may lead to new characterizations of complexity classes by means of families of SNP systems.

Another interesting direction concerns lower bounds and resource trade-offs. Both computability and complexity theory benefit from identifying frontiers: which systems are powerful enough to solve a given class of problems, and which resources are strictly necessary for doing so efficiently.

The dynamics of the structure also deserve special attention. Variants with neuron division, neuron budding, or structural plasticity challenge the assumption that the network is fixed throughout the computation. The interpretation of the speed-up obtained through these types of rules deserves further investigation.

Further work relating SNP systems to spiking neural networks and bioinformatics is also needed, especially in approaches that combine the formal abstraction provided by SNP systems with the dynamics and learning mechanisms of biological neural models.

9 Concluding Remarks

Spiking neural P systems provide a formal framework in which ideas coming from membrane computing and spiking neural computation meet. Although their motivation is biological, their theoretical interest goes beyond biological inspiration. Their mathematical tractability and rigor make it possible to study how computation can arise from the interaction of simple processing units, indistinguishable signals, local rules, temporal coordination, and network structure.

From the computability point of view, SNP systems are expressive enough to have the same computational power as Turing machines and other computationally complete models, and when some resources are bounded, different levels of the Chomsky hierarchy can be characterized. From the computational complexity point of view, they provide a way to revisit questions around the **P** versus **NP** problem from a different perspective, distinguishing them from other classes of membrane systems which use a richer alphabet.

Several variants of SNP systems have been developed in the literature. However, new syntactic and semantic ingredients are introduced in almost every paper. The inspiration can come from different perspectives. On the one hand, biological ingredients such as structural plasticity, refractory period, communication on dendrites and chemical components within neurons, such as calcium, make the systems closer to biological applications. On the other hand, formal ingredients such as new types of rules, alternative derivation modes, restrictions on regular expressions and internal gates in neurons can be used to investigate the mathematical boundaries of the model. In this sense, it would be interesting to integrate all the possible variants under the same formal framework, making it possible to compare different models and, as a further objective, to embed them into a more general formal setting within the framework of membrane computing.

In summary, SNP systems should not be viewed merely as another family of universal devices, but as a formal laboratory for investigating natural computation from the perspectives of computability and computational complexity theory. Further advances in this area may come from research in lower bounds, resource trade-offs, robust notions of uniformity, as well as from the interaction with other areas such as circuit complexity, neural computation and biologically inspired information processing. Spikes, time, synchronization and structure are not merely modelling choices, but also computational ingredients. Understanding their role may contribute to membrane computing, spiking neural networks, and, in general, the theory of computation.

Acknowledgments. This work is part of a project that has received funding from the European Union's Horizon Europe Research and Innovation Programme under the Marie Skłodowska-Curie Actions (MSCA), Grant Agreement No. 101236749.

Disclosure of Interests. The author has no competing interests to declare that are relevant to the content of this article.

References

1. Adleman, L.M.: Molecular computation of solutions to combinatorial problems. Science **266**(5187), 1021–1024 (1994)
2. Alhazov, A., Freund, R., Ivanov, S., Oswald, M., Verlan, S.: Extended spiking neural P systems with white hole rules and their red-green variants. Nat. Comput. **17**(2), 297–310 (2018)
3. Alhazov, A., Freund, R., Oswald, M., Slavkovik, M.: Extended spiking neural P systems. In: Hoogeboom, H.J., Păun, Gh., Rozenberg, G., Salomaa, A. (eds.) Membrane Computing, pp. 123–134. Springer, Heidelberg (2006)
4. Cabarle, F.G.C.: Thinking about spiking neural P systems: some theories, tools, and research topics. J. Membr. Comput. **6**(2), 148–167 (2024)
5. Cabarle, F.G.C., Adorna, H.N., Pérez-Jiménez, M.J., Song, T.: Spiking neural P systems with structural plasticity. Neural Comput. Appl. **26**(8), 1905–1917 (2015)
6. Chen, H., Ionescu, M., Ishdorj, T.: On the efficiency of spiking neural P systems. In: Proceedings of the Fourth Brainstorming Week on Membrane Computing, pp. 195–206. Sevilla, Spain (2006)
7. Eiben, A.E., Smith, J.E.: Introduction to Evolutionary Computing, 2 edn. Springer (2015)
8. Frisco, P., Gheorghe, M., Pérez-Jiménez, M.J. (eds.): Applications of Membrane Computing in Systems and Synthetic Biology. ECC, vol. 7. Springer, Cham (2014). https://doi.org/10.1007/978-3-319-03191-0
9. Haykin, S.: Neural Networks and Learning Machines, 3 edn. Pearson Education India (2009)
10. Hirvensalo, M.: Quantum Computing. Springer (2013)
11. Indiveri, G., Liu, S.C.: Memory and information processing in neuromorphic systems. Proc. IEEE **103**(8), 1379–1397 (2015)

12. Ionescu, M., Păun, Gh., Yokomori, T.: Spiking neural P systems. Fundamenta Informaticae **71**(2–3), 279–308 (2006)
13. Ishdorj, T.O., Leporati, A., Pan, L., Zeng, X., Zhang, X.: Deterministic solutions to QSAT and Q3SAT by spiking neural P systems with pre-computed resources. Theoret. Comput. Sci. **411**(25), 2345–2358 (2010)
14. Istrail, S., De-Leon, S.B.T., Davidson, E.H.: The regulatory genome and the computer. Dev. Biol. **310**(2), 187–195 (2007)
15. Jiménez, M.J.P., Jiménez, Á.R., Caparrini, F.S.: Complexity classes in models of cellular computing with membranes. Nat. Comput. **2**(3), 265–285 (2003)
16. Leporati, A., Zandron, C., Ferretti, C., Mauri, G.: On the computational power of spiking neural P systems. In: Proceedings of the Fifth Brainstorming Week on Membrane Computing, pp. 227–245. Sevilla, Spain (2007)
17. Leporati, A., Mauri, G., Zandron, C.: Spiking neural P systems: main ideas and results. Nat. Comput. **21**(4), 629–649 (2022)
18. Leporati, A., Mauri, G., Zandron, C., Păun, Gh., Pérez-Jiménez, M.J.: Uniform solutions to SAT and Subset Sum by spiking neural P systems. Nat. Comput. **8**(4), 681–702 (2009)
19. Maass, W.: Networks of spiking neurons: the third generation of neural network models. Neural Netw. **10**(9), 1659–1671 (1997)
20. Martín-Vide, C., Păun, Gh., Pazos, J., Rodríguez-Patón, A.: Tissue P systems. Theor. Comput. Sci. **296**(2), 295–326 (2003)
21. McCulloch, W.S., Pitts, W.: A logical calculus of the ideas immanent in nervous activity. Bull. Math. Biophys. **5**, 115–133 (1943)
22. Orellana-Martín, D., Martínez-del Amor, M.A., Valencia-Cabrera, L., Pérez-Hurtado, I., Riscos-Núñez, A., Pérez-Jiménez, M.J.: Dendrite P systems toolbox: representation, algorithms and simulators. Int. J. Neural Syst. **31**(01), 2050071 (2021)
23. Pan, L., Păun, Gh., Pérez-Jiménez, M.J.: Spiking neural P systems with neuron division and budding. SCIENCE CHINA Inf. Sci. **54**(8), 1596–1607 (2011)
24. Pan, L., Wang, J., Hoogeboom, H.J.: Spiking neural P systems with astrocytes. Neural Comput. **24**(3), 805–825 (2012)
25. Păun, Gh.: Computing with Membranes. TUCS Research Report 208, Turku Centre for Computer Science, Turku, Finland (1998)
26. Păun, Gh.: Computing with membranes. J. Comput. Syst. Sci. **61**(1), 108–143 (2000)
27. Păun, Gh.: Computing with membranes: attacking NP-complete problems. In: Antoniou, I., Calude, C.S., Dinneen, M.J. (eds.) Unconventional Models of Computation, UMC'2K, pp. 94–115. Springer, London (2001)
28. Păun, Gh., Pérez-Jiménez, M.J., Riscos-Núñez, A.: Tissue P systems with cell division. Int. J. Comput. Commun. Control **3**(3), 295–303 (2008)
29. Păun, Gh., Rozenberg, G., Salomaa, A. (eds.): The Oxford Handbook of Membrane Computing. Oxford University Press, Oxford (2010)
30. Peng, H., Wang, J.: Advanced Spiking Neural P Systems. Springer (2024)
31. Peng, H., et al.: Spiking neural P systems with multiple channels. Neural Netw. **95**, 66–71 (2017)
32. Pérez-Jiménez, M.J., Graciani, C., Orellana-Martín, D., Riscos-Núñez, A., Romero-Jiménez, Á., Valencia-Cabrera, L.: Fuzzy reasoning spiking neural P systems revisited: a formalization. Theor. Comput. Sci. **701**, 216–225 (2017), at the intersection of computer science with biology, chemistry and physics - In Memory of Solomon Marcus

33. Rozenberg, G., Salomaa, A.: The Mathematical Theory of L Systems, vol. 90. Academic Press (1980)
34. Rumelhart, D.E., Hinton, G.E., Williams, R.J.: Learning representations by back-propagating errors. Nature **323**, 533–536 (1986)
35. Song, B., Li, K., Orellana-Martín, D., Pérez-Jiménez, M.J., PéRez-Hurtado, I.: A survey of nature-inspired computing: membrane computing. ACM Comput. Surv. **54**(1) (2021)
36. Song, T., Pan, L., Păun, Gh.: Spiking neural P systems with rules on synapses. Theoret. Comput. Sci. **529**, 82–95 (2014)
37. Song, T., Rodríguez-Patón, A., Zheng, P., Zeng, X.: Spiking neural P systems with colored spikes. IEEE Trans. Cogn. Dev. Syst. **10**(4), 1106–1115 (2018)
38. Song, X., et al.: Spiking neural P systems with multiple channels and anti-spikes. Biosystems **169–170**, 13–19 (2018)
39. Sosík, P., Paul, P., Ciencialová, L.: A survey on learning models of spiking neural membrane systems. Nat. Comput. **24**(3), 665–677 (2025)
40. Sosík, P., Rodríguez-Patón, A., Ciencialová, L.: Polynomial complexity classes in spiking neural P systems. In: Gheorghe, M., Hinze, T., Păun, Gh., Rozenberg, G., Salomaa, A. (eds.) Membrane Computing, pp. 348–360. Springer, Heidelberg (2011)
41. Wang, J., Hoogeboom, H.J., Pan, L.: Spiking neural P systems with neuron division. In: Gheorghe, M., Hinze, T., Păun, Gh., Rozenberg, G., Salomaa, A. (eds.) Membrane Computing, pp. 361–376. Springer, Heidelberg (2011)
42. Wang, T., Zhang, G., Zhao, J., He, Z., Wang, J., Pérez-Jiménez, M.J.: Application of fuzzy reasoning spiking neural P systems to fault diagnosis. Int. J. Comput. Commun. Control **9**(5), 636–644 (2014)
43. Wolfram, S.: A new kind of science. Appl. Mech. Rev. **56**(2), B18–B19 (2003)
44. Wu, T., Pan, L.: Spiking neural P systems with communication on request and mute rules. IEEE Trans. Parallel Distrib. Syst. **34**(2), 734–745 (2023)
45. Zhang, G., et al.: Membrane Computing Models: Implementations. Springer, Singapore (2021)
46. Zhang, G., et al.: Spiking Neural P Systems. Springer (2024)

The New French School of Philosophy of Computing

Paula Quinon(✉)

Warsaw University of Technology, Poland and University of Gothenburg, Gothenburg, Sweden
paula.quinon@pw.edu.pl

Abstract. This paper proposes the name *The New French School of Philosophy of Computing* for a community of French theoretical computer scientists and philosophers whose work, taken together, advances a distinctive conception of what computation is. Across several interlocking lines of work—Colson's structural theory of primitive recursive algorithms, Berry and Curien's sequential algorithms, Krivine's classical realisability, Coquand's calculus of constructions, Dowek's undecidability results inside subrecursive systems, the programme of implicit computational complexity, and Girard's Geometry of Interaction—a common philosophical sensibility emerges: the algorithm, not the function, is the primary mathematical object, and structural features of formalisms are genuine features of computation rather than artefacts of representation. I argue that these lines admit a candidate unifying framework in Girard's *Geometry of Interaction*. The proposal is offered as the motivation for a special session of the History and Philosophy of Computing (HaPoC) at CiE 2026, where it is meant to be examined and contested.

Keywords: French philosophy of computing · constrained computation · primitive recursion · Curry–Howard correspondence · implicit computational complexity · Geometry of Interaction · intensionality

The objective of this paper is to disclose, define, and describe the specificity of the thought of a community of French theoretical computer scientists and philosophers of computing. Taken together, their work strikes me as more than a loose collection of research lines. For now, my sense that there is such a "thing" rests on a feeling, on the reading I have done for my own philosophical papers on the concept of computing, and on conversations with scholars I believe have contributed to the community. Understanding whether such a school really exists will require much more reading and many more conversations. I do not mean to imply that this is the only such case worth studying—scholars elsewhere may well recognise something similar in the communities they belong to, and those would be different investigations, with their own distinctive findings. The present paper is the presentation of the main motivation for the special session of the

V. Brattka et al. (Eds.): CiE 2026, LNCS 16674, pp. 78–85, 2027.
https://doi.org/10.1007/978-3-032-31348-5_5

History and Philosophy of Computing (HaPoC) at CiE 2026, in which the thesis advanced here is meant to be examined and contested. Still, I want to take the risk and coin the name *The New French School of Philosophy of Computing.*

I will start by looking at several interlocking lines of work that, I think, strongly contribute to *the school's* shape. Each is presented here through its original technical content, with philosophical implications flagged either as already operative in the computer scientists' own work or as drawn out later by the philosophical community.

1 Primitive Recursive Algorithms as a Domain in Their Own Right

Colson (1991) developed a structural theory of primitive recursive algorithms in which primitive recursion is treated not, as in the mainstream recursion-theoretic tradition (Kleene 1952; Rogers 1967; Odifreddi 1989), as a propaedeutic to general recursion, but as an object of study with its own internal logic. Applying denotational semantics with the domain of lazy integers, Colson proved what is now known as the *ultimate obstinacy* property: "once an algorithm begins to consume an argument, it will never switch to another argument" (technical summary in Valarcher–Crolard 2011; cf. Colson 1991). The corollary is sharp: "if a primitive recursive derivation of min(x, y) is expressed faithfully in a programming language, then one of the two computations min(1, 1000) and min(1000, 1) will take at least 1000 steps" (Van den Dries 2003). Coquand (1992), in a paper whose title is literally *Une preuve directe du théorème d'ultime obstination*, supplied a constructive proof of the same result; David (2001) and Colson–Fredholm (1998) extended the analysis to mutual recursion and call-by-value variants. Valarcher (1996) and the subsequent work of Crolard, Polonowski & Valarcher (2011) developed the theory of primitive recursive algorithms further, including a representation theorem that establishes primitive recursive algorithms as mathematical objects with stable identity conditions.

Three philosophical payoffs follow from this. First, Colson's work takes primitive recursion seriously as a formalism with its own structure. The class admits both a denotational and an operational reading, and Colson uses denotational tools to prove an operational fact—that primitive recursive algorithms cannot freely choose which argument to examine first. Second, the result locates the source of inefficiency in the formalism itself. Primitive recursion forces an algorithm to commit to one argument before examining the other, and this commitment is built into the definitional scheme, not into any particular algorithm. The suboptimal complexity of min is structural, not contingent. Third, the result distinguishes the algorithm from the function. The function min is the same mathematical object whether computed by primitive recursion or by general recursion; what differs is the algorithm available to compute it. As I have argued elsewhere (Quinon 2025), Colson's theorem identifies a property of the algorithm, not of the function—a property invisible if one looks only at what is computed.

2 Sequential Algorithms and the Mathematics of Computational Structure

Berry and Curien show, by semantic means, that algorithms can be treated as mathematical objects in their own right. This is the same philosophical move Colson makes proof-theoretically, and the same one Moschovakis makes with recursors—three different routes to the insight that the algorithm is its own thing, in a certain sense "finer" than the function, worth studying directly. Berry and Curien (1982) gave "a sequential denotational semantics [an interpretation of programs as mathematical objects that distinguishes programs by the order in which they compute] for sequential programming languages, based on a new notion of sequential algorithm on the Kahn–Plotkin concrete data structures [data structures whose parts are revealed in a definite order]", in which "an algorithm may be seen either as a concrete object—a 'program' in a simple output-driven language—or as an abstract object—the pair of a sequential function and of a computation strategy for it [the function says what is computed, the strategy says in what order the inputs are examined]" (Berry & Curien 1982, abstract). This was the first category in denotational semantics whose morphisms were "not (presented as) functions, but programs of some sort" (Curien 2022). Berry and Curien were hoping to solve a standard problem in semantics called full abstraction for PCF (Plotkin's small functional language)—roughly, the problem of finding a denotational semantics that matches the language exactly, with no mathematical objects left over that the language cannot express. They did not succeed: Curien notes that "the model SA [sequential algorithms] was shown not to be amenable to a FA [fully abstract] model of PCF" [the SA model contained more mathematical objects than PCF contained programs, so the correspondence could not be established] (Curien 2022). They did, however, achieve full abstraction for a different language, CDS (Concrete Data Structures)—a language together with collaborators they designed so that its programs correspond exactly to the sequential algorithms in their model.

The philosophical contribution does not depend on the full abstraction result: it lies in the demonstration that a sequential algorithm "may be viewed as a sequential input-output function paired with a computation strategy" (Brookes, on Berry–Curien)—a mathematical object whose identity conditions are finer than the extensional equivalence of the function it computes.

3 Abstract Machines and Classical Realisability

The philosophical interest of Krivine's work is that the abstract machine—not the function it computes—is the primary mathematical object of the construction. Logical content and computational behaviour are jointly carried by the structure of the machine itself. As in the Colson and Berry–Curien cases, the algorithmic level of description is treated as mathematically fundamental rather than as an implementation of a function defined elsewhere.

Krivine's work extends the Curry–Howard correspondence—a philosophically important identification of mathematical proofs with computer programs—into classical logic and set theory. The extension is consequential: the correspondence in its original form applied only to *intuitionistic logic*, where every existence proof must produce a concrete witness; classical principles such as the law of excluded middle, which permit reasoning by contradiction without producing witnesses, were thought to lack computational content. The boundary between constructive and non-constructive reasoning was therefore also taken to mark the boundary of what could be given computational meaning. Krivine's "call-by-name lambda-calculus machine" (Krivine 2007) and his subsequent programme of classical realisability (Krivine 2001, 2003, 2009) show otherwise. He builds a framework, now called 'classical realizability', to explore the Curry-Howard (proof-program) correspondence in which "'the programming' side of the correspondence is an imperative low level language, based on a call-by-name lambda-calculus abstract machine [the Krivine machine, a mathematical model of a computing device that processes arguments only when needed], [while] the 'proof' side is usual formalization of Analysis in second-order logic [the standard axiomatic framework for classical mathematics]" (Streicher & Reus 1998, after Krivine). The technical achievement is summarised by Krivine himself: classical realisability "permits to extend the Curry–Howard correspondence between proofs and programs, to Zermelo–Fraenkel set theory and to build new models of ZF, called realizability models" (Krivine 2014).

4 Termination as a Structural Property

The philosophical interest of Coquand's work is that properties of computational processes—that they terminate, that the underlying logic is consistent, that we can mechanically check whether two expressions agree—are made to follow from how definitions are built rather than from anything checked after the fact. The structure of the formalism itself enforces these properties. Coquand's *Calculus of Constructions* (Coquand 1985 thesis; Coquand & Huet 1988)—a formal system in which mathematical proofs and computer programs are written as the same kind of object, and the foundation of modern proof assistants such as Coq and Agda—is strongly normalising, meaning, in Casinghino's gloss, that "there are no infinite reduction sequences from well-typed expressions" (Casinghino 2022): every computation in the system terminates. This single property has two consequences that matter philosophically. First, the calculus is consistent as a logic, which "makes it an attractive target for the formalization of mathematics" (Casinghino 2022). Second, "it implies that there is an algorithm to check whether two expressions are β-convertible [whether one can be transformed into the other by computation steps]. Thus, type checking is decidable and CC provides a practical basis for programming languages" (Casinghino 2022). Termination is therefore not verified externally on a case-by-case basis but follows from the typing discipline itself: 'termination of recursive functions is an important property in proof assistants based on dependent type theories; it implies consistency and decidability of type checking" (Barthe, Grégoire & Pastawski 2006).

In *the school's* broader picture, I take this to be the same pattern visible in the previous sections. Colson's primitive recursion fixed the order in which arguments could be consumed. Berry and Curien's sequential algorithm carried the order of input examination as part of its mathematical identity. Krivine's abstract machine carried logical content in its instructions. Coquand's type system carries the guarantee of termination. In each case, how the computational objects are constructed determines what they can and cannot do—and the formalism, in shaping admissible constructions, shapes admissible computations.

5 Undecidability Inside Subrecursive Systems

The philosophical interest of Dowek's work is that undecidability does not require Turing-completeness [the property of being able to compute every function that a Turing machine can compute]. It can already arise in systems that are much weaker—systems expressive enough only to represent the primitive recursive functions. Dowek (1993) proved this for the *pattern matching problem*: the problem of deciding, given two expressions, whether one can be rewritten into the form of the other. He showed that pattern matching "is undecidable in all the calculi in which primitive recursive functions can be fairly represented" (Dowek 1993), including Girard's system F [a typed lambda calculus that captures higher-order polymorphism] and Gödel's system T [a typed lambda calculus with primitive recursion over natural numbers]. The proof works by reducing Hilbert's tenth problem—a famously undecidable problem about Diophantine equations—to the pattern matching problem. The broader philosophical significance of Dowek's work on the changing relationship between proof and computation has been discussed by Arana (2017).

6 Implicit Computational Complexity

The philosophical interest of implicit computational complexity (ICC) is that it asks whether efficiency—how fast a program runs, how much memory it uses—is best understood as an external constraint imposed on computational processes from the outside, or as an intrinsic feature of certain ways of constructing functions. The standard way to talk about complexity assumes the first: one fixes a machine model, runs a program, counts steps. ICC pursues the second. The programme, developed in particular by Baillot, Marion, Leivant, and their collaborators (Leivant & Marion 1993; Bonfante, Marion & Moyen 2011; Baillot 2010), "aims at studying computational complexity without referring to external measuring conditions or particular machine models, but instead by considering restrictions on programming languages or logical principles implying complexity properties" (Baillot 2010). Equivalently, ICC "characterizes programs by constraints on the way in which they are constructed, without reference to a specific underlying machine model or to explicit bounds on computational resources" (standard survey formulation).

The philosophical move is structurally the same as the one made by Colson at the level of computability and by Coquand at the level of termination: a property classically thought to require external verification (a running clock, a memory monitor) is internalised into the syntax of the formalism. Efficiency becomes a feature of how a function is defined, not of how its execution is observed.

7 Geometry of Interaction as Candidate Unifying Framework

I propose to read the lines surveyed above as broadly unified by Girard's *Geometry of Interaction*—a programme that takes the same idea each of these lines explores in its own way (the algorithm is a mathematical object in its own right, finer than the function it computes) and tries to give it a single mathematical setting. Girard (1989, 1995) introduced *Geometry of Interaction* as a semantics "that accounts for the dynamics of cut-elimination" (Seiller, *Interaction Graphs: Graphings*)—meaning, in plain terms, a semantics that takes seriously the process by which a proof is reduced to its normal form, rather than treating the proof and its normal form as semantically identical. The mathematical content is captured in the slogan that "the interpretation of a proof π and its normal form ρ are not equal, but one has a way of computing the interpretation of the normal form ρ from the interpretation of the proof π" (Seiller, ibid.). This is the same philosophical move encountered throughout §§1–6: the dynamic process of computation, not just its result, is what the mathematics is about. Subsequent developments through Seiller's interaction graphs and graphings "[were] introduced as a general, uniform, construction of dynamic models of linear logic [a logic that tracks resources rather than only truth], encompassing all Geometry of Interaction (GoI) constructions introduced so far" (Seiller, research page).

The *school's* shared concern with the structure of computation is pursued institutionally through Naibo's proof-theoretic work in the ANR projects GoA (Naibo, 2024–2027) and DySCo (Seiller, 2022–2027), which offer a candidate framework for articulating, in mathematically precise terms, what that concern amounts to. The 2023 Dagstuhl meeting *Formalising the Notion of Algorithm* brought together abstract state machines, recursors, and graphing-based models around precisely this question. The unification proposed here is editorial rather than settled: it is a thesis to be examined in the session, not a fact about how the field already sees itself.

References

Baillot, P. (ed.): Proceedings International Workshop on Developments in Implicit Computational Complexity (DICE 2010). EPTCS 23 (2010). https://doi.org/10.4204/EPTCS.23

Barthe, G., Grégoire, B., Pastawski, F.: $\mathrm{CIC}^{\widehat{}}$: type-based termination of recursive definitions in the Calculus of Inductive Constructions. In: LPAR 2006. LNCS, vol. 4246, pp. 257–271. Springer (2006). https://doi.org/10.1007/11916277_18

Berry, G., Curien, P.-L.: Sequential algorithms on concrete data structures. Theor. Comput. Sci. **20**, 265–321 (1982). https://doi.org/10.1016/S0304-3975(82)80002-9

Bonfante, G., Marion, J.-Y., Moyen, J.-Y.: Quasi-interpretations: a way to control resources. Theor. Comput. Sci. **412**(25), 2776–2796 (2011). https://doi.org/10.1016/j.tcs.2011.02.007

Casinghino, C.: Strong normalization for the calculus of constructions. arXiv:2210.11240 (2022)

Colson, L.: About primitive recursive algorithms. Theor. Comput. Sci. **83**, 57–69 (1991). https://doi.org/10.1016/0304-3975(91)90039-5

Colson, L., Fredholm, D.: System T, call-by-value and the minimum problem. Theor. Comput. Sci. **206**, 301–315 (1998). https://doi.org/10.1016/S0304-3975(98)00011-5

Coquand, T.: Une théorie des constructions. Thèse de troisième cycle, Université Paris VII (1985)

Coquand, T.: Une preuve directe du théorème d'ultime obstination. Comptes rendus de l'Académie des sciences, série **I**(314), 389–392 (1992)

Coquand, T., Huet, G.: The calculus of constructions. Inf. Comput. **76**, 95–120 (1988). https://doi.org/10.1016/0890-5401(88)90005-3

Crolard, T., Polonowski, E., Valarcher, P.: Extending the loop language with higher-order procedural variables. In: 2nd International Workshop on Higher-Order Rewriting (HOR 2004)/WRS (2005)

Crolard, T., Polonowski, E., Valarcher, P.: A Representation theorem for primitive recursive algorithms. Fundamenta Informaticae **107**(4), 313–330 (2011). https://doi.org/10.3233/FI-2011-405

Curien, P.-L.: Sequential algorithms "from the source" (talk slides) (2022). https://curien.galene.org/presentations/SA-friends-2022.pdf

David, R.: On the asymptotic behaviour of primitive recursive algorithms. Theor. Comput. Sci. **266**(1–2), 159–193 (2001). https://doi.org/10.1016/S0304-3975(00)00165-1

Dowek, G.: The undecidability of pattern matching in calculi where primitive recursive functions are representable. Theor. Comput. Sci. **107**, 349–356 (1993). https://doi.org/10.1016/0304-3975(93)90175-S

Girard, J.-Y.: Geometry of interaction I: interpretation of system F. In: Logic Colloquium '88, pp. 221–260. North-Holland (1989)

Girard, J.-Y.: Geometry of interaction III: accommodating the additives. In: Advances in Linear Logic, pp. 329–389. Cambridge University Press (1995)

Krivine, J.-L.: Typed lambda-calculus in classical Zermelo–Fraenkel set theory. Arch. Math. Logic **40**(3), 189–205 (2001). https://doi.org/10.1007/s001530000057

Krivine, J.-L.: Dependent choice, "quote" and the clock. Theor. Comput. Sci. **308**(1–3), 259–276 (2003). https://doi.org/10.1016/S0304-3975(02)00776-4

Krivine, J.-L.: A call-by-name lambda-calculus machine. High.-Order Symbolic Comput. **20**(3), 199–207 (2007). https://doi.org/10.1007/s10990-007-9018-9

Krivine, J.-L.: Realizability in classical logic. In: Interactive Models of Computation and Program Behaviour. Panoramas et synthèses 27, pp. 197–229. SMF (2009)

Leivant, D., Marion, J.-Y.: Lambda calculus characterizations of poly-time. Fundamenta Informaticae **19**(1/2), 167–184 (1993). https://doi.org/10.1007/BFb0037112

Quinon, P.: Intensional differences between programming languages: a conceptual and practical analysis. Philosophies **10**(6), 129 (2025). https://doi.org/10.3390/philosophies10060129

Seiller, T.: Interaction graphs: graphings. arXiv:1405.6331 (2014)

Streicher, T., Reus, B.: Classical logic, continuation semantics and abstract machines. J. Funct. Program. **8**(6), 543–572 (1998). https://doi.org/10.1017/S0956796898003141

Valarcher, P.: Contribution à l'étude des algorithmes primitifs récursifs. PhD thesis, Université Paris 7 (1996)

Van den Dries, L.: Generating the greatest common divisor, and limitations of primitive recursive algorithms. Found. Comput. Math. **3**(3), 297–324 (2003). https://doi.org/10.1007/s10208-002-0061-y

Arana, A.: The changing practices of proof in mathematics: Gilles Dowek, Comput., Proof, Mach. Metasci. **26**(1), 131–135 (2017). https://doi.org/10.1007/s11016-016-0150-1

Mathematical Informatics: Algorithms

Thomas Seiller(✉)

CNRS, JFLI – IRL 3527, Tokyo, Japan
thomas.seiller@cnrs.fr
http://www.seiller.org

Abstract. This work continues the development of an intensional approach to computability initiated in previous work [23,24], in which programs and computations, rather than functions, constitute the primary objects of study. In this setting, models of computation are described as monoid actions on a configuration space, and programs as dynamical systems constrained by this action.

Within this framework, we introduce a formal notion of algorithm as a finite directed graph whose edges are labelled by partial maps over an abstract data structure. This definition separates control from data, representing the former as a graph and the latter as an algebra of operations. We then define what it means for a program, in a given model of computation, to implement such an algorithm, by requiring a correspondence between computational steps and labelled transitions that preserves the induced transformations on representations of data. This yields a precise notion of implementation and situates algorithms as abstract partial specifications of computational behaviour.

Keywords: Computability · dynamical systems · models of computation

1 Introduction

The terminology "algorithm" predates the first computers, and does not appear in the earliest works on computability. It originally referred to methods of resolution in mathematics, as reflected for instance in the term algoriste [18]. Its systematic use within computer science appears later, notably within the Russian school [17, 19], even though the term also already occurs in Church's 1936 paper [6]. It was later widely adopted and became central to the discipline, to the point of naming a subfield, algorithmics. More recently, the term has acquired a broader meaning in common usage, where it is used to refer to complex computational systems, in particular machine learning models and platform-based decision systems.

These different usages have been analysed by Airoldi [1], who distinguishes between an analog era, a digital era, and a platform era. While this periodisation is insightful, it should not obscure the fact that these meanings coexist today.

This work was partially funded by the ANR ANR-22-CE48-0003-01 project DySCo.

V. Brattka et al. (Eds.): CiE 2026, LNCS 16674, pp. 86–105, 2027.
https://doi.org/10.1007/978-3-032-31348-5_6

The term "algorithm" is thus used to designate heterogeneous objects, depending on context. This raises a natural question: do these usages refer to a common underlying notion?

The standpoint adopted in this work is that the mathematical (analog) and computer science (digital) notions coincide at an abstract level, and that they admit a common formalisation. The apparent differences arise from constraints related to implementation and physical realisation, rather than from the notion of algorithm itself. In particular, restricting to digital representations does not adequately capture the diversity of computational paradigms, including analog, quantum, or biological ones. This suggests that the notion of algorithm should be studied independently of any specific choice of model of computation, neither any choice of specific representation of data.

I therefore focus on the problem of formalising the notion of algorithm itself, independently of its implementations. This requires identifying a notion which is sufficiently abstract to encompass the various usages discussed above, while remaining mathematically tractable. Founded upon the intensional approach to computability developed in [23,24], I propose a definition of algorithms as finite graphs labelled by operations over an abstract data structure. This definition separates control and data, and integrates naturally with the notions of program and computation introduced in the intensional framework. In particular, it allows for a precise formulation of the notion of implementation, relating algorithms to concrete programs across different models of computation.

By grounding programs in abstract models of computation defined as monoid actions, and algorithms in labelled graphs over abstract data structures, the framework cleanly separates the two notions while making their relationship, implementation via glueing, formally explicit and mathematically tractable. The present proposal therefore satisfies the following four desiderata, which we believe an adequate formalisation of algorithms should satisfy [22]:

1. distinguish algorithms from the programs that implement them;
2. remain independent of any particular model of computation;
3. provide a mathematically precise notion of implementation;
4. allow meaningful comparison between algorithms and implementations.

Comparison to Other Approaches. The question of formalising algorithms has been addressed by several authors. Notably, Moschovakis [20,21] and Gurevich [4,11] have proposed influential frameworks.

The framework proposed here offers several theoretical advantages over existing approaches to formalising the notion of algorithm. Gurevich's approach shares with the current proposal the definition of algorithms as partial specifications. However, despite their generality as a model of computation, abstract state machines do not explicitly distinguish between algorithms and programs, nor do they provide a formal notion of implementation. As such, they do not directly support comparison between different implementations, programs, or algorithms.

Moschovakis' recursors fare better in capturing what we intuitively think of as algorithms, but they are limited to recursive functions and the framework does not satisfactorily addresses the program-level. Indeed, the implementation notion via iterators does not provide explicit treatment of representations and fails to address the diversity of choices of models of computation.

1.1 Contents

The proposed mathematical definition of algorithms will be exposed in three steps. This decomposition is meaningful in itself, and provide a pedagogical approach. The first notion is that of a *syntactic algorithm*. A syntactic algorithm describes the structure of a computation: the collection of elementary actions involved and the way in which they are composed. In particular, it does not prescribe any semantics for these actions, i.e. it does not specify the transformations they induce on data.

As such, syntactic algorithms capture a commonly accepted intuition: an algorithm is given by a finite description of successive operations organised by a control structure. However, this description leaves implicit a number of assumptions regarding the meaning of those operations. Making these assumptions explicit is a central issue in any attempt to formalise the notion of algorithm.

Consider for instance the algorithm shown in Fig. 1b. It can be found in this shape in numerous textbooks on algorithmics. It is unclear, however, how the operations indicated in the text should be understood. What is the meaning of the operations "−" and "mod"? What is the domain of the variables x and y? In standard presentations, one assumes that x and y range over the integers and that these symbols denote the usual arithmetic operations. But nothing in the description itself enforces this interpretation. In fact, much information is usually left implicit: whether the inputs are natural numbers, or polynomials; whether the relation $\neq$ or the operation mod are the usual, intuitive ones, or not.

This observation leads to the idea that the notion of syntactic algorithm should be complemented by a choice of data domain together with interpretations of some primitive operations. This is formalised in the definition of *semantically-specified algorithm*. The introduced distinction between the control and data will form the basis of the formal definitions introduced thereafter.

However, while semantically-specified algorithms solve this first issue, they impose constraints that are arguably too strong. Indeed, let us consider the algorithm shown in Fig. 1a for computing the greatest common divisor. This is what is usually considered as Euclid's algorithm, a name that is used to refer to different interpretations over different domains. For instance, it already appears in Euclide's *Elements* in distinct forms: in Book VII for integers, and in Book X for magnitudes (e.g. line segments) [12–14]. Similarly, Euclid's algorithm can be used to compute the greatest common divisor for polynomials, or for elements of a finite field.

Under the semantically-specified notion, these would constitute distinct algorithms, since the underlying domains and primitive operations differ. Yet one may still wish to regard them as manifestations of a single algorithmic – Euclid's algorithm – which can be instantiated over different structures. I therefore introduce a third notion, that of *logically-specified algorithm*, in which the specification is given in the form of a first-order theory rather than a particular model of that theory. This allows primitive operations to be characterised through their formal properties while remaining independent of a specific semantic realisation.

```
Input: x, y
while y ≠ 0:
    while x ≥ y:
        x := x − y
    z := x
    x := y
    y := z
Output: x
```

(a) GCD 1

```
Input: x, y
while y ≠ 0:
    z := x mod y
    x := y
    y := z
Output: x
```

(b) GCD 2

Fig. 1. Two presentations of Euclid's algorithm.

One key observation is that the proposed approach does not define algorithms as equivalence classes of programs. While complex arguments have been developed to explain how this idea fails [5,7], I propose to approach the question by looking at the two examples of algorithms in Fig. 1. Here are two different presentations of Euclid's algorithm. Are those equal? Are those disjoint? The answer to both questions is negative.

Indeed, programs that do implement the algorithm in Fig. 1a are also implementations of the algorithm in Fig. 1b, implying that they are not disjoint. However, there are implementations of the algorithm in Fig. 1b that are not implementations of the algorithm in Fig. 1a: one could compute the operation mod using a different technique than iterated subtraction.

Thus, implementation induces a non-trivial inclusion structure between algorithms rather than a partition of programs into equivalence classes. Defining algorithms as equivalence classes of programs would therefore fail to capture such asymmetric implementation relations. The relation induced by implementation is consequently better understood as a preorder than as an equivalence relation. This excludes defining algorithms using a *definition by abstraction*; the current approach could be formulated using a more general approach [15]. It also points to an essential difference between the current framework and the approach initiated by Yanofsky [25].

2 Models of Computation and Programs

I first recall the notions of models of computation, programs, and computations introduced in [24], which will serve as the underlying framework in this work. The approach is intensional: programs and computations, rather than functions, are taken as primary objects.

Definition 1 (Model of computation). *A* model of computation *is a monoid action* $\alpha : \mathbb{M}(I) \curvearrowright \mathbf{X}$ *where:*

- $\mathbf{X}$ *is a space, called the* configuration space*;*
- I *is a set of primitive instructions interpreted by a map*[1] $\rho : I \to (\mathbf{X} \to \mathbf{X})$ *extending uniquely*[2] *to the monoid action* α.

The configuration space can be a set, or a more involved structure (topological space, measure space, topological vector space, etc.); it is expected in the latter case that the group action is compatible with the structure. The elements of I are understood as elementary computational steps, while the monoid $\mathbb{M}(I)$ encodes their sequential composition. The action of $\mathbb{M}(I)$ on $\mathbf{X}$ describes the evolution of configurations under composed instructions.

This formulation separates the abstract structure of composition, represented by the monoid $\mathbb{M}(I)$, from its operational realisation on configurations through the action on $\mathbf{X}$.

Example 1. The quintessential example is that of the one-tape Turing machine model, represented as the action $\alpha_{\mathrm{TM}} : \mathbb{M}(\mathrm{Instr}_{\mathrm{TM}}) \curvearrowright \mathbf{X}_{\mathrm{TM}}$. Here the space of configuration $\mathbf{X}_{\mathrm{TM}}$ is defined as the space of $\mathbf{Z}$-indexed sequences in $\{0, 1, \star\}$ which are almost-always equal to $\star$:

$$\{(s_i)_{i \in \mathbf{Z}} \mid s_i \in \{0, 1, \star\}, \mathrm{Card}\{i \in \mathbf{Z} \mid s_i \neq \star\} < \infty\}.$$

The action α_{TM} of the set $\mathrm{Instr}_{\mathrm{TM}} = \{\mathtt{right}, \mathtt{left}\} \cup \{\mathtt{write}_i \mid i \in \{0, 1, \star\}\} \cup \{\mathtt{read}_i \mid i \in \{0, 1, \star\}\}$ of atomic instructions is then defined by:

- $\alpha_{\mathrm{TM}}(\mathtt{right}) : (s_i)_{i \in \mathbf{Z}} \mapsto (s_{i-1})_{i \in \mathbf{Z}}$;
- $\alpha_{\mathrm{TM}}(\mathtt{left}) : (s_i)_{i \in \mathbf{Z}} \mapsto (s_{i+1})_{i \in \mathbf{Z}}$;
- $\alpha_{\mathrm{TM}}(\mathtt{write}_\star) : (s_i)_{i \in \mathbf{Z}} \mapsto (t_i)_{i \in \mathbf{Z}}$ where $t_0 = \star$ and $t_i = s_i$ for $i \neq 0$;
- $\alpha_{\mathrm{TM}}(\mathtt{read}_\star) : (s_i)_{i \in \mathbf{Z}} \mapsto (s_i)_{i \in \mathbf{Z}}$ defined if $s_0 = *$, undefined otherwise.

One can then define the set of programs prescribed by a choice of model of computation.

Definition 2 (Program). *Let* $\alpha : \mathbb{M}(I) \curvearrowright \mathbf{X}$ *be a model of computation. An* α-program *is a tuple* $P = (S, E, s, t, \ell, \mathbf{i}(P), \mathbf{t}(P))$ *where:*

- (S, E, s, t) *is a directed graph, with* $s, t : E \to S$ *the source and target maps;*
- $\ell : E \to I$ *is a labelling of edges by primitive instructions;*
- $\mathbf{i}(P), \mathbf{t}(P) \in S$ *are distinguished initial and terminal states.*

By convention, the set $\{e \in E \mid s(e) = \mathbf{t}(P)\}$ *is empty.*

[1] Here $\mathbf{X} \to \mathbf{X}$ usually denotes the set of partial endomorphisms of $\mathbf{X}$.

[2] Note that $\mathbb{M}(I)$ does not denote here the free monoid, but the monoid generated by I and the relations induced from the interpretation by ρ. In other words, $\mathbb{M}(I)$ is isomorphic to the sub-monoid of $\mathbf{X} \to \mathbf{X}$ generated by the set $\alpha(I)$.

The graph structure of the program describes its control flow, with the vertices being the control states. Edge labels specify the elementary instructions performed during transitions between control states.

An α-program $P = (S, E, s, t, \ell, \mathbf{i}(P), \mathbf{t}(P))$ naturally defines a partial dynamical system on the product space $\mathbf{X} \times S$. Indeed, each edge $e \in E$ determines a partial transition $(x, s(e)) \mapsto (\alpha(\ell(e))(x), t(e))$, combining an evolution of configurations in $\mathbf{X}$ with a transition between control states. The dynamics generated by these transitions describe the execution of the program.

In particular, a computation of the program from a configuration $x_0 \in \mathbf{X}$ is simply an orbit of this partial dynamical system starting from $(x_0, \mathbf{i}(P))$. Equivalently, a computation is given by a sequence $((x_n, s_n))_{n \in \mathbb{N}}$ such that for every $n \in \mathbb{N}$ there exists an edge $e_n \in E$ satisfying $s(e_n) = s_n$, $t(e_n) = s_{n+1}$, and $x_{n+1} = \alpha(\ell(e_n))(x_n)$.

The outcome of a computation, when defined, is determined by the configuration component of a state $(x, \mathbf{t}(P))$ reachable from the initial state. More generally, one may consider observables extracted either from configurations or from complete computation traces.

This framework provides a uniform description of a wide range of computational paradigms. Classical models such as finite automata, Turing machines, or abstract state machines[3] can be expressed through suitable choices of configuration spaces and instruction sets. More generally, the framework accommodates computational systems arising from algebraic, dynamical, geometric, or non-discrete settings.

Within this framework, programs are concrete computational objects defined relative to a fixed model of computation. The notion of algorithm, introduced below, somehow corresponds to replacing the model of computation used in the definition of program by *abstract data structures*, which we will now introduce.

3 Abstract Data Structures

3.1 Definitions

One can already find a notion of "abstract data *type*" in the literature. However, this notion is not operational enough for our purposes, and I will rather work with an abstraction of data *structures*: i.e. a set of *values*, together with a set of allowed operations on those values. Underlying this notion, one can identify a notion of *data representation*. I will therefore define simultaneously two notions: that of data domain and that of data structure, which subsumes a choice of data domain.

Definition 3. *An* abstract data domain $\mathcal{D}$ *is a set* D. *An* abstract data structure $\mathtt{D}$ *over an abstract data domain* $\mathcal{D}$ *is a set of allowed operations* $S_{\mathtt{D}}$ *– called* structural maps *– in* $\sum_{k,k'} \mathcal{D}^k \rightharpoonup \mathcal{D}^{k'}$. *Any element* $s \in S_{\mathtt{D}}$ *belongs to* $\mathcal{D}^k \rightharpoonup \mathcal{D}^{k'}$

[3] A more detailed comparison between those notions and Gurevich's ASM can be found in a paper by Valarcher and Anh-Ton Le [2].

for fixed values of k and k': we will write $dom(s) = k$ and $im(s) = k'$. The maximal arity *of* D *is defined as*[4] $\max\{dom(s), im(s) \mid s \in S_{\mathtt{D}}\}$.

Let us note that once again, no restrictions are imposed on the cardinality of abstract data domains or on the cardinality of the set of allowed maps. A particularly interesting case is when the latter is defined by induction (Example 4). Similarly, in order to define a proper, usable, abstract data structure one would require some maps relating the representations of $\mathcal{D}^k$ for various values of k. But on principle, no such restrictions will be imposed.

This notion is defined in a way which is quite similar to the definition of abstract model of computation: partial maps are used to allow for reading. For instance, a notion of integers with a test to zero could be represented by the set $\mathbf{N}$ together with maps from $\mathbf{N}$ to $\{0, 1\}$ but this would imply the introduction of another data type (booleans), or a particular (ad-hoc) choice of encoding of booleans as specific elements of the set $\mathbf{N}$. Partial functions avoid those by allowing to project on a subset; this can then be used to implement any of these other approaches, and this choice allows for self-contained definitions.

I will now give examples. I note that agreeing on a specific axiomatisation for a given data structure is a complex task, and in fact depends on the expected use. As a consequence, the examples below are proposed axiomatisations but could very well be replaced by others. I am not trying to establish those as *standard* definitions, but rather illustrate how one could manipulate the notion.

Example 2 (Abstract boolean). One can define an abstract data structure representing booleans as follows: the underlying data domain is $\mathcal{B} = \{0, 1\}$, with the structural maps $\mathtt{read}_0$, $\mathtt{read}_1$, and, and, and not defined as:

$$\begin{aligned}
\mathtt{read}_0 :& n \mapsto n \text{ if and only if } n = 0\\
\mathtt{read}_1 :& n \mapsto n \text{ if and only if } n = 1\\
\mathtt{and} :& (m, n) \mapsto m \times n\\
\mathtt{or} :& (m, n) \mapsto m + n - m \times n\\
\mathtt{not} :& n \mapsto 1 - n.
\end{aligned}$$

The following definition captures the standard unary representation of natural numbers, which was used in the early work on computability.

Example 3 (Natural numbers). One can define an abstract data structure representing natural numbers as follows: the underlying abstract data domain is $\mathcal{N} = \mathbf{N}$, with the structural maps $\mathtt{read}_0$, $\mathtt{read}_S$, succ defined by:

$$\begin{aligned}
\mathtt{read}_0 :& n \mapsto n \text{ if and only if } n = 0\\
\mathtt{read}_S :& n \mapsto n \text{ if and only if } n \neq 0\\
\mathtt{succ} :& n \mapsto n + 1.
\end{aligned}$$

[4] By definition, we consider that $\max S = \infty$ when the set S is unbounded.

This data structure can be extended by other operations, such as predecessor, addition and multiplication defined as:

$$\begin{aligned}\texttt{pred} :& n \mapsto n-1 \text{ if and only if } n \neq 0\\ \texttt{add} :& (n,m) \mapsto n+m\\ \texttt{mult} :& (n,m) \mapsto n \times m.\end{aligned}$$

I leave it to the reader to verify that many other abstract structures commonly used in computability and complexity theory, such as lists, trees or graphs, naturally give rise to abstract data structures in the present sense. Additional examples and constructions can be found in [23]. I nevertheless detail one particularly important example, namely that of recursive functions.

The notion of recursive function is sometimes considered as a model of computation. My understanding is that it is not, because the definition of recursive functions is abstracted from computation; it does not determine computation, it only specifies *what* could be computed. As such, I considered for a while that it could be an example of a programming language. But there are no standard operational semantics associated to recursive functions; one may implement the computation of a recursive function in very different ways. This leads to thinking of the notion as something of a more algorithmic nature, separated from consideration on implementations. It finally occurred to me that it was an example of an abstract data structure.

Example 4 (Recursive functions). Primitive recursive and general recursive functions naturally define abstract data structures in the present sense. The underlying data domain is $D = \mathbf{N}$, while the structural maps are given by the corresponding classes of recursive functions. More precisely, one defines inductively a collection Rec of functions over $\mathbf{N}$ generated from constant functions, the successor function, and projection functions, and closed under composition and primitive recursion, together with minimisation in the case of general recursive functions. The resulting collection Rec therefore provides a canonical example of an abstract data structure generated by closure operations on elementary maps.

There are several consequences of this. First, Turing-completeness can be defined as the possibility to implement this data-structure. Second, it provides a new point of view on algebraic characterisation of complexity classes from Implicit complexity (ICC), such as the Bellantoni and Cook result [3]. Those can now be understood as defining abstract data structures that capture time- or space-bounded computations, in the same way linear logic based techniques in ICC capture those same classes through types.

3.2 Representations

Obviously, this notion of data structure is independent of a chosen machine model. To compute on data structures, one needs to implement them. To be precise, one would have to account for the complexity of the implementation, which could lead to overhead in time or space when simulating programs from a model of computation into another. However, questions of complexity will be the subject of a later paper and will not be considered here.

Definition 4. *Suppose given a model of computation* $\alpha : \mathbb{M}(I) \curvearrowright \mathbf{X}$, *and an abstract data domain* $\mathcal{D}$ *with maximal arity* $\mathbf{a}$. *An interpretation of a data domain* $\mathcal{D}$ *is a map*

$$\Delta : \cup_{k \leq \mathbf{a}} D^k \to \mathbf{X}.$$

Let us establish some notations that will be useful below. For each sequence $s \in \{0, 1, \star\}^n$, write $\underline{\star} s \underline{\star}$ the configuration $(t_i)_{i \in \mathbf{Z}}$ defined as $t_i = s_i$ for $i = 0, 1, \dots, n-1$. Given two sequences s_1, s_2, write $s_1 \star s_2$ the concatenation of s_1 and s_2. Lastly, for each natural number n, write $\overline{n}^2$ the sequence in $\{0, 1\}$ corresponding to its binary representation and 1^n the sequence consisting of n occurrences of the symbol 1.

Example 5 (Interpretations of Booleans). Let us consider the Boolean data structure defined above (Example 2). Note that it is has maximal arity 2. We can define the interpretation in $\mathbf{X}_{\text{TM}}$ as follows:

$$\Delta_{\mathcal{B}} : \begin{array}{rl} a \in \mathcal{B} & \mapsto \underline{\star} a \underline{\star} \\ (a, b) \in \mathcal{B}^2 & \mapsto \underline{\star} a \star b \underline{\star} \end{array}$$

Example 6 (Interpretations of natural numbers). One can define two interpretations of the abstract domain of natural numbers in the Turing machines models $amcTM : \mathbb{M}(\text{Instr}_{\text{TM}}) \curvearrowright \mathbf{X}_{\text{TM}}$ given as the following:

$$\Delta_{\mathcal{N}} : (a_0, a_1, \dots, a_n) \mapsto \underline{\star} 1^{a_0} \star 1^{a_1} \star \cdots \star 1^{a_n} \underline{\star}$$
$$\Delta_{\mathcal{N}}^{(2)} : (a_0, a_1, \dots, a_n) \mapsto \underline{\star} \overline{a_0}^2 \star \overline{a_1}^2 \star \cdots \star \overline{a_n}^2 \underline{\star}$$

Definition 5. *Suppose given a model of computation* $\alpha : \mathbb{M}(I) \curvearrowright \mathbf{X}$, *and an abstract data structure* $\mathtt{D} = (\mathcal{D}, S_D)$. *An implementation of* $\mathtt{D}$ *in* α *is an interpretation* Δ *of the underlying data domain* $\mathcal{D}$ *together with a map* $\delta : S_D \to \text{Programs}(\alpha)$ *such that for all* $d \in D^k$, $\exists n, \delta(f)^n(\Delta(d)) = \Delta(f(d))$.

The reader is referred to previous work [23] for concrete examples of implementations of the above data structures, and others, in various abstract models of computation.

3.3 A Bit of History

Recall that the Church-Turing thesis states that the set of *effectively computable functions*[5] (which is an informal notion) is equal to the set of functions computable by Turing machines (or equivalently, are computable in lambda-calculus). A forceful argument for the Church-Turing thesis is that the following

[5] The original statements considered only functions from natural numbers to natural numbers, although some extensions were considered later on [16].

sets of functions (on natural numbers) are equal: functions computed by Turing machines, and lambda-definable functions.

It is interesting to dive into the research work of the time to understand which results are proven exactly. It turns out that the above equivalence is, from a purely formal point of view, not properly established. I am obviously not putting the result into question, but if one reformulates the exact statements of theorem proven in these papers, one obtains the following:

- In his 1937 paper (Computability and λ-Definability), Turing shows[6] that if f is a lambda-definable function (on Church numerals), then there exists a Turing Machine whose output is the sequence

$$\underbrace{11\ldots1}_{f(0)\text{ symbols}}\ 0\ \underbrace{11\ldots1}_{f(1)\text{ symbols}}\ 0\ldots0\ \underbrace{11\ldots1}_{f(n)\text{ symbols}}\ 0\ldots$$

- In the same paper, Turing also shows that if there exists a Turing Machine whose output is the sequence above, then f is general recursive. This is done by means of an encoding of configurations as integers.
- In his 1936 paper (λ-definability and recursiveness), Kleene shows that every general recursive function is λ-definable using a *shifted version of Church numerals*, in which 0 is represented as 1, 2 represented as 3, etc.

The representation of computable functions by Turing shows that the equivalence considered is based on *unary* representations of integers. Moreover, the choice in the definition of what a computable function is leaves open the question of the representation of partial functions. Indeed, the result applies only to total functions[7]. Lastly, one should notice that Kleene uses a different interpretation of (unary) natural numbers.

I will not detail here how the following results can be proved. However, these results can be restated as[8]:

- Lambda-calculus interprets the data structure of partial recursive functions (and can interpret it for both a unary and binary representation of the underlying data domain);
- Turing machines interpret the data structure of partial recursive functions (and can interpret it for both a unary and binary representation of the underlying data domain);

---

[6] Although he writes "No attempt is being made to give a formal proof that this machine has the properties claimed for it. Such a formal proof is not possible unless the ideas of substitution and so forth occurring in the definition of conversion are formally defined, and the best form of such a definition is possibly in terms of machines".

[7] This shows that, even though Turing knows that Turing machines compute partial functions, his notion of *computable function* presupposes totality.

[8] Although binary representations are not considered in these papers, it is easily obtained in a similar manner.

- One can encode the elements of $\mathbf{X}_{\text{TM}}$ as natural numbers and show that, based on this encoding, each instruction is interpreted as a partial recursive function;
- One can encode the elements of Λ as natural numbers and show that, based on this encoding, that a step of β-reduction is interpreted as a partial recursive function.

Based on the encoding of configurations, one can obtain a simulation of lambda-calculus by Turing machines (and conversely). This simulation is, based on defintions from [24], a data-constrained program-wise simulation, which means that the above results could be used as an algorithmic Church-Turing thesis, namely that[9] *the set of effective methods to compute functions is equal to the set of Turing machines (computing functions).*

Let us note however that this alternative statement does not account for complexity. Indeed, the quantitative aspects of the simulations above imply that time and space constrained computations do not coincide.

4 Algorithms

4.1 Syntactic Structure and Glueing

Definition 6. *A syntactic algorithm is a finite labelled graph (with control states)* $A = (S, E, s, t, \mathbf{i}(A), \mathbf{t}(A), L, \ell)$ *where* S *is a finite set of* states *containing the initial and terminal states* $\mathbf{i}(A)$ *and* $\mathbf{t}(A)$, (S, E, s, t) *is a directed graph,* L *a finite set of* labels, *and* $\ell : E \to L$ *is a labelling function.*

The set of all algorithms is written as Algorithms$(*)$.

The guiding intuition behind this definition is that the graph describes a general control structure, while the labels corresponds to the name of operations used in the algorithm. For the moment, the algorithm is *syntactic* because the labels are just arbitrary names. The notions introduced in the next sections refine the definition by associating labels to more restricted classes of operations. Before going through those definitions, we can already define what it means for a program to *implement* a syntactic algorithm.

Definition 7 (Glueing). *Let* $A = (V, S, \mathbf{i}(A), \mathbf{t}(A), E, s, t, L, \ell)$ *be a syntactic algorithm, and an* AMC $\alpha : \mathbb{M}(I) \curvearrowright \mathbf{X}$. *Suppose moreover given a map* $\phi : L \to$ Programs(α). *The pre-glueing of* ϕ *along* A *is the machine* $\mathcal{P}(A, \phi)$ *defined as the disjoint union* $\sum_{e \in E} \phi(\ell(e))$. *The* glueing of ϕ along A *is then defined as the program* $\mathcal{G}(A, \phi)$ *obtained by identifying for all* $v \in V$ *the states* $\{\mathbf{i}(e, \phi(e)) \mid s(e) = v\} \cup \{\mathbf{t}(e, \phi(e)) \mid t(e) = v\}$, *and setting* $\mathbf{i}(\mathcal{G}(A, \phi)) = \mathbf{i}(A)$ *and* $\mathbf{t}(\mathcal{G}(A, \phi)) = \mathbf{t}(A)$.

We illustrate the notion of glueing in Fig. 2.

Definition 8. *A program* G implements *the algorithm* A *when there exists* $\phi : L \to$ Programs(α) *such that* G *is the glueing of* A *along* ϕ.

[9] Here I restrict the discussion to functions from $\mathbf{N}$ to $\mathbf{N}$ for simplicity.

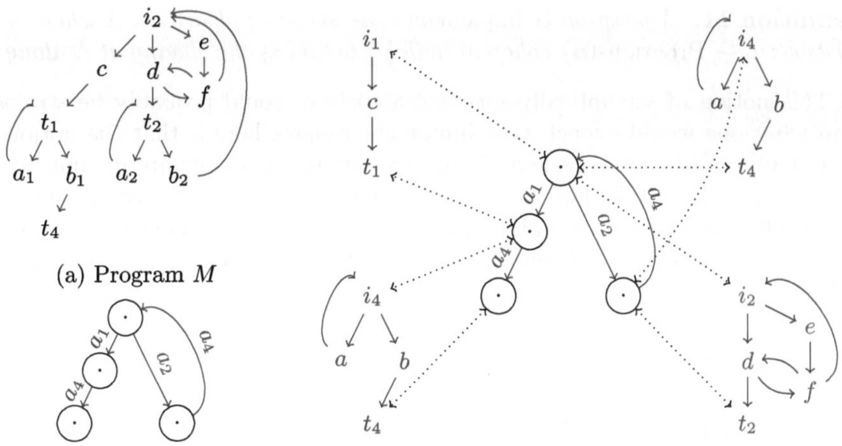

(a) Program M

(b) Algorithm A

(c) M seen as a glueing along an algorithm A

Fig. 2. Example of glueing

4.2 Semantically-Specified Algorithms

The next step is to impose that labels correspond to specific maps defined as part of data structures. In practice, these data structures are implicitly given when writing down algorithm, e.g. using pseudo-code.

Let us note that if an algorithm uses two data structures, say integers and booleans, or simply manipulates several elements of a given data structures, the implicit assumption is that several copies of these data structures are simultaneously interpreted within the implementing model of computation. Formally, this is equivalent to considering products of data structures: if $\mathtt{D} = (\mathcal{D}, S)$ and $\mathtt{D}' = (\mathcal{D}', S')$ are data structures, then their product $\mathtt{D} \times \mathtt{D}'$ is defined as $(\mathcal{D} \times \mathcal{D}', S \bar{\times} S')$ where:

$$S \bar{\times} S' = \{s \times \mathrm{Id}_{\mathcal{D}'} \mid s \in S\} \cup \{\mathrm{Id}_{\mathcal{D}} \times s' \mid s' \in S'\}.$$

As a consequence, one can define algorithms w.r.t. a single data structure.

Definition 9 (Semantically-specified algorithm). *Let* $\mathtt{D} = (\mathcal{D}, S)$ *be a data structure. A* semantically-specified algorithm *with respect to* $\mathtt{D}$ *is defined as a tuple* $A = (V, \mathbf{i}(A), \mathbf{t}(A), E, s, t, L, \ell, [\![\cdot]\!])$ *where:*

- $(V, \mathbf{i}(A), \mathbf{t}(A), E, s, t, L, \ell)$ *is a syntactical algorithm;*
- $[\![\cdot]\!] : L \to S$ *maps the labels to structural maps.*

The set of all specified algorithms w.r.t. $\mathtt{D}$ *is written* $\mathrm{Algorithms}(\mathtt{D})$.

Definition 10 (Coherent labelings). *Suppose given a specified algorithm* $A = (V, S, \mathbf{i}(A), \mathbf{t}(A), E, s, t, L, \ell, [\![\cdot]\!])$ *for a data structure* $\mathtt{D} = (\mathcal{D}, S)$*, and a model of computation* $\alpha : \mathbb{M}(I) \curvearrowright \mathbf{X}$*. A map* $\phi : L \to \mathrm{Programs}(\alpha)$ *is* coherent *with* $[\![\cdot]\!]$ *with respect to an interpretation* Δ *of* $\mathcal{D}$ *when for all* $o \in L$*,* $\phi(o)$ *is an implementation of* $[\![o]\!]$ *w.r.t.* Δ.

Definition 11. *A program G* implements *the specified algorithm A when there exists $\phi : L \to \text{Programs}(\alpha)$ coherent with $[\![\cdot]\!]$ w.t. G is the glueing of A along ϕ.*

This notion of semantically-specified algorithm could probably be stronger than what one would expect. One important remark here is that the notion of data structure is set-theoretic and requires one to fix a concrete domain. As a consequence, the notion does not allow to identify Euclidean division on integers and Euclidean division on polynomials because the underlying abstract data domain is different. The following proposed definition proposes a solution to this problem.

4.3 Logically Specified Data Structures

Definition 12. *A* logical data structure $\mathcal{D}$ *is defined as first-order logical theory over a first-order language* (Var, Fun, Rel).

An abstract data structure $\mathtt{D}$ *is* a model of the logical data structure $\mathcal{D}$ *if it is a model of $\mathcal{D}$, i.e. if* $\mathtt{D} \models \mathcal{D}$.

Definition 13 (Logically-specified algorithm). *Let $\mathcal{D}$ be a logical data structure over a first-order language* (Var, Fun, Rel). *A* logically-specified algorithm *with respect to $\mathcal{D}$ is a tuple $A = (V, \mathbf{i}(A), \mathbf{t}(A), E, s, t, L, \ell, [\![\cdot]\!])$ where:*

- $(V, \mathbf{i}(A), \mathbf{t}(A), E, s, t, L, \ell)$ *is a syntactical algorithm;*
- $[\![\cdot]\!] : L \to \text{Fun} \cup \text{Rel}$ *maps the labels to structural maps.*

The set of all specified algorithms w.r.t. $\mathcal{D}$ is written Algorithms($\mathcal{D}$).

Given a logically-specified algorithm A w.r.t. $\mathcal{D}$ and a model $\mathtt{D}$ of $\mathcal{D}$, one can deduce the unique induced semantically-specified algorithm $A_{\mathcal{D}\models\mathtt{D}}$. Implementing A then boils down to implementing $A_{\mathcal{D}\models\mathtt{D}}$ for a model $\mathtt{D}$ of $\mathcal{D}$.

Definition 14. *A program P implements the logically specified algorithm A if there exists an abstract data structure $\mathtt{D}$ s.t. P implements the induced specified algorithm $A_{\mathcal{D}\models\mathtt{D}}$.*

Note that Euclidean division on integers and on arbitrary Euclidean ring are but different models of the same logical data structure. The corresponding notion of *logically specified algorithm*, intermediate between the syntactical and specified notions considered above, then allows to identify the Euclidean division algorithms defined on different Euclidean rings.

The reader familiar with Gurevich's *abstract state machines* [10] will probably be curious about potential connections with the notion just developed since both are based on first order structures. While there may be a formal relation between the approaches, I note that Gurevich uses the first-order structure to define the states (i.e. the space underlying the AMC) while it is here used to describe instructions (or rather, more precisely, programs) that can be performed. This seems to be a fundamental difference. One consequence of this is that Gurevich modifies the values of the interpretations of functions and relation symbols, while here the interpretation is fixed once and for all.

4.4 Properties

The definition of *glueing* can be adapted to define glueing of algorithms (instead of programs) along another algorithm. This leads to the definition of a preorder on the set of algorithms.

Definition 15 (Glueing of algorithms along a labelled graph). *Suppose given a syntactic algorithm* $A = (V, S, \mathbf{i}(A), \mathbf{t}(A), E, s, t, L, \ell)$, *and an* AMC $\alpha : \mathbb{M}(I) \curvearrowright \mathbf{X}$. *Suppose moreover given a map* $\phi : L \to \mathrm{Algorithms}(*)$. *The pre-glueing of* ϕ *along* A *is defined as the disjoint union* $\mathcal{P}(A, \phi) = \sum_{e \in E} \phi(\ell(e))$. *The* glueing of ϕ along A *is then defined as the algorithm* $\mathcal{G}(A, \phi)$ *obtained by identifying for all* $v \in V$ *the vertices* $\{\mathbf{i}(e, \phi(e)) \mid s(e) = v\} \cup \{\mathbf{t}(e, \phi(e)) \mid t(e) = v\}$, *and setting* $\mathbf{i}(\mathcal{G}(A, \phi)) = \mathbf{i}(A)$ *and* $\mathbf{t}(\mathcal{G}(A, \phi)) = \mathbf{t}(A)$.

If ϕ *maps* L *into* Algorithms(D) *(for* D *either an abstract data structure or a logical data structure), the resulting algorithm belongs to* Algorithms(D).

Proposition 1. *If* B *is obtained as the glueing of* A *along* ϕ, *and* P *is a program implementing* B, *then* P *also implements* A.

Proof. Here is a sketch of the proof, avoiding painful details. If P implements B, then P is a glueing of B along some $\psi : L^B \to \mathrm{Programs}(\alpha)$. Moreover, B is a glueing of A along some $\phi : L^A \to \mathrm{Algorithms}(*)$. Now, from both these maps one can deduce a $\theta : L^A \to \mathrm{Programs}(\alpha)$. It is not difficult to check that P is the glueing of A along this map θ.

The notion of algorithm obtained does not correspond to an equivalence relation on programs. In particular, one program implements many different algorithms. In some way, the notion is almost topological, and one could expect some separation axiom to be satisfied (probably the T_0 axiom: that if P and Q are different programs, there exists an algorithm A such that one of P and Q implement A but not the other).

Among the future research in that direction, one direction seems of particular importance: the definition of a notion of *distance*. A distance between algorithms, together with a notion of *implementation/glueing* up to some error ϵ. This could be used to talk about convergence of programs toward an algorithm for instance.

5 Examples

Variants of the GCD *algorithm.* As a first example, I simply give in Fig. 3 two representations of the GCD algorithms shown in Fig. 1. These algorithms can be considered as *logically specified*, where the data structure is defined as a model of a first order theory of Euclidean rings. Note that these provide an example of the preorder just defined: the algorithm B is obtained as a glueing of some specific algorithm for computing the remainder along the algorithm A.

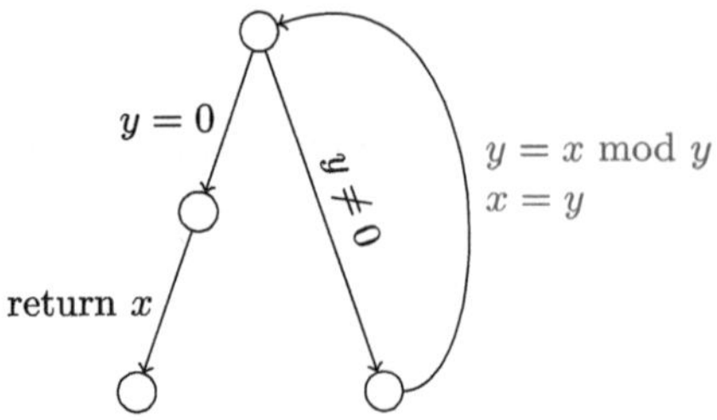

(a) Algorithm A

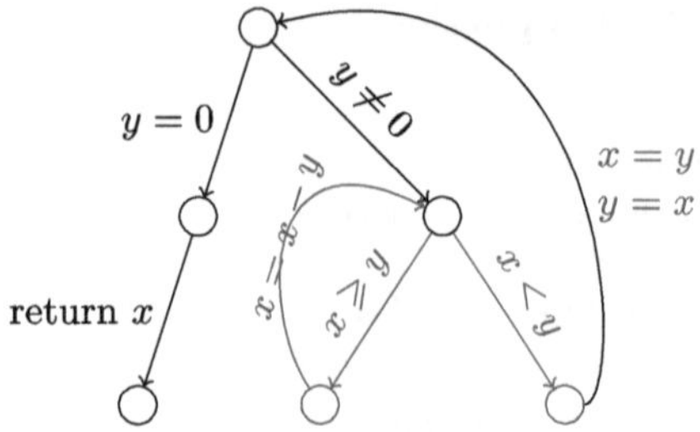

(b) Algorithm B

Fig. 3. Two specified algorithms.

The Merge Sort Algorithm. The merge sort algorithm, and more generally the recursive algorithms, are examples of algorithms which are simultaneously defined with the data structure. I.e. the merge sort algorithm is defined as a specified algorithm with respect to a data structure that contains a sorting algorithm.

More precisely, suppose one wants to define the merge sort algorithm. The first thing is to define the data structure. One will require here a structure that allows for working with several lists, and compare elements (say integers). Defining a recursive algorithm then corresponds to the following: one adds to the data structure the corresponding structural map $\texttt{sort} : \Delta_{\mathcal{L}} \to \Delta_{\mathcal{L}}$. To ease the argument, suppose that there exists a structural map $\texttt{split} : \Delta_{\mathcal{L}} \to \Delta_{\mathcal{L}} \times \Delta_{\mathcal{L}}$ defined as $s_0, s_1, \dots, s_k \mapsto (s_0, s_2, s_4, \dots) \times (s_1, s_3, s_5, \dots)$. It should be clear that if such a map is not given as part of the structure, one can define a corresponding algorithme using more elementary operations on list. One can then define the sorting algorithm shown in Fig. 4a. Note that in the figures, the initial state is represented by an incoming edge and the terminal state by a double circle.

Now if the sort label is interpreted by a program that sorts a list (whichever method is used), a program implementing this algorithm is itself a program that sorts a list. The recursive definition consists in defining the overall algorithm R obtained by glueing A along the label sort within A. This is computing a limit: one starts with A, and then obtains $A[\texttt{sort} \leftarrow A]$, in which one then glues A along the label sort to obtain $A[\texttt{sort} \leftarrow A][\texttt{sort} \leftarrow A]$, etc.

The recursive program R is then defined formally as the limit of this process, which can be represented as an infinite graph that no longer contains the label sort. If one tries to express in more details the recursive structure, one obtains that the obtained algorithm is defined as an implementation of the algorithm shown in Fig. 4b which put forth the recursive structure, which is close to the recursive definition considered by Moschovakis [20]:

$$\mathrm{sort}(x) = \begin{cases} [] & \text{if } x = [] \\ \mathrm{merge}(\mathrm{sort}(\mathrm{split}(x)_1), \mathrm{sort}(\mathrm{split}(x)_2)) & \text{otherwise.} \end{cases}$$

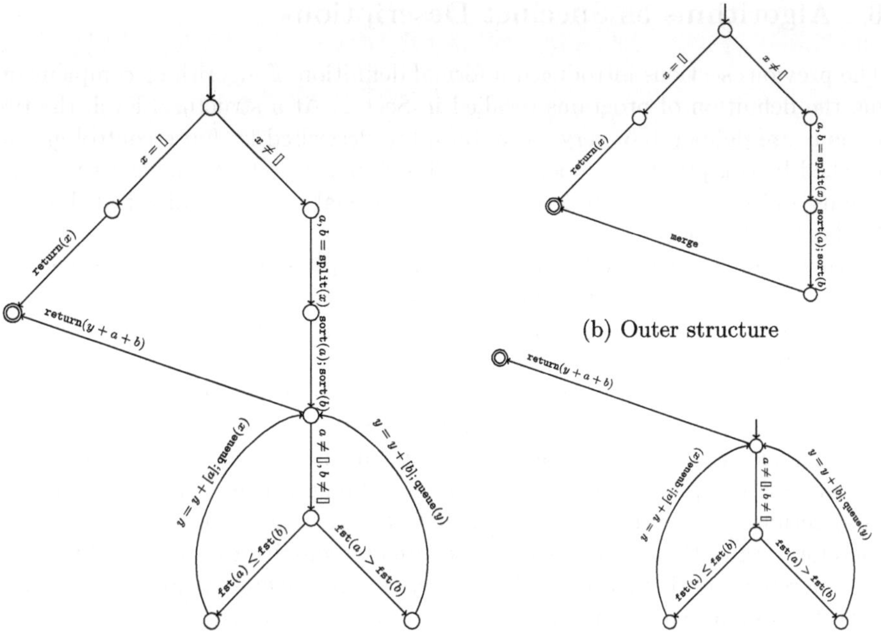

(a) The mergesort algorithm

(b) Outer structure

(c) Structure of the merge algorithm

Fig. 4. The mergesort algorithm.

Here the merge algorithm, shown in Fig. 4c, can also be given a recursive definition:

$$\text{merge}(a, b) = \begin{cases} b & \text{if } a = [] \\ a & \text{if } b = [] \\ \text{fst}(a) + \text{merge}(\text{queue}(a), b) & \text{if } \text{fst}(a) \leq \text{fst}(b) \\ \text{fst}(b) + \text{merge}(a, \text{queue}(b)) & \text{otherwise} \end{cases}$$

Finally, let us note that in the proposed algorithm, no order is enforced on the two subroutines $\texttt{sort}(a)$ and $\texttt{sort}(b)$: the label simply imposes that both operation are performed in this part of the algorithm. As a consequence, an implementation may choose to sort a before b, or b before a, or even sort both in parallel. Variants of this algorithm in which an order is imposed can be considered by replacing the arrow labelled with both sort operations by two consecutive arrows, one sorting a (for instance), the other sorting b. Another variant in which both operations are performed in parallel may be represented by a single arrow labelled with $\texttt{sort}(a) \mid \texttt{sort}(b)$ where $\mid$ indicates explicitly the parallel execution. These algorithms are refinements of the original one, and any implementation of those is an implementation of the algorithm in Fig. 4a. However they are more specific, and any implementation of the original algorithm may not implement one of those, depending on how the subroutine are ordered.

6 Algorithms as Succinct Descriptions

The previous sections introduced a formal definition of algorithms complementing the definition of programs recalled in Sect. 2. At a structural level, the two notions are deliberately very close: both are described by finite control graphs labelled by computational operations. This similarity reflects the fact that algorithms and programs both describe computational processes, albeit at different levels of abstraction.

However, this proximity also raises a conceptual issue. If algorithms and programs are represented by essentially similar mathematical objects, what distinguishes an algorithm from a program? I argue that the distinction cannot rest solely on the formal structure itself, but must also involve the role played by these objects. In particular, an algorithm should provide a succinct and intelligible description of the programs implementing it.

This question becomes especially relevant in the context of neural networks. Indeed, the complete sequence of matrices defining a neural network can itself be described abstractly using an adequate abstract data structure. One might therefore argue that a neural network already constitutes an algorithm, with its concrete execution corresponding merely to an implementation in which each operation is realised directly as a primitive computational step. While this would be coherent with the third usage of the term *algorithm* mentioned in the introduction, it feels incompatible with the intuitive notion of algorithm developed in mathematics and computer science.

I will therefore argue instead that such a fully explicit description should not, in general, be regarded as algorithmically informative. Although it formally specifies a computational process, it somehow fails to do so at a sufficiently abstract level. Algorithms are not merely abstract specifications: they are also intended to provide intelligible and communicable descriptions of computational processes.

This suggests refining the relationship between algorithms and programs by introducing a notion of succinctness. Rather than modifying the definition of algorithm itself, the idea is to distinguish programs that admit genuinely compressed algorithmic descriptions from those that do not.

Definition 16. *Fix a function $f : \mathbb{N} \to \mathbb{N}$ such that $f(n) = o(n)$. Given a program P, we say that P* admits an f-succinct algorithm *if there exists an algorithm A implemented by P satisfying* $\text{size}(A) \leq f(\text{size}(P))$.

The precise choice of the function f determines the degree of compression required from the algorithmic description relative to the implementation. I will not attempt here to determine which asymptotic regimes best reflect practical notions of succinctness. Such questions would require a more detailed study of concrete implementations, for example comparing algorithms with compiled executable code rather than source-level descriptions. The arguments developed below, however, only require the assumption that $f(n) = o(n)$.

This condition admits a natural interpretation in terms of Kolmogorov complexity. Indeed, if a program P admits an f-succinct algorithm A, together with a fixed implementation procedure transforming A into P, then A is a witness that P has low Kolmogorov complexity.

In the context of neural networks, admitting a succinct algorithm may be interpreted as a strong intensional form of explainability. Indeed, the existence of a compressed algorithmic description means that the network is not merely a large collection of parameters, but possesses an underlying computational structure describable at a significantly higher level of abstraction.

The following consequence is immediate from standard counting arguments in Kolmogorov complexity theory.

Proposition 2. *The proportion of programs of size n admitting f-succinct algorithms tends to zero as $n \to \infty$.*

In this sense, most sufficiently large programs do not possess meaningful algorithmic descriptions. Equivalently, most programs are not compressible into significantly smaller abstract computational structures. From the present perspective, they therefore fail to exhibit the kind of higher-level organisation usually associated with algorithms. In layman's term, this theorem states that most programs do not admit f-succinct algorithms.

This observation is particularly relevant in the context of large neural networks. While such systems may implement highly structured behaviours, the proposition suggests that generic parameter configurations should not be expected to admit succinct algorithmic descriptions. Possessing an algorithm, in the sense discussed above, therefore corresponds to a strong intensional form of explainability: it means that the computational behaviour of the network can be described through a substantially more compact abstract structure.

7 Conclusion

The distinction introduced in this work between algorithms, programs, and implementations opens several directions for future research. Since implementations induce a preorder relating abstract algorithms to concrete programs, one may investigate quantitative notions comparing different implementations or measuring the extent to which a program admits an underlying algorithmic structure. This raises the possibility of defining notions analogous to distances, approximations, or convergence between programs and algorithms, potentially allowing one to formalise ideas such as a program "converging" toward an algorithmic description, which could shed some light on learning techniques.

In particular, from the present perspective, an important question is whether such systems possess recoverable intensional structures that may legitimately be regarded as succinct algorithms. If such structure exists, then it should be describable from the training data, independently of the particular underlying model of computation. In a recent collaboration with Jarvis, Gastaldi, and Terilla [8,9], we explore geometric and logical structures arising from data. Reframed in the context of the current discussion, this corresponds to the extraction of an abstract data structure from the training corpus directly, and points to the possibility of mathematically recovering some succinct algorithmic description of the programs obtained through training.

The formal definition of algorithms and implementation could also provide the basis for the development of formal verification and certification techniques at the algorithmic level itself. Rather than verifying individual implementations independently, one could aim to establish properties and correctness results directly for algorithms and then transfer these results systematically to concrete programs through the implementation relation. This would provide a mathematically robust framework for implementation-independent verification and certification, allowing proofs carried out at the abstract algorithmic level to apply uniformly across heterogeneous computational models and implementations.

Finally, the framework also suggests a new approach to computational complexity. Classical complexity theory deliberately abstracts away from concrete implementation details and machine architectures. By contrast, the present approach explicitly relates algorithms, representations, and implementations, making it possible to define complexity measures sensitive to the actual structure of computations. This raises the prospect of developing exact and architecture-aware complexity theories incorporating features such as memory hierarchies, cache sizes, branch prediction, or parallel execution mechanisms, while remaining grounded in abstract algorithmic descriptions.

References

1. Airoldi, M.: Machine Habitus: Toward a Sociology of Algorithms. Polity Press (2022). https://books.google.fr/books?id=sIVdzgEACAAJ
2. Anh-Ton Le, H.L., Valarcher, P.: Completeness of Seiller's Abstract Machine (2025). https://hal.u-pec.fr/hal-05137612, preprint
3. Bellantoni, S., Cook, S.: A new recursion-theoretic characterization of the polytime functions. Comput. Complex. **2** (1992). https://doi.org/10.1007/BF01201998
4. Blass, A., Gurevich, Y.: Algorithms: a quest for absolute definitions. Bull. Eur. Assoc. Theor. Comput. Sci. (2003)
5. Blass, A., Dershowitz, N., Gurevich, Y.: When are two algorithms the same? CoRR abs/0811.0811 (2008). http://arxiv.org/abs/0811.0811
6. Church, A.: An unsolvable problem of elementary number theory. Am. J. Math. **58**(2), 345–363 (1936). http://www.jstor.org/stable/2371045
7. Dean, W.: Algorithms and the mathematical foundations of computer science. In: Horsten, L., Welch, P. (eds.) Gödel's Disjunction: The scope and limits of mathematical knowledge. Oxford University Press (2016). https://doi.org/10.1093/acprof:oso/9780198759591.003.0002
8. Gastaldi, J.L., Jarvis, S., Seiller, T., Terilla, J.: A calculus of types in Isbell nuclei (2026). https://doi.org/10.48550/arXiv.2606.03369
9. Gastaldi, J.L., Jarvis, S., Seiller, T., Terilla, J.: Projective metric geometry of tropical nuclei: gap matrices, event loci, and order chambers (2026). https://doi.org/10.48550/arXiv.2601.07900
10. Gurevich, Y.: Sequential abstract state machines capture sequential algorithms. ACM Trans. Comput. Log. **1**, 77–111 (2000)

11. Gurevich, Y.: What is an algorithm? In: Bieliková, M., Friedrich, G., Gottlob, G., Katzenbeisser, S., Turán, G. (eds.) SOFSEM 2012. LNCS, vol. 7147, pp. 31–42. Springer, Heidelberg (2012). https://doi.org/10.1007/978-3-642-27660-6_3
12. Heath, T.: The Thirteen Books of Euclid's Elements. No. vol. 1, Cambridge University Press (1956)
13. Heath, T.: The Thirteen Books of Euclid's Elements. No. vol. 2, Cambridge University Press (1956)
14. Heath, T.: The Thirteen Books of Euclid's Elements. No. vol. 3, Cambridge University Press (1956)
15. Joinet, J.B., Seiller, T.: From abstraction and indiscernibility to classification and types: revisiting hermann weyl's theory of ideal elements. Kagaku tetsugaku **53**(2), 65–93 (2021). https://doi.org/10.4216/jpssj.53.2_65
16. Kleene, S.C.: Recursive functionals and quantifiers of finite types i. Trans. Am. Math. Soc. **91**(1), 1–52 (1959). http://www.jstor.org/stable/1993145
17. Kolmogorov, A.N., Uspenskii, V.A.: On the definition of an algorithm. Uspekhi Mat. Nauk **13**, 3–28 (1958)
18. Lamassé, S.: Relationships between French "practical arithmetics" and teaching? In: Scientific Sources and Teaching Contexts Throughout History: Problems and Perspectives, pp. 125–153. Springer (2013)
19. Markov, A.A.: The theory of algorithms, Trudy Mat. Inst. Steklov., vol. 42. Acad. Sci. USSR (1954)
20. Moschovakis, Y.: What is an algorithm? In: Mathematics Unlimited — 2001 and beyond (2001)
21. Moschovakis, Y.N.: On founding the theory of algorithms. In: Dales, H.G., Oliveri, G. (eds.) Truth in Mathematics, pp. 71–104. Oxford University Press, USA (1998)
22. Naibo, A., Petrolo, M., Seiller, T.: Goa: the geometry of algorithms. Reasoner **17**(4) (2023). https://riviste.unimi.it/index.php/thereasoner/article/view/24136
23. Seiller, T.: Mathematical informatics (2024). https://theses.hal.science/tel-04616661, habilitation thesis
24. Seiller, T.: Mathematical informatics: models of computation (2026). https://hal.archives-ouvertes.fr/hal-05587108, submitted
25. Yanofsky, N.S.: Towards a definition of an algorithm. J. Log. Comput. **21**(2), 253–286 (2011). https://doi.org/10.1093/logcom/exq016

1. Gurevich, Y.: What is an algorithm? In: Bieliková, M., Friedrich, G., Gottlob, G., Katzenbeisser, S., Turán, G. (eds.) SOFSEM 2012. LNCS, vol. 7147, pp. 31–42. Springer, Heidelberg (2012). https://doi.org/10.1007/978-3-642-27660-6_3
2. Heath, T.: The Thirteen Books of Euclid's Elements, No. vol. 1. Cambridge University Press (1908)
3. Heath, T.: The Thirteen Books of Euclid's Elements, No. vol. 2. Cambridge University Press (1956)
4. Heath, T.: The Thirteen Books of Euclid's Elements, No. vol. 3. Cambridge University Press (1956)
5. Jouhet, J., Seifert, P.: Euclid's straightedge and compass constructions [illegible]
[illegible]

Contributed Papers

Computable Cofinal Fraïssé Limits

Nathanael Ackerman[1], Cameron Freer[2(✉)], and Mostafa Mirabi[3,4]

[1] Harvard University, Cambridge, MA, USA
nate@aleph0.net
[2] Massachusetts Institute of Technology, Cambridge, MA, USA
freer@mit.edu
[3] Wesleyan University, Middletown, CT, USA
mmirabi@wesleyan.edu
[4] The Taft School, Watertown, CT, USA

Abstract. For any collection of finite structures closed under isomorphism (i.e., an *age*) which has the Hereditary Property (HP), the Joint Embedding Property (JEP), and the Cofinal Amalgamation Property (CAP), there is a unique (up to isomorphism) countable structure which is *cofinally ultrahomogeneous* with the given age. Such a structure is called the *cofinal Fraïssé limit* of the age. In this extended abstract, we consider the computational strength needed to construct the cofinal Fraïssé limit of a computable age. We show that this construction can always be done using the oracle $\mathbf{0}'''$, and that there are ages that require $\mathbf{0}''$. In contrast, we show that if one assumes the strengthening of (CAP) known as the Amalgamation Property (AP), then the resulting limit, called the Fraïssé limit, can be constructed from the age using $\mathbf{0}'$. Our results therefore show that the more general case of cofinal Fraïssé limits requires greater computational strength than Fraïssé limits.

1 Introduction

Given a collection of finite structures, an important question is to determine when they can be glued together to get a "canonical infinite limit". In the case where the canonical limit is a structure containing all finite structures and that has the "maximum amount of symmetry" (namely, is *ultrahomogeneous*), this question was answered by Fraïssé. Specifically, he showed that such a limit, called a Fraïssé limit, exists if and only if the collection of structures has the Hereditary Property, the Joint Embedding Property, and the Amalgamation Property.

Fraïssé limits have played a fundamental role in computer science and mathematics, including in database theory, automata theory, model theory, and ergodic theory. Further, there is often a tight connection between properties of the Fraïssé limit (such as having trivial definable closure, $\aleph_0$-categoricity, or having an amenable automorphism group) and corresponding properties of the collection of finite structures used to build it (respectively, having the Strong Amalgamation Property, having only finitely many elements (up to isomorphism) of each size, or having the Ramsey property).

V. Brattka et al. (Eds.): CiE 2026, LNCS 16674, pp. 109–123, 2027.
https://doi.org/10.1007/978-3-032-31348-5_7

Given the important connection between a class of finite structures and its Fraïssé limit, it is worth studying how computably one can construct the Fraïssé limit from the collection of structures. This computability question was first studied by Csima, Harizanov, Miller, and Montalbán, who showed in [14] how to computably build a Fraïssé limit from a computable collection of finite structures along with computable witnesses to the Hereditary Property, the Joint Embedding Property, and the Amalgamation Property.

Of these three properties needed to build a Fraïssé limit from a collection of finite structures, the Amalgamation Property is the most closely related to the symmetry of the limit. However, often the Amalgamation Property is too restrictive and we instead only have the weaker version of this property called the Cofinal Amalgamation Property. The following example shows one of the ways in which the Cofinal Amalgamation Property is not as restrictive as the Amalgamation Property (see Definitions 3.22 and 3.27).

Example 1.1. Let $\mathcal{L}_f = \{f\}$ be the language with a single unary function symbol and $\mathcal{K}_f$ be the Fraïssé class consisting of finite $\mathcal{L}_f$-structures $\mathcal{M}$ such that $\mathcal{M} \models (\forall x)\,(f(x) \neq x)$ and $\mathcal{M} \models (\forall x)\,\Big(\big(f(f(x)) = x\big) \vee \big(f(f(f(x))) = x\big)\Big)$.

Let $\mathcal{L}_R = \{R\}$ be the language with a single binary relation symbol and let $\mathcal{K}_R$ be the collection of $\mathcal{L}_R$-structures $\mathcal{M}$ where $R^{\mathcal{M}}$ is the graph of a function r such that $\mathcal{M} \models (\forall x)\,(r(x) \neq x)$ and $\mathcal{M} \models (\forall x)\,\Big(\big(r(r(x)) = x\big) \vee \big(r(r(r(x))) = x\big)\Big)$. Note that $\mathcal{K}_R$ does not have the Amalgamation Property. To see this, suppose $\mathcal{M}_2$ is the unique element of $\mathcal{K}_R$ with underlying set $\{x, y\}$ and $\mathcal{M}_3$ is the unique element of $\mathcal{K}_R$ with underlying set $\{x, y, z\}$ and $r^{\mathcal{M}_3}(x) = y$ and $r^{\mathcal{M}_3}(y) = z$. If $\mathcal{N}$ were the result of amalgamating $\mathcal{M}_2$ and $\mathcal{M}_3$ over $\{x\}$ then we would have $r^{\mathcal{N}}(r^{\mathcal{N}}(x)) = x$ and $r^{\mathcal{N}}(r^{\mathcal{N}}(r^{\mathcal{N}}(x))) = x$, implying $r^{\mathcal{N}}(x) = x$. However, $\mathcal{K}_R$ does have the Cofinal Amalgamation Property, as it has amalgamation over all subsets that are closed under r.

When studying structures with functions, one often does not care if the function is represented via a function symbol or via a relation symbol encoding the graph of the function. As this example shows, the Cofinal Amalgamation Property, unlike the Amalgamation Property, is not sensitive to the specific way that functions are represented in a structure.

Classes of structures with the Cofinal Amalgamation Property also admit limit objects (called *cofinal Fraïssé limits*), but such limit objects only satisfy the weaker notion of *cofinal ultrahomogeneity*. Here we study how computationally difficult it is to construct a cofinal Fraïssé limit from a collection of finite structures satisfying the Hereditary Property, the Joint Embedding Property, and the Cofinal Amalgamation Property. For proofs and various additional details, please see the full version [2][1] of this extended abstract.

[1] https://arxiv.org/abs/2601.22435.

1.1 Related Work

Fraïssé limits [21], the Amalgamation Property (AP), and ultrahomogeneity are key concepts in classical model theory [22, Chapter 7], which describe an important way in which certain countable structures can be built as a generic limit of an appropriate collection of finite structures [11]. Deep connections between Fraïssé limits and ergodic theory have also been discovered [4,23].

Fraïssé limits and their computability have been considered in the context of verification of database systems [8], and researchers have investigated which Fraïssé limits can be described by automata [25]. Within computer science, Fraïssé limits have also been studied regarding denotational semantics for programming languages [16,17], fuzzy logic [5], and the complexity of constraint satisfaction problems in AI [13] and in phylogenetics and computational linguistics [7].

Here we primarily study the weaker notions of cofinal Fraïssé limits, the Cofinal Amalgamation Property (CAP), and cofinal ultrahomogeneity. Cofinal ultrahomogeneity was introduced by Calais [9,10] (under the name "pseudo-homogeneity") along with the corresponding amalgamation property and limit objects, and was rediscovered by Truss [34] in the context of generic automorphisms. For more on the history of these notions and their relationship to other notions of limit, amalgamation, and ultrahomogeneity, see [27] and [26]. These include the even weaker notions of weak Fraïssé limits, the Weak Amalgamation Property (WAP), and weak ultrahomogeneity, which do generalize the cofinal concepts, but do not yet include many more known examples. For a study of the computability of the connections between weak Fraïssé limits, (WAP), and weak ultrahomogeneity, see [1], which generalizes our results here about cofinal amalgamation; the setting of weak amalgamation often requires substantially more complicated proofs.

For more on the Cofinal Amalgamation Property including several examples, see [28] and [29, Section 3.1]; for a categorical presentation, see [30]. The Cofinal Amalgamation Property has been used in the study of topological dynamics [24] and the Ramsey property [6,15], and has also been studied in [31,32], and [18].

The computability of the Fraïssé construction was studied in a categorical setting in [16,17]. Computability of Fraïssé limits was subsequently studied in [14], and we build on several of their results. Isomorphisms between computable Fraïssé limits were also studied in [20] and [19], and a related computability notion was considered in [3,12]. For more on the computability of certain specific Fraïssé limits, see [33].

2 Summary of Main Results

Given an *age*, i.e., a collection of finitely generated structures closed under isomorphisms,[2] one is often interested in a countable structure which can be thought

[2] Note that some of the model theory literature refers only to the *age of a structure*, in which case an age always has the Hereditary Property and Joint Embedding

of as the limit of the age, i.e., a countable structure whose age is the given one, and where the countable structure is generic in some sense.

In this extended abstract, we will be interested in the computable content of going from an age to its corresponding limit. It turns out that our results on the computable content of these constructions will primarily be in terms of the *embedding information* of the age. The embedding information of a computable age $\mathbb{K}$, which we denote $\mathcal{EI}(\mathbb{K})$, describes when a map between two elements of $\mathbb{K}$ can be extended to an embedding. We observe in Lemma 3.5 that $\mathcal{EI}(\mathbb{K})$ is always computable from the Turing jump of $\mathbb{K}$, and is computable in the case where the language of $\mathbb{K}$ is finite and relational.

The most common instance of the phenomenon of an age having a *limit* is that of the Fraïssé limit. Given an age (with countably many isomorphism classes) satisfying the Hereditary Property (HP), the Joint Embedding Property (JEP), and the Amalgamation Property (AP), there is a unique structure, called the Fraïssé limit, whose age is the given one and which is ultrahomogeneous. In [14] it was shown how to construct a Fraïssé limit from computable witnesses for (HP), (JEP), and (AP). In Lemma 3.15, Proposition 3.18, and Proposition 3.23 we show that such witnesses are always computable from the embedding information. Putting this together, we obtain the following result.

Corollary (Corollary 4.5). *Suppose* $\mathbb{K}$ *is a computable age with* (HP), (JEP), *and* (AP). *Then* $\mathbb{K}$ *has an* $\mathcal{EI}(\mathbb{K})$*-computable Fraïssé limit.*

While Fraïssé limits are the most common type of limit of an age, often we have ages which satisfy (HP) and (JEP), but which do not satisfy (AP), and instead merely satisfy the weaker notion of the Cofinal Amalgamation Property (CAP). In this case the age still has a type of limit, called the cofinal Fraïssé limit, but the limiting object no longer need be ultrahomogeneous and is merely required to satisfy the weaker notion of cofinal ultrahomogeneity.

As with the case of Fraïssé limits, we show that from computable witnesses for (HP), (JEP), and (CAP), we can construct the cofinal Fraïssé limit. We frame this result in terms of (relativized) computable versions of the corresponding properties. (For example, ($\mathfrak{s}$-cHP) denotes the $\mathfrak{s}$-computable version of (HP).)

Theorem (Theorem 4.18). *Let* $\mathfrak{s}$ *be a Turing degree. Suppose*

(a) $\mathbb{K}$ *is an* $\mathfrak{s}$*-computable age,*
(b) $\mathcal{EI}(\mathbb{K}) \leq_T \mathfrak{s}$, *and*
(c) $\mathbb{K}$ *has* ($\mathfrak{s}$-cHP), ($\mathfrak{s}$-cJEP), *and* ($\mathfrak{s}$-cCAP).

Then there is an $\mathfrak{s}$*-computable cofinal Fraïssé limit of* $\mathbb{K}$.

With this result in hand, we reduce the problem of computing the cofinal Fraïssé limit from an age to the problem of computing witnesses to (CAP) from an age (as we have already shown that witnesses to (HP) and (JEP) are computable from the embedding information). We are able to show the following upper bounds on the computability of (CAP) and of cofinal Fraïssé limits.

Property. In this work we use the more general notion, in part so that the definition of *computable age* is computably checkable.

Theorem (Theorem 3.34 (b)). *If* $\mathbb{K}$ *is a computable age with* (CAP), *then it has* $(\mathcal{EJ}(\mathbb{K})''$-cCAP).

Corollary (Corollary 4.19). *Suppose* $\mathbb{K}$ *is a computable age with* (HP), (JEP), *and* (CAP). *Then* $\mathbb{K}$ *has an* $\mathcal{EJ}(\mathbb{K})''$*-computable cofinal Fraïssé limit.*

We then turn our attention to the question of lower bounds for the computability of (CAP). We consider two situations, one where the language is finite and relational (and hence the embedding information is computable), and one where we allow the language to be infinite and to have function symbols. We obtain the following lower bound on the computability of (CAP) for finite relational languages.

Theorem (Theorem 5.1). *There is a computable age* $\mathbb{K}$*, which is the canonical computable age of some structure in a finite relational language, such that* $\mathbb{K}$ *has* (CAP) *but if it has* ($\mathfrak{s}$-cCAP) *for some Turing degree* $\mathfrak{s}$*, then* $\mathbf{0}' \leq_{\mathrm{T}} \mathfrak{s}$.

We also obtain the corresponding lower bound on the computability of cofinal Fraïssé limits for finite relational languages.

Corollary (Corollary 5.2). *There is a computable age* $\mathbb{K}$*, which is the canonical computable age of some structure in a finite relational language, such that if* $\mathfrak{s}$ *is a Turing degree and* $\mathcal{M}$ *is an* $\mathfrak{s}$*-computable structure that is* $\mathfrak{s}$*-computably cofinally ultrahomogeneous and a cofinal Fraïssé limit of* $\mathbb{K}$*, then* $\mathbf{0}' \leq_{\mathrm{T}} \mathfrak{s}$.

Finally, in the general case, we obtain lower bounds on the computability of (CAP) and of cofinal Fraïssé limits.

Theorem (Theorem 5.3). *There is a computable age* $\mathbb{K}$*, which is the canonical computable age of some structure, such that* $\mathbb{K}$ *has* (CAP) *but if it has* ($\mathfrak{s}$-cCAP) *for some Turing degree* $\mathfrak{s}$*, then* $\mathbf{0}'' \leq_{\mathrm{T}} \mathfrak{s}$.

Corollary (Corollary 5.4). *There is a computable age* $\mathbb{K}$*, which is the canonical computable age of some structure, such that if* $\mathfrak{s}$ *is a Turing degree and* $\mathcal{M}$ *is an* $\mathfrak{s}$*-computable structure that is* $\mathfrak{s}$*-computably cofinally ultrahomogeneous and a cofinal Fraïssé limit of* $\mathbb{K}$*, then* $\mathbf{0}'' \leq_{\mathrm{T}} \mathfrak{s}$.

3 Computable Ages

Throughout this extended abstract, we use standard notation from model theory and computable model theory. Note that in particular we call a $\mathcal{L}$**-tuple** a pair $(\mathbf{a}, \mathcal{A})$ where $\mathcal{A}$ is an $\mathcal{L}$-structure and $\mathbf{a} \in \mathcal{A}$.

3.1 Basic Definitions of Computable Ages

Definition 3.1. *An* **age** *for $\mathfrak{L}$ is a collection of finitely generated structures closed under isomorphism. If $\mathfrak{L}$ is a language with a computable representation, a* **computable representation** *of an age $\mathcal{K}$ is a c.e. set $\mathbb{K}$ of objects of the form $(\mathbf{a}, \mathcal{A}, i)$ where*

- *for each $i \in \omega$ there is a unique element of the form $(\mathbf{a}, \mathcal{A}, i) \in \mathbb{K}$,*
- *for each $(\mathbf{a}, \mathcal{A}, i) \in \mathbb{K}$, we have that (I) $i \in \omega$; (II) $(\mathbf{a}, \mathcal{A})$ is an $\mathfrak{L}$-tuple such that $\mathcal{A} = \mathrm{cl}_{\mathcal{A}}(\mathbf{a})$; (III) $\mathcal{A}$ is a computable $\mathfrak{L}$-structure (with respect to the computable representation of $\mathfrak{L}$); and (IV) $\mathcal{A} \in \mathcal{K}$; and*
- *for every $\mathcal{B} \in \mathcal{K}$ there is an $(\mathbf{a}, \mathcal{A}, i) \in \mathbb{K}$ such that $\mathcal{B} \cong \mathcal{A}$.*

We will call a computable representation of an age a **computable age**. *If $\mathbb{K}$ is a computable age and $i \in \omega$, write $(\mathbf{a}_{\mathbb{K}:i}, \mathcal{A}_{\mathbb{K}:i})$ to denote the unique $\mathfrak{L}$-tuple such that $(\mathbf{a}_{\mathbb{K}:i}, \mathcal{A}_{\mathbb{K}:i}, i) \in \mathbb{K}$, and define $\mathbb{I}_{\mathbb{K}}(i) = (\mathbf{a}_{\mathbb{K}:i}, \mathcal{A}_{\mathbb{K}:i}, i)$. We omit the subscript $\mathbb{K}$ from $\mathbb{I}_{\mathbb{K}}$ and omit mention of $\mathcal{K}$ when they are clear from context. We write $\mathcal{A} \in \mathbb{K}$ if $\mathcal{A} = \mathcal{A}_{\mathbb{K}:i}$ for some $i \in \omega$. When i is clear from context we write $\mathcal{A}$ or $(\mathbf{a}, \mathcal{A})$ to denote the triple $(\mathbf{a}, \mathcal{A}, i)$.*

Definition 3.2. *A* **potential embedding** *in $\mathbb{K}$ is a triple of the form $(\mathbb{I}(i), \mathbb{I}(j), \mathbf{c})$ where $i, j \in \omega$ and $\mathbf{c} \in \mathcal{A}_{\mathbb{K}:j}$. We write $E_{\mathbb{I}(i):\mathbf{c}:\mathbb{I}(j)} = (\mathbb{I}(i), \mathbb{I}(j), \mathbf{c})$. Define the* **domain** *of $E_{\mathbb{I}(i):\mathbf{c}:\mathbb{I}(j)}$ to be $\mathrm{dom}(E_{\mathbb{I}(i):\mathbf{c}:\mathbb{I}(j)}) = \mathbb{I}(i)$, the* **codomain** *to be $\mathrm{codom}(E_{\mathbb{I}(i):\mathbf{c}:\mathbb{I}(j)}) = \mathbb{I}(j)$ and the* **range** *of $E_{\mathbb{I}(i):\mathbf{c}:\mathbb{I}(j)}$ to be $\mathrm{range}(E_{\mathbb{I}(i):\mathbf{c}:\mathbb{I}(j)}) = \mathbf{c}$. In diagrams we will write $\mathcal{A} \overset{\mathbf{c}}{\dashrightarrow}_{\mathbb{K}} \mathcal{B}$ to signify that $(\mathcal{A}, \mathcal{B}, \mathbf{c})$ is a potential embedding in $\mathbb{K}$. A potential embedding $(\mathbb{I}(i), \mathbb{I}(j), \mathbf{c})$ in $\mathbb{K}$ is an* **embedding** *if $\mathbf{a}_{\mathbb{K}:i} \simeq \mathbf{c}$. We will write $\mathcal{A} \overset{\mathbf{c}}{\to}_{\mathbb{K}} \mathcal{B}$ to signify that $(\mathcal{A}, \mathcal{B}, \mathbf{c})$ is an embedding in $\mathbb{K}$.*

Note that the collection of potential embeddings is a computable set. However, the collection of embeddings is, in general, only co-c.e. For this reason it will be important to use potential embeddings when building a cofinally ultrahomogeneous structure from its age.

Definition 3.3. *Let F and G be potential embeddings in $\mathbb{K}$ and suppose that $\mathrm{codom}(F) = \mathrm{dom}(G)$. When G is not an embedding, we define $G \circ F = (\mathrm{dom}(F), \mathrm{codom}(G), \mathrm{range}(G))$. When G is an embedding, we define $G \circ F = (\mathrm{dom}(F), \mathrm{codom}(G), \mathfrak{t}_{cl}(\mathbf{a}_j, \mathrm{range}(G))(\mathrm{range}(F)))$, where $j \in \omega$ is such that $\mathbb{I}(j) = \mathrm{dom}(G)$. We write $\mathrm{id}_{\mathbb{I}(i)}$ to denote the triple $(\mathbb{I}(i), \mathbb{I}(i), \mathbf{a}_i)$.*

One can check that this notation gives rise to a category $\mathfrak{Cat}(\mathbb{K})$ whose objects are elements of $\mathbb{K}$, and where the maps between $\mathcal{A}$ and $\mathcal{B}$ are those potential embeddings F with $\mathrm{dom}(F) = \mathcal{A}$ and $\mathrm{codom}(F) = \mathcal{B}$.

The following notion will tell us when a potential embedding between two substructures of an age is in fact an embedding.

Definition 3.4. *Suppose $\mathbb{K}$ is a computable age. Define the* **embedding information** *related to $\mathbb{K}$, denoted $\mathcal{EI}(\mathbb{K})$, to be the collection of tuples $(\mathbb{I}_{\mathbb{K}}(k_0), \mathbf{b}_0, \mathbb{I}_{\mathbb{K}}(k_1), \mathbf{b}_1)$ such that $\mathbf{b}_0 \in \mathcal{A}_{\mathbb{K}:k_0}$ and $\mathbf{b}_1 \in \mathcal{A}_{\mathbb{K}:k_1}$, and $\mathbf{b}_0 \simeq \mathbf{b}_1$.*

We will see that, given two computable cofinally ultrahomogeneous structures that are isomorphic, one can always extend any partial isomorphism (between appropriate elements of the age) to an isomorphism. However, in order to build a *computable* such isomorphism, we will need to be able to computably extend various partial isomorphisms; we will use the embedding information to do so.

Lemma 3.5. *If* $\mathbb{K}$ *is a computable age, then* $\mathcal{EI}(\mathbb{K}) \leq_{\mathrm{T}} \mathbf{0}'$. *Further, if* $\mathbb{K}$ *is a uniformly finite computable age, then* $\mathcal{EI}(\mathbb{K})$ *is computable.*

Definition 3.6. *Let* $\mathbb{K}_0$ *and* $\mathbb{K}_1$ *be computable ages, and let* $\mathfrak{s}$ *be a Turing degree. An* **isomorphism** *from* $\mathbb{K}_0$ *to* $\mathbb{K}_1$ *is a pair of maps* $\boldsymbol{\alpha} = (\alpha_0, \alpha_1)$ *such that* $\alpha_0 : \mathbb{K}_0 \to \mathbb{K}_1$ *and* $\alpha_1 : \mathbb{K}_1 \to \mathbb{K}_0$ *are defined in the obvious way. We write* $\boldsymbol{\alpha}^{-1} = (\alpha_1, \alpha_0)$. *The pair* $\boldsymbol{\alpha}$ *is* $\mathfrak{s}$**-computable** *when* α_0 *and* α_1 *are computable relative to* $\mathfrak{s}$ *under the obvious encoding. We say that* $\mathbb{K}_0$ *and* $\mathbb{K}_1$ *are* $\mathfrak{s}$**-computably isomorphic** *when there is some* $\mathfrak{s}$*-computable isomorphism between them.*

Note that two computable ages $\mathbb{K}_0$ and $\mathbb{K}_1$ are isomorphic if and only if they are both computable representations of the same age, which holds if and only if for some Turing degree $\mathfrak{s}$ they are $\mathfrak{s}$-computably isomorphic.

Definition 3.7. *Let* $\mathbb{K}_0$ *and* $\mathbb{K}_1$ *be computable ages and let* $\boldsymbol{\alpha} = (\alpha_0, \alpha_1)$ *be an isomorphism from* $\mathbb{K}_0$ *to* $\mathbb{K}_1$. *Suppose* $E_{\mathbb{I}(i):\mathbf{c}:\mathbb{I}(j)}$ *is a potential embedding in* $\mathbb{K}_0$. *Define* $\boldsymbol{\alpha}(E_{\mathbb{I}(i):\mathbf{c}:\mathbb{I}(j)}) = \big(\alpha_0(\mathbb{I}(i)),\ \alpha_0(\mathbb{I}(j)),\ \mathfrak{t}_{\mathrm{cl}}(\mathbf{a}_j, \alpha_0(\mathbf{a}_j))(\mathbf{c})\big)$.

Lemma 3.8. *Suppose* $\boldsymbol{\alpha} = (\alpha_0, \alpha_1)$ *is an isomorphism between computable ages* $\mathbb{K}_0$ *and* $\mathbb{K}_1$. *Then* $\boldsymbol{\alpha}$ *extends to an equivalence of categories between* $\mathfrak{Cat}(\mathbb{K}_0)$ *and* $\mathfrak{Cat}(\mathbb{K}_1)$ *with inverse equivalence* $\boldsymbol{\alpha}^{-1}$.

Given a computable structure, the collection of its finitely generated substructures, appropriately enumerated, forms a computable age.

Definition 3.9. *Let* $\mathcal{M}$ *be an* $\mathcal{L}$*-structure. The* **age of** $\mathcal{M}$ *is the collection of all* $\mathcal{L}$*-structures isomorphic to a finitely generated substructure of* $\mathcal{M}$.

Suppose $\mathcal{M}$ *is computable, and let* $(\mathbf{a}_i)_{i\in\omega}$ *be some computable enumeration of all finite tuples in* $\mathcal{M}$ *in which every tuple appears infinitely often. We say that the sequence* $(\mathbf{a}_i,\ \mathrm{cl}_{\mathcal{M}}(\mathbf{a}_i),\ i)_{i\in\omega}$ *is a* **canonical computable age** *of* $\mathcal{M}$. *Let* $\mathbb{K}[\mathcal{M}]$ *be one such canonical computable age. A computable age* $\mathbb{K}$ *is* **computably the age of** $\mathcal{M}$ *if it is computably isomorphic to* $\mathbb{K}[\mathcal{M}]$.

Lemma 3.10. *Let* $\mathcal{M}$ *be a computable* $\mathcal{L}$*-structure. Then (a) any two canonical computable ages of* $\mathcal{M}$ *are computably isomorphic; (b) the canonical computable age of* $\mathcal{M}$ *is computably the age of* $\mathcal{M}$*; and (c) if* $\mathcal{K}$ *is the age of* $\mathcal{M}$*, and* $\mathbb{K}$ *is computably the age of* $\mathcal{M}$*, then* $\mathbb{K}$ *is a computable representation of* $\mathcal{K}$.

Lemma 3.11. *For each* $i \in \{0, 1\}$*, let* $\mathcal{M}_i$ *be a computable structure and suppose that* $\mathbb{K}_i$ *is computably the age of* $\mathcal{M}_i$. *If* $\mathcal{M}_0$ *and* $\mathcal{M}_1$ *are* $\mathfrak{s}$*-computably isomorphic, then* $\mathbb{K}_0$ *and* $\mathbb{K}_1$ *are* $\mathfrak{s}$*-computably isomorphic as well.*

For a computable structure $\mathcal{M}$, we write $\mathcal{EI}(\mathcal{M})$ to mean $\mathcal{EI}(\mathbb{K}\mathcal{M})$. The following result (which is immediate from [14, Lemma 2.9]) will allow us to convert a computable sequence of compatible embeddings into a structure.[3]

Proposition 3.12. *Let* $\mathbb{K}$ *be a computable age and suppose* $(F_i)_{i\in\omega}$ *is a computable sequence of embeddings where for each* $i \in \omega$ *there is a* $k_i \in \omega$ *such that* $\mathrm{dom}(F_i) = \mathbb{I}_{\mathbb{K}}(k_i)$ *and* $\mathrm{codom}(F_i) = \mathbb{I}_{\mathbb{K}}(k_{i+1})$. *Then there is a computable age* $\mathbb{K}^*$ *which is computably isomorphic to* $\mathbb{K}$, *and a computable sequence* $(\ell_i)_{i\in\omega}$ *such that for* $i \in \omega$, *we have* $\mathbf{a}_{\mathbb{K}^*:\ell_i} \simeq \mathbf{a}_{\mathbb{K}:k_i}$ *and* $\mathcal{A}_{\mathbb{K}^*:\ell_i} \subseteq \mathcal{A}_{\mathbb{K}^*:\ell_{i+1}}$, *and* $\mathfrak{t}_{\mathrm{cl}}(\mathbf{a}_{\mathbb{K}:k_{i+1}}, \mathbf{a}_{\mathbb{K}^*:\ell_{i+1}}) \circ \mathfrak{t}_{\mathrm{cl}}(\mathbf{a}_{\mathbb{K}:k_i}, \mathrm{range}(F_i)) = \mathfrak{t}_{\mathrm{cl}}(\mathbf{a}_{\mathbb{K}:k_i}, \mathbf{a}_{\mathbb{K}^*:\ell_i})$.

Corollary 3.13. *Let* $\mathbb{K}$ *be a computable age and suppose* $(F_i)_{i\in\omega}$ *are as in Proposition 3.12. Then there is a structure* $\mathcal{D}_\omega$ *along with embeddings* $G_i : \mathcal{A}_{\mathbb{K}:k_i} \to \mathcal{D}_\omega$ *(for* $i \in \omega$*) such that for any* $i < j$, *we have* $G_i = G_j \circ F_{j-1} \circ F_{j-2} \circ \cdots \circ F_i$. *Further, both* $\mathcal{D}_\omega$ *and* $(G_n)_{n\in\omega}$ *are computable.*

3.2 Properties of Computable Ages

Definition 3.14. *An age* $\mathcal{K}$ *has the* **hereditary property**, *written* (HP), *if for all* $\mathcal{B} \in \mathcal{K}$, *we have* $\mathcal{A} \subseteq \mathcal{B}$ *implies* $\mathcal{A} \in \mathcal{K}$. *Suppose that* $\mathfrak{s}$ *is a Turing degree. We say a computable age* $\mathbb{K}$ *has the* $\mathfrak{s}$**-computable hereditary property**, *written* ($\mathfrak{s}$-cHP), *if there is an* $\mathfrak{s}$*-computable function that takes a pair* $(\mathbb{I}(i), \mathbf{b})$ *where* $\mathbf{b} \in \mathcal{A}_i$ *and returns some* $\mathbb{I}(j)$ *such that* $\mathbf{a}_j \simeq \mathbf{b}$. *When* $\mathfrak{s}$ *is* $\mathbf{0}$, *we speak of the* **computable hereditary property**, *and write* (cHP).

Lemma 3.15. *Suppose* $\mathbb{K}$ *is a computable age with* (HP). *Then* $\mathbb{K}$ *has* ($\mathcal{EI}(\mathbb{K})$-cHP).

Not every computable age with (HP) has (cHP), but every computable age with (HP) is isomorphic to one with (cHP), as shown in [14].

Proposition 3.16 ([14, **Theorem 2.8**]). *Suppose* $\mathbb{K}$ *is a computable age with* (HP). *Then there is a computable age* $\mathbb{K}^*$ *with* (cHP) *that is isomorphic to* $\mathbb{K}$.

Definition 3.17. *An age* $\mathcal{K}$ *has the* **joint embedding property**, *written* (JEP), *if for all* $\mathcal{A}, \mathcal{B} \in \mathcal{K}$ *there is a* $\mathcal{C} \in \mathcal{K}$ *for which there are embeddings* $\alpha_{\mathcal{A}} : \mathcal{A} \to \mathcal{C}$ *and* $\alpha_{\mathcal{B}} : \mathcal{B} \to \mathcal{C}$. *Suppose* $\mathfrak{s}$ *is a Turing degree. We say a computable age* $\mathbb{K}$ *has the* $\mathfrak{s}$**-computable joint embedding property**, *written* ($\mathfrak{s}$-cJEP), *if there is an* $\mathfrak{s}$*-computable function* f *with domain* $\mathbb{K} \times \mathbb{K}$ *such that for all* $\mathbb{I}(\ell_0), \mathbb{I}(\ell_1) \in \mathbb{K}$, *we have* $f(\mathbb{I}(\ell_0), \mathbb{I}(\ell_1)) = (F_0, F_1)$ *where (i)* F_0 *and* F_1 *are embeddings; (ii)* $\mathrm{dom}(F_0) = \mathbb{I}(\ell_0)$ *and* $\mathrm{dom}(F_1) = \mathbb{I}(\ell_1)$; *and (iii)* $\mathrm{codom}(F_0) = \mathrm{codom}(F_1)$. *When* $\mathfrak{s}$ *is* $\mathbf{0}$, *we speak of the* **computable joint embedding property**, *and write* (cJEP).

[3] Note that [14] refers to the canonical computable age of $\mathcal{M}$ as "the canonical representation of the age of $\mathcal{M}$", and speaks of "a canonical representation of the age of $\mathcal{M}$" to mean any computable representation that is computably isomorphic to the canonical one.

Proposition 3.18. *Suppose $\mathbb{K}$ is a computable age with* (JEP). *Then $\mathbb{K}$ also has* ($\mathcal{E}\mathcal{I}(\mathbb{K})$-cJEP).

The following two results are due to [14].

Proposition 3.19 ([14, Lemma 2.6]). *If $\mathcal{M}$ is a computable structure then* $\mathbb{K}[\mathcal{M}]$ *has* (cHP) *and* (cJEP).

Proposition 3.20 ([14, Theorem 2.10]). *If $\mathbb{K}$ is a computable age with* (cHP) *and* (cJEP), *then it is canonically the age of some computable structure.*

The *span* is a key technical notion relevant to amalgamation properties.

Definition 3.21. *A* **potential span** *in a computable age $\mathbb{K}$ is a pair (F_0, F_1) of potential embeddings with* $\mathrm{dom}(F_0) = \mathrm{dom}(F_1)$. *We say that a potential span (F_0, F_1) is* **over** $\mathrm{dom}(F_0)$. *A potential span is a* **span** *if in addition, F_0 and F_1 are both embeddings. An* **amalgamation diagram** *over a potential span (F_0, F_1) is a tuple (G_0, G_1) such that*

- *G_0 is an embedding and G_1 is a potential embedding,*
- $\mathrm{dom}(G_i) = \mathrm{codom}(F_i)$ *for* $i \in \{0, 1\}$,
- $\mathrm{codom}(G_0) = \mathrm{codom}(G_1)$, *and*
- *if (F_0, F_1) is a span then G_1 is an embedding and $G_0 \circ F_0 = G_1 \circ F_1$.*

Definition 3.22. *An age $\mathcal{K}$ has the* **amalgamation property**, *written* (AP), *if whenever $\mathcal{A}, \mathcal{B}, \mathcal{C} \in \mathcal{K}$ and whenever $f_{\mathcal{B}} : \mathcal{A} \to \mathcal{B}$ and $f_{\mathcal{C}} : \mathcal{A} \to \mathcal{C}$ are embeddings, there is a $\mathcal{D} \in \mathcal{K}$ such that there are embeddings $\alpha_{\mathcal{B}} : \mathcal{B} \to \mathcal{D}$ and $\alpha_{\mathcal{C}} : \mathcal{C} \to \mathcal{D}$ where $\alpha_{\mathcal{B}} \circ f_{\mathcal{B}} = \alpha_{\mathcal{C}} \circ f_{\mathcal{C}}$. Suppose $\mathfrak{s}$ is a Turing degree. We say a computable age $\mathbb{K}$ has the* **$\mathfrak{s}$-computable amalgamation property**, *written* ($\mathfrak{s}$-cAP), *if there is an $\mathfrak{s}$-computable function which maps each potential span to an amalgamation diagram over it.*

Proposition 3.23. *Any computable age $\mathbb{K}$ with* (AP) *must have* ($\mathcal{E}\mathcal{I}(\mathbb{K})$-cAP).

The focus of this extended abstract is a weakening of the amalgamation property, known as the cofinal amalgamation property; we will see that this is computationally more complicated than the amalgamation property.

Definition 3.24. *Suppose $\mathbb{K}$ is a computable age. A* **collection of distinguished extensions**, *written* $\mathrm{DE}_{\mathbb{K}}$, *is a collection of embeddings such that*

- *for $F \in \mathrm{DE}_{\mathbb{K}}$ we have* $\mathrm{dom}(F), \mathrm{codom}(F) \in \mathbb{K}$,
- *for $F \in \mathrm{DE}_{\mathbb{K}}$ and isomorphisms G with* $\mathrm{codom}(G) = \mathrm{dom}(F)$ *and* $\mathrm{dom}(G) \in \mathbb{K}$ *we have $F \circ G \in \mathrm{DE}_{\mathbb{K}}$,*
- *for $F \in \mathrm{DE}_{\mathbb{K}}$ and isomorphisms H with* $\mathrm{dom}(H) = \mathrm{codom}(F)$ *and* $\mathrm{codom}(H) \in \mathbb{K}$ *we have $H \circ F \in \mathrm{DE}_{\mathbb{K}}$,*
- *for all $\mathbb{I}(i) \in \mathbb{K}$ there is an $F \in \mathrm{DE}_{\mathbb{K}}$ with* $\mathrm{dom}(F) = \mathbb{I}(i)$.

Such a collection $\mathrm{DE}_{\mathbb{K}}$ is $\mathfrak{s}$-computable when it is $\mathfrak{s}$-computable as a set.

Definition 3.25. *Suppose* $\boldsymbol{\alpha} = (\alpha_0, \alpha_1) : \mathbb{K}_0 \to \mathbb{K}_1$ *is a computable isomorphism and* $\mathrm{DE}_{\mathbb{K}_0}$ *is a collection of distinguished extensions in* $\mathbb{K}_0$*. Let* $\boldsymbol{\alpha}(\mathrm{DE}_{\mathbb{K}_0})$ *be the collection of embeddings* F *in* $\mathfrak{Cat}(\mathbb{K}_1)$ *such that* $\boldsymbol{\alpha}^{-1}(F) \in \mathrm{DE}_{\mathbb{K}_0}$*.*

Lemma 3.26. *Suppose* $\boldsymbol{\alpha} = (\alpha_0, \alpha_1) : \mathbb{K}_0 \to \mathbb{K}_1$ *is a computable isomorphism and* $\mathrm{DE}_{\mathbb{K}_0}$ *is a computable collection of distinguished extensions in* $\mathbb{K}_0$*. Then* $\boldsymbol{\alpha}(\mathrm{DE}_{\mathbb{K}_0})$ *is a computable collection of distinguished extensions in* $\mathbb{K}_1$*.*

Definition 3.27. *An age* $\mathcal{K}$ *has the* **cofinal amalgamation property**, *written* (CAP), *if for every* $\mathcal{A} \in \mathcal{K}$ *there is an* $\mathcal{A}' \in \mathcal{K}$, *called an* **amalgamation base**, *such that (a) there is an embedding* $\beta : \mathcal{A} \to \mathcal{A}'$ *and (b) whenever* $\mathcal{B}, \mathcal{C} \in \mathcal{K}$ *and* $f_{\mathcal{B}} : \mathcal{A}' \to \mathcal{B}$ *and* $f_{\mathcal{C}} : \mathcal{A}' \to \mathcal{C}$ *are embeddings, there is a* $\mathcal{D} \in \mathcal{K}$ *and there are embeddings* $\alpha_{\mathcal{B}} : \mathcal{B} \to \mathcal{D}$ *and* $\alpha_{\mathcal{C}} : \mathcal{C} \to \mathcal{D}$ *where* $\alpha_{\mathcal{B}} \circ f_{\mathcal{B}} = \alpha_{\mathcal{C}} \circ f_{\mathcal{C}}$*.*

Definition 3.28. *Suppose* $\mathbb{K}$ *is a computable age. We define a* **witness to** (CAP) *to be a pair* $(\mathrm{AB}_{\mathbb{K}}, f_{\mathrm{AB}_{\mathbb{K}}})$ *satisfying the following.*

(a) $\mathrm{AB}_{\mathbb{K}}$ *is a collection of distinguished extensions such that if* $F \in \mathrm{AB}_{\mathbb{K}}$*, then* $\mathrm{codom}(F)$ *is an amalgamation base, and*
(b) $f_{\mathrm{AB}_{\mathbb{K}}}$ *takes as input tuples* (F, G_0, G_1) *such that (i)* $F \in \mathrm{AB}_{\mathbb{K}}$*, (ii)* (G_0, G_1) *is a potential span, and (iii)* $\mathrm{dom}(G_0) = \mathrm{dom}(G_1) = \mathrm{codom}(F)$*, and outputs an amalgamation diagram over* (G_0, G_1)*.*

Such a witness is $\mathfrak{s}$**-computable** *if* $\mathrm{AB}_{\mathbb{K}}$ *is* $\mathfrak{s}$*-c.e. and* $f_{\mathrm{AB}_{\mathbb{K}}}$ *is an* $\mathfrak{s}$*-computable partial function. We say that* $\mathbb{K}$ *has* ($\mathfrak{s}$-cCAP) *if it has an* $\mathfrak{s}$*-computable witness to* (CAP)*.*

Definition 3.29. *Let* $\boldsymbol{\alpha} = (\alpha_0, \alpha_1) : \mathbb{K}_0 \to \mathbb{K}_1$ *and suppose* $(\mathrm{AB}_{\mathbb{K}_0}, f_{\mathrm{AB}_{\mathbb{K}_0}})$ *is a witness to* (CAP) *for* $\mathbb{K}_0$*. Let* $\boldsymbol{\alpha}(f_{\mathrm{AB}_{\mathbb{K}_0}})$ *be the map which takes as input those tuples* (F, G_0, G_1) *such that (i)* $F \in \boldsymbol{\alpha}(\mathrm{AB}_{\mathbb{K}_0})$*, (ii)* (G_0, G_1) *is a potential span in* $\mathbb{K}_1$*, and (iii)* $\mathrm{dom}(G_0) = \mathrm{dom}(G_1) = \mathrm{codom}(F)$*, and outputs* $(\boldsymbol{\alpha}(H_0), \boldsymbol{\alpha}(H_1))$ *where* $(H_0, H_1) = f_{\mathrm{AB}_{\mathbb{K}_0}}(\boldsymbol{\alpha}^{-1}(F), \boldsymbol{\alpha}^{-1}(G_0), \boldsymbol{\alpha}^{-1}(G_1))$*.*

Lemma 3.30. *Let* $\boldsymbol{\alpha} = (\alpha_0, \alpha_1) : \mathbb{K}_0 \to \mathbb{K}_1$ *be a computable isomorphism, and suppose that* $(\mathrm{AB}_{\mathbb{K}_0}, f_{\mathrm{AB}_{\mathbb{K}_0}})$ *is a witness to* (CAP) *for* $\mathbb{K}_0$ *which is computable. Then* $(\boldsymbol{\alpha}(\mathrm{AB}_{\mathbb{K}_0}), \boldsymbol{\alpha}(f_{\mathrm{AB}_{\mathbb{K}_0}}))$ *is a witness to* (CAP) *for* $\mathbb{K}_1$ *which is computable.*

The following corollary is immediate.

Corollary 3.31. *Let* $\mathfrak{s}$ *be a Turing degree, and suppose* $\mathbb{K}_0$ *has* ($\mathfrak{s}$-cCAP) *and* $\mathbb{K}_1$ *is computably isomorphic to* $\mathbb{K}_0$*. Then* $\mathbb{K}_1$ *has* ($\mathfrak{s}$-cCAP)*.*

Proposition 3.32. *The following are equivalent for a computable age* $\mathbb{K}$*.*

- $\mathbb{K}$ *has* (CAP)*.*
- $\mathbb{K}$ *has* ($\mathfrak{s}$-cCAP) *for some Turing degree* $\mathfrak{s}$*.*

In an age with (AP), every element is an amalgamation base, yielding the following.

Lemma 3.33. *If an age has* (AP), *then it also has* (CAP). *If a computable age has* (cAP), *then it also has* (cCAP).

Earlier we saw how to obtain bounds on the computational power needed for the effective versions of each of (HP), (JEP), and (AP). We now do analogously with (CAP), but the bounds are higher.

Theorem 3.34. *Let* $\mathbb{K}$ *be a computable age, and let* AB *be the collection of all embeddings between elements of* $\mathbb{K}$ *whose codomain is some amalgamation base in* $\mathbb{K}$. *Then (a)* $\mathrm{AB} \leq_{\mathrm{T}} \mathcal{EI}(\mathbb{K})''$, *and (b) when* $\mathbb{K}$ *has* (CAP), *it has* ($\mathcal{EI}(\mathbb{K})''$-cCAP).

4 Upper Bounds

4.1 Ultrahomogeneous Structures

Definition 4.1. *A structure* $\mathcal{M}$ *is* **ultrahomogeneous** *if for every finitely generated structure* $\mathcal{A} \subseteq \mathcal{M}$ *and every isomorphism* $f : \mathcal{A} \to \mathcal{A}^*$ *with* $\mathcal{A}^* \subseteq \mathcal{M}$ *there is an automorphism* $g : \mathcal{M} \to \mathcal{M}$ *such that* $g\restriction_{\mathcal{A}} = f$. *The structure* $\mathcal{M}$ *is a* **Fraïssé limit of** $\mathcal{K}$ *if the age of* $\mathcal{M}$ *is* $\mathcal{K}$ *and* $\mathcal{M}$ *is ultrahomogeneous.*

Definition 4.2. *A computable structure* $\mathcal{M}$ *is* **computably ultrahomogeneous** *if there is a function* g *which takes in pairs of finite tuples* $(\mathbf{a}, \mathbf{b})$ *of equal length from* $\mathcal{M}$ *and returns a code for some bijection of the underlying set of* $\mathcal{M}$ *with itself, such that if* $\mathfrak{t}_{\mathrm{cl}}(\mathbf{a}, \mathbf{b})$ *is an isomorphism from* $\mathrm{cl}_{\mathcal{M}}(\mathbf{a})$ *to* $\mathrm{cl}_{\mathcal{M}}(\mathbf{b})$, *then* $g(\mathbf{a}, \mathbf{b})$ *is an automorphism of* $\mathcal{M}$ *extending* $\mathfrak{t}_{\mathrm{cl}}(\mathbf{a}, \mathbf{b})$. *We call such a map* g *a* **witness** *to the computable ultrahomogeneity of* $\mathcal{M}$. *The structure* $\mathcal{M}$ *is a* **computable Fraïssé limit of** $\mathbb{K}$ *if* $\mathcal{M}$ *is computably ultrahomogeneous and* $\mathbb{K}$ *is a computable representation of the age of* $\mathcal{M}$.

Our next result strengthens one direction of [14, Theorem 3.9].[4]

Proposition 4.3. *Suppose* $\mathcal{M}$ *is an arbitrary countable structure with* $\mathfrak{s}$*-computable domain which is* $\mathfrak{s}$*-computably ultrahomogeneous. Then* $\mathbb{K}[\mathcal{M}]$ *has* ($\mathfrak{s}$-cAP).

The other direction of [14, Theorem 3.9] is equivalent to the following result.

Proposition 4.4 ([14, **Theorem 3.9**]). *Suppose* $\mathbb{K}$ *is a computable age that has* (cHP), (cJEP), *and* (cAP). *Then* $\mathbb{K}$ *has a computable Fraïssé limit.*

The next result follows from Lemma 3.15 and Propositions 3.18, 3.23, and 4.4.

[4] Note that [14] defines a notion of "computable homogeneity" where instead of taking in two $\simeq$-equivalent tuples and returning an automorphism it takes in two $\simeq$-equivalent tuples $\mathbf{a}, \mathbf{b}$ and an extra point x and returns a new point y such that $\mathbf{a}y \simeq \mathbf{b}x$. Note that their notion of computable homogeneity is equivalent to our notion of computable ultrahomogeneity via a standard back-and-forth construction of an automorphism from one-point extension axioms.

Corollary 4.5. *Suppose $\mathbb{K}$ is a computable age with* (HP), (JEP), *and* (AP). *Then $\mathbb{K}$ has an $\mathcal{EI}(\mathbb{K})$-computable Fraïssé limit.*

The next result uses a standard back-and-forth construction to build the appropriate automorphism.

Proposition 4.6. *Let $\mathcal{M}$ be a computable structure, and suppose $\mathcal{M}$ is ultrahomogeneous. Then $\mathcal{M}$ is $\mathcal{EI}(\mathcal{M})$-computably ultrahomogeneous.*

4.2 Cofinally Ultrahomogeneous Structures

Definition 4.7. *Let $\mathcal{M}$ be a structure. A set E is a* **cofinal collection** *(in $\mathcal{M}$) when the following hold.*

- *E is a collection of finite tuples in $\mathcal{M}$.*
- *For every finite tuple $\mathbf{a}$ in $\mathcal{M}$ there is some $\mathbf{b} \in E$ such that $\mathbf{a} \subseteq \mathrm{cl}_{\mathcal{M}}(\mathbf{b})$.*
- *If $\mathbf{b} \in E$ and if $\mathbf{a} \subseteq \mathrm{cl}_{\mathcal{M}}(\mathbf{b})$ is a finite tuple, then $\mathbf{ab} \in E$.*
- *For all finite tuples $\mathbf{a}$ and $\mathbf{b}$ in $\mathcal{M}$, if $\mathbf{a} \in E$ and $\mathbf{a} \simeq \mathbf{b}$, then $\mathbf{b} \in E$.*

Definition 4.8. *A structure $\mathcal{M}$ is* **cofinally ultrahomogeneous** *if for every finite set $A \subseteq \mathcal{M}$ there is a finitely generated $\mathcal{B} \subseteq \mathcal{M}$ with $A \subseteq \mathcal{B}$ such that whenever $f : \mathcal{B} \to \mathcal{B}^*$ is an isomorphism with $\mathcal{B}^* \subseteq \mathcal{M}$ then there is an automorphism $f^* : \mathcal{M} \to \mathcal{M}$ such that $f^* \restriction_{\mathcal{B}} = f$. Given a cofinal collection E in $\mathcal{M}$, we say that $\mathcal{M}$ is* **cofinally ultrahomogeneous for** *E if every isomorphism between two elements of E can be extended to an automorphism of $\mathcal{M}$.*

In the full paper [2], we provide an example of a cofinally ultrahomogeneous structure that is not ultrahomogeneous, but which encodes (in a relational language) a particular bijection of an infinite set.

Definition 4.9. *Let $\mathcal{M}$ be a computable structure and suppose E is an $\mathfrak{s}$-computable cofinal collection in $\mathcal{M}$. We say that $\mathcal{M}$ is* **$\mathfrak{s}$-computably cofinally ultrahomogeneous** *(with respect to E) if there is an $\mathfrak{s}$-computable g such that*

- $\mathrm{dom}(g)$ *is the collection of pairs $(\mathbf{a}, \mathbf{b})$ of tuples in $\mathcal{M}$ where $\mathbf{a} \in E$ and $\mathbf{a} \sim \mathbf{b}$,*
- *if $(\mathbf{a}, \mathbf{b}) \in \mathrm{dom}(g)$, then $g(\mathbf{a}, \mathbf{b})$ is the index of an $\mathfrak{s}$-computable bijection $g_{\mathbf{a},\mathbf{b}} : \mathcal{M} \to \mathcal{M}$, where (i) $g_{\mathbf{a},\mathbf{b}}(\mathbf{a}) = \mathbf{b}$; and (ii) if $\mathbf{a} \simeq \mathbf{b}$ then $g_{\mathbf{a},\mathbf{b}}$ is an automorphism of $\mathcal{M}$.*

Proposition 4.10. *Let $\mathcal{M}$ be a computable structure. Then $\mathcal{M}$ is cofinally ultrahomogeneous if and only if there is some Turing degree $\mathfrak{s}$ such that $\mathcal{M}$ is $\mathfrak{s}$-computably cofinally ultrahomogeneous.*

Proposition 4.11. *Let $\mathcal{M}$ be a cofinally ultrahomogeneous structure for some cofinal collection E. If $\mathbf{a} \in E$, then $\mathrm{cl}_{\mathcal{M}}(\mathbf{a})$ is an amalgamation base in $\mathbb{K}[\mathcal{M}]$.*

Proposition 4.12. *Let $\mathcal{M}$ be a cofinally ultrahomogeneous structure. Let $\mathcal{A}_0, \mathcal{A}_1 \subseteq \mathcal{M}$ be amalgamation bases in the age of $\mathcal{M}$ and suppose $k : \mathcal{A}_0 \to \mathcal{A}_1$ is an isomorphism. Then there is an automorphism $g : \mathcal{M} \to \mathcal{M}$ such that $g \restriction_{\mathcal{A}_0} = k$.*

Proposition 4.13. *Let $\mathfrak{s}$ be a Turing degree, and let $\mathcal{M}$ be an $\mathfrak{s}$-computable structure that is $\mathfrak{s}$-computably cofinally ultrahomogeneous. If $\mathcal{EI}(\mathcal{M}) \leq_T \mathfrak{s}$ then $\mathbb{K}[\mathcal{M}]$ has ($\mathfrak{s}$-cCAP).*

We already knew, by Proposition 3.19, that for any $\mathfrak{s}$-computable $\mathfrak{s}$-cofinally ultrahomogeneous structure $\mathcal{M}$, its canonical $\mathfrak{s}$-computable age $\mathbb{K}[\mathcal{M}]$ has ($\mathfrak{s}$-cHP) and ($\mathfrak{s}$-cJEP). Proposition 4.13 shows that it also has ($\mathfrak{s}$-cCAP) if $\mathcal{EI}(\mathcal{M}) \leq_T \mathfrak{s}$.

Definition 4.14. *Let $\mathfrak{s}$ be a Turing degree, and suppose E is a cofinal collection. An $\mathfrak{s}$-computable structure $\mathcal{M}$ has the* **$\mathfrak{s}$-computable cofinal extension property** *(with respect to E) if there is an $\mathfrak{s}$-computable partial function p which takes tuples $(\mathbf{a}, \mathbf{b}, \mathbf{c})$ with (i) $\mathbf{a}, \mathbf{b}, \mathbf{c} \in \mathcal{M}$, (ii) $\mathbf{a}, \mathbf{c} \in E$, (iii) $\mathbf{a} \subseteq \mathrm{cl}_{\mathcal{M}}(\mathbf{c})$, and (iv) $\mathbf{a} \sim \mathbf{b}$, and which returns a tuple $\mathbf{d}$ where $\mathbf{ac} \sim \mathbf{bd}$ and whenever $\mathbf{a} \simeq \mathbf{b}$, then (I) $\mathbf{d} \simeq \mathbf{c}$, (II) $\mathbf{b} \in \mathrm{cl}_{\mathcal{M}}(\mathbf{d})$, and (III) $\mathfrak{t}_{\mathrm{cl}}(\mathbf{d}, \mathbf{c}) \restriction_{\mathbf{b}} = \mathfrak{t}_{\mathrm{cl}}(\mathbf{b}, \mathbf{a})$.*

The next result uses the cofinal extension property to build an automorphism extending any isomorphism of elements in E, generalizing the proof that a structure is ultrahomogeneous iff its age has the 1-point extension property.

Proposition 4.15. *Let $\mathfrak{s}$ be a Turing degree, and $\mathcal{M}$ an $\mathfrak{s}$-computable structure. Suppose that $\mathcal{EI}(\mathcal{M}) \leq_T \mathfrak{s}$ and let E be an $\mathfrak{s}$-computable collection of cofinal pairs. Then $\mathcal{M}$ is $\mathfrak{s}$-computably cofinally ultrahomogeneous with respect to E if and only if $\mathcal{M}$ has the $\mathfrak{s}$-computable cofinal extension property with respect to E.*

Definition 4.16. *Let $\mathcal{K}$ be an age in $\mathcal{L}$. A countable $\mathcal{L}$-structure $\mathcal{M}$ is a* **cofinal Fraïssé limit** *of $\mathcal{K}$ if $\mathcal{M}$ is cofinally ultrahomogeneous and the age of $\mathcal{M}$ is $\mathcal{K}$.*

Definition 4.17. *Suppose $\mathbb{K}$ is a computable age. An $\mathfrak{s}$-computable structure $\mathcal{M}$ is an $\mathfrak{s}$-computable cofinal Fraïssé limit if $\mathcal{M}$ is $\mathfrak{s}$-computably cofinally ultrahomogeneous and $\mathbb{K}$ is an $\mathfrak{s}$-computable representation of the age of $\mathcal{M}$.*

Theorem 4.18. *Let $\mathfrak{s}$ be a Turing degree. Suppose that $\mathbb{K}$ is an $\mathfrak{s}$-computable age, $\mathcal{EI}(\mathbb{K}) \leq_T \mathfrak{s}$, and $\mathbb{K}$ has ($\mathfrak{s}$-cHP), ($\mathfrak{s}$-cJEP), and ($\mathfrak{s}$-cCAP). Then there is an $\mathfrak{s}$-computable cofinal Fraïssé limit of $\mathbb{K}$.*

The next three results follow from the above.

Corollary 4.19. *Suppose $\mathbb{K}$ is a computable age with (HP), (JEP), and (CAP). Then $\mathbb{K}$ has an $\mathcal{EI}(\mathbb{K})''$-computable cofinal Fraïssé limit.*

Theorem 4.20. *Suppose $\mathcal{M}$ is a cofinally ultrahomogeneous structure. Then $\mathcal{M}$ is $\mathcal{EI}(\mathcal{M})''$-computably cofinally ultrahomogeneous.*

Corollary 4.21. *Suppose $\mathcal{M}$ is a computable cofinally ultrahomogeneous structure. Then $\mathcal{M}$ is $\mathbf{0}'''$-computably cofinally ultrahomogeneous.*

5 Lower Bounds

We conclude with two lower bounds, Theorem 5.1 (for finite relational languages) and Theorem 5.3 (for countable languages which may be infinite or include function symbols), and an associated corollary for each in combination with Proposition 4.13.

Theorem 5.1. *There is a computable age* $\mathbb{K}$*, which is the canonical computable age of some structure in a finite relational language, such that* $\mathbb{K}$ *has* (CAP) *but if it has* ($\mathfrak{s}$-cCAP) *for some* $\mathfrak{s}$*, then* $\mathbf{0}' \leq_{\mathrm{T}} \mathfrak{s}$*.*

Corollary 5.2. *There is a computable age* $\mathbb{K}$*, which is the canonical computable age of some structure in a finite relational language, such that if* $\mathfrak{s}$ *is a Turing degree and* $\mathcal{M}$ *is an* $\mathfrak{s}$*-computable structure that is* $\mathfrak{s}$*-computably cofinally ultrahomogeneous and a cofinal Fraïssé limit of* $\mathbb{K}$*, then* $\mathbf{0}' \leq_{\mathrm{T}} \mathfrak{s}$*.*

Theorem 5.3. *There is a computable age* $\mathbb{K}$*, which is the canonical computable age of some structure, such that* $\mathbb{K}$ *has* (CAP) *but if it has* ($\mathfrak{s}$-cCAP) *for some* $\mathfrak{s}$*, then* $\mathbf{0}'' \leq_{\mathrm{T}} \mathfrak{s}$*.*

Corollary 5.4. *There is a computable age* $\mathbb{K}$*, which is the canonical computable age of some structure, such that if* $\mathfrak{s}$ *is a Turing degree and* $\mathcal{M}$ *is an* $\mathfrak{s}$*-computable structure that is* $\mathfrak{s}$*-computably cofinally ultrahomogeneous and a cofinal Fraïssé limit of* $\mathbb{K}$*, then* $\mathbf{0}'' \leq_{\mathrm{T}} \mathfrak{s}$*.*

References

1. Ackerman, N., Freer, C., Mirabi, M.: Computable weak Fraïssé limits. In preparation (2026)
2. Ackerman, N., Freer, C., Mirabi, M.: On the computability of cofinal Fraïssé limits (2026). https://arxiv.org/abs/2601.22435
3. Adams, F., Cenzer, D.: Computability and categoricity of weakly ultrahomogeneous structures. Computability **6**(4), 365–389 (2017)
4. Angel, O., Kechris, A.S., Lyons, R.: Random orderings and unique ergodicity of automorphism groups. J. Eur. Math. Soc. (JEMS) **16**(10), 2059–2095 (2014)
5. Badia, G., Noguera, C.: Fraïssé classes of graded relational structures. Theor. Comput. Sci. **737**, 81–90 (2018)
6. Bartoš, A., Bice, T., Barbosa, K.D., Kubiś, W.: The weak Ramsey property and extreme amenability (2021). https://arxiv.org/abs/2110.01694
7. Bodirsky, M., Jonsson, P., Van Pham, T.: The complexity of phylogeny constraint satisfaction problems. ACM Trans. Comput. Log. **18**(3), Art. 23, 42 (2017)
8. Bojańczyk, M., Segoufin, L., Toruńczyk, S.: Verification of database-driven systems via amalgamation. In: Hull, R., Fan, W. (eds.) Proceedings of 32nd Symposium Principles of Database Systems (PODS 2013), pp. 63–74. ACM (2013)
9. Calais, J.: Relation et multirelation pseudo-homogènes. C. R. Acad. Sci. Paris Sér. A-B **265**, A2–A4 (1967)
10. Calais, J.: Relations pseudo-homogènes, application à la théorie des arbres. C. R. Acad. Sci. Paris Sér. A-B **266**, A324–A326 (1968)

11. Cameron, P.J.: The age of a relational structure. Disc. Math. **95**(1–3), 49–67 (1991)
12. Cenzer, D., Adams, F., Ng, K.M.: Computability and categoricity of weakly homogeneous Boolean algebras and p-groups. In: Aspects of Computation and Automata Theory with Applications, pp. 141–158. World Scientific (2024)
13. Cristani, M., Hirsch, R.: The complexity of constraint satisfaction problems for small relation algebras. Artif. Intell. **156**(2), 177–196 (2004)
14. Csima, B.F., Harizanov, V.S., Miller, R., Montalbán, A.: Computability of Fraïssé limits. J. Symb. Logic **76**(1), 66–93 (2011)
15. Dasilva Barbosa, K.: Ramsey Degree Theory of Ordered and Directed Sets. Ph.D. thesis, University of Toronto (2022)
16. Droste, M., Göbel, R.: Universal domains in the theory of denotational semantics of programming languages. In: Proceedings of 5th Symposium Logic in Computer Science (LICS '90), pp. 19–34. IEEE Computer Society (1990)
17. Droste, M., Göbel, R.: Universal domains and the amalgamation property. Math. Struct. Comput. Sci. **3**(2), 137–159 (1993)
18. Drzewiecka, M., Ivanov, A., Mokry, B.: Generics in invariant subsets of some highly homogeneous permutation groups (2023)
19. Fokina, E., Harizanov, V., Turetsky, D.: Computability-theoretic categoricity and Scott families. Ann. Pure Appl. Logic **170**(6), 699–717 (2019)
20. Fouché, W.L.: Martin-Löf randomness, invariant measures and countable homogeneous structures. Theory Comput. Syst. **52**(1), 65–79 (2013)
21. Fraïssé, R.: Sur l'extension aux relations de quelques propriétés des ordres. Ann. Sci. École Norm. Sup. **71**(4), 363–388 (1954)
22. Hodges, W.: Model Theory, Encyclopedia of Mathematics and its Applications, vol. 42. Cambridge University Press, Cambridge (1993)
23. Kechris, A.S., Pestov, V.G., Todorcevic, S.: Fraïssé limits, Ramsey theory, and topological dynamics of automorphism groups. Geom. Funct. Anal. **15**(1), 106–189 (2005)
24. Kechris, A.S., Rosendal, C.: Turbulence, amalgamation, and generic automorphisms of homogeneous structures. Proc. Lond. Math. Soc. (3) **94**(2), 302–350 (2007)
25. Khoussainov, B., Nies, A., Rubin, S., Stephan, F.: Automatic structures: richness and limitations. In: Proceedings of 19th IEEE Symposium Logic in Computer Science (LICS 2004), pp. 44–53. IEEE Computer Society (2004)
26. Krawczyk, A., Kruckman, A., Kubiś, W., Panagiotopoulos, A.: Examples of weak amalgamation classes. MLQ Math. Log. Q. **68**(2), 178–188 (2022)
27. Krawczyk, A., Kubiś, W.: Games with finitely generated structures. Ann. Pure Appl. Logic **172**(10), 103016, 13 (2021)
28. Kruckman, A.: Notes on generalized Fraïssé theory (2015). https://math.berkeley.edu/~kruckman/fraisse.pdf
29. Kruckman, A.: Infinitary Limits of Finite Structures. Ph.D. thesis, University of California, Berkeley (2016)
30. Kubiś, W.: Weak Fraïssé categories. Theory Appl. Categ. **38**(2), 27–63 (2022)
31. Malicki, M.: Remarks on weak amalgamation and large conjugacy classes in non-archimedean groups. Arch. Math. Logic **61**(5–6), 685–704 (2022)
32. Panagiotopoulos, A., Tent, K.: Universality vs genericity and C_4-free graphs. Eur. J. Combin. **106**, 103590, 12 (2022)
33. Steiner, R.M.: Reducibility, Degree Spectra, and Lowness in Algebraic Structures. Ph.D. thesis, City University of New York (2012)
34. Truss, J.K.: Generic automorphisms of homogeneous structures. Proc. London Math. Soc. (3) **65**(1), 121–141 (1992)

Bishop's (Up)Crossing Inequality and Lower Semicomputable Random Reals Revisited

Mikhail Andreev[1] and Alexander Shen[2(✉)]

[1] CRITEO, Berlin, Germany
mikhail.andreev.57@gmail.com
[2] LIRMM, Univ Montpellier, CNRS, Montpellier, France
sasha.shen@gmail.com, alexander.shen@lirmm.fr

Abstract. In this paper we provide an easy proof of Barmpalias–Lewis-Pye result saying that all computable increasing sequences converging to random reals converge at the same speed by noting that it easily follows from Bishop's upcrossing inequality. We also provide a simple derivation of this inequality.

Keywords: Upcrossing inequalities · Lower semicomputable random reals · Barmpalias–Lewis-Pye theorem

1 Crossing Inequalities

We start by a symmetric version of Bishop's upcrossing inequality [2] (so we name it 'crossing inequality' instead of 'upcrossing' or 'downcrossing' one).

Consider the following "anti-slalom" game. A player starts at some point (x, y) and moves to the right. She wants to *miss* as many gates (vertical segments $[m_i, M_i]$ located on the vertical line with first coordinate x_i) as possible. The restriction is that the slope of her trajectory should be between α and β all the time.

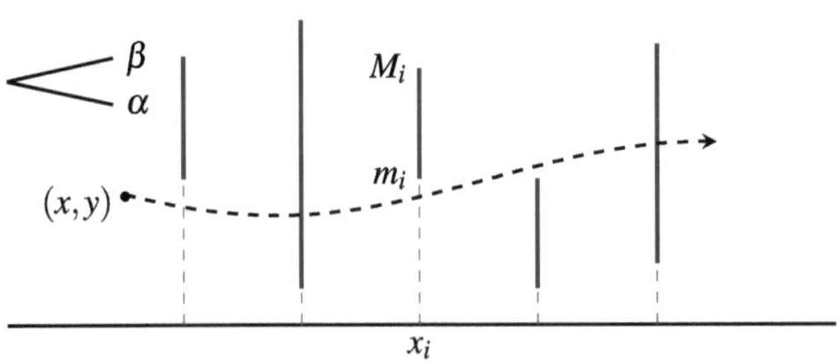

The game is called *anti-slalom* since the player wants to miss the gates as much as possible. If there were no restrictions on the slope ($\alpha = -\infty$, $\beta = \infty$),

To the memory of Vladimir V'yugin.

V. Brattka et al. (Eds.): CiE 2026, LNCS 16674, pp. 124–134, 2027.
https://doi.org/10.1007/978-3-032-31348-5_8

she could avoid all the gates. The smaller is the interval $[\alpha, \beta]$, the more restricted is the movement. When the interval is very small, the player essentially moves along some straight line, and crosses all the gates that intersect this straight line.

Fix the positions of the gates (i.e., x_i, m_i and M_i) as well as $\alpha < \beta$. Define the function $t(x, y)$ that is the minimal number of crossings if player starts at the point (x, y). Then consider the function $T(x) = \max_y t(x, y)$.

For example, if x is greater than all x_i, we have $T(x) = 0$ since there are no gates on the right of x_i. We also have $T(x) = 0$ if x is much smaller than all x_i: if we start far away, small changes in the slope (that should be between α and β) are enough to avoid all the gates. So $T(x)$ is a function with finite support (and integer values).

Theorem 1 (Crossing inequality).

$$\int_{\mathbb{R}} T(x)\, dx \leq \frac{1}{\beta - \alpha} \sum_i (M_i - m_i).$$

For example, if there is only one gate, then $t(x, y) = 1$ inside the triangle (see the picture) and 0 elsewhere, so T equals 1 on an interval of length $l/(\beta - \alpha)$ where l is the length of the gate, and the inequality we claim becomes an equality.

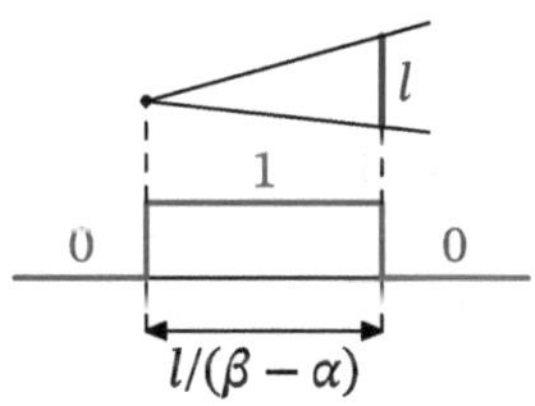

If there are two gates, the situation is more complicated; for example, on the left picture the inequality is still an equality (since two marked intervals are equal), while on the right picture the inequality is strict.

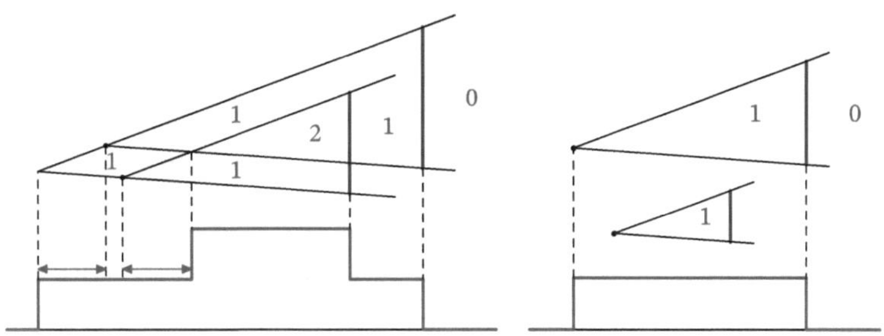

Proof (of Theorem 1). To understand the behavior of the function t, we use the moving line approach. For every x we consider the function $y \mapsto t(x, y)$ and look how this function changes when x decreases.

- When x is greater than all x_i (positions of all gates), the function is zero everywhere.
- When x crosses the coordinate x_i of ith gate (changing from $x_i + 0$ to $x_i - 0$, so to say), the indicator function of the interval $[m_i, M_i]$ is added (new gate crossing).
- When x decreases from $x + \Delta x$ to x not crossing any of x_i, the function $t(x, y)$ can be computed as follows:

$$t(x,y) = \min_{\Delta y \in [\alpha \Delta x, \beta \Delta x]} t(x + \Delta x, y + \Delta y).$$

  Indeed, the player could move (without crossing any gates) from (x, y) to the point $(x + \Delta x, y + \Delta y)$ for arbitrary $\Delta y \in [\alpha \Delta x, \beta \Delta x]$. She needs to choose the point that minimizes the number of subsequent crossings.

The last formula means that the widths of the peaks of the function $t(x, \cdot)$ (marked by arrows in the diagram) decrease with speed $\beta - \alpha$ when x decreases

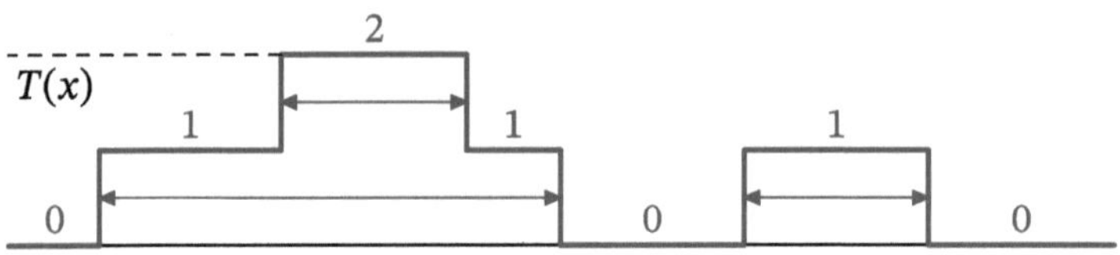

(and does not cross any x_i). Therefore, the integral $I(x) = \int_y t(x, y)\, dy$ of this function decreases with speed at least $T(x)(\beta - \alpha)$.

The total decrease of the integral cannot exceed the sum of its increases (that happen due to gates), therefore

$$(\beta - \alpha) \int_x T(x) \leq \sum_i (M_i - m_i),$$

as required. □

2 Crossing a Curve

Now we want to get an upper bound for the number of crossings for the case where we cross some (polygonal) line instead of a set of vertical gates.

A simple trick reduces this case to the case of vertical gates. Indeed, if the trajectory with slope in $[\alpha, \beta]$ crosses a non-vertical segment AB (with horizontal size Δx and vertical size Δy), then it has to cross also the vertical segment $A'B$ that is a union of all intervals reachable (with slope in $[\alpha, \beta]$) from all points between A and B.

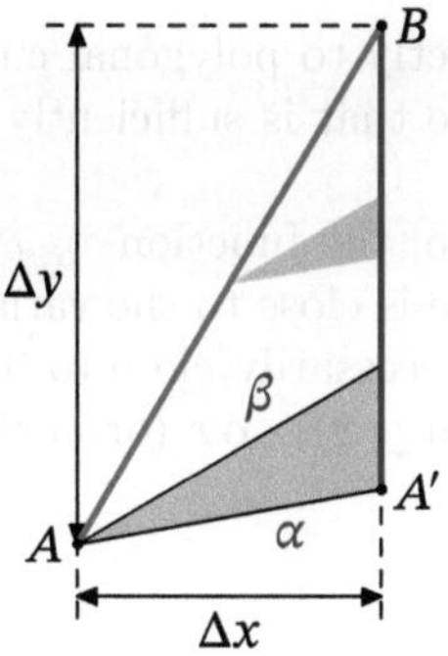

What is the length of this new interval? In our example $|A'B| = \Delta y - \alpha\Delta x$. In the general case, the length of the resulting vertical segment depends on the slope of AB (whether it is less than α, between α and β, or greater than β). Assuming $\Delta x > 0$ (while Δy can be positive or negative), we get the following answer for the length of the replacement vertical segment:

$$\text{vertical length} = \begin{cases} \Delta y - \alpha\Delta x \text{ , if } \Delta y/\Delta x \geq \beta \text{ ;} \\ (\beta - \alpha)\Delta x \text{ , if } \alpha \leq \Delta y/\Delta x \leq \beta \text{ ;} \\ \beta\Delta x - \Delta y \text{ , if } \Delta y/\Delta x \leq \alpha \text{ .} \end{cases}$$

Note that the dependence on Δy is continuous, and any of the two expressions can be used in border cases. Another way to say the same thing: the length of the replacement vertical segment is $\tau_{\alpha,\beta}(\Delta y/\Delta x)\Delta x$, where $\tau_{\alpha,\beta}$ is the following function:

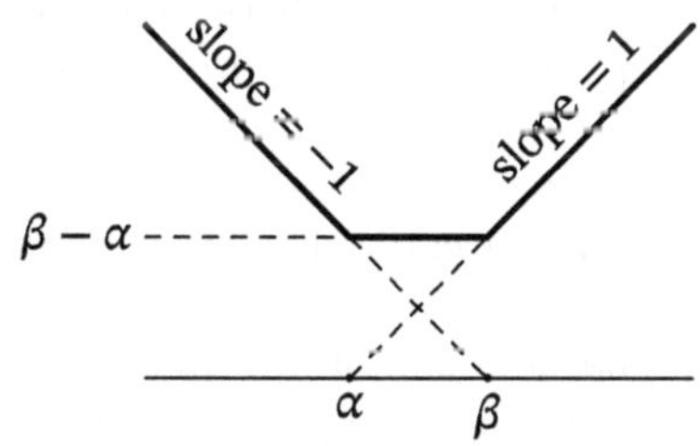

Now we may consider a polygonal curve $y = y(x)$ for $x \in [l, r]$ as a union of segments, and define $t(x, y)$ as the minimal number of intersections with the curve for a player who starts at x, y and keeps her slope in $[\alpha, \beta]$. Then $T(x)$ is defined as $\max_y t(x, y)$ (as before). Replacing each segment by a vertical one as described, we get the following bound:

Theorem 2 (Curve crossing inequality).

$$\int_{\mathbb{R}} T(x)\, dx \leq \frac{1}{\beta - \alpha} \int_l^r \tau_{\alpha,\beta}(y'(x))\, dx.$$

Our argument applies directly to polygonal curves; we may use approximations to extend it to any curve that is sufficiently well-behaved, but this is not needed for the sequel.

If α and β are close to zero, the function $\tau_{\alpha,\beta}$ is close to the absolute value function, so the right hand size is close to the variation of the function y. If α is close to β (but they are not necessarily close to 0), the right hand side is close to the variation of the function $y(x) - \alpha x$ (or $y(x) - \beta x$).

3 Crossing a Gap

We switched to curves because in this way we can get a bound for the up- and downcrossings. Let us explain what it means. Consider again some curve C (again we assume that C is a graph of a piecewise linear function, this is enough for us), some point (x, y) and slopes $\alpha < \beta$. We count how many times the curve crosses up and down the gap between slopes α and β, i.e., an angle formed by α- and β-rays starting at (x, y).

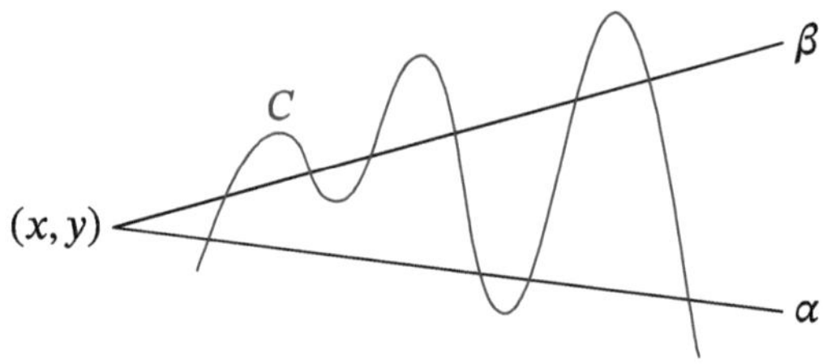

In other words, we are in (x, y) and watch the movement of a point along C (from left to right). In this example (see picture) we see that initially the point is at an angle smaller than α, then the angle becomes greater than β, then it goes down below β but does not reach α and turns back exceeding β again, then becomes less than α, than again greater than β and finally less than α. We say that there are two *upcrossings* and two *downcrossings* of the same gap (between α and β) in this picture. The numbers of up- and downcrossings differ at most by 1 (since they alternate), and we consider the sum of them.

More formally, we consider sequences of points $c_0, \ldots, c_m$ on the curve (going from left to right) that alternate being below α-ray and above β-ray, and take the maximal m in all those sequences. This maximal m is the number of crossings for the gap and is denoted $t(x, y)$. So for every fixed curve that is the graph of the function $y(x)$ for $x \in [a, b]$, and for every fixed $\alpha < \beta$ we get some non-negative integer function $t(x, y)$ that depends on the starting point (x, y). Then we let $T(x) = \max_y t(x, y)$.

Theorem 3 (Bishop's crossings inequality).

$$\int_{\mathbb{R}} T(x)\,dx \le \frac{1}{\beta - \alpha} \int_a^b \tau_{\alpha,\beta}(y'(x))\,dx.$$

This inequality is the same as in Theorem 2, and this result is an immediate corollary of that theorem. Indeed, any trajectory that goes from (x, y) with slope between α and β, is confined in the gap and therefore has at least one intersection point for each part of the curve that crosses the gap.

Note that $t_{\alpha,\beta}(s) \leq |s| + O(|\alpha| + |\beta|)$, and the integral of $|y'|$ is the variation of the function $y(x)$, therefore we get

$$\int_{\mathbb{R}} T(x)\, dx \leq \frac{O(|\alpha| + |\beta|)(b - a) + \mathrm{Var}(y)}{\beta - \alpha} \tag{$*$}$$

and this will be enough for our application.

4 Barmpalias–Lewis-Pye Theorem

The Bishop's crossing inequality implies (almost immediately) the following result of Barmpalias and Lewis-Pye [1, Theorem 1.7]; see also [7]:

Theorem 4 (Barmpalias, Lewis-Pye). *Let a_n and b_n be two increasing computable sequences of rational numbers that converge to real numbers A and B. Assume that A is Martin-Löf random. Then the sequence*

$$\frac{B - b_n}{A - a_n}$$

has a limit.

(See, e.g., [9] for the definition of Martin-Löf randomness and basic results about monotone converging sequences of rational numbers and their limits, lower semicomputable reals.)

The argument below proves that there exists a finite or infinite limit, but in fact the second case is impossible (and this was known much before the result of Barmpalias and Lewis-Pye, see [6]; we reproduce the argument below in Sect. 5 for reader's convenience).

Proof (of Theorem 4). If the sequence $(B - b_n)/(A - a_n)$ has no finite or infinite limit, there is a gap $\alpha < \beta$ that is crossed infinitely many times (there are infinitely many terms strictly below α and infinitely many terms strictly above β). We need to show that A is non-random.

Consider the curve connecting (a_1, b_1), (a_2, b_2), ... and converging to (A, B).

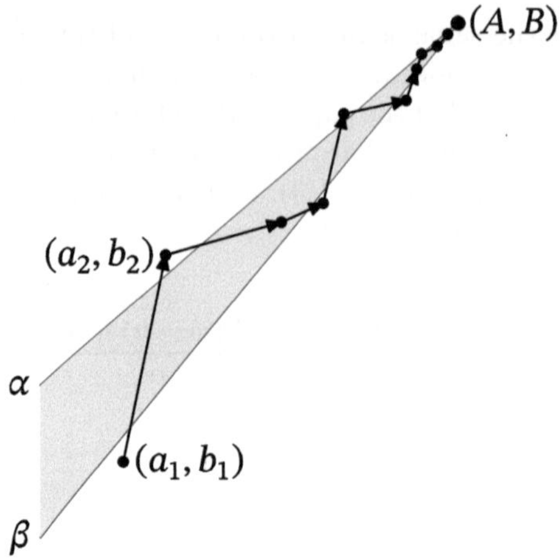

We know that this curve has infinitely many crossings with the angle that has vertex (A, B) and slopes α and β. Now we can apply Bishop's inequality (rotated by 180°): let $t(a, b)$ be number of crossings, considered as a function of the angle vertex (a, b) and let $T(a)$ be the maximum of $t(a, b)$ over all b. Then $T(A) = +\infty$ while $\int_a T(a)$ is finite (the curve is monotone and bounded, so the variation is finite). It remains to note that function T is lower semicomputable (for the finite part of the curve it is computable, and it increases when we add new segments to the curve), so we get an integral test that rejects A.

Formally speaking, we proved Theorem 3 for finite polygonal curves and now we have infinite one. To avoid problems, we consider the bound for the finite part of the curve and note that all integrals are uniformly bounded. Then we apply the monotone convergence theorem.

5 Remarks

$(B - b_n)/(A - a_n)$ is bounded

For reader's convenience, let us provide a proof of a classical result from [6] saying that $(B{-}b_n)/(A{-}a_n)$ is bounded (under the assumptions of Theorem 4). Assume that for every $\varepsilon > 0$ there exists some n such that $A - a_n < \varepsilon(B - b_n)$. Let us show how to get an effectively open cover for A of arbitrarily small measure.

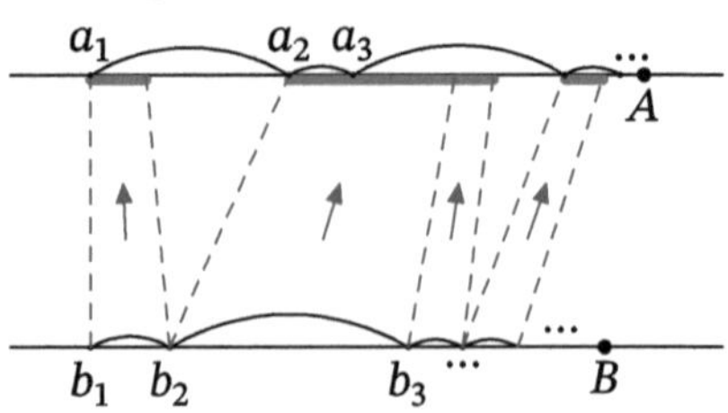

For every interval (b_i, b_{i+1}) on the bottom line consider its "counterpart" on the top line, an interval whose length is multiplied by ε, i.e., equals $\varepsilon(b_{i+1} - b_i)$, that starts from b_i or from the right endpoint of the counterpart of the previous interval, *whichever is bigger*.

The total length of all counterparts is at most $\varepsilon(B - b_1)$, and can be made arbitrarily small if ε is small enough. On the other hand, they cover A (so A is non-random). Why? Since $A - a_n < \varepsilon(B - b_n)$ for some n, the counterparts of (b_n, b_{n+1}), $(b_{n+1}, b_{n+2}), \ldots$ cannot be all below A (note that they are disjoint). So some of them should go above A. On the other hand, A cannot get into the gaps between the counterparts, since all gaps are below some a_i, and A is greater than all a_i. This finishes the argument.

Zero Limit and Non-random *B*

If B is random, the limit of the inverse fraction $(A - a_n)/(B - b_n)$ exists (and is finite, as we have seen), therefore both limits are non-zero. Therefore, if the limit $(B - b_n)/(A - a_n)$ is zero, then B is non-random.

On the other hand, if a_n and b_n are computable increasing sequences converging to A and B, and $A - a_n = \Theta(B - b_n)$ (i.e., $A - a_n$ and $B - b_n$ differ at most by some constant factor), then A and B are either both random or both non-random. This is another classical result that goes back to Solovay (see [5, Theorem 4.9]). The direct argument may go as follows. Let us assume that A is non-random; we need to show that B is also non-random. Assume that we have an enumeration of intervals that together cover A and have small total measure. How can we cover B by some other enumerable family of intervals with small measure? When some interval (p, q) appears in the cover of A, we can wait until some a_i turns out to be in this interval. (If this never happens, the interval does not cover A and is useless.) If some a_i appears in (p, q), we take an interval of c times bigger length, where c is the constant in $B - b_i = O(A - a_i)$, with left endpoint b_i. We do not know whether (p, q) covers A, but if it does, then the new interval covers B (due to the choice of c). Since some interval in the enumeration covers A by assumption, the new family covers B and it at most c times longer. This shows that if $B - b_i = O(A - a_i)$ and A is non-random, then B is non-random, too.

These arguments show that the limit in Theorem 4 is always finite and equals zero if and only if B is non-random.

Only *A* and *B* Matter

One can note that the limit in Theorem 4 does not depend on the choice of computable increasing sequences that converge to A and B (and depends on A and B only). Indeed, for non-random B this limit is always 0. For random B it is enough to show that two increasing sequences a_n and a'_n that converge to the same random A, have the same convergence speed, i.e. $(A - a_n)/(A - a'_n)$ converges to 1. Indeed, we know already that the limit exists, assume that it is not 1. Then one sequence converges faster than the other; we may assume without loss of generality that $A - a'_n < c(A - a_n)$ for some $c < 1$ and for all n (discarding some initial parts of the sequences). This implies that A is computable. Indeed, if we know some approximation a to A from below, we can wait until a_i becomes greater than a. Then a'_i is a better approximation to A that a (by factor c). Repeating this trick, we can compute A with arbitrary precision.

Extension to Reals that are Not Left-c.e. and Finite Variation Functions

Ivan Titov [10] proved the following generalization of the Barmpalias–Lewis-Pye result where A is not assumed to be a left-c.e. real:

Theorem 5 (Titov). *Let A be a Martin-Löf random real, and let g be a partial computable non-decreasing bounded function that is defined on all rational numbers smaller than A (and, maybe, on some other rational numbers). Let $B = \lim_{r \to A-0} g(r)$. Then the ratio*

$$\frac{B - g(r)}{A - r}$$

has limit as $r \to A - 0$.

This result can be obtained as a corollary of crossing inequalities in the same way. The only difference is that here we have to deal not with a polygonal line, but with a graph of a partial function defined on a countable set. But the definitions of the number of crossings and the variation can be naturally extended to this case, and our main tool, the inequality (∗) (see the end of Sect. 3) remain valid for this case. Indeed, it is true for a function with finite domain (we can connect the dots by a polygonal line), and then we can take the limit and prove it for countable domains.

Note that in this way we prove only that finite or infinite limit exists (and to exclude an infinite limit another argument is needed; it is provided in [10]).

This argument extends (without any changes) to functions g that have finite variation (but are not necessarily monotone)—a result proven by Titov in another paper [11, Theorem 20].

Historical Remarks

Two results (of Bishop and Barmpalias–Lewis-Pye) we are reproving here have a quite complicated history. The upcrossing inequality appeared in Bishop's paper [2] in slightly different, but essentially equivalent form; Lemma 1 in this paper corresponds to our Theorem 1, and Theorem 1 in Bishop's paper roughly corresponds to Theorem 3. The main technical innovation in our proof is to introduce the "anti-slalom" game that provides an upper bound for the number of upcrossings and is easier to analyze than the number of upcrossings itself.

Bishop used these results to provide a new proof of Birkhoff ergodic theorem (a more constructive one—he was interested in constructive versions of different mathematical results, and this was his main motivation to revisit the ergodic theorem). In his book [2] he reproves the constructive ergodic theorem, this time essentially using some other (and somehow weaker) version of the upcrossing inequality. The difference is that in this weaker version we consider not the maximum of $t(x, y)$ for given x and for all y, but consider only points (x, y) on the curve. This is enough for Bishop's proof of ergodic theorem, but not for our application. Finally, Bishop [4] revisits ergodic theorem in more general setting, but does not state explicitly any version of upcrossing inequality for plane curves.

The upcrossing inequalities already appeared in algorithmic randomness theory for a completely different reason: V'yugin used them in [12] to prove the ergodic theorem for Martin-Löf random sequences (under some computability assumptions on the measure). For this he used the weaker version of upcrossing inequality from [3]. In fact, neither Bishop in [3] nor V'yugin in [12] explicitly use this weaker version of the upcrossing inequality; instead, they incorporate its proof in the proof of the ergodic theorem, but all the tools are there. See [8] for the details (the weak version of the inequality appears there as Lemma 1, and the Bishop–V'yugin potential argument is reproduced in Sect. 3). Note also that a bound for crossings appears also in the proof of the martingale convergence theorem.

Theorem 4 was proven by Barmpalias and Lewis-Pye [1] (without any connection to ergodic theorem for Martin-Löf randomness or upcrossing inequalities). It appears in their paper as Theorem 1.7, and the crucial step in the proof is Lemma 2.1 that says essentially that the number of (α, β)-crossings is finite for every gap (α, β). This Lemma is proven in Sect. 3 using quite complicated construction with movable markers.

Another proof of Theorem 4 was given by Miller [7]; as the author states, one of the main purposes of his note is "to give relatively short, self-contained proofs of the results of Barmpalias and Lewis-Pye". The statement of Theorem 4 appears as Lemma 1.2 there, and the proof is relatively short (though a bit technical). Of course, this is a matter of taste, but we hope that our proofs are more intuitive and maybe even can be considered as "proofs from The Book". The other results mentioned in this section are also well known; we included them only to provide easy direct proofs.

Acknowledgments. The authors thank Ivan Titov, Laurent Bienvenu and Wolfgang Merkle who tried hard to explain to us the results about convergence speed for lower semicomputable reals. Ivan Titov kindly informed us that some of his results also follow from crossing inequalities. We also thank Ilya Brodsky without whom this paper could not appear. We are grateful to the anonymous referees of our submission to CiE 2026 conference whose critical remarks helped us to provide more historical context and make clear that we just reproduce the proofs of known results for readers' convenience (and to correct language errors).

During the preparation of this paper one of us (A. Shen) was supported by ANR-21-CE48-0023 FLITTLA and ANR-24-CE48-3758 CADO grants. Part of the work was done while visiting Neapolis University, Paphos, Cyprus.

This paper is dedicated to the memory of our late colleague Vladimir V'yugin, a great mathematician and a kind man, who first used Bishop's upcrossing inequalities in the algorithmic randomness theory.

Disclosure of Interests. Authors declare no competing interests.

References

1. Barmpalias, G., Lewis-Pye, A.: Differences of halting probabilities. J. Comput. Syst. Sci. **89**, 349–360 (2017). https://doi.org/10.1016/j.jcss.2017.06.002. https://arxiv.org/abs/1604.00216v2
2. Bishop, E.: An upcrossing inequality with application. Michigan Math. J. **13**(1), 1–13 (1966). https://doi.org/10.1307/mmj/1028999474
3. Bishop, E.: Foundations of Constructive Analysis. Mc-Graw Hill (1967). https://archive.org/details/foundationsofcon0000bish
4. Bishop, E.: A constructive ergodic theorem. J. Math. Mech. **17**(7), 631–639 (1968). http://www.jstor.org/stable/24901875
5. Calude, C.S., Hertling, P.H., Khoussainov, B., Wang, Y.: Recursively enumerable reals and Chaitin Ω numbers. Theor. Comput. Sci. **255**(1–2), 125–149 (2001). https://doi.org/10.1016/S0304-3975(99)00159-0
6. Kučera, A., Slaman, T.A.: Randomness and recursive enumerability. SIAM J. Comput. **31**(1), 199–211 (2001). https://doi.org/10.1137/S0097539799357441
7. Miller, J.S.: On work of Barmpalias and Lewis-Pye: a derivation on the D.C.E. reals. In: Day, A., Fellows, M., Greenberg, N., Khoussainov, B., Melnikov, A., Rosamond, F. (eds.) Computability and Complexity. LNCS, vol. 10010, pp. 644–659. Springer, Cham (2017). https://doi.org/10.1007/978-3-319-50062-1_39
8. Shen, A.: Ergodic theorem and algorithmic randomness (following V.V'yugin). Inf. Comput. **307**, 105375 (2025). https://doi.org/10.1016/j.ic.2025.105375. https://arxiv.org/abs/2306.12769
9. Shen, A., Uspensky, V.A., Vereshchagin, N.: Kolmogorov complexity and algorithmic randomness. In: Mathematical Surveys and Monographs, vol. 220. American Mathematical Society, Providence (2018). https://www.lirmm.fr/~ashen/kolmbook-eng-scan.pdf
10. Titov, I.: Extending the Barmpalias–Lewis-Pye limit theorem to all reals. J. Symb. Logic (2025). https://doi.org/10.1017/jsl.2025.10157. Accessed 28 Oct 2025
11. Titov, I.: Relative randomness and continuous translation functions. In: 50th International Symposium on Mathematical Foundations of Computer Science (MFCS 2025), Warsaw, Poland, August 2025, p. 91 (2025). https://doi.org/10.4230/LIPIcs.MFCS.2025.91. https://drops.dagstuhl.de/entities/document/10.4230/LIPIcs.MFCS.2025.91
12. Vladimir, V.: V'yugin, Ergodic theorems for individual random sequences. Theor. Comput. Sci. **207**(2), 343–361 (1998). https://doi.org/10.1016/S0304-3975(98)00072-3

What is a *Polynomial-Time* Computable Square-Integrable Function?

Aras Bacho[1], Svetlana Selivanova[2], and Martin Ziegler[3](✉)

[1] California Institute of Technology, Pasadena, USA
[2] Novosibirsk State University and A.P. Ershov Institute of Informatics Systems SB RAS, Novosibirsk, Russia
[3] KAIST, Daejeon, Republic of Korea
ziegler@kaist.ac.kr

Abstract. We give two natural definitions of polynomial-time computability for L^2 (=square-integrable) functions; and show them incomparable (unless $\#P_1 \subseteq FP_1$), with applications to partial differential equations.

A good Definition is worth a thousand Lemmas!
(Variation of Doron Zeilberger's *Opinion 82)*

A mathematician with a definition knows what to build upon. A mathematician with two *definitions is never sure.*
(Variation of *Segal's Law)*

1 Introduction

The computational cost of an algorithm $\mathcal{A}$ receiving binary strings as input is commonly measured in dependence on the length n of said input, more precisely: When $t_\mathcal{A}(x) < \infty$ denotes the number of steps that $\mathcal{A}$ makes on input $x \in \{0,1\}^*$, one usually considers its worst-case runtime $t_\mathcal{A}(n) = \max\{t_\mathcal{A}(x) : |x| \leq n\}$, which is obviously well-defined (namely pointwise finite) and a non-decreasing function of type $\mathbb{N} \nearrow \mathbb{N}$, its asymptotic growth being the core subject of classical complexity theory. This well-definition of worst-case cost

$$t_\mathcal{A}(n) \;=\; \max\{t_\mathcal{A}(x) : x \in X_n\} \tag{1}$$

holds trivially as well for computations over any countable domain X, when equipped with a monotone cover X_n by *finite* subsets of X: for instance $X_n = \{\text{graph } G = (V, E) \text{ with } |V| \leq n \text{ vertices}\}$.

Perhaps less known, but folklore in Computable Analysis [37, Theorems 7.1.5 +7.2.7]: Equation (1) remains well-defined when generalizing 'finite' to 'compact'; specifically when the domain $X = \bigcup_n X_n$ is a suitably encoded [30] σ-compact topological space with compact cover $X_n \subseteq X_{n+1}$. For instance $t_\mathcal{A}(n) = \max\{t_\mathcal{A}(x, n) : 0 \leq x \leq 1\}$ is still well-defined when $\mathcal{A}$ computes a

V. Brattka et al. (Eds.): CiE 2026, LNCS 16674, pp. 135–149, 2027.
https://doi.org/10.1007/978-3-032-31348-5_9

continuous total function $f : [0;1] \to \mathbb{C}$ in the sense of [20, Definition 2.11] or [37, Example 7.2.14], namely pointwise approximating $x \mapsto f(x)$ up to error 2^{-n}. This well-definition is crucial to the rich complexity theory of real (and complex) continuous functions [10,20], including investigations on the complexity blow-up [5,6,8,12,15,17,21] incurred by popular operators and functionals on suitable spaces of (uniformly) continuous functions.

Indeed, Computable Analysis has long focused on continuous functions only [13]: because any terminating computation $\mathcal{A}(x)$ by itself depends continuously on its input x, namely $\mathcal{A}$ can read only a finite initial part of the infinite information comprising $x \in X$. Integrable functions generalize continuous ones. They arise naturally as (weak derivatives of) solutions to Partial Differential Equations (PDEs). A p-integrable function $f : [0;1] \to \mathbb{C}$ is technically an equivalence class of mappings which differ on some subset of $[0;1]$ of measure zero. One cannot reasonably define computing such generalized f in terms of algorithmic evaluation $x \mapsto f(x)$. Various better definitions of qualitative computability are well-known, and equivalent [9,22,28,37]. The present work continues previous explorations of quantitative/complexity-theoretic definitions [4,23,24,33]. Recall that $\#\mathsf{P} \supseteq \mathsf{FP}$ denotes the class of polynomial witness counting problems

$$\varphi(\vec{v}) \;=\; \operatorname{Card}\left\{\vec{w} \in \{\mathtt{0},\mathtt{1}\}^* : |\vec{w}| \leq \operatorname{poly}(|\vec{v}|), \vec{w} \in P\right\} \geq 0, \quad P \in \mathsf{P} \tag{2}$$

and $\mathsf{GapP} \supseteq \#\mathsf{P}$ its closure under negation/subtraction [14]. Intuitively $\#\mathsf{P}$ lies between classes NP and PSPACE (although technically the latter two consist of decision and the former of function problems). Theorems 5, 6 and 11 below suppose $\#\mathsf{P}_1 \not\subseteq \mathsf{FP}_1$, where subscript 1 indicates restriction to unary inputs [35, §5]. A weaker variant of the *Millennium Prize Problem* asks whether $\mathsf{P}_1 = \mathsf{NP}_1$, equivalently: whether $\mathsf{EXP} = \mathsf{NEXP}$ [2, Exercises 2.30+2.32].

1.1 Overview

Subsection 1.2 recalls common both computability and complexity theoretical notions of (polynomial-time, $\#\mathsf{P}$) computation for continuous functions. Section 2 formalizes two plus one natural versions (a,b,c) for L^2 functions: (c) pertains to *average-case* complexity, as elaborated in Subsect. 2.1. Subsection 2.2 compares these notions, exhibiting both implications and counterexamples (unless $\#\mathsf{P}_1 \subseteq \mathsf{FP}_1$): also in relation to the continuous case from Subsect. 1.2. Subsection 2.3 collects various closure properties of (a,b,c), such as under shifts and differentiation and indefinite integration. Due to page limitations, several proofs had to be omitted; selected representative proofs are collected in Sect. 3.

As applications (Sect. 4), Subsect. 4.1 treats the Heat Equation from our new perspective: With respect to Fourier computation (a), it turns out to map polynomial-time initial conditions to polynomial-time solutions. This contrasts the classical setting, see Remark 17. The Wave Equation in Subsect. 4.2 exhibits a similar effect: preserving (a) Fourier polynomial-time computability, while $\#\mathsf{P}_1$-hard w.r.t. classical (=continuous/ly differentiable) solutions.

1.2 Computability and Complexity of Continuous Functions

Recall that Gaussian dyadic rationals are of the form $(x + iy)/2^n$ where $x, y, n \in \mathbb{Z}$. An equivalent definition of computability for continuous functions $f \in \mathcal{C}[0;1]$, based on the Weierstraß approximation theorem, has been generalized successfully: algorithmic output only of (the coefficient vectors of) a sequence of Gaussian dyadic rational polynomials p_n approximating f in the ∞-norm up to error 2^{-n} [28, §I.0.3]. Indeed, the set of such polynomials is countable, and dense in $\mathcal{C}[0;1]$ by Stone-Weierstraß. This suggests [37, Definition 8.1.2] defining computability of a point x in a fixed separable metric space (X, d) in terms of computably approximating x up to error 2^{-n} by a suitable sequence of elements from some fixed countable dense subset of X. Unfortunately already for $\mathcal{C}[0;1]$ this alternative definition, although qualitatively equivalent, can differ exponentially in quantitative computational cost [20, §8]. This leaves us kind of in the dark how to *define* polynomial-time computability for square-integrable functions $f : [0;1] \to \mathbb{C}$ [23]. The following alternative characterization of polynomial-time computation for $\mathcal{C}[0;1]$ from [20, Theorem 2.22] does offer some guidance:

Definition 1. Let $\mathcal{C}[0;1]$ denote the set of total continuous complex functions $f : [0;1] \to \mathbb{C}$. Call such f *computable in polynomial time* iff there exists a Gaussian integer double sequence $\bar{v} = (v_{n,L,\ell}) \in \mathbb{Z} + i\mathbb{Z}$, indexed by $n \in \mathbb{N}$ and $0 \leq \ell < L \in \mathbb{N}$ such that (i), (ii), and (iii) hold:

i) $\forall n, \ell, L : \ \left|f(\ell/L) - v_{n,L,\ell}/2^n\right| \leq 2^{-n}$
ii) $(n, L, \ell) \mapsto \operatorname{Re} v_{n,L,\ell}, \operatorname{Im} v_{n,L,\ell} \in \mathbb{Z}$ are computable in time $\operatorname{poly}(n + \log L)$.
iii) f has a *polynomial* modulus of continuity $\mu \in \mathbb{N}[N]$ [4, Fact 3]:

$$|t - t'| \leq 2^{-\mu(m)} \Rightarrow |f(t) - f(t')| \leq 2^{-m} \ . \tag{3}$$

Call f computable *in* #P if (i), (iii), and (ii$'$) hold:

ii$'$) Both $\{0,1\}^* \ni 1^n \, 0 \operatorname{bin}(L) \operatorname{bin}(\ell) \mapsto \operatorname{Re} v_{n,L,\ell}, \operatorname{Im} v_{n,L,\ell} \in \mathbb{Z}$ are in GapP.

Call f *computable* if (i), (ii$''$), and (iii$''$) hold:

ii$''$) Both $(n, L, \ell) \mapsto \operatorname{Re} v_{n,L,\ell}, \operatorname{Im} v_{n,L,\ell} \in \mathbb{Z}$ are recursive, i.e. computable in the discrete sense.
iii$''$) f has a recursive modulus of continuity μ. □

Other than [21, Definitions 13+14], we here do not allow (repeated and adaptive and post-processable) oracle access to #P: for sharper (pun) complexity claims without climbing up the Counting Hierarchy [36], such as in Theorems 4+7+9 below. Note that $\mathcal{O}(n + \log L)$ is the input length in (ii$'$), which thus generalizes (ii) meaning that both mappings $1^n \, 0 \operatorname{bin}(L) \operatorname{bin}(\ell) \mapsto \operatorname{Re} v_{n,L,\ell}, \operatorname{Im} v_{n,L,\ell} \in \mathbb{Z}$ belong to FP. Reasonably enough, weakening polynomial complexity conditions (ii+iii) to merely recursive (ii$''$+iii$''$) makes Definition 1 boil down to the many equivalent conceptions of a computable continuous function [28, §I.0].

2 Candidate Definitions of Polynomial-Time for L^2

Let us consider the separable space $L^2[0;1]$ of complex-valued square-integrable functions on $[0;1]$. Recall that this set comes equipped with the inner product $\langle f,g\rangle = \int_0^1 f(t)\bar{g}(t)\,dt$ and induced norm $\|f\|_2 = \sqrt{\langle f,f\rangle}$, which makes $L^2[0;1]$ a Hilbert space. $L^2[0;1]$ is thus a special—and arguably the best-behaved—case of the hierarchy $L^p[0;1]$ of Banach spaces, $1 \le p \le \infty$, which in turn generalize and contain the space $\mathcal{C}[0;1]$ with norm $\|\cdot\|_\infty$ from Subsect. 1.2.

Calculus knows (at least) two common choices of countable dense subsets of $L^2[0;1]$: a) Fourier series with Gaussian dyadic rational coefficients and b) piecewise constant functions comprised of equidistant blocks of Gaussian dyadic rational heights. This suggests the following:

Definition 2. **a)** Call $f \in L^2[0;1]$ *Fourier*-computable in polynomial time iff there exists a Gaussian integer double sequence $\bar{c} = (c_{k,n}) \in \mathbb{Z} + i\mathbb{Z}$, indexed by $n \in \mathbb{N}$ and $k \in \mathbb{Z}$, such that (a i), (a ii), and (a iii) hold:

a i) $\forall n,k: \; \left|\tilde{f}_k - c_{k,n}/2^n\right| \le 2^{-n}$

a ii) Both $(n,k) \mapsto \operatorname{Re} c_{k,n}, \operatorname{Im} c_{k,n} \in \mathbb{Z}$ are computable in time $\operatorname{poly}(n+\log|k|)$.

a iii) It holds $\left\|f - \mathcal{F}_K(f)\right\|_2 \le 2^{-m}$ for $K = 2^{\operatorname{poly}(m)}$, where $\mathcal{F}_K(f,t) := \sum_{|k|\le K} \tilde{f}_k \cdot \exp(2\pi i k t)$ denotes the first $2K+1$ terms of the Fourier series with coefficients $\tilde{f}_k = \int_0^1 \exp(-2\pi i k t)\cdot f(t)\,dt$.

a′) Call f Fourier-computable in #P iff (a i), (a iii), and (a′ ii) hold:

a′ ii) Both $\{0,1\}^* \ni 1^n\,0\,\operatorname{bin}(k) \mapsto \operatorname{Re} c_{k,n}, \operatorname{Im} c_{k,n} \in \mathbb{Z}$ belong to GapP.

b) Call $f \in L^2[0;1]$ *Step*-computable in polynomial time iff there exists a Gaussian integer multi sequence $\bar{h} = (h_{S,s,n})$, indexed by $n \in \mathbb{N}$ and $0 \le s < S \in \mathbb{N}$, such that (b i), (b ii), and (b iii) hold:

b i) $\forall n,S,s: \; \left|\hat{f}_{S,s} - h_{S,s,n}/2^n\right| \le 2^{-n}$

b ii) $(S,s,n) \mapsto \operatorname{Re} h_{S,s,n}, \operatorname{Im} h_{S,s,n} \in \mathbb{Z}$ are computable in time $\operatorname{poly}(n+\log S)$.

b iii) It holds $\left\|f - \mathcal{S}_S(f)\right\|_2 \le 2^{-m}$ for $S = 2^{\operatorname{poly}(m)}$,where $\mathcal{S}_S(f,t) := \hat{f}_{S,s}$ for $\frac{s}{S} \le t < \frac{s+1}{S}$ denotes the piecewise constant function with S equidistant steps of complex heights $\hat{f}_{S,s} := S \cdot \int_{s/S}^{(s+1)/S} f(t)\,dt$.

b′) Call $f \in L^2[0;1]$ Step-computable in #P iff (b i), (b iii), and (b′ ii) hold:

b′ ii) $\{0,1\}^* \ni 1^n\,0\,\operatorname{bin}(S)\operatorname{bin}(s) \mapsto \operatorname{Re} h_{S,s,n}, \operatorname{Im} h_{S,s,n} \in \mathbb{Z}$ belong to GapP.

c) Call f Step-computable in polynomial time on *average* iff there exists a Gaussian integer multi sequence $\bar{h} = (h_{S,s,n})$ s.t. (c i), (b ii), (b iii) hold:

c i) $\sqrt{\sum_{s=0}^{S(n)-1} \left|\hat{f}_{S(n),s} - h_{S(n),s,n}/2^n\right|^2 / S(n)} \le 2^{-n}$ for $S(n) = 2^{\operatorname{poly}(n)}$.

c′) Call f Step-computable in #P on *average* iff (c i), (b′ ii), and (b iii) hold.

The sequence $(\tilde{f}_k)$ of Fourier coefficients constitutes the well-known Parseval isometry between $L^2[0;1]$ and ℓ^2. And (a iii) means that said Fourier series converges no slower than exponentially; similarly, (b iii) means that piecewise averaged step functions converge no slower than exponentially. Note that (iii) and (a iii) and (b iii) are mathematical conditions, while (ii)/(ii′) and variants (a ii)/(a′ ii)/(b ii)/(b′ ii) pertain to computation. In fact (a iii) and (b iii) have been shown equivalent in [4, Theorem 17c+a]; and (iii) implies (a/b iii) [4, Fact 12d]. Record also that $\tilde{f}_0 = \hat{f}_{1,0}$.

Similarly to Definition 1(ii′′+iii′′), weakening complexity conditions (a ii+iii)/ (b ii+iii) to merely recursive makes Definition 2 boil down to the many equivalent conceptions of *computable* L^2 function [22,28,39, §II.1]; see Corollary 10.

2.1 Step-Computability on Average

See [4, §1.3.4] for a discussion of *binary* 2^{-n} versus *unary* $1/n$ error bounds. Items (b i) and (c i) in Definition 2 both require that error $\leq 2^{-n}$ of the S-step approximation be attainable within runtime polynomial in n, but in (b) for every $S \in \mathbb{N}$ independently of n while in (c) only for $S(n) = 2^{p(n)}$ with some fixed $p \in \mathbb{N}[N]$. Put differently, (b i) requires approximating of each step #s individually up to error $\leq 2^{-n}$, whereas (c i) amounts to approximation with residual sum of squares $\leq 2^{-n}$; similarly for (b′) and (c′). In particular (c) does not require polynomial-time computability of $\hat{f}_{1,0}$, but (b) does: an observation we will employ frequently, such as in Theorem 6.

In the L^2 setting, Euclidean error norm (c i) is arguably more natural than maximum error norm (b i). Theorems 4 and 6 below apply to (c) and (c′) rather than to (b) and (b′), respectively. Note that, when (c i) holds for $S(n) = 2^{p(n)}$, then it holds also for $S(n) = 2^{p(q(n))}$ with any non-constant $q \in \mathbb{N}[N]$. *Leonid Levin*'s notion of average-case complexity [2, Definition 18.4] arises frequently in Real Computation [18,31]. Average polynomial runtime in dependence on the worst-case precision is equivalent to average polynomial precision in dependence on the worst-case runtime:

Lemma 3. *Let $\mathcal{A}$ denote an algorithm computing on input $s \in \mathbb{N}$ an infinite sequence of Gaussian rational approximations $\big(h_{s,p}\big)_{p=0}^{\infty}$ up to absolute error 2^{-p} to the s-th entry of some finite or infinite complex sequence $(h_s)_{s\geq 0}$.*

Let $\mathrm{time}_{\mathcal{A}}(\mathrm{s},\mathrm{p})$ denote the number of steps made by $\mathcal{A}$ to produce said p-th binary approximation; for $\mathrm{t} \in \mathbb{N}$ let $\mathrm{prec}_{\mathcal{A}}(\mathrm{s},\mathrm{t})$ denote the (exponent n to base $1/2$ of the) approximation error attained by $\mathcal{A}$ within t steps. Then, for $k,n,t \in \mathbb{N}$:

$$\big\{s : \mathrm{time}_{\mathcal{A}}^{1/k}(s,n) \leq t\big\} \;=\; \big\{s : \mathrm{prec}_{\mathcal{A}}(s,t^k) \geq n\big\} \ .$$

In particular, for any probability distribution w.r.t. s, expected values $\mathbb{E}_s$ satisfy:

$$\mathbb{E}_s\big[\mathrm{time}_{\mathcal{A}}^{1/k}(s,n)\big] \;\leq\; \mathcal{O}(n) \quad\Longleftrightarrow\quad \mathbb{E}_s\big[\mathrm{prec}_{\mathcal{A}}(s,t^k)\big] \;\geq\; \Omega(t) \ .$$

2.2 Comparing Notions of Polynomial-Time Computability

This section collects both relations and separations among the above notions of computational complexity for L^2-functions.

Theorem 4. *A function $f \in \mathrm{L}^2[0;1]$ is Fourier-computable in $\#\mathsf{P}$ (a′) iff f is Step-computable in $\#\mathsf{P}$ on average (c′).*

Theorem 5. *There exists a function $f \in \mathrm{L}^2[0;1]$ which is Fourier-computable in polynomial time (a), but $\hat{f}_{S,0} \in \mathbb{C}$ is not computable in polynomial time (b) for $S(n) = 2^{n^2+n}$ unless $\#\mathsf{P}_1 \subseteq \mathsf{FP}_1$.*

Theorem 6. *There exists a function $f \in \mathrm{L}^2[0;1]$ which is polynomial time Step-computable on average (c), but $\tilde{f}_0 = \hat{f}_{1,0} \in \mathbb{C}$ is not computable in polynomial time (a+b) unless $\#\mathsf{P}_1 \subseteq \mathsf{FP}_1$.*

Theorem 7. *If $f \in \mathcal{C}[0;1]$ is computable in $\#\mathsf{P}$ according to Definition 1ii′), then it is also both Fourier-computable in $\#\mathsf{P}$ (a′) and Step-computable in $\#\mathsf{P}$ (b).*

[20, Theorem 5.32de] exhibits a polynomial-time computable $f \in \mathcal{C}^\infty[0;1]$ such that $\tilde{f}_0 = \hat{f}_{1,0}$ is not computable in polynomial time (a+b) unless $\#\mathsf{P}_1 \subseteq \mathsf{FP}_1$.

Theorem 8. *If $f \in \mathcal{C}[0;1]$ is Step-computable in polynomial time (b) and has a polynomial modulus of continuity (iii), then f is computable in polynomial time.*

Recall that a convergent Fourier series need not be continuous, and the Fourier series of a continuous function need not converge pointwise.

Theorem 9. *If $f \in \mathcal{C}[0;1]$ is Fourier-computable in $\#\mathsf{P}$ (a′) and has a polynomial modulus of continuity (iii), then f is computable in $\#\mathsf{P}$.*

Corollary 10. *Theorems 4, 7 and 9 together show: If $f \in \mathcal{C}[0;1]$ has a polynomial modulus of continuity (Definitions 1iii), then all four conceptions of computability in $\#\mathsf{P}$ are equivalent: (ii′) and (a′) and (b′) and (c′).*

Theorem 11. *There exists $f \in \mathcal{C}^\infty[0;1]$ which is Fourier-computable in polynomial time (a) but $f(0)$ is not computable in polynomial time unless $\#\mathsf{P}_1 \subseteq \mathsf{FP}_1$.*

2.3 Closure Properties

Naturally all the above notions are closed under pointwise binary sum and by scaling with a polynomial-time computable constant. The following is well-known about continuous functions:

Fact 12. *(1) If $t > 0$ is polynomial-time computable and $f \in \mathcal{C}[0;1]$ is polynomial-time computable with $f(0) = f(1)$, then its cyclic shift $\tau_t f : x \mapsto f(x + t \bmod 1)$ is again polynomial-time computable.*
(2) There exists a polynomial-time computable $f \in \mathcal{C}^1[0;1]$ whose derivative f' is not computable [20, Theorem 6.4].
(3) If $f \in \mathcal{C}^1[0;1]$ is polynomial-time computable and its derivative f' satisfies (iii), then f' is again polynomial-time computable [20, Theorem 6.2b].
(4) If $f \in \mathcal{C}[0;1]$ is computable in $\#\mathsf{P}$, then so is its anti-derivative aka indefinite Riemann integral $\int f : [0;1] \ni x \mapsto \int_0^x f(y)\,dy \in \mathcal{C}^1[0;1]$.

(5) There exists a polynomial-time computable $f \in C^\infty[0;1]$ *such that* $\int f$ *is not polynomial-time computable unless* $\#\mathsf{P} \subseteq \mathsf{FP}$ [20, Theorem 5.33].

The Fourier setting is additionally closed under indefinite integration:

Theorem 13. *(1) If* $t > 0$ *is polynomial-time computable and* $f \in \mathrm{L}^2[0;1]$ *is Fourier-computable in polynomial time, then its cyclic shift* $\tau_t f \in \mathrm{L}^2[0;1]$ *is again Fourier-computable in polynomial time.*
(2) There exists $f \in \mathrm{L}^2[0;1]$ *Fourier-computable in polynomial time, whose weak derivative* f' *belongs to* $\mathrm{L}^2[0;1]$ *but is not Fourier-computable.*
(3) If $f \in \mathrm{L}^2[0;1]$ *is Fourier-computable in polynomial time and* f' *belongs to* $\mathrm{L}^2[0;1]$ *and satisfies (a iii), then* f' *is Fourier-computable in polynomial time.*
(4) If $f \in \mathrm{L}^2[0;1]$ *is Fourier-computable in polynomial time, then its indefinite integral* $\int f$ *is Fourier-computable in polynomial time as well.*

Recall that $\int f$ is Hölder-continuous for $f \in \mathrm{L}^2[0;1]$; but the *Fundamental Theorem of Calculus* "$f = \partial \int f$" may fail on points of discontinuity. Regarding Theorem 13(4), we cannot hope for polynomial-time computability of $\int f$ as a continuous function in view of Theorem 11.

Theorem 14. *(1) If* $t > 0$ *is polynomial-time computable and* $f \in \mathrm{L}^2[0;1]$ *is Step-computable in polynomial time, then its cyclic shift* $\tau_t f \in \mathrm{L}^2[0;1]$ *is again Step-computable in polynomial time.*
(2) If $f \in \mathrm{L}^2[0;1]$ *is Step-computable in polynomial time, then its indefinite integral* $\int f$ *is Step-computable in polynomial time as well.*

3 Selected Proofs

Due to page constraints, we can only include few selected proofs. Record that GapP is closed under polynomial products and exponential sums [11, Lemma 3.1].

3.1 Proof of Theorem 4: #P-Equivalence in L^2

Recall that (a iii) and (b iii) are equivalent by [4, Theorem 17c+a]. Let $e_{S,s,k} :=$

$$S \cdot \int_{\frac{s}{S}}^{\frac{s+1}{S}} \exp(2\pi i k t)\, dt \;=\; \big(\exp(2\pi i k \tfrac{s+1}{S}) - \exp(2\pi i k \tfrac{s}{S})\big) \cdot S/(2\pi i k) \;. \quad (4)$$

These complex coefficients are polynomial-time computable [37, §7.3] [4, Example 4]; more precisely: There exist Gaussian integers $e_{S,s,k,n} \in \mathbb{Z}+i\mathbb{Z}$, computable in time $\mathrm{poly}(n + \log k + \log S)$, such that $|e_{S,s,k} - e_{S,s,k,n}/2^n| \leq 2^{-n}$; for convenience let $e_{S,s,0} := 1$ and $e_{S,s,0,n} := 2^n$, so that $|e_{S,s,k}| \leq 1$ for all s, k.

First suppose (a) that $f \in \mathrm{L}^2[0;1]$ is Fourier-computable in #P: $|\tilde{f}_k - c_{k,n}/2^n| \leq 2^{-n}$ for both $\{0,1\}^* \ni 1^n\,0\,\mathrm{bin}(k) \mapsto \mathrm{Re}\, c_{k,n}, \mathrm{Im}\, c_{k,n} \in \mathbb{Z}$ in GapP.

Record that $\hat{f}_{S,s} = \tilde{f}_0 + \sum_{k\neq 0} \tilde{f}_k \cdot e_{S,s,k}$. Now let $h_{S,s,K,n} := \sum_{|k|\leq K} c_{k,m} \cdot e_{S,s,k,m'}$ for $m := n + \log K + \mathcal{O}(1)$ and $m' := n + \log K + \log F + \mathcal{O}(1)$, where $F \in \mathbb{N}$ is some constant (depending only on f) such that $\max_k |\tilde{f}_k| \leq F$. Then closure of GapP under binary multiplication and under exponential sums implies that $h_{S,s,K,n}$ belongs to GapP, since $\log S + \log K \leq \text{poly}(n)$ by (a iii) and by [4, Theorem 17ac]: establishing (b′ii). [4, Theorem 17ac] with Parseval also yield

$$\sum_{s=0}^{S-1} \left|\hat{f}_{S,s} - h_{S,s,K}\right|^2 / S \;=\; \left\|\mathcal{S}_S(\mathcal{F}_K(f)) - \mathcal{F}_K(f)\right\|_2^2 \;\leq\; 4^{-n}$$

for $h_{S,s,K} := \sum_{|k|\leq K} \tilde{f}_k \cdot e_{S,s,k}$. Moreover

$$\begin{aligned}|h_{S,s,K} - h_{S,s,K,n}/2^{m+m'}| &\leq \sum_{|k|\leq K} |\tilde{f}_k - c_{k,m}/2^m| \cdot |e_{S,s,k}| \\ &\quad + \sum_{|k|\leq K} |c_{k,m}/2^m| \cdot |e_{S,s,k} - e_{S,s,k,m'}/2^{m'}| \\ &\leq (2K+1)\cdot 2^{-m}\cdot 1 \;+\; (2K+1)\cdot(F+1)\cdot 2^{-m'}\end{aligned}$$

$\leq 2^{-n}$ by the above choices of K, S and m, m': establishing (c i), and thus Step-computability of f in #P on average (c′).

Now suppose conversely (c′) that $f \in \mathrm{L}^2[0;1]$ is Step-computable in #P on average with $h_{S,s,n}$ in GapP for $S = 2^{\text{poly}(n)}$ according to (c i) and (b′ii). Then $c_{k,S} := \widetilde{\mathcal{S}_S(f)}_k = \sum_{s=0}^{S-1} \hat{f}_{S,s} \cdot e_{S,s,-k}$ has

$$|\tilde{f}_k - c_{k,S}|^2 \;\leq\; \sum_k |\tilde{f}_k - c_{k,S}|^2 \;=\; \|f - \mathcal{S}_S(f)\|_2^2 \;\leq\; 4^{-n}$$

by Parseval and (b iii). And $g_{S,n} \in \mathrm{L}^2[0;1]$, defined by $[\frac{s}{S}; \frac{s+1}{S}) \mapsto h_{S,s,n}/2^n$, has

$$|c_{k,S} - \widetilde{(g_{S,n})}_k|^2 \;\leq\; \sum_k |c_{k,S} - \widetilde{(g_{S,n})}_k|^2 \;=\; \|\mathcal{S}_S(f) - g_{S,n}\|_2^2 \;\leq\; 4^{-n}$$

by (c i) and Parseval. Note $\widetilde{(g_{S,n})}_k = \sum_{s=0}^{S-1} h_{S,s,n}/2^n \cdot e_{S,s,-k}$.
Finally $c_{k,S,n,m'} := \sum_{s=0}^{S-1} h_{S,s,n} \cdot e_{S,s,-k,m'}$ belongs to GapP and has

$$|\widetilde{(g_{S,n})}_k - c_{k,S,n,m'}/2^{n+m'}| \;\leq\; \sum_{s=0}^{S-1} |h_{S,s,n}/2^n| \cdot |e_{S,s,-k} - e_{S,s,-k,m'}/2^{m'}|$$

$\leq S \cdot \|g_{S,n}\|_1 \cdot 2^{-m'} \leq 2^{-n}$ for $m' := n + \log S + \mathcal{O}(1) \leq \text{poly}(n)$ since $\|g_{S,n}\|_1 \leq \|g_{S,n}\|_2 \leq \|f\|_2 + 1$: showing (a i) and Fourier-computability of f in #P (a). □

3.2 Encoding Discrete Problems Into the Reals

Our proof of Theorem 5 below (and of 6, omitted) encodes discrete decision and counting problems into real numbers and functions such that efficient approximations of the latter allow to recover the former. Similar techniques are well-known [15,19,26,32]. For the unfamiliar reader we collect some subtleties of such encodings. Recall [20, §2.2] that computing a real number (in polynomial time) means approximating by dyadic rationals up to any given error bound 2^{-n} (in time polynomial in n).

Remark 15. Fix discrete decision problem $P \subseteq \mathbb{N}$. Let $r_P := \sum_{N \in P} 2^{-N} \in [0;2]$ denote the real number with characteristic function of P as binary expansion.

(1) P is decidable iff r_P is computable [20, Theorem 2.3], [37, Lemma 4.2.1].
(2) The equivalence (1) is not uniform: No algorithm can decide P from given approximations to r_P [34], [37, Theorem 4.1.13.3].
(3) The equivalence (1) does not extend to computational complexity: There exist polynomial-time computable r_P such that $P \notin \mathsf{P}$, regardless of the Millennium Prize Problem [20, Theorem 2.8].
(4) Proceeding from binary to ternary encoding, let $r'_P := \sum_{N \in P} 3^{-N}$. Then r'_P is polynomial-time computable iff $P \in \mathsf{P}$. This equivalence holds uniformly.

To see (4), note that

$$0 \in P \iff r'_P \geq 1, \qquad 0 \notin P \iff r'_P \leq 1/2 \ :$$

the two cases can be distinguished by approximating r'_P up to error $< \frac{1}{2}$. Forcing $0 \notin P$ by replacing r'_P with $r'_P - 1$ in case $0 \in P$, "$1 \in P$?" can be recovered next. Indeed it holds

$$N \in P \Leftrightarrow r'_P \geq 3^{-N}, \quad N \notin P \Leftrightarrow r'_P \leq 3^{-N}/2 = \sum\nolimits_{n>N} 3^{-n} \tag{5}$$

provided that $0, 1, \ldots, N-1 \notin P$. The latter (at most N-fold) repeated conditional subtractions of $3^{-0}, 3^{-1}, \ldots, 3^{-N+1}$ from r'_P are performed with precision 4^{-N} each: such that the two cases in Equation (5) remain numerically 'well-separated' in spite of error propagation. □

Equation (5) says that all n-th terms, $n > N$, can sum up to at most half the N-th term; which in turn affects only the first poly(N) bits i.e. up to error $2^{-\text{poly}(N)}$. The next subsection encodes a discrete counting problem into a L^2-function with similar numerical 'well-separation' (*).

3.3 Proof of Theorem 5: Fourier but Not Step Polytime

$$\hat{f}_{S,0} \;=\; \tilde{f}_0 + \sum\nolimits_{k \neq 0} \tilde{f}_k \cdot e_{S,0,k}, \quad e_{S,0,k} \;=\; \big(\exp(2\pi i k/S) - 1\big) \cdot \tfrac{S}{2\pi i k} \tag{6}$$

according to Equation (4). We also record $|e_{S,0,k}| \leq 1$ and, for $|k/S| \ll 1$, $e_{S,0,k} \approx 1$ by L'Hôpital; more precisely

$$|k/S| \leq 1 \quad \Longrightarrow \quad |e_{S,0,k} - 1| \leq |k/S| \ . \tag{7}$$

Also, $\sum_{n \geq N} 2^{-n^2} \;=\; \sum_{m \geq 0} 2^{-(N^2+2mN+m^2)} \;\leq\; 2^{-N^2+\mathcal{O}(1)}$. To construct f as in Theorem 5, fix some $\#\mathsf{P}_1$-complete $\varphi : \mathbb{N}_+ \to \mathbb{N}$ w.l.o.g. of the form

$$\varphi(m) \;=\; \mathrm{Card}\big\{\vec{w} \in \{0,1\}^m : \vec{w} \in P\big\}, \quad P \in \mathsf{P} \tag{8}$$

and let f be the Fourier series with coefficients

$$\tilde{f}_k := 2^{-m^2} \cdot \begin{Bmatrix} 1 \ \mathrm{bin}_m(w) \in P \\ 0 \ \mathrm{bin}_m(w) \notin P \end{Bmatrix}, \quad k = 2^m + w,\ w \in \{0, 1, \ldots, 2^m - 1\} \tag{9}$$

where $\mathrm{bin}_m\big(\sum_{j=0}^{m-1} b_j 2^j\big) := (b_0, \ldots, b_{m-1}) \in \{0,1\}^m$ and $\tilde{f}_k :\equiv 0$ for $k \leq 1$. Said coefficient sequence $\tilde{f}_k$ satisfies Definition 2a iii), and in fact is polynomial-time computable (Definition 2a i+ii) since $P \in \mathsf{P}$ and $\log|k| = \mathcal{O}(m)$; moreover it is square summable, thus giving rise to an L^2-function f.

Intuitively, in view of Equations (6) and (7) and (9), $\hat{f}_{S,0} \approx \sum \tilde{f}_k$ should allow for recovering $\varphi(N)$ from $\hat{f}_{S,0}$ within time polynomial in N for all sufficiently large N, and to thus conclude the proof. To make this idea rigorous note that, in $\sum_k \tilde{f}_k$, the case $\varphi(N) \geq 1$ is 'well-separated' (Subsect. 3.2) from the case $\varphi(N) = 0$; namely the weight $\tilde{f}_{2^N+W}$ of one witness W to $N \in P$ is larger than the weight of all possible witnesses w to all $n > N$ combined: For $K := 2^N + W$ with $0 \leq W < 2^N$,

$$\mathrm{bin}_N(W) \in P \quad \Rightarrow \quad \tilde{f}_K = 2^{-N^2} \overset{(*)}{\gg} 2^{-N^2-N+\mathcal{O}(1)}$$
$$= \sum_{m>N} 2^m \cdot 2^{-m^2} \geq \sum_{m>N} \sum_{w=0}^{2^m-1} \tilde{f}_{2^m+w} = \sum_{k \geq 2^{N+1}} \tilde{f}_k .$$

Moreover the two cases are also well-separated in $\hat{f}_{S,0}$ for $S := 2^{N^2+N}$:

$$\Big|\hat{f}_{S,0} - \sum_{k<2^{N+1}} \tilde{f}_k\Big| \leq \sum_{k<2^{N+1}} \underbrace{\tilde{f}_k}_{\leq 1} \cdot \underbrace{|e_{S,0,k} - 1|}_{\leq k/S \text{ by } (7)} + \sum_{k \geq 2^{N+1}} |\tilde{f}_k| \cdot \underbrace{|e_{S,0,k}|}_{\leq 1}$$
$$\leq 2^{N+1} \cdot (2^{N+1}/2^{N^2+N}) + \sum\nolimits_{k \geq 2^{N+1}} |\tilde{f}_k| \overset{(*)}{\leq} 2^{-N^2-N+\mathcal{O}(1)} .$$

4 Applications

In this section we investigate applications of Definition 2 to two important PDE examples: Heat (parabolic) and Wave (hyperbolic) equations.

4.1 Heat Equation

Consider an initial-value problem for the heat equation on the unit interval with periodic boundary conditions:

$$\begin{cases} u_t = u_{xx}, \ x \in [0, 1) \bmod 1, \ t > 0, \\ u(x, 0) = f(x), \end{cases} \tag{10}$$

where $f \in \mathrm{L}^2[0,1]$. As is well known, this problem has a unique solution u, and for any fixed $t > 0$, $u(\cdot, t) \in C^\infty[0, 1]$ (hence also in $\mathrm{H}^s[0,1]$ for all $s > 0$). Recall that $r \in \mathbb{R}$ is *polynomial-time computable* if there is an algorithm that, given $n \in \mathbb{N}$, outputs, within time polynomial in n, a sequence of $r_n \in \mathbb{Z}$ such that $|r - r_n/2^n| \leq 2^{-n}$.

Theorem 16. *Let $t > 0$ be a polynomial-time computable real number. If the initial function $f \in \mathrm{L}^2[0,1]$ is Fourier-computable in polynomial time, then the solution $u(\cdot, t)$ at time t is also Fourier-computable in polynomial time.*

Proof. The solution of (10) is given by $u(x,t) = \sum_{k \in \mathbb{Z}} \tilde{f}_k e^{-4\pi^2 k^2 t} e^{2\pi i k x}$. Thus, the Fourier coefficients of $u(\cdot, t)$ are $\tilde{u}_k(t) = \tilde{f}_k e^{-4\pi^2 k^2 t}$.

Since f is Fourier-computable in polynomial time, there exist a polynomial p and a double sequence $(c_{k,n})$ of Gaussian integers, such that a i), a ii) of Definition 2 hold, and $\left\|f - \mathcal{F}_K(f)\right\|_2 \leq 2^{-n}$ with $K = 2^{p(n)}$.

By Parseval's identity,

$$\left\|u(\cdot,t) - \mathcal{F}_K(u(\cdot,t))\right\|_2^2 = \sum_{|k|>K} |\tilde{u}_k(t)|^2 = \sum_{|k|>K} |\tilde{f}_k|^2 e^{-8\pi^2 k^2 t} \leq \sum_{|k|>K} |\tilde{f}_k|^2 \leq 2^{-2n},$$

hence $\left\|u(\cdot,t) - \mathcal{F}_K(u(\cdot,t))\right\|_2 \leq 2^{-n}$, therefore (a iii) for $u(\cdot,t)$ holds for the same $K = 2^{p(n)}$.

Now we need to construct a double sequence $(d_{k,n})$ of Gaussian integers, such that (a i) and (a ii) hold for $u(\cdot,t)$. Since t is polynomial-time computable, we can compute, within polynomial time, an integer sequence t_m such that $|t - t_m/2^m| \leq 2^{-m}$. It is also possible to compute, within polynomial time, an integer sequence $b_{k,m}$ (i.e., a sequence of Gaussian integers with zero imaginary part) such that $\left|e^{-4\pi^2 k^2 t} - b_{k,m}/2^m\right| \leq 2^{-m}$, since the exponential function is polynomial-time computable, as well as the argument $4\pi^2 k^2 t$. We also have $\left|\tilde{f}_k - c_{k,m}/2^m\right| \leq 2^{-m}$.

From a iii) with $n = 0$ we have $K_0 = 2^{p(0)}$ such that $\left\|f - \mathcal{F}_{K_0}(f)\right\|_2 \leq 1$, hence $\left\|f\right\|_2 \leq \left|\mathcal{F}_{K_0}(f)\right|_2 + 1$. Since K_0 is a constant, we can compute a rational number M with $\left\|f\right\|_2 \leq M$ in constant time (by approximating the finitely many coefficients $\tilde{f}_k$ for $|k| \leq K_0$ using the given $c_{k,1}$).

Define $\alpha_n = \left\lceil \sqrt{\frac{\ln M + n \ln 2}{4\pi^2 t}} \right\rceil$. For $|k| \geq \alpha_n$ we have $|\tilde{u}_k(t)| \leq M e^{-4\pi^2 k^2 t} \leq 2^{-n}$; set $d_{k,n} = 0$ for these k.

For $|k| < \alpha_n$ choose an auxiliary precision parameter $m = n + \lceil \log_2(M + 2) \rceil + 1$.

For each such k, let $P_{k,m} = c_{k,m} \cdot b_{k,m} \in \mathbb{Z}[i]$ and define $d_{k,n}$ by rounding $P_{k,m}/2^{2m-n}$ to the nearest Gaussian integer: $d_{k,n} = \left\lfloor \frac{P_{k,m}}{2^{2m-n}} \right\rceil$. Then $\left|\tilde{u}_k(t) - d_{k,n}/2^n\right| \leq 2^{-n}$.

Finally, let us check Condition (a ii) for u. For $|k| \geq \alpha_n$, $d_{k,n} = 0$. For $|k| < \alpha_n$, note that $\alpha_n = O(\sqrt{n})$, hence $\log|k| = O(\log n)$. Computing $c_{k,m}$ takes time polynomial in $m + \log|k| = \mathrm{poly}(n)$. Computing $b_{k,m}$ takes time polynomial in m. The multiplication and rounding are polynomial in the bit-length of $P_{k,m}$, which is $O(m)$. Thus $(n,k) \mapsto d_{k,n}$ is computable in time polynomial in $n + \log|k|$. □

Remark 17. In contrast to Theorem 16, [21, Theorem 4e/§4.3] has constructed a polynomial-time computable initial condition $f \in \mathcal{C}^\infty$ such that $u(\cdot, 1)$ is continuous but not polynomial-time computable as a continuous function unless $\#\mathsf{P}_1 \subseteq \mathsf{FP}_1$; cmp. also [3].

4.2 Wave Equation

The d-dimensional Wave Equation

$$\begin{cases} u_{tt}(\vec{x},t) \;=\; \Delta u(\vec{x},t) := \sum_{j=1}^{d} u_{jj}, & t>0, \\ u(\vec{x},0) \;=\; f(\vec{x}), \quad u_t(\vec{x},0) \;=\; g(\vec{x}), & \vec{x}\in\Omega \end{cases} \tag{11}$$

is notorious for purportedly violating the Church-Turing Hypothesis, namely as a physical system evolving from computable initial state (f, g) at $t = 0$ into an uncomputable state $u(\cdot, t)$ at a computable time $t > 0$ [7,27,29,38]. In dimension $d = 1$, the explicit solution to (11) is given by d'Alembert:

$$u(x,t) \;=\; \Big(f(x+t) \,+\, f(x-t) \,+\, \int_{x-t}^{x+t} g(y)\,dy\Big)/2 \;. \tag{12}$$

As in Subsect. 4.1, we consider the compact unit interval with periodic boundary conditions $\Omega := [0;1) \bmod 1$.

Theorem 18. *(1) If $f, g \in \mathcal{C}(\Omega)$ are computable in $\#\mathsf{P}$, then so is $u(\cdot, t)$ for every polynomial-time computable $t > 0$.*
(2) There exist polynomial-time computable $f, g \in \mathcal{C}^\infty(\Omega)$ such that $u(0, \frac{1}{2})$ is not polynomial-time computable unless $\#\mathsf{P}_1 \subseteq \mathsf{FP}_1$.
(3) If $f, g \in \mathrm{L}^2(\Omega)$ are Fourier-computable in polynomial time (a), then so is $u(\cdot, t)$ for every polynomial-time computable $t > 0$.

Proof (Theorem 18).

(1) follows from Eq. (12) and Fact 12.
(2) Take $f :\equiv 0$ and g from [20, Theorem 5.32de].
(3) follows from Equation (12) and Theorem 13. □

5 Conclusion

Choosing the 'right' basis is well-known computationally beneficial [1]. In d-dimensional Linear Algebra, say, matrix-vector multiplication in a basis of eigenvectors reduces the arithmetic cost from quadratic to linear. The present work rigorously confirms a much larger efficiency gain. We introduced two possible definitions of polynomial time computability of L^2 functions, which we have shown to be not equivalent unless $\#\mathsf{P}_1 \subseteq \mathsf{FP}_1$. One of the possible insights to the question which of them is "more suitable" could be provided by PDEs, which are a large source of applications of generalized (including integrable) functions.

For both Heat and Wave equations Fourier-polynomial computability turned out to be preserved by the solution operator (these results can be similarly extended to higher dimensions and some more general cases of boundary conditions), which is not the case for the "classical" definition of polynomial-time computability for continuous functions, unless $\#\mathsf{P}_1 \subseteq \mathsf{FP}_1$ [21]. Step-polynomial computability (and, in particular, Step-polynomial computability on average), which is well compatible with grid or finite element methods, is potentially applicable to an even broader class of PDEs.

Further systematic studies of computational complexity of various (classes of systems of) PDEs for the case of generalized solutions, as well as developing corresponding Exact Real Computation methods [25], are of interest for future work. In particular, according to [21], computing "classical" solutions to well posed evolutionary systems of PDEs with variable coefficients is PSPACE-complete, unless the coefficients and initial data are analytic functions; how about generalized solutions w.r.t. the definitions of the present paper? Theorems 16 and 18(3) give a hope for efficiency of Exact Real Computation methods for some examples of computing generalized solutions of PDEs. Also, further investigations could extend the presented L^2 polynomial-time computability notions to the case of operators, as in [16] for classical computability of continuous functions.

Remark 19. Some of our proofs/algorithms are non-uniform, namely require 'knowing' an integer upper bound F to $\|f\|_2$: such as in Theorem 4$\mathcal{F} \Rightarrow \mathcal{S}$ or in Theorem 14(2)

Question 20. (1) If $f \in \mathrm{L}^2[0;1]$ is Step-computable in polynomial time (b), is f then also Fourier-computable in polynomial time (a)?

(2) Suppose $f \in \mathrm{L}^2[0;1]$ is Step-computable in polynomial time and $g \in \mathcal{C}[0;1]$ is polynomial-time computable. Is $f \cdot g \in \mathrm{L}^2[0;1]$ then necessarily Step-computable in polynomial time again?

(3) Is there a definition of polynomial-time computability for L^2 functions with closure properties (even) better than our Fourier (a)?

Note that (2) is related to (1) via Theorem 14(2) since $g_k(t) := \exp(-2\pi ikt) \in \mathcal{C}[0;1]$ is polynomial-time computable.

Acknowledgements. A. Bacho acknowledges support from the Air Force Office of Scientific Research (AFOSR) under awards FA9550-20-1-0358 (MURI: "Machine Learning and Physics-Based Modeling and Simulation") and FA9550-24-1-0237, the U.S. Department of Energy (DOE), Office of Science, Office of Advanced Scientific Computing Research (ASCR) under award DE-SC0023163 (SEA-CROGS: "Scalable, Efficient and Accelerated Causal Reasoning Operators, Graphs and Spikes for Earth and Embedded Systems"), and the Office of Naval Research (ONR) under award N00014-25-1-2035. S. Selivanova is supported by the Mathematical Center in Akademgorodok under the agreement No 075-15-2025-349 with the Ministry of Science and Higher Education of the Russian Federation. M. Ziegler is supported by KAIST in-house grant KC30. We thank Donghyun Lim for promoting this research direction http://youtu.be/KAQi8kdwJM0

References

1. Alpert, B.K.: A class of bases in l2 for the sparse representations of integral operators. SIAM J. Math. Anal. **24**(1), 246–262 (1993)
2. Arora, S., Barak, B.: Computational Complexity: A Modern Approach. Cambridge University Press (2009)
3. Bacho, A., Boche, H., Kutyniok, G.: Complexity blowup for solutions of the laplace and the diffusion equation (2023). arXiv:2212.00693
4. Bacho, A., Ziegler, M.: Second-order parameterizations for the complexity theory of integrable functions. In: Boulier, F., Mou, C., Sadykov, T.M., Vorozhtsov, E.V. (eds.) Computer Algebra in Scientific Computing - Proc. 27th International Workshop, CASC 2025, Dubai, volume 16235 of Lecture Notes in Computer Science, pp. 27–46. Springer (2025)
5. Boche, H., Grigorescu, A., Schaefer, R.F., Vincent Poor, H.: Characterization of the complexity of computing the capacity of colored Gaussian noise channels. IEEE Trans. Commun. **72**(8), 4844–4856 (2024)
6. Boche, H., Pohl, V.: Complexity blowup in simulating analog linear time-invariant systems on digital computers. IEEE Trans. Signal Process. **69**, 5005–5020 (2021)
7. Boche, H., Pohl, V.: On non-detectability of non-computability and the degree of non-computability of solutions of circuit and wave equations on digital computers. IEEE Trans. Inf. Theory **68**(8), 5561–5578 (2022)
8. Boche, H., Volker Pohl, H., Poor, V.: Characterization of the complexity of computing the minimum mean square error of causal prediction. IEEE Trans. Inf. Theory **70**(9), 6627–6638 (2024)
9. Brattka, V., Dillhage, R.: Computability of compact operators on computable Banach spaces with bases. Math. Log. Q. **53**(4–5), 345–364 (2007)
10. Braverman, M., Cook, S.A.: Computing over the reals: foundations for scientific computing. Not. AMS **53**(3), 318–329 (2006)
11. Fortnow, L.: Counting Complexity, pp. 81–107. Springer-Verlag, Berlin, Heidelberg (1998)
12. Friedman, H.: The computational complexity of maximization and integration. Adv. Math. **53**, 80–98 (1984)
13. Grzegorczyk, A.: On the definitions of computable real continuous functions. FM **44**, 61–71 (1957)
14. Hemachandra, L.A., Ogiwara, M.: Is #P closed under subtraction? Curr. Trends Theor. Comput. Sci., 523–536 (1993)
15. Kawamura, A.: Lipschitz continuous ordinary differential equations are polynomial-space complete. Comput. Complex. **19**(2), 305–332 (2010)
16. Kawamura, A., Cook, S.: Complexity theory for operators in analysis. In: Proceedings of the 42nd ACM Symposium on Theory of Computing, STOC '10, pp. 495–502. ACM, New York (2010)
17. Kawamura, A., Steinberg, F., Ziegler, M.: On the computational complexity of the Dirichlet problem for Poisson's equation. Math. Struct. Comput. Sci. **27**(8), 1437–1465 (2017)
18. Kawamura, A., Thies, H., Ziegler, M.: Average-case polynomial-time computability of Hamiltonian dynamics. In: 43rd International Symposium on Mathematical Foundations of Computer Science, volume 117 of LIPIcs. Leibniz International Proceedings in Informatics, pp. 30:17. Schloss Dagstuhl Leibniz-Zent. Inform, Wadern (2018)

19. Ko, K.-I.: The maximum value problem and NP real numbers. JCSS **24**, 15–35 (1982)
20. Ko, K.-I.: Complexity Theory of Real Functions. Progress in Theoretical Computer Science. Birkhäuser, Boston (1991)
21. Ivan, K., Gleb, P., Svetlana, S., Martin, Z.: Bit-complexity of classical solutions of linear evolutionary systems of partial differential equations. J. Complexity, 101727 (2023)
22. Kunkle, D.: Type-2 computability on spaces of integrable functions. MLQ **50**(4,5), 417–430 (2004)
23. Lim, D., Selivanova, S., Ziegler, M.: What is a polynomial-time computable L^2 function? In: Proceedings of the 17th International Conference Computability and Complexity in Analysis (CCA), Bologna (Italy), pp. 41–42 (2020). http://cca-net.de/cca2020/CCA2020Abstracts.pdf
24. Lim, D., Ziegler, M.: Quantitative coding and complexity theory of continuous data. J. ACM **72**(1), 1–39 (2025)
25. Park, S., et al.:. Semantics, specification logic, and Hoare logic of exact real computation. Logical Methods Comput. Sci. **20**(2) (2024)
26. Pour-El, M.B., Richards, J.I.: A computable ordinary differential equation which possesses no computable solution. Ann. Math. Logic **17**, 61–90 (1979)
27. Pour-El, M.B., Richards, J.I.: The wave equation with computable initial data such that its unique solution is not computable. Adv. Math. **39**, 215–239 (1981)
28. Pour-El, M.B., Richards, J.I.: Computability in Analysis and Physics. Perspectives in Mathematical Logic. Springer, Berlin (1989)
29. Pour-El, M.B., Zhong, N.: The wave equation with computable initial data whose unique solution is nowhere computable. MLQ **43**(4), 499–509 (1997)
30. Schröder, M.: Spaces allowing type-2 complexity theory revisited. MLQ **50**(4,5), 443–459 (2004)
31. Schröder, M., Steinberg, F., Ziegler, M.: Average-case bit-complexity theory of real functions. In: Kotsireas, I.S., Rump, S.M., Yap, C.K. (eds.) MACIS 2015. LNCS, vol. 9582, pp. 505–519. Springer, Cham (2016). https://doi.org/10.1007/978-3-319-32859-1_43
32. Specker, E.: Nicht konstruktiv beweisbare Sätze der Analysis. JSL **14**(3), 145–158 (1949)
33. Florian Steinberg. Complexity theory for spaces of integrable functions. Logical Methods Comput. Sci. **13**(3), Paper No. 21, 39 (2017)
34. Turing, A.M.: On computable numbers, with an application to the Entscheidungsproblem. LMS **42**(2), 230–265 (1937)
35. Valiant, L.G.: The complexity of enumeration and reliability problems. SIAM J. Comput. **8**(3), 410–421 (1979)
36. Wagner, K.W.: Some observations on the connection between counting and recursion. Theoret. Comput. Sci. **47**, 131–147 (1986)
37. Weihrauch, K.: Computable Analysis. Springer, Berlin (2000)
38. Weihrauch, K., Zhong, N.: Is wave propagation computable or can wave computers beat the Turing machine? LMS **85**(2), 312–332 (2002)
39. Zhong, N., Zhang, B.-Y.: L^p-computability. MLQ **45**(4), 449–456 (1999)

On Computability of Ideal Lattices

Nikolay Bazhenov[1,2,3](✉), Manat Mustafa[3], and Stanislav Yun[3]

[1] Novosibirsk State University, Novosibirsk, Russia
[2] Kazan Federal University, Kazan, Russia
[3] Nazarbayev University, Astana, Kazakhstan
nickbazh@yandex.ru, {manat.mustafa,stanislav.yun}@nu.edu.kz

Abstract. Investigations of ideal lattices play an important role in classical lattice theory. For a given lattice L, its ideal lattice $\mathrm{Id}(L)$ consists of all ideals of L, ordered by set-theoretic inclusion. In this paper, we study ideal lattices from the point of view of computable structure theory. We focus on the Turing complexity of $\mathrm{Id}(L)$ for the case when the lattice L is computable and L has only countably many ideals.

For a computable lattice L, we show that either the ideal lattice $\mathrm{Id}(L)$ is uncountable, or else $\mathrm{Id}(L)$ has a hyperarithmetical copy. We establish that, in general, the lattice $\mathrm{Id}(L)$ can have an arbitrary level of hyperarithmetical complexity. We prove that the index set of computable lattices M having computably presentable $\mathrm{Id}(M)$ is Π^1_1-hard. We also discuss connections with effective presentations of domains.

Keywords: Lattice · Ideal lattice · Computable structure · Index set · Effective domain

1 Introduction

The paper investigates ideal lattices from the point of view of computable structure theory. Consider a join-semilattice $\mathcal{U} = (U, \vee)$, where $a \vee b$ is the supremum of $\{a, b\}$. Recall that a non-empty set $I \subseteq U$ is an *ideal* of $\mathcal{U}$ if I is a down-set that is closed with respect to $\vee$. The set of all ideals of $\mathcal{U}$ forms a lattice with respect to set-theoretic inclusion of ideals. We denote this *ideal lattice* by $\mathrm{Id}(\mathcal{U})$.

Studies of ideal lattices play important role in classical lattice theory [12,13]. For example, algebraic lattices are characterized as precisely those lattices that are isomorphic to $\mathrm{Id}(\mathcal{U})$ for some join-semilattice $\mathcal{U}$ having a least element (see Section I.3.15 of [12])

In this paper we focus on ideal lattices $\mathrm{Id}(\mathcal{L})$ induced by *computable lattices* $\mathcal{L}$. Here we view lattices as structures in the signature $\{\vee, \wedge\}$.

Our motivation is twofold. First, countable lattices constitute a classical topic in computable structure theory [2,9]. For example, there are a lot of known results on Boolean lattices and Boolean algebras (see, e.g., monographs [8,11]). Furthermore, the paper [15] established that the class of countable (non-distributive) lattices has a strong computability-theoretic universality property:

V. Brattka et al. (Eds.): CiE 2026, LNCS 16674, pp. 150–161, 2027.
https://doi.org/10.1007/978-3-032-31348-5_10

in particular, within this class one can realize arbitrary degree spectrum (of a structure) and arbitrary computable dimension. In recent years, there has been a renewed interest in algorithmic properties of computable distributive lattices (see, e.g., [3,5,27])

Second, we are naturally interested in the familiar class of complete lattices. Recall that a lattice $\mathcal{A}$ is *complete* iff any subset $X \subseteq \mathcal{A}$ has supremum $\sup_{\mathcal{A}} X$ and infimum $\inf_{\mathcal{A}} X$. The following fact is well-known: if a lattice $\mathcal{L}$ has a least element, then its ideal lattice $\mathrm{Id}(\mathcal{L})$ is complete (see, e.g., Section I.3.14 in [12]).

Complete lattices naturally arise in lattice theory and its applications. In particular, the concept lattices (of formal contexts) that appear in Formal Concept Analysis [10] are complete lattices. The computability-theoretic properties of concept lattices were extensively studied by Morozov and his students (see [17–20]). The recent paper [4] provided analysis of some concept lattices that arise in the theory of numberings.

Intuitively speaking, our general (informal) question can be stated as follows. How (recursion-theoretically) 'far' can the complete lattice $\mathrm{Id}(\mathcal{L})$ take us from the original *computable* lattice $\mathcal{L}$?

The paper is arranged as follows. Section 2 contains the necessary preliminaries. We work with computable lattices $\mathcal{L}$. In Sect. 3 we study the (Turing) complexity of *countable* ideal lattices $\mathrm{Id}(\mathcal{L})$. Proposition 1 shows that such $\mathrm{Id}(\mathcal{L})$ always has a hyperarithmetical copy. In Theorem 1, for a given computable ordinal α, we construct $\mathcal{L}$ such that $\mathrm{Id}(\mathcal{L})$ has no copies computable in $\mathbf{0}^{(\alpha)}$. This implies that $\mathrm{Id}(\mathcal{L})$ can have an arbitrary level of hyperarithmetical complexity. Theorem 2 shows that the index set of computable lattices $\mathcal{M}$ having computably presentable $\mathrm{Id}(\mathcal{M})$ is Π^1_1-hard.

In Sect. 4, we provide a finer analysis of lattices $\mathrm{Id}(\mathcal{L})$ that have no computable copies. We construct a countable $\mathrm{Id}(\mathcal{L})$ such that $\mathrm{Id}(\mathcal{L})$ is not computably presentable, but every proper principal ideal of $\mathrm{Id}(\mathcal{L})$, viewed as a sublattice of $\mathrm{Id}(\mathcal{L})$, admits a computable copy (Theorem 3). Our Sects. 3 and 4 use ideas from Sect. 4 of [18]. We note that every lattice $\mathcal{L}$ constructed in Theorems 1–3 have a least element, and hence, $\mathrm{Id}(\mathcal{L})$ is a complete lattice.

Finally, Sect. 5 discusses connections with effective presentations of (not necessarily countable) domains. We show that there exists a computable algebraic domain (in the sense of [6,26]) that is not homeomorphic to the space of ideals of an arbitrary join-semilattice (Proposition 3).

2 Preliminaries

As usual, $(\varphi_e)_{e\in\omega}$ denotes the standard list of all unary partial computable functions. If σ and τ are elements from $\omega^{<\omega}$, then $\sigma\widehat{\ }\tau$ denotes the concatenation of σ and τ. By $|\sigma|$ we denote the length of σ. Preliminaries on computability theory can be found, e.g., in [21,23,24].

We refer to the monographs [2,8,9] for the background on computable structure theory. Without loss of generality, we may assume that every infinite computable structure has domain ω.

For a tree $T \subseteq \omega^{<\omega}$, by $[T]$ we denote the set of all (infinite) paths through T. For two trees $S, T \subseteq \omega^{<\omega}$, we define their *product tree*

$$V := S * T \tag{1}$$

as follows. A string $\sigma \in \omega^{<\omega}$ belongs to V if and only if

$$\sigma(0)^\frown\sigma(2)^\frown\sigma(4)^\frown\dots^\frown\sigma(2\cdot[(|\sigma|-1)/2]) \in S, \text{ and}$$
$$\sigma(1)^\frown\sigma(3)^\frown\sigma(5)^\frown\dots^\frown\sigma(2\cdot[|\sigma|/2]-1) \in T,$$

where $[m/2]$ denotes the integer part of the rational number $m/2$.

If $\sigma \in \omega^{<\omega}$ and $\theta \in \omega^{<\omega} \cup \omega^\omega$, then $\sigma \subset \theta$ means that σ is a proper initial segment of θ.

If $\mathcal{L} = (L, \vee, \wedge)$ is a lattice, then its standard ordering $\leq_\mathcal{L}$ is recovered as follows: $x \leq_\mathcal{L} y$ if and only if $x \vee y = y$.

For an element $a \in \mathcal{L}$, by $\downarrow a$ we denote the principal ideal

$$\downarrow a = \{x \in \mathcal{L} : x \leq_\mathcal{L} a\}.$$

It is well-known that the map $a \mapsto \downarrow a$ is an isomorphic embedding from $\mathcal{L}$ into $\mathrm{Id}(\mathcal{L})$ (see, e.g., Corollary 7 in Section I.3.4 of [12]).

Recall that a lattice $\mathcal{L}$ is *bounded* if it has a least element $0_\mathcal{L}$ and a greatest element $1_\mathcal{L}$. Sometimes (when it is clear from the context) we view bounded lattices as structures in the signature $\{\vee, \wedge, 0, 1\}$.

By ① we denote the one-element lattice. For further background on lattices, we refer to the monograph [12].

2.1 Sums of Lattices

Let $\mathcal{M}$ be a partial order, and let $(\mathcal{P}_i)_{i\in\mathcal{M}}$ be a family of partial orders. The $\mathcal{M}$-*sum* of the sequence $(\mathcal{P}_i)_{i\in\mathcal{M}}$ is a partial order $\mathbf{Sum}(\mathcal{P}_i;\mathcal{M})$ that is defined as follows.

- The domain of $\mathbf{Sum}(\mathcal{P}_i;\mathcal{M})$ is equal to $\{(x,i) : i \in \mathcal{M},\ x \in \mathcal{P}_i\}$.
- We have $(x,i) \leq (y,j)$ if and only if either $i <_\mathcal{M} j$, or $i = j$ and $x \leq_{\mathcal{P}_i} y$.

The following fact will be useful for us:

Lemma 1 (folklore). *Suppose that $\mathcal{M}$ is a lattice, and every $\mathcal{P}_i$, $i \in \mathcal{M}$, is a bounded lattice. Then the sum $\mathbf{Sum}(\mathcal{P}_i;\mathcal{M})$ is also a lattice. In addition, if $\mathcal{M} = (M, \vee, \wedge)$ is a computable lattice and $\mathcal{P}_i = (P_i, \vee_i, \wedge_i, 0_{\mathcal{P}_i}, 1_{\mathcal{P}_i})$, $i \in M$, are uniformly computable bounded lattices, then the lattice $\mathbf{Sum}(\mathcal{P}_i;\mathcal{M})$ has a computable copy.*

Proof. Let $a = (x,i)$ and $b = (y,j)$ be incomparable elements from $\mathbf{Sum}(\mathcal{P}_i;\mathcal{M})$. If $i \neq j$, then the indices i and j are $\leq_\mathcal{M}$-incomparable. We choose elements $k = \sup_\mathcal{M}(i,j)$ and $\ell = \inf_\mathcal{M}(i,j)$. Then the element $(0_{\mathcal{P}_k}, k)$ is supremum of $\{a,b\}$, and $(1_{\mathcal{P}_\ell}, \ell)$ is infimum of $\{a,b\}$.

If $i = j$, then inside $\mathcal{M}_i$, we take the elements $u = x \vee_i y$ and $v = x \wedge_i y$. The elements (u, i) and (v, i) are supremum and infimum of $\{a, b\}$, respectively.

An analysis of the argument given above shows that computability of the lattice $\mathcal{M}$ and uniform computability of bounded lattices $\mathcal{P}_i$ implies that one can construct a computable copy of the lattice $\mathbf{Sum}(\mathcal{P}_i; \mathcal{M})$. □

If we choose our $\mathcal{M}$ as the linear order ω, then the sum $\mathbf{Sum}(\mathcal{P}_i; \omega)$ will be also denoted by

$$\sum_{i \in \omega} \mathcal{P}_i = \mathcal{P}_0 \oplus \mathcal{P}_1 \oplus \mathcal{P}_2 \oplus \dots .$$

We also use similar notations for the case of a finite linear order $\mathcal{M}$.

Remark 1. Selivanov [25] has constructed a strongly computable family $(\mathcal{R}_i)_{i \in \omega}$ of finite distributive lattices such that for all $i \in \omega$, the lattice $\mathcal{R}_i$ is not isomorphically embeddable (via a lattice embedding) into any finite sum

$$\mathcal{R}_{j_0} \oplus \mathcal{R}_{j_1} \oplus \dots \oplus \mathcal{R}_{j_k}, \text{ where } i \notin \{j_0, j_1, \dots, j_k\}$$

(see also the proof of Lemma 3.1 in [3] for further discussion). In what follows, we will use $\mathcal{R}_i$ only to denote Selivanov's i-th lattice.

3 Hyperarithmetical Ideal Lattices

Firstly, we show that every lattice of the form $\mathrm{Id}(\mathcal{L})$, where $\mathcal{L}$ is computable, either has a hyperarithmetical copy, or is uncountable.

Proposition 1. *Suppose that $\mathcal{L}$ is a computable lattice such that its ideal lattice* $\mathrm{Id}(\mathcal{L})$ *is countable. Then* $\mathrm{Id}(\mathcal{L})$ *has a hyperarithmetical copy.*

Proof. Without loss of generality, we assume that the domain of $\mathcal{L}$ is equal to ω. Then the set of all ideals of $\mathcal{L}$ is a Δ^0_2 class (in Cantor space 2^ω). Indeed, $A \subseteq \omega$ is an ideal if and only if

$$\exists x (x \in A) \,\&\, \forall x \forall y (x \leq y \,\&\, y \in A \to x \in A) \,\&\, \forall x \forall y (x, y \in A \to x \vee y \in A). \quad (2)$$

By Perfect Set Theorem (see Theorem III.6.3 in [24]), if $P \subset 2^\omega$ is a non-empty countable Σ^1_1 class, then there exists a computable ordinal α such that $X \leq_T \mathbf{0}^{(\alpha)}$ for all $X \in P$. Hence, since $\mathrm{Id}(\mathcal{L})$ is countable, we can choose a computable α such that $A \leq_T \mathbf{0}^{(\alpha)}$ for all $A \in \mathrm{Id}(\mathcal{L})$. Using Eq. (2), it is straightforward to show that there exists a total $\mathbf{0}^{(\alpha+2)}$-computable function $f(x)$ such that the list $(\varphi^{\emptyset^{(\alpha)}}_{f(i)})_{i \in \omega}$ enumerates the characteristic functions of all sets from $\mathrm{Id}(\mathcal{L})$, without repetitions. Hence, the lattice $\mathrm{Id}(\mathcal{L})$ has a $\mathbf{0}^{(\alpha+3)}$-computable copy. □

Our next result shows that, in general, countable lattices $\mathrm{Id}(\mathcal{L})$ can have arbitrary level of hyperarithmetical complexity.

Theorem 1. *Let α be a computable ordinal. There exists a computable lattice $\mathcal{L}$ such that the ideal lattice* $\mathrm{Id}(\mathcal{L})$ *is countable and* $\mathrm{Id}(\mathcal{L})$ *has no* $\mathbf{0}^{(\alpha)}$*-computable copies.*

Proof. Recall that a function $f \in \omega^\omega$ is a Π^0_1 *function singleton* if there exists a computable tree $S \subset \omega^{<\omega}$ such that $[S] = \{f\}$, that is, f is a unique (infinite) path through the tree S.

It is well-known that for every computable ordinal β, there exists a Π^0_1 function singleton f such that $f \equiv_T \mathbf{0}^{(\beta)}$ (we refer to Corollary II.4.3 in [24] and to [16], see also Exercise XII.2.14 in [22]). Hence, we choose a computable tree S such that S has a unique path $f \equiv_T \mathbf{0}^{(\alpha+1)}$.

In a standard way, we transform the tree $S \subset \omega^{<\omega}$ into a computable lattice $\mathcal{S}^1$ as follows:

- we add an external greatest element 1_S,
- for $x, y \in S$, we have $x <_S y$ if and only if x is an ancestor of y in the tree (i.e., the string x is a proper initial segment of y).

Note the following. If x and y are incomparable elements from S, then

$$\sup_{\mathcal{S}^1}(x, y) = 1_S$$

and $\inf_{\mathcal{S}^1}(x, y)$ is the longest string z such that $z <_S x$ and $z <_S y$.

Without loss of generality, we may identify the domain of the lattice $\mathcal{S}^1$ with the set ω, and we assume that 1 is the greatest element in $\mathcal{S}^1$. Recall that $\mathcal{R}_i$, $i \in \omega$, are Selivanov's finite lattices from Remark 1. We set

$$\tilde{\mathcal{R}}_i := \begin{cases} \mathcal{R}_i, & \text{if } i \neq 1, \\ \textcircled{1}, & \text{if } i = 1. \end{cases} \tag{3}$$

By Lemma 1, the lattice $\mathcal{L} := \mathbf{Sum}(\tilde{\mathcal{R}}_i; \mathcal{S}^1)$ has a computable copy.

Consider the ideal lattice $\mathrm{Id}(\mathcal{L})$. Every ideal $I \in \mathrm{Id}(\mathcal{L})$ satisfies precisely one of the following three cases:

(a) I is a principal ideal of the form $\downarrow (x, i)$ for some $i \neq 1$ and $x \in \mathcal{R}_i$, and hence I is a finite set.
(b) I contains two elements (x, i) and (y, j) such that i and j are $\leq_{\mathcal{S}^1}$-incomparable. Then I also contains the top element $(1_{\tilde{\mathcal{R}}_1}, 1)$, and hence, $I = \mathrm{dom}(\mathcal{L})$.
(c) Otherwise, the ideal I is equal to the set

$$I_f = \{(x, \sigma) : x \in \mathcal{R}_\sigma,\ \sigma \subset f\}, \tag{4}$$

where f is the unique path through the tree S.

We deduce that the lattice $\mathrm{Id}(\mathcal{L})$ is countable.

Towards a contradiction, assume that $\mathcal{N}$ is a $\mathbf{0}^{(\alpha)}$-computable copy of the lattice $\mathrm{Id}(\mathcal{L})$. Consider the element $a_0 \in \mathcal{N}$ such that a_0 is the isomorphic image of the ideal I_f from Eq. (4). Then the set

$$X := \{\sigma : \mathcal{R}_\sigma \text{ is isomorphically embeddable into the lattice } (\downarrow a_0, \vee_\mathcal{N}, \wedge_\mathcal{N})\}$$

is c.e. with respect to $\mathbf{0}^{(\alpha)}$. On the other hand, the main property of Selivanov's lattices (see Remark 1) implies that

$$X = \{\sigma : \sigma \subset f\}.$$

We deduce that $\mathbf{0}^{(\alpha+1)} \equiv_T f \leq_T \mathbf{0}^{(\alpha)}$, and this gives a contradiction. Therefore, the lattice $\mathrm{Id}(\mathcal{L})$ has no $\mathbf{0}^{(\alpha)}$-computable copies. Theorem 1 is proved. □

Fix a (standard) computable list $(\mathcal{M}_e)_{e\in\omega}$ of all computable structures in the signature $\{\vee, \wedge\}$. We use the construction of Theorem 1 to obtain the following further result.

Theorem 2. *The index set*

$$C = \{e \in \omega : \mathcal{M}_e \text{ is a computable lattice such that its ideal lattice } \mathrm{Id}(\mathcal{M}_e) \text{ has a computable copy}\}$$

is Π^1_1-hard (i.e., every Π^1_1 set $A \subseteq \omega$ is many-one reducible to C).

Proof. For the ordinal $\alpha = 0$, fix the computable tree $S \subset \omega^{<\omega}$ from the proof of Theorem 1. Let $f_0 \in \omega^\omega$ be the unique path through S. Note that $f_0 \equiv_T \mathbf{0}'$.

Let $A \subseteq \omega$ be an arbitrary Π^1_1 set. It is well-known (see, e.g., Theorem 16.XX in [23]) that there exists a computable sequence of computable trees $(U_e)_{e\in\omega}$ such that $U_e \subseteq \omega^{<\omega}$ and for all e,

$$e \in A \Leftrightarrow U_e \text{ is well-founded.} \tag{5}$$

For an index $e \in \omega$, we define a new computable tree $V_e := S \ast U_e$ (recall Eq. (1)).

For the tree V_e, we apply the construction of Theorem 1, and obtain computable lattices $\mathcal{V}^1_e$ and $\mathcal{L}_e = \mathbf{Sum}(\tilde{\mathcal{R}}_i; \mathcal{V}^1_e)$.

If an index e does not belong to the set A, then by Eq. (5), the tree U_e has an infinite path g. The tree V_e has an infinite path f such that $f(2k) = f_0(k)$ and $f(2k+1) = g(k)$ for all $k \in \omega$. Notice that $\mathbf{0}' \equiv_T f_0 \leq_T f$. Then the family $\mathrm{Id}(\mathcal{L}_e)$ contains the ideal I_f from Eq. (4). An argument similar to that of Theorem 1 shows that the lattice $\mathrm{Id}(\mathcal{L}_e)$ cannot have a computable copy.

Suppose that $e \in A$. Then both trees U_e and V_e are well-founded. An analysis of the ideals belonging to $\mathrm{Id}(\mathcal{L}_e)$ then shows that every ideal of $\mathcal{L}_e$ is principal, and hence, the lattice $\mathrm{Id}(\mathcal{L}_e)$ is isomorphic to the computable structure $\mathcal{L}_e$.

We conclude that $e \in A$ if and only if the lattice $\mathrm{Id}(\mathcal{L}_e)$ has a computable copy. Therefore, the map $e \mapsto \mathcal{L}_e$ induces a many-one reduction from A to our index set C. Theorem 2 is proved. □

The proof of Theorem 2 implies the following:

Corollary 1. *The index set*

$$\widehat{C} = \{e \in \omega : \mathcal{M}_e \text{ is a computable lattice such that its ideal lattice } \mathrm{Id}(\mathcal{M}_e) \text{ is countable}\}$$

is Π^1_1-complete (i.e., $\widehat{C}$ is a Π^1_1 set, and $\widehat{C}$ is Π^1_1-hard).

Proof (sketch). First, we establish that $\widehat{C}$ is Π^1_1-hard. In the proof of Theorem 2, we need to replace the tree $V_e = S * U_e$ with $\widehat{V}_e = \omega^{<\omega} * U_e$. If $e \notin A$, then the tree $\widehat{V}_e$ has continuum many paths, and hence the corresponding ideal lattice $\mathrm{Id}(\mathcal{L}_e)$ has size continuum. If $e \in A$, then $\widehat{V}_e$ is well-founded and we have $\mathrm{Id}(\mathcal{L}_e) \cong \mathcal{L}_e$.

Second, we show that $\widehat{C}$ belongs to the class Π^1_1. The proof of Proposition 1 implies that $e \in \widehat{C}$ if and only if $\mathcal{M}_e$ is a computable lattice satisfying the following condition:

$$(\exists X \in \mathrm{HYP})\, \forall A\, (A \text{ is an ideal of } \mathcal{M}_e \rightarrow A \leq_T X). \tag{6}$$

By the Spector–Gandy Theorem (see Theorem III.3.5 in [24]), Eq. (6) is equivalent to a Π^1_1-condition. Hence, $\widehat{C}$ is a Π^1_1-set. □

We leave open whether the index set C from Theorem 2 is Π^1_1-complete.

4 Lattices with All Proper Principal Ideals Computable

We note that the ideal lattice $\mathrm{Id}(\mathcal{L})$ constructed in Theorem 1 has the following property:

(∗) There exists an element $I_f \in \mathrm{Id}(\mathcal{L})$ such that $I_f \neq \mathcal{L}$ and the corresponding principal ideal (viewed as a sublattice of $\mathrm{Id}(\mathcal{L})$) $(\downarrow I_f, \vee, \wedge)$ does not have a computable copy.

Hence, it is natural to ask the following question:

> Does there exist a computable lattice $\mathcal{L}$ such that $\mathrm{Id}(\mathcal{L})$ has no computable copies, *but* every proper principal ideal of $\mathrm{Id}(\mathcal{L})$ has a computable copy?

We give a positive answer to this question:

Theorem 3. *There exists a computable lattice $\mathcal{L}$ such that:*

- *its ideal lattice* $\mathrm{Id}(\mathcal{L})$ *is countable,*
- $\mathrm{Id}(\mathcal{L})$ *has no computable copies,*
- *every proper principal ideal of* $\mathrm{Id}(\mathcal{L})$, *viewed as a sublattice of* $\mathrm{Id}(\mathcal{L})$, *has a computable copy.*

Proof. Essentially following [14], we say that a tree $T \subset \omega^{<\omega}$ is a *generalized PAC tree* (where PAC stands for 'paths all computable') iff:

- T is a c.e. tree with no dead ends (i.e., for every $\sigma \in T$ at least one of the elements $\sigma^\frown k$, $k \in \omega$, belongs to T), and
- every path in $[T]$ is computable.

Let $\mathcal{X}$ be a countable subset of ω^ω. A *computable listing* of $\mathcal{X}$ is a uniformly computable list $(f_i)_{i \in \omega}$ of computable functions such that $\mathcal{X} = \{f_i : i \in \omega\}$. Note that the list may contain repetitions: that is, it could be the case that $f_i = f_j$ for some $i \neq j$. Morozov [18] proved the following:

Lemma 2 (Lemma 2 of [18]). *There exists a generalized PAC tree $T \subset \omega^{<\omega}$ such that the family $[T]$ has no computable listing.*

Similarly to the proof of Theorem 1, we transform the c.e. tree T from Lemma 2 into a lattice $\mathcal{T}^1 = (T \cup \{1_T\}, \vee, \wedge)$ by appending the external greatest element 1_T. As usual, we choose a computable bijection θ acting from ω onto the set $\mathrm{dom}(\mathcal{T}^1)$. By identifying a number $x \in \omega$ with $\theta(x) \in \mathcal{T}^1$, we may assume that $\mathcal{T}^1$ is a computable lattice having domain ω.

Consider the finite lattices $\tilde{\mathcal{R}}_i$ from Eq. (3). By Lemma 1, the lattice $\mathcal{L} := \mathbf{Sum}(\tilde{\mathcal{R}}_i; \mathcal{T}^1)$ has a computable copy.

Consider the corresponding ideal lattice $\mathrm{Id}(\mathcal{L})$. Every element $I \in \mathrm{Id}(\mathcal{L}) \setminus \{\mathcal{L}\}$ satisfies one of the following two cases:

(a) I is a finite set, and hence, the lattice $(\downarrow I, \vee, \wedge)$ obviously has a computable copy.
(b) Otherwise, I is equal to $\{(x, \sigma) : x \in \mathcal{R}_\sigma,\ \sigma \subset f\}$ for some path f through the tree T. Since every $f \in [T]$ is computable, it is straightforward to build a computable copy of the lattice

$$(\downarrow I, \vee, \wedge) \cong (\mathcal{R}_{f(0)} \oplus \mathcal{R}_{f(0)^\frown f(1)} \oplus \mathcal{R}_{f(0)^\frown f(1)^\frown f(2)} \oplus \dots) \oplus \textcircled{1}.$$

Therefore, $\mathrm{Id}(\mathcal{L})$ is a countable lattice, and every proper principal ideal of $\mathrm{Id}(\mathcal{L})$ has a computable copy.

Now, towards a contradiction, assume that $\mathcal{N}$ is a computable copy of the lattice $\mathrm{Id}(\mathcal{L})$. Given an arbitrary element $a \neq 1_{\mathcal{N}}$ from $\mathcal{N}$, we construct a c.e. set $Y_a \subset \omega^{<\omega}$ as follows.

At stage 0, put $b[0] := a$. Consider a stage $s+1$. Suppose that there exist a string $\sigma \in \omega^{<\omega}$ and elements $d_0, d_1, \dots, d_k$ such that:

- $d_i \leq s+1$,
- $d_i \leq_{\mathcal{N}} b[s]$, and
- the elements d_i, $i \leq k$, form a finite sublattice of $\mathcal{N}$ that is isomorphic to $\mathcal{R}_\sigma$.

Then we enumerate (every) such σ into the set Y_a.

If there exists (the least) $u \leq s+1$ such that $b[s] <_{\mathcal{N}} u$ and $u \neq 1_{\mathcal{N}}$, then we put $b[s+1] := u$. Otherwise, set $b[s+1] := b[s]$. This concludes the construction of Y_a.

We show that for every $a \neq 1_{\mathcal{N}}$, the set Y_a enumerates an infinite path through the tree T. Fix an isomorphism $\Psi \colon \mathcal{N} \to \mathrm{Id}(\mathcal{L})$.

First, suppose that there exists $s \in \omega$ such that the ideal $I := \Psi(b[s])$ satisfies Case (b) discussed above. Then it is clear that the set Y_a enumerates precisely the path f (induced by I) discussed in Case (b).

Otherwise, for every $s \in \omega$, the ideal $J_s := \Psi(b[s])$ satisfies Case (a) above. This implies that J_s is a principal ideal of the form $\downarrow (x_s, \sigma_s)$ for some $\sigma_s \in T$ and $x_s \in \mathcal{R}_{\sigma_s}$. Since the tree T has no dead ends, for every σ_s there exists a string $\tau \in T$ such that τ extends σ_s. This fact implies that for every $s \in \omega$, there exists a stage $s' > s$ such that $b[s'] >_{\mathcal{N}} b[s]$ and $\sigma_s \subset \sigma_{s'}$. We choose a sequence of stages $s_0 < s_1 < s_2 < \dots$ such that $\sigma_{s_k} \subset \sigma_{s_{k+1}}$ for all $k \in \omega$. Then one can deduce that the set Y_a is equal to $\{\tau : \tau \subset \sigma_{s_k}$ for some $k \in \omega\}$, and hence Y_a enumerates a path through T.

Given $a \neq 1_{\mathcal{N}}$, we put $f_a := \bigcup_{\sigma \in Y_a} \sigma$. Then the list $(f_a)_{a \neq 1_{\mathcal{N}}}$ induces a computable listing of the family $[T]$, which contradicts our original choice of the tree T. We deduce that the lattice $\mathrm{Id}(\mathcal{L})$ has no computable copies. Theorem 3 is proved. □

5 Connections to Effective Domains

The previous two sections have discussed in detail how to deal with effective presentations of *countable* ideal lattices $\mathrm{Id}(\mathcal{L})$. On the other hand, it is also natural to ask the following question: can one effectively present uncountable lattices $\mathrm{Id}(\mathcal{L})$? In the current section, we briefly discuss a possible approach to this question—here we view lattices $\mathrm{Id}(\mathcal{L})$ as *domains*.

We refer to [1] for the preliminaries on domain theory. Following [6], we recall some background on effective presentations of domains.

Let $\prec_A$ be a transitive relation on a set A. Recall that a non-empty subset $B \subseteq A$ is *directed* iff for any $a, b \in B$ there exists an element $c \in B$ such that $a \prec_A c$ and $b \prec_A c$.

A poset $(A, \leq_A)$ is *directed-complete* iff every directed (with respect to $\leq_A$) subset $B \subseteq A$ has supremum $\sup B$.

Let $\mathbb{X}$ be a T_0-space. The *specialization order* $\leq_{\mathbb{X}}$ is defined as follows: points $x, y \in \mathbb{X}$ satisfy $x \leq_{\mathbb{X}} y$ if and only if for every open set $U \subseteq \mathbb{X}$, $x \in U$ implies $y \in U$. An element $x \in \mathbb{X}$ is *compact* if the set

$$\uparrow x := \{y : x \leq_{\mathbb{X}} y\}$$

is open. By $K(\mathbb{X})$ we denote the set of all compact elements of $\mathbb{X}$.

Definition 1. *A T_0-space $\mathbb{X}$ is an* algebraic domain *if the poset $(\mathbb{X}, \leq_{\mathbb{X}})$ is directed-complete, and the sets $\uparrow b$, where $b \in K(\mathbb{X})$, form a base of the topology in $\mathbb{X}$. An ω-*algebraic domain *is an algebraic domain with countably many compact elements.*

Definition 2 ([26])**.** *A* computable algebraic domain *is a pair $(\mathbb{X}, \nu)$, where $\mathbb{X}$ is an ω-algebraic domain and ν is a surjective map from ω onto $K(\mathbb{X})$ such that the relation*

$$\{(i, j) : \nu(i) \leq_{\mathbb{X}} \nu(j)\}$$

is computable.

Next, we consider spaces of ideals:

Definition 3 ([7])**.** *Let $\prec$ be a transitive relation on the set ω. A subset $I \subseteq \omega$ is an* ideal *(for $\prec$) if:*

(a) $I \neq \emptyset$;
(b) *if* $b \in I$ *and* $a \prec b$, *then* $a \in I$;
(c) *the set* I *is directed.*

The family $\mathbf{I}(\prec)$ *of all ideals is viewed as a topological space, where the topology is generated by basic open sets of the form*

$$[n]_{\prec} := \{I \in \mathbf{I}(\prec) : n \in I\}, \ n \in \omega.$$

Remark 2. Suppose that $\mathcal{U} = (U, \vee)$ is a join-semilattice. Then the notion of an ideal (from Definition 3) for the relation $\leq_{\mathcal{U}}$ coincides with the classical notion of an ideal of a join-semilattice. Hence, one can view the family $\mathrm{Id}(\mathcal{U})$ as the topological space $\mathbf{I}(\leq_{\mathcal{U}})$.

We will use the following fact connecting Definitions 1 and 2 to spaces $\mathbf{I}(\prec)$:

Proposition 2 ([26], see also Sect. 5.1 of [6]). (1) *An infinite topological space* $\mathbb{X}$ *is an* ω*-algebraic domain if and only if* $\mathbb{X}$ *is homeomorphic to the space* $\mathbf{I}(\sqsubseteq)$ *for some partial order* $\sqsubseteq$ *on* ω.

(2) *If* $\sqsubseteq$ *is a computable partial order on* ω, *then the pair* $(\mathbf{I}(\sqsubseteq), \nu)$, *where*

$$\nu(i) = \downarrow i = \{j \in \omega : j \sqsubseteq i\}, \tag{7}$$

is a computable algebraic domain.

Proposition 2.(2) immediately implies the following:

Corollary 2. *Let* $\mathcal{U} = (\omega, \vee)$ *be a computable join-semilattice. Then the pair* $(\mathrm{Id}(\mathcal{U}), \nu) = (\mathbf{I}(\leq_{\mathcal{U}}), \nu)$, *where* ν *is taken from Eq. (7) is a computable algebraic domain.*

We note that in a certain strict sense, Proposition 2.(2) characterizes *all* computable algebraic domains, up to effective homeomorphisms (see Proposition 4 in [6] for a precise statement). On the other hand, we show that the algebraic domains induced by join-semilattices (as in Corollary 2) cannot give such a characterization.

Proposition 3. *Suppose that* $\sqsubseteq$ *is a computable partial order on* ω *such that the set* ω *is not directed with respect to* $\sqsubseteq$. *Then* $(\mathbf{I}(\sqsubseteq), \nu)$ *is a computable algebraic domain such that the space* $\mathbf{I}(\sqsubseteq)$ *is not homeomorphic to the space* $\mathrm{Id}(\mathcal{A})$ *for any join-semilattice* $\mathcal{A}$.

Proof. Since the set ω is not directed with respect to $\sqsubseteq$, ω does not belong to $\mathbf{I}(\sqsubseteq)$. Therefore, for every ideal $I \in \mathbf{I}(\sqsubseteq)$, there exists $n_I \in \omega$ such that $n_I \notin I$ and hence, I does not belong to the basic open set $[n_I]_{\sqsubseteq}$. Notice that $[n_I]_{\sqsubseteq}$ is non-empty, since the principal ideal $\downarrow n_I$ belongs to $[n_I]_{\sqsubseteq}$. We conclude that for every element $I \in \mathbf{I}(\sqsubseteq)$, there exists a non-empty open set $V \subset \mathbf{I}(\sqsubseteq)$ satisfying $I \notin V$.

On the other hand, for an arbitrary join-semilattice $\mathcal{A} = (A, \vee)$, the set A is always an ideal of $\mathcal{A}$. Therefore, the element A belongs to *all* non-empty open sets $V \subseteq \mathrm{Id}(\mathcal{A})$. Hence, we deduce that the topological spaces $\mathbf{I}(\sqsubseteq)$ and $\mathrm{Id}(\mathcal{A}) = \mathbf{I}(\leq_{\mathcal{A}})$ cannot be homeomorphic. □

Acknowledgments. We are grateful to the anonymous reviewers for their helpful comments. The work was funded by the Science Committee of the Ministry of Science and Higher Education of the Republic of Kazakhstan (Grant No. AP26198617). The work was also supported by Nazarbayev University Faculty Development Competitive Research Grants 201223FD8823. The work of Bazhenov was supported by the Russian Science Foundation (project no. 24-11-00227).

Disclosure of Interests. The authors have no competing interests to declare that are relevant to the content of this article.

References

1. Abramsky, S., Jung, A.: Domain theory. In: Abramsky, S., Gabbay, D.M., Maibaum, T.S.E. (eds.) Handbook of Logic in Computer Science, vol. 3, pp. 1–168. Oxford University Press (1994)
2. Ash, C.J., Knight, J.F.: Computable structures and the hyperarithmetical hierarchy. Stud. Logic Found. Math. **144** (2000)
3. Bazhenov, N.A., Frolov, A.N., Kalimullin, I.S., Melnikov, A.G.: Computability of distributive lattices. Sib. Math. J. **58**(6), 959–970 (2017). https://doi.org/10.1134/S0037446617060052
4. Bazhenov, N., Mustafa, M., Nurakunov, A.: On concept lattices for numberings. Tsinghua Sci. Technol. **29**(6), 1642–1650 (2024). https://doi.org/10.26599/TST.2023.9010102
5. Bazhenov, N., Mustafa, M., Yamaleev, M.: Computable isomorphisms of distributive lattices. In: Gopal, T.V., Watada, J. (eds.) TAMC 2019. LNCS, vol. 11436, pp. 28–41. Springer, Cham (2019). https://doi.org/10.1007/978-3-030-14812-6_3
6. de Brecht, M., Kihara, T., Selivanov, V.: Ideal presentations and numberings of some classes of effective quasi-Polish spaces. Computability **13**(3–4), 325–348 (2024). https://doi.org/10.3233/COM-230442
7. de Brecht, M., Pauly, A., Schröder, M.: Overt choice. Computability **9**(3–4), 169–191 (2020). https://doi.org/10.3233/COM-190253
8. Downey, R.G., Melnikov, A.: Computable Structure Theory: A Unified Approach. Springer, Cham (2026)
9. Ershov, Y.L., Goncharov, S.S.: Constructive Models. Kluwer Academic, New York (2000)
10. Ganter, B., Wille, R.: Formal Concept Analysis: Mathematical Foundations. Springer, Berlin (1999)
11. Goncharov, S.S.: Countable Boolean Algebras and Decidability. Consultants Bureau, New York (1997)
12. Grätzer, G.: Lattice Theory: Foundation. Springer, Basel (2011)
13. Grätzer, G., Wehrung, F. (eds.): Lattice Theory: Special Topics and Applications, vol. 1. Springer, Basel (2014)
14. Hirschfeldt, D.R.: Computable trees, prime models, and relative decidability. Proc. Amer. Math. Soc. **134**(5), 1495–1498 (2006). https://doi.org/10.1090/S0002-9939-05-08097-4
15. Hirschfeldt, D.R., Khoussainov, B., Shore, R.A., Slinko, A.M.: Degree spectra and computable dimensions in algebraic structures. Ann. Pure Appl. Logic **115**(1–3), 71–113 (2002). https://doi.org/10.1016/S0168-0072(01)00087-2

16. Jockusch, C.G., McLaughlin, T.G.: Countable retracing functions and Π_2^0 predicates. Pacific J. Math. **30**(1), 67–93 (1969). https://doi.org/10.2140/pjm.1969.30.67
17. Levental, S.M.: Study of a universal formal context. Sib. Math. J. **53**(5), 810–820 (2012). https://doi.org/10.1134/S0037446612050072
18. Morozov, A.S.: On effective presentations of formal concept lattices. Sib. Math. J. **50**(3), 481–494 (2009). https://doi.org/10.1007/s11202-009-0054-z
19. Morozov, A.S.: On computable automorphisms in formal concept analysis. Sib. Math. J. **51**(2), 289–295 (2010). https://doi.org/10.1007/s11202-010-0029-0
20. Morozov, A.S., L'vova, M.A.: On computable formal concepts in computable formal contexts. Sib. Math. J. **48**(5), 871–878 (2007). https://doi.org/10.1007/s11202-007-0089-y
21. Odifreddi, P.: Classical recursion theory. Stud. Logic Found. Math. **125** (1992)
22. Odifreddi, P.: Classical recursion theory. Volume II. Stud. Logic Found. Math. **143** (1999)
23. Rogers, H.: Theory of Recursive Functions and Effective Computability. McGraw-Hill, New York (1967)
24. Sacks, G.E.: Higher Recursion Theory. Springer, Berlin (1990)
25. Selivanov, V.L.: Algorithmic complexity of algebraic systems. Math. Notes **44**(6), 944–950 (1988). https://doi.org/10.1007/BF01158034
26. Selivanov, V.L.: On degree spectra of topological spaces. Lobachevskii J. Math. **41**(2), 252–259 (2020). https://doi.org/10.1134/S1995080220020146
27. Turlington, A.: Computability of Heyting algebras and distributive lattices. Ph.D. thesis, University of Connecticut (2010)

Glivenko's Theorem Underneath Structure

Riccardo Borsetto[1], Giulio Fellin[1,(✉)], Tarmo Uustalu[2,3], and Cheng-Syuan Wan[3]

[1] Università degli Studi di Verona, Verona, Italy
{riccardo.borsetto,giulio.fellin}@univr.it
[2] Reykjavik University, Reykjavik, Iceland
tarmo@ru.is
[3] Tallinn University of Technology, Tallinn, Estonia
{tarmo,cswan}@cs.ioc.ee

Abstract. Glivenko's theorem has been studied in an abstract setting, where double negation is replaced by arbitrary nuclei, and classical and intuitionistic propositional logics are generalised to abstract entailment relations. This paper extends that framework to substructural logics, focusing on cases where structural rules such as weakening, contraction, and exchange are restricted. We apply our findings to Full Lambek logic and its extensions, in order to derive a few Glivenko-style theorems. All results are developed within an intuitionistic meta-theory, allowing a computational interpretation and direct formalisation in Agda.

Keywords: Glivenko's theorem · Full Lambek calculus · Substructural logics · Nuclei · Structural proof theory · Agda

1 Introduction

Glivenko's theorem is one of the earliest and most influential results relating classical and intuitionistic logic. Established in the setting of propositional logic, it states that the classical provability of a formula implies the intuitionistic provability of its double negation [14,15]:

$$\mathbf{Cla} \vdash U \rhd a \iff \mathbf{Int} \vdash U \rhd \neg\neg a,$$

where **Cla** and **Int** denote classical and intuitionistic propositional logic, respectively.

Results of this kind are often referred to as *conservation* or *translation* results between logical systems. Their importance lies in the fact that they allow classical reasoning to be studied within a constructive framework. Rather than treating classical and intuitionistic mathematics as entirely separate enterprises, such results make it possible to identify which parts of classical reasoning can be faithfully interpreted in constructive terms. This is particularly significant because intuitionistic proofs often admit computational interpretations—for

V. Brattka et al. (Eds.): CiE 2026, LNCS 16674, pp. 162–177, 2027.
https://doi.org/10.1007/978-3-032-31348-5_11

instance through the Curry–Howard correspondence—so that translating classical arguments into intuitionistic ones may reveal their underlying computational content.

Over time, a number of generalisations of Glivenko's theorem have been investigated, often by replacing **Int** with weaker logics. In 1968, Segerberg [35] introduced the *Glivenko logic* **Gli**, defined as Johansson's minimal logic extended by the double negation of *ex falso quodlibet sequitur*, namely $\neg\neg(\bot \to a)$. In the same paper, he claimed—without proof or citation—a characterisation result that was later formally established by Odintsov [27,28]: the Glivenko logic is the smallest logic for which Glivenko's theorem holds. More precisely, for any logic **L** such that $\mathbf{Min} \subseteq \mathbf{L} \subseteq \mathbf{Cla}$, the following are equivalent:

1. $\mathbf{Cla} \vdash U \rhd a$ if and only if $\mathbf{L} \vdash U \rhd \neg\neg a$;
2. **L** is an extension of **Gli**, i.e. it satisfies the condition $\mathbf{L} \vdash \neg\neg(\bot \to a)$.

Extensions of this result to substructural logics were developed by Cignoli and Torrens [6,7], and by Galatos and Ono [12,13]. Subsequently, Ono [30] established an analogue of Segerberg's characterisation for the Full Lambek logic with exchange, $\mathbf{FL_e}$—that is, intuitionistic linear logic without exponentials—and its *involutive* extension $\mathbf{InFL_e}$, obtained by adding the stability axiom $\neg\neg a \rhd a$ to $\mathbf{FL_e}$. Ono's theorem states that for any logic **L** such that $\mathbf{FL_e} \subseteq \mathbf{L} \subseteq \mathbf{InFL_e}$, the following conditions are equivalent:

1. $\mathbf{InFL_e} \vdash U \rhd a$ if and only if $\mathbf{L} \vdash U \rhd \neg\neg a$;
2. The following conditions are satisfied:
   - $\mathbf{L} \vdash \neg(a \wedge b) \rhd \neg(\neg\neg a \wedge \neg\neg b)$;
   - $\mathbf{L} \vdash \neg(a \to b) \rhd \neg(\neg\neg a \to \neg\neg b)$.

Double negation in intuitionistic logic is a specific instance of a *nucleus* [2,17, 18,25,33,36,37]—also referred to as a strong monad [20] or lax modality [1]—defined as a map j on formulae satisfying the following [4]:

$$(a \to j(b)) \lhd\rhd (j(a) \to j(b)).$$

Fellin and Schuster [9,10] generalised Glivenko's theorem by replacing double negation with an arbitrary nucleus, thereby moving from syntactic provability in a particular calculus to the more general setting of inductively generated abstract consequence relations, and extending the framework from propositional formulas to arbitrary objects. They provided precise criteria under which a classical-style generalisation yields a conservative extension of its intuitionistic counterpart via a nucleus.

The present work extends the framework of [9,10] to entailment relations that may lack structural rules such as weakening, contraction, and exchange [31,32]. In the traditional structural setting, the presence of exchange together with implication ensures that j remains a nucleus over all inductive extensions of **L**. In the more general substructural context, however, this property no longer holds automatically. As a consequence, the assumption that j is a nucleus over the initial relation cannot in general be propagated to its extensions, and the role of the nucleus must be analysed more carefully. This makes it possible, in particular, to dispense with the initial nucleus assumption that is typically required in the structural case.

Another important feature of the main theorem in the present paper is that it further investigates the inclusion relationships between the various extensions induced by nuclei, thereby generalising and expanding characterisations that have appeared in the literature.

The paper is organised as follows. Section 2 introduces substructural entailment relations and establishes the abstract Conservation theorem (Theorem 1). Section 3 analyses left, right and bi-nuclei. Section 4 specialises the framework to the substructural *Full Lambek logic* **FL** and its extensions [21–23], where we prove the Conservation theorem for **FL** (Theorem 3); from it we derive Corollary 2, which, by instantiating the Glivenko nucleus $\neg\neg$ [4,8], recovers the substructural results of Ono and of Galatos–Ono [12,13,30], and Corollary 3 for the *open nucleus*, which remains meaningful in the noncommutative setting. Section 5 outlines directions for future work.

All the results in this paper are developed within an intuitionistic metatheory. Working constructively at the meta-level ensures that the arguments themselves admit a computational interpretation and avoids relying on classical principles that would be incompatible with the constructive perspective underlying much of the work. An additional benefit of this approach is that the main constructions and proofs can be directly formalised in a proof assistant. In particular, we have implemented the results in Agda [26][1], providing a machine-checked formalisation of the main definitions and theorems. The source code is available at [5]. This formalisation is partially based on the Agda development by Altenkirch [3]. Throughout the paper, the symbol ۞ next to a definition or theorem header is a clickable link to the corresponding formalised statement in the HTML rendering of the Agda source.

2 Substructural Entailment Relations and Conservation

Let S be a set and S^* be the set of finite lists of elements of S. Given $\mathbf{L} \subseteq S^* \times S$, we write $\mathbf{L} \vdash U \rhd b$ in place of $(U, b) \in \mathbf{L}$. Expressions "$U \rhd b$" are called *sequents*.

A *(single-conclusion) substructural entailment relation* ۞ is a relation $\mathbf{L} \subseteq S^* \times S$ satisfying closure conditions

$$\frac{}{a \rhd a}\,\text{Refl} \qquad\qquad \frac{U \rhd a \qquad V_1, a, V_2 \rhd b}{V_1, U, V_2 \rhd b}\,\text{Trans}$$

We call such closure conditions rules. If such a condition is satisfied by $\mathbf{L}$, we say that the rule holds or is admissible. These two rules are the *structural rules* of *reflexivity* and *transitivity*, respectively.

---

[1] The formalisation is built on top of the `agda/cubical` library [38].

Optional structural rules include *commutativity* (or *exchange*), *monotonicity* (or *weakening*) and *contraction*:

$$\frac{U_1, a_1, a_2, U_2 \rhd b}{U_1, a_2, a_1, U_2 \rhd b}\,\mathrm{Comm} \qquad \frac{U_1, U_2 \rhd b}{U_1, a, U_2 \rhd b}\,\mathrm{Mono} \qquad \frac{U_1, a, a, U_2 \rhd b}{U_1, a, U_2 \rhd b}\,\mathrm{Contr}$$

Quite often an entailment relation **L** is inductively generated from a given set $\mathcal{R}$ of rules by closing up with respect to the structural rules above. These *non-structural rules* have the form

$$\frac{U_1 \rhd b_1 \quad \dots \quad U_n \rhd b_n}{U \rhd b}\,r \qquad \text{or, more compactly,} \qquad \frac{\{U_i \rhd b_i \colon 1 \leq i \leq n\}}{U \rhd b}\,r$$

We call *axioms* the rules that are premissless, i.e. have the form

$$\frac{}{U \rhd b}\,r$$

If **L** is inductively generated by a set $\mathcal{R}$ of rules, we write $\mathbf{L} = \langle \mathcal{R} \rangle$.

An inductively generated entailment relation $\rhd$ can have multiple generating rule sets $\mathcal{R}$. Despite this, it is common to refer to an inductively generated entailment relation $\rhd$ without specifying a set of generating rules $\mathcal{R}$, unless statements about $\rhd$ depend on $\mathcal{R}$.

By an *extension* $\mathbf{L}'$ of an entailment relation **L** we mean in general an entailment relation such that $\mathbf{L} \subseteq \mathbf{L}'$. We also say that $\mathbf{L}' = \langle \mathcal{R}, \mathcal{R}' \rangle$ is an *inductive extension* of $\mathbf{L} = \langle \mathcal{R} \rangle$, and we call $\mathcal{R}'$ the set of *additional rules*.

Definition 1 (⚙). *Consider the following rules:*

$$\frac{U, a, V \rhd j(b)}{U, j(a), V \rhd j(b)}\,\mathrm{L}j \qquad\qquad \frac{U \rhd b}{U \rhd j(b)}\,\mathrm{R}j$$

Given a set S endowed with an entailment relation **L**, *we say that a function* $j \colon S \to S$ *is*

- expansive (over **L**) *if* Rj *holds in* **L**;
- *a* bi-nucleus (over **L**) *if both* Lj *and* Rj *hold in* **L**.

Definition 2 (⚙). *Let* $j \colon S \to S$. *Given a rule r, its j*-version r_j *is obtained by putting j in front of every succedent:*

$$\frac{\{U_i \rhd b_i \colon i \in I\}}{U \rhd b}\,r \quad \rightsquigarrow \quad \frac{\{U_i \rhd j(b_i) \colon i \in I\}}{U \rhd j(b)}\,r_j$$

We write $\mathcal{R}_j$ *for* $\{r_j \colon r \in \mathcal{R}\}$. *Given a rule r that is admissible in* **L**, *we say that* r survives after j (over **L**) *if* r_j *is admissible in* **L**. *We also define the* stability axiom *for j:*

$$\frac{}{j(a) \rhd a}\,j\text{-stab}$$

Stability for j is an abstract version of the *double negation elimination* axiom, which defines classical logic.

Remark 1 (⚙). We observe the following:

1. Rj and Refl_j are equivalent in **L**.
2. Lj and Trans_j are equivalent in **L**.
3. j-stab and the following are equivalent in **L**

$$\frac{U, a, V \rhd b}{U, j(a), V \rhd b}\,\mathrm{L}j^{+}$$

In particular, j-stab entails Lj.

Definition 3 (⚙). *Given an inductively generated entailment relation* $\mathbf{L} = \langle \mathcal{R} \rangle$ *and an expansive function* j *on* **L**, *we define the following extensions:*

1. *the* Kleisli j-extension $\mathrm{K}_j(\mathbf{L})$ *of* **L** *by*
$$\mathrm{K}_j(\mathbf{L}) := \{(U, a) \colon \mathbf{L} \vdash U \rhd j(a)\},$$
2. *the* Glivenko j-extension $\mathrm{G}_j(\mathbf{L})$ *of* **L** *as*
$$\mathrm{G}_j(\mathbf{L}) := \langle \mathcal{R}, \mathcal{R}_j, \{\mathrm{L}j\} \rangle,$$
3. *the* Kleisli-stable j-extension $\mathrm{M}_j(\mathbf{L})$ *of* **L** *as*
$$\mathrm{M}_j(\mathbf{L}) := \langle \mathcal{R}, \{j\text{-stab}\} \rangle.$$

Intuitively, the Glivenko j-extension corresponds to an abstract version of the Glivenko logic, and the Kleisli-stable j-extension corresponds to an abstract version of classical logic. Although the definitions of $\mathrm{G}_j(\mathbf{L})$ and $\mathrm{M}_j(\mathbf{L})$ are given in terms of the generating set $\mathcal{R}$ rather than **L** itself, we always consider $\mathcal{R}$ fixed as the generator of **L**.

Remark 2. Clearly, j is a bi-nucleus over $\mathrm{G}_j(\mathbf{L})$.

Proposition 1 (⚙). *Let* S *be a set endowed with an inductively generated entailment relation* $\mathbf{L} = \langle \mathcal{R} \rangle$, *and let* j *be expansive over* **L**. *Then the following inclusions hold:*

$$\begin{array}{ccc} \mathbf{L} & \hookrightarrow & \mathrm{G}_j(\mathbf{L}) \\ \downarrow & & \downarrow \\ \mathrm{K}_j(\mathbf{L}) & \hookrightarrow & \mathrm{M}_j(\mathbf{L}) \end{array}$$

Proof.

- $\mathbf{L} \subseteq \mathrm{K}_j(\mathbf{L})$ is nothing but Rj.
- $\mathbf{L} \subseteq \mathrm{G}_j(\mathbf{L})$ follows from the definition of $\mathrm{G}_j(\mathbf{L})$.
- $\mathrm{G}_j(\mathbf{L}) \subseteq \mathrm{M}_j(\mathbf{L})$: It suffices to show that the latter admits all rules in the inductive definition of the former. Rule Lj is guaranteed by j-stab, see Remark 1. The only non-trivial remaining cases are the r_j where $r \in \mathcal{R}$. Let us consider

$$\frac{\{U_i \rhd b_i : i \in I\}}{U \rhd b}\,r \quad \rightsquigarrow \quad \frac{\{U_i \rhd j(b_i) : i \in I\}}{U \rhd j(b)}\,r_j$$

  We have the following in $\mathrm{M}_j(\mathbf{L})$:

$$\cfrac{\cfrac{\cdots \qquad \cfrac{U_i \rhd j(b_i) \qquad \cfrac{}{j(b_i) \rhd b_i}\,j\text{-stab}}{U_i \rhd b_i}\,\mathrm{Trans} \qquad \cdots}{U \rhd b}\,r}{U \rhd j(b)}\,\mathrm{R}_j$$

- $\mathrm{K}_j(\mathbf{L}) \subseteq \mathrm{M}_j(\mathbf{L})$:

$$\cfrac{\cfrac{\cfrac{\mathrm{K}_j(\mathbf{L}) \vdash U \rhd b}{\mathbf{L} \vdash U \rhd j(b)}\,\mathrm{defn}}{\mathrm{M}_j(\mathbf{L}) \vdash U \rhd j(b)}\,\mathrm{defn} \qquad \cfrac{}{\mathrm{M}_j(\mathbf{L}) \vdash j(b) \rhd b}\,j\text{-stab}}{\mathrm{M}_j(\mathbf{L}) \vdash U \rhd b}\,\mathrm{Trans}$$

□

Theorem 1 (Conservation). *Let* $\mathbf{L} = \langle \mathcal{R} \rangle$, *let* $\mathbf{L}' = \langle \mathcal{R}, \mathcal{R}' \rangle$, *and let* j *be expansive over* $\mathbf{L}$. *Then:*

$$\mathrm{K}_j(\mathbf{L}') \supseteq \mathbf{L}' \tag{1}$$

$$\mathrm{K}_j(\mathbf{L}') \subseteq \mathbf{L}' \iff \mathbf{L}' \supseteq \mathrm{M}_j(\mathbf{L}) \tag{2}$$

$$\mathrm{K}_j(\mathbf{L}') \subseteq \mathrm{M}_j(\mathbf{L}) \iff \mathbf{L}' \subseteq \mathrm{M}_j(\mathbf{L}) \tag{3}$$

$$\mathrm{K}_j(\mathbf{L}') \supseteq \mathrm{M}_j(\mathbf{L}) \iff \mathbf{L}' \supseteq \mathrm{G}_j(\mathbf{L}) \tag{4}$$

Proof.

(1) A particular case of Proposition 1.

(2) If $\mathbf{L}' \supseteq \mathrm{M}_j(\mathbf{L})$, then in particular $\mathbf{L}' \vdash j(a) \rhd a$. Let us suppose $\mathrm{K}_j(\mathbf{L}') \vdash U \rhd a$, i.e. $\mathbf{L}' \vdash U \rhd j(a)$. Then by Trans we conclude $\mathbf{L}' \vdash U \rhd a$. We have shown $\mathrm{K}_j(\mathbf{L}') \subseteq \mathbf{L}'$.
Now suppose that $\mathrm{K}_j(\mathbf{L}') \subseteq \mathbf{L}'$. From $\mathbf{L}' \vdash j(a) \rhd j(a)$, i.e. $\mathrm{K}_j(\mathbf{L}') \vdash j(a) \rhd a$, we get $\mathbf{L}' \vdash j(a) \rhd a$ by assumption. This means that j-stab holds in $\mathbf{L}'$, which suffices for $\mathbf{L}' \supseteq \mathrm{M}_j(\mathbf{L})$.

(3) The direction "$\Rightarrow$" easily follows from $\mathbf{L}' \subseteq \mathrm{K}_j(\mathbf{L}')$ (Proposition 1).
Now suppose that $\mathbf{L}' \subseteq \mathrm{M}_j(\mathbf{L})$. Since $\mathbf{L} \subseteq \mathbf{L}'$, we have $\mathrm{M}_j(\mathbf{L}) \subseteq \mathrm{M}_j(\mathbf{L}')$. On the other hand, as stability and all the rules of $\mathbf{L}'$ are admissible in $\mathrm{M}_j(\mathbf{L})$, we also have $\mathrm{M}_j(\mathbf{L}') \subseteq \mathrm{M}_j(\mathbf{L})$. Therefore $\mathrm{M}_j(\mathbf{L}) = \mathrm{M}_j(\mathbf{L}')$. From this and $\mathrm{K}_j(\mathbf{L}') \subseteq \mathrm{M}_j(\mathbf{L}')$ (Proposition 1), we conclude $\mathrm{K}_j(\mathbf{L}') \subseteq \mathrm{M}_j(\mathbf{L})$.

(4) Assume $\mathbf{L}' \supseteq \mathrm{G}_j(\mathbf{L})$, and suppose $\mathrm{M}_j(\mathbf{L}) \vdash U \rhd a$. We prove that $\mathrm{G}_j(\mathbf{L}) \vdash U \rhd ja$ by induction on the last rule r used in the derivation of $\mathrm{M}_j(\mathbf{L}) \vdash U \rhd a$. The case in which $r \in \mathcal{R} \cup \{\text{Trans}, \text{Refl}\}$ is converted in $\mathrm{G}_j(\mathbf{L})$ to an application of the rule r_j, while the case in which r is j-stab is converted to an instance of Refl. So we have proved $\mathrm{G}_j(\mathbf{L}) \vdash U \rhd ja$, which by hypothesis implies $\mathbf{L}' \vdash U \rhd ja$, i.e. $\mathrm{K}_j(\mathbf{L}') \vdash U \rhd a$ as desired.
Assume $\mathrm{K}_j(\mathbf{L}') \supseteq \mathrm{M}_j(\mathbf{L})$. Suppose that $\mathrm{G}_j(\mathbf{L}) \vdash U \rhd a$. We prove $\mathbf{L}' \vdash U \rhd a$ by induction on the last rule used in the derivation of $\mathrm{G}_j(\mathbf{L}) \vdash U \rhd a$.

(a) Case $r \in \mathcal{R}$: We have

$$\frac{\mathrm{G}_j(\mathbf{L}) \vdash U_1 \rhd a_1 \quad \dots \quad \mathrm{G}_j(\mathbf{L}) \vdash U_n \rhd a_n}{\mathrm{G}_j(\mathbf{L}) \vdash U \rhd a}\, r$$

which leads to

$$\frac{\dfrac{\mathrm{G}_j(\mathbf{L}) \vdash U_1 \rhd a_1}{\mathbf{L}' \vdash U_1 \rhd a_1}\,\text{ind. hp.} \quad \dots \quad \dfrac{\mathrm{G}_j(\mathbf{L}) \vdash U_n \rhd a_n}{\mathbf{L}' \vdash U_n \rhd a_n}\,\text{ind. hp.}}{\mathbf{L}' \vdash U \rhd a}\, r$$

(b) Case $r = \mathrm{L}j$ or r_j (where $r \in \mathcal{R}$): We have $a = j(b)$ for some $b \in S$, and hence we can write $\mathrm{G}_j(\mathbf{L}) \vdash U \rhd j(b)$. Since $\mathrm{G}_j(\mathbf{L}) \subseteq \mathrm{M}_j(\mathbf{L}) \subseteq \mathrm{K}_j(\mathbf{L}')$, it follows that $\mathrm{K}_j(\mathbf{L}') \vdash U \rhd j(b)$. By the definition of $\mathrm{K}_j(\mathbf{L}')$, this is equivalent to $\mathbf{L}' \vdash U \rhd j(j(b))$. On the other hand, we observe that $\mathrm{M}_j(\mathbf{L}) \vdash j(j(b)) \rhd b$ by applying $\mathrm{L}j^+$ twice. Then $\mathrm{M}_j(\mathbf{L}) \subseteq \mathrm{K}_j(\mathbf{L}')$ implies $\mathrm{K}_j(\mathbf{L}') \vdash j(j(b)) \rhd b$, which is nothing but $\mathbf{L}' \vdash j(j(b)) \rhd j(b)$. By an application of transitivity between $\mathbf{L}' \vdash U \rhd j(j(b))$ and $\mathbf{L}' \vdash j(j(b)) \rhd j(b)$, we conclude $\mathbf{L}' \vdash U \rhd j(b)$. □

In particular, we observe that $\mathrm{M}_j(\mathbf{L})$ is the smallest extension $\mathbf{L}'$ of $\mathbf{L}$ for which $\mathrm{K}_j(\mathbf{L}') = \mathbf{L}'$. We also notice that $\mathrm{K}_j(\mathbf{L}') = \mathrm{K}_j\,\mathrm{M}_j(\mathbf{L})$ if and only if $\mathbf{L}'$ is intermediate between $\mathrm{G}_j(\mathbf{L})$ and $\mathrm{M}_j(\mathbf{L})$. Visually, we can summarise some of the main cases as follows:

$$\begin{array}{ccccccccccc}
\mathbf{L} & \hookrightarrow & \mathbf{L}_1 & \hookrightarrow & \mathrm{G}_j(\mathbf{L}) & \hookrightarrow & \mathbf{L}_2 & \hookrightarrow & \mathrm{M}_j(\mathbf{L}) & \hookrightarrow & \mathbf{L}_3 \\
\downarrow & & \downarrow & & \downarrow & & \downarrow & & \| & & \| \\
\mathrm{K}_j(\mathbf{L}) & \hookrightarrow & \mathrm{K}_j(\mathbf{L}_1) & \hookrightarrow & \mathrm{K}_j\,\mathrm{G}_j(\mathbf{L}) & = & \mathrm{K}_j(\mathbf{L}_2) & = & \mathrm{K}_j\,\mathrm{M}_j(\mathbf{L}) & \hookrightarrow & \mathrm{K}_j(\mathbf{L}_3)
\end{array}$$

3 Left and Right Nuclei

In the previous section, we introduced the notion of *bi-nucleus*, which allows rule Lj to be applied to any element on the left-hand side of the sequent. This is sufficient for the main conservation theorem. However, in concrete cases many functions j satisfy only a weaker version of the rule Lj, which can be applied only to elements occurring in specific positions. For this reason, we introduce the notions of *left nucleus* and *right nucleus.*

Definition 4 (⚙). *Consider the following rules:*

$$\frac{U, a \rhd j(b)}{U, j(a) \rhd j(b)}\,\mathrm{L}j_{\mathrm{left}} \qquad\qquad \frac{a, U \rhd j(b)}{j(a), U \rhd j(b)}\,\mathrm{L}j_{\mathrm{right}}$$

Given a set S endowed with an entailment relation **L**, *we say that an expansive function $j\colon S \to S$ is*

- *a* right nucleus (over **L**) *if* $\mathrm{L}j_{\mathrm{right}}$ *holds in* **L**;
- *a* left nucleus (over **L**) *if* $\mathrm{L}j_{\mathrm{left}}$ *holds in* **L**.

Remark 3 (⚙).

1. Every bi-nucleus is both a left and right nucleus.
2. If Comm is admissible in **L**, then the definitions of right, left and bi-nucleus coincide.

Remark 4 (⚙). Let j be a left/right nucleus over **L**. Then an easy consequence of Lj_* (where $* \in \{\mathrm{left}, \mathrm{right}\}$), the context-free version of transitivity survives after j:

$$\frac{a \rhd b \quad b \rhd c}{a \rhd c}\,\mathrm{trans} \quad\rightsquigarrow\quad \frac{a \rhd j(b) \quad \dfrac{b \rhd j(c)}{j(b) \rhd j(c)}\,\mathrm{L}j_*}{a \rhd j(c)}\,\mathrm{trans}$$

4 Full Lambek Logic and Its Structural Extensions

Fix a set S of formulae, which contains constants $0, 1$ and is closed under connectives $\wedge, \vee, \cdot, \multimapinv, \multimap$. As a basis, we consider *Full Lambek logic* **FL**, which is inductively defined by the set $\mathcal{FL}$ of rules in Table 1.

Two unary connectives are defined as follows:

$$\sim a := a \multimap 0, \qquad\qquad -a := 0 \multimapinv a.$$

Table 1. The set $\mathcal{FL}$ of rules that generate **FL**, adapted from [12, §2.1]. To sometimes have a slicker notation, we subsume under L∧ and R∨ the two left rules $\mathrm{L}\wedge_1$ and $\mathrm{L}\wedge_2$ and the two right rules $\mathrm{R}\vee_1$ and $\mathrm{R}\vee_2$, respectively.

$$\frac{U,a,V \rhd c \quad U,b,V \rhd c}{U,a \vee b,V \rhd c}\,\mathrm{L}\vee \qquad \frac{U \rhd a}{U \rhd a \vee b}\,\mathrm{R}\vee_1 \qquad \frac{U \rhd b}{U \rhd a \vee b}\,\mathrm{R}\vee_2$$

$$\frac{U,a,V \rhd c}{U,a \wedge b,V \rhd c}\,\mathrm{L}\wedge_1 \qquad \frac{U,b,V \rhd c}{U,a \wedge b,V \rhd c}\,\mathrm{L}\wedge_2 \qquad \frac{U \rhd a \quad U \rhd b}{U \rhd a \wedge b}\,\mathrm{R}\wedge$$

$$\frac{U,V \rhd c}{U,1,V \rhd c}\,\mathrm{L}1 \qquad \frac{}{\rhd 1}\,\mathrm{R}1$$

$$\frac{U,a,b,V \rhd c}{U,a \cdot b,V \rhd c}\,\mathrm{L}\cdot \qquad \frac{U \rhd a \quad V \rhd b}{U,V \rhd a \cdot b}\,\mathrm{R}\cdot$$

$$\frac{U \rhd a \quad W,b,V \rhd c}{W,U,a \multimap b,V \rhd c}\,\mathrm{L}\!\multimap \qquad \frac{a,U \rhd b}{U \rhd a \multimap b}\,\mathrm{R}\!\multimap$$

$$\frac{U \rhd a \quad W,b,V \rhd c}{W,b \multimapinv a,U,V \rhd c}\,\mathrm{L}\!\multimapinv \qquad \frac{U,a \rhd b}{U \rhd b \multimapinv a}\,\mathrm{R}\!\multimapinv$$

Remark 5 ( ). Let **L** be an inductive extension of **FL**.

1. If **L** admits Mono and Contr, then ∧ and · coincide, i.e. $a \wedge b \lhd\!\rhd a \cdot b$.
2. If **L** admits Comm, then $\multimapinv$ and $\multimap$ coincide, i.e. $b \multimapinv a \lhd\!\rhd a \multimap b$.
3. R∧, R$\multimapinv$ and R$\multimap$ are invertible in **L**.

In the *Full Lambek logic with exchange* $\mathbf{FL_e} := \langle \mathcal{FL}, \{\mathrm{Comm}\}\rangle$ [12, §2.1] and its extensions, we use the usual symbol → for symmetric implication and the usual symbol ¬ for symmetric negation.

An important structural extension is *Minimal logic* **Min**, which can be generated by the same rules as **FL** plus all the structural rules Comm, Mono, Contr. We also define, as usual, *Intuitionistic logic* **Int** as **Min** plus

$$\frac{}{U,0,V \rhd c}\,\mathrm{L}0$$

For the rest of this section, let **L** be an inductive extension of **FL**, and let j be an expansive function on formulae. By the *j-shift* for the logical operator $*$ we understand the corresponding axiom as in the following table:

$*$	j-shift for $*$
1	$1 \rhd j(1)$
$\wedge$	$j(a) \wedge j(b) \rhd j(a \wedge b)$
$\cdot$	$j(a) \cdot j(b) \rhd j(a \cdot b)$

$*$	j-shift for $*$
$\vee$	$j(a) \vee j(b) \rhd j(a \vee b)$
$\multimap$	$a \multimap jb \rhd j(a \multimap b)$
$\multimapinv$	$jb \multimapinv a \rhd j(b \multimapinv a)$

Theorem 2 ( ). *Let* **L** *be an inductive extension of* **FL**. *The right rule* R$*$ *is admissible for the Kleisli extension* $\mathrm{K}_j(\mathbf{L})$ *if and only if the j-shift for $*$ holds for* **L**.

Proof. Let us consider the case $* = \wedge$; the other cases are similar.

Suppose first that the j-shift for $\wedge$ holds in $\mathbf{L}$. If $\mathrm{K}_j(\mathbf{L}) \vdash U \rhd a$ and $\mathrm{K}_j(\mathbf{L}) \vdash U \rhd b$, i.e. $\mathbf{L} \vdash U \rhd j(a)$ and $\mathbf{L} \vdash U \rhd j(b)$, then $\mathrm{R}\wedge$ for $\mathbf{L}$ gives $\mathbf{L} \vdash U \rhd j(a) \wedge j(b)$, which by transitivity and the j-shift for $\wedge$ implies $\mathbf{L} \vdash U \rhd j(a \wedge b)$, i.e. $\mathrm{K}_j(\mathbf{L}) \vdash U \rhd a \wedge b$.

As for the converse, assume that $\mathrm{R}\wedge$ is admissible in $\mathrm{K}_j(\mathbf{L})$. By reflexivity, $\mathbf{L} \vdash j(a) \rhd j(a)$ and $\mathbf{L} \vdash j(b) \rhd j(b)$. By $\mathrm{L}\wedge_1$ and $\mathrm{L}\wedge_2$, we get $\mathbf{L} \vdash j(a) \wedge j(b) \rhd j(a)$ and $\mathbf{L} \vdash j(a) \wedge j(b) \rhd j(b)$, i.e. $\mathrm{K}_j(\mathbf{L}) \vdash j(a) \wedge j(b) \rhd a$ and $\mathrm{K}_j(\mathbf{L}) \vdash j(a) \wedge j(b) \rhd b$. Then, $\mathrm{R}\wedge$ for $\mathrm{K}_j(\mathbf{L})$ entails $\mathrm{K}_j(\mathbf{L}) \vdash j(a) \wedge j(b) \rhd a \wedge b$, i.e. $\mathbf{L} \vdash j(a) \wedge j(b) \rhd j(a \wedge b)$.
□

Lemma 1 (⚙). *Let* $\mathbf{L} = \langle \mathcal{R} \rangle$ *be an inductive extension of* $\mathbf{FL}$*, and let* j *be a (right / left / bi-) nucleus on* $\mathbf{L}$*. Then*

1. *The* j*-shifts for* $1, \vee$ *hold in* $\mathbf{L}$*.*
2. *The* j*-shift for* $\cdot$ *holds in* $\langle \mathcal{R}, \{\mathrm{Comm}\} \rangle$*.*
3. *The* j*-shift for* $\wedge$ *holds in* $\langle \mathcal{R}, \{\mathrm{Comm}, \mathrm{Mono}, \mathrm{Contr}\} \rangle$*.*
4. *If the* j*-shift for* $\cdot$ *holds, then* j *is a bi-nucleus.*
5. *If* j *is a bi-nucleus, then* $\mathrm{L}\multimapinv$ *and* $\mathrm{L}\multimap$ *survive after* j*.*
6. *If the* j*-shift for* $\cdot$ *holds, then* $\mathrm{L}\multimapinv$*,* $\mathrm{L}\multimap$ *and* Trans *survive after* j*.*

Proof. Straightforward. □

Lemma 2 (⚙). *Let* $\mathbf{L} = \langle \mathcal{FL}, \mathcal{R}, \mathcal{S} \rangle$*, where* $\mathcal{S} \subseteq \{\mathrm{Comm}, \mathrm{Mono}, \mathrm{Contr}\}$*, and let* j *be a (right / left / bi-) nucleus on* $\mathbf{L}$*. Then*

$$\mathrm{G}_j(\mathbf{L}) = \langle \mathcal{FL}, \mathcal{R}, \mathcal{R}_j, \{j\text{-shift}\cdot, j\text{-shift}\wedge, j\text{-shift}\multimapinv, j\text{-shift}\multimap\} \rangle.$$

If, moreover, $\mathrm{Comm} \in \mathcal{S}$*, then* $\mathrm{G}_j(\mathbf{L}) = \langle \mathcal{FL}, \mathcal{R}, \mathcal{R}_j, \{j\text{-shift}\wedge, j\text{-shift} \to\} \rangle$*. If, moreover,* $\mathcal{S} = \{\mathrm{Comm}, \mathrm{Mono}, \mathrm{Contr}\}$*, then* $\mathrm{G}_j(\mathbf{L}) = \langle \mathcal{FL}, \mathcal{R}, \mathcal{R}_j, \{j\text{-shift} \to\} \rangle$*.*

Proof. We show that all rules generating $\mathrm{G}_j(\mathbf{L})$ are admissible in $\langle \mathcal{FL}, \mathcal{R}, \mathcal{R}_j, \{j\text{-shift}\cdot, j\text{-shift}\wedge, j\text{-shift}\multimapinv, j\text{-shift}\multimap\} \rangle$. In fact, it is straightforward that rules Comm, Mono, Contr, if in $\mathcal{S}$, survive after j; rules L1, $\mathrm{L}\vee$, $\mathrm{L}\wedge_1$, $\mathrm{L}\wedge_2$, $\mathrm{L}\cdot$ survive after j since they do not change succedents; rules R1, $\mathrm{R}\vee_1$, $\mathrm{R}\vee_2$ survive after j by Theorem 2 since their j-shifts hold by Lemma 1(1). Observe that admissibilities of $\mathrm{R}\wedge_j$, $\mathrm{R}\cdot_j$, $\mathrm{R}\multimapinv_j$ and $\mathrm{R}\multimap_j$ are equivalent to admissibilities of j-shift$\wedge$, j-shift$\cdot$, j-shift$\multimapinv$, j-shift$\multimap$, respectively, by Theorem 2. By Lemma 1(6), we know that j-shift$\cdot$ suffices also for $\mathrm{L}\multimap_j$, $\mathrm{L}\multimapinv_j$ and Trans_j, which is tantamount to $\mathrm{L}j$ by Remark 1. This proves our claim.

If, moreover, $\mathrm{Comm} \in \mathcal{S}$, then j-shift$\cdot$ holds by Lemma 1; and $\multimap$ and $\multimapinv$ coincide by Remark 5. If, moreover, $\mathcal{S} = \{\mathrm{Comm}, \mathrm{Mono}, \mathrm{Contr}\}$, then j-shift$\wedge$ holds by Lemma 1. □

Theorem 3 (Conservation for FL ⚙). *Let* j *be a (right / left / bi-) nucleus over* $\mathbf{FL}$*, and let* $\mathbf{L}$ *be an inductive extension of* $\mathbf{FL}$ *such that* $\mathbf{L} \subseteq \mathrm{M}_j(\mathbf{FL})$*. Then the following are equivalent:*

1. ***Generalised Glivenko's theorem:*** $\mathrm{M}_j(\mathbf{FL}) \vdash U \rhd a$ *if and only if* $\mathbf{L} \vdash U \rhd ja$.
2. $\mathrm{G}_j(\mathbf{FL}) \subseteq \mathbf{L}$.
3. *The following conditions hold:*
   - $\mathbf{L} \vdash ja \cdot jb \rhd j(a \cdot b)$,
   - $\mathbf{L} \vdash ja \wedge jb \rhd j(a \wedge b)$,
   - $\mathbf{L} \vdash a \multimap jb \rhd j(a \multimap b)$,
   - $\mathbf{L} \vdash jb \multimapinv a \rhd j(b \multimapinv a)$.

Proof. The equivalence between the first two items is just an instance of Theorem 1(4). The equivalence between the last two items follows directly from the characterisation of $\mathrm{G}_j(\mathbf{FL})$ given in Lemma 2. □

Remark 6 (⚙). A common definition of a nucleus, as given by Johnstone [19, §C1.1], consists of the following conditions:

1. $a \rhd j(a)$,
2. $j(a) \lhd\!\rhd j(j(a))$,
3. $j(a \cdot b) \lhd\!\rhd j(a) \cdot j(b)$.

Conditions 1 and 2 follow directly from the rules Lj and Rj. Moreover, as a consequence of Lemma 2, we can observe that if all structural rules are satisfied, then condition 3 also holds. However, if some structural rules are omitted, condition 3 may not hold. Thus, in the substructural setting, these two notions of nucleus may differ.

In the final part of this section, we examine a few examples of nuclei and show how the general framework developed above allows us to recover a number of results from the literature. To keep the exposition focused on the main ideas, most of the technical proof details are omitted here. We start by presenting two versions of the *Glivenko nucleus* [4,8], i.e. $a \mapsto \neg\neg a$, that work in substructural logic.

Remark 7 (⚙).

1. The *left Glivenko nucleus* $a \mapsto {\sim}{-}a$ is a left nucleus over **FL**;
2. The *right Glivenko nucleus* $a \mapsto {-}{\sim}a$ is a right nucleus over **FL**.

Corollary 1 (⚙).

1. *Let* $\mathbf{L}$ *be an inductive extension of* $\mathbf{FL}$ *such that* $\mathbf{L} \subseteq \mathrm{M}_{\sim-}(\mathbf{FL})$. *Then the following are equivalent:*
   (a) $\mathrm{M}_{\sim-}(\mathbf{FL}) \vdash U \rhd a$ *if and only if* $\mathbf{L} \vdash U \rhd {\sim}{-}a$.
   (b) $\mathrm{G}_{\sim-}(\mathbf{FL}) \subseteq \mathbf{L}$.
   (c) The following conditions hold:
   - $\mathbf{L} \vdash ({\sim}{-}a) \cdot ({\sim}{-}b) \rhd {\sim}{-}(a \cdot b)$,
   - $\mathbf{L} \vdash ({\sim}{-}a) \wedge ({\sim}{-}b) \rhd {\sim}{-}(a \wedge b)$,
   - $\mathbf{L} \vdash a \multimap {\sim}{-}b \rhd {\sim}{-}(a \multimap b)$,

- $\mathbf{L} \vdash {\sim}{-}b \multimapinv a \rhd {\sim}{-}(b \multimapinv a)$.

2. *Let* **L** *be an inductive extension of* **FL** *such that* $\mathbf{L} \subseteq \mathrm{M}_{-\sim}(\mathbf{FL})$. *Then the following are equivalent:*
   (a) $\mathrm{M}_{-\sim}(\mathbf{FL}) \vdash U \rhd a$ *if and only if* $\mathbf{L} \vdash U \rhd -{\sim}a$.
   (b) $\mathrm{G}_{-\sim}(\mathbf{FL}) \subseteq \mathbf{L}$.
   (c) *The following conditions hold:*
   - $\mathbf{L} \vdash (-{\sim}a) \cdot (-{\sim}b) \rhd -{\sim}(a \cdot b)$,
   - $\mathbf{L} \vdash (-{\sim}a) \wedge (-{\sim}b) \rhd -{\sim}(a \wedge b)$,
   - $\mathbf{L} \vdash a \multimap -{\sim}b \rhd -{\sim}(a \multimap b)$,
   - $\mathbf{L} \vdash -{\sim}b \multimapinv a \rhd -{\sim}(b \multimapinv a)$.

Proof. Direct application of Theorem 3. □

If we consider $\mathbf{FL_e}$, then the *Glivenko nucleus* $\neg\neg$ is a nucleus over $\mathbf{FL_e}$. We notice that the extension $\mathrm{M}_{\neg\neg}(\mathbf{FL_e})$, corresponds to the one known in the literature as the *Involutive Full Lambek logic with exchange* $\mathbf{InFL_e}$ [12, §7.3]. If, moreover, we consider **Min**, we obtain two well-known extensions: $\mathbf{Cla} := \mathrm{M}_{\neg\neg}(\mathbf{Min})$ is *Classical logic* and $\mathbf{Gli} := \mathrm{G}_{\neg\neg}(\mathbf{Min})$ is the *Glivenko logic* [12,29, 35].

Corollary 2 (⚙).

1. (Ono *[30]*) *Let* **L** *be an inductive extension of* $\mathbf{FL_e}$ *such that* $\mathbf{L} \subseteq \mathbf{InFL_e}$. *Then the following are equivalent:*
   (a) $\mathbf{InFL_e} \vdash U \rhd a$ *if and only if* $\mathbf{L} \vdash U \rhd \neg\neg a$.
   (b) $\mathrm{G}_{\neg\neg}(\mathbf{FL_e}) \subseteq \mathbf{L}$.
   (c) *The following conditions hold:*
   - $\mathbf{L} \vdash \neg\neg a \wedge \neg\neg b \rhd \neg\neg(a \wedge b)$,
   - $\mathbf{L} \vdash a \to \neg\neg b \rhd \neg\neg(a \to b)$.

2. (Segerberg *[35]*, Odintsov *[27, 28]*) *Let* **L** *be an inductive extension of* **Min** *such that* $\mathbf{L} \subseteq \mathbf{Cla}$. *Then the following are equivalent:*
   (a) $\mathbf{Cla} \vdash U \rhd a$ *if and only if* $\mathbf{L} \vdash U \rhd \neg\neg a$.
   (b) $\mathbf{Gli} \subseteq \mathbf{L}$.
   (c) $\mathbf{L} \vdash a \to \neg\neg b \rhd \neg\neg(a \to b)$.
3. (Glivenko *[14, 15]*) $\mathbf{Cla} \vdash U \rhd a$ *if and only if* $\mathbf{Int} \vdash U \rhd \neg\neg a$.

Proof. Provided that $\mathbf{FL_e} \vdash (\neg\neg a) \cdot (\neg\neg b) \rhd \neg\neg(a \cdot b)$, these become particular cases of Corollary 2. The only thing left to show for item 3 is $\mathbf{Int} \vdash a \to \neg\neg b \rhd \neg\neg(a \to b)$, which is well-known (see, e.g. [10]). □

Lastly, we present a version of the *open nucleus* [4,8], i.e. $a \mapsto (0 \to a)$, that works in substructural logic.

Remark 8 (⚙).

1. The *right open nucleus* $o_R : a \mapsto 0 \multimap 0 \cdot a$ is a right nucleus over **FL**;
2. The *left open nucleus* $o_L : a \mapsto a \cdot 0 \mathbin{\circ\!-} 0$ is a left nucleus over **FL**.

Corollary 3 (⚙).

1. Let $\mathbf{L}$ *be an inductive extension of* $\mathbf{FL}$ *such that* $\mathbf{L} \subseteq \mathrm{M}_{o_R}(\mathbf{FL})$. *Then the following are equivalent:*
   (a) $\mathrm{M}_{o_R}(\mathbf{FL}) \vdash U \rhd a$ *if and only if* $\mathbf{L} \vdash 0, U \rhd 0 \cdot a$.
   (b) $\mathrm{G}_{o_R}(\mathbf{FL}) \subseteq \mathbf{L}$.
   (c) The following conditions hold:
   - $\mathbf{L} \vdash 0, (0 \multimap 0 \cdot a) \cdot (0 \multimap 0 \cdot b) \rhd 0 \cdot (a \cdot b)$,
   - $\mathbf{L} \vdash 0, (0 \multimap 0 \cdot a) \wedge (0 \multimap 0 \cdot b) \rhd 0 \cdot (a \wedge b)$,
   - $\mathbf{L} \vdash 0, a \multimap (0 \multimap 0 \cdot b) \rhd 0 \cdot (a \multimap b)$,
   - $\mathbf{L} \vdash 0, (0 \multimap 0 \cdot b) \mathbin{\circ\!-} a \rhd 0 \cdot (b \mathbin{\circ\!-} a)$,

2. Let $\mathbf{L}$ *be an inductive extension of* $\mathbf{FL}$ *such that* $\mathbf{L} \subseteq \mathrm{M}_{o_L}(\mathbf{FL})$. *Then the following are equivalent:*
   (a) $\mathrm{M}_{o_L}(\mathbf{FL}) \vdash U \rhd a$ *if and only if* $\mathbf{L} \vdash U, 0 \rhd a \cdot 0$.
   (b) $\mathrm{G}_{o_L}(\mathbf{FL}) \subseteq \mathbf{L}$.
   (c) The following conditions hold:
   - $\mathbf{L} \vdash (a \cdot 0 \mathbin{\circ\!-} 0) \cdot (b \cdot 0 \mathbin{\circ\!-} 0), 0 \rhd (a \cdot b) \cdot 0$,
   - $\mathbf{L} \vdash (a \cdot 0 \mathbin{\circ\!-} 0) \wedge (b \cdot 0 \mathbin{\circ\!-} 0), 0 \rhd (a \wedge b) \cdot 0$,
   - $\mathbf{L} \vdash a \multimap (b \cdot 0 \mathbin{\circ\!-} 0), 0 \rhd (a \multimap b) \cdot 0$,
   - $\mathbf{L} \vdash (b \cdot 0 \mathbin{\circ\!-} 0) \mathbin{\circ\!-} a, 0 \rhd (b \mathbin{\circ\!-} a) \cdot 0$,

Proof. This is Theorem 3 with invertibility of R$\multimap$ and R$\mathbin{\circ\!-}$ (Remark 5). □

5 Future Work

In this paper, our approach has remained *proof-irrelevant* in that it treats derivability relations extensionally, without distinguishing between different proofs of the same judgment. A natural direction for future work is to explore the same framework from a *proof-relevant* perspective. This would involve enriching the underlying entailment relations to retain information about the structure or identity of proofs, such as labelling derivations, formalising proof terms, or working within a categorical semantics where morphisms represent proofs explicitly.

Several questions arise in this setting:

- How do Glivenko extensions interact with proof equivalences? Given an equivalence relation on proofs in an entailment relation $\mathbf{L}$, can one define a canonical transformation between proofs of $\mathbf{L} \vdash U \rhd a$ and proofs of $\mathrm{G}_j(\mathbf{L}) \vdash U \rhd j(a)$ in a way that preserves the equivalence relation? Moreover, if such a transformation exists, does it have an inverse? This question is related to the decorations studied in Schellinx's PhD thesis [34], where the codomain of the translation is classical linear logic, while here we are in a more abstract setting.

- In what sense are j-shifts invertible or conservative at the level of proofs, not just derivability? What kinds of monads arise categorically?
- Can we extend our results to a setting like type theory or logical frameworks (e.g., LF [16], FOLDS [24]), where proofs themselves are first-class objects?

Adopting a proof-relevant viewpoint could also facilitate connections with categorical logic and type theory, and yield new insights into normalisation, proof transformation, or computational interpretations of substructural logics.

We also note that it would be natural to extend our approach beyond the propositional setting considered here. In particular, one could investigate substructural logics with quantifiers and (co)modalities, such as the exponentials of linear logic (e.g. ! and ?). The interaction between nuclei (such as double negation) and linear structure has already been studied in specific settings; see, for instance, [11] for translations between classical, intuitionistic, and linear logic. Moreover, modalities such as $\Box$ and ! suggest considering the dual notion of a *co-nucleus*, and whether a corresponding Glivenko-type theorem exists, along with its interaction with the structural features of the logic.

We also note that the structural rule of associativity may be relevant in this context [39].

Acknowledgments. We are thankful to Thorsten Altenkirch, Francesco Ciraulo, Hajime Ishihara, Sara Negri, Peter Schuster, Niccolò Veltri and Joshua Wrigley for their interest and suggestions. G.F. is a member of the "Gruppo Nazionale per le Strutture Algebriche, Geometriche e le loro Applicazioni" (GNSAGA) of the Istituto Nazionale di Alta Matematica (INdAM). The collaboration of G.F. and T.U. was supported by the Iceproof project of the Collaboration Fund of the Icelandic Ministry of Culture, Innovation and Higher Education and by "Programma di Internazionalizzazione di Ateneo (PIA)" of Università degli Studi di Verona. T.U. and C.-S.W. were supported by the Estonian Research Council grants PRG1210 and PSG749 respectively.

References

1. Aczel, P.: The Russell-Prawitz modality. Math. Struct. Comput. Sci. **11**(4), 541–554 (2001). https://doi.org/10.1017/s0960129501003309
2. Aczel, P.: Aspects of general topology in constructive set theory. Ann. Pure Appl. Logic **137**(1–3), 3–29 (2006). https://doi.org/10.1016/j.apal.2005.05.016
3. Altenkirch, T.: Nuclear — Agda formalisation (2024). https://github.com/txa/nuclear
4. Berg, B.v.d.: A Kuroda-style j-translation. Arch. Math. Log. **58**(5), 627–634 (2019). https://doi.org/10.1007/s00153-018-0656-x
5. Borsetto, R.: Glivenko's theorem underneath structure — Agda formalisation (2026). https://github.com/eapiova/nuclear
6. Cignoli, R., Torrens, A.: Hájek basic fuzzy logic and Łukasiewicz infinite-valued logic. Arch. Math. Log. **42**(4), 361–370 (2003). https://doi.org/10.1007/s001530200144
7. Cignoli, R., Torrens, A.: Glivenko like theorems in natural expansions of BCK-logic. Math. Log. Q. **50**, 111–125 (2004). https://doi.org/10.1002/malq.200310082

8. Escardó, M., Oliva, P.: The Peirce translation. Ann. Pure Appl. Logic **163**, 681–692 (2012). https://doi.org/10.1016/j.apal.2011.11.002
9. Fellin, G., Schuster, P.: A general Glivenko–Gödel theorem for nuclei. In: Sokolova, A. (ed.) Proc. of 37th Conf. on the Mathematical Foundations of Programming Semantics, MFPS 2021, Electronic Proceedings in Theoretical Computer Science, vol. 351, pp. 51–66. Open Publishing Assoc. (2021). https://doi.org/10.4204/eptcs.351.4
10. Fellin, G., Schuster, P.: Conservation as translation. Rev. Symb. Log. **18**(1), 316–348 (2025). https://doi.org/10.1017/s1755020324000170
11. Ferreira, G., Oliva, P., Protin, C.L.: On the various translations between classical, intuitionistic and linear logic. Studia Logica (2024). https://arxiv.org/abs/2409.02249, accepted for publication
12. Galatos, N., Jipsen, P., Kowalski, T., Ono, H.: Residuated Lattices: An Algebraic Glimpse at Substructural Logics, Studies in Logic and the Foundations of Mathematics, vol. 151. Elsevier (2007)
13. Galatos, N., Ono, H.: Glivenko theorems for substructural logics over FL. J. Symb. Log. **71**(4), 1353–1384 (2006). https://doi.org/10.2178/jsl/1164060460
14. Glivenko, V.: Sur la logique de M. Brouwer. Acad. Roy. Belg. Bull. Cl. Sci. **5**(14), 225–228 (1928)
15. Glivenko, V.: Sur quelques points de la logique de M. Brouwer. Acad. Roy. Belg. Bull. Cl. Sci. **5**(15), 183–188 (1929)
16. Harper, R., Honsell, F., Plotkin, G.D.: A framework for defining logics. J. ACM **40**(1), 143–184 (1993). https://doi.org/10.1145/138027.138060
17. Haykazyan, L.: More on a curious nucleus. J. Pure Appl. Algebra **224**, 860–868 (2020). https://doi.org/10.1016/j.jpaa.2019.06.014
18. Johnstone, P.T.: Stone Spaces, Cambridge Studies in Advanced Mathematics, vol. 3. Cambridge University Press (1982)
19. Johnstone, P.T.: Sketches of an Elephant: A Topos Theory Compendium. Vol. 2, Oxford Logic Guides, vol. 44. Clarendon Press (2002)
20. Kock, A.: Synthetic Differential Geometry, London Math. Soc. Lecture Note Ser., vol. 333. Cambridge University Press, 2nd edn. (2006). https://doi.org/10.1017/cbo9780511550812
21. Lambek, J.: The mathematics of sentence structure. Am. Math. Mon. **65**, 154–170 (1958). https://api.semanticscholar.org/CorpusID:123801856
22. Lambek, J.: On the calculus of syntactic types. In: Jakobson, R. (ed.) Structure of Language and its Mathematical Aspects (1961). https://api.semanticscholar.org/CorpusID:118284222
23. Lambek, J.: Some lattice models of bilinear logic. Algebra Universalis **34**, 541–550 (1995). https://api.semanticscholar.org/CorpusID:119782781
24. Makkai, M.: Towards a categorical foundation of mathematics. In: Makowsky, J.A., Ravve, E.V. (eds.) Logic Colloquium '95. Lecture Notes in Logic, vol. 11, pp. 153–190. Cambridge University Press (1998)
25. Negri, S.: Continuous domains as formal spaces. Math. Struct. Comput. Sci. **12**(1), 19–52 (2002). https://doi.org/10.1017/s0960129501003450
26. Norell, U.: Towards a Practical Programming Language Based on Dependent Type Theory. Ph.D. thesis, Chalmers University of Technology (2007)
27. Odintsov, S.P.: Logic of classical refutability and class of extensions of minimal logic. Logic Logical Philos. **9**(9), 91–107 (2001). https://doi.org/10.12775/llp.2001.006
28. Odintsov, S.P.: Negative equivalence of extensions of minimal logic. Stud. Logica. **78**(3), 417–442 (2004). https://doi.org/10.1007/s11225-004-6043-0

29. Odintsov, S.P.: Constructive Negations and Paraconsistency, Trends in Logic, vol. 26. Kluwer (2008). https://doi.org/10.1007/978-1-4020-6867-6
30. Ono, H.: Glivenko theorems revisited. Ann. Pure Appl. Log. **161**(2), 246–250 (2009). https://doi.org/10.1016/j.apal.2009.05.006
31. Paoli, F.: Substructural Logics: A Primer, Trends in Logic, vol. 13. Kluwer (2002). https://doi.org/10.1007/978-94-017-3179-9
32. Restall, G.: Substructural Logics. In: Zalta, E.N., Nodelman, U. (eds.) The Stanford Encyclopedia of Philosophy. Metaphysics Research Lab, Stanford University, Fall 2024 edn. (2024)
33. Rosenthal, K.I.: Quantales and their Applications, Pitman Research Notes in Mathematics, vol. 234. Longman Scientific & Technical, Essex (1990)
34. Schellinx, H.: The Noble Art of Linear Decorating. Ph.D. thesis, University of Amsterdam (1994). https://eprints.illc.uva.nl/id/eprint/1964
35. Segerberg, K.: Propositional logics related to Heyting's and Johansson's. Theoria **34**(1), 26–61 (1968). https://doi.org/10.1111/j.1755-2567.1968.tb00337.x
36. Simmons, H.: A framework for topology. In: Macintyre, A., Pacholski, L., Paris, J. (eds.) Logic Colloquium '77, Studies in Logic and the Foundations of Mathematics, vol. 96, pp. 239–251. North-Holland, Amsterdam (1978). https://doi.org/10.1016/S0049-237x(08)72007-x
37. Simmons, H.: A curious nucleus. J. Pure Appl. Algebra **214**, 2063–2073 (2010). https://doi.org/10.1016/j.jpaa.2010.02.011
38. Vezzosi, A., Mörtberg, A., Abel, A.: Cubical Agda: a dependently typed programming language with univalence and higher inductive types. J. Funct. Program. **31**, e8 (2021)
39. Wan, C.S.: Semi-substructural logics à la lambek. Electr. Proc. Theor. Comput. Sci. **415**, 195–213 (2024). https://doi.org/10.4204/eptcs.415.18

Complexities of Effective Reductions with Ordinal Turing Machines

Merlin Carl(✉)

Institut für Mathematik, Europa-Universität Flensburg, Flensburg, Germany
merlin.carl@uni-flensburg.de

Abstract. In [2] and [3], we introduced a notion of effective reducibility between set-theoretical Π_2-statements; in [4], this was extended to statements of arbitrary (potentially even infinite) quantifier complexity. We also considered a corresponding notion of Weihrauch reducibility, which allows only one call to the effectivizer of ψ in a reduction of φ to ψ. In Stammes [17], a considerably refined analysis was proposed that interpolates between these two notions: Namely, one can ask how many calls to an effectivizer for ψ are required for effectivizing φ. This allows us to make formally precise questions such as "how many ordinals does one need to check for being cardinals in order to compute the cardinality of a given ordinal?". In this paper, we will (partially) answer several questions of this type. Many of these answers turn out to be independent of ZFC.

1 Introduction and Basic Notions

Imagine you are given an ordinal α and you want to know its cardinality. You have a black box, which, given an ordinal β, will answer "yes" or "no", depending on whether or not β is a cardinal. Then you can solve your problem by checking all ordinals below α and then taking the supremum of all those that turn out to be cardinals. This method requires α many (in terms of the order type) accesses to the black box. Can you do any better?

In classical computability theory, such questions are treated under the name of "bounded queries", see, e.g., Martin and Gasarch, [7], Gasarch and Stefan [8], and Gasarch [9].[1]

To answer questions such as the one just formulated, this needs to be generalized to the transfinite. In line with [2,3] and [4], we will use Koepke's ordinal Turing machines ([13]) to model the idea of a transfinite "method".

More generally, in this paper, we want to measure the relative complexity – which we call "reduction complexity" – of certain functions embodying natural set-theoretical principles (such as "every set is equivalent to a cardinal") by the number of calls to one function that one needs in order to compute (on an

[1] We thank Vasco Brattka for pointing out this reference to us.

V. Brattka et al. (Eds.): CiE 2026, LNCS 16674, pp. 178–192, 2027.
https://doi.org/10.1007/978-3-032-31348-5_12

ordinal Turing machine) another. In the transfinite, such a notion was proposed by Stammes in his Master's thesis [17], Definition 4.2.[2]

Our main results are the following: Whether one call to the power set operator is enough to compute the continuum function is independent of ZFC (Proposition 5; to compute the cardinal successor α^+ of an ordinal α, one in general needs to check $\alpha^+ + 1$ many ordinals for being cardinals (Theorem 6), while the number of calls required to compute the cardinality of α depends heavily on the set-theoretical background and varies at least between being always finite and having no constant bound at all (Proposition 7 and Theorem 11); and to compute the subset of a set x of all elements satisfying a certain formula from a truth predicate, one cannot get away with fewer than the cardinality of x many calls to the truth predicate.

1.1 Basic Definitions and Notation

The model of computation underlying this paper are Koepke's ordinal Turing machines (OTMs), introduced in [13]. A pair (P, ρ) of an OTM-program and an ordinal parameter ρ is called a *parameter-program.* We encode arbitrary sets as sets of ordinals in the usual way by fixing, for a set x, an ordinal α and a bijection $f : \alpha \to \mathrm{tc}(\{x\})$ (where tc denotes the transitive closure) with $f(0) = x$ and letting $c_f(x) := \{p(\iota, \xi) : \iota, \xi < \alpha \wedge f(\iota) \in f(\xi)\}$, where p is Cantor's ordinal pairing function. If s is an ordered pair (a, b), we write $(s)_0 := a$, $(s)_1 := b$. By $\mathfrak{P}(X)$, we denote the power set of the class X, i.e., the class of all sub*sets* of X (which is a set in case X is). If M is a set or a class and x is a set, then $\mathrm{card}^M(x)$ denotes the cardinality of x in M, and, for an ordinal α, we denote by $\mathrm{cf}^M(\alpha)$ the M-cofinality of α (where the superscript M is dropped when M is V or clear from the context). For a set a, we denote by $L[a]$ the constructible universe relativized to a, see, e.g., [12], p. 192f. When $X \subseteq Y$ are sets or classes and $k \in \omega$, we write $X \prec_{\Sigma_k} Y$ to indicate that X is a Σ_k-submodel of Y.

Although the function types considered in this paper can be regarded as types of "effectivizers" (in the sense of [4]) of certain set-theoretical statements that were considered in [4], it saves some technical details to define them directly here.

- A Pot[3] function is a class function $F : \mathfrak{P}(\mathrm{On}) \to \mathfrak{P}(\mathrm{On})$ that maps every encoding of a set to an encoding of its power set.
- A PowerCard function is a class function $F : \mathfrak{P}(\mathrm{On}) \to \mathrm{On}$ that maps every encoding of a set to the cardinality of its power set.
- A NextCard function is a class function $F : \mathrm{On} \to \mathrm{On}$ that maps every ordinal α to its cardinal successor α^+.
- An OrdCard function is a class function $F : \mathrm{On} \to \mathrm{On}$ that maps every ordinal α to its cardinality $\mathrm{card}(\alpha)$.

[2] We only became aware of Stammes' work after finishing this paper and thank one of the anonymous referees for pointing it out to us.

[3] *Pot* here abbreviates "Potenzmenge", which is German for "power set".

- DecCard denotes the class function $F : \mathrm{On} \to \{0,1\}$ that is defined by

$$F(\alpha) = \begin{cases} 1, \text{ if } \alpha \text{ is a cardinal,} \\ 0, \text{ otherwise} \end{cases}.$$

- For $n \in \omega$, a Σ_n-Sep function is a class function $F : \mathfrak{P}(\mathrm{On}) \times \omega \times \mathfrak{P}(\mathrm{On}) \to \mathfrak{P}(\mathrm{On})$ that maps a triple $(c(S), k, c(\boldsymbol{p}))$ consisting of an encoding of a set S, an index k for a Σ_n $\in$-formula φ_k and an encoding of a finite tuple $\boldsymbol{p}$ of sets to an encoding of the set $\{x \in S : \varphi_k(x, \boldsymbol{p})\}$ if $\boldsymbol{p}$ has the right length, and to $\emptyset$, otherwise.
- For $n \in \omega$, a Σ_n-Truth function is a class function $F : \omega \times \mathfrak{P}(\mathrm{On}) \to \{0,1\}$ that maps a pair $(k, c(\boldsymbol{p}))$ consisting an index k for a Σ_n $\in$-formula φ_k and an encoding of a finite tuple $\boldsymbol{p}$ of sets to 0 or 1, according to the following condition:

$$F(k, c(\boldsymbol{p})) = \begin{cases} 1, \text{ if } \varphi_k(\boldsymbol{p}) \\ 0, \text{ otherwise} \end{cases}.$$

Note that, due to the possibility of different encodings, these are function types rather than particular functions, although for the types OrdCard, NextCard, PowerCard and DecCard, there is only one (class) function that belongs to them. Since the functions we consider are proper classes, these types cannot be introduced as objects in ZFC, as, taken at face value, they are classes of classes. One way to formalize the work below in ZFC is via talking about the formulas defining these functions instead.[4] We will not go into the details of the formalization.

In agreement with the definitions in [4], we say that one function type A is OTM-reducible to another function type B, written $A \leq_{\mathrm{OTM}} B$ if and only if there is a parameter-OTM-program (P, ρ) such that, for each function F of type B, $P^F(\rho)$ computes a function of type A. If there are two such programs P and Q, such that applying F once to the output of P and then applying Q to the output of F computes an A-function for each input, we say that A is ordinal Weihrauch reducible to B and write $A \leq_{\mathrm{oW}} B$.

The gap between OTM-reducibility and oW-reducibility is rather wide: In the case of the former, we allow any number of calls to the external function (i.e., the "effectivizer"), while in the latter, only a single one is allowed. In this paper, we work towards a more refined notion, differentiating reductions by the required number of calls to the extra function. To this end, we fix the following definition (which was first considered by Stammes, see [17], Definition 4.2):[5]

Definition 1 *(Cf. Stammes, [17], Definition 4.2)*
Let Φ and Ψ be types of (class) functions, and let $f : V \to On$ be a (class) function. We say that Φ is f-OTM-reducible to Ψ if and only if there is a

[4] Thus, one could, e.g., say that a pair $(\varphi, \boldsymbol{p})$ of a formula and a set parameter is a *PowerCard-formula* if $\varphi(x, y, \boldsymbol{p})$ defines a PowerCard function.

[5] Stammes only seems to consider *constant* bounds on the number of calls to the extra function; thus, the definition below is a slight generalization of his.

parameter-program (P,ρ) which OTM-reduces Φ to Ψ and, for any instance a and any F of type Ψ, the order type of calls to F in the computation $P^F(a,\rho)$ is at most $f(a)$. We denote this by $\Phi \leq^f_{OTM} \Psi$. In particular, if f is constant with value ξ, we write $\Phi \leq^{\xi}_{OTM} \Psi$.

If the order type of the set of times at which calls to the extra function take place is strictly below $f(x)$ for all but set many x, we write $\Phi \leq^{<f}_{OTM} \Psi$ and say that f is an upper bound to the reduction complexity of Φ to Ψ. If, on the other hand, the number of calls to the extra function is at least $f(x)$ for all but set many inputs x, we write $\Phi \leq^{\geq f}_{OTM} \Psi$ and say that f is a lower bound for the reduction complexity of Φ to Ψ.

2 Basic Tools

In this section, we gather some basic observations about reduction complexity, along with some lemmata that will help in proving concrete results later on.

Proposition 1. *(Cf. Stammes, [17], Proposition 4.5) If $\Phi \leq^f_{OTM} \Psi \leq^g_{OTM} \Gamma$, then $\Phi \leq^{f\cdot g}_{OTM} \Gamma$. (Here, $f \cdot g$ denotes the product of the functions $f, g : V \to On$.)*

While upper bounds for reduction complexities can be read off from concrete constructions, lower bounds require more work. Currently, our best tool for lower bounds is Proposition 4 below.

Notation 1 *For any (F-)OTM-program P, any set a and any sequence $\boldsymbol{s} := (s_\iota : \iota < \xi)$ of sets of ordinals, denote by $P^{F\to \boldsymbol{s}}(a)$ the computation that is obtained when, for any $\iota < \xi$, the ι-th call that P makes to F is answered with $\boldsymbol{s}(\iota)$.*

Thus, if ξ is the order type of all calls that $P^F(a)$ makes to F and $\boldsymbol{s}$ is the sequence of values that F returns at these calls, then $P^{F\to \boldsymbol{s}}(a)$ is the same computation (as a sequence of machine states) as $P^F(a)$. In particular, the computation of $P^F(a)$ can be OTM-effectively obtained from $\boldsymbol{s}$ and a.

We recall the definition of Σ_k-stable ordinals (see, e.g., [1], p. 178):

Definition 2 *For $k \in \omega$ and a a set, let $\sigma_k(a)$ be the minimal ordinal α such that $L_\alpha[a] \prec_{\Sigma_k} L[a]$.*

Remark 2. If a is transitive, let H be the Σ_n-hull of $a \cup \{a\}$ in $L[a]$; forming the transitive collapse $\overline{H}$ of the result will leave a fixed, so that, by the condensation lemma, we have $\overline{H} = L_\xi[a]$ for some ordinal ξ; by definition, $L_\xi[a] \prec_{\Sigma_n} L[a]$. Thus, whenever a is transitive, $\sigma_k(a)$ exists (and these are the only cases we care about in this paper). Moreover, the cardinality of H in $L[a]$ is bounded above by $\text{card}^{L[a]}(a) \times \aleph_0$, so, when a is infinite, we will have $\text{card}^{L[a]}(\xi) = \text{card}^{L[a]}(a)$ and in particular $\xi < (\text{card}(a))^+$.

The idea of the next two results is to show that, under certain conditions, a program needs to make many calls to the extra function in order to be able to run for a long time (but still halt afterwards).

Lemma 3. *Let P be an OTM-program, $\rho \in On$ a parameter, let F be a class function and t an initial tape content (i.e., an input, which we assume to be set-sized, i.e., bounded). Let $(\tau_\iota : \iota < \xi)$ be the sequence of times at which F is called, and let $\boldsymbol{v} := (v_\iota : \iota < \xi)$ be the sequence of values returned by F at these times. Then we have the following statements:*

1. *$\tau_\gamma < \sigma_1(\{t\} \cup (\rho+1) \cup tc(\{\boldsymbol{v} \restriction \gamma\}))$.*
2. *Thus, in particular, $\tau_\gamma < \sigma_1(\{t\} \cup (\rho+1) \cup tc(\{\boldsymbol{v}\}))$ for all $\gamma < \xi$.*
3. *If $P^F(t,\rho)$ halts, then its halting time is also strictly smaller than $\sigma_1(\{t\} \cup (\rho+1) \cup tc(\{\boldsymbol{v}\}))$.*

Proof. 1. Let $\gamma < \xi$, and let z be the computation state (tape content and inner state) of $P^F(t,\rho)$ at time $\tau_\gamma + 1$, i.e., right after the γ-th call to F. Then z is Σ_1-definable in t, ρ and γ, and, since γ is Σ_1-definable from $\boldsymbol{v} \restriction \gamma$, it is an element of $L_{\sigma_1(\{t\}\cup(\rho+1)\cup\mathrm{tc}(\{\boldsymbol{v}\}))}[t, \mathrm{tc}(\{\boldsymbol{v}\})]$.
Now, the statement that P, when run on the initial configuration z, makes at least one call to the extra function is Σ_1, and (by assumption) true in V and thus in $L[t, \mathrm{tc}(\{\boldsymbol{v} \restriction \gamma\})]$. Consequently, it must then also be true in $L_{\sigma_1(\{t\}\cup(\rho+1)\cup\mathrm{tc}(\{\boldsymbol{v}\restriction\gamma\}))}$. But then, the γ-th call must take place before time $\sigma_1(\{t\} \cup (\rho+1) \cup \mathrm{tc}(\{\boldsymbol{v} \restriction \gamma\}))$, which is what we wanted.
2. The second claim now follow immediately.
3. For the third claim, just note that, if all calls to F in the computation of $P^F(t,\rho)$ are contained in $\boldsymbol{v}$, then the statement "There is a time at which $P^F(t,\rho)$ halts" is Σ_1 in $\boldsymbol{v}$, t and ρ, and thus the same argument as above shows that the halting time must be below $\sigma_1(\{t\} \cup (\rho+1) \cup \mathrm{tc}(\{\boldsymbol{v}\}))$.

Proposition 4. *Suppose that $F : V \to V$ is a class function such that, for all sets x, we have $card(tc(F(x))) \leq card(tc(x))$ (i.e., in the terminology of Hodges [11], F does not "raise cardinalities"), let P be an OTM-program, $\rho \in On$, $t \subseteq On$ a set of ordinals (encoding the initial tape content) and $\kappa > max\{\rho, sup(t)\}$ an uncountable cardinal.*

1. *Then, if $P^F(t,\rho)$ halts in more than κ many steps, it makes κ many calls to F before time κ.*
2. *Moreover, if $P^F(t,\rho)$ computes for at least κ many steps and makes less than κ many calls to F before time κ, then it will diverge without making any further calls to F at or after time κ.*

Proof. 1. The computation clearly cannot make more than κ many calls to F before time κ; we thus only need to show that it cannot make less than that many calls.
First, let us assume that κ is regular. Suppose for a contradiction that $P^F(t,\rho)$ makes $\gamma < \kappa$ many calls to F before time κ. As in Lemma 3, let $(\tau_\iota : \iota < \gamma)$ be the times before κ at which calls to F were made during this computation and let $\boldsymbol{v} := (v_\iota : \iota < \gamma)$ be the values returned by F at these requests. By regularity of κ, we have $\delta := \sup\{\tau_\iota : \iota < \gamma\} < \kappa$. By induction, we have $\mathrm{card}(\mathrm{tc}(\{v_\iota\})) < \kappa$ for all $\iota < \gamma$: At successor levels, this follows from the

assumption on F, while at limit levels δ, this follows from the regularity of κ: If tape portions written on before time δ are always of length strictly smaller than κ, then the tape portion written on at time δ, being bounded above by the union of the lengths of these tape portions, will, as a union of strictly less than κ many ordinals strictly smaller than κ, also be strictly smaller than κ. It follows that $\text{card}(\text{tc}(\{\boldsymbol{v}\})) < \kappa$, so $\boldsymbol{v} \in H_\kappa^{L[\boldsymbol{v}]} = L_\kappa[\boldsymbol{v}]$. Now, if no further calls to F are made at all after time δ (including times greater than or equal to κ), then it follows from the remark above that $\sigma_1(\text{tc}(\{t\}) \cup \{\boldsymbol{v}\} \cup (\rho+1)) < \kappa$, so, by Lemma 3, the computation will halt in less than κ many steps, a contradiction. Thus, in this case, there must be at least one call to F taking place at time $\tau \geq \kappa$.

Let $\varphi(t, \delta, \boldsymbol{v}, \rho)$ be the statement "There is a time strictly above δ at which $P^{F \to \boldsymbol{v}}(t, \rho)$ makes a call to F". Then φ is Σ_1 (note that the value returned by F at this call is irrelevant to the truth of this statement). Since this statement is by assumption true in $L[\text{tc}(\{t\}), \text{tc}(\{\boldsymbol{v}\})]$, it must then also be true in $L_{\sigma_1(\text{tc}(\{t\}) \cup \{\boldsymbol{v}\} \cup (\rho+1))}[\text{tc}(\{t\}), \text{tc}(\{\boldsymbol{v}\})]$ (which contains all occurring parameters). But then, there must be a call to F between times δ and $\sigma_1(\text{tc}(\{t\}) \cup \text{tc}(\{\boldsymbol{v}\}) \cup (\rho + 1 \cup (\delta + 1))) < \kappa$, contradicting the definition of δ.

Now, if κ is singular, we can write it as a union of an increasing sequence $(\kappa_\iota : \iota < \gamma)$ of regular cardinals. Since $\kappa > \max\{\rho, \sup(v)\}$, there is $\xi < \gamma$ such that $\kappa_\iota > \max\{\rho, \sup(v)\}$ for $\iota \geq \xi$; without loss of generality, assume that $\xi = 0$. If $P^F(t, \rho)$ halts in more than κ many steps, then, in particular, for every $\iota < \gamma$, it halts in $> \kappa_\iota$ many steps and thus, before time κ_ι, makes at least κ_ι many calls to F. Since this is true for all $\iota < \gamma$, it will make κ many calls to F before time κ.

2. We now show the second claim. Suppose first that $P^F(t, \rho)$ makes actually less than $\text{cf}(\kappa) \leq \kappa$ many calls to F. From what we just showed, it follows that $P^F(t, \rho)$ will not halt. To see that there will be no calls to F at or after time κ, let δ be the supremum of times at which such calls are made before time κ; by assumption, we have $\delta < \kappa$. Let z be the computation state of $P^F(t, \rho)$ at time δ. Consider a slightly modified version Q of the program P which terminates when P makes a call to F. Thus, Q is an ordinary OTM-program that makes no calls to an extra function. Consequently, if Q is started on the initial configuration z, it will either halt in less than $\sigma_1(z) < \kappa$ many steps or not at all. However, if P made calls to F after time δ, then Q would, by assumption, terminate at or after time κ, a contradiction.
   This implies the second claim immediately if κ is regular. If κ is singular, let ξ be the order type of the calls made to F in the computation of $P^F(t, \rho)$ before time κ, and pick a regular cardinal $\lambda \in (\xi, \kappa)$. Before time λ, the computation has made $\leq \xi < \lambda = \text{cf}(\lambda)$ many calls to F, so the above implies it will in fact not make any further calls to F at or after time λ, and in particular not at or after time κ.

3 Several Reduction Complexities

We will now apply the above tools to various concrete cases.

3.1 PowerCard and Pot

We start by considering the question how many applications of Pot are necessary for computing PowerCard (that such a reduction is possible was observed in [4]).

It is easy to see that PowerCard becomes effective when two uses of Pot are allowed: One for computing the power set, and another one for computing the power set of the power set, which can then be searched for the (code for a) well-ordering of minimal length (note that a well-ordering of any given set x is implicit in its encoding as a set of ordinals). If only one application is allowed, the answer is less obvious. Indeed, the proof that one application of Pot is in general insufficient is considerably more technical.

Proposition 5. *PowerCard$\leq^1_{OTM}$Pot is independent of ZFC.*

Proof. If $V = L$, then we have PowerCard$\leq_{OTM}$Pot: Given $\alpha \in \mathrm{On}$, use the Pot-function to obtain $\mathfrak{P}(\alpha)$. Now use Koepke's algorithm to enumerate L on an OTM[6] until the first L-level $L_\gamma \ni \mathfrak{P}(\alpha)$ is found. The first such level will be $L_{\mathrm{card}^L(\mathfrak{P}(\alpha))+1}$ (since $L_{\mathrm{card}^L(\mathfrak{P}(\alpha))}$ is the first L-level that contains all constructible subsets of α and over this level, $\mathfrak{P}(\alpha)$ is definable), so γ is guaranteed to be a successor ordinal $\gamma = \beta + 1$ and we can return β.

To see that this does not follow from ZFC, let M be a transitive model of ZFC which satisfies $2^{\aleph_{\omega\alpha+1}} = \aleph_{\omega\alpha+4}$ for all ordinals α. Such a model can be obtained by Easton forcing (see [14], S. 265). We will obtain a model of ZFC in which PowerCard$\not\leq_{\mathrm{OTM}}$Pot by an iterated (class) forcing which successively sabotages all parameter-programs (P, ρ) that might be candidates for witnessing the reduction. To this end, let such a pair (P_k, ρ) be encoded as $\omega\rho + k$. This induces an ordering on these pairs; we will take care of these pairs in this induced ordering.

We now explain how to obtain, starting in a transitive model N of ZFC in which $2^{\aleph_{\omega\alpha+1}} = \aleph_{\omega\alpha+4}$ for all $\alpha \geq \omega(\omega\rho + k)$, a generic extension $N[G]$ of N in which (i) $2^{\aleph_{\omega\beta+1}} = \aleph_{\omega\beta+4}$ for all $\beta > \omega(\omega\rho + k)$ and (ii) (P_k, ρ) does not witness the $\leq^1_{\mathrm{OTM}}$-reduction between PowerCard and Pot. Let $\alpha := \omega\rho + k$. Let F be a Pot-function in N. If $P_k^F(\aleph_{\omega\alpha+1}, \rho)$ does not halt with output $\aleph^N_{\omega\alpha+4}$, we take the trivial generic extension $N[G] = N$.

Otherwise, we need to modify N to ensure that (P_k, ρ) no longer works. To this end, define $\mathbb{P}_{\rho,k}$ to be the standard forcing for collapsing $\aleph_{\omega\alpha+4}$ to $\aleph_{\omega\alpha+3}$, i.e., the set of partial functions from $\aleph_{\omega\alpha+3}$ to $\aleph_{\omega\alpha+4}$ of size $< \aleph_{\omega\alpha+2}$. As a successor cardinal, $\aleph_{\omega\alpha+2}$ is regular, so, by [14], Lemma 6.13, $\mathbb{P}_{\rho,k}$ is $\aleph_{\omega\alpha+2}$-closed for all ρ, k. Let $N[G]$ be a $\mathbb{P}_{\rho,k}$-generic extension of N. By [14], Theorem 6.14, $N[G]$ contains no subsets of $\aleph_{\omega\alpha+1}$ that are not in N, so that $\mathfrak{P}^N(\aleph_{\omega\alpha+1}) = \mathfrak{P}^{N[G]}(\aleph_{\omega\alpha+1})$. Moreover, the forcing will collapse $\aleph^N_{\omega\alpha+4}$ to $\aleph^N_{\omega\alpha+3}$.

Now, all elements of $\mathbb{P}_{\rho,k}$ have size $\leq \aleph_{\omega\alpha+1}$, so $\aleph_{\omega\alpha+4} \leq |\mathbb{P}_{\rho,k}| \leq \aleph_{\omega\alpha+4}^{\aleph_{\omega\alpha+1}}$. Using the Hausdorff formula ([12], p. 57), the fact that $\kappa^\lambda = 2^\lambda$ for $\kappa \leq \lambda$ for infinite cardinals κ and λ ([12], Lemma 5.20) and the assumption that $2^{\aleph_{\omega\alpha+1}} = \aleph_{\omega\alpha+4}$, we compute $\aleph_{\omega\alpha+4}^{\aleph_{\omega\alpha+1}} = \aleph_{\omega\alpha+3}^{\aleph_{\omega\alpha+1}} \cdot \aleph_{\omega\alpha+4} = \aleph_{\omega\alpha+2}^{\aleph_{\omega\alpha+1}} \cdot \aleph_{\omega\alpha+3}\aleph_{\omega\alpha+4} =$

[6] Cf, e.g., [6], Lemma 3.5.3.

$\aleph_{\omega\alpha+1}^{\aleph_{\omega\alpha+1}}\aleph_{\omega\alpha+2}\aleph_{\omega\alpha+3}\aleph_{\omega\alpha+4} = 2^{\aleph_{\omega\alpha+1}}\aleph_{\omega\alpha+4} = \aleph_{\omega\alpha+4}\cdot\aleph_{\omega\alpha+4} = \aleph_{\omega\alpha+4}$. Since an antichain cannot have more elements than the whole partial order, $\mathbb{P}_{\rho,k}$ satisfies the $\aleph_{\omega\alpha+5}$-cc and thus does not change the cardinals above $\aleph_{\omega\alpha+4}$ ([14], Lemma 6.9). Moreover, the continuum function above $\aleph_{\omega\alpha+4}$ is also not changed, for, if $\kappa \geq \aleph_{\omega\alpha+4}$, then the number of nice names for subsets of κ is bounded by the number of maximal antichains in $\mathbb{P}_{\rho,k}$ to the power of κ,[7] which, by the above, is bounded above by $(\text{card}(\mathbb{P}_{\rho,k})^{\aleph_{\omega\alpha+4}})^{\kappa} \leq (\aleph_{\omega\alpha+4}^{\aleph_{\omega\alpha+4}})^{\kappa} = \aleph_{\omega\alpha+4}^{\kappa} \leq (2^{\aleph_{\omega\alpha+4}})^{\kappa} = 2^{\kappa}$.

We now show that, in $N[G]$, (P_k, ρ) does not oW-reduce PowerCard to Pot. So let F be a Pot-function in $N[G]$. We consider $P_k^F(\aleph_{\omega\alpha+1}, \rho)$. Prior to the application of F in this computation, the cardinality of the number of computation steps is bounded by $\aleph_{\omega\alpha+1}$ (note that, as α is chosen so that $\aleph_{\omega\alpha+1}$ is guaranteed to be strictly larger than ρ, the computation will either halt in $< \aleph_{\omega\alpha+1}$ many steps or not at all). Moreover, the cardinality of the transitive closure of the input is also $\aleph_{\omega\alpha+1}$. Thus, the set S to which F is applied in the course of the computation also has the property that its transitive closure has cardinality $\leq \aleph_{\omega\alpha+1}$. Since such sets can be encoded as subsets of $\aleph_{\omega\alpha+1}$, and the forcing does not add any subsets of $\aleph_{\omega\alpha+1}$, we have $\mathfrak{P}^{N[G]}(S) = \mathfrak{P}^N(S)$. Thus, F will return (a code for) the same set that we would have obtained had the computation instead been performed with a Pot-function in N. Now, by assumption, in N, the result of the computation was $\aleph_{\omega\alpha+4}^N$. However, in $N[G]$, this is not even a cardinal, and certainly not the cardinality of the power set of $\aleph_{\omega\alpha+1}$. Thus, (P_k, ρ) does not witness the oW-reduction of PowerCard to Pot in $N[G]$.

We note that the step just described ensures that (P_k, ρ) gets the cardinality of the power set of $\aleph_{\omega\alpha+1}$ wrong. Since no new subsets of $\aleph_{\omega\alpha+1}$ are added by this step, it will preserve the fact that (P_l, ξ) gets the result for $\aleph_{\omega(\omega\xi+l)+1}$ wrong for all (P_l, ξ) that preceed (P_k, ρ) in the ordering defined above.

This explains one step of the iteration. We use iterated forcing with Easton support ([12], p. 395) to iterate it through all ordinals; since the iteration is progressively closed, it follows from [16], Lemma 117 and Theorem 98, that this iteration yields a model of ZFC.

What we have just seen thus means that PowerCard$\leq^2_{\text{OTM}}$Pot, where it is consistent with ZFC that this bound is optimal (but it is also consistent with ZFC that it is not).

3.2 Reductions to DecCard

In this section, we consider how many ordinals need to be checked for being cardinals for finding cardinal successors and for computing cardinalities of ordinals.

Theorem 6. *Let $f : On \to On$ be the (class) function $f(\alpha) = card(\alpha)^+ + 1$. Then f is the reduction complexity of NextCard to DecCard.*

[7] Cf. [14], p. 209f.

Proof. 1. A reduction from NextCard to DecCard works as follows (see [5], Proposition 14): Given $\alpha \in \mathrm{On}$, apply DecCard successively to all ordinals, starting with $\alpha+1$, until the answer is positive for some ordinal β; then return β. This works, and it clearly works within the required bound on the number of calls to DecCard.

2. Clearly, DecCard satisfies the assumption of Proposition 4. To see that f is optimal, let (P, ρ) witness the reduction, let F be DecCard, and let $\alpha > \rho$ be infinite, but not a cardinal. Let $\kappa := \mathrm{card}(\alpha)^{+}$. Now assume for a contradiction that $P^F(\alpha, \rho)$ makes less than $\kappa + 1$ many calls to F. This means that the number of calls to F it makes is at most κ, and we already know from Proposition 4 that many calls are made prior to time κ. Thus, all calls to F are made before time κ. But this means that all calls to F evaluate F at ordinals strictly less than κ; in particular, if F is applied to an ordinal greater than α during the computation, then it always returns 0 (since there are no cardinals between α, $\kappa = \alpha^{+}$).
   Let us modify (P, ρ) a bit to work as follows: On input α, it starts by successively calling F for all $\xi \leq \alpha$ and storing the results on some extra tape. After that, F is never used again; instead, we use the stored information to evaluate F for ordinals $< \alpha$, while, if $F(\xi)$ is requested for some $\xi > \alpha$, we always return 0. Using this, we can now simulate the computation of $P^F(\alpha, \rho)$ without actually using F ever again.
   Now, this modified computation makes $\alpha + 1 < \kappa$ many calls to F and thus, by Proposition 4, must halt in less than κ many steps or will not halt at all. But, by assumption, it outputs κ, which means that it runs for at least κ many steps before halting (for the read-write-head cannot reach cell κ in fewer than κ many steps), a contradiction.

The naive approach to reducing OrdCard to DecCard explained in the proof of Proposition 14(4) in [5] takes $\alpha + 1$ many steps in input α (in the worst case that α itself is a cardinal). A slight improvement would be to first check whether α is finite; if it is, return α; and if it is not, start by applying DecCard to α and then to the ordinals strictly below α, which would give us the new upper bound α. One might conjecture that this is optimal. Surprisingly, it is consistent with ZFC that it is not at all:

Proposition 7. *If $V = L$, then OrdCard$\leq^{<\omega}_{OTM}$DecCard.*

Proof. Given an ordinal α, the reduction works as follows: Use Koepke's algorithm to enumerate L on an OTM.[8] Whenever a new L-level L_β with $\beta > \alpha$ is produced, compute $\mathrm{card}^{L_\beta}(\alpha)$ and store it on some extra tape. If that value is not already present on that tape, check it with DecCard. If the answer is positive, return that value; otherwise, continue with the next β.

This clearly yields the right result: If some L-level contains a bijection between some ordinal γ and α, and γ is in fact an L-cardinal, then γ is the L-cardinality of α.

[8] Cf. the proof of Proposition 5 above.

Moreover, the sequence of values checked with DecCard is a strictly decreasing sequence of ordinals, and hence finite.

The above algorithm will work in general when $V = L[a]$ in the oracle a when a is a set of ordinals. If V is very much unlike L, and no extra oracle is given, however, this will not be true:

Proposition 8. *If $0^\sharp$ exists, then OrdCard$\not\leq^{<\omega}_{OTM}$DecCard.*

Proof. If $0^\sharp$ exists, then the V-cardinals are order indiscernibles for L.[9] Assume for a contradiction that (P, ρ) is a parameter-program that witnesses the reduction OrdCard$\leq^{<\omega}_{\text{OTM}}$DecCard. Pick a Silver indiscernible $\xi > \rho, \aleph_1$ which is not a V-cardinal, and let F be DecCard. By assumption, $P^F(\xi, \rho)$ computes card(ξ) and uses F only finitely often along the way. This will in particular reveal only finitely many cardinals; let us say that $s := (\kappa_i : i < n)$ is the sequence of cardinals found along the way, where $n \in \omega$. Then we can view the computation as running relative to a function that returns 1 on elements of s and 0 everywhere else; thus, the fact that $P^F(\xi, \rho) \downarrow= \text{card}(\xi)$ can be expressed as a first-order formula $\varphi(\kappa_0, ..., \kappa_{n-1}, \text{card}(\xi))$. Due to absoluteness of computations, this formula will be absolute between L and V. However, since card(ξ) is an uncountable cardinal, the class of Silver indiscernibles is unbounded in card(ξ); thus, there will be a Silver indiscernible β such that $\text{card}(\xi) > \beta > \max(\{\kappa \in s : \kappa < \text{card}(\xi)\})$. It follows that $L \models \varphi(\kappa_0, ..., \kappa_{n-1}, \beta)$; but the computation $P^F(\xi, \rho)$ cannot halt with two different outputs, a contradiction.

Remark 9. We note that, in $L[0^\sharp]$, regardless of the input, we can get away with $< \aleph_\omega + \omega$ calls to DecCard. This works by first running through the first $\aleph_\omega + 1$ ordinals, checking all of them with DecCard until $\omega + 1$ many cardinals have been found (which will be the cardinals $\aleph_0, \aleph_1, ..., \aleph_\omega$). Then $(\aleph_i : i < \omega)$ is an infinite set of Silver indiscernibles, and we have $L_{\aleph_\omega} \prec L$. Thus, by computing a code for $L_{\aleph_\omega}$ from $\aleph_\omega$ and evaluating formulas with parameter $\aleph_1, \aleph_2, ...$ in $L_{\aleph_\omega}$, we can compute $0^\sharp$. But then, as described above, relative to $0^\sharp$, we only need finitely many extra calls to DecCard to compute the cardinality of any given ordinal. Thus, in $L[0^\sharp]$, we still have a constant upper bound to the number of necessary calls.

Note that this construction also works relative to iterated sharps: For example, in order to evaluate OrdCard in $L[(0^\sharp)^\sharp]$, we first determine $\aleph_1, ..., \aleph_\omega$, as above; then, on the basis of this, we compute $0^\sharp$ by evaluating the truth predicate in $L_{\aleph_\omega}$; and then, we compute $(0^\sharp)^\sharp$ by evaluating the truth predicate in $L_{\aleph_\omega}[0^\sharp]$.

We do not know whether this bound is optimal, but conjecture that it is not.

Question 10. Is there a reduction of OrdCard to DecCard which provably in ZFC improves on the naive approach explained above in the sense that, for some $\alpha \in \text{On}$, the cardinality of the number of calls required on input $\xi > \alpha$ will be strictly smaller than that of ξ?

[9] See, e.g., [12], Theorem 18.1(ii).

At least consistently, there need not be a constant bound on the reduction complexity of OrdCard to DecCard:

Theorem 11. *There is a class forcing extension of L which satisfies ZFC such that OrdCard$\not\leq^{<\alpha}_{OTM}$DecCard for every ordinal α.*

Proof. For each triple (P, ρ, α) consisting of a parameter-program (P, ρ) and an ordinal α, we sabotage the claim that (P, ρ) reduces OrdCard to DecCard with complexity bounded above by α. Let T be the class of all such triples (P, ρ, α), and let $\leq^T$ be the ordering on T induced by Cantor's pairing function (used twice to encode triples); this is a linear ordering of order type On.

With each triple $t = (P, \rho, \alpha) \in T$, we associate an ordinal $\kappa(t)$ so that $\kappa(t)$ is an uncountable limit cardinal in L and such that

$$\mathrm{cf}(\kappa(t)) > \lambda(t) := (\sup(\{\kappa(t')^+ : t' <^T t\}) \cup \rho + 1 \cup \alpha + 1)^+.$$

Note that, by Proposition 4, if F is DecCard (which is clearly definable and does not raise cardinals, as it only outputs 0 or 1), if $P^F(\kappa(t), \rho)$ stopped after at least $\kappa(t)^+$ many steps, it would have made at least $\kappa(t)^+ > \alpha$ many calls to F at time $\kappa(t)^+$, and thus have violated the supposed upper bound α to the number of these calls. Thus, we only need to take care of cases in which $P^F(\kappa(t), \rho)$ halts in less than $\kappa(t)^+$ many steps – in all other cases, it is either guaranteed to make more calls to the extra function, or it is guaranteed not to halt.

The desired target model will arise as an iterated forcing of class length On using Easton support. Suppose that, for some $t \in T$, an intermediate model $M_{<t}$ has been obtained that takes care of all $t' < t = (P, \rho, \alpha) \in T$. The forcing will be set up in a way that taking care of t does not change cardinals or cofinalities $\leq \kappa(t')^+$ for all $t' < t$. (*) It will, moreover, not collapse cardinals greater than $\kappa(t)$. (**) Thus, in particular, if κ satisfies the definition of $\kappa(t)$ in the ground model, it will continue to do so in the generic extension: All forcings for $t' <^T t$ will leave $\lambda(t)$ intact.

Let F be DecCard in M. Our aim is to ensure that, in the target model, the cardinality of $\kappa(t)$ is not computed correctly with less than α many calls to F. We define $G : \mathrm{On} \to \mathrm{On}$ by

$$G(\xi) := \begin{cases} 0, \text{ if } \xi = \kappa(t) \\ F(\xi), \text{ otherwise} \end{cases}.$$

We now consider the computation of $P^G(\kappa(t), \rho)$. We distinguish the following cases:

1. Before time $\kappa(t)^+$, the computation $P^G(\kappa(t), \rho)$ contains fewer than α many calls to G and has not halted.
2. Before time $\kappa(t)^+$, the computation $P^G(\kappa(t), \rho)$ contains fewer than α many calls to G and it has halted.
3. The computation $P^G(\kappa(t), \rho)$ has made at least α many calls to G before time $\kappa(t)^+$.

Note that, since $\kappa(t) > \rho$ and because F does not raise cardinals, the program can write on at most one extra cell on each tape per time step, so that, whatever it can write in less than $\kappa(t)^+$ many steps in inputs ρ and $\kappa(t)$ will have cardinality less than $\kappa(t)^+$; in particular, all calls to G the computation will make before time $\kappa(t)^+$ will concern ordinals less than $\kappa(t)^+$.

We now let

$$\delta := \sup(\{\iota < \kappa(t) : \text{Among the first } \alpha \text{ calls to } G \text{ before time } \kappa(t)^+ \text{ in the computation } P^G(\kappa(t), \rho) \text{ one concerns } \iota\}). \quad (1)$$

Since $\mathrm{cf}(\kappa(t)) > \alpha$ by definition, we have $\delta < \kappa(t)$. We now let λ be the smallest regular $M_{<t}$-cardinal in $((\max(\delta, \lambda(t)^+), \kappa(t))$ which is not equal to the output of $P^G(\kappa(t), \rho)$. If there is no such output, this is trivial; if there is, $\kappa(t)$ being a limit cardinal guarantees the existence of λ.

We now force with the Levy collapse forcing $\mathrm{Coll}(\kappa(t), \lambda)$, which consists of the partial functions from $\kappa(t)$ to λ of cardinality $< \lambda$, ordered by $\supseteq$.

Since this is λ-closed, no cardinals below λ will be changed; thus, $(*)$ is satisfied. Moreover, it satisfies the κ^+-cc, and thus does not collapse cardinals $\geq \kappa(t)^+$, so that $(**)$ is satisfied as well.

Let M_t be the generic extension, and let F' be DecCard in M_t. We now show for each of the three cases that, in M_t, it is not true that $P^{F'}(\kappa(t), \rho) \downarrow= \mathrm{card}(\kappa(t))$ with less than α many calls to F'.

Let γ be the time before $\kappa(t)^+$ at which the α-th call to G takes place in the computation of $P^G(\kappa(t), \rho)$ if such a time exists, and $\gamma := \kappa(t)^+$, otherwise. The crucial observation is that, as seen above, (i) this forcing does not change cardinals $\leq \delta$, (ii) no calls concerning ordinals in $(\delta, \kappa(t))$ are made to G before time γ in the computation of $P^G(\kappa(t), \rho)$, (iii) we have $F'(\kappa(t)) = 0$ since $\kappa(t)$ is collapsed and (iv) both F' and G return 0 for every ordinal in $(\kappa(t), \kappa(t)^+)$. Thus, F' will return the same value as G for any ordinal to which it is applied in the computation before time γ. Consequently, the computation of $P^G(\kappa(t), \rho)$ and $P^{F'}(\kappa(t), \rho)$ in fact agree up to time γ.

Now, in case (1), we know that $P^G(\kappa(t), \rho)$ will not halt and it will make no calls to G at or after time $\kappa(t)^+$. Thus, the computations $P^G(\kappa(t), \rho)$ and $P^{F'}(\kappa(t), \rho)$ in fact agree entirely in this case, so $P^{F'}(\kappa(t), \rho) \uparrow$.

In case (2), the output of $P^{F'}(\kappa(t), \rho)$ is the same as that of $P^G(\kappa(t), \rho)$ (since the computations agree). But in M_t, we have $\mathrm{card}(\kappa(t)) = \lambda$, which, by choice of λ, is not the output of $P^G(\kappa(t), \rho)$.

In case (3), $P^{F'}$ has made at least α many calls to F' before time γ and thus does not adhere to the supposed bound on the number of calls.

This sequence of forcings is progressively closed. Thus, again by Reitz [16] Lemma 117 and Theorem 98, the iteration yields a model of ZFC.

Remark 12. The above proof only invokes rather general properties of DecCard; it thus applies at least to every class function instead of DecCard mapping to $\{0, 1\}$, and in fact to a considerably wider range of class functions.

3.3 On Separation and Truth

In Lemma 11 of [5], we showed that Σ_n-separation is reducible to Σ_n-truth using at least[10] card(a) applications in input a, using the obvious idea: Run through the given set and test each element with the truth predicate for satisfying the property in question. But are that many calls to truth really necessary? This is the question we will treat in this section.

We begin by noting that a finite number of calls to any Σ_k-truth predicate – and thus, in particular, a single such call – is not enough.

Proposition 13. *There is no $n \in \omega$ such that Σ_1-Sep$\leq^{<\omega}_{OTM}$ Σ_n-truth.*

Proof. Assume for a contradiction that (P, ρ) witnesses such a reduction of Σ_1-Sep to Σ_n-truth, for some $n \in \omega$. Let $h_\rho := \{i \in \omega : P_i(\rho)\downarrow\}$ be the OTM-halting problem in the parameter ρ. Then h_ρ is a subset of ω. Let $\varphi(\rho)$ be the Σ_1-formula that defines h_ρ as a subset of ω in the parameter ρ, and let F be a Σ_n-truth-function. Now, by assumption, only finitely many calls are made to F in the computation of $P^F(\omega, \rho)$. But this means that the sequence $\boldsymbol{s}$ of the finitely many outcomes can be hardcoded in a variant Q of P that, when P calls F for the j-th time, just uses the j-th bit of $\boldsymbol{s}$ as the result to continue. Thus, Q is an OTM-program which, in the parameter ρ, computes h_ρ, i.e., solving the halting problem for OTM-programs in the parameter ρ, a contradiction.

The following lemma summarizes the main idea behind the argument:

Lemma 14. *Let F, G be functions mapping sets of ordinals to sets of ordinals. Assume that there is a parameter-program (P, ρ) such that, for some set a, $P^F(a, \rho)$ computes $G(a)$ and makes only finitely many calls to F.*

1. *If $F(x) \in L$ for all x (i.e., $F(x)$ is parameter-OTM-computable), then $G(a) \in L$.*
2. *If $F(x)$ is OTM-computable in the parameter ρ for all x, then $G(a)$ is OTM-computable in the parameter ρ and the input a, for every input a.*

Proof. We only show (2); the proof for (1) is completely analogous.

Let $\boldsymbol{v} = (v_1, ..., v_k)$ be the sequence of values that F returns in the finitely many calls to F. By assumption, let $Q_1, ..., Q_k$ be OTM-programs that compute $v_1, ..., v_k$ in the parameter ρ, respectively. Then we can modify P to work as follows: For $i \leq k$, in the i-th call to F, it runs $Q_i(\rho)$ and uses the output as the return value of F. The computation will be (as a sequence of computational states) identical to that of $P^F(a, \rho)$, and thus have the same output; but it is also a computation that only uses the input a and the parameter ρ.

Intuitively, one should expect that there is no better way to obtain separated sets from truth than considering each element of the given set separately. Under a moderate extra assumption, we can indeed prove this:

[10] Note that, since the computation works through a given code for a, which may order a in a non-minimal way, it may well make more such calls in terms of the order-type.

Lemma 15. *Assume that there is a definable global well-ordering $\leq^*$ of V which is compatible with the $\in$-relation.*[11] *Then the following is true: Let $f: V \rightarrow On$ be a class function such that, for all but set many values of x, we have $f(x) <$ card(x). Then, for no $m \in \omega$ we have Σ_4-Sep$\leq^f_{OTM} \Sigma_m$-truth.*

Proof. Let f be as in the assumption. We can assume without loss of generality[12] that $\leq^*$ is Σ_2-definable. Assume for a contradiction that, for some $m \in \omega$, some parameter-program (P, ρ) witnesses Σ_4-Sep$\leq^f_{\text{OTM}} \Sigma_m$-truth. Pick an uncountable cardinal $\kappa > \rho$ such that $2^{<\kappa} = \kappa$ and let $\leq'$ be the $\leq^*$-smallest well-ordering of $\mathfrak{P}^{<\kappa}(\kappa)$ in order type κ.[13] Let $g : \kappa \rightarrow \mathfrak{P}^{<\kappa}(\kappa)$ be the enumeration induced by $\leq'$, and define $h : \kappa \rightarrow \omega \times \mathfrak{P}^{<\kappa}(\kappa)$ as $h(\omega\iota + k) = (k, f(\iota))$ for $\iota < \kappa$, $k \in \omega$.

Define a subset $S \subseteq \kappa$ as follows: For $\iota < \kappa$, we have $\iota \in S :\Leftrightarrow \neg P^{F \rightarrow h(\iota)_1}_{h(\iota)_0}(\rho) \downarrow = 1$. S is clearly definable as a subset of κ, and the definition is Σ_4.

Now, by assumption, $P^F(\rho)$ computes S, making $\xi < \kappa$ many calls to F. Let $\boldsymbol{v} := (i_\iota : \iota < \xi)$ be the sequence of values returned by F to these requests. Thus, $P^{F \rightarrow \boldsymbol{v}}(\rho)$ computes S as well. Clearly, $\boldsymbol{v}$ can be regarded as (corresponding to) an element of $\mathfrak{P}^{<\kappa}(\kappa)$. We can modify P to a program Q – in the same parameters – which, rather than writing S to the tape and halting, takes as an additional input some $\iota < \kappa$ and decides whether $\iota \in S$. Let k be the index of Q in the enumeration of programs, and let $\alpha < \kappa$ be the pre-image of $(k, \boldsymbol{v})$ under h. Then $P^{F \rightarrow h(\alpha)_1}_{h(\alpha)_0}(\rho) \downarrow = 1 \Leftrightarrow \alpha \in S \Leftrightarrow \neg P^{F \rightarrow h(\iota)_1}_{h(\iota)_0}(\rho) \downarrow = 1$, a contradiction.

Question 16. Can the assumption of a definable global well-ordering be eliminated from the last result? We conjecture that this is the case.

4 Conclusion and Further Work

Clearly, there are many other principles that could be meaningfully investigated with respect to reduction complexity.

While most results in this paper should be conceptually stable under changes of the underlying model of computation, some of them might allow for refinements that are more to the point. The reducibility concept defined and applied in this paper allows formalizing of intuitively meaningful and natural questions such as "how many applications of power set are needed in order to calculate the cardinality of power sets?". However, there are some questions of this kind for which the answer given by our formalization is not quite satisfying. A typical example would be "how many applications of power set are needed in order to calculate cardinalities?". The answer given in [5], Proposition 14 – that one application is enough – depends heavily on the fact that, since sets need to be

[11] This is equivalent to assuming $V = \text{HOD}$.

[12] Cf., e.g., Hamkins, [10].

[13] To see that there are unboundedly many such κ, note that, defining $\alpha_0 := \aleph_1$, $\alpha_{\iota+1} := 2^{\alpha_\iota}$ and $\alpha_\lambda := \bigcup_{\iota<\lambda} \alpha_\iota$ for a limit ordinal λ, each fixed point of the normal function $\iota \mapsto \alpha_\iota$ will have this property.

encoded before an OTM can operate on them, every set given to an OTM comes with a well-ordering. For such questions, it should be interesting to study similar reducibility notions on models of transfinite computability that can compute directly on sets, rather than on encodings of sets; Passmann's "Set Register Machines" introduced in [15] would be an example of such a notion.

Acknowledgements. The author is grateful to the three anonymous referees of this paper for pointing out a number of typos and also several ways to improve the presentation and the correctness of the paper.

References

1. Barwise, J.: Admissible Sets and Structures. Cambridge University Press, Cambridge (2017)
2. Carl, M.: Generalized effective reducibility. In: Beckmann, A., et al. (eds.) Pursuit of the Universal. 12th Conference on Computability in Europe (2016)
3. Carl, M.: Effectivity and reducibility with ordinal turing machines. Computability **10**(4) (2021)
4. Carl, M.: Full generalized effective reducibility. In: A. Beckmann et al. (eds.) Crossroads of Computability and Logic: Insights, Inspirations, and Innovations, 21st Conference on Computability in Europe. Proceedings, CiE 2025 Proceedings (2025)
5. Carl, M.: Effective reducibility for statements of arbitrary quantifier complexity with ordinal turing machines. Preprint: arXiv:2411.19386 (2025)
6. Carl, M.: Ordinal Computability. An Introduction to Infinitary Machines. De Gruyter Berlin Boston (2019)
7. Gasarch, W., Martin, G.: Bounded Queries in Recursion Theory. Progress in Computer Science and Applied Logic, vol. 16. Birkhäuser Boston Basel (1998)
8. Gasarch, W.: Bounded queries in recursion theory: a survey. In: Proceedings of the Sixth Annual Structure in Complexity Theory Conference (1991). https://doi.org/10.1109/SCT.1991.160245
9. Gasarch, W., Stephan, F.: A techniques oriented survey of bounded queries. In: Cooper, S., Truss, J. (eds.) Models and Computability. London Mathematical Society Lecture Note Series. Cambridge University Press (1999)
10. Hamkins, J.D.: Complexity of definable global choice functions. https://mathoverflow.net/users/1946/joel-david-hamkins, https://mathoverflow.net/q/472862. Accessed 08 June 2024
11. Hodges, W.: On the effectivity of some field constructions. In: Proceedings of the London Mathematical Society, vol. s3-32(1) (1976)
12. Jech, T.: Set Theory. Third Millenium Edition. Springer, Berlin Heidelberg New York (2003)
13. Koepke, P.: Turing computations on ordinals. Bull. Symbolic Logic **11**(3)
14. Kunen, K.: Set Theory. An Introduction to Independence Proofs, Elsevier, Amsterdam (1980)
15. Passmann, R.: The first-order logic of CZF is intuitionistic first-order logic. J. Symbolic Logic **89**(1) (2022)
16. Reitz, J.: The Ground Axiom. PhD Thesis, City University of New York (2024). https://arxiv.org/pdf/math/0609064
17. Stammes, J.: Ordinal Reductions on Choice Principles. Master Thesis, Universiteit van Amsterdam (2024). https://dspace.uba.uva.nl/server/api/core/bitstreams/82b5ccde-8cc8-464b-a968-74d7ea88b41a/content

Weihrauch Reducibility Between Ramsey-Type Theorems and Well-Ordering Principles at the Level of Σ_2^0-Induction

Lorenzo Carlucci(✉) and Giordano Celli

Department of Mathematics, Sapienza University,
Piazzale Aldo Moro 5, 00185 Roma, Italy
lorenzo.carlucci@uniroma1.it, celli.1873376@studenti.uniroma1.it

Abstract. We study the relations under Weihrauch reducibility of the well-ordering preservation principle for the operator $X \mapsto X^\omega$ and the Ordered Ramsey Theorem. Both are known to be equivalent to Σ_2^0-induction in Reverse Mathematics.

We show that the Ordered Ramsey Theorem is Weihrauch-equivalent to the parallel product of the well-ordering preservation principle for the operator $X \mapsto X^\omega$ and the Eventually Constant Tail principle.

By previous work from Pauly, Pradic and Soldà, the Ordered Ramsey Theorem is known to be Weihrauch-equivalent to the parallel product of the Eventually Constant Tail principle and the parallelization of the jump of the Limited Principle of Omniscience. We show that the latter principle and the well-ordering preservation principle for $X \mapsto X^\omega$ are Weihrauch-incomparable.

1 Introduction and Motivation

A *well-ordering (preservation) principle* – WOP for short – states that some operator on linear orderings preserves well-ordering. A *Ramsey-type theorem* states that some class of colorings is partition-regular. Both families of principles are well investigated in Reverse Mathematics [5]. Many prominent systems can be characterized in terms of WOPs (see [9,11]) and of Ramsey-type theorems (see [5]). Usually the equivalence of the latter two is obtained as an indirect by-product. We refer to [2] for examples and discussion.

In the present paper we study from the viewpoint of Weihrauch reducibility the well-ordering principle for the operator $X \mapsto X^\omega$ and the Ordered Ramsey Theorem. This is one way of making precise the question from [3] whether *direct proofs* exist, in one direction or the other, between Ramsey-type principles and WOPs.

The motivation for the present work is twofold. First, we continue the line of research initiated in [2] concerning the possibility of establishing Weihrauch reductions from WOPs to combinatorial theorems from Ramsey Theory. For

V. Brattka et al. (Eds.): CiE 2026, LNCS 16674, pp. 193–207, 2027.
https://doi.org/10.1007/978-3-032-31348-5_13

example it is shown in [2] that the WOP for the operator $X \mapsto \omega^X$ is Weihrauch-reducible to Ramsey's Theorem for 2-colorings of 3-elements subsets of $\mathbb{N}$ as well as to Hindman's Theorem restricted to finite sums of 3 terms. The initial motivation for the present work comes from the observation in [2] that a similar proof shows that the WOP for the operator $X \mapsto X^\omega$ reduces to Ramsey's Theorem for finite colorings of 2-elements sets. The latter result is not optimal from the point of view of Reverse Mathematics, since the WOP in question is equivalent to Σ^0_2-induction [12], while Ramsey's Theorem for pairs and arbitrarily many colors implies the stronger principle of Σ^0_3-bounding [7]. It was therefore natural to ask whether the same proof-technique from [2] could be applied to reduce the WOP for $X \mapsto X^\omega$ to a Ramsey-type statement equivalent to Σ^0_2-induction. The Ordered Ramsey Theorem, introduced in [8], is one such candidate. Its Weihrauch complexity is studied in [10] by Pauly, Pradic and Soldà, whose work served both as a starting point and as a guide for ours. Note that reductions of Ramsey-type theorems to WOPs are not to be found in previous literature.

2 Background

We introduce the needed technical notions, the main principles of interest and the relevant results from previous literature.

The principles considered in this paper can be expressed as *instance-solution problems*, i.e. as non-empty partial functions $\mathsf{P} : A \to \mathcal{P}(B) \setminus \{\emptyset\}$ for some sets A and B. The elements of the domain of P are called *instances* of P. If $a \in A$ is an instance of P then each $b \in \mathsf{P}(a) \subseteq B$ is called a *solution of* P *to* a, or a P-solution to a. It is well-known that $\forall\exists$ theorems of the form $\forall x(F(x) \to \exists y G(x, y))$ of interest in Reverse Mathematics can be naturally expressed as problems (see [5]).

2.1 Weihrauch Reducibility

Weihrauch reducibility has become in the last years of major interest in Computable Mathematics and in Reverse Mathematics. We give the basic definitions.

Definition 1 ((Strong) Weihrauch reducibility). *Let* P *and* Q *be problems.*

1. P *is Weihrauch-reducible to* Q *(denoted* $\mathsf{P} \leq_{\mathrm{W}} \mathsf{Q}$*), if there exist Turing functionals* H *and* K *such that: if* X *is any* P*-instance then* $H(X)$ *is a* Q*-instance, and if* $\hat{Y}$ *is any* Q*-solution to* $H(X)$ *then* $K(X \oplus \hat{Y})$ *is a* P*-solution to* X*.*
2. P *is strongly Weihrauch-reducible to* Q *(denoted* $\mathsf{P} \leq_{\mathrm{sW}} \mathsf{Q}$*), if there exist Turing functionals* H *and* K *such that: if* X *is any* P*-instance then* $H(X)$ *is a* Q*-instance, and if* $\hat{Y}$ *is any* Q*-solution to* $H(X)$ *then* $K(\hat{Y})$ *is a* P*-solution to* X*.*

We abbreviate by $\mathsf{P} \equiv_{\mathrm{W}} \mathsf{Q}$ the fact that $\mathsf{P} \leq_{\mathrm{W}} \mathsf{Q}$ and $\mathsf{Q} \leq_{\mathrm{W}} \mathsf{P}$. Similarly for $\mathsf{P} \equiv_{\mathrm{sW}} \mathsf{Q}$. We abbreviate by $\mathsf{P} <_W \mathsf{Q}$ the fact that $\mathsf{P} \leq_{\mathrm{W}} \mathsf{Q}$ and $\mathsf{Q} \not\leq_{\mathrm{W}} \mathsf{P}$. Similarly for $<_{\mathrm{sW}}$.

Some of our results are expressed in terms of a number of structural operations on problems, which preserve Weihrauch degrees.

Definition 2 (Parallel product and finite parallelization). *Let* P *and* Q *be problems.*

1. *The parallel product of* P *and* Q*, denoted by* $\mathsf{P} \times \mathsf{Q}$*, is the problem whose instances are all pairs* (X_0, X_1) *where* X_0 *is a* P*-instance and* X_1 *is a* Q*-instance, with the solutions being all pairs* (Y_0, Y_1) *such that* Y_0 *is a* P*-solution to* X_0 *and* Y_1 *is a* Q*-solution to* X_1*.*
2. *The finite parallelization of* P*, denoted by* P^**, is the problem whose instances are all sequences* $\langle n, X_0, \ldots, X_{n-1}\rangle$*, where* $n \geq 1$ *and* $X_0, \ldots, X_{n-1}$ *are* P*-instances, with the solutions being all sequences* $\langle Y_0, \ldots, Y_{n-1}\rangle$ *such that* Y_i *is a* P*-solution to* X_i *for all* $i < n$*.*

For every problem P and $n \geq 1$ we write P^n to mean $\underbrace{\mathsf{P} \times \cdots \times \mathsf{P}}_{n \text{ times}}$.

The following problems have emerged as useful benchmarks in Weihrauch Complexity.

1. LPO (Limited Principle of Omniscience) takes as input an infinite binary sequence p, and outputs 0 if $p = 0^{\mathbb{N}}$ and 1 otherwise.
2. LPO' (Jump[1] of Limited Principle of Omniscience) takes as input (a code for) an infinite sequence $\langle p_0, p_1, \ldots\rangle$ of binary sequences such that the function $p(i) = \lim_{s\to\infty} p_i(s)$ is defined for every $i \in \mathbb{N}$, and outputs $\mathsf{LPO}(p)$.
3. $\mathsf{IsFinite}$ (Finiteness Problem) takes as input a binary sequence $p \in 2^{\mathbb{N}}$, and outputs 1 if p contains finitely many occurrences of 1 and 0 otherwise.

As noted in [10], both a call to LPO' and a call to $\mathsf{IsFinite}$ amount to a binary Σ^0_2-question. We will only use the fact that $\mathsf{IsFinite} \leq_{\mathrm{W}} \mathsf{LPO}'$ (a witnessing reduction can be given as in the proof of [10, Lemma 9]), although it is fairly easy to see that the strong equivalence $\mathsf{IsFinite} \equiv_{\mathrm{sW}} \mathsf{LPO}'$ holds.

4. $\mathsf{TC}_{\mathbb{N}}$ (Totalization of Closed Choice on $\mathbb{N}$) takes as input an enumeration e of any subset of $\mathbb{N}$ (we allow the possibility that $\mathrm{ran}(e) = \mathbb{N}$) and outputs a natural number m such that either $m \notin \mathrm{ran}(e)$ or $\mathrm{ran}(e) = \mathbb{N}$.
5. ECT (Eventually Constant Tail) takes as input a pair (n, f), where $n \geq 1$ and $f : \mathbb{N} \to \{0, \ldots, n-1\}$, and outputs a natural number n_0 such that for all $x \geq n_0$ there exists $y > x$ satisfying $f(y) = f(x)$. A number such as n_0 is called a bound for f.

From the point of view of Reverse Mathematics the $\forall\exists$ theorem corresponding to ECT, stating that every finite coloring of the set of natural numbers admits a bound, is proved equivalent to Σ^0_2-induction in [4, Theorem 7]. In [4, Theorem 9] ECT is characterized as the finite parallelization of $\mathsf{TC}_{\mathbb{N}}$.

Lemma 1 (Davis et alii). $\mathsf{ECT} \equiv_{\mathrm{W}} (\mathsf{TC}_{\mathbb{N}})^*$.

[1] As defined in [1].

2.2 Well-Ordering Principles

A well-ordering preservation principle states that some operator on linear orders preserves well-orderings. In the present paper we are interested in the WOP based on the operator $X \mapsto X^\omega$, which corresponds to ordinal exponentiation to the power ω. When referring to a linear ordering $\mathcal{X} = (X, \leq_X)$ we systematically use X also for $(X, \leq_X)$. The following definition generalizes exponentiation of well-orders ([6, Definition 2.1]) to linearly ordered bases (as in [12, Definition 4]). We are only interested in exponent ω.

Definition 3 (The operator $X \mapsto X^\omega$). *Let X be a linear ordering. Define X^ω as the set of all finite sequences of the form $\langle (b_0, a_0), \dots, (b_n, a_n) \rangle$, where $(b_i, a_i) \in \mathbb{N} \times X$ for all $i \leq n$ and $b_0 > b_1 > \cdots > b_n$. If X has a minimal element 0_X we require that $a_i \neq 0_X$ for all $i \leq n$. Let $<_{X^\omega}$ be the lexicographic ordering on X^ω: for any two distinct sigma, au such that $\sigma = \langle (b_0, a_0), \dots, (b_n, a_n) \rangle$, $\tau = \langle (b'_0, a'_0), \dots, (b'_m, a'_m) \rangle \in X^\omega$, $\sigma <_{X^\omega} \tau$ if and only if either τ extends σ or there exists $j \geq 0$ such that $(b_i, a_i) = (b'_i, a'_i)$ for all $i < j$ and either $b_j < b'_j$ or both $b_j = b'_j$ and $a_j <_X a'_j$. Otherwise $\tau >_{X^\omega} \sigma$.*

If X is an ordinal α, the element $\langle (b_0, a_0), \dots, (b_n, a_n) \rangle$ of α^ω is intended to correspond to the ordinal $\alpha^{b_0} \cdot a_0 + \cdots + \alpha^{b_n} \cdot a_n$.

In the Reverse Mathematics literature $\mathsf{WOP}(X^\omega)$ denotes the statement

$$\forall X(\mathrm{WO}(X) \to \mathrm{WO}(X^\omega)),$$

where $\mathrm{WO}(X)$ is any canonical Π^1_1-formula expressing that X is well-ordered. Uftring [12, Theorem 5] proved that this principle is equivalent to Σ^0_2-induction over RCA_0.

We will also make use of the following simpler operator.

Definition 4 (The operator $X \mapsto X^n_{lex}$). *Let X be a linear ordering, and n a positive integer. Define X^n_{lex} as the set of sequences of length n of elements of X, ordered lexicographically.*

Notice that if F is one of the above operators then any element of $F(X)$ can be regarded as a word over an alphabet containing X. We can then make the WOP for F into an instance-solution problem through the following definition.

Definition 5 (WOP Problem for F). *Define the problem $\mathsf{WOP}(F(X))$ as follows. An instance of $\mathsf{WOP}(F(X))$ is an infinite $<_{F(X)}$-descending sequence $\sigma = \langle \sigma_n \rangle_{n \geq 0}$ in $F(X)$, and a corresponding solution is an infinite $<_X$-descending sequence $\rho = \langle x_i \rangle_{i \geq 0}$ in X contained in σ in the following sense: there exists a weakly increasing sequence of indices $\langle n_i \rangle_{i \geq 0} \subseteq \mathbb{N}$ such that x_i occurs in the word σ_{n_i}.*

2.3 Ramsey-Type Problems

We define the Ramsey-type principles of interest to the present paper: the Ordered and the Additive Ramsey Theorems. We formulate both as problems.

Fix a linear order X. For every poset $(P, <_P)$, we call a coloring $c : [X]^2 \to P$ *right-ordered* if we have $c(x, y) \leq_P c(x, y')$ whenever $x <_X y \leq_X y'$. A set $Y \subseteq X$ is called *c-monochromatic* if c is constant on $[Y]^2$.

Definition 6 (Ordered Ramsey Theorem). *The problem* $\mathsf{ORT}_{\mathbb{N}}$ *takes as input a finite poset* $(P, \leq_P)$ *and a right-ordered coloring* $c : [\mathbb{N}]^2 \to P$*, and outputs an infinite c-monochromatic subset of* $\mathbb{N}$.

A coloring $c : [X]^2 \to S$ is called *additive* with respect to a semigroup structure $(S, \cdot)$ if we have $c(x, z) = c(x, y) \cdot c(y, z)$ whenever $x <_X y <_X z$.

Definition 7 (Additive Ramsey Theorem). *The problem* $\mathsf{ART}_{\mathbb{N}}$ *takes as input a finite semigroup S and an additive coloring* $c : [\mathbb{N}]^2 \to S$*, and outputs an infinite c-monochromatic subset of* $\mathbb{N}$.

In Reverse Mathematics the $\forall\exists$ theorems naturally corresponding to the problems $\mathsf{ORT}_{\mathbb{N}}$ and $\mathsf{ART}_{\mathbb{N}}$ in Second-Order Arithmetic were introduced in [8] and proved equivalent to Σ^0_2-induction over RCA_0. The Weihrauch complexity of the problems $\mathsf{ORT}_{\mathbb{N}}$ and $\mathsf{ART}_{\mathbb{N}}$ is studied in [10] (see Theorem 1 below).

2.4 Known Results

As mentioned above, in Reverse Mathematics the principles corresponding to $\mathsf{ORT}_{\mathbb{N}}$, $\mathsf{ART}_{\mathbb{N}}$ and ECT are all equivalent to Σ^0_2-induction [4].

The following results about their Weihrauch complexity were recently proved by Pauly, Pradic and Soldà [10].

Theorem 1 (Pauly, Pradic, Soldà). $\mathsf{ORT}_{\mathbb{N}} \equiv_W \mathsf{ECT} \times (\mathsf{LPO}')^* \equiv_W \mathsf{ART}_{\mathbb{N}}$.

Moreover, the two factors in the product characterizing $\mathsf{ORT}_{\mathbb{N}}$ in the above theorem are Weihrauch-incomparable. This follows from [10, Lemma 9], which proves that $(\mathsf{LPO}')^*$ and $(\mathsf{TC}_{\mathbb{N}})^*$ are Weihrauch-incomparable, and Lemma 1 above.

Lemma 2 (Pauly, Pradic, Soldà; Davis et alii). $(\mathsf{LPO}')^*$ *and* ECT *are Weihrauch-incomparable.*

Thus, both ECT and $(\mathsf{LPO}')^*$ are strictly weaker than their product, and the splitting in Theorem 1 is non-trivial.

2.5 Overview of the Paper

The paper is organized as follows. In Sect. 3 we prove the Weihrauch equivalence of the Ordered Ramsey Theorem with the parallel product of the Eventually Constant Tail principle and the WOP for $X \mapsto X^\omega$:

Theorem 2. $\mathsf{ORT}_{\mathbb{N}} \equiv_W \mathsf{ECT} \times \mathsf{WOP}(X^\omega)$.

In Sect. 4 we compare the problems featuring in Theorems 1 and 2. First, we prove that ECT is strictly weaker than $\mathsf{WOP}(X^\omega)$. We then prove that $\mathsf{WOP}(X^\omega)$ is Weihrauch-incomparable with $(\mathsf{LPO}')^*$. This ensures that Theorem 2 gives a different splitting of $\mathsf{ORT}_{\mathbb{N}}$ than the one from Theorem 1 above. A second consequence is that $\mathsf{WOP}(X^\omega)$ is strictly weaker than $\mathsf{ECT} \times \mathsf{WOP}(X^\omega)$. Therefore both ECT and $\mathsf{WOP}(X^\omega)$ are indispensable for our characterization of $\mathsf{ORT}_{\mathbb{N}}$.

3 The Ordered Ramsey Theorem and $\mathsf{WOP}(X^\omega)$

In this section we establish the Weihrauch equivalence of the Ordered Ramsey Theorem with the parallel product of the Eventually Constant Tail principle and the WOP for the operator $X \mapsto X^\omega$. The lower bound on the Ordered Ramsey Theorem is based on an idea from [2]. The upper bound is, as far as we know, the first example of a reduction of a Ramsey-type principle to a well-ordering principle.

3.1 Lower Bound: $\mathsf{ORT}_{\mathbb{N}} \geq_W \mathsf{ECT} \times \mathsf{WOP}(X^\omega)$

Let X be a linearly ordered set. Let $\tau = \langle (b_0, a_0), \ldots, (b_m, a_m) \rangle$ with $m \geq 0$ be an element of X^ω. Define the *weight* of τ as $\|\tau\| := 2m + 2$. For each s such that $0 \leq s < \|\tau\|$, define the s-th *letter* of τ as

$$\tau(s) := \begin{cases} b_t \in \mathbb{N}, & \text{if } s = 2t; \\ a_t \in X, & \text{if } s = 2t + 1. \end{cases}$$

Notice that $\|\tau\|$ is the least $s \geq 0$ such that $\tau(s)$ is undefined. Using the above terminology, we can add an order structure to the coloring devised in [2] to establish a reduction from RT^2_∞ to $\mathsf{WOP}(X^\omega)$. We use $\prec$ to denote the prefix relation on sequences, and we identify each natural number with the set of its predecessors.

Proposition 1. $\mathsf{WOP}(X^\omega) \leq_W \mathsf{ORT}_{\mathbb{N}}$.

Proof. Let $\sigma = \langle \sigma_i \rangle_{i \geq 0}$ be an instance of $\mathsf{WOP}(X^\omega)$. If

$$\sigma_i = \langle (b_0^i, a_0^i), \ldots, (b_{m_i}^i, a_{m_i}^i) \rangle,$$

then $\|\sigma_i\| = 2m_i + 2$ and $m_i \leq b_0^i \leq b_0^0 =: k_\sigma$. Hence $\|\sigma_i\| \leq 2k_\sigma + 2$ for every $i \geq 0$.

Define a coloring $c : [\mathbb{N}]^2 \to 2k_\sigma + 2$ as follows: for $i < j$ in $\mathbb{N}$,

$$c(i,j) := \begin{cases} \|\sigma_j\| & \text{if } \sigma_j \prec \sigma_i; \\ \text{least } s < \|\sigma_i\| \text{ such that } \sigma_i(s) \neq \sigma_j(s) & \textit{otherwise}. \end{cases}$$

If $\sigma_j \not\prec \sigma_i$, since $\sigma_j <_{X^\omega} \sigma_i$ there exists $0 \leq s < \|\sigma_i\|$ such that $\sigma_i(s) \neq \sigma_j(s)$. Also, in both cases of the definition of c, $c(i,j) < \|\sigma_i\|$ and $\sigma_j(s) = \sigma_i(s)$ for all $0 \leq s < c(i,j)$.

Notice that for all i such that $0 \leq i < j < k$ we have $c(i,k) \leq c(i,j)$. Suppose otherwise, as witnessed by i, j, k such that $0 \leq i < j < k$ and $c(i,j) < c(i,k)$. Then $\sigma_k(s) = \sigma_i(s)$ for all s such that $0 \leq s \leq c(i,j)$. On the other hand, $\sigma_j(s) = \sigma_i(s)$ for all s such that $0 \leq s < c(i,j)$ and $\sigma_j(c(i,j))$ is either undefined or strictly less than $\sigma_i(c(i,j))$, either in $\mathbb{N}$ or in X, depending on the parity of $c(i,j)$. It follows that $\sigma_k >_{X^\omega} \sigma_j$, which is impossible because σ is decreasing.

Let $P := \{0, 1, \ldots, 2k_\sigma + 1\}$, and let $\leq_P$ be the inverse of the natural order.

Then the above remarks show that $c : [\mathbb{N}]^2 \to P$ and that c is right-ordered with respect to $\leq_P$. Hence (P, c) is an instance of $\mathsf{ORT}_{\mathbb{N}}$, uniformly computable from σ. Let $H = \{h_0 < h_1 < \cdots\}$ be a solution to $\mathsf{ORT}_{\mathbb{N}}$ for (P, c). Denote by c^* the c-color of $[H]^2$. Notice that $c^* = c(h_i, h_{i+1}) < \|\sigma_{h_i}\|$ for all $i \geq 0$, hence $\sigma_{h_i}(c^*)$ is defined for all $i \geq 0$.

Suppose that c^* is even. Then

$$b^{h_0}_{\frac{c^*}{2}} >_{\mathbb{N}} b^{h_1}_{\frac{c^*}{2}} >_{\mathbb{N}} \cdots$$

which is impossible. Therefore c^* is odd, and

$$a^{h_0}_{\frac{c^*-1}{2}} >_X a^{h_1}_{\frac{c^*-1}{2}} >_X \cdots ,$$

so that $\langle a^{h_i}_{\frac{c^*-1}{2}} \rangle_{i \geq 0}$ is the desired solution to $\mathsf{WOP}(X^\omega)$ for σ. □

Lemma 3. $(\mathsf{ORT}_{\mathbb{N}})^* \equiv_{\mathrm{sW}} \mathsf{ORT}_{\mathbb{N}}$.

Proof. Obviously $\mathsf{ORT}_{\mathbb{N}} \leq_{\mathrm{sW}} (\mathsf{ORT}_{\mathbb{N}})^*$. The opposite inequality can be proved by a straightforward product coloring and product ordering argument. □

Proposition 2. $\mathsf{ECT} \times \mathsf{WOP}(X^\omega) \leq_{\mathrm{W}} \mathsf{ORT}_{\mathbb{N}}$.

Proof. From [10, Lemma 52] it follows that $\mathsf{ECT} \leq_{\mathrm{sW}} \mathsf{ORT}_{\mathbb{N}}$. From this and Proposition 1 we have $\mathsf{ECT} \times \mathsf{WOP}(X^\omega) \leq_{\mathrm{W}} (\mathsf{ORT}_{\mathbb{N}})^2$. Clearly $(\mathsf{ORT}_{\mathbb{N}})^2 \leq_{\mathrm{sW}} (\mathsf{ORT}_{\mathbb{N}})^*$ and, by Lemma 3, $(\mathsf{ORT}_{\mathbb{N}})^* \leq_{\mathrm{sW}} \mathsf{ORT}_{\mathbb{N}}$. □

3.2 Upper Bound: $\mathsf{ORT}_{\mathbb{N}} \leq_{\mathrm{W}} \mathsf{ECT} \times \mathsf{WOP}(X^\omega)$

In this subsection we prove that $\mathsf{ORT}_{\mathbb{N}} \leq_{\mathrm{W}} \mathsf{ECT} \times \mathsf{WOP}(X^\omega)$. We obtain the result from a parametrized version. For every $n \geq 1$, let $\mathsf{ORT}_{\mathbb{N}}[n]$ denote the restriction of $\mathsf{ORT}_{\mathbb{N}}$ to instances (P, c) with $|P| = n$.

Proposition 3. *For all $n \geq 1$, $\mathsf{ORT}_{\mathbb{N}}[n] \leq_{\mathrm{sW}} (\mathsf{ECT})^n \times \mathsf{WOP}(X^n_{\mathsf{lex}})$.*

Proof. Let (P, c) be an instance of $\mathsf{ORT}_{\mathbb{N}}[n]$. Fix a linear ordering $p_1 < p_2 < \cdots < p_n$ of P extending $\leq_P$.

The idea is to first define, for each value $p_i \in P$ an instance $f_i : \mathbb{N} \to 2$ of ECT which keeps track of the (repeated) use of the color p_i in c.

For each j such that $1 \leq j \leq n$ we define a binary sequence $\mathtt{improve}_j$ and an auxiliary sequence $\mathtt{previous}_j$ as follows. First, set

$$\mathtt{improve}_j(0) := 1\,,\ \ \mathtt{previous}_j(0) = \mathtt{previous}_j(1) := 0.$$

Now let $i \geq 1$. If there exists k such that $\mathtt{previous}_j(i) \leq k < i$ and such that $c(k, i) = p_j$, then call k^i_j the least such k and set $\mathtt{improve}_j(i) := 1$, $\mathtt{previous}_j(i+1) := i$. Otherwise, set $\mathtt{improve}_j(i) := 0$, $\mathtt{previous}_j(i+1) = \mathtt{previous}_j(i)$. We say that p_j improves at stage $i \geq 0$ if and only if $\mathtt{improve}_j(i) = 1$.

When interpreted as 2-colorings of $\mathbb{N}$, the sequences $\mathtt{improve}_j$, for $1 \leq j \leq n$, constitute an instance of $(\mathsf{ECT})^n$. Let n_0^j, for j such that $1 \leq j \leq n$, be corresponding solutions, and set n_0 to be their maximum. Then, by definition of ECT, for all j such that $1 \leq j \leq n$ either p_j doesn't improve after stage n_0 or p_j improves infinitely many times after stage n_0.

Next, we define the fitting instance $\sigma = \langle \sigma_i \rangle_{i \geq 0}$ of $\mathsf{WOP}(X^n_{\mathsf{lex}})$. The major difficulty is to define a linear ordering $(X, \leq_X)$ in such a way as to encode in a decreasing sequence of n-tuples of elements from X associated to the above-defined n instances of ECT and their respective solutions sufficient information for extracting a c-monochromatic set.

We choose $X := \{1, \ldots, n\} \times \mathbb{N} \times \mathbb{Q} \times \{-n, \ldots, -1\}$. Each factor is equipped with its natural order, so that the resulting lexicographic order on X is linear. We now proceed to define an infinite $<_{\mathsf{lex}}$-decreasing sequence σ.

Stage 0. Set $\sigma_0 := \langle a_n^0, \ldots, a_1^0 \rangle$, where $a_j^0 := \langle j,\ 1,\ 2, -1 \rangle$ for all j such that $n \geq j \geq 1$.

Stage $i \geq 1$. Consider the set I^i of all j-s such that p_j improves at stage i. Notice that I^i is non-empty for every $i \geq 1$, as it contains the index of $c(i-1, i)$. For every $j \in I^i$, let m_j^i be the largest m such that $n \geq m \geq j$ and p_m has improved at least once in the time-frame $(\mathtt{previous}_j(i), i]$.
Suppose that $j \in I^i$. Set

$$a_j^i := \begin{cases} \langle j,\ a_j^{i-1}(2),\ \frac{1}{k_j^i+1},\ -1 \rangle & \text{if } m_j^i = j \\ \langle j,\ a_j^{i-1}(2) + 1,\ \frac{1}{k_j^i+1},\ -1 \rangle & \text{if } m_j^i > j \end{cases}$$

Suppose instead that $j \notin I^i$. Set

$$a_j^i := \begin{cases} \langle j,\ a_j^{i-1}(2),\ a_j^{i-1}(3),\ a_j^{i-1}(4) - 1 \rangle \text{ if } j = \max\{m_l^i \mid l \in I^i\}, \\ a_j^{i-1} \text{ otherwise.} \end{cases}$$

Finally, set $\sigma_i := \langle a_n^i, \ldots, a_1^i \rangle$. Notice that σ_i is well-defined as an element of X: if $a_j^i(4)$ were to fall below $-n$ for a certain j, then $\mathtt{previous}_j(i)$ would have to be sitting between two consecutive improvements of p_l for at least n different indices $l < j$, and this is impossible. Moreover, if $i \geq 1$ then $\sigma_i <_{\mathsf{lex}} \sigma_{i-1}$ because, as observed above, I^i is non-empty. This completes the definition of σ.

Let $\rho = \langle x_s \rangle_{s \geq 0}$ be a sequence in X that is a solution to $\mathsf{WOP}(X^n_{\mathsf{lex}})$ for σ. We aim to compute, uniformly from n_0 and ρ, an increasing sequence $\gamma = \langle y_s \rangle_{s \geq 0}$ of natural numbers enumerating a c-monochromatic set. The x_s-s are drawn from the matrix

$$\Sigma := \langle a_j^i \rangle_{n \geq j \geq 1}^{i \geq 0},$$

where i is interpreted as a row index and j as a (decreasing) column index. We are able to locate the x_s-s more precisely in the matrix Σ by reasoning as follows. Let j^* be the largest j such that $n \geq j \geq 1$ and p_j improves infinitely many times. Let i^* be the least $i \geq 0$ such that p_{j*} improves at stage i and no color p_j, for j such that $n \geq j > j^*$, improves after stage i. Then:

1. If p_j improves finitely many times, then $a^i_j(1)$, $a^i_j(2)$, and $a^i_j(3)$ are constant for $i \geq n_0$; also, by construction $a^i_j(4) > a^{i+1}_j(4)$ for at most $(n-1)$ many $i \geq n_0$. Hence column j contains at most $(n_0+1)+(n-1) = n_0+n$ distinct coefficients.
2. Together, columns $n \geq j \geq j^*+1$ contain at most $M := n(n_0+n)$ distinct coefficients. This follows immediately from point (1) and the choice of j^*.
3. ρ can't move from a given column to a column on its left. This is true because by construction $a^i_j >_X a^{i'}_{j'}$ for all j such that $n \geq j > j' \geq 1$ and for all $i, i' \geq 0$.
4. If $j^*-1 \geq j \geq 1$, then for every $i \geq 0$ column j contains finitely many distinct coefficients smaller than a^i_j. This follows trivially from (1) if p_j doesn't improve indefinitely. Suppose otherwise: since p_{j^*} improves at some stage $i' > i$ and $j^* > j$, $a^i_j(2)$ is eventually increased.
5. ρ can't visit columns $j^*-1 \geq j \geq 1$. This follows from points (3) and (4).
6. x_s lies in column j^* for all $s \geq M+1$. This follows from points (2) and (5).
7. Let $z := a^{i^*}_{j^*}(2)$. Then $a^i_{j^*}(2) < z$ for all i such that $0 \leq i < i^*$ and $a^i_{j^*}(2) = z$ for all $i \geq i^*$. Hence if $i < i^*$ then σ contains finitely many distinct coefficients smaller than $a^i_{j^*}$, meaning that ρ can't intercept the cell $a^i_{j^*}$.
8. For $s \geq M+1$, the x_s-s are drawn in increasing order of rows from column j^* (so that $x_{s+n}(3) < x_s(3) < 2$ for all $s \geq M+2$). This follows from points (6) and (7), since $x_s(2) = z$ is fixed for $s \geq M+1$.

Set

$$y_s := \frac{1}{x_{f(s)}(3)} - 1 \quad \forall s \geq 0,$$

where $f(s) := (M+2) + (s \times n)$. Point (8) above proves that γ is an increasing sequence of natural numbers.

It only remains to show that γ is c-monochromatic. We fix $0 \leq s < t$, and show that $c(y_s, y_t) = p_{j^*}$. Let i be least such that $a^i_{j^*}(3) = x_{f(s)}(3)$, so that $y_s = k^i_{j^*}$ and $c(k^i_{j^*}, i) = p_{j^*}$ by definition of $k^i_{j^*}$. Similarly, let i' be least such that $a^{i'}_{j^*}(3) = x_{f(t)}(3)$, so that $y_t = k^{i'}_{j^*}$. By construction we have $i^* \leq k^i_{j^*} < i \leq k^{i'}_{j^*}$. Since c is right-ordered, $c(k^i_{j^*}, k^{i'}_{j^*}) \geq_P c(k^i_{j^*}, i) = p_{j^*}$. Now if $c(k^i_{j^*}, k^{i'}_{j^*}) = p_j >_P p_{j^*}$ then $j > j^*$ and p_j improves after i^*, contradicting the choices of i^* and j^*. Therefore $c(y_s, y_t) = c(k^i_{j^*}, k^{i'}_{j^*}) = p_{j^*}$, as required. □

Let $\mathsf{WOP}(X^\omega)[\leq n]$ denote the restriction of $\mathsf{WOP}(X^\omega)$ to those sequences $\tau = \langle \tau_i \rangle_{i \geq 0}$ satisfying $|\tau_i| \leq n$ for all $i \geq 0$. The following is obvious.

Lemma 4. *For all* $n \geq 1$, $\mathsf{WOP}(X^n_{\mathsf{lex}}) \leq_{\mathrm{sW}} \mathsf{WOP}(X^\omega)[\leq 2n]$.

Corollary 1. $\mathsf{ORT}_{\mathbb{N}} \leq_{\mathrm{W}} \mathsf{ECT} \times \mathsf{WOP}(X^\omega)$.

Proof. From Proposition 3 and Lemma 4 we have $\mathsf{ORT}_{\mathbb{N}}[n] \leq_{\mathrm{sW}} (\mathsf{ECT})^n \times \mathsf{WOP}(X^\omega)[\leq 2n]$. Hence $\mathsf{ORT}_{\mathbb{N}} \leq_{\mathrm{W}} (\mathsf{ECT})^* \times \mathsf{WOP}(X^\omega)$ (the argument in Proposition 2 doesn't establish a strong Weihrauch reduction, since f depends on $n = |P|$). From Lemma 1 it follows that $(\mathsf{ECT})^* \equiv_{\mathrm{W}} \mathsf{ECT}$, since $(\mathsf{P}^*)^* \equiv_{\mathrm{W}} \mathsf{P}^*$ for every problem P. Therefore the desired inequality $\mathsf{ORT}_{\mathbb{N}} \leq_{\mathrm{W}} \mathsf{ECT} \times \mathsf{WOP}(X^\omega)$ holds. □

4 Completing the Picture

In the previous section we proved that $\mathsf{ORT}_{\mathbb{N}}$ is Weihrauch-equivalent to $\mathsf{ECT} \times \mathsf{WOP}(X^\omega)$. Since $\mathsf{ORT}_{\mathbb{N}}$ is also equivalent to $\mathsf{ECT} \times (\mathsf{LPO}')^*$ by Theorem 1, it is natural to ask how $\mathsf{WOP}(X^\omega)$ compares with $(\mathsf{LPO}')^*$ and with ECT. We show that that ECT is strictly weaker than $\mathsf{WOP}(X^\omega)$ and that $\mathsf{WOP}(X^\omega)$ and $(\mathsf{LPO}')^*$ are incomparable.

4.1 ECT Is Strictly Weaker Than $\mathsf{WOP}(X^\omega)$

We first observe that $\mathsf{WOP}(X^\omega) \not\leq_{\mathrm{W}} \mathsf{ECT}$. Otherwise, $\mathsf{ORT}_{\mathbb{N}} \leq_{\mathrm{W}} \mathsf{ECT}$ by Corollary 2 and thus $(\mathsf{LPO}')^* \leq_{\mathrm{W}} \mathsf{ECT}$ by Theorem 1, contradicting Lemma 2. We mention without proof that the stronger non-reduction result $\mathsf{WOP}(X^2_{\mathsf{lex}}) \not\leq_{\mathrm{W}} \mathsf{ECT}$ can be obtained by a variation on the proof of [10, Lemma 9].

In order to prove that $\mathsf{ECT} \leq_{\mathrm{W}} \mathsf{WOP}(X^\omega)$ we use the following proposition.

Proposition 4. $(\mathsf{TC}_{\mathbb{N}})^n \leq_{\mathrm{sW}} \mathsf{WOP}(X^{2^n}_{\mathsf{lex}})$ *for all* $n \geq 1$.

For our proof of Proposition 4 we need the following technical definition and lemma. Below, $[n]$ denotes the set $\{1, \ldots, n\}$.

Definition 8. *Let* $(U_i)_{i\geq 0}$ *be any sequence of subsets of* $[n]$. *Define a set of sequences* $\mathtt{reset}_A$, $A \subseteq [n]$, *as follows. Start by setting* $\mathtt{reset}_A(0) := -1$ *for all* $A \subseteq [n]$. *Let* $i \geq 0$ *and assume that, for all* $A \subseteq [n]$, $\mathtt{reset}_A(i)$ *has been defined and* $\mathtt{reset}_A(i) < i$. *We say that* $B \subseteq [n]$ *completes a cycle at stage* i *if*

$$\bigcup_{\mathtt{reset}_B(i) < h \leq i} U_h = B.$$

For all $A \subseteq [n]$, *set* $\mathtt{reset}_A(i+1) := i$ *if some* $B \supseteq A$ *completes a cycle at stage* i, *in which case we say that* A *is reset at stage* i; $\mathtt{reset}_A(i+1) := \mathtt{reset}_A(i)$ *otherwise.*

Lemma 5. *For all* $i \geq 0$,

$$V_i := \bigcup_{\mathtt{reset}_{U_i}(i) < h \leq i} U_h.$$

is the unique subset of $[n]$ *completing a cycle at stage* i.

Proof. It's clear that $V_0 = U_0$ is the unique subset of $[n]$ completing a cycle at stage 0.

Let $i \geq 1$. Assume by induction that, for all $0 \leq h < i$, V_h is the unique subset of $[n]$ completing a cycle at stage h. This implies that $A \subseteq [n]$ is reset at stage $0 \leq h < i$ if and only if $A \subseteq V_h$.

First, we show that if $\mathtt{reset}_{U_i}(i) \geq 0$ then $V_i \subseteq V_{\mathtt{reset}_{U_i}(i)}$. Suppose otherwise, and let $\mathtt{reset}_{U_i}(i) < h \leq i$ be least such that $U_h \not\subseteq V_{\mathtt{reset}_{U_i}(i)}$. Since U_i is reset at stage $\mathtt{reset}_{U_i}(i)$, it must be contained in $V_{\mathtt{reset}_{U_i}(i)}$ by the inductive hypothesis. Hence $h < i$. By the choice of h, $U_l \subseteq V_{\mathtt{reset}_{U_i}(i)}$ is also reset at stage $\mathtt{reset}_{U_i}(i)$ for all $\mathtt{reset}_{U_i}(i) < l < h$. Then, for each $\mathtt{reset}_{U_i}(i) < k < h$, $V_k \subseteq \bigcup_{\mathtt{reset}_{U_i}(i)<l\le k} U_l \subseteq V_{\mathtt{reset}_{U_i}(i)}$, so that $U_h \nsubseteq V_k$ is not reset at stage k by the inductive hypothesis. We infer that $\mathtt{reset}_{U_h}(h) < \mathtt{reset}_{U_i}(i)$. This implies that $U_{\mathtt{reset}_{U_i}(i)} \subseteq V_h$, which in turn implies that $\mathtt{reset}_{V_h}(\mathtt{reset}_{U_i}(i)) \le \mathtt{reset}_{U_{\mathtt{reset}_{U_i}(i)}}(\mathtt{reset}_{U_i}(i))$. Since $\mathtt{reset}_{V_h}(\mathtt{reset}_{U_i}(i)) = \mathtt{reset}_{V_h}(h)$, it follows that $V_h \supseteq V_{\mathtt{reset}_{U_i}(i)}$ ($\supsetneq$, actually). In particular, $U_i \subseteq V_{\mathtt{reset}_{U_i}(i)}$ is reset at stage $h \in (\mathtt{reset}_{U_i}(i), i)$, which contradicts the definition of $\mathtt{reset}_{U_i}(i)$.

We show that V_i completes a cycle at stage i. This is clearly the case if $\mathtt{reset}_{U_i}(i) = -1$. Suppose instead that $\mathtt{reset}_{U_i}(i) \ge 0$. Surely $\mathtt{reset}_{V_i}(i) \le \mathtt{reset}_{U_i}(i)$, as $V_i \supseteq U_i$. It then suffices to show that V_i is reset at stage $\mathtt{reset}_{U_i}(i)$. By the inductive hypothesis, this is the same as saying that $V_i \subseteq V_{\mathtt{reset}_{U_i}(i)}$, which was proven above.

Lastly, we prove that V_i is unique. Suppose by way of contradiction that $W \ne V_i$ completes a cycle at stage i. Then $W \supseteq U_i$, hence $\mathtt{reset}_W(i) \le \mathtt{reset}_{U_i}(i) = \mathtt{reset}_{V_i}(i)$. Since both W and V_i complete a cycle at stage i, this implies that $W \supseteq V_i$, or rather that $W \supsetneq V_i$, by our assumption. We must then have $\mathtt{reset}_W(i) < \mathtt{reset}_{U_i}(i)$ and $W - V_i \subseteq \bigcup_{\mathtt{reset}_W(i)<h\le\mathtt{reset}_{U_i}(i)} U_h$. The former gives $W \supseteq U_{\mathtt{reset}_{U_i}(i)}$, so that $\mathtt{reset}_W(i) = \mathtt{reset}_W(\mathtt{reset}_{U_i}(i)) \le \mathtt{reset}_{U_{\mathtt{reset}_{U_i}(i)}}(\mathtt{reset}_{U_i}(i))$. It follows that

$$V_{\mathtt{reset}_{U_i}(i)} \subseteq \bigcup_{\mathtt{reset}_W(i)<h\le\mathtt{reset}_{U_i}(i)} U_h.$$

Therefore

$$W = (W - V_i) \cup V_i \subseteq (W - V_i) \cup V_{\mathtt{reset}_{U_i}(i)} \subseteq \bigcup_{\mathtt{reset}_W(i)<h\le\mathtt{reset}_{U_i}(i)} U_h \subseteq W.$$

We conclude that W completes a cycle at some stage $h \in (\mathtt{reset}_W(i), i)$, which contradicts the definition of $\mathtt{reset}_W(i)$. □

Proof (Proof of Proposition 4). Let $\eta = \langle e_1, \dots, e_n\rangle$ be an instance of $(\mathsf{TC}_{\mathbb{N}})^n$.

For a fixed $m \in [n]$, $g_m(-1) := 0$ is the default, optimistic guess for a solution to $\mathsf{TC}_{\mathbb{N}}$ for e_m. At each stage $i \ge 0$, the $(i-1)$-th guess is updated to $g_m(i) := \min\left(\mathbb{N} - \mathrm{ran}(e_m \upharpoonright \{0, \dots, i\})\right)$. Let γ be the sequence of n-tuples $\langle g_1(i), \dots, g_n(i)\rangle$, $i \ge -1$. Then the set of components of γ that are updated at each stage $i \ge 0$ is

$$U_i := \{m \in [n] \ : \ e_m \text{ enumerates } g_m(i-1) \text{ at stage } i\}.$$

Both γ and the cycle structure associated to its update sets $(U_i)_{i\ge0}$ can be encoded into a sequence $\sigma = \langle \sigma_i\rangle_{i\ge0}$ in $[0,\infty]^{2^n}$. To define σ, first fix a total ordering $A_1 = \emptyset < A_2 < \cdots < A_{2^n} = [n]$ of $\mathcal{P}([n])$ extending $\subseteq$ and, for each j such that $2^n \ge j \ge 1$, initialize a sequence $\mathtt{count}_j$ by setting $\mathtt{count}_j(0) := 2^n - |A_j|$. We define σ in stages.

Stage 0. Set:

$$a^0_{j_0} := \mathtt{count}_{j_0}(0) + \frac{1}{p_1^{g_1(0)} p_2^{g_2(0)} \dots p_n^{g_n(0)}},$$

where $A_{j_0} = U_0$ and p_m denotes the m-th prime number; set

$$a^0_j := \infty \text{ for all } j \neq j_0;$$

and set $\sigma_0 := \langle a^0_{2^n}, \dots , a^0_1\rangle$. For all j such that $2^n \geq j \geq 1$, set

$$\mathtt{count}_j(1) := \begin{cases} \mathtt{count}_{j_0}(0) + |A_{j_0} - A_j|, & \text{if } A_j \subsetneq A_{j_0} \\ \mathtt{count}_j(0), & \text{otherwise.} \end{cases}$$

Stage $i \geq 1$. Let j_i satisfy $A_{j_i} = V_i$, where the latter is defined as in Lemma 5. Set

$$a^i_j := a^{i-1}_j \text{ for all } j > j_i;$$

$$a^i_{j_i} := \mathtt{count}_{j_i}(i) + \frac{1}{p_1^{g_1(i)} p_2^{g_2(i)} \dots p_n^{g_n(i)} p^i_{n+1}};$$

$$a^i_j := \infty \text{ for all } j \text{ such that } j_i > j \geq 1.$$

Finally set $\sigma_i := \langle a^i_{2^n}, \dots, a^i_1\rangle$. For all j such that $2^n \geq j \geq 1$, set

$$\mathtt{count}_j(i+1) := \begin{cases} \max_{0 \leq h \leq i} \mathtt{count}_{j_h}(h) + |A_{j_i} - A_j|, & \text{if } A_j \subsetneq A_{j_i} \\ \mathtt{count}_j(i), & \text{otherwise.} \end{cases}$$

This completes the definition of σ, which is then uniformly computable from η. In order to show that σ is an instance of $\mathsf{WOP}(X^{2^n}_{\mathsf{lex}})$ (where $X = [0, \infty]$) i.e., that it's descending with respect to $<_{\mathsf{lex}}$, we only have to check that $a^i_{j_i} < a^{i-1}_{j_i}$ for all $i \geq 1$. This is obviously the case if $a^{i-1}_{j_i} = \infty$. Now $a^{i-1}_{j_i} < \infty$ if and only if there exists i' such that $0 \leq i' < i$ such that $j_{i'} = j_i$ and $A_{j_h} \not\supsetneq A_{j_i}$ for all h such that $i' < h < i$. This implies that $\mathtt{count}_{j_i}(i') = \mathtt{count}_{j_i}(i)$. Moreover, $g_m(i') \leq g_m(i)$ for all $m \in [n]$ (with $g_m(i') < g_m(i)$ if and only if $m \in A_{j_i}$), and $p^{i'}_{n+1} < p^i_{n+1}$. The claim follows.

Let y be such that $2^n \geq y \geq 1$ and such that the following equation holds:

$$A_y = \{m \in [n] \mid \mathrm{ran}(e_m) = \mathbb{N}\}.$$

Notice that, for all $m \in [n]$, $m \in A_y$ if and only if $m \in U_i$ for infinitely many i-s. We make the following observations.

1. Obviously if $m \in A_y$ then any natural number is a solution to $\mathsf{TC}_{\mathbb{N}}$ for e_m. Let $i_* \geq -1$ be least such that $U_i \subseteq A_y$ for all $i > i_*$. Then, for all $m \notin A_y$, $g_m(i_*)$ is a (the least) solution to $\mathsf{TC}_{\mathbb{N}}$ for e_m and $g_m(i) = g_m(i_*)$ for all $i \geq i_*$.

2. Let $i_{**} \geq -1$ be least such that $A_{j_i} \subseteq A_y$ for all $i > i_{**}$. If $i_* = -1$ then $i_{**} = -1$ as well. Otherwise $U_{i_*} \nsubseteq A_y$, hence $A_{j_{i_*}} \nsubseteq A_y$: in this case, i_{**} is the unique $i \geq i_*$ such that $A_{j_i} \supsetneq A_y$.
3. Let i_0 be the least $i > i_{**}$ such that $y = j_i$. For all $k \geq 1$, let i_k be the least $i > i_{k-1}$ such that $y = j_i$. Then $a_y^i = \infty$ for all (possible) i such that $i_{**} < i < i_0$, and

$$\infty > a_y^{i_0} = \ldots = a_y^{i_1 - 1} > a_y^{i_1} = \ldots > a_y^{i_{k-1}} = \cdots = a_y^{i_k - 1} > a_y^{i_k} = \ldots .$$

4. Suppose that $i_{**} \geq 0$. Let i be such that $0 \leq i \leq i_{**}$ and j be such that $2^n \geq j \geq 1$. By point 2 , either $a_j^i = \infty$ or $a_j^i < a_y^{i_k}$ for all $k \geq 0$.
5. For all j such that $y > j \geq 1$ and for all $i > i_{**}$, either $a_j^i = \infty$ or $a_j^i = a_{j_h}^h$ for some h such that $i_{**} < h \leq i$. In both cases, $a_j^i > a_y^{i_0}$.
6. For all j such that $2^n \geq j \geq 1$ and such that $A_j \nsubseteq A_y$ (hence for all j such that $2^n \geq j > y$) and for all $i > i_{**}$, $a_j^i = a_j^{i'_{**}}$, where $i'_{**} = i_{**}$ or $i'_{**} = 0$.

Let ρ be a solution to $\mathsf{WOP}(X^\omega)$ for σ. It follows from points 3, 4, 5, 6 above that

$$\rho = \alpha\beta\delta,$$

where: α is either (∞) or the empty sequence; β is a (necessarily finite, and possibly empty) descending sequence contained in $\{a_{j_i}^i : (i > i_{**}) \,\&\, (y > j_i)\}$; δ is a subsequence of $\langle a_y^{i_k} \rangle_{k \geq 0}$. Pick any finite term of ρ, and compute the corresponding guess $\langle g_1(i), \ldots, g_n(i) \rangle$, $i > i_{**} \geq i_*$. This is a solution to $(\mathsf{TC}_\mathbb{N})^n$ for η by point 1 above. □

Corollary 2. $\mathsf{ECT} \leq_\mathrm{W} \mathsf{WOP}(X^\omega)$.

Proof. By Proposition 4, $(\mathsf{TC}_\mathbb{N})^n \leq_{\mathrm{sW}} \mathsf{WOP}(X_{\mathsf{lex}}^{2^n}) \leq_\mathrm{W} \mathsf{WOP}(X^\omega)$ for all $n \geq 1$. Hence $(\mathsf{TC}_\mathbb{N})^* \leq_\mathrm{W} \mathsf{WOP}(X^\omega)$. The desired conclusion follows from the fact that $\mathsf{ECT} \equiv_\mathrm{W} (\mathsf{TC}_\mathbb{N})^*$ (Lemma 1). □

Note that, since the proof of $(\mathsf{TC}_\mathbb{N})^* \equiv_\mathrm{W} \mathsf{ECT}$ in [4] shows $(\mathsf{TC}_\mathbb{N})^n \equiv_\mathrm{W} \mathsf{ECT}_{n+1}$, where ECT_{n+1} is the restriction of ECT to colorings in $(n+1)$ colors, Proposition 4 also shows $\mathsf{ECT}_{n+1} \leq_\mathrm{W} \mathsf{WOP}(X_{\mathsf{lex}}^{2^n})$ for all $n \geq 1$.

4.2 $\mathsf{WOP}(X^\omega)$ and $(\mathsf{LPO}')^*$ are Weihrauch Incomparable

The proof of the following proposition is a variation on the proof of [10, Lemma 8], which establishes that ECT doesn't reduce the weaker version $\mathsf{IsFinite}_\mathrm{S}$ of $\mathsf{IsFinite}$.

Proposition 5. $\mathsf{IsFinite} \nleq_\mathrm{W} \mathsf{WOP}(X^\omega)$.

Proof. Suppose by way of contradiction that $\mathsf{IsFinite} \leq_\mathrm{W} \mathsf{WOP}(X^\omega)$, as witnessed by the functionals H (sending instances to instances) and K (sending solutions to solutions).

Consider $s^0 := 0^{\mathbb{N}}$ and any solution $\langle y_0^0, y_1^0, \dots \rangle$ to $\mathsf{WOP}(X^\omega)$ for $H(s^0)$. Since $\mathsf{IsFinite}(s^0) = 1$, $K\langle\langle y_0^0, \dots, y_{m_0}^0 \rangle, 0^{n_0}\rangle \downarrow = 1$ for some $m_0, n_0 \geq 0$. By choosing a large enough n_0, we can make sure that all the terms $y_0^0, \dots, y_{m_0}^0$ are mentioned in $H(0^{n_0})$. We proceed inductively to obtain, for each $k \geq 1$: a binary sequence $s^k := 0^{n_0}10^{n_1} \dots 10^{n_{k-1}}10^{\mathbb{N}}$; a solution $\langle y_0^k, y_1^k, \dots \rangle$ to $\mathsf{WOP}(X^\omega)$ for $H(s^k)$; two integers $m_k, n_k \geq 0$ such that $K\langle\langle y_0^k, \dots, y_{m_k}^k \rangle, 0^{n_0}10 \dots 10^{n_{k-1}}10^{n_k}\rangle \downarrow = 1$ and such that $y_0^k, \dots, y_{m_k}^k$ are all mentioned in $H(0^{n_0}10 \dots 10^{n_{k-1}}10^{n_k})$.

Now, let $p := 0^{n_0}10^{n_1}10^{n_2}10^{n_3}1\dots$, and let $\langle x_0, x_1, \dots \rangle$ be a solution to $\mathsf{WOP}(X^\omega)$ for $H(p)$. Since $\mathsf{IsFinite}(p) = 0$, then $K\langle\langle x_0, \dots, x_m \rangle, 0^{n_0}1 \dots 10^{n_k}\rangle \downarrow = 0$ for some $m, k \geq 0$, and again we can assume without loss of generality that $x_0, \dots, x_m$ are all mentioned in $H(0^{n_0}1 \dots 10^{n_k})$. The latter, which is an initial segment of both $H(s^k)$ and $H(p)$, also contains the terms $y_0^k, \dots, y_{m_k}^k$, and thus decides on the relation between x_m and $y_{m_k}^k$ in the linear ordering $\leq_X$.

If $x_m \geq_X y_{m_k}^k$, then $\beta := \langle x_0, \dots, x_m, y_{m_k+1}^k, y_{m_k+2}^k, \dots \rangle$ is a solution to $\mathsf{WOP}(X^\omega)$ for $H(s^k)$. Hence $K\langle \beta, s^k \rangle = 1$. But

$$K\langle \beta, s^k \rangle = K\langle\langle x_0, \dots, x_m \rangle, 0^{n_0}1 \dots 10^{n_k}\rangle = 0,$$

which is impossible.

Similarly, if $x_m \leq_X y_{m_k}^k$ then $\gamma := \langle y_0^k, \dots, y_{m_k}^k, x_{m+1}, x_{m+2}, \dots \rangle$ is a solution to $\mathsf{WOP}(X^\omega)$ for $H(p)$, hence $K\langle \gamma, p \rangle = 0$. But

$$K\langle \gamma, p \rangle = K\langle\langle y_0^k, \dots, y_{m_k}^k \rangle, 0^{n_0}1 \dots 10^{n_k}\rangle = 1,$$

which is again impossible. This gives the desired contradiction. □

Corollary 3. $(\mathsf{LPO}')^* \not\leq_W \mathsf{WOP}(X^\omega)$.

Proof. Combine Proposition 5 with the fact that $\mathsf{IsFinite} \leq_W \mathsf{LPO}'$. □

Corollary 4. $\mathsf{WOP}(X^\omega) <_W \mathsf{ECT} \times \mathsf{WOP}(X^\omega)$.

Proof. We argue by way of contradiction. If $\mathsf{ECT} \times \mathsf{WOP}(X^\omega) \leq_W \mathsf{WOP}(X^\omega)$ implies $(\mathsf{LPO}')^* \leq_W \mathsf{WOP}(X^\omega)$ in view of Theorems 1 and 2. But this contradicts Corollary 3. □

Proposition 6. $\mathsf{WOP}(X^2_{\mathsf{lex}}) \not\leq_W (\mathsf{LPO}')^*$.

Proof. From Proposition 4 with $n = 1$ have that $\mathsf{TC}_{\mathbb{N}} \leq_{sW} \mathsf{WOP}(X^2_{\mathsf{lex}})$. On the other hand, $(\mathsf{LPO}')^*$ does not compute $\mathsf{TC}_{\mathbb{N}}$ (see [10, Lemma 9]). □

Corollary 5. $\mathsf{WOP}(X^\omega) \not\leq_W (\mathsf{LPO}')^*$.

Proof. This follows immediately from Proposition 6, by recalling from Lemma 4 that $\mathsf{WOP}(X^n_{\mathsf{lex}}) \leq_{sW} \mathsf{WOP}(X^\omega)[\leq 2n]$ for every $n \geq 1$. □

In [10] it is shown that $(\mathsf{LPO}')^*$ is Weihrauch-equivalent to the variants of $\mathsf{ORT}_{\mathbb{N}}$ and $\mathsf{ART}_{\mathbb{N}}$ in which the output is the color of some infinite monochromatic set for the instance coloring (these variants are denoted $c\mathsf{ORT}_{\mathbb{N}}$ and $c\mathsf{ART}_{\mathbb{N}}$ in [10]. By the results of the present section we have the obvious corollary that $\mathsf{WOP}(X^\omega)$ is incomparable with these principles as well.

5 Conclusion and Perspectives

We have compared in terms of Weirauch reducibility the Ordered Ramsey Theorem $\mathsf{ORT}_{\mathbb{N}}$ and the WOP for ω-exponentiation $\mathsf{WOP}(X^\omega)$. The fact that the Ordered Ramsey Theorem is reducible to the parallel product of ECT and $\mathsf{WOP}(X^\omega)$ is the first example of a Weihrauch reduction of a combinatorial principle to a well-ordering preservation principle. While we obtained a complete picture concerning the principles studied in this paper we consider the present work as a specimen of a wider study of Weihrauch reducibilities between WOPs and Ramsey-type theorems. We believe that the reductions from WOPs can be useful to obtain analogous results also at higher level of logical complexity, e.g., for the principles studied in [2].

Acknowledgment. We thank the referees for helpful comments.

References

1. Brattka, V., Gherardi, G., Pauly, A.: Weihrauch complexity in computable analysis. In: Brattka, V., Hertling, P. (eds.) Handbook of Computability and Complexity in Analysis. Theory and Applications of Computability, pp. 367–417. Springer, Cham (2021)
2. Carlucci, L., Mainardi, L., Zdanowski, K.: Reductions of well-ordering principles to combinatorial theorems. To appear in Notre Dame Journal of Formal Logic. **67**(2), 199–218 (2026)
3. Carlucci, L., Zdanowski, K.: A note on Ramsey theorems and Turing jumps. In: Cooper, S.B., Dawar, A., Löwe, B. (eds.) CiE 2012. LNCS, vol. 7318, pp. 89–95. Springer, Heidelberg (2012). https://doi.org/10.1007/978-3-642-30870-3_10
4. Davis, C., Hirschfeldt, D.R., Hirst, J.L., Pardo, J., Pauly, A., Yokoyama, K.: Combinatorial principles equivalent to weak induction. Computability **9**(3–4), 219–229 (2020)
5. Dzhafarov, D., Mumert, C.: Reverse Mathematics. Problems, Reductions, and Proofs. Springer, Cham (2022)
6. Hirst, J.L.: Reverse mathematics and ordinal exponentiation. Ann. Pure Appl. Logic **66**, 1–18 (1994)
7. Hirst, J.L.: Combinatorics in Subsystems of Second Order Arithmetic. Ph.D. thesis, The Pennsylvania State University, ProQuest LLC, Ann Arbor, MI (1987)
8. Kolodziejczyk, L.A., Michalewski, H., Pradic, C., Skrzypczak, M.: The logical strength of Büchi's decidability theorem. Logical Methods Comput. Sci. **15**(2), 1–31 (2019)
9. Marcone, A., Montalbàn, A.: The Veblen function for computability theorists. J. Symbolic Logic **76**(2), 575–602 (2011)
10. Pauly, A., Pradic, C., Soldà, G.: On the Weihrauch degree of the additive Ramsey theorem. Computability **13**(3–4), 459–483 (2024)
11. Rathjen, M., Weiermann, A.: Reverse mathematics and well-ordering principles. In: Cooper, S.B., Sorbi, A. (eds.) Computability in context, pp. 351–370. Imperial College Press (2011)
12. Uftring, P.: Weak and strong versions of effective transfinite recursion. Ann. Pure Appl. Logic **174**, 4 (2023)

Free Sets, Thin Sets and Rainbows for Barriers

Lorenzo Carlucci[1(✉)] and Oriola Gjetaj[2]

[1] Department of Mathematics, Sapienza University, Piazzale Aldo Moro 5, 00185 Roma, Italy
lorenzo.carlucci@uniroma1.it

[2] Department of Computer, Control, and Management Engineering, Sapienza University, Via Ariosto 25, 00185 Roma, Italy

Abstract. We formulate and prove the generalizations of Friedman's free set and thin set theorems and of the rainbow Ramsey theorem to colorings of barriers. We analyze the strength of these theorems from the point of view of computability theory proving some upper and lower bounds on the complexity of solutions for computable instances and some uniform computable reductions. We obtain as corollaries some proof-theoretical results on the logical strength of the theorems, in the spirit of reverse mathematics.

Keywords: Ramsey Theory · Barriers · Hyperarithmetical Hierarchy

1 Introduction and Motivation

Many variants or consequences of Ramsey's theorem have been extensively investigated in computability, proof theory and reverse mathematics. Among these, the free set theorem, the thin set theorem and the rainbow Ramsey theorem have attracted significant interest in recent decades (see, e.g., [6,7,15,19,25]).

The free set theorem (FS^n) states that, for every coloring of the n-subsets of the natural numbers in unboundedly many colors, there exists an infinite set $H \subseteq \mathbf{N}$ such that for all n-subsets s of H, the color of s is either not in H or else is in s itself. Such a set is called *free* for the coloring. The thin set theorem, denoted TS^n_ω (or just TS^n) is a weak variant of the free set theorem asserting that for any coloring of the n-subsets of $\mathbf{N}$ there is an infinite set $H \subseteq \mathbf{N}$ such that the n-subsets of H avoid at least one color. Such a set is called *thin* for the coloring. Both the free set theorem and the thin set theorem are consequences of Ramsey's theorem. Cholak et al. [7] started the study of the logico-computational content of these theorems. Contrary to Ramsey's theorem, for any $n \geq 3$, none of these principles for colorings of n-subsets codes the halting set. This surprising result was established by Wang [25].

The rainbow Ramsey theorem (RRT^n_k) is an "anti-Ramsey" principle that guarantees, for any coloring of the n-subsets of the natural numbers such that no color is used more than k times, the existence of an infinite set $H \subseteq \mathbf{N}$ whose

V. Brattka et al. (Eds.): CiE 2026, LNCS 16674, pp. 208–223, 2027.
https://doi.org/10.1007/978-3-032-31348-5_14

n-subsets are assigned distinct colors (i.e., the coloring is injective on the n-subsets). Such a set is called a *rainbow* for the coloring. Galvin [13] observed that the rainbow Ramsey theorem follows from the classical Ramsey's theorem. This theorem is studied by Csima and Mileti [10] and Wang [25] from the perspective of computability and reverse mathematics. The rainbow Ramsey theorem for pairs (RRT^2_2) is strictly weaker than Ramsey's theorem for pairs (RT^2_2) over RCA_0 and for no finite dimension n and no bound k the rainbow Ramsey theorem for k-bounded colorings of n-sets (RRT^n_k) does code the halting set [25].

In joint work with Le Houérou and Patey [2] we generalized these principles to colorings of objects of unbounded dimensions, in particular to colorings of sets $s \subseteq \mathbf{N}$ such that $|s| = \min(s) + 1$. These sets are known as *exactly large sets* or *ω-size* sets in Logic literature (see [3,4,17]).

Ramsey's theorem for colorings of exactly large sets is computationally and proof-theoretically stronger than the usual Ramsey theorem for fixed finite dimensions (RT^n_k) and even stronger than Ramsey's theorem for all finite dimensions ($\forall n \mathsf{RT}^n_k$). From the point of view of computability, the theorem corresponds to (closure under) the ω-th Turing jump (see [8] and [4]). Generalizations to exactly large sets of the free set, thin set and rainbow Ramsey theorems were studied in [2]. The free set and thin set theorems for exactly ω-large sets turn out to be significantly stronger than their versions for sets of fixed finite size. In particular, they can code the ω-th Turing jump and imply the system ACA_0^+. By contrast, the rainbow Ramsey theorem for exactly ω-large sets satisfies cone avoidance and therefore does *not* code the first Turing jump.

The family of exactly large sets forms a barrier in the sense of Nash-Williams better quasi-ordering theory and is known as the Schreier barrier in Ramsey theory and Banach space theory literature (see, e.g., [24]). Indeed, Nash-Williams proved that Ramsey's theorem holds for arbitrary barriers. The computational strength of Ramsey's theorem for barriers is investigated in [8]. It is thus quite natural to inquire into the generalization of the free set, thin set and rainbow Ramsey theorems to general barriers.

In the present paper we formulate and prove these theorems and establish some bounds on the computational complexity of their solutions. All principles P studied in this paper have the form $\forall X\, (I(X) \rightarrow \exists Y\, S(X, Y))$, where I and S are arithmetical formulas. We call such statements $\forall\exists$-principles. An X satisfying I is called an *instance* of P, and a Y satisfying $S(X, Y)$ is called a *solution* to P for X. Our results are naturally stated in terms of strong uniform computable reductions, known as strong Weihrauch reductions (see [12]). For $\forall\exists$-principles P and Q, we say that Q is strongly Weihrauch reducible to P, written $\mathsf{Q} \leq_{\mathrm{sW}} \mathsf{P}$, if there are Turing functionals Φ and Ψ such that for every instance X of Q, $\Phi(X)$ is an instance of P, and every solution $\hat{Y}$ to $\Phi(X)$ yields a solution $\Psi(\hat{Y})$ to Q for X. We occasionally also mention results in reverse mathematics. We refer to [12] for background and definitions.

The paper is organized as follows. In Sect. 2 we introduce barriers and Nash-Williams' Ramsey theorem for barriers. In Sect. 3 we introduce and prove the free set, thin set and rainbow Ramsey theorems for barriers. In Sect. 4 we prove some lower bounds on their solutions.

2 Ramsey Theory on Barriers

We start by fixing some notation. We use $\mathbf{N}$ to denote the set of natural numbers and $\mathbf{N}^+$ to denote the set of positive integers. We use uppercase letters X, Y, Z to denote subsets of $\mathbf{N}$ which may be finite or infinite. We use lowercase letters s, t to denote finite subsets of $\mathbf{N}$. If $X \subseteq \mathbf{N}$ we denote by $[X]^{<\omega}$ (respectively $[X]^\omega$) the set of finite (respectively, infinite) subsets of X. For $n \in \mathbf{N}$ we denote by $[X]^n$ the set of subsets of X with exactly n elements (which we call the n-subsets of X). We identify a subset X of $\mathbf{N}$ with the strictly increasing sequence (finite or infinite) which enumerates it and we denote by $X(i)$ or X_i the element in position i in the sequence. If s is such a finite sequence then $|s|$ denotes its length (which coincides with the cardinality of the range of s since s is increasing) and we write $s = (s_0, \ldots, s_{|s|-1})$. Given $s, X \subseteq \mathbf{N}$, by $s \sqsubseteq X$ we mean that s, as a sequence, is an initial segment (or prefix) of X. This is stronger than $s \subseteq X$, which denotes set-theoretic inclusion as usual. The irreflexive versions of the previous relations are denoted by $\sqsubset$ and $\subset$. By $s < X$ we mean that each element of s is strictly smaller than each element of X or equivalently (when s and X are nonempty) $\max s < \min X$. We write $s^\frown X$ for the concatenation of the sequences s and X. Notice that in our setting $s^\frown X$ does make sense only if $s < X$ and, set-theoretically, coincides with $s \cup X$. Let $F \subseteq [\mathbf{N}]^{<\omega}$ and $X \in [\mathbf{N}]^\omega$ then $F|X = \{s \in F : s \subseteq X\}$. Thus, $F|X$ is an abbreviation for $[X]^{<\omega} \cap F$. Let $X \in [\mathbf{N}]^\omega$ and $n \in \mathbf{N}$, then we denote by X/n (the n-tail of X) the set $X \setminus [0, n]$.

We now introduce the basic definitions concerning barriers. Barriers are combinatorial objects first introduced by Nash-Williams [18] to define better quasi orderings.

Definition 1 (Barriers). *A set $B \subseteq [\mathbf{N}]^{<\omega}$ is a barrier on $X \in [\mathbf{N}]^\omega$ if the following conditions are met:*

1. *(Base) $\bigcup B = X$;*
2. *(Sperner property) For all $s, t \in B$, $s \not\subset t$;*
3. *(Density) For all $Y \in [X]^\omega$ there exists an $s \in B$ such that $s \sqsubset Y$.*

The set $X = \bigcup B$ is called the base of B and is denoted by $b(B)$. The simplest examples of barriers are the families $[\mathbf{N}]^n$ for $n \in \mathbf{N}$. A trivial barrier often refers to the barrier of singletons, $[\mathbf{N}]^1 = \{\{n\} : n \in \mathbf{N}\}$. The family $\{s \subset \mathbf{N} : |s| = \min(s) + 1\}$ is a barrier, known as the Schreier barrier or as the barrier of exactly ω-large sets (also called ω-size sets). For $X \subseteq \mathbf{N}$ the family $\{s \subset X : |s| = \min(s) + 1\}$ is known as the Schreier barrier on X [24].

Marcone [16] started the study of barriers in the context of reverse mathematics. Among other things he showed that RCA_0 proves that if B is a barrier then $b(B)$ exists as a set and B is isomorphic to a barrier B' with $b(B') = \mathbf{N}$.

A fundamental fact, proved by Pouzet [20], is that the lexicographic order on barriers is always a well-ordering. Given a barrier $B \subseteq [\mathbf{N}]^{<\omega}$ and $s, t \in B$, let

$s <_{\text{lex}} t$ if $s(i) < t(i)$ for the least $i \in \mathbf{N}$ such that $s(i) \neq t(i)$, if it exists. Based on this, the order type of a barrier is the order type of its lexicographic order. If B is a barrier we denote its order type by $ot(B)$. Assous [1] characterized the order types of barriers, and proved that they are either of the form ω^n for some $n \in \mathbf{N}^+$, or $\omega^\alpha \cdot k$ for some $\alpha \geq \omega$ and $k \in \mathbf{N}^+$. The Schreier barrier is an example of a barrier of order type ω^ω.

If B is a barrier on $X \in [\mathbf{N}]^\omega$, then for any infinite subset $Y \in [X]^\omega$, the set $B|Y = \{s \in B \mid s \subseteq Y\}$ forms a barrier on Y, which we call a sub-barrier of B. Additionally, if B is a non-trivial barrier on X and $n \in X$ then $B_n = \{s \in [\mathbf{N}/n]^{<\omega} : \{n\} \cup s \in B\}$ is a barrier on X/n. Note that B_n is not in general a sub-barrier of B.

Proposition 1. *Let B be a barrier on $X \in [\mathbf{N}]^\omega$. Then $B' \subseteq B$ is a barrier if and only if there exists $Y \in [X]^\omega$ such that $B|Y = B'$.*

Proof. See the proof of Proposition 2.9 in [5]. Note that both directions between B' and Y are B-computable □

One reason for interest in barriers is that Ramsey's theorem holds for finite colorings of barriers as the following well-known result by Nash-Williams [18] shows.

Theorem 1 (Barrier Ramsey Theorem[1]). *Fix a barrier B on $X \in [\mathbf{N}]^\omega$. For every $f : B \to k$, where k is a positive integer, there exists an infinite set $H \in [X]^\omega$ such that f is constant on $B|H$. We say that H is monochromatic for f. We denote this principle by RT^B_k.*

More succinctly, in the light of Proposition 1, the previous theorem asserts that for all barriers B and for all k-coloring f of B there exists a sub-barrier $B' \subseteq B$ such that f is constant on B'. We call B' monochromatic for f.

Ramsey's theorem for the Schreier barrier, studied in [4], is a particular case of Nash-Williams' barrier Ramsey theorem, which in turn is a consequence of the clopen Ramsey theorem (see [22]). The barrier Ramsey theorem *for all barriers* is equivalent to the system ATR_0 (see Theorem 4.9 in [16]). Thus, the barrier Ramsey theorem is proof-theoretically much stronger than Ramsey's theorem for the Schreier barrier, which stands at the level of ACA_0^+ as proved in [4].

Clote [8] showed that the order-type of barriers is relevant to the computability-theoretic complexity of the barrier Ramsey theorem. He obtained a level-wise analysis of the complexity of solutions to computable instances of RT^B_k in the hyperarithmetic hierarchy as a function of the order type of the barrier B. He proved in particular the following theorem. The Σ^0_α sets are defined as in [21].

Theorem 2 (Clote [8]). *Fix $k \in \mathbf{N}^+$.*

[1] Note that in the recent [17] the "barrier Ramsey theorem" denotes a finitary theorem generalizing the Paris-Harrington principle.

1. *For every computable ordinal α, for every computable barrier B of order type at most $\omega^{1+\alpha}$ and every computable coloring $f : B \to k$, there is an infinite f-monochromatic set computable in $\emptyset^{(\alpha)}$.*
2. *For every computable limit ordinal α there exists a computable barrier B of order type ω^α and a computable coloring $f : B \to 2$ such that every infinite f-monochromatic set computes $\emptyset^{(\alpha)}$.*
3. *For every computable ordinal α there exists a computable barrier B of order type ω^α and a computable coloring $f : B \to 2$ that omits all $\Sigma^*_{\omega^\alpha} = \bigcup_{\beta<\omega^\alpha} \Sigma^0_\beta$-sets.*

3 Free Sets, Thin Sets and Rainbows for Barriers

In this section we introduce and prove the natural generalizations of the free set, thin set and rainbow Ramsey theorems to arbitrary barriers.

Definition 2 (Barrier Free Set Theorem). *Fix a barrier B on $X \in [\mathbf{N}]^\omega$. For every $f : B \to \mathbf{N}$ there exists a barrier $B' \subseteq B$ such that for all $s \in B'$ if $f(s) \in b(B')$ then $f(s) \in s$. We say that B' is free for f. If $B' = B|Y$ for some $Y \in [X]^\omega$, we call Y a free set for f. We denote this principle by FS^B.*

Definition 3 (Barrier Thin Set Theorem). *Fix a barrier B on $X \in [\mathbf{N}]^\omega$. For every $f : B \to \mathbf{N}$ there exists a barrier $B' \subseteq B$ such that the image of f restricted to B' is different from $\mathbf{N}$. We say that B' is thin for f. If $B' = B|Y$ for some $Y \in [X]^\omega$, we call Y a thin set for f. We denote this principle by TS^B.*

Let $k \in \mathbf{N}$. A coloring $f : X \to Y$ is called k-bounded if for all $x \in Y$ $|f^{-1}(x)| \leq k$.

Definition 4 (Barrier Rainbow Ramsey Theorem). *Fix a barrier B on $X \in [\mathbf{N}]^\omega$ and let $k \in \mathbf{N}$. For every k-bounded $f : B \to \mathbf{N}$, there exists a barrier $B' \subseteq B$ such that f is injective on B'. We say that B' is a rainbow for f. If $B' = B|Y$ for some $Y \in [X]^\omega$, we call Y a rainbow for f. We denote this principle by RRT^B_k.*

The proofs of the thin set and rainbow Ramsey theorems for barriers are straightforward generalizations of the proofs for the finite dimensional version of these theorems, see Sects. 3.2 and 3.3. This observation was indeed the initial motivation for the present project. The case of the free set theorem for barriers is more interesting. We treat this case in the next subsection.

3.1 Barrier Free Set Theorem

We prove the barrier free set theorem from the barrier Ramsey theorem by generalizing the proof for the Schreier barrier versions of these theorems from [2]. Interestingly, some operations that *prima facie* looked specific to exactly ω-large sets naturally generalize to arbitrary barriers. Let $X \in [\mathbf{N}]^\omega$ and let S_X be the Schreier barrier on X. The following facts are easy to check.

(1) If $s \in S_X$ and $k \in X$ and $k < \min(s)$ then $\{k\} \cup \{s_0, s_1, \dots, s_{k-1}\} \in S_X$.
(2) If $s = \{s_0, \dots, s_{s_0}\} \in S_X$ and $k \in X$ and $s_i < k < s_{i+1}$ for some $i \in [0, s_0 - 1]$ then $\{s_0, s_1, \dots, s_i, k, s_{i+1}, \dots s_{s_0-1}\} \in S_X$.

We jointly generalize the above two properties to general barriers in the following lemma.

Lemma 1. *Let B be a barrier with base $X \in [\mathbf{N}]^\omega$ and let $s = \{s_0, \dots, s_n\} \in B$ with $s_0 < \dots < s_n$. Fix $i \in [-1, n-1]$ (with $s_{-1} = -1$) and let $k \in (s_i, s_{i+1})$ such that $k \in X$. Then there exists a (unique) element of B of the form*

$$\{s_0, \dots, s_i, k, s_{i+1}, \dots, s_j\}$$

with $j < n$. We call such an element the k-variant of s and denote it by $s[k]$.

Proof. Such an element must exist by the Density and Sperner properties by considering the infinite set

$$\{s_0, \dots, s_i\} \cup \{k\} \cup \{s_{i+1}, \dots, s_n\} \cup X/s_n.$$

Necessarily the desired initial segment must stop not earlier than $\{s_0, \dots, s_i\} \cup \{k\}$ (otherwise it would be a subset of s) and not later than $\{s_0, \dots, s_i\} \cup \{k\} \cup \{s_{i+1}, \dots, s_{n-1}\}$ (otherwise s would be a subset of it). □

Note that, whenever defined, $s[k] <_{\text{lex}} s$. If $s \in B$ and $k \in X$ is such that $s_{n-1} < k < s_n$, then $(s_0, \dots, s_{n-1}, k) \in B$. For $k < \min(s)$ the notion of k-variant is closely related with the definition of the relation $\lhd$ on a barrier that is a key step in the definition of a better quasi order. For $s, t \in [\mathbf{N}]^{<\omega}$ the relation $s \lhd t$ is defined to hold if there exists a $u \in [\mathbf{N}]^{<\omega}$ such that $s \sqsubseteq u$ and $t \sqsubseteq u[1, |u|]$. If s is the k-variant of t for some $k < \min(t)$ then $s \lhd t$ holds. If s and t are members of a barrier and $s \lhd t$, then the reverse implication also holds; s is the k-variant of t for some $k < \min(t)$.

We next observe that barriers are preserved under translations. We use the following notation. If $X = \{x_1, x_2, \dots\} \subseteq \mathbf{N}$ we define $X^+ = \{x_i + 1 : i \in \mathbf{N}\}$.

Lemma 2. *Let B be a barrier with base $X \in [\mathbf{N}]^\omega$. Let*

$$B^+ = \{s^+ \cup \{m\} : s \in B \text{ and } m \in X^+/(\max(s) + 1)\}$$

where $s^+ = \{x + 1 : x \in s\}$. B^+ is barrier with base X^+.

Proof. We verify that B^+ satisfies the Base, Sperner and Density properties:

(Base) We need to show that the base of B^+ is infinite and is equal to X^+. Notice that every element of $s' \in B^+$ is either $s_i + 1$ with $s_i \in X$, or the last coordinate $m \in X^+$. Thus, $b(B^+) \subseteq X^+$. Conversely, let $x \in X^+$. Then $x = x_k + 1$ for some $x_k \in X$. Since X is the base of B there is some $s \in B$ such that $x_k \in s$. Let $m \in X^+/\max(s^+)$. Then $s^+ \cup \{m\} \in B^+$ and contains $x_k + 1$.

(Sperner property) For all $s', t' \in B^+$ such that $s' \neq t'$, $s' \not\subseteq t'$. Let $s' = s^+ \cup \{a\}$ and $t' = t^+ \cup \{b\}$ for some $s, t \in B$, $a > \max(s) + 1$ and $b > \max(t) + 1$

and such that $s' \neq t'$. Obviously if $s^+ = t^+$ then $a \neq b$ and therefore $s^+ \cup \{a\} \not\subseteq t^+ \cup \{b\}$. So suppose that $s^+ \neq t^+$, i.e. $s \neq t$. By the Sperner Property on B we have $s \not\subseteq t$. Suppose $s^+ \cup \{a\} \subseteq t^+ \cup \{b\}$. If $s^+ \cup \{a\} \subseteq t^+$ then $s \subseteq t$, which is impossible.

(Density) We want to show that for all $Y \in [X^+]^\omega$ there exists an $s' \in B^+$ such that $s' \sqsubset Y$. By definition we have that $Y = Z^+$ for some $Z \in [X]^\omega$. By the density of B on X there exists an $s \in B$ such that $s \sqsubset Z$. Thus $s^+ \sqsubset Z^+$. Let z be the first element of Z larger than $\max(s^+)$. Then $s^+ \cup \{z\} \sqsubset Z^+$ and $s^+ \cup \{z\} \in B^+$. □

In the next theorem we prove the free set theorem for barriers using Ramsey's theorem for barriers. We express this result (and later positive results) in terms of strong Weihrauch reductions, which we denote by $\leq_{\mathrm{sW}}$ (see [12] for definitions).

Theorem 3. *For each barrier B with base* $\mathbf{N}$, $\mathsf{FS}^B \leq_{\mathrm{sW}} \mathsf{RT}_2^{B^+}$.

Proof. Let B be a barrier with base $\mathbf{N}$ and let $f : B \to \mathbf{N}$ be an instance of FS^B. For $s = (s_0, \dots, s_n) \in B^+$, let $s \ominus 1 = (s_0 - 1, \dots, s_{n-1} - 1)$. Then $s \ominus 1 \in B$. Define g by recursion on $<_{\mathrm{lex}}$ as follows.

$$g(s) = \begin{cases} 0 & \text{if } f(s \ominus 1) = s_i - 1 \textit{ for some } i < n, \\ 1 - g(s[f(s \ominus 1) + 1]) & \text{if } f(s \ominus 1) < s_0 - 1, \text{or, for some } i < n - 1, \\ & f(s \ominus 1) \in (s_i - 1, s_{i+1} - 1), \\ 0 & \text{if } f(s \ominus 1) \in (s_{n-1} - 1, s_n - 1), \\ 1 & \text{otherwise } (\textit{if } f(s \ominus 1) \geq s_n) \end{cases}$$

Note that g is well-defined: if $f(s\ominus 1) < s_0 - 1$ then $f(s\ominus 1) + 1 < s_0 = \min(s)$, so that $s[f(s \ominus 1) + 1]$ is well-defined, is in B^+ and is $<_{\mathrm{lex}}$-smaller than s. If $f(s \ominus 1) \in (s_i - 1, s_{i+1} - 1)$ for some $i < n - 1$, then $f(s \ominus 1) + 1 \in (s_i, s_{i+1})$ for some $i < n - 1$ and $s[f(s\ominus 1) + 1]$ is defined, is in B^+ and is $<_{\mathrm{lex}}$-smaller than s. Let $H = \{x_0, x_1, \dots\} \subseteq \mathbf{N}^+$ be infinite such that $B^+|H$ is g-monochromatic as given by $\mathsf{RT}_2^{B^+}$. We claim that $B|H^-$ is f-free, where $H^- = \{x_0 - 1, x_1 - 1, \dots\}$. Consider some $(s_0 - 1, \dots, s_{n-1} - 1) \in B|H^-$. Then $\{s_0, \dots, s_{n-1}\} \subseteq H$ and for every $k \in H/s_{n-1}$, $(s_0, \dots, s_{n-1}, k) \in B^+|H$. To show that $B|H^-$ is f-free it is sufficient to show that for some k, writing $s = (s_0, \dots, s_{n-1}, k) \in B^+|H$, we have that if $f(s \ominus 1) \in H^-$ then $f(s \ominus 1) \in s \ominus 1$, since all elements of $B|H^-$ are of the form $s \ominus 1$ for some $s \in B^+|H$. There are two cases:

Case 1: $B^+|H$ is monochromatic for the color 0. Take s_n to be the next element of H after s_{n-1} and write $s = (s_0, \dots, s_n)$, then $g(s) = 0$. We consider three subcases:

Subcase 1.1: $f(s \ominus 1) = s_i - 1$ for some $i < n$. Then $f(s \ominus 1) \in s \ominus 1$, so we are done.

Subcase 1.2: $f(s\ominus 1) \in (s_{n-1} - 1, s_n - 1)$. Then $f(s\ominus 1) \notin H^-$ by definition of s_n as the next element of H after s_{n-1}.

Subcase 1.3: $f(s\ominus 1) < s_0-1$ or $f(s\ominus 1) \in (s_i-1, s_{i+1}-1)$ for some $i < s_0-1$ and $g(s[f(s\ominus 1)+1]) = 1$. Since $f(s\ominus 1)+1 \in H$, we have $s[f(s\ominus 1)+1] \in B^+|H$. But then $g(s[f(s\ominus 1)+1]) = 1$ contradicts the fact that $B^+|H$ is g-monochromatic for the color 0.

Case 2: $B^+|H$ is monochromatic for the color 1. Take $s_n \in H$ bigger than s_{n-1} and write $s = (s_0, \ldots, s_n)$, then $s \in B^+|H$ and $g(s) = 1$. We consider two subcases:

Subcase 2.1: $f(s\ominus 1) \geq s_n$. This case is impossible, indeed, as H is infinite, there exists some element $x \in H$ such that $x > f(s\ominus 1)+1$ and therefore $f(s\ominus 1) \in (s_n-1, x-1)$, which leads to $g(s_0, \ldots, s_{n-1}, x) = 0$ contradicting the fact that $B^+|H$ is g-monochromatic for the color 1. Note that by definition of B^+, $(s_0, \ldots, s_{n-1}, x) \in B^+$.

Subcase 2.2: $f(s\ominus 1) < s_0-1$ or $f(s\ominus 1) \in (s_i-1, s_{i+1}-1)$ for some $i < n-1$ and $g(s[f(s\ominus 1)+1]) = 0$. Since $f(s\ominus 1)+1 \in H$, we have that $s[f(s\ominus)+1] \in B^+|H$. But $g(s[f(s\ominus 1)+1]) = 0$ contradicts the fact that $B^+|H$ is g-monochromatic for the color 1.

Therefore $B|H^-$ is f-free. Notice that g is uniformly computable in f and that H^- is uniformly computable in H. Hence, $\mathsf{FS}^B \leq_{\mathrm{sW}} \mathsf{RT}_2^{B^+}$. □

In terms of reverse mathematics, the above proof shows that FS^B follows from $\mathsf{RT}_2^{B^+}$ over $\mathsf{RCA}_0 + \mathrm{WO}(ot(B^+))$, where $ot(B^+)$ is the order type of B^+. In terms of computability FS^B inherits the upper bound on solutions to $\mathsf{RT}_2^{B^+}$ from Theorem 2. It is easy to show that, for every barrier B,

$$ot(B^+) = \omega \cdot ot(B),$$

since decomposes B^+ into intervals $I_s := (s^+, m)$ with $m > \max(s^+)$ indexed by $s \in B$ such that each interval is ordered according m (so that $ot(I_s) = \omega$) and the relative order of the intervals is determined by the order of their index s. Since RCA_0 proves $\forall\alpha(\mathrm{WO}(\alpha) \rightarrow \mathrm{WO}(\omega\cdot\alpha))$, the proof of Theorem 3 shows that $\mathsf{RT}_2^{B^+}$ implies FS^B over $\mathrm{WO}(ot(B))$.

3.2 Barrier Thin Set Theorem

The proof of the following proposition is a straightforward adaptation from the analogous result for the finite dimensional versions of the thin set theorem (Proposition 5.2 in [11]).

Proposition 2. *For all barriers B, $\mathsf{TS}^B \leq_{\mathrm{sW}} \mathsf{RT}_2^B$.*

Proof. Let $f : B \rightarrow \mathbf{N}$ be an instance of TS^B. Define $g : B \rightarrow 2$ as follows:

$$g(s) = \begin{cases} f(s) & \text{if } f(s) = 0, \\ 1 & \text{otherwise.} \end{cases}$$

Any g-monochromatic sub-barrier B' of B is obviously thin for f. □

The barrier thin set theorem also follows from the barrier free set theorem. The proof of the next theorem is the same as the proof for the finite dimensional case (Theorem 3.2 in [7]). The observation that the argument works for barriers started the investigations on FS and TS for barriers.

Theorem 4. *For all barriers B, $\mathsf{TS}^B \leq_{\mathrm{sW}} \mathsf{FS}^B$.*

Proof. Let $f : B \to \mathbf{N}$. Let X be an infinite set such that $X \subseteq b(B)$ and $B|X$ is free for f. Let Y be a non-empty subset of X such that $X \setminus Y$ is infinite. We claim that $B|(X \setminus Y)$ is thin for f. Assume, by way of contradiction, that for all $n \in \mathbf{N}$ there exists an $s_n \in B|(X \setminus Y)$ such that $f(s_n) = n$. Take $n \in Y$. Thus, $n \in X$. Since $B|X$ is free for f, it must be the case that $n \in s_n$, contradicting the fact that s_n was chosen in $X \setminus Y$. □

As a corollary of the above reductions and Theorem 2 we get the following.

Corollary 1. *For every computable ordinal α, for every computable barrier B of order type at most $\omega^{1+\alpha}$ and every computable coloring $f : B \to \mathbf{N}$, there is an infinite f-thin set computable in $\emptyset^{(\alpha)}$.*

3.3 Barrier Rainbow Ramsey Theorem

Galvin's proof of rainbow Ramsey theorem from Ramsey's theorem (see the proof of [10, Theorem 1.6]) adapts with minor modifications to the barrier version.

Proposition 3. *For all $k \in \mathbf{N}^+$, for all barriers B, $\mathsf{RRT}^B_k \leq_{\mathrm{sW}} \mathsf{RT}^B_k$.*

Proof. Let $f : B \to \mathbf{N}$ be k-bounded. Fix a B-computable bijection $c : B \to \mathbf{N}$. Define $g : B \to k$ as follows:

$$g(s) = |\{t \in B : c(t) < c(s) \text{ and } f(s) = f(t)\}|.$$

The set $\{t \in B : c(t) < c(s)\}$ is finite and finite coded. The fact that g is a k-coloring follows from the hypothesis that f is k-bounded.

Let $H \subseteq \mathbf{N}$ be an infinite subset of $b(B)$ such that g is constant on $B|H$, as given by RT^B_k. We claim that $B|H$ is a rainbow for f.

Let $s, t \in B|H$. Since $g(s) = g(t)$ and either $c(s) < c(t)$ or $c(t) < c(s)$, we have that $f(s) \neq f(t)$. Thus $B|H$ is a rainbow for f. □

As is the case for the Schreier barrier (see [2, Proposition 3.11]) and for $[\mathbf{N}]^n$ (see [25, Theorem 4.2]), the rainbow Ramsey theorem for 2-bounded colorings of barriers follows from the free set theorem for barriers. We don't know if the same holds for $k > 2$.

Proposition 4. *For all barriers B, $\mathsf{RRT}^B_2 \leq_{\mathrm{sW}} \mathsf{FS}^B$.*

Proof. Let B be a barrier. Fix a B-computable bijection $c : B \to \mathbf{N}$ so that for each s the set $\{t \in B : c(t) < c(s)\}$ is finite and finite coded. Let $f : B \to \mathbf{N}$ be 2-bounded. For $s \in B$ define $g(s)$ as follows:

$$g(s) = \begin{cases} \min(t \setminus s) & \text{if there exists a } t \in B \text{ s.t. } c(t) < c(s) \text{ and } f(s) = f(t), \\ 0 & \text{otherwise.} \end{cases}$$

Since f is 2-bounded, if t exists in the definition of g then it is unique. If t and s are distinct elements in the barrier B then $(t \setminus s) \neq \emptyset$, since $t \subseteq s$ is impossible. Let A be an infinite subset of the base of B such that $B' = B|A$ is g-free. We claim that B' is a rainbow for f. Suppose otherwise, by way of contradiction, as witnessed by $s, t \in B'$ such that $f(s) = f(t)$. Without loss of generality we can assume $c(t) < c(s)$. Then $g(s) = \min(t \setminus s) \in A \setminus s$, contradicting that B' is g-free. □

As a corollary of the above reductions and Theorem 2 we get the following.

Corollary 2. *For every computable ordinal α, for every computable barrier B of order type at most $\omega^{1+\alpha}$ and every computable coloring $f : B \to \mathbf{N}$, there is an infinite f-rainbow computable in $\emptyset^{(\alpha)}$.*

4 Lower Bounds

The argument used by Jockusch to establish the seminal result that RT^2_2 omits Σ^0_2-solutions (Theorem 3.1 in [14]) turned out to be extremely versatile. It stands out as the model whose variations allowed to establish a large number of similar omission results for variants of Ramsey's theorem. Among those are the increasing polarized Ramsey theorem, Hindman's finite sums theorem, the free set theorem, the thin set theorem and the rainbow Ramsey theorem. The fertility of Jockusch's construction is not limited to the lower levels of the arithmetical hierarchy. Indeed, Clote [8] showed how the scheme of Jockusch's proof can be lifted to obtain weak anti-basis results with respect to the hyperarithmetical hierarchy. In particular Clote used this idea to obtain lower bounds for the barrier Ramsey theorem RT^B_2 as a function of the order-type of the barrier B as shown in Theorem 2 above.

It is thus natural to inquire whether the same scheme of argument applies to the barrier versions of other variants of Ramsey's theorem for which Jockusch's argument works in the finite case. In this section we verify that this is the case for the barrier thin set theorem and for the barrier rainbow Ramsey theorem.

4.1 Clote's Canonical Barriers and Generalized Limit Lemma

Clote's lifting of Jockusch's argument relies on a generalization of Shoenfield's limit lemma proved in [9]. Clote's proof uses limits taken along barrier elements rather than along the natural numbers. The barriers used in the construction are called canonical barriers, and they are defined using notations for computable ordinals in Kleene's notation system $\mathcal{O}$ (see [21]).

We need the following definition. For barriers B and B' let $B * B' = \{s \cup t : s \in B, t \in B' \text{ and } \max(s) < \min(t)\}$. Note that if barriers B and B' have the same base, then $B * B'$ is a barrier on that base.

Definition 5 (Canonical Barriers). *For each ordinal notation $a \in \mathcal{O}$, the canonical barrier B_a is defined by induction as follows.*

1. $B_1 = \{()\}$,
2. $B_2 = \{(n) : n \in \mathbf{N}\}$,
3. $B_{2^a} = B_2 * B_a$,
4. $B_{3\cdot 5^z} = \bigcup_{n\in\mathbf{N}} \{(n)\} * B_{\varphi_z(n)} * B_{\varphi_z(n-1)} * \cdots * B_{\varphi_z(0)}$.

Recall that if $3 \cdot 5^z$ is a notation then $\varphi_z(0), \varphi_z(1), \varphi_z(2), \ldots$ are notations for an increasing sequence of ordinals whose limit is the ordinal denoted by $3\cdot 5^z$. Each B_a has base $\mathbf{N}$ and order type $\omega^{|a|}$ (see [8]). Note that Clote's canonical barriers are close (but not identical) to so-called *exactly* α-large sets (a.k.a. α-size sets).

For an ordinal notation $a \in \mathcal{O}$, limits along a canonical barrier are defined inductively as as follows. Let g be a partial computable function, then

$$\lim_{s\in B_1} g(x,s) = g(x,()),$$

$$\lim_{s\in B_2} g(x,s) = y \text{ if and only if } \exists m \forall n \geq m \;\; g(x,(n)) = y,$$

$$\lim_{s\in B_{2^a}} g(x,s) = \lim_{(n)\in B_2} \lim_{\substack{t\in B_a \\ n<\min(t)}} g(x,(n)^\frown t),$$

$$\lim_{s\in B_{3\cdot 5^z}} g(x,s) = \lim_{\langle n\rangle\in B_2} \lim_{\substack{t_n\in B_{\varphi_z(n)} \\ n<\min(t_n)}} \cdots \lim_{\substack{t_0\in B_{\varphi_z(0)} \\ \max(t_1)<\min(t_0)}} g(x,(n)^\frown t_n^\frown \cdots ^\frown t_0).$$

The limits along a canonical barrier correspond, as noted in [9], to the notion of limit with respect to a generalized Fréchet filter.

The proof in [9] of the following generalization of the limit lemma uses effective transfinite induction on ordinal notations. The set H_a for $a \in \mathcal{O}$ is defined as usual and captures the a-iteration of the Turing jump (see, e.g., [21] for a definition). By a result of Spector [23] the sets H_a only depend on the ordinal denoted by a, up to Turing equivalence.

Theorem 5 (Clote [9]). *There exists a partial computable function g such that for all ordinal notations $a \in \mathcal{O}$, for all $e, x \in \mathbf{N}$ and $s \in \mathbf{N}^{<\omega}$, $g(a,e,x,s)$ is defined and if $\varphi_e^{H_a}(x) = y$ then $\lim_{s\in B_a} g(a,e,x,s) = y$.*

We will need the following corollary of the generalized limit lemma (it is proved as a Claim in the proof of Theorem 9 in [8]).

Lemma 3 (Clote, [8]). *For all infinite $X \subseteq Nat$, for all $a \in \mathcal{O}$, for all $e, k \in \mathbf{N}$, there exists $s \in B_a|X$ such that for all $x \leq k$, $g(a,e,x,s) = \varphi_e^{H_a}(x)$, if $\varphi_e^{H_a}(x)$ is defined.*

From the proof in [9] it is clear that in the above lemma the element $s \in B_a$ can be chosen so as to satisfy $k < \min(s)$. This will be useful in some later proofs.

Clote [8] used the above results to prove anti-basis results for Ramsey's theorem for barriers. With the same technique, we will obtain (weak) anti-basis results for the free set, thin set and rainbow Ramsey theorems for barriers.

4.2 Omission Results

We start with the barrier thin set theorem. The proof of the following theorem by Cholak et al. (Theorem 4.1 in [7]) is a variation on Jockusch's proof of the analogous omitting theorem for Ramsey for pairs (Theorem 3.1 in [14]).

Theorem 6 (Cholak et al. [7]). *There is a computable function $f : [\mathbf{N}]^2 \to \mathbf{N}$ such that no infinite Σ_2^0 set is thin for f.*

Using Clote's generalized limit lemma (Theorem 5 above) we obtain the following.

Theorem 7. *For all $a \in \mathcal{O}$ there is a computable function $f : B_2 * B_a \to \mathbf{N}$ with no infinite thin set computable in H_a.*

Proof. We want to define a computable function $f : B_2 * B_a \to \mathbf{N}$ such that for all $e, i \in \mathbf{N}$, if $X \subseteq \mathbf{N}$ is infinite and $X = \varphi_e^{H_a}$ then $i \in f([X]^{<\omega} \cap (B_2 * B_a))$, i.e. $(B_2 * B_a)|X$ is not thin for f.

From Lemma 5 we have that for all $a \in \mathcal{O}$, for all $e, x \in \mathbf{N}$ and for all $s \in [\mathbf{N}]^{<\omega}$, $g(a,e,x,s)$ is defined, $\{0,1\}$-valued and, moreover,

$$\varphi_e^{Ha}(x) \downarrow \Longrightarrow \lim_{s \in B_a} g(a,e,x,s) = 1.$$

For each $e \in \mathbf{N}$, define A_e as the set with characteristic function $\varphi_e^{H_a}$ if the latter is a characteristic function, and let A_e be undefined otherwise. Denote by $F_{e,i}$ the set of least $\langle e,i\rangle + 1$ elements of A_e, if A_e is defined and has more than $\langle e,i\rangle$ elements. Let $F_{e,i,s}$ be the natural approximation of $F_{e,i}$ at stage s where $s = (s_1, \ldots, s_n) \in B_a$; that is if there are more than $\langle e,i\rangle$ numbers $x < s_1$ with $g(a,e,x,s) = 1$ then $F_{e,i,s}$ consists of the first $\langle e,i\rangle + 1$ such numbers, otherwise let $F_{e,i,s}$ be undefined.

The construction of f is carried out in stages $s = (s_1 \ldots, s_n) \in B_a$ and $f(m,s)$ is defined at stage s for each $m < s_1$. At stage $s = (s_1, \ldots, s_n) \in B_a$, with $s_1 < \ldots < s_n$, there are substages $0, 1, \ldots, \langle e,i\rangle, \ldots, s_1$ and f is defined on at most one new argument at each substage.

Substage $\langle e,i\rangle < s_1$ of stage s: if $F_{e,i,s}$ is not defined, continue to the next substage without doing anything. If $F_{e,i,s}$ is defined, let $m_{e,i,s}$ be the smallest element m of $F_{e,i,s}$ with $f(m,s)$ not yet defined, and set $f(m_{e,i,s}, s) := i$.

Substage s_1 of stage s: let $f(m, s) := 1$ for all $m < s_1$, for which $f(m, s)$ is not yet defined. This completes the construction.

Now let $X \subseteq \mathbf{N}$ be an infinite, H_a-computable set. Let $\varphi_e^{H_a}$ be its characteristic function. We show that $(B_2 * B_a)|X$ is not thin for f.

If X is as above then by Lemma 3, for all $i \in \mathbf{N}$, letting $k = \max(F_{e,i})$, there exists $s \in B_a$ such that

$$(\forall x \leq k)(k < \min(s) \text{ and } g(a, e, x, s) = \varphi_e^{H_a}(x)).$$

By construction there exists $m \in X$ with $m < s_1$ where $f(m, s) = i$. Then $i \in f([X]^{<\omega} \cap (B_2 * B_a))$, thus X is not thin for f relative to the barrier $B_2 * B_a$. □

We now turn to the case of the barrier Ramsey theorem. We proceed as for the thin set theorem, using the proof of the following results by Csima and Mileti ([10], Theorem 2.4) as a template.

Theorem 8 (Csima and Mileti). *There exists a computable 2-bounded function $f : [\mathbf{N}]^2 \to \mathbf{N}$ such that no infinite Σ_2^0 set is a rainbow for f.*

We extend this result to barriers in the spirit of Clote [8].

Theorem 9. *For all $a \in \mathcal{O}$ there is a 2-bounded computable $f : B_2 * B_a \to \mathbf{N}$ with no infinite rainbow computable in H_a.*

Proof. We want to show that for all e, if $\varphi_e^{H_a}$ is the characteristic function of an infinite subset X of $b(B_a)$[2] then X is not a rainbow for f, in the sense that f is not injective on $(B_2 * B_a)|X$. To this aim it is sufficient to show that there are $\ell, m \in X$ and $s \in B_a$ such that $m < \ell < \min(s)$ for which $f(m, s) = f(\ell, s)$.

Let g be the partial computable function given by Theorem 5.

We define a 2-bounded computable $f : B_2 * B_a \to \mathbf{N}$ in stages, where the stages are elements of the barrier B_a. For each $s = (s_1, \ldots, s_n) \in B_a$ at stage s there are substages $0, \ldots, e, \ldots, s_1$.

At substage $e < s_1$, if there exist $m < \ell < s_1$ such that $g(a, e, m, s) = g(a, e, \ell, s) = 1$ which have not been claimed during stage s then claim the least such m and ℓ and set

$$f(m, s) := \langle m, s \rangle =: f(\ell, s).$$

Otherwise, do nothing at this substage. Continue to the next substage.

At substage s_1 set $f(\ell, s) := \langle \ell, s \rangle$ for all $\ell < s_1$ for which $f(\ell, s)$ is not yet defined.

Notice that f is computable and 2-bounded by construction. The construction above is such that, for $s \in B_a$, the value $f(\ell, s)$ is defined for all $\ell < s_1$ at stage s. Thus, overall, f is defined on $B_2 * B_a$.

Now let $X \subseteq \mathbf{N}$ be an infinite H_a-computable subset of $b(B_a)$ with characteristic function $\varphi_e^{H_a}$.

[2] By definition it should be a subset of the base of $B_2 * B_a$ but the base of B_2 is $\mathbf{N}$.

From Theorem 3 we can deduce the following. Let D_e denote the first $2e+2$ elements of X. Then there exists a $s \in B_a$ such that $\max(D_e) < \min(s)$ and for all $x \leq \max(D_e)$, we have

$$g(a, e, x, s) = \varphi_e^{H_a}(x).$$

We say that s *is correct for* D_e.

By choice of s and since X is infinite there are exactly $2e+2$ numbers $x \leq \max(D_e) < \min(s)$ such that $g(a, e, x, s) = 1$ (since s is correct for D_e). Since at most $2e$ elements are claimed at stage s before substage e, the construction at substage e of stage s claims the least unclaimed numbers $m, \ell < \min(s)$ with $g(a, e, m, s) = g(a, e, \ell, s) = 1$ and defines $f(m, s) = f(\ell, s)$.

By the minimality of m and ℓ and by the correctness of s for D_e it must be the case that $m, \ell \in D_e \subseteq X$. But then $m, \ell \in X$ and $s \in B_a$ and $f(m, s) = f(\ell, s)$, so X is not a rainbow for f relative to the barrier $B_2 * B_a$ (i.e. it is not the case that f is injective on $(B_2 * B_a)|X$). □

We briefly observe how some non-provability corollaries follow from the omitting theorems proved in the previous sections. Let α be a notation for a computable ordinal. Let Π^0_α-CA_0 be the system axiomatized by RCA_0 augmented with the well-ordering of α and the axiom of closure under the α-th jump. Note that Π^0_α-CA_0 is strictly stronger than Π^0_β-CA_0 if and only if $\alpha \geq \omega \cdot \beta$. For a notation $a \in \mathcal{O}$ for an ordinal $\alpha \geq \omega \cdot \beta$ the existence of a computable coloring of $B_2 * B_a$ with no thin set computable in H_a (Theorem 7) implies that the barrier thin set theorem is not provable in Π^0_β-CA_0, even if restricted to barriers of order type $\omega^{1+\alpha}$. The same applies to the barrier free set theorem and to the barrier rainbow Ramsey theorem.

5 Conclusions and Future Work

We formulated and proved natural generalizations of the free set theorem, the thin set theorem and the rainbow Ramsey theorem to colorings of barriers and we initiated the study of their effective and logical strength.

We showed that the barrier free set theorem, the barrier thin set theorem and the barrier rainbow Ramsey theorem are consequences of the barrier Ramsey theorem. Moreover, the barrier thin set theorem and the barrier rainbow Ramsey theorem (for 2-bounded colorings) are reducible to the barrier free set theorem as is the case in the fixed-dimension versions.

We then showed that for each infinite computable ordinal α, the free set, thin set and rainbow Ramsey theorems for barriers of order type ω^α have computable instances with no solution computable in $\emptyset^{(\alpha)}$. This was previously known only for $\alpha = \omega$ (see [2]). It remains to investigate whether computable instances of the free set, thin set or rainbow Ramsey theorems can code transfinite jumps. We plan to address this issue in future work.

Acknowledgements. We thank the referees for helpful suggestions.

References

1. Assous, M.: Caractérisation du type d'ordre des barrières de Nash-Williams. Publications du Département de mathématiques (Lyon) **11**(4), 89–106 (1974)
2. Carlucci, L., Gjetaj, O., Le Houérou, Q., Patey, L.L.: Ramsey-like theorems for the Schreier barrier. J. Symb. Log. 1–29 (2025)
3. Carlucci, L., Volpi, A., Zdanowski, K.: The strength of Ramsey's theorem for coloring. α-large sets (2026). arXiv:2603.22579
4. Carlucci, L., Zdanowski, K.: The strength of Ramsey's theorem for coloring relatively large sets. J. Symb. Log. **79**(1), 89–102 (2014)
5. Carroy, R., Pequignot, Y.: From well to better, the space of ideals. Fundam. Math. **227**(3), 247–270 (2014)
6. Cholak, P., Patey, L.: Thin set theorems and cone avoidance. Trans. Amer. Math. Soc. **373**(4), 2743–2773 (2020)
7. Cholak, P.A., Giusto, M., Hirst, J.L., Jockusch, Jr. C.G.: Free sets and reverse mathematics. In: Reverse mathematics 2001, volume 21 of Lecture Notes in Logic, pp. 104–119. Assoc. Symbol. Logic, La Jolla (2005)
8. Clote, P.: A recursion theoretic analysis of the Clopen Ramsey theorem. J. Symb. Log. **49**(2), 376–400 (1984)
9. Clote, P.: A generalization of the limit lemma and Clopen games. J. Symb. Log. **51**(2), 273–291 (1986)
10. Csima, B.F., Mileti, J.R.: The strength of the rainbow Ramsey theorem. J. Symb. Log. **74**(4), 1310–1324 (2009)
11. Dorais, F., Dzhafarov, D., Hirst, J., Mileti, J., Shafer, P.: On uniform relationships between combinatorial problems. Trans. Amer. Math. Soc. **368**(2), 1321–1359 (2016)
12. Dzhafarov, D.D., Mummert, C.: Reverse Mathematics: Problems, Reductions, and Proofs. Theory and Applications of Computability. Springer International Publishing, Cham (2022)
13. Galvin, F., Prikry, K.: Borel sets and Ramsey's theorem. J. Symb. Log. **38**(2), 193–198 (1973)
14. Jockusch, C.G.: Ramsey's theorem and recursion theory. J. Symb. Log. **37**(2), 268–280 (1972)
15. Liu, L., Patey, L.: The reverse mathematics of the Thin set and Erdős-Moser theorems. J. Symb. Log. **87**(1), 313–346 (2022)
16. Marcone, A.: Wqo and Bqo theory in subsystems of second order arithmetic. In: Reverse Mathematics 2001, volume 21 of Lect. Notes Log., pp. 303–330. Assoc. Symbol. Logic, La Jolla (2005)
17. Marcone, A., Montalbán, A., Volpi, A.: The barrier Ramsey theorem (2025). arXiv:2505.02544
18. Nash-Williams, C.S.J.A.: On better-quasi-ordering transfinite sequences. Proc. Cambridge Philos. Soc. **64**, 273–290 (1968)
19. Patey, L.: The weakness of being cohesive, thin or free in reverse mathematics. Israel J. Math. **216**(2), 905–955 (2016)
20. Pouzet, M.: Sur les prémeilleurordres. Ann. Inst. Four. **22**(2), 1–19 (1972)
21. Rogers, H.: Theory of Recursive Functions and Effective Computability. MIT Press, Cambridge, 1st mit press edition (1987)
22. Simpson, S.G.: Subsystems of Second Order Arithmetic. Perspectives in Logic. Cambridge University Press, Cambridge; Association for Symbolic Logic, Poughkeepsie, NY, second edition (2009)

23. Spector, C.: Recursive well-orderings. J. Symb. Log. **20**(2), 151–163 (1955)
24. Todorcevic, S.: Introduction to Ramsey Spaces. Princeton University Press, Princeton (2010)
25. Wang, W.: Some logically weak Ramseyan theorems. Adv. Math. **261**, 1–25 (2014)

Adaptive Multi-Head Finite-State Gamblers

Julianne Cruz[1], Sho Glashausser[1], Xiaoyuan Li[2], and Neil Lutz[1](✉)

[1] Swarthmore College, Swarthmore, PA 19081, USA
{jcruz2,sglasha1,nlutz1}@swarthmore.edu
[2] Iowa State University, Ames, IA 50011, USA
forbesii@iastate.edu

Abstract. Multi-head finite-state dimensions and predimensions quantify the predictability of a sequence by a gambler with trailing heads acting as "probes to the past". These additional heads allow the gambler to exploit patterns that are simple but non-local, such as in a sequence S with $S[n] = S[2n]$ for all n. In the original definitions of Huang, Li, Lutz, and Lutz (2025), the head movements were required to be *oblivious* (i.e., data-independent). Here, we introduce a model in which head movements are *adaptive* (i.e., data-dependent) and compare it to the oblivious model. We establish that for each $h \geq 2$, adaptivity enhances the predictive power of h-head finite-state gamblers, in the sense that there are sequences whose oblivious h-head finite-state predimensions strictly exceed their adaptive h-head finite-state predimensions. We further prove that adaptive finite-state predimensions admit a strict hierarchy as the number of heads increases, and in fact that for all $h \geq 1$ there is a sequence whose adaptive $(h+1)$-head finite-state predimension is strictly less than its adaptive h-head predimension.

1 Introduction

This paper initiates the study of adaptive multi-head finite-state dimensions and predimensions, showing that the predictive power of multi-head finite-state gamblers depends both on the adaptivity of their head movements and on the number of heads.

Finite-state dimensions were introduced by Dai, Lathrop, Lutz, and Mayordomo [4] and Athreya, Hitchcock, Lutz, and Mayordomo [1]. As with other effective dimensions [11,12], they measure the unpredictability of an infinite sequence S by considering the rates at which a gambler placing bets on successive symbols of S can amass capital. For finite-state dimensions, the gambler's betting function must be implementable by a finite-state machine. There is a large body of literature on these dimensions and their applications, especially to topics related to Borel's theory of normal numbers; see [9] for some of these applications.

Due to the strict computational constraint on the betting function, finite-state gamblers are only able to exploit highly localized patterns in the sequence.

V. Brattka et al. (Eds.): CiE 2026, LNCS 16674, pp. 224–239, 2027.
https://doi.org/10.1007/978-3-032-31348-5_15

For example, given a sequence S where $S[n] = S[2n]$ for all $n \in \mathbb{N}$, an unconstrained gambler can quickly amass capital by remembering the whole sequence and placing "all-in" bets on the even-numbered symbols, but a finite-state gambler has no way to take advantage of this simple pattern more than a constant number of times.

Inspired by the work of Becher, Carton, and Heiber [2] on *finite-state independence*, Huang, Li, Lutz, and Lutz [8] recently introduced *multi-head finite-state gamblers* that are still strictly constrained but are well-suited to this type of non-local pattern. These gamblers are still required to implement their betting functions with a finite-state machine, but they operate under a different data access model: rather than accessing transitory symbols only once, this model permits the gambler to have additional read heads acting as "probes to the past," accessing symbols from earlier in the sequence. Using this model, [8] defined, for each positive integer h, the *h-head finite-state predimension* of a sequence, where the original definition of finite-state dimension [4] is the $h = 1$ case.

The term *predimension* was used in [8] because these quantities lack a standard stability property expected of dimension notions; that work then defined the stable quantity *multi-head finite state dimension*, as well as versions of these notions that are *strong* in the sense of [1]. Lutz [13] subsequently showed that these predimensions and dimensions admit equivalent characterizations in terms of lossless compression by multi-head finite-state transducers.

The main result of [8] was a hierarchy theorem, showing that for all $h \geq 1$, there is a sequence whose $(h+1)$-head finite-state predimension is strictly less than its h-head finite-state predimension. This is reminiscent of the classic hierarchy theorem by Yao and Rivest [15] showing that "$k+1$ heads are better than k" in the context of language recognition by multi-head automata, and the study of multi-head finite-state gamblers is broadly analogous to the long line of work on multi-head automata.

There is an important distinction, however, in the rules governing head movement. In the definitions of [8], the read heads are required to move in an *oblivious* (or *data-independent*) way, meaning they move right according to a predetermined schedule that does not depend on the input string. By contrast, most work on multi-head automata—including the hierarchy theorem of Yao and Rivest [15]—concerns *adaptive* (or *data-dependent*) head movements [7], although oblivious variants have also been studied (e.g., [6]). Hence, in this work, we define and study adaptive multi-head finite-state gamblers, predimensions, and dimensions.

Our first main result is a separation theorem, showing that for all $h \geq 2$, there is a sequence whose adaptive h-head finite-state predimension is strictly less than its oblivious h-head finite-state predimension. We prove this by constructing a family of sequences in which the "correct" relative position for one of the read heads changes over time in a way described by the sequence itself. An adaptive gambler can adjust that head's speed accordingly to keep it in the correct position, but an oblivious gambler can make no such adjustment. We prove an upper bound on the adaptive h-head finite-state predimension of this sequence by constructing a suitable gambler, and we use a Kolmogorov complexity argument to derive a lower bound on the sequence's oblivious h-head finite-state predimension.

Our second main result is a hierarchy theorem, directly analogous to the hierarchy theorem of Yao and Rivest [15]: we show that for all $h \geq 1$, there is a sequence whose adaptive $(h+1)$-head finite-state predimension is strictly less than its adaptive h-head finite-state predimension. We prove this adaptive hierarchy using the same family of sequences used in [8] to prove an oblivious hierarchy, and our proof tracks their Kolmogorov complexity arguments closely. It is surprising, in light of our separation theorem and the additional flexibility granted by adaptivity, that the same techniques can be so straightforwardly applied in this setting.

The rest of the paper is organized as follows. In Sect. 2, we define adaptive and oblivious multi-head finite-state gamblers. In Sect. 3, we present the algorithmic information ideas used in our later definitions and proofs, including Kolmogorov complexity and gales, and we give a general lemma for translating Kolmogorov complexity bounds (which concern compressibility) into gale bounds (which concern unpredictability) in the context of multi-head finite-state gamblers. In Sect. 4, we use gales to define adaptive and oblivious multi-head finite-state predimensions and dimensions. In Sect. 5, we define the family of sequences used in our separation theorem, and then we prove adaptive upper bounds and oblivious lower bounds on this sequence family in Sects. 6 and 7, respectively. In Sect. 8, we prove our strict hierarchy theorem, and we conclude with several directions for future work in Sect. 9.

2 Adaptive Multi-Head Finite-State Gamblers

Throughout this paper, an alphabet is a finite set Σ such that $|\Sigma| \geq 2$. Sequences and strings over an alphabet Σ are indexed starting from 0, and for each sequence S and $m \leq n$, we write $S[m..n]$ for the $(n-m+1)$-symbol string $S[m]S[m+1]\ldots S[n]$. The empty string is denoted λ, and $S[m..n] = \lambda$ when $m > n$. We denote the length of a string x by $|x|$ and write $\sqsubseteq$ for the prefix relation.

We will often consider finite subsequences of infinite sequences, and it will be convenient to retain the position of each symbol in the subsequence, for which we use placeholder symbols: Given a fixed "horizon" $n \in \mathbb{N}$, a sequence S over some alphabet Σ, and a set $A \subseteq [0, n]$, we let $S[A]$ denote the $(n+1)$-symbol string such that $S[A][i]$ is $S[i]$ if $i \in A$ and 0 otherwise. Here 0 indicates the lexicographically first symbol in Σ; in particular, the placeholder symbol will be 0^L when the alphabet is $\{0,1\}^L$. Note that $S[A] = S[A \cap \mathbb{N}]$ for all A and that $S[[m,n]]$ is not the $(n-m+1)$-symbol string $S[m..n]$; instead, it is the $(n+1)$-symbol string consisting of m 0s followed by $S[m..n]$.

2.1 Adaptive Multi-Head Finite-State Gamblers

We now define adaptive multi-head finite-state gamblers as a generalization of the multi-head finite-state gamblers defined in [8], which we reclassify below as *oblivious* multi-head finite-state gamblers. Note that while *adaptive* often means *not oblivious* in informal language, our formal definitions treat oblivious gamblers as a special case of adaptive gamblers.

Definition 2.1. *Let h be a positive integer. An* adaptive h-head finite-state gambler *(*adaptive h-FSG*) is a 6-tuple $G = (Q, \Sigma, \delta, \beta, q_0, c_0)$, where*

- *Q is a finite, non-empty state space;*
- *Σ is an alphabet;*
- *$\delta : Q \times \Sigma^h \to Q \times \{0,1\}^{h-1}$ is the* transition function*;*
- *$\beta : Q \to \Delta_{\mathbb{Q}}(\Sigma)$, where $\Delta_{\mathbb{Q}}(\Sigma)$ denotes the set of all rational-valued discrete probability distributions over Σ, is the* betting function*;*
- *$q_0 \in Q$ is the* initial state*; and*
- *$c_0 \in [0, \infty)$ is the* initial capital.

There are h *heads*: $h-1$ *trailing heads*, numbered 1 through $h-1$, and a *leading head*, numbered h. All heads start at position 0, and the gambler proceeds through an input sequence $S \in \Sigma^\omega$ in discrete time steps. In each step, when the leading head is at some position $S[n-1]$ and the gambler is in state q, the following events happen simultaneously.

- The gambler places bets on each symbol in Σ according to its current state q; allocating to each symbol $b \in \Sigma$ the fraction $\beta(q)(b)$ of its current capital.
- The current state is updated to the first component of $\delta(q, (b_1, \ldots, b_h))$, where each b_i is the symbol in S at the position of head i.
- The leading head moves to the right, and some subset of the trailing heads also move. Specifically, each head i will move right if and only if the ith bit of the second component of $\delta(q, (b_1, \ldots, b_h))$ is 1.

When the sequence S is fixed, for each n we write q_n for the state of the gambler and $\pi_i(n)$ for the position of head i after n of these time steps.

2.2 Oblivious Multi-Head Finite-State Gamblers

Here we characterize oblivious multi-head finite-state gamblers, as defined in [8], formally as a special case of adaptive finite-state gamblers. In particular, the parts of the gambler that govern its movements need to be decoupled from the parts that read symbols in the input sequence.

Definition 2.2. *Let h be a positive integer. An adaptive h-head finite-state gambler $G = (Q, \Sigma, \delta, \beta, q_0, c_0)$ is an* oblivious h-head finite-state gambler *if there are sets P and T and functions $\delta_P : P \times \Sigma^h \to P$, $\delta_T : T \to T$, and $\mu : T \to \{0,1\}^{h-1}$ such that $Q = P \times T$ and, for all $(p,t) \in Q$ and $(b_1, \ldots, b_h) \in \Sigma^h$,*

$$\delta((p,t), (b_1, \ldots, b_h)) = \big((\delta_P(p, (b_1, \ldots, b_h)), \delta_T(t)), \mu(t)\big).$$

Thus, the component of the state from T governs the head movements, while only the component from P depends on the symbols that are read, and the head movements are therefore independent of the input sequence.

As observed in [8], the obliviousness condition implies that each trailing head moves at what is essentially a constant speed, in the following sense.

Observation 2.3 ([8]). *In an oblivious h-FSG, for each trailing head i, there is a speed parameter η_i such that, for all n, $\eta_i n - |T| \leq \pi_i(n) \leq \eta_i n + |T|$. Since heads with speed parameters 0 or 1 can be simulated by the leading head using finitely many states, one can assume without loss of generality that $\eta_i \in (0, 1)$.*

3 Algorithmic Information Preliminaries

Fix an alphabet Σ. We are primarily concerned with how predictable sequences are to multi-head finite-state gamblers, and we will quantify this predictability using s-gales. Predictability is closely tied to compressibility, so we will also reason about the compressibility of strings in these sequences, formalized using Kolmogorov complexity.

3.1 s-Gales and Martingales

Definition 3.1. *For $s \in [0, \infty)$, an* s-gale *is a function $d : \Sigma^* \to [0, \infty)$ such that, for all $w \in \Sigma^*$,*

$$d(w) = |\Sigma|^{-s} \sum_{b \in \Sigma} d(wb).$$

A martingale *is a 1-gale.*

Intuitively, we regard an s-gale as representing the capital of a gambler who places successive bets on symbols in a sequence. After gambling on some string w, the gambler has capital $d(w)$, and it now places another bet, allocates portions of its capital according to its beliefs about the next symbol. If the gambler allocates some portion $\alpha d(w)$ to the symbol b and b is the actual next symbol, the gambler's new capital is $d(wb) = |\Sigma|^s \alpha d(w)$. A martingale is considered the "fair" version of this setup; in a martingale, if the gambler makes a uniform bet, allocating $1/|\Sigma| d(w)$ to each symbol b, then it is guaranteed to maintain constant capital. The lower s is, the less favorable the environment is for the gambler. The gambler's goal is to have unbounded capital, either infinitely often or cofinitely often; these outcomes characterize *success* and *strong success*, respectively.

Definition 3.2. *An s-gale d* succeeds *on a sequence $S \in \Sigma^\omega$, and we write $S \in S^\infty(d)$, if $\limsup_{n\to\infty} d(S[0..n-1]) = \infty$. We say d* succeeds strongly *on S, and write $S \in S^\infty_{\mathrm{str}}(d)$, if $\liminf_{n\to\infty} d(S[0..n-1]) = \infty$.*

As in [8], given a positive integer h and an oblivious or adaptive h-head finite-state gambler G, the *martingale of* G is given by $d_G(\lambda) = c_0$ and, for each string $w \in \Sigma^*$ and symbol $b \in \Sigma$, $d_G(wb) = |\Sigma| d_G(w) \beta(q_{|w|})(b)$, where $q_{|w|}$ is the gambler's state after reading w. Given any non-negative real s, the *s-gale of G* is defined, for each string $w \in \Sigma^*$, by $d_G^{(s)}(w) = |\Sigma|^{(s-1)|w|} d_G(w)$. Note that the gambler does not have access to its martingale (or other s-gale) value, which might involve an unbounded amount of information.

3.2 Kolmogorov Complexity and Martin-Löf Randomness

Fix a universal prefix Turing machine $\mathcal{U}$. For all strings $\sigma, \tau \in \Sigma^*$, the *conditional (prefix) Kolmogorov complexity* of σ given τ is

$$K(\sigma \mid \tau) = \min\{|\pi| \mid \pi \in \{0,1\}^* \text{ and } \mathcal{U}(\pi, \tau) = \sigma\},$$

and the *(prefix) Kolmogorov complexity* of σ is $K(\sigma) = K(\sigma \mid \lambda)$.

We will use only a few basic properties of Kolmogorov complexity; more details and background can be found in [10]. First, the complexity of a string is never much more than its length multiplied by the log of the alphabet size: for all $x \in \Sigma^*$, $K(x) \leq |x| \log |\Sigma| + O(\log |x|)$. Second, we have *symmetry of information*: for all $x, y \in \Sigma^*$,

$$K(x,y) - O(1) \leq K(x \mid y) + K(y) \leq K(x,y) + O(\log |y|).$$

Third, computable functions cannot significantly increase complexity: if $f : \Sigma^* \to \Sigma^*$ is a computable function, then for all $\sigma, \tau \in \Sigma^*$, applying f cannot significantly add to the information content of x, in the sense that $K(f(\sigma)) \leq K(\sigma) + O(1)$, and by similar reasoning, $K(\tau \mid f(\sigma)) \geq K(\tau \mid \sigma) - O(1)$.

A sequence $S \in \Sigma^\omega$ is *Martin-Löf random* [14] if $K(S[0..n-1]) \geq n \log |\Sigma| - O(1)$ holds for all n. Such a sequence is essentially *incompressible*, and it is well-known that almost all sequences have this property [5].

3.3 Using Kolmogorov Complexity to Bound Martingales

Intuitively, if a string is difficult to compress, then it is also difficult to predict. Furthermore, if it is difficult to compress *even given the information accessible to the trailing heads*, then it is also difficult to predict with access to that information. This idea is formalized in the following lemma, which will be our main tool for proving upper bounds on the martingales of multi-head finite-state gamblers.

Lemma 3.3. *Let Σ be an alphabet, $h \geq 1$, G be any adaptive or oblivious h-head finite-state gambler over Σ, $\alpha, \delta \in (0,1) \cap \mathbb{Q}$, $S \in \Sigma^\omega$, $m, n \in \mathbb{N}$, and let U be any set that includes all positions strictly less than m of the trailing heads while the leading head reads $S[m..n]$, i.e.,*

$$U \supseteq [0, m-1] \cap \bigcup_{i=1}^{h-1} [\pi_i(m), \pi_i(n)].$$

If $n - m$ is sufficiently large and $K(S[[m,n]] \mid S[U]) \geq (1 - \alpha + \delta)(n-m) \log |\Sigma|$, then, letting $x = S[m..n]$ and $y = S[0..m-1]$,

$$\max_{w \sqsubseteq x} \frac{d_G(yw)}{d_G(y)} \leq |\Sigma|^{\alpha \cdot (n-m)}. \tag{1}$$

This is a direct generalization of Lemma 5.7 in [8], which was stated only for oblivious gamblers and for specific choices of Σ, S, m, U, α, and δ. We closely follow that proof, which uses a standard simulation technique to relate incompressibility and unpredictability; the more general form stated here requires no major changes to the original argument. See the arXiv version [3] for a full proof.

4 Adaptive Multi-Head Finite-State Dimensions

To define adaptive multi-head finite predimensions and dimensions, we follow the standard template for defining effective dimension notions with s-gales, just as Huang, Li, Lutz, and Lutz [8] did in defining oblivious multi-head finite predimensions and dimensions. We deviate from the terminology and notation of [8] by including *oblivious* explicitly and prepending an o to the dimension notations defined in [8] to emphasize that these are the oblivious varieties.

Definition 4.1 ([8]). *Let S be a sequence. For each positive integer h, the* oblivious h-head finite-state predimension *of S is*

$$\mathrm{odim}_{\mathrm{FS}}^{(h)}(S) = \inf\left\{ s \in [0,\infty) \;\middle|\; \exists \text{ oblivious } h\text{-FSG } G \text{ with } \left(S \in S^{\infty}\left(d_G^{(s)}\right)\right)\right\},$$

and the oblivious h-head finite-state strong predimension *of S is*

$$\mathrm{oDim}_{\mathrm{FS}}^{(h)}(S) = \inf\left\{ s \in [0,\infty) \;\middle|\; \exists \text{ oblivious } h\text{-FSG } G \text{ with } \left(S \in S^{\infty}_{\mathrm{str}}\left(d_G^{(s)}\right)\right)\right\}.$$

The oblivious multi-head finite-state dimension *and* oblivious multi-head finite-state strong dimension *of S are*

$$\mathrm{odim}_{\mathrm{FS}}^{\mathrm{MH}}(S) = \inf_{h\geq 1} \mathrm{odim}_{\mathrm{FS}}^{(h)}(S) \text{ and } \mathrm{oDim}_{\mathrm{FS}}^{\mathrm{MH}}(S) = \inf_{h\geq 1} \mathrm{oDim}_{\mathrm{FS}}^{(h)}(S).$$

The adaptive versions $\mathrm{adim}_{\mathrm{FS}}^{(h)}(S)$, $\mathrm{aDim}_{\mathrm{FS}}^{(h)}(S)$, $\mathrm{adim}_{\mathrm{FS}}^{\mathrm{MH}}(S)$, and $\mathrm{aDim}_{\mathrm{FS}}^{\mathrm{MH}}(S)$ are defined identically, except with adaptive h-FSGs replacing the oblivious h-FSGs. Since oblivious h-FSGs are a special case of adaptive h-FSGs, it is immediate that the adaptive versions of these quantities are bounded above by the oblivious versions.

5 Defining Our Family of Sequences

In this section we define, for each integer $h \geq 2$, sequences with patterns that can be exploited more thoroughly by adaptive h-FSGs than by oblivious h-FSGs; we will prove this separation in Sects. 6 and 7.

Fix any constant $L \geq 1$. For each integer $h \geq 2$, we now define the function

$$\Phi_h : (\{0,1\}^L)^\omega \to (\{0,1\}^L \cup \{\$\})^\omega.$$

Here \$ is a special symbol outside the original alphabet.

Given any sequence $S \in (\{0,1\}^L)^\omega$, we construct the sequence $\Phi_h(S)$ by modifying S in the following way for each $k \in \mathbb{N}$. Let $s_k = (h+1)^{2k}h$ and $t_k = (h+1)^{2k+1}$, and let $\oplus$ denote the bitwise parity (XOR) operation. First, set $\Phi_h(S)[s_k] = \Phi_h(S)[t_k] = \$$. Second, for each integer $n \in (s_k, t_k)$ such that $n \bmod (h+1) = 0$, set

$$\Phi_h(S)[n] = \left(\bigoplus_{i=1}^{h-2} \Phi_h(S)\left[\frac{i}{h+1}n\right]\right) \oplus \Phi_h(S)\left[\frac{h-(k \bmod 2)}{h+1}n\right]. \tag{2}$$

For all other indices n, set $\Phi_h(S)[n] = S[n]$.

The sequence $\Phi_h(S)$ thus oscillates between three modes, with the symbol \$ indicating a switch between modes. From each t_k to s_{k+1}, the sequence is in one mode: it is identical to S, with no additional structure. From each s_k to t_k, a $\frac{1}{h+1}$ fraction of the symbols are replaced by the bitwise parity of $h-1$ symbols from earlier in the sequence. This comprises two modes: the relative positions of the first $h-2$ of these earlier symbols are always the same, but the relative position of the rightmost of the symbols differs according to whether k is even or odd. Intuitively, the fact that this last piece of pertinent information moves between two different relative positions is precisely what makes this sequence more suitable for an adaptive gambler than an oblivious one.

6 Upper Bound for Adaptive Multi-Head Finite-State Predimensions

The self-referential structure imposed by the function Φ_h defined in Sect. 5 makes every sequence $\Phi_h(S)$ somewhat predictable to an adaptive h-head finite-state gambler. In this section, we construct such gamblers to prove the following upper bounds on the adaptive h-head finite-state predimensions of such sequences.

Theorem 6.1. *For every $h \geq 2$ and every sequence $S \in (\{0,1\}^L)^\omega$,*

$$\mathrm{adim}_{\mathrm{FS}}^{(h)}(\Phi_h(S)) \leq \frac{L}{\log(2^L+1)}\left(1 - \frac{1}{(h+2)h}\right)$$

and

$$\mathrm{aDim}_{\mathrm{FS}}^{(h)}(\Phi_h(S)) \leq \frac{L}{\log(2^L+1)}\left(1 - \frac{1}{(h+2)(h+1)h^2}\right).$$

Let $h \geq 2$, $S \in (\{0,1\}^L)^\omega$, $X = \Phi_h(S)$, and $\varepsilon \in (0,1) \cap \mathbb{Q}$. To prove Theorem 6.1, we construct an adaptive h-head finite-state gambler G over the alphabet $\Gamma = \{0,1\}^L \cup \{\$\}$ for the sequence X.

The heads of G will be numbered $1, \ldots, h$, where h is the leading head. Each trailing head $i \leq h-2$ moves at a fixed speed of $\frac{i}{h+1}$, and the speed of the last trailing head $h-1$ will range between $\frac{h-1}{h+1}$ and $\frac{h}{h+1}$. The gambler G has five modes whose behaviors can be implemented with finitely many states. These modes will dictate the betting pattern and the speed of head $h-1$. Two modes are for intervals between positions s_k and t_k, where a $\frac{1}{h+1}$ fraction of the sequence's symbols are bitwise parities of earlier symbols. In these modes, the trailing heads access the relevant earlier symbols, and G bets accordingly. During two other modes, G is repositioning head $h-1$, either by slowing down or speeding up, to be prepared for the next s_k-to-t_k interval. The fifth mode (which we number -1 to make the transition rule simpler) is a one-time start mode.

To define the gambler in more detail, we first define probability distributions that G will use for betting. Define $\nu : \Gamma \to [0,1]$ by $\nu(\$) = \varepsilon$ and $\nu(b) = (1-\varepsilon)/2^L$ for all $b \in \{0,1\}^L$. For each symbol $a \in \Gamma$, define the distribution $\chi_a : \Gamma \to [0,1]$ by $\chi_a(b) = 1 - \varepsilon$ if $b = a$ and $\chi_a(b) = \varepsilon/2^L$ otherwise.

The gambler always "hedges" for the possibility that the symbol $ will appear rather than attempting to predict its appearance. Because the instances of this symbol are exponentially sparse in X, this hedge can safely be arbitrarily small.

- In Mode -1, head $h-1$ moves right in the first $h-1$ out of every $h+1$ steps, and G bets according to ν.
- In Mode 0, head $h-1$ moves right in the first $h-1$ out of every $h+1$ steps. If the current position is not a multiple of $h+1$, then G bets according to ν. If the current position is a multiple of $h+1$, let $b_1, \ldots, b_h$ be the symbols observed by the h heads. If $b_{h-1} = \$$, then G bets according to $\chi_\$$; otherwise, it bets according to $\chi_{\bigoplus_{i=1}^{h-1} b_i}$.
- In Mode 1, head $h-1$ *speeds up*, moving right in the first h^3+h^2-h+1 out of every h^3+2h^2-1 steps, and G bets according to ν.
- In Mode 2, head $h-1$ moves right in the first $h-1$ out of every $h+1$ steps, and the betting behavior of G is identical to Mode 0.
- In Mode 3, head $h-1$ *slows down*, moving right in the first h^3-2h out of every h^3+2h^2-1 steps, and G bets according to ν.

The gambler G starts in mode -1. Whenever the leading head reads a $ in mode m, the gambler moves into mode $(m+1) \bmod 4$; mode -1 is never revisited.

See the arXiv version [3] for a proof of the following lemma.

Lemma 6.2. *For $n \in \mathbb{N}$, let*

$$\rho(n) = \frac{1}{n}\left|\left\{ i < n \;\middle|\; \frac{d_G(X[0..i])}{d_G(X[0..i-1])} = (1-\varepsilon)|\Gamma| \right\}\right|.$$

Informally, this is the fraction of symbols in $X[0..n-1]$ on which G makes a full winning bet. Let $\rho_1 = \limsup_{n\to\infty} \rho(n)$ and $\rho_2 = \liminf_{n\to\infty} \rho(n)$. Then

$$\rho_1 = \frac{1}{h(h+2)} \text{ and } \rho_2 = \frac{1}{h^2(h+1)(h+2)}.$$

To prove Theorem 6.1, it suffices to show, for all $\varepsilon > 0$, that the $(L/\log(2^L+1)(1-\rho_1)+\varepsilon)$-gale of G succeeds on X and the $(L/\log(2^L+1)(1-\rho_2)+\varepsilon)$-gale of G succeeds strongly on X. In the arXiv version [3], we use Lemma 6.2 to show this.

7 Lower Bound for Oblivious Multi-Head Finite-State Predimensions

In this section we complete our separation between adaptive and oblivious multi-head finite-state gamblers by showing that for all $h \geq 2$, no oblivious h-FSG can match the performance of the adaptive h-FSG defined in Sect. 6 on the same family of sequences. Formally, we use Kolmogorov complexity arguments to establish the following lower bounds on the oblivious h-head finite-state predimension and strong predimension of these sequences.

Theorem 7.1. *For every $h \geq 2$ and every Martin-Löf random sequence $R \in (\{0,1\}^L)^\omega$,*

$$\mathrm{odim}_{\mathrm{FS}}^{(h)}(\Phi_h(R)) \geq \frac{L}{\log(2^L+1)}\left(1 - \frac{(h+1)^2}{(h+1)^4-1}\right)$$

and

$$\mathrm{oDim}_{\mathrm{FS}}^{(h)}(\Phi_h(R)) \geq \frac{L}{\log(2^L+1)}\left(1 - \frac{1}{(h+1)h\cdot((h+1)^4-1)}\right).$$

To prove this theorem, let $h \geq 2$, let $R \in (\{0,1\}^L)^\omega$ be Martin-Löf random, and let $X = \Phi_h(R)$. Let $\Gamma = \{0,1\}^L \cup \{\$\}$, consider any oblivious h-FSG G over the alphabet Γ, and let $H = \{\eta_1, \dots, \eta_h\}$ be the set of head speeds, where $0 < \eta_1 < \dots < \eta_h = 1$. For each $i \in \{1, \dots, h\}$ and $m, n \in \mathbb{N}$, let $a_i(m) = \lfloor \eta_i m \rfloor - |T|$ and $b_i(n) = \lceil \eta_i n \rceil + |T|$, and define the set

$$U(m,n) = \bigcup_{i=1}^{h-1} [a_i(m), b_i(n)].$$

By Observation 2.3, this $U(m,n)$ is a superset of the positions of the trailing heads while the leading head reads $X[m..n]$, suitable for use in Lemma 3.3. In order to apply that lemma, we now establish conditional Kolmogorov complexity bounds.

Lemma 7.2. *If $j \in \{1, \dots, h-1\}$ satisfies $\frac{j}{h+1} \notin H$, then there is some $\gamma \in (0,1)$ satisfying the following property for all sufficiently large k: For all integers $n \in (s_k, s_{k+1}]$ and $m \in [\gamma n, n]$, we have $\max U(m,n) < m$ and, if any of $j \leq h-2$, $j = h - (k \bmod 2)$, or $m \geq t_k$ hold, then*

$$K(X[[m,n]] \mid X[U(m,n)]) \geq (n-m)L - O(\log n).$$

To see why this lemma is true, let $\eta = \frac{j}{h+1}$ and assume $\eta \notin H$. Intuitively, during the interval where the gambler bets on symbols in some string $X[m..n]$, it might be helpful to access some of the symbols in $X[\eta m..\eta n]$. This will be the case if the index range $[m,n]$ includes part of the sequence that uses parity calculations involving symbols with indices in the range $[\eta m, \eta n]$—that is, some index range $[s_k, t_k]$ where either $j \leq h-2$ or $j = h - (k \bmod 2)$.

Each trailing head i will only read symbols within the string $X[a_i(m)..b_i(n)]$, though. We will choose m and n such that, for each trailing head i, there is no overlap between $[a_i(m), b_i(n)]$ and $[\eta m, \eta n]$ or $[m, n]$, so that head i does not access the pertinent information during this interval. We will further argue that beyond this non-overlap condition, each $X[a_i(m)..b_i(n)]$ also does not tell the gambler anything about $X[\eta m..\eta n]$ in any subtler ways. Informally, this is because the structure of X dictates that $X[\eta m..\eta n]$ consists entirely of random symbols directly from R, each of which is copied at most once in the sequence, and all these copies appear in $X[m..n]$, with which $X[a_i(m)..b_i(n)]$ does not overlap. A complete proof appears in the arXiv version [3].

Even if the trailing heads are optimally positioned for some subsequence, at least an $\frac{h}{h+1}$ fraction of the symbols in the sequence are "fresh" random symbols from R, which the trailing heads' observations say nothing about. The following lemma formalizes this intuition; the complete proof appears in the arXiv version [3].

Lemma 7.3. *Let γ be as in Lemma 7.2. If k is sufficiently large, then for all integers $n \in (s_k, s_{k+1}]$ and $m \in [\gamma n, n]$,*

$$K(X[[m,n]] \mid X[U(m,n)]) \geq \frac{h}{h+1}(n-m)L - O(\log n).$$

With these conditional Kolmogorov complexity bounds in hand, we are now ready to complete the proof of Theorem 7.1. There are two cases to consider.

Case 1: Suppose $\frac{h}{h+1} \notin H$. Define the function $\beta : \mathbb{N} \to [0,1]$ by

$$\beta(n) = \frac{1}{n} \sum_{\text{odd } k} |[s_k..t_k] \cap [0..n-1]| .$$

Informally, this is the density in $X[0..n-1]$ of intervals on which G *might* gamble more successfully, in case $\frac{h-1}{h+1} \in H$. We prove in the arXiv version [3] that

$$\delta_1 := \limsup_{n\to\infty} \frac{\beta(n)}{h+1} = \frac{(h+1)^2}{(h+1)^4 - 1}, \tag{3}$$

$$\delta_2 := \liminf_{n\to\infty} \frac{\beta(n)}{h+1} = \frac{1}{(h+1)h \cdot ((h+1)^4 - 1)}. \tag{4}$$

Let $\varepsilon < 1 - \max\{L/\log|\Gamma|, \gamma\}$ be rational. In the arXiv version [3], we use Lemmas 3.3, 7.2, and 7.3 to prove that, for all sufficiently large n,

$$\log d_G(X[0..n-1]) \leq \left(\log|\Gamma| - \left(1 - \frac{\beta(n)}{h+1}\right)L + (L+4)\varepsilon\right)n. \tag{5}$$

As we prove in the arXiv version [3], this implies

$$\limsup_{n\to\infty} d_G^{((1-\delta_1)L/\log|\Gamma| - (L+5)\varepsilon)}(X[0..n-1]) = 0, \tag{6}$$

$$\liminf_{n\to\infty} d_G^{((1-\delta_2)L/\log|\Gamma| - (L+5)\varepsilon)}(X[0..n-1]) = 0. \tag{7}$$

Thus, the $((1-\delta_1)L/\log|\Gamma|-(L+5)\varepsilon)$-gale of G does not succeed on X, and the $((1-\delta_2)L/\log|\Gamma|-(L+5)\varepsilon)$-gale of G does not succeed strongly on X.

Case 2: If $\frac{h}{h+1} \in H$, then since $|H| = h$ and $h \in H$, there must be some $j \leq h-1$ such that $\frac{j}{h+1} \notin H$. The rest of the argument is symmetric to Case 1, and we reach the same conclusion about the $((1-\delta_1)L/\log|\Gamma|-(L+5)\varepsilon)$-gale and $((1-\delta_2)L/\log|\Gamma|-(L+5)\varepsilon)$-gale of G.

Every oblivious h-FSG G falls into one of the two cases above. Letting ε approach 0, this implies

$$\mathrm{odim}_{\mathrm{FS}}^{(h)}(X) \geq \frac{(1-\delta_1)L}{\log|\Gamma|} \text{ and } \mathrm{oDim}_{\mathrm{FS}}^{(h)}(X) \geq \frac{(1-\delta_2)L}{\log|\Gamma|},$$

and applying (3) and (4) completes the proof of Theorem 7.1.

Combined with Theorem 6.1, this establishes a strict separation between adaptive and oblivious h-head predimensions.

8 A Hierarchy of Adaptive Predimensions

In this section we prove our hierarchy theorem, describing, for each $h \geq 1$, sequences whose adaptive $(h+1)$-head predimensions are strictly less than their adaptive h-head predimensions. In fact, even the *oblivious* $(h+1)$-head predimensions of these sequences are strictly less than their adaptive h-head predimensions.

Theorem 8.1. *For each $h \geq 1$ there is a sequence $Y \in \{0,1\}^\omega$ such that*

$$\mathrm{oDim}_{\mathrm{FS}}^{(h+1)}(Y) < \mathrm{adim}_{\mathrm{FS}}^{(h)}(Y).$$

Since adaptive gamblers generalize oblivious gamblers, this theorem has the following immediate corollary.

Corollary 8.2. *For each $h \geq 1$ there is a sequence $Y \in \{0,1\}^\omega$ such that*

$$\mathrm{adim}_{\mathrm{FS}}^{(h+1)}(Y) < \mathrm{adim}_{\mathrm{FS}}^{(h)}(Y) \text{ and } \mathrm{aDim}_{\mathrm{FS}}^{(h+1)}(Y) < \mathrm{aDim}_{\mathrm{FS}}^{(h)}(Y).$$

This hierarchy is analogous to the oblivious predimension hierarchy theorem of Huang, Li, Lutz, and Lutz [8]. In fact, we prove Theorem 8.1 using the same family of sequences used in [8], which is surprising in light of the separation between adaptive and oblivious predimensions established by Theorems 6.1 and 7.1. We now describe that family of sequences, which is constructed by starting from Martin-Löf random sequences and then imposing structures of backward-looking self-reference onto these sequences. Prime numbers play an important role in keeping these structures tree-like, so that paths of self-reference are unique. For each positive integer k, let p_k denote the kth prime number, and define $\nu_k : \mathbb{Z} \to \mathbb{N}$, where $\nu_k(n)$ is the multiplicity of p_k in n. Let $p_0 = 1$.

Definition 8.3 ([8]). *For each positive integer h, define the function $F_{h+1} : \{0,1\}^\omega \to \{0,1\}^\omega$ by, for each sequence $S \in \{0,1\}^\omega$, $F_{h+1}(S)[0] = 0$, and, for all $q, r \geq 0$ such that $r < p_{h+1}$,*

$$F_{h+1}(S)[qp_{h+1} + r] = \begin{cases} \bigoplus_{k=1}^{h} F_{h+1}(S)[qp_k] & \text{if } r = 0 \\ S[q \cdot (p_{h+1} - 1) + r] & \text{if } r > 0, \end{cases}.$$

Now let h be any positive integer, R be any Martin-Löf random binary sequence, and $Y = F_{h+1}(R)$. Informally, every bit $Y[n]$ in the sequence is either the parity of exactly h bits from earlier in the sequence (when n is a multiple of p_{h+1}) or a "fresh" random bit from R that no effective gambler can hope to predict. In the former case, $Y[n]$ is always *sensitive* to each of those h earlier bits, so the gambler would need information about all h of them to successfully predict $Y[n]$. Intuitively, $h+1$ heads (including h trailing heads) are sufficient to access that information and make confident predictions about these parity bits, but with only h heads (including $h-1$ trailing heads), the gambler is always missing something important.

An upper bound with $h+1$ heads comes directly from Lemma 5.4 of [8], which constructed an oblivious $(h+1)$-FSG to establish that $\mathrm{oDim}_{\mathrm{FS}}^{(h+1)}(Y) \leq 1 - 1/p_{h+1}$. Hence, to prove Theorem 8.1, it suffices to show that $\mathrm{adim}_{\mathrm{FS}}^{(h)}(Y) = 1$. Huang, Li, Lutz, and Lutz [8] proved the oblivious analog of by arguing that for every oblivious h-head finite-state gambler, there is some fixed $j \in \{1, \ldots, h\}$ and some threshold n_0 such that, for nearly all $n > n_0$, the gambler has no information about the jth of these h bits, and that without this j^{th} piece of pertinent information, it has very little ability to predict bits in the sequence. The recursive structure of the sequence makes the details require care (e.g., the parameterization of "nearly all"), but [8] formalized this idea with a conditional Kolmogorov complexity lower bound on substrings of Y, from which they inductively derived $\mathrm{odim}_{\mathrm{FS}}^{(h)}(Y) = 1$.

Our proof that $\mathrm{adim}_{\mathrm{FS}}^{(h)}(Y) = 1$ is based on reversing the order of the quantifiers on j and n_0 in the argument of [8]. With adaptive head movements, we cannot guarantee that there is a *single* j for which the jth pertinent piece of information is not available to the gambler, but intuitively, it is enough to say that there is nearly always *some* j for which the gambler lacks that information.

Dropping the assumption of oblivious head movements slightly complicates the argument of [8], which is cast in terms of the fixed speeds of oblivious heads. Those fixed speeds guarantee that the gap between the trailing heads and the leading head grows linearly, which enables reasoning about linear-length strings that have been visited by the leading head but not yet by any trailing heads. In the adaptive setting, this gap can be sublinear and even sublogarithmic, which makes the directly analogous argument too delicate for the logarithmic error terms incurred in our Kolmogorov complexity calculations. This pushes us to instead reason about strings that may also have been visited by trailing heads. Nevertheless, portions of the argument from [8] hold unchanged in this setting.

For all $m, n \in \mathbb{N}$ such that $m \leq n$, define

$$U(m,n) = [0, m-1] \cap \bigcup_{i=1}^{h-1} [\pi_i(m-1), \pi_i(m-1) + n - m + 1],$$

which includes all possible trailing head positions less than m while the leading head reads $n - m + 1$ bits after $Y[0..m-1]$.

Fix a positive integer d to bound the recursion depth we consider. For each $1 \leq i \leq h$ and all $m, n \in \mathbb{N}$ such that $m \leq n$, define

$$V_i(m,n) = \left\{ (p_i/p_{h+1})^{\nu_{h+1}(k)} k \mid k \in [m..n] \text{ and } \nu_{h+1}(k) \leq d \right\}.$$

Informally, each index in $V_i(m,n)$ is a leaf in the tree of parity operations that calculate some bit in $Y[m..n]$; in particular, it is the leaf reached by taking the ith child at each level. The indices in $V_i(m,n)$ are thus weakly "upstream" of $Y[m..n]$. Note that $V_i(m,n)$ and $[m..n]$ will generally not be disjoint, as most indices ℓ have $\nu_{h+1}(\ell) = 0$. We also define the *closure* of V_i as

$$\overline{V}_i(m,n) = \left\{ \ell \cdot \frac{p_{h+1}^{a_1+\ldots+a_h}}{p_1^{a_1} \cdots p_h^{a_h}} \mid \ell \in V_i \text{ and } a_1, \ldots, a_h \in \mathbb{N} \right\}.$$

An index k belongs to this closure if it is weakly "downstream" of some index $\ell \in V_i(m,n)$, in the sense that $Y[\ell]$ appears *somewhere* in the tree of parity operations used to compute $Y[k]$. This includes the root, so $V_i(m,n) \subseteq \overline{V}_i(m,n)$.

The following Lemma, which we prove in the arXiv version [3], formalizes the intuition that an h-FSG is missing some key piece of information.

Lemma 8.4. *There is a constant $\gamma \in (1/2, 1) \cap \mathbb{Q}$ such that whenever n is sufficiently large and $m \in [\gamma n, n]$, there is some $j(m,n) \in \{1, \ldots, h\}$ satisfying*

$$U(m,n) \cap \overline{V}_{j(m,n)}(m,n) = \emptyset.$$

Due to this missing information, the string $Y[[m,n]]$ is nearly incompressible given information from the trailing heads, in the sense of the following lemma, which we prove in the arXiv version [3].

Lemma 8.5. *If n is sufficiently large and $m \in [\gamma n, \sqrt{\gamma} n]$, then*

$$K(Y[[m,n]] \mid Y[U(m,n)]) \geq (n-m)(1 - p_{h+1}^{-d-2}).$$

Using this Kolmogorov complexity bound, we now bound the value of the gambler's s-gale for s arbitrarily close to 1; the proof appears in the arXiv version [3].

Lemma 8.6. *For all $\varepsilon \in \mathbb{Q} \cap (0,1)$, the set $\left\{ d_G^{(1-\varepsilon/\gamma)}(Y[0..n-1]) \mid n \in \mathbb{N} \right\}$ is bounded by a constant.*

Since Lemma 8.6 holds for every adaptive h-head finite-state gambler and for arbitrarily small ε, we conclude that $\mathrm{adim}_{\mathrm{FS}}^{(h)}(Y) = 1$, completing the proof of Theorem 8.1.

9 Conclusion

By establishing a separation between the predictive power of oblivious and adaptive multi-head finite-state gamblers, we have shown that adaptive multi-head finite-state dimensions and predimensions are independently interesting objects of study. The statement and proof of our hierarchy theorem demonstrate that this separation is subtle and that some of the pre-existing techniques for analyzing oblivious multi-head finite-state predimensions can be modified to apply in the adaptive setting. This opens several natural directions for further work, some of which we now highlight. First, what is the maximum separation? The separation given by Theorems 6.1 and 7.1 is quantitatively quite small. Second, are adaptive multi-head finite-state predimensions stable under unions? The technique used by Huang, Li, Lutz, and Lutz [8] to prove instability in the oblivious setting relies heavily on obliviousness. Third, does the compression characterization of Lutz [13] hold in the adaptive setting? We conjecture that it does. Finally, and more broadly, what would be the effects of allowing two-way head movement or non-determinism? These modifications have been studied in the context of multi-head automata for language recognition [7], but not for multi-head gamblers.

References

1. Athreya, K.B., Hitchcock, J.M., Lutz, J.H., Mayordomo, E.: Effective strong dimension in algorithmic information and computational complexity. SIAM J. Comput. **37**(3), 671–705 (2007)
2. Becher, V., Carton, O., Heiber, P.A.: Finite-state independence. Theory Comput. Syst. **62**, 1555–1572 (2018)
3. Cruz, J., Glashausser, S., Li, X., Lutz, N.: Adaptive multi-head finite-state gamblers (2026). https://arxiv.org/abs/2603.16034
4. Dai, J.J., Lathrop, J.I., Lutz, J.H., Mayordomo, E.: Finite-state dimension. Theoret. Comput. Sci. **310**(1), 1–33 (2004)
5. Downey, R.G., Hirschfeldt, D.R.: Algorithmic Randomness and Complexity. Springer (2010)
6. Ďuriš, P., Královič, R., Pardubská, D.: Tight hierarchy of data-independent multi-head automata. J. Comput. Syst. Sci. **114**, 126–136 (2020)
7. Holzer, M., Kutrib, M., Malcher, A.: Multi-head finite automata: characterizations, concepts and open problems. In: Electronic Proceedings in Theoretical Computer Science (EPTCS), vol. 1, pp. 93–107 (2009)
8. Huang, X., Li, X., Lutz, J.H., Lutz, N.: Multi-head finite-state dimension (2026). https://arxiv.org/abs/2509.22912
9. Kozachinskiy, A., Shen, A.: Automatic Kolmogorov complexity, normality, and finite-state dimension revisited. J. Comput. Syst. Sci. **118**, 75–107 (2021)
10. Li, M., Vitányi, P.M.B.: An Introduction to Kolmogorov Complexity and Its Applications, 4th edn. Springer, Berlin (2019)
11. Lutz, J.H.: Dimension in complexity classes. SIAM J. Comput. **32**(5), 1236–1259 (2003)
12. Lutz, J.H.: The dimensions of individual strings and sequences. Inf. Comput. **187**(1), 49–79 (2003)

13. Lutz, N.: Multi-head finite-state compression. In: LATIN 2026: Theoretical Informatics (to appear)
14. Martin-Löf, P.: The definition of random sequences. Inf. Control **9**(6), 602–619 (1966)
15. Yao, A.C., Rivest, R.L.: $k + 1$ heads are better than k. J. ACM **25**(2), 337–340 (1978)

A Note on Computable Étale Spaces

Matthew de Brecht(✉)

Graduate School of Human and Environmental Studies, Kyoto University, Kyoto, Japan
matthew@i.h.kyoto-u.ac.jp

Abstract. An étale space over a topological space Y is defined as a local homeomorphism from a topological space X into Y. They often come up in topos theory because of the equivalence between sheaves and étale spaces over a space. In this note, we define computable étale spaces over a computable topological space Y within the TTE framework of computable topology, and show they are naturally equivalent to computable functions from Y to ODS, the effective quasi-Polish category of overt-discrete quasi-Polish spaces. More generally, if $\mathcal{C}$ is a computable category (or groupoid), then there is an equivalence between computable functors from $\mathcal{C}$ to ODS, and computable étale spaces equipped with a computable action by $\mathcal{C}$.

1 Introduction

We recommend [26] for an intuitive explanation of the connections between Grothendiek toposes and geometric logic. Briefly, frames (complete lattices satisfying a certain infinitary distributive law) correspond to the Lindenbaum-algebras of propositional geometric theories, and Grothendiek toposes in a sense correspond to the "Lindenbaum algebras" of predicate geometric theories. A famous result by Joyal and Tierney [15] shows that every Grothendiek topos is equivalent to the category of continuous actions of a localic groupoid.

An analogous connection between ω_1-pretoposes and ω_1-coherent theories was established by R. Chen in [1,2]. An ω_1-pretopos is a generalization of a Grothendiek topos where only countable coproducts are required to exist, rather than all small coproducts. Similarly, ω_1-coherent logic is the restriction of geometric logic that requires infinitary disjunctions to be countable. R. Chen showed that the syntactic ω_1-pretopos (the "Lindenbaum algebra") of a countable ω_1-coherent theory is equivalent to the category $\mathbf{Act}(\mathcal{C})$ of (countably based) continuous actions of a quasi-Polish groupoid $\mathcal{C}$. This can be viewed as a generalization (from propositional logic to predicate logic) of R. Heckmann's spatiality

The author thanks Ruiyuan Chen for helpful discussions. This note contains proofs of the main results of the author's presentation at the 2025 RIMS workshop on proof theory and computability theory, which was organized by T. Kihara. This work was supported by JSPS Bilateral Program Number JPJSBP120265001.

V. Brattka et al. (Eds.): CiE 2026, LNCS 16674, pp. 240–255, 2027.
https://doi.org/10.1007/978-3-032-31348-5_16

result [12], which implies the Lindenbaum algebra of a countable ω_1-coherent propositional theory is isomorphic to the frame of opens of a quasi-Polish space.

This note is a contribution towards effectivizing R. Chen's result and other aspects of Grothendiek topos theory. We define computable étale spaces over a computable topological space Y within the TTE framework of computable topology [27], and show they are equivalent to computable functions from Y to ODS, the effective quasi-Polish category of overt discrete quasi-Polish spaces. More generally, we show that for any computable category $\mathcal{C}$ there is a computability preserving equivalence between the category $\mathbf{Act}(\mathcal{C})$ of (countably based) étale spaces with continuous actions by $\mathcal{C}$, and the functor category $\mathsf{ODS}^{\mathcal{C}}$ of (topologically) continuous functors from $\mathcal{C}$ to ODS. It follows that the syntactic ω_1-pretopos of a countable ω_1-coherent theory is equivalent to a functor category $\mathsf{ODS}^{\mathcal{C}}$ for some quasi-Polish category $\mathcal{C}$. These syntactic ω_1-pretoposes are analogous to classifying (Grothendiek) toposes of geometric theories, and the categorically equivalent functor categories $\mathsf{ODS}^{\mathcal{C}}$ are suitable for computability theory because they are internal categories of $\mathbf{QCB}_0$.

See [5,10,11,19] for related work applying computability theory to category theoretical structures. See also [16] for connections between Lawvere-Tierney topologies and computability theory. We mainly work with computable topological spaces[1], but will need $\mathbf{QCB}_0$-spaces [24] for functor categories. One could also generalize further to represented spaces [22], which appears necessary to internalize some categories of interest [5]. We will also emphasize the constructions for quasi-Polish spaces, because they are exactly what is needed to establish the connections with ω_1-coherent logic. By encoding quasi-Polish spaces as spaces of ideals and using the techniques of [20], most of our results can be formalized within the subsytem ACA_0 of second order arithmetic (or even weaker systems).

2 Preliminaries

2.1 Computable Topological Spaces

Let $\prec$ be a transitive relation on $\mathbb{N}$. A subset $I \subseteq \mathbb{N}$ is an *ideal* if and only if:

1. $I \neq \emptyset$, (*I is non-empty*)
2. $(\forall a \in I)(\forall b \in \mathbb{N})\,(b \prec a \Rightarrow b \in I)$, (*I is a lower set*)
3. $(\forall a, b \in I)(\exists c \in I)\,(a \prec c \,\&\, b \prec c)$. (*I is directed*)

The collection $\mathbf{I}(\prec)$ of all ideals has the topology generated by basic open sets of the form $[a]_\prec = \{I \in \mathbf{I}(\prec) \mid a \in I\}$.

A topological space is *quasi-Polish* if it is homeomorphic to a space of the form $\mathbf{I}(\prec)$ for some transitive relation $\prec$ on $\mathbb{N}$. A represented space is *effectively quasi-Polish* if it is computably isomorphic to a space of the form $\mathbf{I}(\prec)$ for some

[1] Readers uninterested in computability can interpret "computable topological space" as "countably based space", and "computable function" as "continuous function".

computably enumerable (c.e.) transitive relation $\prec$ on $\mathbb{N}$. An effectively quasi-Polish space is *computably overt* if furthermore the corresponding relation $\prec$ is such that $E_{\prec} = \{a \in \mathbb{N} \mid [a]_{\prec} \neq \emptyset\}$ is c.e. A computably overt effective quasi-Polish space is called a *computable quasi-Polish space.*

Quasi-Polish spaces were introduced in [3], and the equivalent characterization in terms of spaces of ideals that we use here was first shown in [8]. Effective quasi-Polish spaces have been studied in [7,8,13,17,25].

An admissible representation [23] of $\mathbf{I}(\prec)$ represents each ideal by an enumeration of its elements. A function $f\colon \mathbf{I}(\prec_1) \to \mathbf{I}(\prec_2)$ is *computable* iff there is a Type Two Turing Machine [27] which when given any enumeration of the elements of any $I \in \mathbf{I}(\prec_1)$ outputs an enumeration of $f(I) \in \mathbf{I}(\prec_2)$.

A *c.e. open subset* (or *Σ^0_1-subset*) of $\mathbf{I}(\prec)$ is an open subset of the form $\bigcup_{a \in S}[a]_{\prec}$ for some c.e. subset $S \subseteq \mathbb{N}$. A Π^0_2-subset is a subset of the form

$$\{I \in \mathbf{I}(\prec) \mid (\forall i \in \mathbb{N})[I \in U_i \Rightarrow I \in V_i]\}$$

for some c.e. sequence $(U_i, V_i)_{i \in \mathbb{N}}$ of c.e. open subsets of $\mathbf{I}(\prec)$. Effective quasi-Polish spaces are closed under Π^0_2-subspaces (this is a well-known result, but see the proof of Theorem 3 in [4] for how to prove this in ACA_0).

A pair $(\prec, X)$, where $\prec$ is a c.e. transitive relation and $X \subseteq \mathbf{I}(\prec)$, is called a *computable topological space*[2]. Computable functions, c.e. open subsets, and Π^0_2-subsets of computable topological spaces are defined by restricting the definition for effective quasi-Polish spaces to their subspaces. Computable topological spaces are countably based, and conversely every countably based space is homeomorphic to a computable topological space for some suitable $\prec$.

2.2 Computable Categories

A *computable category* is an internal category [14] in the category of computable topological spaces and computable functions. More concretely, a computable category is a tuple $\mathcal{C} = (\mathcal{C}_{\mathsf{Obj}}, \mathcal{C}_{\mathsf{Mor}}, \sigma, \tau, \iota, \circ)$ consisting of the following data:

- $\mathcal{C}_{\mathsf{Obj}}$ (objects) and $\mathcal{C}_{\mathsf{Mor}}$ (morphisms) are computable topological spaces.
- $\sigma\colon \mathcal{C}_{\mathsf{Mor}} \to \mathcal{C}_{\mathsf{Obj}}$ (source), $\tau\colon \mathcal{C}_{\mathsf{Mor}} \to \mathcal{C}_{\mathsf{Obj}}$ (target), and $\iota\colon \mathcal{C}_{\mathsf{Obj}} \to \mathcal{C}_{\mathsf{Mor}}$ (identity) are computable functions.
- $\circ :\subseteq \mathcal{C}_{\mathsf{Mor}} \times \mathcal{C}_{\mathsf{Mor}} \to \mathcal{C}_{\mathsf{Mor}}$ (composition) is a partial computable function with domain
$$dom(\circ) = \{\langle g, f\rangle \in \mathcal{C}_{\mathsf{Mor}} \times \mathcal{C}_{\mathsf{Mor}} \mid \sigma(g) = \tau(f)\}$$

subject to the following:

- $\sigma(g \circ f) = \sigma(f)$ and $\tau(g \circ f) = \tau(g)$,
- $\sigma(\iota(c)) = c$ and $\tau(\iota(c)) = c$,
- $(h \circ g) \circ f = h \circ (g \circ f)$ when the compositions $h \circ g$ and $g \circ f$ are defined,
- if $\sigma(f) = c$ and $\tau(f) = d$ then $f \circ \iota(c) = f = \iota(d) \circ f$.

[2] See Sect. 4.2 of [4] for a proof that this definition is computably equivalent to the more common definition of a *computable (or effective) topological space.*

We write $f\colon c \to d$ to denote that f is a morphism with $\sigma(f) = c$ and $\tau(f) = d$.

We call $\mathcal{C}$ an *effective quasi-Polish category* if $\mathcal{C}_{\mathsf{Obj}}$ and $\mathcal{C}_{\mathsf{Mor}}$ are effective quasi-Polish spaces. This implies $\mathcal{C}$ is an internal category in the category of effective quasi-Polish spaces and computable functions, because effective quasi-Polish spaces are closed under finite limits (in particular, the pullback $dom(\circ)$ is an effective quasi-Polish space). $\mathcal{C}$ is *locally compact* if $\mathcal{C}_{\mathsf{Mor}}$ and $\mathcal{C}_{\mathsf{Obj}}$ are locally compact spaces.

As a useful example, note that every computable topological space X can be viewed as a computable category $\mathcal{C}_X$ by defining $(\mathcal{C}_X)_{\mathsf{Mor}} = (\mathcal{C}_X)_{\mathsf{Obj}} = X$, and defining σ, τ, and ι to be the identity functions. The only morphisms in $\mathcal{C}_X$ are the identity morphisms, so composition $\circ :\subseteq X \times X \to X$ is the (partial) inverse of the diagonal embedding $\Delta\colon X \to X \times X$, $x \mapsto \langle x, x\rangle$. Then $\mathcal{C}_X$ is a discrete category, although not necessarily discrete in the topological sense.

2.3 Computable Functors

Let $\mathcal{C} = (\mathcal{C}_{\mathsf{Obj}}, \mathcal{C}_{\mathsf{Mor}}, \sigma_{\mathcal{C}}, \tau_{\mathcal{C}}, \iota_{\mathcal{C}}, \circ_{\mathcal{C}})$ and $\mathcal{D} = (\mathcal{D}_{\mathsf{Obj}}, \mathcal{D}_{\mathsf{Mor}}, \sigma_{\mathcal{D}}, \tau_{\mathcal{D}}, \iota_{\mathcal{D}}, \circ_{\mathcal{D}})$ be computable categories. A *computable functor* from $\mathcal{C}$ to $\mathcal{D}$ is a pair $F = (F_{\mathsf{Obj}}, F_{\mathsf{Mor}})$ of computable functions $F_{\mathsf{Obj}}\colon \mathcal{C}_{\mathsf{Obj}} \to \mathcal{D}_{\mathsf{Obj}}$ and $F_{\mathsf{Mor}}\colon \mathcal{C}_{\mathsf{Mor}} \to \mathcal{D}_{\mathsf{Mor}}$ satisfying

1. $F_{\mathsf{Obj}} \circ \sigma_{\mathcal{C}} = \sigma_{\mathcal{D}} \circ F_{\mathsf{Mor}}$,
2. $F_{\mathsf{Obj}} \circ \tau_{\mathcal{C}} = \tau_{\mathcal{D}} \circ F_{\mathsf{Mor}}$,
3. $F_{\mathsf{Mor}} \circ \iota_{\mathcal{C}} = \iota_{\mathcal{D}} \circ F_{\mathsf{Obj}}$, and
4. $F_{\mathsf{Mor}}(g \circ_{\mathcal{C}} f) = F_{\mathsf{Mor}}(g) \circ_{\mathcal{D}} F_{\mathsf{Mor}}(f)$ for all composable $f, g \in \mathcal{C}_{\mathsf{Mor}}$.

A computable functor F from $\mathcal{C}$ to $\mathcal{D}$ will be denoted $F\colon \mathcal{C} \to \mathcal{D}$. It is clear that the identity functor $1_{\mathcal{C}}\colon \mathcal{C} \to \mathcal{C}$ (which is the identity on objects and morphisms) is computable, and that computable functors are closed under composition.

Given computable functors $F, G\colon \mathcal{C} \to \mathcal{D}$, a *computable natural transformation* from F to G is given by a computable function $\eta\colon \mathcal{C}_{\mathsf{Obj}} \to \mathcal{D}_{\mathsf{Mor}}$ satisfying:

- $\sigma_{\mathcal{D}} \circ \eta = F_{\mathsf{Obj}}$ and $\tau_{\mathcal{D}} \circ \eta = G_{\mathsf{Obj}}$
  (i.e., $\eta(c)\colon F_{\mathsf{Obj}}(c) \to G_{\mathsf{Obj}}(c)$ for each $c \in \mathcal{C}_{\mathsf{Obj}}$),
- $\eta(\tau_{\mathcal{C}}(f)) \circ_{\mathcal{D}} F_{\mathsf{Mor}}(f) = G_{\mathsf{Mor}}(f) \circ_{\mathcal{D}} \eta(\sigma_{\mathcal{C}}(f))$ for each $f \in \mathcal{C}_{\mathsf{Mor}}$.

Usually we will write $\eta\colon F \to G$ if it will not cause any confusion.

A *continuous functor*[3] is a functor that is computable with respect to an oracle. Similarly, a *continuous natural transformation* is a natural transformation that is computable with respect to an oracle. We will need these relativized versions when defining functor categories in the next section.

[3] We use *continuous* in the topological sense, which conflicts with the established usage of *continuous functor* as a functor that preserves all small limits. This will not cause a problem because we do not use the latter notion in this note.

2.4 Functor Categories

Since the category of computable topological spaces (or quasi-Polish spaces) is not cartesian closed, when we discuss functor categories we will potentially need to step into the larger cartesian closed category $\mathbf{QCB}_0$ [24].

The functor category $\mathcal{D}^{\mathcal{C}}$ has continuous functors from $\mathcal{C}$ to $\mathcal{D}$ as objects and continuous natural transformations as morphisms. It is formally defined as:

- $(\mathcal{D}^{\mathcal{C}})_{\mathsf{Obj}}$ is the subspace of $(\mathcal{D}_{\mathsf{Obj}})^{\mathcal{C}_{\mathsf{Obj}}} \times (\mathcal{D}_{\mathsf{Mor}})^{\mathcal{C}_{\mathsf{Mor}}}$ of continuous functors.
- $(\mathcal{D}^{\mathcal{C}})_{\mathsf{Mor}}$ is the subspace of $(\mathcal{D}_{\mathsf{Mor}})^{\mathcal{C}_{\mathsf{Obj}}} \times (\mathcal{D}^{\mathcal{C}})_{\mathsf{Obj}} \times (\mathcal{D}^{\mathcal{C}})_{\mathsf{Obj}}$ of tuples $\langle \eta, F, G \rangle$ where η is a continuous natural transformation from F to G.
- $\sigma_{\mathcal{D}^{\mathcal{C}}}(\langle \eta, F, G \rangle) = F$, $\tau_{\mathcal{D}^{\mathcal{C}}}(\langle \eta, F, G \rangle) = G$, $\iota_{\mathcal{D}^{\mathcal{C}}}(F) = \langle \iota_{\mathcal{D}} \circ F_{\mathsf{Obj}}, F, F \rangle$.
- $\langle \theta, G, H \rangle \circ_{\mathcal{D}^{\mathcal{C}}} \langle \eta, F, G \rangle = \langle \zeta, F, H \rangle$, where $\zeta(c) = \theta(c) \circ_{\mathcal{D}} \eta(c)$.

The function spaces $(\mathcal{D}_{\mathsf{Obj}})^{\mathcal{C}_{\mathsf{Obj}}}$ and $(\mathcal{D}_{\mathsf{Mor}})^{\mathcal{C}_{\mathsf{Mor}}}$ are constructed as in $\mathbf{QCB}_0$, and the topology is the sequentialization of the compact-open topology.

If X and Y are quasi-Polish, and X is locally compact, then Y^X is quasi-Polish and has the compact-open topology (see Theorem 16.3 of [6]). Therefore, if $\mathcal{C}$ and $\mathcal{D}$ are quasi-Polish categories and $\mathcal{C}$ is locally compact, then $(\mathcal{D}_{\mathsf{Obj}})^{\mathcal{C}_{\mathsf{Obj}}}$ and $(\mathcal{D}_{\mathsf{Mor}})^{\mathcal{C}_{\mathsf{Mor}}}$ are quasi-Polish. Thus $\mathcal{D}^{\mathcal{C}}$ is a quasi-Polish category, because $(\mathcal{D}^{\mathcal{C}})_{\mathsf{Obj}}$ and $(\mathcal{D}^{\mathcal{C}})_{\mathsf{Mor}}$ can be constructed as equalizers of continuous functions between quasi-Polish spaces, and $\sigma_{\mathcal{D}^{\mathcal{C}}}$, $\tau_{\mathcal{D}^{\mathcal{C}}}$, and $\iota_{\mathcal{D}^{\mathcal{C}}}$ are continuous.

2.5 Computable Étale Spaces

An *étale space* over a topological space Y is a local homeomorphism $p\colon X \to Y$, which means that each $x \in X$ has an open neighborhood U such that the restriction $p|_U$ is a homeomorphism with an open subset of Y. A morphism between étale spaces $p\colon X \to Y$ and $q\colon Z \to Y$ is a continuous function $f\colon X \to Z$ such that $q \circ f = p$. It is well-known that the category of étale spaces over Y is equivalent to the category of sheaves of sets over Y (see [18]).

We only consider countably based étale spaces in this note. Having a countable basis forces the fibers to be countable, hence these can be viewed as sheaves of countable sets over a space. Note that a countably based étale space over a quasi-Polish space is automatically quasi-Polish (Lemma 2.2 of [1]).

Let X and Y be computable topological spaces. A *computable étale space* consists of a computable function $p\colon X \to Y$ along with the following:

1. a computable enumeration of c.e. open subsets $(U_n)_{n\in\mathbb{N}}$ that cover X,
2. a computable enumeration of c.e. open subsets $(V_n)_{n\in\mathbb{N}}$ of Y,
3. a computable enumeration of computable functions (sections) $s_n\colon V_n \to U_n$ ($n \in \mathbb{N}$) satisfying
   - $s_n(p(x)) = x$ for each $x \in U_n$, and
   - $p(s_n(y)) = y$ for each $y \in V_n$.

   (i.e., each s_n is a computable inverse of the restriction $p|_{U_n}\colon U_n \to V_n$).

In this case, we call $p\colon X \to Y$ a *computable local homeomorphism.* We will sometimes call X a computable étale space if omitting the computable local homeomorphism p and the computable sections does not cause confusion.

We have the following easy fact that we write here as a lemma for convenience.

Lemma 1. *If $p\colon X \to Y$ is a local homeomorphism with sections $s_n\colon V_n \to U_n$ ($n \in \mathbb{N}$) and $p(x) = y$, then $x = s_n(y)$ if and only if $x \in U_n$.* □

2.6 Computable Actions

The definitions in this section are a minor generalization of the continuous groupoid actions in [1] to continuous actions by topological categories.

Let $\mathcal{C}$ be a computable category, and $p\colon X \to \mathcal{C}_{\mathsf{Obj}}$ be a computable function from a computable topological space X to the space of objects of $\mathcal{C}$. A *computable action* by $\mathcal{C}$ on X is a partial computable function $\alpha :\subseteq \mathcal{C}_{\mathsf{Mor}} \times X \to X$ with domain

$$dom(\alpha) = \{\langle f, x\rangle \in \mathcal{C}_{\mathsf{Mor}} \times X \mid \sigma(f) = p(x)\}$$

that satisfies the following conditions (where we abbreviate $\alpha(f, x)$ by $f \cdot x$):

1. $\tau(f) = p(f \cdot x)$
2. $\iota(p(x)) \cdot x = x$
3. $(g \circ f) \cdot x = g \cdot (f \cdot x)$.

Intuitively, α interprets (in a computable way) each morphism $f\colon c \to d$ in $\mathcal{C}$ as a function $\alpha(f, -)\colon p^{-1}(c) \to p^{-1}(d)$ between the fibers of c and d, while preserving identity morphisms and composition.

A *computable equivariant function* from $p\colon X \to \mathcal{C}_{\mathsf{Obj}}$ to $q\colon Y \to \mathcal{C}_{\mathsf{Obj}}$ (both equipped with computable actions by $\mathcal{C}$) is a computable function $f\colon X \to Y$ such that $p = q \circ f$ and $f(g \cdot x) = g \cdot f(x)$.

If $\mathcal{C}$ is a computable category, then a computable étale space over $\mathcal{C}_{\mathsf{Obj}}$ equipped with a computable action by $\mathcal{C}$ is called a *computable $\mathcal{C}$-set.*

These definitions relativize to countably based étale spaces, continuous actions, and continuous equivariant functions. We write $\mathbf{Act}(\mathcal{C})$ for the category of countably based $\mathcal{C}$-sets and continuous equivariant functions.

2.7 The Category of Overt Discrete Quasi-polish Spaces

Let $\mathcal{P}(\mathbb{N})$ be the powerset of the natural numbers with the Scott-topology. If $(F_i)_{i\in\mathbb{N}}$ is a computable enumeration of all finite subsets of $\mathbb{N}$ and we define $i \prec j \iff F_i \subseteq F_j$, then $\mathcal{P}(\mathbb{N})$ is computably homeomorphic to $\mathbf{I}(\prec)$. The space $\mathcal{P}(\mathbb{N} \times \mathbb{N})$ of sets of pairs of natural numbers is defined similarly, and is computably homeomorphic to $\mathcal{P}(\mathbb{N})$.

A symmetric transitive relation is called a *partial equivalence relation* [14]. The effective quasi-Polish category ODS of overt discrete quasi-Polish spaces is defined as follows (compare with Sect. 4 of [1]):

– $\mathsf{ODS}_{\mathsf{Obj}}$ is the subspace of $\mathcal{P}(\mathbb{N} \times \mathbb{N})$ of all partial equivalence relations.

- $\mathsf{ODS_{Mor}}$ is the subspace of $\mathcal{P}(\mathbb{N} \times \mathbb{N}) \times \mathsf{ODS_{Obj}} \times \mathsf{ODS_{Obj}}$ of tuples $\langle G, \equiv_\sigma, \equiv_\tau \rangle$ satisfying (for all $a, a', b, b' \in \mathbb{N}$):
  1. $G(a,b)$ implies $a \equiv_\sigma a \,\&\, b \equiv_\tau b$.
  2. $G(a,b) \,\&\, a \equiv_\sigma a'$ implies $G(a',b)$.
  3. $G(a,b) \,\&\, b \equiv_\tau b'$ implies $G(a,b')$.
  4. $G(a,b) \,\&\, G(a,b')$ implies $b \equiv_\tau b'$.
  5. $a \equiv_\sigma a$ implies $(\exists b \in \mathbb{N})\, G(a,b)$.
- $\sigma(\langle G, \equiv_\sigma, \equiv_\tau \rangle) = \equiv_\sigma$.
- $\tau(\langle G, \equiv_\sigma, \equiv_\tau \rangle) = \equiv_\tau$.
- $\iota(\equiv) = \langle \equiv, \equiv, \equiv \rangle$.
- Composition $\circ$ is defined as

$$\langle G', \equiv_\rho, \equiv_\tau \rangle \circ \langle G, \equiv_\sigma, \equiv_\rho \rangle = \langle \{\langle a, c \rangle \mid (\exists b \in \mathbb{N})\, G(a,b) \,\&\, G'(b,c)\}, \equiv_\sigma, \equiv_\tau \rangle,$$

  which is easily seen to be well-defined and computable.

Clearly, $\mathsf{ODS_{Obj}}$ and $\mathsf{ODS_{Mor}}$ are effective quasi-Polish spaces because they are defined as Π^0_2-subspaces of effective quasi-Polish spaces.

An effective quasi-Polish space $X \cong \mathbf{I}(\prec)$ is *computably discrete* if there is a c.e. set $D_\prec \subseteq \mathbb{N} \times \mathbb{N}$ such that $I = J$ in $\mathbf{I}(\prec)$ if and only if there exists $\langle a, b \rangle \in D_\prec$ with $a \in I$ and $b \in J$.

Let $\mathbb{S} = \{\bot, \top\}$ be the Sierpinski space. Then an effective quasi-Polish space X is computably discrete if and only if equality $(=)\colon X \times X \to \mathbb{S}$, defined as $=(x,y) = \top \iff x = y$, is computable. Furthermore, X is computably overt if and only if $(\exists_X)\colon \mathbb{S}^X \to \mathbb{S}$, defined as $\exists_X(U) = \top \iff U \neq \emptyset$, is computable. Therefore, our definitions are equivalent to the ones in [9,22].

The next theorem is closely related to Theorem 1 of [21], which includes several other important results on overt discrete quasi-Polish spaces within the general context of represented spaces.

Theorem 1. ODS *is equivalent to the category of overt discrete quasi-Polish spaces. Under this equivalence, the computable points of* $\mathsf{ODS_{Obj}}$ *correspond to the computably overt computably discrete effective quasi-Polish spaces, and the computable points of* $\mathsf{ODS_{Mor}}$ *correspond to the computable functions between computably overt computably discrete effective quasi-Polish spaces.*

Proof. We construct a full and faithful and essentially surjective functor $\mathcal{E}$ from ODS to the category of overt discrete quasi-Polish spaces. Given $\equiv$ in $\mathsf{ODS_{Obj}}$, define $\mathcal{E}_{\mathsf{Obj}}(\equiv) = \mathbf{I}(\equiv)$, the space of ideals of $\equiv$. This is quasi-Polish by definition, and discrete because if $I, J \in \mathcal{E}_{\mathsf{Obj}}(\equiv)$ then $I = J$ if and only if $(\exists a \in \mathbb{N})[a \in I \,\&\, a \in J]$. Finally, $\mathcal{E}_{\mathsf{Obj}}(\equiv)$ is overt (relative to $\equiv$) because $[a]_\prec \neq \emptyset$ if and only if $a \equiv a$. Clearly, if $\equiv$ is a computable point then $\mathcal{E}_{\mathsf{Obj}}(\equiv)$ is a computably overt computably discrete effective quasi-Polish space.

$\mathcal{E}_{\mathsf{Mor}}$ maps $\langle G, \equiv_\sigma, \equiv_\tau \rangle$ in $\mathsf{ODS_{Mor}}$ to the function $f_G\colon \mathcal{E}_{\mathsf{Obj}}(\equiv_\sigma) \to \mathcal{E}_{\mathsf{Obj}}(\equiv_\tau)$ defined as

$$f_G(I) = \{b \in \mathbb{N} \mid (\exists a \in I)\, G(a,b)\}$$

for each $I \in \mathcal{E}_{\mathsf{Obj}}(\equiv_\sigma)$. It follows from the definition of ideals and the five axioms defining elements of $\mathsf{ODS}_{\mathsf{Mor}}$ that $f_G(I)$ is an ideal of $\equiv_\tau$, hence f_G is a well-defined function. Clearly, f_G is computable from $\langle G, \equiv_\sigma, \equiv_\tau \rangle$. Note that a computable point of $\mathsf{ODS}_{\mathsf{Mor}}$ has computable source and target because σ and τ are computable.

It is easy to see that $\mathcal{E}$ is a full and faithful functor. Next, we show that each computably overt computably discrete effective quasi-Polish space is computably homeomorphic to $\mathcal{E}_{\mathsf{Obj}}(\equiv)$ for some computable point $\equiv$ of $\mathsf{ODS}_{\mathsf{Obj}}$.

Assume $\prec$ is a c.e. transitive relation, that $E_\prec = \{a \in \mathbb{N} \mid [a]_\prec \neq \emptyset\}$ is c.e., and that $D_\prec \subseteq \mathbb{N} \times \mathbb{N}$ is c.e. and such that $I = J$ in $\mathbf{I}(\prec)$ if and only if there is $\langle a, b \rangle \in D_\prec$ with $a \in I$ and $b \in J$. Without loss of generality, we can assume that $\langle a, b \rangle \in D_\prec$ implies $a \in E_\prec$ and $b \in E_\prec$.

Define

$$S = \{c \in E_\prec \mid (\exists \langle a, b \rangle \in D_\prec)[a \prec c \,\&\, b \prec c]\}.$$

Note that for each $c \in S$ there is a unique $I \in \mathbf{I}(\prec)$ such that $\{I\} = [c]_\prec$, and conversely for each $I \in \mathbf{I}(\prec)$ there is at least one $c \in S$ with $\{I\} = [c]_\prec$. If $[a]_\prec = [b]_\prec = \{I\}$ then a and b must have a $\prec$-upper bound in I, so by defining

$$a \equiv b \iff a, b \in S \,\&\, (\exists c \in S)[a \prec c \,\&\, b \prec c],$$

we obtain that $a \equiv b$ if and only if $a, b \in S$ and $[a]_\prec = [b]_\prec = \{I\}$ for a unique $I \in \mathbf{I}(\prec)$. Now define $g \colon \mathbf{I}(\prec) \to \mathcal{E}_{\mathsf{Obj}}(\equiv)$ and $h \colon \mathcal{E}_{\mathsf{Obj}}(\equiv) \to \mathbf{I}(\prec)$ as

$$g(I) = \{a \in I \mid a \equiv a\}$$
$$h(J) = \{b \in S \mid (\exists a \in J)\, b \prec a\}.$$

Then $g(I)$ is the $\equiv$-ideal composed of all $a \in S$ with $[a]_\prec = \{I\}$. Furthermore, $h(J)$ is a $\prec$-ideal because each $\equiv$-ideal is $\prec$-directed. Therefore, g and h are well-defined, and easily seen to be computable inverses of each other. Therefore, $\mathcal{E}_{\mathsf{Obj}}(\equiv)$ is computably homeomorphic to $\mathbf{I}(\prec)$. Since the proof relativizes, it follows that $\mathcal{E}$ is essentially surjective. □

3 Main Result

Our main result is the following:

Theorem 2. *Let $\mathcal{C}$ be a computable category. The category $\mathbf{Act}(\mathcal{C})$ of countably based $\mathcal{C}$-sets and continuous equivariant functions and the category $\mathsf{ODS}^{\mathcal{C}}$ of (topologically) continuous functors from $\mathcal{C}$ to ODS are equivalent. The equivalence preserves the computability of both objects and morphisms.* □

As a special case, if X is a computable topological space then there is an equivalence between computable étale spaces over X and computable functions from X to $\mathsf{ODS}_{\mathsf{Obj}}$. Thus, the functor category $\mathsf{ODS}^{\mathcal{C}_X}$ is equivalent to the category of countably based étale spaces over X.

We prove our main result in the next four sections by constructing functors $\mathcal{F} \colon \mathsf{ODS}^{\mathcal{C}} \to \mathbf{Act}(\mathcal{C})$ and $\mathcal{G} \colon \mathbf{Act}(\mathcal{C}) \to \mathsf{ODS}^{\mathcal{C}}$ that preserve computable objects and morphisms, and then show that that $\mathcal{G} \circ \mathcal{F}$ and $\mathcal{F} \circ \mathcal{G}$ are naturally isomorphic to the identity functors.

3.1 From $\mathsf{ODS}^{\mathcal{C}}$ to $\mathbf{Act}(\mathcal{C})$

The next two subsections define the functor $\mathcal{F}\colon \mathsf{ODS}^{\mathcal{C}} \to \mathbf{Act}(\mathcal{C})$.

Objects (Functors to $\mathcal{C}$-Sets)

$\mathcal{F}_{\mathsf{Obj}}$ maps a functor $F\colon \mathcal{C} \to \mathsf{ODS}$ to an étale space $p_F\colon X_F \to \mathcal{C}_{\mathsf{Obj}}$ equipped with an action α_F defined as follows.

Assume $F\colon \mathcal{C} \to \mathsf{ODS}$ is a computable functor. For $c \in \mathcal{C}_{\mathsf{Obj}}$ we write $\equiv_c$ for the partial equivalence relation $F_{\mathsf{Obj}}(c)$, and for $n \in \mathbb{N}$ we write $[n]_c = \{m \in \mathbb{N} \mid n \equiv_c m\}$ for the equivalence class of n with respect to $\equiv_c$. Let $X_F \subseteq \mathcal{C}_{\mathsf{Obj}} \times \mathcal{P}(\mathbb{N})$ be the subspace defined as

$$X_F = \{\langle c, [n]_c\rangle \in \mathcal{C}_{\mathsf{Obj}} \times \mathcal{P}(\mathbb{N}) \mid n \equiv_c n\}.$$

Define $p_F\colon X_F \to \mathcal{C}_{\mathsf{Obj}}$ to be the projection on to the first coordinate. Define open sets

$$\begin{aligned} U_n &= \{\langle c, [n]_c\rangle \in X_F \mid n \equiv_c n\}, \text{ and} \\ V_n &= \{c \in \mathcal{C}_{\mathsf{Obj}} \mid n \equiv_c n\}. \end{aligned}$$

Define $s_n\colon V_n \to U_n$ as $s_n(c) = \langle c, [n]_c\rangle$. Then s_n is the inverse for $p|_{U_n}$, which shows that p is a computable local homeomorphism.

Define the action $\alpha_F :\subseteq \mathcal{C}_{\mathsf{Mor}} \times X_F \to X_F$ as

$$\alpha_F(f, \langle c, [n]_c\rangle) = \langle \tau(f), F_{\mathsf{Mor}}(f)(n)\rangle$$

where for $F_{\mathsf{Mor}}(f) = \langle G_f, \equiv_\sigma, \equiv_\tau\rangle$ we set $F_{\mathsf{Mor}}(f)(n) = \{m \in \mathbb{N} \mid G_f(n,m)\}$. Using the assumption that F is a computable functor, it is straightforward to check that α_F is a computable action.

Morphisms (Natural Transformations to Equivariant Functions)

$\mathcal{F}_{\mathsf{Mor}}$ maps a natural transformation η to an equivariant function h as follows.

Assume $F, F'\colon \mathcal{C} \to \mathsf{ODS}$ are computable functors, and that $\eta\colon \mathcal{C}_{\mathsf{Obj}} \to \mathsf{ODS}_{\mathsf{Mor}}$ corresponds to a computable natural transformation from F to F'. Let $p\colon X \to \mathcal{C}_{\mathsf{Obj}}$ and $q\colon Y \to \mathcal{C}_{\mathsf{Obj}}$ be the $\mathcal{C}$-sets $\mathcal{F}_{\mathsf{Obj}}(F)$ and $\mathcal{F}_{\mathsf{Obj}}(F')$, respectively (we omit the subscripts). Define $h\colon X \to Y$ as

$$h(\langle c, [n]_c\rangle) = \langle c, \eta(c)(n)\rangle$$

where for $\eta(c) = \langle G, F_{\mathsf{Obj}}(c), F'_{\mathsf{Obj}}(c)\rangle$ we set $\eta(c)(n) = \{m \in \mathbb{N} \mid G(n,m)\}$.

It is clear that h is computable. We check that it is equivariant. Since p and q project the first coordinate and h does not change it, it is clear that $p = q \circ h$.

For $f\colon c \to d$ in $\mathcal{C}_{\mathsf{Mor}}$, we have (with a slight abuse of notation in the second and third lines)

$$\begin{aligned} h(f \cdot \langle c, [n]_c \rangle) &= h(\langle d, F_{\mathsf{Mor}}(f)(n) \rangle) \\ &= \langle d, \eta(d)(F_{\mathsf{Mor}}(f)(n)) \rangle \\ &= \langle d, F'_{\mathsf{Mor}}(f)(\eta(c)(n)) \rangle \\ &= f \cdot \langle c, \eta(c)(n) \rangle \\ &= f \cdot h(\langle c, [n]_c \rangle), \end{aligned}$$

hence h is equivariant.

3.2 From $\mathbf{Act}(\mathcal{C})$ to $\mathsf{ODS}^{\mathcal{C}}$

The next two subsections define the functor $\mathcal{G}\colon \mathbf{Act}(\mathcal{C}) \to \mathsf{ODS}^{\mathcal{C}}$.

Objects ($\mathcal{C}$-Sets to Functors)

$\mathcal{G}_{\mathsf{Obj}}$ maps an étale space $p\colon X \to \mathcal{C}_{\mathsf{Obj}}$ equipped with an action α to a functor $F\colon \mathcal{C} \to \mathsf{ODS}$ as follows.

Assume $p\colon X \to \mathcal{C}_{\mathsf{Obj}}$ is a computable étale space with sections $s_n\colon V_n \to U_n$ ($n \in \mathbb{N}$), and computable action $\alpha :\subseteq \mathcal{C}_{\mathsf{Mor}} \times X \to X$. Define $F_{\mathsf{Obj}}\colon \mathcal{C}_{\mathsf{Obj}} \to \mathsf{ODS}_{\mathsf{Obj}}$ as

$$F_{\mathsf{Obj}}(c) = \{\langle n, m \rangle \in \mathbb{N} \times \mathbb{N} \mid c \in V_n \cap V_m \,\&\, s_n(c) = s_m(c)\}$$

and define $F_{\mathsf{Mor}}\colon \mathcal{C}_{\mathsf{Mor}} \to \mathsf{ODS}_{\mathsf{Mor}}$ as

$$F_{\mathsf{Mor}}(f) = \langle G_f, \equiv_\sigma, \equiv_\tau \rangle,$$

where

$$\begin{aligned} &G_f = \{\langle n, m \rangle \in \mathbb{N} \times \mathbb{N} \mid n \equiv_\sigma n \,\&\, m \equiv_\tau m \,\&\, \alpha(f, s_n(\sigma(f))) = s_m(\tau(f))\} \\ &\equiv_\sigma = F_{\mathsf{Obj}}(\sigma(f)) \\ &\equiv_\tau = F_{\mathsf{Obj}}(\tau(f)). \end{aligned}$$

We show that $F\colon \mathcal{C} \to \mathsf{ODS}$ is a computable functor. It is clear that $F_{\mathsf{Obj}}(x)$ is a partial equivalence relation, hence $F_{\mathsf{Obj}}\colon \mathcal{C}_{\mathsf{Obj}} \to \mathsf{ODS}_{\mathsf{Obj}}$ is well-defined and easily seen to be computable by using Lemma 1.

$F_{\mathsf{Mor}}\colon \mathcal{C}_{\mathsf{Mor}} \to \mathsf{ODS}_{\mathsf{Mor}}$ is also easily seen to be computable by using Lemma 1 and the fact that α being an action implies $p(\alpha(f, s_n(\sigma(f)))) = \tau(f)$, but we must check that $F_{\mathsf{Mor}}(f)$ satisfies the 5 conditions to be in $\mathsf{ODS}_{\mathsf{Mor}}$. Condition 1 holds by definition. For condition 2, if $G_f(n, m)$ and $n \equiv_\sigma n'$, then $s_n(\sigma(f)) = s_{n'}(\sigma(f))$ by definition of $\equiv_\sigma$, hence $G_f(n', m)$ follows. Condition 3 is similar. For condition 4, if $G_f(n, m)$ and $G_f(n, m')$ then $\tau(f) \in V_m \cap V_{m'}$ and $s_m(\tau(f)) = s_{m'}(\tau(f))$ because α is single-valued, hence $m \equiv_\tau m'$. For condition 5, if $n \equiv_\sigma n$, then $\sigma(f) \in V_n$, hence $p(s_n(\sigma(f))) = \sigma(f)$, thus $\langle f, s_n(\sigma(f)) \rangle \in dom(\alpha)$. Choose

$m \in \mathbb{N}$ with $\alpha(f, s_n(\sigma(f))) \in U_m$. Then $\tau(f) = p(\alpha(f, s_n(\sigma(f))))$ because α is an action. Thus $\tau(f) \in V_m$, which implies $m \equiv_\tau m$, and it follows from Lemma 1 that $\alpha(f, s_n(\sigma(f))) = s_m(\tau(f))$. Therefore, $G_f(n, m)$, and condition 5 is satisfied.

Finally, it is straightforward to verify that F_{Obj} and F_{Mor} together satisfy the 4 conditions required to be a functor: the first two conditions hold by definition, and the last two follow from the properties of actions.

Morphisms (Equivariant Functions to Natural Transformations)

$\mathcal{G}_{\mathsf{Mor}}$ maps an equivariant function h to a natural transformation η_h as follows.

Assume $h \colon X \to Y$ is a computable equivariant function from $p \colon X \to \mathcal{C}_{\mathsf{Obj}}$ to $q \colon Y \to \mathcal{C}_{\mathsf{Obj}}$. Let $s_n \colon V_n \to U_n$ $(n \in \mathbb{N})$ be the sections of p, and $s'_n \colon V'_n \to U'_n$ the sections of q. Let $F, F' \colon \mathcal{C} \to \mathsf{ODS}$ be the computable functors corresponding to p and q, respectively.

Define $\eta_h \colon \mathcal{C}_{\mathsf{Obj}} \to \mathsf{ODS}_{\mathsf{Mor}}$ as

$$\eta_h(c) = \langle G, \equiv_\sigma, \equiv_\tau \rangle$$

where

- $G = \{\langle n, m\rangle \in \mathbb{N} \times \mathbb{N} \mid n \equiv_\sigma n \,\&\, m \equiv_\tau m \,\&\, h(s_n(c)) = s'_m(c)\}$
- $\equiv_\sigma = F_{\mathsf{Obj}}(c)$
- $\equiv_\tau = F'_{\mathsf{Obj}}(c)$

It is straightforward to see that η_h is well-defined, and since $q(h(s_n(c))) = p(s_n(c)) = c$, Lemma 1 implies $h(s_n(c)) = s'_m(c)$ if and only if $h(s_n(c)) \in U'_m$, hence η_h is computable.

We check that it is in fact a natural transformation $\eta_h \colon F \to F'$. For $f \colon c \to d$ in $\mathcal{C}_{\mathsf{Mor}}$,

$$\begin{aligned}
&\eta_h(d) \circ F_{\mathsf{Mor}}(f)\\
&= \langle \{\langle n, m\rangle \mid (\exists k)[\cdots] \,\&\, f \cdot s_n(c) = s_k(d) \,\&\, h(s_k(d)) = s'_m(d)\}, F_{\mathsf{Obj}}(c), F'_{\mathsf{Obj}}(d)\rangle\\
&= \langle \{\langle n, m\rangle \mid [\cdots] \,\&\, h(f \cdot s_n(c)) = s'_m(d)\}, F_{\mathsf{Obj}}(c), F'_{\mathsf{Obj}}(d)\rangle\\
&= \langle \{\langle n, m\rangle \mid [\cdots] \,\&\, f \cdot h(s_n(c)) = s'_m(d)\}, F_{\mathsf{Obj}}(c), F'_{\mathsf{Obj}}(d)\rangle\\
&= \langle \{\langle n, m\rangle \mid (\exists k)[\cdots] \,\&\, h(s_n(c)) = s'_k(c) \,\&\, f \cdot s'_k(c) = s'_m(d)\}, F_{\mathsf{Obj}}(c), F'_{\mathsf{Obj}}(d)\rangle\\
&= F'_{\mathsf{Mor}}(f) \circ \eta_h(c)
\end{aligned}$$

where $[\cdots]$ abbreviates $n \equiv n$, $m \equiv m$, $k \equiv k$ (when relevant, and with the appropriate subscripts). The other condition for being a natural transformation is trivial.

3.3 Natural Isomorphism (from $\mathsf{ODS}^{\mathcal{C}}$ and Back)

We show that $\mathcal{G} \circ \mathcal{F}$ is actually equal to the identity functor on $\mathsf{ODS}^{\mathcal{C}}$.

Assume $F\colon \mathcal{C} \to \mathsf{ODS}$ is a computable functor, and let $\widehat{F} = \mathcal{G}_{\mathsf{Obj}}(\mathcal{F}_{\mathsf{Obj}}(F))$. Then $\widehat{F}_{\mathsf{Obj}}\colon \mathcal{C}_{\mathsf{Obj}} \to \mathsf{ODS}_{\mathsf{Obj}}$ is as follows:

$$\begin{aligned}\widehat{F}_{\mathsf{Obj}}(c) &= \{\langle n,m\rangle \in \mathbb{N}\times\mathbb{N} \mid c \in V_n \cap V_m \,\&\, s_n(c) = s_m(c)\}\\ &= \{\langle n,m\rangle \in \mathbb{N}\times\mathbb{N} \mid n \equiv_c n \,\&\, m \equiv_c m \,\&\, \langle c,[n]_c\rangle = \langle c,[m]_c\rangle\}\\ &= \{\langle n,m\rangle \in \mathbb{N}\times\mathbb{N} \mid n \equiv_c m\}\\ &= F_{\mathsf{Obj}}(c),\end{aligned}$$

where $\equiv_c$ and $[\cdot]_c$ refer to $F_{\mathsf{Obj}}(c)$, and $s_n\colon V_n \to U_n$ are the sections of $\mathcal{F}_{\mathsf{Obj}}(F)$.

For morphisms, assuming $F_{\mathsf{Mor}}(f) = \langle G_f, \equiv_\sigma, \equiv_\tau\rangle$, we have that $\widehat{F}_{\mathsf{Mor}}(f) = \langle \widehat{G}_f, \equiv_\sigma, \equiv_\tau\rangle$, where $\equiv_\sigma = F_{\mathsf{Obj}}(\sigma(f)) = \widehat{F}_{\mathsf{Obj}}(\sigma(f))$, and $\equiv_\tau = F_{\mathsf{Obj}}(\tau(f)) = \widehat{F}_{\mathsf{Obj}}(\tau(f))$. Furthermore, assuming $n \equiv_\sigma n$ and $m \equiv_\tau m$, we have

$$\begin{aligned}\widehat{G}_f(n,m) &\iff \alpha_F(f, s_n(\sigma(f))) = s_m(\tau(f))\\ &\iff \alpha_F(f, \langle \sigma(f), [n]_{\sigma(f)}\rangle) = \langle \tau(f), [m]_{\tau(f)}\rangle\\ &\iff \langle \tau(f), F_{\mathsf{Mor}}(f)(n)\rangle = \langle \tau(f), [m]_{\tau(f)}\rangle\\ &\iff F_{\mathsf{Mor}}(f)(n) = [m]_{\tau(f)}\\ &\iff G_f(n,m),\end{aligned}$$

where α_F is the action of $\mathcal{F}_{\mathsf{Obj}}(F)$. Therefore, $\widehat{F} = F$.

Next, assume $F, F'\colon \mathcal{C} \to \mathsf{ODS}$ are functors and that $\eta\colon \mathcal{C}_{\mathsf{Obj}} \to \mathsf{ODS}_{\mathsf{Mor}}$ corresponds to a natural transformation from F to F'. Let $\widehat{\eta} = \mathcal{G}_{\mathsf{Mor}}(\mathcal{F}_{\mathsf{Mor}}(\eta))$. Fix $c \in \mathcal{C}_{\mathsf{Obj}}$ and assume $\eta(c) = \langle G, F_{\mathsf{Obj}}(c), F'_{\mathsf{Obj}}(c)\rangle$ and $h = \mathcal{F}_{\mathsf{Mor}}(\eta)$. Then $\widehat{\eta}(c) = \langle \widehat{G}, F_{\mathsf{Obj}}(c), F'_{\mathsf{Obj}}(c)\rangle$, where for $\langle n,n\rangle \in F_{\mathsf{Obj}}(c)$ and $\langle m,m\rangle \in F'_{\mathsf{Obj}}(c)$, we have

$$\begin{aligned}\widehat{G}(n,m) &\iff h(s_n(c)) = s'_m(c)\\ &\iff h(\langle c,[n]_c\rangle) = \langle c,[m]'_c\rangle\\ &\iff \langle c, \eta(c)(n)\rangle = \langle c,[m]'_c\rangle\\ &\iff \eta(c)(n) = [m]'_c\\ &\iff \{b \in \mathbb{N} \mid G(n,b)\} = \{b \in \mathbb{N} \mid b \equiv'_c m\}\\ &\iff G(n,m),\end{aligned}$$

where s_n refers to $\mathcal{F}_{\mathsf{Obj}}(F)$, and $\equiv'_c$, $[\cdot]'_c$, and s'_m refer to $\mathcal{F}_{\mathsf{Obj}}(F')$. Thus, $\widehat{\eta} = \eta$.

3.4 Natural Isomorphism (from $\mathbf{Act}(\mathcal{C})$ and Back)

We show that $\mathcal{F} \circ \mathcal{G}$ is naturally isomorphic to the identity functor on $\mathbf{Act}(\mathcal{C})$.

Assume $p\colon X \to \mathcal{C}_{\mathsf{Obj}}$ is a computable étale space with sections $s_n\colon V_n \to U_n$ ($n \in \mathbb{N}$), and computable action $\alpha :\subseteq \mathcal{C}_{\mathsf{Mor}} \times X \to X$. Let $F\colon \mathcal{C} \to \mathsf{ODS}$ be the functor obtained from applying $\mathcal{G}_{\mathsf{Obj}}$ to X (see Sect. 3.2).

We calculate the étale space $p_F\colon X_F \to \mathcal{C}_{\mathsf{Obj}}$ with action α_F corresponding to $\mathcal{F}_{\mathsf{Obj}}(F)$ (see Sect. 3.1). We have

$$\begin{aligned} X_F &= \{\langle c, [n]_c\rangle \in \mathcal{C}_{\mathsf{Obj}} \times \mathcal{P}(\mathbb{N}) \mid n \equiv_c n\} \\ &= \{\langle c, [n]_c\rangle \in \mathcal{C}_{\mathsf{Obj}} \times \mathcal{P}(\mathbb{N}) \mid c \in V_n\}, \end{aligned}$$

where for $c \in V_n$,

$$\begin{aligned} [n]_c &= \{m \in \mathbb{N} \mid n \equiv_c m\} \\ &= \{m \in \mathbb{N} \mid c \in V_m \;\&\; s_n(c) = s_m(c)\} \\ &= \{m \in \mathbb{N} \mid s_n(c) \in U_m\} \end{aligned}$$

and $p_F\colon X_F \to \mathcal{C}_{\mathsf{Obj}}$ is the projection on to the first coordinate.

After unwinding the definitions, the action $\alpha_F :\subseteq \mathcal{C}_{\mathsf{Mor}} \times X_F \to X_F$ is

$$\begin{aligned} \alpha_F(f, \langle c, [n]_c\rangle) &= \langle \tau(f), \{m' \in \mathbb{N} \mid \tau(f) \in V_{m'} \;\&\; \alpha(f, s_n(c)) = s_{m'}(\tau(f))\}\rangle \\ &= \langle \tau(f), \{m' \in \mathbb{N} \mid \alpha(f, s_n(c)) \in U_{m'}\}\rangle \\ &= \langle \tau(f), \{m' \in \mathbb{N} \mid s_m(p(\alpha(f, s_n(c)))) \in U_{m'}\}\rangle \\ &= \langle \tau(f), \{m' \in \mathbb{N} \mid s_m(\tau(f)) \in U_{m'}\}\rangle \\ &= \langle \tau(f), [m]_{\tau(f)}\}\rangle \end{aligned}$$

where by assumption $\sigma(f) = c$ and $c \in V_n$, and $m \in \mathbb{N}$ is any number satisfying $\alpha(f, s_n(c)) \in U_m$.

Define $\theta_X\colon X \to X_F$ as $\theta_X(x) = \langle p(x), [n]_{p(x)}\rangle$, where $n \in \mathbb{N}$ satisfies $x \in U_n$ (if $x \in U_n \cap U_m$ then $[n]_{p(x)} = [m]_{p(x)}$ so the definition of $\theta_X(x)$ does not depend on the choice of n). It is clear that θ_X is computable, and it is equivariant because

$$\begin{aligned} \theta_X(\alpha(f, x)) &= \langle p(\alpha(f, x)), [m]_{p(\alpha(f,x))}\rangle \quad \text{(where } m \in \mathbb{N} \text{ satisfies } \alpha(f, x) \in U_m\text{)} \\ &= \langle \tau(f), [m]_{\tau(f)}\rangle \\ &= \alpha_F(f, \langle p(x), [n]_{p(x)}\rangle) \\ &\qquad \text{(as shown above, because } \sigma(f) = p(x) \text{ and } p(x) \in V_n\text{)} \\ &= \alpha_F(f, \theta_X(x)). \end{aligned}$$

Next, define $\theta'_X\colon X_F \to X$ as $\theta'_X(\langle c, [n]_c\rangle) = s_n(c)$. Note that θ'_X is well-defined because $s_n(c) = s_m(c)$ for each $m \in [n]_c$. It is clear that θ'_X is computable, and it is equivariant because:

$$\begin{aligned} \theta'_X(\alpha_F(f, \langle c, [n]_c\rangle)) &= \theta'_X(\langle \tau(f), [m]_{\tau(f)}\rangle) \\ &\qquad \text{(where } m \in \mathbb{N} \text{ satisfies } \alpha(f, s_n(c)) \in U_m\text{)} \\ &= s_m(\tau(f)) \\ &= \alpha(f, s_n(c)) \quad \text{(by Lemma 1 because } \alpha(f, s_m(c)) \in U_m\text{)} \\ &= \alpha(f, \theta'_X(\langle c, [n]_c\rangle)). \end{aligned}$$

Furthermore, θ_X and θ'_X are inverses because:

$$\begin{aligned}\theta'_X(\theta_X(x)) &= \theta'_X(\langle p(x), [n]_{p(x)}\rangle) \quad \text{(where } n \in \mathbb{N} \text{ satisfies } x \in U_n \text{)}\\ &= s_n(p(x))\\ &= x,\end{aligned}$$

and

$$\begin{aligned}\theta_X(\theta'_X(\langle c, [n]_c\rangle)) &= \theta_X(s_n(c))\\ &= \langle p(s_n(c)), [n]_{p(s_n(c))}\rangle \quad \text{(because } s_n(c) \in U_n \text{)}\\ &= \langle c, [n]_c\rangle.\end{aligned}$$

Therefore, X and X_F are computably isomorphic $\mathcal{C}$-sets.

Finally, we prove that $\theta\colon 1_{\mathbf{Act}(\mathcal{C})} \to \mathcal{F} \circ \mathcal{G}$ is a natural transformation (it automatically follows that the inverse θ' is also natural). Assume $h\colon X \to Y$ is a continuous equivariant function from $p\colon X \to \mathcal{C}_{\mathsf{Obj}}$ to $q\colon Y \to \mathcal{C}_{\mathsf{Obj}}$. Let $s_n\colon V_n \to U_n$ ($n \in \mathbb{N}$) be the sections of p, and $s'_n\colon V'_n \to U'_n$ the sections of q. We write $\eta_h\colon F \to F'$ for $\mathcal{G}_{\mathsf{Mor}}(h)$ and $h'\colon X_F \to Y_{F'}$ for $\mathcal{F}_{\mathsf{Mor}}(\eta_h)$. Then for $c \in \mathcal{C}_{\mathsf{Obj}}$ with $c \in V_n$ we have

$$h'(\langle c, [n]_c\rangle) = \langle c, \eta_h(c)(n)\rangle$$

where

$$\begin{aligned}m \in \eta_h(c)(n) &\iff \langle n, n\rangle \in F_{\mathsf{Obj}}(c) \,\&\, \langle m, m\rangle \in F'_{\mathsf{Obj}}(c) \,\&\, h(s_n(c)) = s'_m(c)\\ &\iff c \in V_n \cap V'_m \,\&\, h(s_n(c)) = s'_m(c)\\ &\qquad\qquad \text{(by definition of } F_{\mathsf{Obj}} \text{ and } F'_{\mathsf{Obj}})\\ &\iff c \in V_n \cap V'_m \,\&\, h(s_n(c)) \in U'_m\\ &\qquad\qquad \text{(by Lemma 1 because } q(h(s_n(c))) = c \text{).}\end{aligned}$$

Assume $x \in X$, and fix $n \in \mathbb{N}$ with $x \in U_n$. Since $s_n(p(x)) = x$, if $m \in \eta_h(p(x))(n)$ then $h(x) = h(s_n(p(x))) \in U'_m$. Conversely, if $h(x) \in U'_m$, then since $p(x) \in V_n$ and $p(x) = q(h(x)) \in V'_m$ and $h(s_n(p(x))) = h(x) \in U'_m$, it follows that $m \in \eta_h(p(x))(n)$. Therefore, $m \in \eta_h(p(x))(n)$ if and only if $h(x) \in U'_m$.

Now fix any $m \in \mathbb{N}$ with $h(x) \in U'_m$. Since $s'_m(p(x)) = s'_m(q(h(x))) = h(x)$, we have $m' \in [m]'_{p(x)}$ if and only if $h(x) = s'_m(p(x)) \in U'_{m'}$.

It follows that $\eta_h(p(x))(n) = [m]'_{p(x)}$, hence

$$\begin{aligned}h'(\theta_X(x)) &= h'(\langle p(x), [n]_{p(x)}\rangle)\\ &= \langle p(x), \eta_h(p(x))(n)\rangle\\ &= \langle p(x), [m]'_{p(x)}\rangle\\ &= \langle q(h(x)), [m]'_{q(h(x))}\rangle\\ &= \theta_Y(h(x)).\end{aligned}$$

Therefore, $\theta\colon 1_{\mathbf{Act}(\mathcal{C})} \to \mathcal{F} \circ \mathcal{G}$ is a natural isomorphism.

References

1. Chen, R.: Borel functors, interpretations, and strong conceptual completeness for $\mathcal{L}_{\omega_1\omega}$. Trans. Am. Math. Soc. **372**, 8955–8983 (2019)
2. Chen, R.: Étale structures and the Joyal-Tierney representation theorem in countable model theory. Bull. Symb. Logic (2025). arxiv:2310.11539
3. de Brecht, M.: Quasi-Polish spaces. Ann. Pure Appl. Logic **164**, 356–381 (2013)
4. de Brecht, M.: Some notes on spaces of ideals and computable topology. In: Anselmo, M., Della Vedova, G., Manea, F., Pauly, A. (eds.) CiE 2020. LNCS, vol. 12098, pp. 26–37. Springer, Cham (2020). https://doi.org/10.1007/978-3-030-51466-2_3
5. de Brecht, M.: The category of quasi-Polish spaces as a represented space. In: The 68th Topology Symposium of the Topology Section of the Mathematical Society of Japan (2021). https://www.mathsoc.jp/section/topology/topsymp.html
6. de Brecht, M., Goubault-Larrecq, J., Jia, X., Lyu, Z.: Domain-complete and LCS-complete spaces. Electron. Notes Theor. Comput. Sci. **345**, 3–35 (2019)
7. de Brecht, M., Kihara, T., Selivanov, V.: Ideal presentations and numberings of some classes of effective quasi-Polish spaces. Computability **13**, 325–348 (2024)
8. de Brecht, M., Pauly, A., Schröder, M.: Overt choice. Computability **9**, 169–191 (2020)
9. Escardó, M.: Synthetic topology of data types and classical spaces. Electron. Notes Theor. Comput. Sci. **87**, 21–156 (2004)
10. Franklin, J., Neumann, E., Pauly, A., Pradic, C., Valenti, M., Represented spaces of represented spaces. In: Beckmann, A., Oitavem, I., Manea, F. (eds.) CiE 2025. LNCS, vol. 15764, pp. 47–61. Springer, Cham (2025). https://doi.org/10.1007/978-3-031-95908-0_4
11. Harrison-Trainor, M., Melnikov, A., Miller, R., Montalbán, A.: Computable functors and effective interpretability. J. Symb. Log. **82**, 77–97 (2017)
12. Heckmann, R.: Spatiality of countably presentable locales (proved with the Baire category theorem). Math. Struct. in Comp. Science **25**, 1607–1625 (2015)
13. Hoyrup, M., Rojas, C., Selivanov, V., Stull, D.M.: Computability on quasi-polish spaces. In: Hospodár, M., Jirásková, G., Konstantinidis, S. (eds.) DCFS 2019. LNCS, vol. 11612, pp. 171–183. Springer, Cham (2019). https://doi.org/10.1007/978-3-030-23247-4_13
14. Jacobs, B.: Categorical Logic and Type Theory. Elsevier Science (1999)
15. Joyal, A., Tierney, M.: An extension of the Galois theory of Grothendieck. Memoirs of the American Mathematical Society, vol. 309 (1984)
16. Kihara, T.: Lawvere-Tierney topologies for computability theorists. Trans. Amer. Math. Soc. Ser. B **10**, 48–85 (2023)
17. Korovina, M., Kudinov, O.: On higher effective descriptive set theory. In: Kari, J., Manea, F., Petre, I. (eds.) CiE 2017. LNCS, vol. 10307, pp. 282–291. Springer, Cham (2017). https://doi.org/10.1007/978-3-319-58741-7_27
18. Lane, S.M., Moerdijk, I.: Sheaves in Geometry and Logic. Springer (1992)
19. Miller, R., Poonen, B., Schoutens, H., Shlapentokh, A.: A computable functor from graphs to fields. J. Symb. Log. **83**(1), 326–348 (2018)
20. Mummert, C.: On the reverse mathematics of general topology. Ph.D. thesis, Pennsylvania State University (2005)
21. Neumann, E., Pauly, A., Pradic, C., Valenti, M.: Computably discrete represented spaces. In: Beckmann, A., Oitavem, I., Manea, F. (eds) CiE 2025. LNCS, vol. 15764, pp. 349–364. Springer, Cham (2025). https://doi.org/10.1007/978-3-031-95908-0_25

22. Pauly, A.: On the topological aspects of the theory of represented spaces. Computability **5**(2), 159–180 (2016)
23. Schröder, M.: Extended admissibility. Theoret. Comput. Sci. **284**(2), 519–538 (2002)
24. Schröder, M.: Admissibly represented spaces and Qcb-spaces. In: Brattka, V., Hertling, P. (eds.) Handbook of Computability and Complexity in Analysis. TAC, pp. 305–346. Springer, Cham (2021). https://doi.org/10.1007/978-3-030-59234-9_9
25. Selivanov, V.: Towards the effective descriptive set theory. In: Beckmann, A., Mitrana, V., Soskova, M. (eds.) CiE 2015. LNCS, vol. 9136, pp. 324–333. Springer, Cham (2015). https://doi.org/10.1007/978-3-319-20028-6_33
26. Vickers, S.: Locales and toposes as spaces. In: Aiello, M., Pratt-Hartman, I., van Bentham, J. (eds.) Handbook of Spatial Logics, pp. 429–496. Springer, Heidelberg (2007). https://doi.org/10.1007/978-1-4020-5587-4_8
27. Weihrauch, K.: Computable Analysis. Springer, Cham (2000)

Placing Green Bridges Optimally for Robust Habitat Reconnection

Gero Ellmies and Till Fluschnik(✉)

Department of Computer Science, Algorithm Engineering Group, Humboldt-Universität zu Berlin, Berlin, Germany
{gero.ellmies,till.fluschnik}@hu-berlin.de

Abstract. We study the problem of robustly reconnecting habitats via the placement of green bridges at minimum total cost. Habitats are fragmented into patches and we seek to reconnect each habitat such that it remains connected even if any of its patches becomes unavailable. Formally, we are given an undirected graph with edge costs, a set of fixed green bridges represented as a subset of the graph's edges, a set of habitats represented as vertex subsets, and some budget. We decide whether there exists a subset of the graph's edges containing all fixed green bridges such that, for each habitat, the induced subgraph on the solution edges is 2-vertex-connected, and the total cost does not exceed the budget. We also study the 2-edge-connectivity variant, modeling the case where any single reconnecting green bridge may fail. We analyze the computational complexity of these problems, focusing on the boundary between NP-hardness and polynomial-time solvability when the maximum habitat size and maximum vertex degree are bounded by constants. We prove that for each constant maximum habitat size of at least four there exists a small constant maximum degree for which the problems are NP-hard, and complement this with polynomial-time algorithms yielding partial dichotomies for bounded habitat size and degree.

Keywords: Wildlife crossings · NP-hardness · 2-connectivity · Basic habitat graphs

1 Introduction

When reconnecting fragmented habitats via green bridges, one is concerned with how robustly a habitat is connected. Robustness here refers to the scenario in which either a habitat patch or a green bridge becomes unavailable. A habitat patch can become unavailable due to forest fires, flooding, or invasive species. A green bridge can become unavailable due to defects or maintenance. In graph-theoretic terms, the unavailability of habitat patches corresponds to vertex-connectivity, while the unavailability of green bridges corresponds to edge-connectivity. It is well-known that vertex-connectivity is stronger in the sense that any k-vertex-connected graph is also k-edge connected, while the converse does not hold. We investigate the cases of 2-vertex- and 2-edge-connectivity,

V. Brattka et al. (Eds.): CiE 2026, LNCS 16674, pp. 256–271, 2027.
https://doi.org/10.1007/978-3-032-31348-5_17

modelling the simplest case where a single failure does not disconnect the habitat. We present the case of 2-vertex-connectivity and address the consequences for the 2-edge-connectivity case mostly in a full version [5] due to the space limits.

We model the problem in line with the literature [8,12,15] via an undirected graph $G = (V, E)$. The habitat patches are represented by the vertex set V, and the edges in E represent possible green bridges. Since the cost of green bridges can differ, we equip each edge with a cost $c\colon E \to \mathbb{N}_0$. Moreover, to represent already built bridges, we provide a set $F_{\text{in}} \subseteq E$ of so-called *forced* edges. We next state our central problem. Herein, for a graph $G = (V, E)$, vertex subset $H \subseteq V$ and edge subset $F \subseteq E$, we denote by $G[H, F] = (H, \{\{u, v\} \in F \mid u, v \in H\})$ the graph with vertex set H and all edges from F with both endpoints in H.

Problem 1 (Robust Green Bridges Placement (RGBP)). **Given** an undirected graph $G = (V, E)$ with edge costs $c\colon E \to \mathbb{N}_0$, a set $F_{\text{in}} \subseteq E$, a set $\mathcal{H}$ of subsets of V, and an integer $k \in \mathbb{N}$ there exists a subset $F \subseteq E$ with $F_{\text{in}} \subseteq F$ and $\sum_{e \in F} c(e) \leq k$ such that for every $H \in \mathcal{H}$ we have that $G[H, F]$ is 2-vertex-connected

This work aims to study the boundary between NP-hardness and polynomial-time solvability when the maximum habitat size $\eta = \max_{H \in \mathcal{H}} |H|$ and maximum vertex degree Δ are bounded by constants. Concretely, for each fixed constant value of η and Δ, we wonder whether RGBP is polynomial-time solvable or NP-hard.

1.1 Our Contributions

By B-RGBP we denote the special case of RGBP with unit costs and no forced edges, i.e., where $c(e) = 1$ for all $e \in E$ and $F_{\text{in}} = \emptyset$ (hence, we drop c and F_{in} from the input). Our main contribution is the following (depicted in Fig. 1).

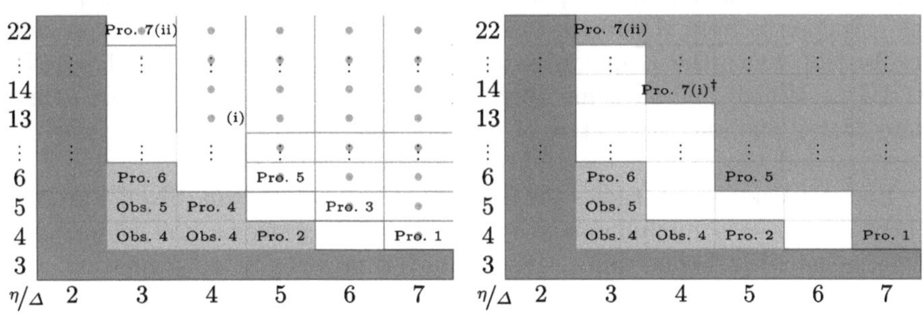

Fig. 1. Overview of our results for (left) RGBP and (right) the edge-connectivity version of RGBP. Green corresponds to polynomial-time solvability, red to NP-hardness even with unit costs and no forced edges. (†: explained in a full version [5])

Theorem 1. RGBP *is polynomial-time solvable if* $\Delta \leq 2$ *or* $\eta \leq 3$ *Moreover:*

(i) RGBP *is polynomial-time solvable if* $\eta \leq 4$ *and* $\Delta \leq 5$*, but* B-RGBP *is* NP*-hard even if* $\eta \geq 4$ *and* $\Delta \geq 7$*.*
(ii) RGBP *is polynomial-time solvable if* $\eta \leq 5$ *and* $\Delta \leq 4$*, but* B-RGBP *is* NP*-hard even if* $\eta \geq 5$ *and* $\Delta \geq 6$*.*
(iii) RGBP *is polynomial-time solvable if* $\eta \leq 6$ *and* $\Delta \leq 3$*, but* B-RGBP *is* NP*-hard even if* $\eta \geq 6$ *and* $\Delta \geq 5$*.*
(iv) B-RGBP *is* NP*-hard even if* $\eta \geq 13$ *and* $\Delta \geq 4$ *or if* $\eta \geq 22$ *and* $\Delta \geq 3$*.*

Because of the space limits, we present the case of size-four habitats (Theorem 1(i)) in more detail and sketch or defer most details of the other results (details to results marked with ★ are deferred to a full version [5]).

1.2 Related Work

Placing Green Bridges Optimally. Fluschnik and Kellerhals [8] introduced the problem of cost-optimal placement of green bridges under habitat constraints and studied d-Reach GBP (connectivity), d-Diam GBP (diameter d), and d-Closed GBP (clique-structure). Herkenrath et al. [12] focus on 1-Reach GBP with small habitats in degree bounded planar graphs, and provide experiments on real-world landscapes fragmented by highways. Wallisch et al. [15] study 2-Diam GBP with the same focus as ours, i.e., on small habitats and bounded degrees. For every constant Δ there exist a constant η_Δ such that 2-Diam GBP on graphs of maximum degree Δ is trivially polynomial-time solvable if a habitat has size at least η_Δ; we prove that this does not hold for RGBP.

Spanning 2-Connected Subgraph and Survivable Network Design. RGBP with only one habitat is equivalent to the Minimum 2-connected Subgraph (M2cS) problem, where one seeks to select a minimum-cost (spanning) edge subset from a graph while keeping it 2-connected. M2cS is NP-hard [9], and the reduction is from Hamiltonian Cycle (HC), which is NP-hard on cubic graphs [10]. Since the graph in reduction remains unchanged, it follows that RGBP is already NP-hard for only one habitat and maximum degree three. Note that if the only habitat is of constant size, then the problem is trivial. Cheriyan and Végh [2] proves a 6-approximation for the problem.

Ene and Vakilian [6] studies the Survivable Network Design (SNDP) problem where one seeks a minimum-cost subgraph fulfilling degree and connectivity requirements. For 2-edge-connectivity, Jain [14] proves a 2-approximation.

Community-Aware Network Sparsification. The green bridge placement framework of Fluschnik and Kellerhals [8] is closely related to community-aware network sparsification [1,11,13], where habitats correspond to communities and constraints include connectivity [1,11,13], star containment [11,13], and density [11]. Note that density constraints, even for large $\alpha < 1$, do not guarantee 2-connectivity. For a comprehensive overview, see [13]. Cohen et al. [3] study the cost-free clique case, implying polynomial-time solvability for $\eta = 4$ but NP-hardness for $\eta > 4$. In contrast, we consider bounded-degree graphs, complementing their clique-based setting.

2 Preliminaries

An undirected graph $G = (V, E)$ consists of a vertex set V and edge set $E \subseteq \binom{V}{2}$. We also write $V(G)$ and $E(G)$ to refer to the vertex and edge set of G, respectively. For a vertex subset $W \subseteq V$, we denote by $G[W] := (W, E \cap \binom{W}{2})$ the graph G induced by W. For a vertex subset $W \subseteq V$ or edge subset $E' \subseteq E$, we denote the deletion from G by $G - W := G[V \setminus W]$ and by $G - E' := (V, E \setminus E')$ (for singletons, we often write $G - v$ and $G - e$). By $N_G(v) = \{w \in V \mid \{v, w\} \in E\}$ we define the neighborhood of vertex $v \in V$. The degree of vertex v is $\deg_G(v) = |N_G(v)|$. The maximum degree of G is $\max_{v \in V} \deg_G(v)$. Two different vertices $s, t \in V$ are connected if there is a sequence $(v_1, \ldots, v_q)$ of pairwise different vertices in V with $s = v_1$ and $t = v_q$ such that $\{v_i, v_{i+1}\} \in E$ for all $i \in \{1, \ldots, q-1\}$. Such a sequence is called a *path* (a path on n vertices is also denoted by P_n). A graph is connected if every pair of vertices is connected. A subset $W \subseteq V$ forms a *(connected) component* in G if every pair of vertices in W is connected and no vertex $v \in V \setminus W$ is connected to a vertex in W. A graph is 2-vertex-connected (2-connected for short) if it has at least three vertices, and for every vertex $v \in V$, $G - v$ is connected. A graph is 2-edge-connected if it has at least two edges, and for every edge $e \in E$, $G - e$ is connected.

2.1 Decision Problems

We use the following Problem 2 and Problem 3, which are NP-complete [4,7], in our polynomial-time many-one reductions.

*Problem 2 (*Hamiltonian Cubic Vertex Cover*(HCVC)).* **Given** an undirected, cubic graph $G = (V, E)$, a Hamiltonian cycle C in G, and an integer $p \in \mathbb{N}$ there is vertex subset $W \subseteq V$ of size at most p such that every edge in E has an endpoint in W

Problem 3 ((2,2)-3 Satisfiability*((2,2)-3SAT)).* **Given** a set X of variables and a (2,2)-3-CNF formula $\phi = \bigwedge_{i=1}^{M} C_i$ of clauses over X, i.e., a 3-CNF formula of clauses over X each of size exactly three such that each variable appears exactly twice negated and twice unnegated there is a satisfying truth assignment $\alpha \colon X \to \{\top, \bot\}$

2.2 Preprocessing

To solve an instance of RGBP, it will be useful to preprocess the instance using the following Reduction Rules 1 to 5. We call an instance of RGBP *reduced* if and only if none of them is applicable. Since their correctness and polynomial-time applicability are immediate, the corresponding brief proofs are deferred to full version of the paper. Our first rule detects trivial no-instances.

Reduction Rule 1. *If $G[H]$ is not 2-connected for some $H \in \mathcal{H}$, return* **no**.

We can safely assume that there are no duplicated habitats.

Reduction Rule 2. *If there are $H, H' \in \mathcal{H}$ with $H = H'$, delete one of them.*

Every edge required for $G[H]$ to be 2-connected can be safely enforced.

Reduction Rule 3. *If for a habitat H there is an edge $e \in E(G[H])$ such that $G[H] - e$ is not 2-connected, then add e to F_{in}.*

Note that this implies that if for a habitat H the graph $G[H]$ contains a vertex v of degree two, then its two adjacent edges in $G[H]$ are added to F_{in}. In particular, all edges of a 2-connected size-3 habitat (which is a triangle) are forced.

Since $F_{\text{in}} \subseteq F$ for every solution F, we have:

Reduction Rule 4. *If $G[H, F_{\text{in}}]$ is 2-connected for some habitat H, delete H.*

An edge induced by no habitat can be deleted (budget update may be required).

Reduction Rule 5. *If there is an edge $e \in E$ not contained in $G[H]$ for every $H \in \mathcal{H}$, then delete e and if $e \in F_{\text{in}}$, then set $k := k - c(e)$.*

2.3 Basic Habitat Graph

Most of our polynomial-time algorithms rely on the following graph that captures habitat intersections in the input graph (see Fig. 2 for an example).

Fig. 2. Illustrative example to Definition 1. (Left) A graph G, a habitat set $\mathcal{H}$ with a blue, orange, magenta, and green size-4 habitat (contained vertices are encircled), and the set F_{in} of forced edges consisting of the thick/red edges. (Right) The corresponding basic habitat graph $\mathcal{G}_{G,\mathcal{H},F_{\text{in}}}$ with two components each being a path of length one.

Definition 1. *Given a graph $G = (V, E)$ with forced edges $F_{\text{in}} \subseteq E$ and a set $\mathcal{H}$ of habitats, the* basic habitat graph *$\mathcal{G}_{G,F_{\text{in}},\mathcal{H}} = (\mathcal{H}, Y)$ is the graph with vertex set $\mathcal{H}$ and edge set $Y = \{\{H, H'\} \mid (E(G[H]) \cap E(G[H'])) \setminus F_{\text{in}} \neq \emptyset\}$.*

Wallisch et al. [15] introduced the *habitat (intersection) graph*, which is similar to our basic habitat graph but also requires that no third habitat "lies in between" two adjacent habitats. We prove, in analogue to [15, Lemma 12], that the problem essentially decomposes across connected components of the habitat graph.

Observation 1 (★). *Let* $I = (G, c, F_{\text{in}}, \mathcal{H}, k)$ *be an instance of* RGBP. *Let* $C_1, \ldots, C_q$ *form the connected components in* $\mathcal{G}_{G,F_{\text{in}},\mathcal{H}}$. *For each* $p \in \{1, \ldots, q\}$, *let* $G_p := (\bigcup_{H \in C_p} H, \bigcup_{H \in C_p} E(G[H]))$. *Then,* F *is a minimum-cost solution for* I *if and only if* $F \cap E(G_p)$ *is a minimum cost solution for* $I_p := (G_p, c, F_{\text{in}} \cap E(G_p), C_p, k)$ *for every* $p \in \{1, \ldots, q\}$.

Next we show that if the basic habitat graph of an instance is a path or a cycle, we can use dynamic programming to solve it. The dynamic program essentially enumerates all feasible edge sets for each habitat and combines them with solutions of its at most two neighboring habitats.

Lemma 1 (★). *Let* $I = (G, c, F_{\text{in}}, \mathcal{H}, k)$ *be an instance of* RGBP *with* $\mathcal{G}_{G,F_{\text{in}},\mathcal{H}}$ *being (i) a path or (ii) a cycle. Then, we can either report that* I *is a* no*-instance or output a minimum-cost solution* F *for* I *in (i)* $O(2^{\eta \cdot \Delta} \cdot \text{poly}(|I|))$ *time or (ii)* $O(2^{3\eta \cdot \Delta/2} \cdot \text{poly}(|I|))$ *time.*[1]

We also introduce the following notation that will simplify our proofs. For a set $\mathcal{H}' \subseteq \mathcal{H}$, we denote by $\partial_{\mathcal{G}_{G,F_{\text{in}},\mathcal{H}}}(\mathcal{H}')$ the set of all habitats $H \in \mathcal{H} \setminus \mathcal{H}'$ with $H \cap \bigcup_{H' \in \mathcal{H}'} H' \neq \emptyset$, $H \setminus \bigcup_{H' \in \mathcal{H}'} H' \neq \emptyset$, and $\{H, H'\} \in \mathcal{G}_{G,F_{\text{in}},\mathcal{H}}$ for some $H' \in \mathcal{H}'$. Intuitively, $\partial_{\mathcal{G}_{G,F_{\text{in}},\mathcal{H}}}(\mathcal{H}')$ denotes all habitats that lie directly around $\mathcal{H}'$ in $\mathcal{G}_{G,F_{\text{in}},\mathcal{H}}$, none of which are part of $\mathcal{H}'$. We often write $\partial(\mathcal{H}')$ when the subscript is clear from the context.

3 Habitats of Size Four

We first settle that for the case of size-4 habitats, we need not distinguish between 2-vertex- and 2-edge-connectivity.

Observation 2 (★). *For every graph* G *on four vertices, it holds that* G *is 2-connected if and only if it is 2-edge-connected.*

Proposition 1. B-RGBP *is* NP*-hard even if* $\eta = 4$ *and* $\Delta \geq 7$.

We give a polynomial-time many-one reduction from HCVC (Problem 2).

Intuition 1. Each vertex of the vertex cover instance is represented by a K_4-habitat. All but three edges are forced into every solution (via auxiliary habitats), leaving a choice between selecting either one or two of the remaining edges to achieve 2-connectivity. Selecting two edges corresponds to choosing the vertex into the vertex cover; selecting only one corresponds to excluding it. If two edges are selected, then in the incident edge gadget only two further edges are needed to make each of its three K_4-habitats 2-connected.

Construction 1. Let $I = (G, p, C)$ be an instance of HCVC with $G = (V, E)$ and let $V = \{v_1, \ldots, v_n\}$ be enumerated such that $C = (V, \bigcup_{i=1}^{n-1} \{\{v_i, v_{i+1}\}\} \cup \{\{v_n, v_1\}\})$ forms a Hamiltonian cycle in G, and let $m := |E|$. Let $X = \{(i, j) \mid \{v_i, v_j\} \in E \wedge i < j\} \setminus \{(1, n)\} \cup \{(n, 1)\}$ (note that $|X| = |E|$). Construct an

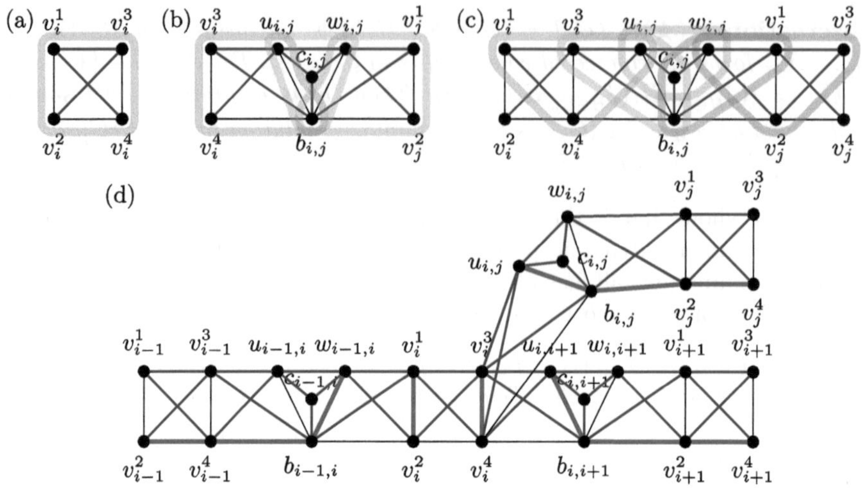

Fig. 3. Illustration to Construction 1. (a) V_i and habitat H_i (cyan). (b) $V_{i,j}$ with its neighbors in V_i and V_j, as well as the habitats $H_{i,j}$ (orange), $H'_{i,j}$ (cyan), and $H''_{i,j}$ (magenta). (c) V_i, V_j, and $V_{i,j}$, as well as the habitats $H^1_{i,j}$ (cyan), $H^2_{i,j}$ (orange), $H^3_{i,j}$ (orange), $H^4_{i,j}$ (teal), and $H^5_{i,j}$ (blue). (d) An excerpt from G' around V_i, where green edges indicate additional edges in a solution. (Every red edge is contained in E'_2).

instance $I' = (G', \mathcal{H}, k)$ with $G' = (V', E')$ of B-RGBP as follows (see Fig. 3 for an illustration).

Let $V' = \bigcup_{i=1}^n V_i \cup \bigcup_{(i,j)\in X} V_{i,j}$ with $V_i = \{v_i^1, v_i^2, v_i^3, v_i^4\}$ and $V_{i,j} = \{u_{i,j}, w_{i,j}, c_{i,j}, b_{i,j}\}$. Next, construct E'. For each $i \in \{1,\dots,n\}$, let V_i form a clique, i.e., for each $x, y \in V_i$, add the edge $\{x, y\}$. Moreover, for each $(i,j) \in X$, let each of $V_{i,j}$, $\{v_i^3, v_i^4, u_{i,j}, b_{i,j}\}$, and $\{v_j^1, v_j^2, w_{i,j}, b_{i,j}\}$ form a clique. This finishes the construction of G'. Denote by $E'_1 := \bigcup_{i=1}^n E_i^* \cup \bigcup_{(i,j)\in X} E^*_{i,j}$, where $E_i^* := \{\{v_i^1, v_i^2\}, \{v_i^2, v_i^4\}, \{v_i^3, v_i^4\}\}$ for each $i \in \{1,\dots,n\}$ and $E^*_{i,j} := \{\{v_i^4, b_{i,j}\}, \{b_{i,j}, v_j^2\}, \{u_{i,j}, b_{i,j}\}, \{w_{i,j}, b_{i,j}\}\}$ for each $(i,j) \in X$. Denote by $E'_2 := E' \setminus E'_1$. We introduce this notation to distinguish between the edges E'_1 a solution can select from, and the edges E'_2 which are fully contained in every solution. Construct the habitat set $\mathcal{H}$ as follows. Add the habitat $H_i = V_i$ for each $i \in \{1,\dots,n\}$ (intuitively, H_i encodes whether v_i is part of the vertex cover). For each $(i,j) \in X$, add the habitats $H_{i,j} = V_{i,j}$, $H'_{i,j} = \{v_i^3, u_{i,j}, b_{i,j}, v_i^4\}$, and $H''_{i,j} = \{v_j^1, w_{i,j}, b_{i,j}, v_j^2\}$ (intuitively, these three habitats encode that edge $\{v_i, v_j\}$ must be covered). Moreover, for each $(i,j) \in X$, add the auxiliary habitats $H^1_{i,j} = \{v_i^1, v_i^3, u_{i,j}, v_i^4\}$, $H^2_{i,j} = \{v_i^3, u_{i,j}, c_{i,j}, b_{i,j}\}$, $H^3_{i,j} = \{u_{i,j}, c_{i,j}, w_{i,j}\}$, $H^4_{i,j} = \{v_j^1, w_{i,j}, c_{i,j}, b_{i,j}\}$, and $H^5_{i,j} = \{v_j^1, v_j^3, w_{i,j}, v_j^2\}$. Set $k = |E'_2| + 2m + n + p$. ◇

The following two statements can be directly derived from the construction.

[1] We denote by poly($|I|$) a polynomial in the input size $|I|$.

Observation 3 (★). *The maximum degree of G' is seven.*

Lemma 2 (★). *Let I' be a* yes*-instance. Then every solution contains E'_2.*

Next, we show that in case of a yes-instance, there is always a solution that matches our Intuition 1; the proof essentially uses edge switching arguments.

Lemma 3 (★). *Let I' be a* yes*-instance. Then there is a solution F where (i) for every $(i,j) \in X$, F contains exactly two edges in $E^*_{i,j}$, and (ii) for every $i \in \{1,\dots,n\}$, F contains no two adjacent edges in E^*_i.*

Proof (of Proposition 1). Let I' be the instance obtained from input instance $I = (G, p, C)$ of HCVC via Construction 1 in polynomial-time. We prove that I is a yes-instance if and only if I' is a yes-instance.

($\Rightarrow$) Let $W \subseteq V$ be a vertex cover of size p. Let $\mathcal{I}_W := \{i \in \{1,\dots,n\} \mid v_i \in W\}$ and $\mathcal{J}_W := \{1,\dots,n\} \setminus I_W$. Let $F' := \{\{v_i^1, v_i^2\}, \{v_i^3, v_i^4\} \mid i \in \mathcal{I}_W\} \cup \{\{v_i^2, v_i^4\} \mid i \in \mathcal{J}_W\}$ and $F'' := \bigcup_{(i,j)\in X} F_{i,j}$, where $F_{i,j} := \{\{u_{i,j}, b_{i,j}\}, \{b_{i,j}, v_j^2\}\}$ if $i \in \mathcal{I}_W$, and $F_{i,j} := \{\{w_{i,j}, b_{i,j}\}, \{b_{i,j}, v_i^4\}\}$ if $i \in \mathcal{J}_W$. Let $F = E'_2 \cup F' \cup F''$. Note that $|F| = |E'_2| + 2m + n + p$. We claim that F is a solution to I'. For every $(i,j) \in X$, we have that $G[H^x_{i,j}, F]$ is 2-connected for every $x \in \{1,\dots,5\}$ since $E'_2 \subseteq F$. Also, it holds that $G[H_{i,j}, F]$ is 2-connected since either $\{u_{i,j}, b_{i,j}\}$ or $\{w_{i,j}, b_{i,j}\}$ is contained in $F_{i,j}$. By the definitions of F' and $F_{i,j}$, we also have that $G[H'_{i,j}, F]$ is 2-connected since either $\{b_{i,j}, v_i^4\}$ or $\{\{u_{i,j}, b_{i,j}\}, \{v_i^3, v_i^4\}\}$ is contained in F. Moreover, for every $i \in \{1,\dots,n\}$, it holds that $G[H_i, F]$ is 2-connected. Suppose towards a contradiction that there is $(i,j) \in X$ such that $G[H''_{i,j}, F]$ is not 2-connected. Then, neither $\{b_{i,j}, v_j^2\}$ nor $\{\{w_{i,j}, b_{i,j}\}, \{v_j^1, v_j^2\}\}$ is contained in F. This implies that $i \in \mathcal{J}_W$ and that $j \in \mathcal{J}_W$. Hence, edge $\{v_i, v_j\} \in E$ is not covered by W, a contradiction.

($\Leftarrow$) Let F be a solution where for every $(i,j) \in X$, F contains exactly two edges in $E^*_{i,j}$ (let $F'' \subseteq F$ denote the set of all these edges), and for every $i \in \{1,\dots,n\}$, F contains no two adjacent edges in E^*_i (which exists due to Lemma 3). We know that $E'_2 \subseteq F$ (see Lemma 2). We claim that $W = \{v_i \in V \mid |F \cap E^*_i| = 2\}$ is a vertex cover of G. Suppose towards a contradiction that there is an $(i,j) \in X$ such that $\{v_i, v_j\} \cap W = \emptyset$. Thus, $F \cap E^*_i = \{v_i^2, v_i^4\}$ and $F \cap E^*_j = \{v_j^2, v_j^4\}$. Since $G[H'_{i,j}, F]$ and $G[H''_{i,j}, F]$ are 2-connected, F contains both $\{v_i^4, b_{i,j}\}$ and $\{v_j^2, b_{i,j}\}$. Since F contains exactly two edges in $E^*_{i,j}$, $G[H_{i,j}, F]$ is not 2-connected; contradicting the fact that F is a solution. Finally, we have that $k \geq |F| = |E'_2| + |F''| + |\{i \in \{1,\dots,n\} \mid |F \cap E^*_i| = 1\}| + 2 \cdot |\{i \in \{1,\dots,n\} \mid |F \cap E^*_i| = 2\}| = |E'_2| + 2m + n + |W|$. Hence, $p \geq |W|$. □

If I is reduced, then every habitat induces a K_4. In a K_4, every vertex has degree 3; hence, we get the following.

Observation 4 (★). RGBP *is polynomial-time solvable if $\eta = 4$ and $\Delta \leq 4$.*

Proposition 2. RGBP *is polynomial-time solvable if $\eta = 4$ and $\Delta = 5$.*

We first show the following.

Lemma 4. *Let $I = (G, c, F_{\text{in}}, \mathcal{H}, k)$ be reduced. If there is $\{H, H'\} \in E(\mathcal{G}_{G,F_{\text{in}},\mathcal{H}})$ such that $G[H \cap H']$ is a triangle, then the union of all habitats in the component containing H and H' in $\mathcal{G}_{G,F_{\text{in}},\mathcal{H}}$ is of cardinality six.*

Proof. Let H and H' intersect in u, v, w and let $x \in H \setminus H'$ and $x' \in H' \setminus H$. Note that $\deg_{G[H \cup H']}(y) = 4$ for each $y \in \{u, v, w\}$. We distinguish how a third habitat $H'' \in \partial(\{H, H'\})$ intersects $H \cup H'$ (see Fig. 4).

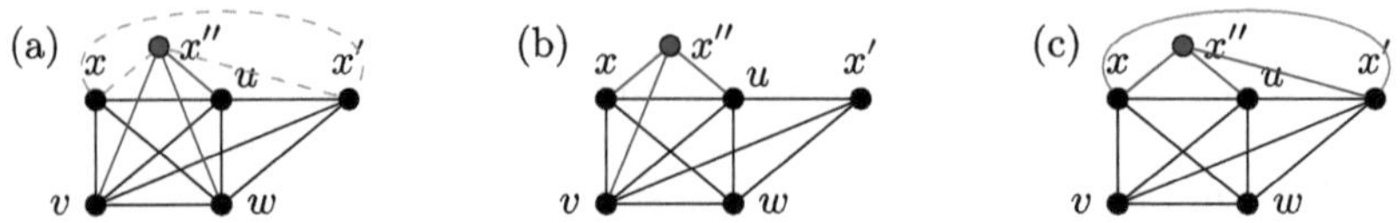

Fig. 4. Illustration to the three cases (a), (b), and (c) in the proof of Lemma 4.

Note that H'' cannot be intersecting and have two vertices $x'', y'' \in H'' \setminus (H \cup H')$ since H and H' have no adjacent vertices of degree at most three. Let $\{x''\} = H'' \setminus (H \cup H')$.

Case (a): $H \cap H'' = \{u, v, w\}$ Then, $\deg_{G[H \cup H' \cup H'']}(y) = 5$ for each $y \in \{u, v, w\}$. We claim that $\partial(\{H, H', H''\}) = \emptyset$. Suppose that there is $H^* \in \partial(\{H, H', H''\})$ with $z \in H^* \setminus (H \cup H' \cup H'')$. Vertex z cannot be adjacent with all of x, x', x'' in $G[H^*]$, since either two of them are not adjacent, or they have degree six. Hence, H^* must contain a vertex from $\{u, v, w\}$, none of which is adjacent with z; contradicting the fact that I is reduced.

Case (b): H'' contains x and two vertices from $\{u, v, w\}$ W.l.o.g. let $H \cap H'' = \{x, u, v\}$. Then, $\deg_{G[H \cup H' \cup H'']}(x) = 4$ and $\deg_{G[H \cup H' \cup H'']}(y) = 5$ for each $y \in \{v, u\}$. Overall, in $G[H \cup H' \cup H'']$ only the vertices x, x', w, x'' can be of degree smaller than five, where x', x'' are of degree at least 3, and x, w are of degree at least 4. We claim that $\partial(\{H, H', H''\}) = \emptyset$. Suppose towards a contradiction that there is $H^* \in \partial(\{H, H', H''\})$. Note that $|H^* \setminus (H \cup H' \cup H'')| < 2$ since $G[H^*]$ must contain an edge from $G[H]$, $G[H']$, or $G[H'']$. Let $\{z\} = H^* \setminus (H \cup H' \cup H'')$. Then, H^* must contain a vertex $a \in \{x', x''\}$ and one vertex $b \in \{x, w\}$ where $\{a, b\} \notin E(G[H \cup H' \cup H''])$; contradicting the fact that I is reduced.

Case (c): H'' contains x, x' and one vertex from $\{u, v, w\}$ In this case, we have $\{x, x'\} \in E$ since otherwise H'' does not form a K_4. W.l.o.g. let $u = H'' \cap \{u, v, w\}$. Then, only the vertices in $\{x'', v, w\}$ can have degree at most four in $G[H \cup H' \cup H'']$. We claim that $\partial(\{H, H', H''\}) = \emptyset$. Suppose towards a contradiction that there is $H^* \in \partial(\{H, H', H''\})$. Note that $|H^* \setminus (H \cup H' \cup H'')| < 2$ since x'' is the only vertex in $G[H \cup H' \cup H'']$ that can have degree three. Let $z \in H^* \setminus (H \cup H' \cup H'')$. Then, H^* must contain $\{v, w\}$, each of which is then of degree two in $G[H^*]$; contradicting the fact that I is reduced. ☐

Lemma 4 implies that the following is correct and executable in polynomial-time.

Reduction Rule 6. *Let $\{H, H'\} \in E(\mathcal{G}_{G,F_{\text{in}},\mathcal{H}})$ such that $G[H \cap H']$ is a triangle. Let $\mathcal{H}'$ be the set of all habitats in the connected component in $\mathcal{G}_{G,F_{\text{in}},\mathcal{H}}$ containing H and H', and let $V' = \bigcup_{H'' \in \mathcal{H}'} H''$. Decide instance $I' = (G[V'], c, F_{\text{in}} \cap E(G[V']), \mathcal{H}', k)$. If I' is a* `no`*-instance, then output a trivial* `no`*-instance. If I' is a* `yes`*-instance, then compute a minimum-cost solution F for I', remove $\mathcal{H}'$ from $\mathcal{H}$, and add F to F_{in}.*

Lemma 5 (★). *If $I = (G, c, F_{\text{in}}, \mathcal{H}, k)$ is reduced and Reduction Rule 6 is inapplicable, then every connected component in $\mathcal{G}_{G,F_{\text{in}},\mathcal{H}}$ is a path or a cycle.*

Proof (of Proposition 2). Let $I = (G, c, F_{\text{in}}, \mathcal{H}, k)$ be a reduced instance such that Reduction Rule 6 is inapplicable. Compute the basic habitat graph $\mathcal{G}_{G,F_{\text{in}},\mathcal{H}}$ in polynomial time. Let $\mathcal{C}$ denote the set of all connected components in $\mathcal{G}_{G,F_{\text{in}},\mathcal{H}}$. Due to Lemma 5, we know that each component is a path or a cycle. For each connected component $C \in \mathcal{C}$, let $\mathcal{H}_C$ denote the set of all habitats in C and $G_C = (\bigcup_{H \in \mathcal{H}_C} H, \bigcup_{H \in \mathcal{H}_C} E(G[H]))$. Let $I_C = (G_C, c, F_{\text{in}} \cap E(G_C), \mathcal{H}_C, k)$ denote the subinstance of I restricted to C. Due to Lemma 1, we can compute a minimum-cost solution F_C for I_C in polynomial time (or report that I is a `no`-instance). Then, let $F = F_{\text{in}} \cup \bigcup_{C \in \mathcal{C}} F_C$. Output `yes` if $c(F) \leq k$ (which is correct due to Observation 1), and `no` otherwise. □

4 Habitats of Size Five

Proposition 3 (★). B-RGBP *is* NP*-hard even if $\eta = 5$ and $\Delta \geq 6$.*

We give a polynomial-time many-one reduction from (2,2)-3SAT (Problem 3).

Intuition 2. Each variable is represented by a vertex gadget inducing a wheel W_5 (a size 4 cycle C_4 with an additional *center* vertex adjacent to all cycle vertices), and each clause by three overlapping W_5-habitats sharing the same center. A variable gadget is connected via an additional W_5-habitat to each clause gadget in which it appears. In every W_5-habitat, all but two non-adjacent edges can be forced, so that selecting either of them suffices for 2-connectivity. For a variable gadget, this choice encodes a truth assignment. For a clause gadget, a non-selected edge corresponds to a satisfied literal, and ensures that the connecting W_5-habitat is already 2-connected via the choice made in the variable gadget.

Construction 2. Let $(X, \phi = \bigwedge_{i=1}^{M} C_i)$ be an instance of (2,2)-3SAT with $X = \{x_1, \ldots, x_N\}$. Construct an instance $I' = (G, \mathcal{H}, k)$ with $G = (V, E)$ of B-RGBP as follows (see Fig. 5 for an illustration).

First, construct the graph G. For each $i \in \{1, \ldots, N\}$, add the vertex set $V_i = \{v_i\} \cup \{t_i, t_i^1, t_i^2\} \cup \{f_i, f_i^1, f_i^2\}$, and the edge set $E_{x_i} = E'_{x_i} \cup E''_{x_i} \cup E^*_{x_i}$ with $E'_{x_i} = \{\{v_i, z\} \mid z \in \{t_i^1, t_i^2, f_i^1, f_i^2\}\} \cup \{\{t_i^1, f_i^1\}, \{t_i^2, f_i^2\}\}$, $E''_{x_i} = \{\{t_i, t_i^1\}, \{t_i, t_i^2\}, \{f_i, f_i^1\}, \{f_i, f_i^2\}\}$, and edge set $E^*_{x_i} = \{\{t_i^1, t_i^2\}, \{f_i^1, f_i^2\}\}$ (intuitively, a solution selects exactly one of $E^*_{x_i}$, encoding whether x_i is set to true, selecting $\{t_i^1, t_i^2\}$, or to false, selecting $\{f_i^1, f_i^2\}$). For each $j \in \{1, \ldots, M\}$, add the vertex set $W_j = \{w_j\} \cup U_j$ with $U_j = \{u_{j,p}^q \mid p \in$

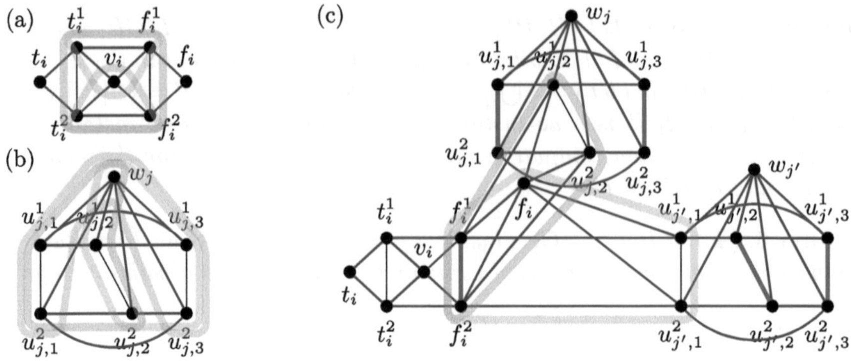

Fig. 5. Illustration to Construction 2 (red edges are in every solution). (a) V_{x_i} and the habitats $H_{x_i}^1$ (orange), $H_{x_i}^2$ (magenta), and H_{x_i} (cyan). (b) W_j and the habitats $H_{C_j}^{-1}$ (orange), $H_{C_j}^{-2}$ (magenta), and $H_{C_j}^{-3}$ (cyan) (c) An excerpt around V_i according to x_i's negated appearances with habitats H_{x_i,C_j} (cyan) and $H_{x_i,C_{j'}}$ (orange). Green edges indicate a part of a solution that corresponds to setting x_i to false.

$\{1,2,3\}, q \in \{1,2\}\}$ and the edge set $E_{C_j} = E'_{C_j} \cup E''_{C_j} \cup E^*_{C_j}$ with $E'_{C_j} = \{\{w_j, u\} \mid u \in U_j\}$, $E''_{C_j} = \bigcup_{q=1}^{2} \{\{u_{j,1}^q, u_{j,2}^q\}, \{u_{j,2}^q, u_{j,3}^q\}, \{u_{j,1}^q, u_{j,3}^q\}\}$, and $E^*_{C_j} = \{\{u_{j,p}^1, u_{j,p}^2\} \mid p \in \{1,2,3\}\}$ (intuitively, a solution selects exactly two from $E^*_{C_j}$, leaving out the one for which the literal is evaluated to true). If variable x_i appears unnegated in C_j as the p-th literal, $p \in \{1,2,3\}$, then add the edge set $E_{x_i,C_j} = \{\{u_{j,p}^1, t_i\}, \{u_{j,p}^2, t_i\}, \{u_{j,p}^1, t_i^1\}, \{u_{j,p}^2, t_i^2\}\}$. If variable x_i appears negated in C_j as the p-th literal, $p \in \{1,2,3\}$, then add the edge set $E_{x_i,C_j} = \{\{u_{j,p}^1, f_i\}, \{u_{j,p}^2, f_i\}, \{u_{j,p}^1, f_i^1\}, \{u_{j,p}^2, f_i^2\}\}$. Next, we construct the habitat set $\mathcal{H}$. For each $i \in \{1,\ldots,N\}$, add the habitats $H_{x_i}^1 = \{v_i, t_i^1, f_i^1\}$, $H_{x_i}^2 = \{v_i, t_i^2, f_i^2\}$, and $H_{x_i} = H_{x_i}^1 \cup H_{x_i}^2$ (intuitively, H_{x_i} encodes the truth assignment of x_i). For each $j \in \{1,\ldots,M\}$, and each $p \in \{1,2,3\}$, add the habitat $H_{C_j}^{-p} = W_j \setminus \{u_{j,p}^1, u_{j,p}^2\}$ (intuitively, $H_{C_j}^{-1}$, $H_{C_j}^{-2}$, and $H_{C_j}^{-3}$ together encode whether and which literal of C_j is evaluated to true), and for each $p' \in \{1,2,3\}$ with $p' > p$ and each $\ell \in \{1,2\}$, add $H_{C_j,\ell}^{p,p'} = \{w_j, u_{j,p}^\ell, u_{j,p'}^\ell\}$. If variable x_i appears unnegated in C_j as the p-th literal, $p \in \{1,2,3\}$, then add the habitats $H_{x_i,C_j} = \{u_{j,p}^1, u_{j,p}^2, t_i, t_i^1, t_i^2\}$, $H'_{x_i,C_j} = \{u_{j,p}^1, t_i, t_i^1\}$, and $H''_{x_i,C_j} = \{u_{j,p}^2, t_i, t_i^2\}$. If variable x_i appears negated in C_j as the p-th literal, $p \in \{1,2,3\}$, then add the habitats $H_{x_i,C_j} = \{u_{j,p}^1, u_{j,p}^2, f_i, f_i^1, f_i^2\}$, $H'_{x_i,C_j} = \{u_{j,p}^1, f_i, f_i^1\}$, and $H''_{x_i,C_j} = \{u_{j,p}^2, f_i, f_i^2\}$. (Intuitively, H_{x_i,C_j} makes sure that the truth assignment is correctly propagated to the clause gadget.) Set $k = 27N + 14M$. ◇

Figure 6 depicts all pairwise non-isomorphic 2-connected size-5 habitats with degree at most 3. In a reduced instance, only size-5 habitats isomorphic to (b) can appear. Since all but x_1 have degree 3, no other habitat can intersect in an edge. Thus, each component of the basic habitat graph is small (cf. Observation 4).

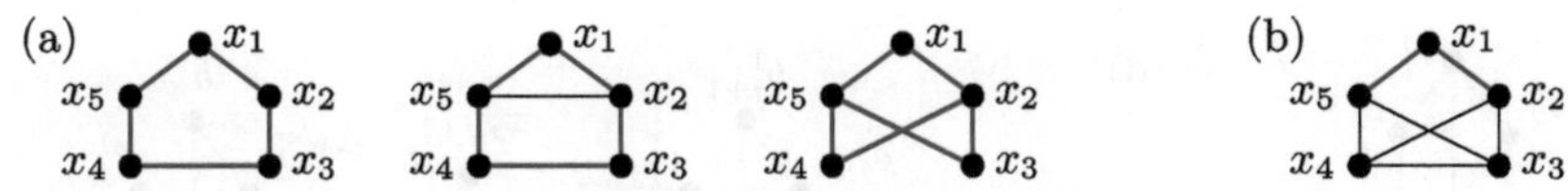

Fig. 6. All pairwise non-isomorphic 2-vertex- and 2-edge-connected habitats of size five with degree at most three. Red edges are forced in every reduced instance. (Color figure online)

Observation 5 (★). RGBP *is polynomial-time solvable if* $\eta = 5$ *and* $\Delta = 3$.

For $\Delta = 4$, the proof is more involved, consisting of an exhaustive case distinction over all possible ways a size-5 habitat can intersect with another habitat.

Proposition 4 (★). RGBP *is polynomial-time solvable if* $\eta = 5$ *and* $\Delta = 4$.

5 Habitats of Size Six

Proposition 5 (★). B-RGBP *is* NP*-hard even if* $\eta = 6$ *and* $\Delta \geq 5$.

We give a polynomial-time many-one reduction from HCVC (Problem 3). The intuition for the reduction is similar to Intuition 1.

Construction 3. Let $I = (G, p, C)$ be an instance of HCVC with $G = (V, E)$ and let $V = \{v_1, \ldots, v_n\}$ be enumerated such that $C = (V, \bigcup_{i=1}^{n-1} \{\{v_i, v_{i+1}\}\} \cup \{\{v_n, v_1\}\})$ forms a Hamiltonian cycle in G, and let $m := |E|$. Let $X = X' \cup X''$ with $X' = \{(i, i+1) \mid i \in \{1, \ldots, n-1\}\} \cup \{(n, 1)\}$ and $X'' = \{(i, j) \mid \{v_i, v_j\} \in E \wedge i < j\} \setminus (X' \cup \{(1, n)\})$ (note that $|X| = |E|$). Construct an instance $I' = (G', \mathcal{H}, k)$ with $G' = (V', E')$ of B-RGBP as follows (see Fig. 7 for an illustration).

For each $i \in \{1, \ldots, n\}$, add the vertex set $V_i = \{t_i^z, b_i^z, \ell_i^z \mid z \in \{1, 2\}\}$ and the edge set $E_i = E'_i \cup E_i^{\text{in}} \cup E_i^{\text{out}}$, where $E'_i = \{\{\ell_i^2, t_i^1\}, \{\ell_i^1, b_i^1\}, \{t_i^1, b_i^1\}, \{b_i^2, t_i^2\}\}$, $E_i^{\text{in}} = \{\{t_i^1, \ell_i^2\}, \{b_i^2, b_i^1\}, \{\ell_i^1, \ell_i^2\}\}$ and $E_i^{\text{out}} = \{\{\ell_i^1, b_i^2\}, \{\ell_i^2, t_i^2\}\}$. For each $(i, j) \in X'$, add two vertices $y_{i,j}$ and $z_{i,j}$ and the edge set $E_{i,j} = \{\{y_{i,j}, z_{i,j}\}\} \cup E'_{i,j}$ with $E'_{i,j} := \{\{b_i^1, \ell_j^1\}, \{b_i^1, z_{i,j}\}, \{z_{i,j}, \ell_j^1\}\} \cup \{\{b_i^2, \ell_j^2\}, \{b_i^2, y_{i,j}\}, \{y_{i,j}, \ell_j^2\}\}$. For each $(i, j) \in X''$, add three vertices $y_{i,j}, z_{i,j}, d_{i,j}$ and the edge set $E_{i,j} = E'_{i,j} \cup \{\{y_{i,j}, z_{i,j}\}, \{y_{i,j}, d_{i,j}\}, \{d_{i,j}, z_{i,j}\}\}$ with $E'_{i,j} := \{\{t_i^1, t_j^2\}, \{t_i^1, y_{i,j}\}, \{y_{i,j}, t_j^2\}\} \cup \{\{t_i^2, t_j^1\}, \{t_i^2, z_{i,j}\}, \{z_{i,j}, t_j^1\}\}$. Add the edges $\{z_{i,j}, y_{i,i+1}\}$ and $\{y_{i,j}, y_{j,j'}\}$, where $j' = j + 1$ if $j < n$, and $j' = 1$ otherwise. For each $i \in \{1, \ldots, n\}$, add the habitat $H_i = V_i$. For each $(i, j) \in X'$, add the habitats $H_{i,j}^y = \{b_i^2, y_{i,j}, \ell_j^2\}$, $H_{i,j}^z = \{b_i^1, z_{i,j}, \ell_j^1\}$, and $H_{i,j} = H_{i,j}^y \cup H_{i,j}^z$, as well as $H_j^\dagger = \{y_{i,j}, z_{i,j}, \ell_j^1, b_j^1, t_j^1, \ell_j^2\}$. For each $(i, j) \in X''$, add $H_{i,j}^y = \{t_i^1, y_{i,j}, t_j^2\}$, $H_{i,j}^z = \{t_i^2, z_{i,j}, t_j^1\}$, $H_{i,j}^d = \{y_{i,j}, z_{i,j}, d_{i,j}\}$ and $H_{i,j} = H_{i,j}^y \cup H_{i,j}^z$, as well as $H_i^\ddagger = \{t_i^2, z_{i,j}, y_{i,i+1}, b_i^2\}$ and $H_j^\ddagger = \{t_j^2, y_{i,j}, y_{j,j'}, b_j^2\}$, where $j' = j + 1$ if $j < n$, and $j' = 1$ otherwise. Set $k = (7n + \frac{11}{2}n) + (4n + 2n + p)$. ◇

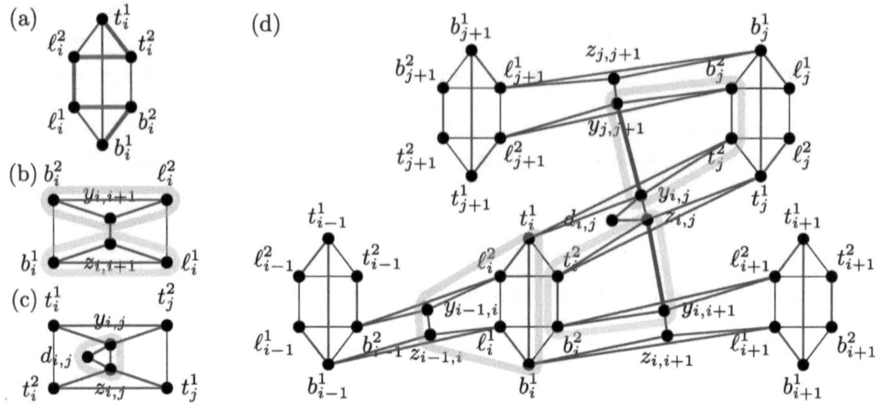

Fig. 7. Illustration to Construction 3 (red edges are in every solution). (a) The two minimal solutions for H_i (either green or blue). (b) Habitats $H^y_{i,j}$ (cyan) and $H^z_{i,j}$ (orange) for some $(i,j) \in X'$. (c) Habitat $H^d_{i,j}$ (magenta) for some $(i,j) \in X''$. (d) An excerpt from G' around V_i, with habitats $H^\dagger_i$ (cyan), $H^\ddagger_i$ (orange), and $H^\ddagger_j$ (magenta).

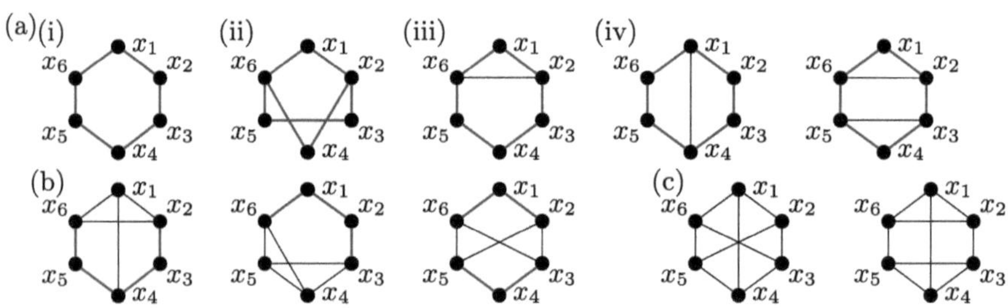

Fig. 8. All pairwise non-isomorphic 2-vertex- and 2-edge-connected habitats of size six with degree at most three. Red edges are forced in every reduced instance. (Color figure online)

Figure 8 shows all pairwise non-isomorphic 2-vertex- and 2-edge-connected size-6 habitats of degree at most 3.

In a reduced instance, no habitat in (a) appears. Thus, the habitats in (b) and (c) remain. We call them *(6,3)-habitats.*

Proposition 6 (★). RGBP *is polynomial-time solvable if* $\eta = 6$ *and* $\Delta = 3$.

To prove Proposition 6, we essentially combine Observation 1 and 5 with the following.

Lemma 6 (★). *Let* $I = (G, c, F_{\text{in}}, \mathcal{H}, k)$ *be reduced and* $H \in \mathcal{H}$ *be a (6,3)-habitat. If* $\{H, H'\} \in E(\mathcal{G}_{G,F_{\text{in}},\mathcal{H}})$ *for some* $H' \in \mathcal{H}$*, then* $H' \subseteq H$.

6 Degree Three and Four

Proposition 7 (★). B-RGBP *is* NP*-hard even if (i)* $\eta = 13$ *and* $\Delta \geq 4$*, or (ii)* $\eta = 22$ *and* $\Delta \geq 3$*.*

Both reductions follow the same conceptional idea by Herkenrath et al. [12, Constr. 3]. We only present the construction to Proposition 7(ii) due to the space constraints. We give a polynomial-time many-one reduction from HCVC (Problem 2).

Construction 4. Let $I = (G, p, C)$ be an instance of HCVC with $G = (V, E)$ and let $V = \{v_1, \ldots, v_n\}$ be enumerated such that $C = (V, \bigcup_{i=1}^{n-1}\{\{v_i, v_{i+1}\}\} \cup \{\{v_n, v_1\}\})$ forms a Hamiltonian cycle in G. Construct an instance $I' = (G', \mathcal{H}, k)$ of B-RGBP as follows (see Fig. 9 for an illustration).

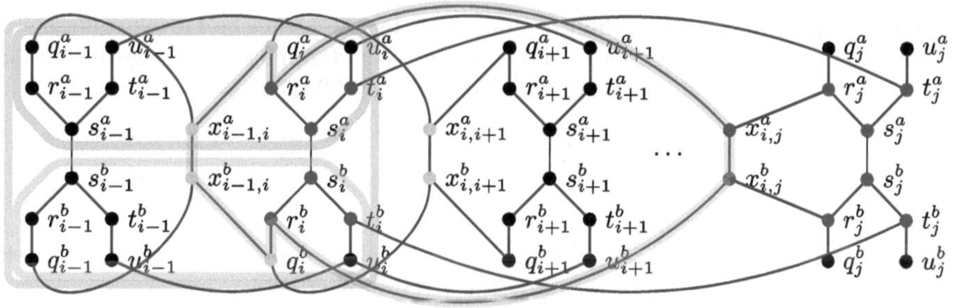

Fig. 9. Illustration to Construction 4: An excerpt around V_i where $(i, j) \in X''$, with habitats $H^a_{i,i+1}$ (cyan), $H^b_{i,i+1}$ (orange), and $H_{i,i+1}$ (magenta), as well as $H^\dagger_{i,j}$ (green), $H^a_{i,j}$ (blue vertices), $H^b_{i,j}$ (magenta vertices), and $H^\dagger_i$ (yellow vertices).

For each $i \in \{1, \ldots, n\}$, add the vertex set $V_i = V_i^a \cup V_i^b$ with $V_i^c = \{q_i^c, r_i^c, s_i^c, t_i^c, u_i^c\}$ for $c \in \{a, b\}$ and the edge set $E_i = E_i^a \cup E_i^b \cup \{e_i^*\}$ with $E_i^c = \{\{q_i^c, r_i^c\}, \{r_i^c, s_i^c\}, \{s_i^c, t_i^c\}, \{t_i^c, u_i^c\}\}$ for $c \in \{a, b\}$, and $e_i^* = \{s_i^a, s_i^b\}$. Let $X = X' \cup X''$ with $X' = \{(i, i+1) \mid i \in \{1, \ldots, n-1\}\} \cup \{(n, 1)\}$ and $X'' = \{(i, j) \mid \{v_i, v_j\} \in E \wedge i < j\} \setminus (X' \cup \{(1, n)\})$ (note that $|X| = |E|$). For each $(i, j) \in X$, add vertices $x^a_{i,j}$ and $x^b_{i,j}$, and the edge $\{x^a_{i,j}, x^b_{i,j}\}$. For each $(i, j) \in X'$, add the edge set $E_{i,j} = E^a_{i,j} \cup E^b_{i,j}$ with $E^c_{i,j} = \{\{q_i^c, x^c_{i,j}\}, \{q_j^c, x^c_{i,j}\}, \{u_i^c, u_j^c\}\}$ for $c \in \{a, b\}$; add the habitats $H^c_{i,j} = V_i^c \cup V_j^c \cup \{x^c_{i,j}\}$ for each $c \in \{a, b\}$, and the habitat $H_{i,j} = H^a_{i,j} \cup H^b_{i,j}$. For each $(i, j) \in X''$, add the edge set $E_{i,j} = E^a_{i,j} \cup E^b_{i,j}$ with $E^c_{i,j} = \{\{r_i^c, x^c_{i,j}\}, \{r_j^c, x^c_{i,j}\}, \{t_i^c, t_j^c\}\}$ for $c \in \{a, b\}$; add the habitats $H^c_{i,j} = (V_i^c \setminus \{q_i^c, u_i^c\}) \cup (V_j^c \setminus \{q_j^c, u_j^c\}) \cup \{x^c_{i,j}\}$ for each $c \in \{a, b\}$, and the habitat $H_{i,j} = H^a_{i,j} \cup H^b_{i,j}$. Moreover, add the habitat $H^\dagger_{i,j} = \{x^1_{i',i}, q_i^a, r_i^a, x^a_{i,j}, x^b_{i,j}, r_i^b, q_i^b, x^b_{i',i}\}$, where $i' = i-1$ if $i > 1$ and $i' = n$ otherwise. Finally, for each $i \in \{1, \ldots, n\}$, add habitat $H^\dagger_i = \{x^a_{i',i}, q_i^a, x^a_{i,i''}, x^b_{i,i''}, q_i^b, x^b_{i',i}\}$, where $i' = i - 1$ if $i > 1$ and $i' = n$ otherwise, and $i'' = i + 1$ if $i < n$ and $i'' = 1$ otherwise. Set $k = |E'| - n + p$. ◇

7 Conclusion

Our results suggest that some real-world instances of our problem could be tractable. For example, graphs derived from Manhattan-like street networks are likely of maximum degree at most 4 (each bounded face has at most four adjacent faces). Thus, when habitats are of size at most 5, these instances are tractable. Even if the maximum degree increases from 4 to 5, for small habitats of size at most 4 we are still in the tractable regime. Overall, since all of our NP-hardness results hold without edge costs or forced edges, this suggests that these two aspects play at most a limited role in the problem's intractability.

Natural directions for future work include (i) closing the gaps for η and Δ in our table and (ii) κ-connected habitats with $\kappa \geq 3$. As to (i), the largest gap is for maximum degree 3: we prove NP-hardness for habitats of size 22, but the problem is polynomial-time solvable for habitats of size 6. In the edge-variant of RGBP, the remaining gap for size-5 habitats is particularly interesting.

Acknowledgments. T. Fluschnik is supported by the Deutsche Forschungsgemeinschaft (DFG, German Research Foundation), project "Parameterized Algorithmics in Computational Sustainability (PACS)"—FL 1247/1-1, 522475669.

References

1. Angluin, D., Aspnes, J., Reyzin, L.: Network construction with subgraph connectivity constraints. J. Comb. Optim. **29**(2), 418–432 (2015)
2. Cheriyan, J., Végh, L.A.: Approximating minimum-cost k-node connected subgraphs via independence-free graphs. In: Proceedings of the 54th FOCS, pp. 30–39. FOCS '13. IEEE Computer Society, USA (2013)
3. Cohen, N., Havet, F., Mazauric, D., Sau, I., Watrigant, R.: Complexity dichotomies for the minimum F-overlay problem. J. Discrete Algorithms **52–53**, 133–142 (2018)
4. Darmann, A., Döcker, J.: On simplified NP-complete variants of Monotone 3-Sat. Discret. Appl. Math. **292**, 45–58 (2021)
5. Ellmies, G., Fluschnik, T.: Placing green bridges optimally for robust habitat reconnection. CoRR **abs/2602.19834** (2026)
6. Ene, A., Vakilian, A.: Improved approximation algorithms for degree-bounded network design problems with node connectivity requirements. In: Proceedings of the 46th STOC, pp. 754–763. STOC '14. Association for Computing Machinery, New York, NY, USA (2014)
7. Fleischner, H., Sabidussi, G., Sarvanov, V.I.: Maximum independent sets in 3- and 4-regular Hamiltonian graphs. Discret. Math. **310**(20), 2742–2749 (2010)
8. Fluschnik, T., Kellerhals, L.: Placing green bridges optimally, with a multivariate analysis. Theory Comput. Syst. **68**(5), 1312–1338 (2024)
9. Garey, M.R., Johnson, D.S.: Computers and Intractability: A Guide to the Theory of NP-Completeness. Freeman, W. H (1979)
10. Garey, M.R., Johnson, D.S., Tarjan, R.E.: The planar Hamiltonian circuit problem is NP-complete. SIAM J. Comput. **5**(4), 704–714 (1976)
11. Gionis, A., Rozenshtein, P., Tatti, N., Terzi, E.: Community-aware network sparsification. In: Chawla, N.V., Wang, W. (eds.) Proceedings of the 2017 SDM, pp. 426–434. SIAM (2017)

12. Herkenrath, M., Fluschnik, T., Grothe, F., Kellerhals, L.: Placing green bridges optimally, with habitats inducing cycles. In: Proceedings of the 31st IJCAI, pp. 3825–3831 (2022)
13. Herrendorf, E., Komusiewicz, C., Morawietz, N., Sommer, F.: On the complexity of community-aware network sparsification. In: Královic, R., Kucera, A. (eds.) Proceedings of the 49th MFCS. LIPIcs, vol. 306, pp. 60:1–60:18. Schloss Dagstuhl - Leibniz-Zentrum für Informatik (2024)
14. Jain, K.: A factor 2 approximation algorithm for the generalized Steiner network problem. Combinatorica **21**(1), 39–60 (2001)
15. Wallisch, C., Fluschnik, T., Kellerhals, L.: Placing green bridges optimally, with close-range habitats in sparse graphs. CoRR **abs/2510.21540** (2025). To appear in the proceedings of AAMAS'26.

Nuclear Shifts for Conservation

Giulio Fellin[1(✉)], Sara Negri[2], and Peter Schuster[1]

[1] Dipartimento di Informatica, Università degli Studi di Verona, strada le Grazie 15, 37134 Verona, Italy
{giulio.fellin,peter.schuster}@univr.it

[2] Dipartimento di Matematica (Dipartimento di Eccellenza 2023–2027), Università degli Studi di Genova, via Dodecaneso 35, 16146 Genova, Italy
sara.negri@unige.it

Abstract. Many a Glivenko-style conservation theorem bears on a variant of Kuroda's double negation shift (DNS), e.g. in first-order, infinitary, and modal logic. A unified proof-theoretic explanation of this phenomenon is made possible: by abstracting provability to inductively generated entailment relations and double negation to nuclei. Conservation is characterised in terms of the admissibility of inference rules in the Kleisli extension induced by a nucleus. While the left rules typically carry over to this extension without much ado, the right rules correspond exactly to DNS-type principles, of which the ones for implication, the universal quantifier, infinitary conjunction, box and diamond are the critical cases.

Keywords: Intuitionistic logic · Minimal logic · Infinitary logic · Glivenko's theorem · Double negation

1 Introduction

The double negation shift (DNS) is indispensable for Kuroda's generalisation to first-order logic of Glivenko's theorem for propositional logic: up to double negation, classical first-order predicate logic is conservative over intuitionistic logic—with DNS added to the latter in the predicate case. Besides Kuroda's result, quite a few conservation theorems in the vein of Glivenko's have turned out to hinge upon variants of DNS. In the present note, we shed light on this phenomenon in the borderland of classical and constructive reasoning.

Abstracting Glivenko's theorem from double negation and provability to a nucleus over an inductively generated entailment relation requires that all generating rules of the latter remain admissible in the Kleisli extension, i.e. when the nucleus is applied to the succedent of every premiss and the conclusion. This conservation criterion clearly is automatic for every rule which is right-invariant, i.e. such that all premisses and the conclusion have the same succedent. Axioms formulated as zero-premiss rules are trivially right-invariant. In the single-succedent calculi for propositional, infinitary, first-order and basic modal minimal logic, one can express all rules as right-invariant rules except for the right rules for implication, infinite conjunction, the universal quantifier and the

V. Brattka et al. (Eds.): CiE 2026, LNCS 16674, pp. 272–288, 2027.
https://doi.org/10.1007/978-3-032-31348-5_18

two modalities. The conservation criteria for the critical right rules have proved to be counterparts of DNS.

2 Preliminaries

We use uppercase Latin letters $A, B, \ldots$ for formulas and uppercase Greek letters $\Gamma, \Delta, \ldots$ for sets of formulas. Additionally, we set $\neg\neg\Gamma = \{\neg\neg B \colon B \in \Gamma\}$ and $\forall\Gamma = \{\forall B \mid B \in \Gamma\}$ where $\forall B$ stands for the universal closure of B.

Recall that from provability in *minimal logic* $\vdash_{\mathrm{m}}$ one obtains provability first in *intuitionistic logic* $\vdash_{\mathrm{i}}$ and then in *classical logic* $\vdash_{\mathrm{c}}$ by allowing as additional assumptions (in first-order logic, the universal closures of) the axiom schemes of *ex falso quodlibet* EFQ : $\bot \to B$ and *tertium non datur* TND : $B \vee \neg B$, respectively. [1] In other words,

$$\Gamma \vdash_{\mathrm{i}} A \equiv \Gamma, \mathrm{EFQ} \vdash_{\mathrm{m}} A\,, \qquad \Gamma \vdash_{\mathrm{c}} A \equiv \Gamma, \mathrm{TND} \vdash_{\mathrm{i}} A\,.$$

Here we identify an axiom scheme with the set of all its instances; for example, by $\Gamma, \mathrm{EFQ} \vdash_{\mathrm{m}} A$ we mean $\Gamma, \bot \to B_1, \ldots, \bot \to B_k \vdash_{\mathrm{m}} A$ for some formulas $B_1, \ldots, B_k$ with $k \geqslant 0$.

As is well known, EFQ and TND together amount in minimal logic to *reductio ad absurdum* RAA : $\neg\neg B \to B$, by which it is a single step to classical logic:

$$\Gamma \vdash_{\mathrm{c}} A \equiv \Gamma, \mathrm{RAA} \vdash_{\mathrm{m}} A\,.$$

An established tool is what we want to dub *Brouwer's lemma*:

$$\Delta, D \vdash_{*} \neg\neg C \implies \Delta, \neg\neg D \vdash_{*} \neg\neg C \qquad (* \in \{\mathrm{m}, \mathrm{i}\})\,. \tag{1}$$

This rests on $D \to E \vdash_{\mathrm{m}} \neg E \to \neg D$, especially $D \to E \vdash_{\mathrm{m}} \neg\neg D \to \neg\neg E$, and $\vdash_{\mathrm{m}} \neg\neg\neg F \to \neg F$, the latter of which goes back to Brouwer [8]. [2]

3 Glivenko's Theorem and Double Negation Shifts

At first sight, derivability in classical logic is stronger than in intuitionistic logic, and stronger in the latter than in minimal logic:

$$\Gamma \vdash_{\mathrm{m}} A \implies \Gamma \vdash_{\mathrm{i}} A \implies \Gamma \vdash_{\mathrm{c}} A\,.$$

---

1 For any given derivation, finitely many instances of these axiom schemes suffice.

2 This argument factors through auxiliary principles, such as $D \to E \vdash \neg\neg D \to \neg\neg E$, close to the interpretation of double negation as a nucleus. A reviewer has kindly pointed out the following alternative only using implication introduction and elimination. Assume Δ and $\neg\neg D$ but $\neg C$. If D, then $\neg\neg C$ by Δ and $\Delta, D \vdash \neg\neg C$, and thus $\bot$ by $\neg C$. Hence $\neg D$, which by $\neg\neg D$ entails $\bot$. In all, $\neg\neg C$.

Among the earliest results which may be regarded as converses of this implication is *Glivenko's theorem* [18,19] in *propositional* logic: [3] [4]

$$\Gamma \vdash_{c} A \Longrightarrow \Gamma \vdash_{i} \neg\neg A \tag{2}$$

Odintsov [33, Theorem 5.2] and Cignoli–Torrens [9, Theorem 3.4] have shown that Glivenko's theorem can be extended to minimal logic augmented with $\neg\neg$EFQ, and to bounded BCK-algebras equipped with the counterpart of $\neg\neg$RAA. The latter equally works in minimal logic, and Glivenko's theorem can be restated:

$$\Gamma \vdash_{c} A \Longrightarrow \Gamma, \neg\neg\,\mathrm{EFQ} \vdash_{m} \neg\neg A\,; \qquad \Gamma \vdash_{c} A \Longrightarrow \Gamma, \neg\neg\,\mathrm{RAA} \vdash_{m} \neg\neg A\,.$$

The first of these two forms of Glivenko's theorem was noticed by Segerberg [41, page 46]. Unlike $\neg\neg$ TND, which is an instance of non-contradiction $\vdash_{m} \neg(C \wedge \neg C)$, one cannot prove $\neg\neg$ RAA with minimal logic, as seen by Johansson [20]. Cignoli–Torrens [9, Lemma 3.3] have noticed the analogue for BCK-algebras of the equivalence in minimal logic between $\neg\neg$ RAA and what we want to call the *double negation shift for* $\rightarrow$:

$$\mathrm{DNS}_{\rightarrow}: \quad (B \rightarrow \neg\neg C) \rightarrow \neg\neg(B \rightarrow C)\,.$$

This $\mathrm{DNS}_{\rightarrow}$ is $\vdash_{m}$-equivalent [13, Lemma 5.5] to $\neg\neg$ EFQ. Hence Glivenko's theorem can be equivalently formulated with $\mathrm{DNS}_{\rightarrow}$ on top of $\vdash_{m}$:

$$\Gamma \vdash_{c} A \quad \Longrightarrow \quad \Gamma, \mathrm{DNS}_{\rightarrow} \vdash_{m} \neg\neg A\,.$$

To sum up the literature cited above:

Proposition 1. *In minimal logic,* $\neg\neg$ EFQ, $\neg\neg$ RAA *and* $\mathrm{DNS}_{\rightarrow}$ *are equivalent.*

In the light of this it is clear why $\neg\neg$ EFQ, $\neg\neg$ RAA and $\mathrm{DNS}_{\rightarrow}$ do not show up in the original formulation of Glivenko's theorem with intuitionistic logic $\vdash_{i}$, in which these add-ons are derivable anyway. Removing any of them from the above variants of Glivenko's theorem with minimal logic $\vdash_{m}$ would make this theorem fail, e.g. for any instance $A \equiv \neg\neg P \rightarrow P$ of EFQ, as noticed by Johansson [20].

We insist on $\mathrm{DNS}_{\rightarrow}$ because this is analogous to other double negation shifts:

1. the *double negation shift for* $\forall$

$$\mathrm{DNS}_{\forall}: \quad \forall x \neg\neg Cx \rightarrow \neg\neg \forall x Cx$$

needed for Kuroda's extension [25] of Glivenko's theorem to *first-order* logic

$$\Gamma \vdash_{c} A \quad \Longrightarrow \quad \Gamma, \mathrm{DNS}_{\forall} \vdash_{i} \neg\neg A\,;$$

---

[3] A different converse had been obtained earlier by Kolmogorov [24].

[4] We will not make use of the following equivalent formulations of Glivenko's theorem:

$$\Gamma \vdash_{c} A \Longrightarrow \neg\neg\Gamma \vdash_{i} \neg\neg A\,; \quad \Gamma \vdash_{c} \neg A \Longrightarrow \Gamma \vdash_{i} \neg A\,; \quad \Gamma \vdash_{c} \bot \Longrightarrow \Gamma \vdash_{i} \bot\,.$$

The last one is equiconsistency of classical and intuitionistic propositional logic.

2. the *double negation shift for* $\bigwedge$

$$\mathrm{DNS}_{\bigwedge}: \quad \bigwedge_{n\in\mathbb{N}} \neg\neg C_n \to \neg\neg \bigwedge_{n\in\mathbb{N}} C_n$$

required for Tesi's [3,43] counterpart of Kuroda's extension in *infinitary* logic

$$\Gamma \vdash_{\mathrm{c}} A \implies \Gamma, \mathrm{DNS}_{\bigwedge} \vdash_{\mathrm{i}} \neg\neg A\,.$$

Related phenomena are present in substructural logics; see e.g. [7,15,16,34].

Why Double Negation Shifts? A quick proof of Glivenko's theorem in propositional logic goes as follows:

$$\begin{aligned} \Gamma &\vdash_{\mathrm{c}} A \\ \Gamma &\vdash_{\mathrm{c}} \neg\neg A \\ \Gamma, \mathrm{RAA} &\vdash_{\mathrm{m}} \neg\neg A && \text{by the definition of } \vdash_{\mathrm{c}} \text{ from } \vdash_{\mathrm{m}} \\ \Gamma, \neg\neg\,\mathrm{RAA} &\vdash_{\mathrm{m}} \neg\neg A && \text{by Brouwer's lemma} \\ \Gamma &\vdash_{\mathrm{i}} \neg\neg A && \text{because} \vdash_{\mathrm{i}} \neg\neg\,\mathrm{RAA} \end{aligned}$$

To show $\mathrm{DNS}_{\to}$ in action we refine the very last step:

$$\begin{aligned} \Gamma, \neg\neg\,\mathrm{RAA} &\vdash_{\mathrm{m}} \neg\neg A \\ \Gamma, \{\neg\neg(\widetilde{C} \to C)\} &\vdash_{\mathrm{m}} \neg\neg A \quad \text{with } \widetilde{C} \equiv \neg\neg C \\ \Gamma, \{\widetilde{C} \to \neg\neg C\}, \{(\widetilde{C} \to \neg\neg C) \to \neg\neg(\widetilde{C} \to C)\} &\vdash_{\mathrm{m}} \neg\neg A \\ \Gamma, \mathrm{DNS}_{\to} &\vdash_{\mathrm{m}} \neg\neg A \\ \Gamma &\vdash_{\mathrm{i}} \neg\neg A \quad \text{because } \vdash_{\mathrm{i}} \mathrm{DNS}_{\to} \end{aligned}$$

Where $\mathrm{DNS}_{\forall}$ is needed for Kuroda's extension can be seen as follows:

$$\begin{aligned} \Gamma &\vdash_{\mathrm{c}} A \\ \Gamma &\vdash_{\mathrm{c}} \neg\neg A \\ \Gamma, \forall\,\mathrm{RAA} &\vdash_{\mathrm{m}} \neg\neg A && \text{by the definition of } \vdash_{\mathrm{c}} \text{ from } \vdash_{\mathrm{m}} \\ \Gamma, \neg\neg\forall\,\mathrm{RAA} &\vdash_{\mathrm{m}} \neg\neg A && \text{by Brouwer's lemma} \\ \Gamma, \forall\neg\neg\,\mathrm{RAA}, \mathrm{DNS}_{\forall} &\vdash_{\mathrm{m}} \neg\neg A \\ \Gamma, \mathrm{DNS}_{\forall} &\vdash_{\mathrm{i}} \neg\neg A && \text{because } \vdash_{\mathrm{i}} \neg\neg\,\mathrm{RAA} \end{aligned}$$

Writing x for the list of free variables of a formula Bx, we detail the last two steps:

$$\begin{aligned} \Gamma, \neg\neg\forall\,\mathrm{RAA} &\vdash_{\mathrm{m}} \neg\neg A \\ \Gamma, \{\neg\neg\forall x Cx\} &\vdash_{\mathrm{m}} \neg\neg A \quad \text{with } Cx \equiv \neg\neg Bx \to Bx \\ \Gamma, \{\forall x \neg\neg Cx\}, \{\forall x \neg\neg Cx \to \neg\neg\forall x Cx\} &\vdash_{\mathrm{m}} \neg\neg A \\ \Gamma, \mathrm{DNS}_{\forall} &\vdash_{\mathrm{i}} \neg\neg A \quad \text{because } \vdash_{\mathrm{i}} \neg\neg Cx \end{aligned}$$

The case of $\mathrm{DNS}_{\bigwedge}$ for Tesi's analogue in infinitary logic goes in parallel.

4 Nuclei over Entailment Relations

4.1 Entailment Relations

We consider relations $\rhd$ between finite subsets $U, V, W, \ldots$ and elements $a, b, c, \ldots$ of any given set S. In this context we write $U \rhd b$ rather than $(U, b) \in \rhd$, and abbreviate $U_1 \cup \ldots \cup U_k \cup \{a_1, \ldots, a_\ell\} \rhd b$ by $U_1, \ldots, U_k, a_1, \ldots, a_\ell \rhd b$ where $k, \ell \geqslant 0$; in particular, $\rhd b$ stands for $\emptyset \rhd b$. We say that an expression of the form $U \rhd b$ is a *sequent* with *succedent* b, and that $U \rhd b$ *holds* if the pair $(U, b) \in \rhd$.

We recall from [32, page 21] that a rule (of inference) is called an *admissible rule*, or is said *admissible* in a sequent calculus if derivability is closed under the rule: that is, whenever the premisses of the rule are derivable in the system, the conclusion also is. While admissibility thus is a notion which rather pertains to the realm of inductively defined relations, we extend it to arbitrary relations $\rhd$ of the above type as follows: a *rule*

$$\frac{U_1 \rhd c_1 \quad \ldots \quad U_n \rhd c_n \quad \ldots}{V \rhd d} \quad \text{or, more compactly,} \quad \frac{\{U_i \rhd c_i \colon i \in I\}}{V \rhd d} \tag{3}$$

with $I \subseteq \mathbb{N}$ is *admissible* for $\rhd$ if the *conclusion* $V \rhd d$ holds whenever all the *premisses* $U_i \rhd c_i$ with $i \in I$ hold.

A relation $\rhd$ as above is a *(single-succedent) entailment relation* on S if the following three *structural rules* are admissible in $\rhd$:

$$\begin{array}{ccc} \textit{reflexivity} & \textit{monotonicity} & \textit{transitivity} \\ \dfrac{}{a \rhd a} & \dfrac{U \rhd a}{U, V \rhd a} & \dfrac{U \rhd a \quad V, a \rhd b}{V, U \rhd b} \end{array}$$

The least entailment relation is inverse membership $\ni$. In sequent calculi, $\rhd$ is a derivability relation, and S the set of formulas of a formal language.

A sequent $U \rhd b$ holds in an entailment relation $\rhd$ if and only if the corresponding *zero-premiss rule*

$$\frac{}{U \rhd b} \tag{4}$$

is admissible in $\rhd$. Hence sequents may be identified with zero-premiss rules, e.g. the sequent $a \rhd a$ with the rule of reflexivity, and we may subsume sequents under rules unless we need to distinguish them.

Following [13,14,35,36], by an *extension* $\rhd'$ of an entailment relation $\rhd$ on a set S we understand an entailment relation $\rhd'$ on the same set S such that $\rhd \subseteq \rhd'$. In the same vein we say that an extension $\rhd'$ of $\rhd$ is *conservative* if also $\rhd \supseteq \rhd'$ and thus altogether $\rhd = \rhd'$.

4.2 Inductive Generation

We consider rules for entailment relations $\rhd$ on a set S. Let $\mathcal{R}$ be a set of rules. A sequent φ is $\mathcal{R}$-*derivable* from a set Γ of sequents if there is an $\mathcal{R}$-*derivation* of φ from Γ: that is, a partially labelled finite tree in which

- every non-leaf node is labelled by a sequent;
- the root is labelled by φ;
- every branching is done according to either a structural rule or a rule in $\mathcal{R}$;
- the *unlabelled* leaves accommodate instances of reflexivity and of zero-premiss rules in $\mathcal{R}$; and
- every *labelled* leaf has as label one of the sequents in Γ.

A rule r is *derivable* from $\mathcal{R}$ if there is an $\mathcal{R}$-derivation of the conclusion of r from the set of premisses of r. For example, every $r \in \mathcal{R}$ is derivable from $\mathcal{R}$ with r itself as its own $\mathcal{R}$-derivation.

Any given sequent is $\mathcal{R}$-derivable from $\emptyset$ precisely when the corresponding zero-premiss rule is derivable from $\mathcal{R}$. If *every* sequent of $\rhd$ is $\mathcal{R}$-derivable from $\emptyset$, then $\rhd$ is *inductively generated* by $\mathcal{R}$, and $\mathcal{R}$ is a set of *generating rules* for $\rhd$. In this case, $\rhd$ is the least entailment relation on S for which, in addition to the structural rules, all rules in $\mathcal{R}$ are admissible. An *axiom* is a sequent the corresponding zero-premiss rule of which belongs to the set of generating rules.

Example 1. The inverse membership relation $\ni$, the least entailment relation at all, is inductively generated by no rules apart from the structural rules.

Example 2. Any entailment relation $\rhd$ whatsoever is inductively generated by the set $\mathcal{Z}$ consisting of the zero-premiss rules corresponding to the elements of $\rhd$. In this case, a sequent $U \rhd a$ holds precisely when it is $\mathcal{Z}$-derivable from $\emptyset$; in fact $\mathcal{Z}$ is closed under the structural rules, as $\rhd$ is an entailment relation.

Albeit perfectly licit, Example 2 is of course not the intended case of inductive generation. Typically, much fewer and more characteristic generating rules suffice, see e.g. the cases of in Sect. 6 with finitely or countably many generating rules. [5] In fact, an inductively generated entailment relation $\rhd$ can have many sets of generating rules $\mathcal{R}$; for example, one can enlarge any such $\mathcal{R}$ by adding rules derivable from $\mathcal{R}$ while thus generating the same $\rhd$.

In spite of this it is quite common to speak of an inductively generated entailment relation $\rhd$ without specifying any set of generating rules $\mathcal{R}$ for $\rhd$. This is alright unless what one says about $\rhd$ depends on $\mathcal{R}$, which eventuality one needs to exclude beforehand; see e.g. Remark 5. To be more precise, we suppose throughout that *each inductively generated entailment relation* $\rhd$ *be tacitly accompanied by a set of generating rules* $\mathcal{R}$; we thus say that a rule (a sequent) is *derivable* in $\rhd$ if it can be derived from $\mathcal{R}$ (is $\mathcal{R}$-derivable from $\emptyset$).

Remark 1. Let $\rhd$ be an inductively generated entailment relation.

1. The generating rules of $\rhd$ are derivable in $\rhd$, and so are the structural rules.
2. Every rule derivable in $\rhd$ is admissible in $\rhd$.

---

5 Unless $\rhd$ is required to be a set from the outset, which we do require in the present paper, the cases in which $\rhd$ can be inductively generated by a *set* of rules are of particular relevance in predicative set theories such as **CZF**, **BCST**, and **ECST** [2].

3. If a zero-premiss rule r is admissible in $\rhd$, then r is even derivable in $\rhd$.

In particular, a sequent is derivable if and only if the corresponding zero-premiss rule is admissible. Also, an arbitrary rule is admissible precisely when derivability is closed under the rule, which is to say that for inductively generated entailment relations admissibility of rules essentially is the same as for sequent calculi.

While an admissible rule need not be derivable, [6] admissibility is often proved by induction on the last rules applied in the derivation of the premisses.

4.3 Inductive Extension

If an entailment relation $\rhd$ is inductively generated by a set of rules $\mathcal{R}$, then the *inductive extension* [13] of $\rhd$ by a set $\mathcal{R}^+$ of *additional rules* is the entailment relation $\rhd$ *plus* $\mathcal{R}^+$ inductively generated by $\mathcal{R} \cup \mathcal{R}^+$.

Example 3. Every inductively generated entailment relation is an inductive extension of $\ni$ inductively generated by $\emptyset$ (Example 1).

Proposition 2. *Let $\rhd$ and $\rhd'$ be entailment relations on the same set S and inductively generated by the sets of rules $\mathcal{R}$ and $\mathcal{R}'$. Each of the following assertions implies the next:*

1. *$\rhd'$ is an inductive extension of $\rhd$, i.e. $\mathcal{R} \subseteq \mathcal{R}'$.*
2. *Every rule derivable from $\mathcal{R}$ is derivable from $\mathcal{R}'$.*
3. *Every rule derivable from $\mathcal{R}$ is admissible in $\rhd'$.*
4. *$\rhd'$ is an extension of $\rhd$, i.e. $\rhd \subseteq \rhd'$.*

Moreover, statements 2 and 3 are respectively equivalent to the following:

2′. Every rule in $\mathcal{R}$ is derivable from $\mathcal{R}'$.
3′. Every rule in $\mathcal{R}$ is admissible in $\rhd'$.

Corollary 1. *The sets of rules $\mathcal{R}$ and $\mathcal{R}'$ inductively generate the same entailment relation whenever every element of $\mathcal{R}$ is derivable from $\mathcal{R}'$ and vice versa.*

Remark 2. None of the implications in Proposition 2 can be reversed. In particular, except for the case of sequents as zero-premiss rules, a rule admissible in an inductively generated entailment relation $\rhd$ need not remain admissible in every inductive extension of $\rhd$.

---

[6] Common examples in logic include the substitution rule:

$$\frac{U \rhd c}{U[y/x] \rhd c[y/x]}\,\text{Subs},$$

where, supposing that there is no variable collision, $[y/x]$ denotes that every free occurrence of the variable x is replaced by y. In fact, given any derivation of $U \rhd c$, by replacing with y every occurrence of x in the derivation and after renaming bound variables to avoid variable collision, one gets a derivation of $U[y/x] \rhd c[y/x]$, but there is no derivation from $U \rhd c$ to $U[y/x] \rhd c[y/x]$. See e.g. [31, Lemma 4.1.2].

4.4 Nuclei

Let $\rhd$ be an entailment relation on a set S.

Definition 1. *A* nucleus *over* $\rhd$ *is a map* $j\colon S \to S$ *satisfying*

$$\frac{U, a \rhd jb}{U, ja \rhd jb}\,\mathrm{L}j \qquad\qquad \frac{U \rhd b}{U \rhd jb}\,\mathrm{R}j$$

If only Rj *is admissible in* $\rhd$*, then* j *is* expansive *with respect to* $\rhd$*.*

By reflexivity and transitivity, the rule Rj can be expressed as the sequent $b \rhd jb$, and Rj is equivalent to the inverse of Lj; whence the two rules Lj and Rj can be subsumed by the following double rule: [7]

$$\frac{U, a \rhd jb}{\overline{U, ja \rhd jb}} \tag{5}$$

Nuclei thus are extension systems [29], algebraic theories in extension form [28, page 32] or Kleisli triples [30]. Nuclei [42] are also known as strong monads [23], lax modalities [1,17] and local operators [26]; they correspond to sublocales [22, Theorem II.2.3] and Lawvere–Tierney topologies [27, Corollary IX.5.6].

Remark 3. Let $\rhd$ come with a *relative pseudo-complement* $\to$ characterised by

$$\frac{U, a \rhd b}{\overline{U \rhd a \to b}}\ .$$

Then "j is a nucleus" can be expressed by a sequent together with its converse: [8]

$$a \to jb \lhd\!\rhd ja \to jb\,. \tag{6}$$

Here $\lhd\!\rhd$ is to be read as the (meta-)conjunction of the sequent and its converse.

Incidentally, for $\neg\neg$ as j and $\vdash_{\mathrm{m}}$ or $\vdash_{\mathrm{i}}$ as $\rhd$ the Condition (5) is nothing but Brouwer's Lemma (1). Besides $\neg\neg$, examples of nuclei in logic include:

$(\_\_ \to A) \to A$	continuation nucleus
$(\_\_ \to A) \to \_\_$	selection nucleus, or Peirce monad [11]
$\_\_ \vee A$	closed nucleus
$A \to \_\_$	open nucleus

[7] As usual, a rule with a double line is to be understood both up and down.

[8] This is well-known in locale theory [21, Remark C.1.1.14], as $a \to jb = ja \to jb$.

5 Extensions Induced by Nuclei

Let now $\rhd$ be an *inductively generated* entailment relation on a set S. We follow [12,13]—but unlike the latter start from any map $j\colon S \to S$ whatsoever—when now considering the following two relations.[9]

Definition 2. *Let $j\colon S \to S$ be an arbitrary map.*

1. *The* Kleisli extension $\rhd_j$ *is defined by* $U \rhd_j a \equiv U \rhd ja$.
2. *The* stable extension $\rhd^j$ *is the inductive extension of* $\rhd$ *with the additional axiom of* stability $ja \rhd a$.

Remark 4. $\rhd_j \subseteq \rhd^j$, and stability is admissible in the Kleisli extension $\rhd_j$ too.

Proof. If $U \rhd_j a$, i.e. $U \rhd ja$, then $U \rhd^j ja$ by (inductive) extension; applying transitivity to the latter and stability for the stable extension $ja \rhd^j a$ yields $U \rhd^j a$. On the other hand, stability for the Kleisli extension $ja \rhd_j a$ is nothing but the instance $ja \rhd ja$ of reflexivity for $\rhd$. □

As an inductive extension of $\rhd$, the strong extension $\rhd^j$ is an entailment relation and an extension of $\rhd$ (Proposition 2). To have these properties for the Kleisli extension $\rhd_j$ as well requires j to be a nucleus, as we detail in the sequel.

Proposition 3. *Let $j\colon S \to S$ be an arbitrary map.*

1. *Monotonicity is admissible in the Kleisli extension $\rhd_j$.*
2. *The following are equivalent:*
   (a) *j is expansive, i.e. $\mathrm{R}j$ is admissible in $\rhd$.*
   (b) *Reflexivity is admissible in the Kleisli extension $\rhd_j$.*
   (c) *The Kleisli extension of $\rhd$ is an extension of $\rhd$, that is, $\rhd \subseteq \rhd_j$.*
   (d) *Every zero-premiss rule admissible in $\rhd$ is admissible in $\rhd_j$ too.*
3. *$\mathrm{L}j$ is admissible in $\rhd$ if and only if transitivity is admissible in $\rhd_j$.*

Proof. Items 1 and 2 are straightforward. As for item 3, suppose first that $\mathrm{L}j$ is admissible in $\rhd$. Let $V \rhd_j c$, i.e. $V \rhd jc$, and $W, c \rhd_j d$, i.e. $W, c \rhd jd$. By $\mathrm{L}j$, the latter entails $W, jc \rhd jd$. Applying transitivity of $\rhd$ gives $V, W \rhd jd$, i.e. $V, W \rhd_j d$. Conversely, if $W, c \rhd jd$, i.e. $W, c \rhd_j d$, then the admissibility of transitivity for $\rhd_j$ yields $W, jc \rhd_j d$, i.e. $W, jc \rhd jd$, since also $jc \rhd_j c$ by stability of $\rhd_j$ (Remark 4). □

Corollary 2. *The Kleisli extension $\rhd_j$ is an entailment relation precisely when j is a nucleus; in this case $\rhd_j$ is indeed an extension of $\rhd$.*

By Remark 4 we always have *extension*, namely $\rhd_j \subseteq \rhd^j$, whereas by *conservation* we mean $\rhd_j \supseteq \rhd^j$. In the prime example, with $\rhd \equiv \vdash_{\mathrm{i}}$ and $j \equiv \neg\neg$, conservation is Glivenko's theorem, as the stable extension $\rhd^j$ of intuitionistic provability $\vdash_{\mathrm{i}}$ is classical provability $\vdash_{\mathrm{c}}$.

All (generating) axioms and rules of $\rhd$ are admissible in the stable extension $\rhd^j$, once more by the latter's very inductive definition. In the Kleisli extension $\rhd_j$, on the other hand, *a priori* only the sequents of $\rhd$ seen as zero-premiss rules (4) are admissible whenever j is expansive.

Sequents aside, what does admissibility for the Kleisli extension mean?

---

[9] We have noticed a related construction in arrow algebras [6, Proposition 3.15].

Theorem 1. [12,13] *Let $j\colon S \to S$ be any map. The following are equivalent:*

1. *The stable extension $\rhd^j$ is a conservative extension of the Kleisli extension $\rhd_j$, that is, $\rhd_j = \rhd^j$.*
2. *The Kleisli extension $\rhd_j$ is an entailment relation for which all generating rules of $\rhd$ are admissible.*

Similar Glivenko-style conservation theorems [13] apply to the translations due to Kolmogorov, Gentzen and Kuroda rather than double negation.

There is no natural choice of a neat set of generators for the Kleisli extension unless this already coincides with the stable extension. In particular, the notion of derivability for rules does hardly apply to the Kleisli extension.

In view of Corollary 2 and Theorem 1, that j be a nucleus is *necessary* for conservative extension:

Corollary 3. *$\rhd^j$ is a conservative extension of $\rhd_j$* only if *$j\colon S \to S$ is a nucleus.*

Remark 5. When applying the conservation criterion of Theorem 1, namely that all generating rules of $\rhd$ be admissible in the Kleisli extension $\rhd_j$, Proposition 2 guarantees us that it is irrelevant which generating rules we check as long as they inductively generate the same $\rhd$.

5.1 Right-Invariant Rules

We say that a rule is *right-invariant* whenever it has the form

$$\frac{U_1 \rhd c \quad \dots \quad U_n \rhd c}{V \rhd c} \qquad (n \geqslant 0) \tag{7}$$

where all the premisses and the conclusion are sequents with *the very same but arbitrary* succedent c. Right-invariant rules align with the *left rule schemata* [31] for sequent calculi: all active formulas are in the antecedent, and the succedent remains unchanged.[10]

Example 4. With arbitrary succedent c, the following rules are right-invariant:

1. sequents $V \rhd c$ seen as a zero-premiss rules $\dfrac{}{V \rhd c}$;
2. rules of the form [11] $\dfrac{U, b_1 \rhd c \quad \dots \quad U, b_\ell \rhd c}{U, a_1, \dots, a_k \rhd c}$.

The subsequent lemma and corollary have been noticed before [13] but only for sequents rather than for the wider class of right-invariant rules.

[10] We refer to [31, §6.3] for a detailed discussion of different approaches to extending calculi for pure logic by axioms.

[11] Any such rule represents in terms of $\rhd$ the sequent $a_1, \dots, a_k \vdash b_1, \dots, b_\ell$ of any given Scott-style [38–40] multi-succedent extension $\vdash$ of $\rhd$ taken up recently [14,35–37], provided that the comma is understood *à la* Gentzen, Lorenzen and Scott: as $\wedge$ in the antecedent and as $\vee$ in the succedent.

Lemma 1. *Every* right-invariant *rule admissible for* $\rhd$ *is admissible for* $\rhd_j$ *too.*

Proof. Specify c to jc throughout any given right-invariant rule (7). □

Corollary 4. $\rhd_j = \rhd^j$ *whenever* $\rhd$ *is inductively generated by* right-invariant *rules only.*

Needless to say this applies e.g. if $\rhd$ is inductively generated only by axioms.

6 Provability as Entailment

By S we mean a set of formulas, of the logic under consideration, which contains $\bot, \top$ and is closed under the connectives $\wedge, \vee, \to$. For extensions of propositional logic, we further require S to be closed under the additional logical operators: under the quantifiers $\forall, \exists$ for predicate logic; under the infinitary connectives $\bigwedge, \bigvee$ for infinitary logic; and under the modal operators $\Box, \Diamond$ for modal logic.

Let first $\rhd$ be $\vdash_{\mathrm{m}}$. For propositional logic this entailment relation can be inductively generated by the following rules, on top of the structural rules: [12]

$$\frac{\Gamma, A \rhd C}{\Gamma, A \wedge B \rhd C}\mathrm{L}\wedge_1 \qquad \frac{\Gamma, B \rhd C}{\Gamma, A \wedge B \rhd C}\mathrm{L}\wedge_2 \qquad \frac{\Gamma \rhd A \quad \Gamma \rhd B}{\Gamma \rhd A \wedge B}\mathrm{R}\wedge$$

$$\frac{\Gamma, A \rhd C \quad \Gamma, B \rhd C}{\Gamma, A \vee B \rhd C}\mathrm{L}\vee \qquad \frac{\Gamma \rhd A}{\Gamma \rhd A \vee B}\mathrm{R}\vee_1 \qquad \frac{\Gamma \rhd B}{\Gamma \rhd A \vee B}\mathrm{R}\vee_2$$

$$\frac{\Gamma \rhd A \quad \Gamma, B \rhd C}{\Gamma, A \to B \rhd C}\mathrm{L}\to \qquad \frac{\Gamma, A \rhd B}{\Gamma \rhd A \to B}\mathrm{R}\to \qquad \frac{}{\Gamma \rhd \top}\mathrm{R}\top$$

The subsequent technical remark is well-known:

Remark 6. The following are derivably equivalent by the structural rules only:

$$\frac{\Gamma \rhd A \quad \Gamma, B \rhd C}{\Gamma, A \to B \rhd C}\mathrm{L}\to \qquad \frac{}{A \to B, A \rhd B}\mathrm{m.p.} \qquad \frac{\Gamma \rhd A \to B}{\Gamma, A \rhd B}\mathrm{R}\to^{\mathrm{inv}}$$

Therefore $\to$ is a relative pseudo-complement for minimal logic. By Remark 3, a map $j : S \to S$ thus is a nucleus if and only if both directions of the counterpart of (6) are derivable sequents:

$$A \to jB \lhd\rhd jA \to jB$$

In this case j remains a nucleus in all inductive extensions of $\vdash_{\mathrm{m}}$ (Proposition 2).

From minimal propositional logic as above one obtains as inductive extensions:

[12] These are variants of the rules of the sequent calculus **G3mp**, adapted from [31].

1. *minimal first-order logic* by adding the rules

$$\frac{\Gamma, A[t/x] \rhd C}{\Gamma, \forall x\, A \rhd C}\mathrm{L}\forall \qquad \frac{\Gamma \rhd A[y/x]}{\Gamma \rhd \forall x\, A}\mathrm{R}\forall \quad (y \text{ fresh})$$

$$\frac{\Gamma, A[y/x] \rhd C}{\Gamma, \exists x\, A \rhd C}\mathrm{L}\exists \quad (y \text{ fresh}) \qquad \frac{\Gamma \rhd A[t/x]}{\Gamma \rhd \exists x\, A}\mathrm{R}\exists$$

2. *minimal infinitary logic* by adding the rules

$$\frac{\Gamma, A_n \rhd C}{\Gamma, \bigwedge_{i\in\mathbb{N}} A_i \rhd C}\mathrm{L}{\textstyle\bigwedge_n} \ (n \in \mathbb{N}) \qquad \frac{\{\Gamma \rhd A_n : n \in \mathbb{N}\}}{\Gamma \rhd \bigwedge_{i\in\mathbb{N}} A_i}\mathrm{R}{\textstyle\bigwedge}$$

$$\frac{\{\Gamma, A_n \rhd C : n \in \mathbb{N}\}}{\Gamma, \bigvee_{i\in\mathbb{N}} A_i \rhd C}\mathrm{L}{\textstyle\bigvee} \qquad \frac{\Gamma \rhd A_n}{\Gamma \rhd \bigvee_{i\in\mathbb{N}} A_i}\mathrm{R}{\textstyle\bigvee_n} \ (n \in \mathbb{N})$$

3. *minimal basic modal logic* by adding the rules [13]

$$\frac{\Gamma, A \rhd B}{\Box\Gamma, \Diamond A \rhd \Diamond B}\mathrm{R}\Diamond \qquad \frac{\Gamma \rhd A}{\Box\Gamma \rhd \Box A}\mathrm{R}\Box$$

For simplicity's sake, we uniformly write $\vdash_{\mathrm{m}}$, $\vdash_{\mathrm{i}}$, and $\vdash_{\mathrm{c}}$ for provability in all these variants (propositional, first-order, infinitary, and modal) of minimal, intuitionistic, and classical logic. [14] Also, we subsume under L$\wedge$ the two left rules L$\vee_1$ and L$\vee_2$; under R$\vee$ the two right rules R$\vee_1$ and R$\vee_2$; under L$\bigwedge$ all the left rules L$\bigwedge_n$ with $n \in \mathbb{N}$; and under R$\bigvee$ all the right rules R$\bigvee_n$ with $n \in \mathbb{N}$.

Up to the end of Section 6, let $\rhd$ *be provability in minimal propositional logic or one of the three inductive extensions listed above;* in particular, $\rhd$ is inductively generated. *Let also* $j\colon S \to S$ *be any map where* S *is a set of formulas as before.*

In the case of first-order logic, j needs to be compatible with substitution [5]:[15]

$$j(A[t/x]) = (jA)[t/x]\,.$$

For this reason, we will simply write $jA[t/x]$.

[13] These rules (and their variations) are usually called P ('possibility') and N ('necessitation') in the literature, and they work simultaneously as left and right rules from a bottom-to-top perspective [4]. We decided to classify them as right rules since this aligns better to their properties (compare Theorem 2). It is worth noting that there is no consensus in the literature on how to define the sub-classical companion of modal logics such as the minimal basic modal logic **mK**. A detailed discussion lies beyond the scope of this note. We follow the approach of [4], since this aligns with our method and gives a representative example of how modal rules can be translated into j-shifts and yield conservation. The same treatment can also be adapted to alternative notions of sub-classical modal logic.

[14] In particular, we do so for the basic modal companions of minimal, intuitionistic, and classical logic **mK**, **iK**, and **K** [4,10,44].

[15] This request is related to the concept of an internal nucleus [45, Section 6].

Remark 7. All left rules listed above but L$\rightarrow$ are admissible in the Kleisli extension $\rhd_j$. If j is expansive, then L$\rightarrow$ too is admissible in $\rhd_j$.

Proof. All left rules but L$\rightarrow$ are right-invariant, and Lemma 1 applies, whereas L$\rightarrow$ in the form of m.p. (Remark 6) is derivable in $\rhd_j$ by L$\rightarrow$ for $\rhd$ and Rj. $\square$

By the j-*shift* for the logical operator $*$ we understand the sequent as follows:

$*$	j-shift for $*$
$\top$	$\top \rhd j\top$
$\wedge$	$jA \wedge jB \rhd j(A \wedge B)$
$\vee$	$jA \vee jB \rhd j(A \vee B)$
$\rightarrow$	$A \rightarrow jB \rhd j(A \rightarrow B)$
$\forall$	$\forall x\, jA \rhd j\forall x\, A$

$*$	j-shift for $*$
$\exists$	$\exists x\, jA \rhd j\exists x\, A$
$\bigwedge$	$\bigwedge jA_i \rhd j \bigwedge A_i$
$\bigvee$	$\bigvee jA_i \rhd j \bigvee A_i$
$\square$	$\square jA \rhd j\square A$
$\Diamond$	$\Diamond jA \rhd j\Diamond A$

Remark 8. If j is a nucleus, then by (6) the j-shift for $\rightarrow$ can equally be written as $jA \rightarrow jB \rhd j(A \rightarrow B)$.

If j is $\neg\neg$, then the j-shifts for $\rightarrow$, $\forall$ and $\bigwedge$ are nothing but the double negation shifts $\mathrm{DNS}_{\rightarrow}$, $\mathrm{DNS}_{\forall}$ and $\mathrm{DNS}_{\bigwedge}$ discussed before.

While the left rules are settled by Remark 7, for the right rules R$*$ the admissibility criterion given in [13, Lemma 5.1] for a nucleus j in the cases $* \in \{\rightarrow, \forall, \bigwedge\}$ carries over to an arbitrary map j and all relevant logical operators:

Theorem 2. *The right rule* R$*$ *is admissible in the Kleisli extension* $\rhd_j$ *if and only if the* j*-shift for* $*$ *is derivable in* $\rhd$.

Proof. Note first that every j-shift is a sequent, for which derivability amounts to the admissibility of the corresponding zero-premiss rule (Remark 1). We only treat the operators $\square$ and $\Diamond$.

R$\square$ Assume that the j-shift for $\square$ is derivable in $\rhd$. If $\Gamma \rhd_j A$, i.e. $\Gamma \rhd jA$, then R$\square$ yields $\square\Gamma \rhd \square jA$, which by transitivity and the j-shift for $\square$ entails $\square\Gamma \rhd j\square A$, i.e. $\square\Gamma \rhd_j \square A$.
Conversely, assume that R$\square$ is admissible in $\rhd_j$. By reflexivity, $jA \rhd jA$, i.e. $jA \rhd_j A$, which by R$\square$ gives $\square jA \rhd_j \square A$, i.e. $\square jA \rhd j\square A$.

R$\Diamond$ Assume that the j-shift for $\Diamond$ is derivable in $\rhd$. If $\Gamma, A \rhd_j B$, i.e. $\Gamma, A \rhd jB$, then R$\Diamond$ yields $\square\Gamma, \Diamond A \rhd \Diamond jB$, which by transitivity and the j-shift for $\Diamond$ entails $\square\Gamma, \Diamond A \rhd j\Diamond B$, i.e. $\square\Gamma, \Diamond A \rhd_j \Diamond B$. Conversely, assume that R$\Diamond$ is admissible in $\rhd_j$. By reflexivity, $jA \rhd jA$, i.e. $jA \rhd_j A$, which by R$\Diamond$ gives $\Diamond jA \rhd_j \Diamond A$, i.e. $\Diamond jA \rhd j\Diamond A$. $\square$

Remark 9. Let j be a nucleus over $\rhd$, and let $* \in \{\wedge, \vee, \top, \to, \exists, \forall, \bigvee, \bigwedge, \Box, \Diamond\}$.

1. The converse $\lhd$ of the j-shift for $*$ is derivable in $\rhd$.
2. If $* \in \{\wedge, \vee, \top, \exists, \bigvee\}$, then the j-shift for $*$ is derivable in $\rhd$.

Proof. We only deal with the case in which $*$ is $\wedge$.

1. By reflexivity, $A \rhd A$, which by L$\wedge_1$ entails $A \wedge B \rhd A$. Applying first Rj and then Lj gives $j(A \wedge B) \rhd jA$. Similarly, $j(A \wedge B) \rhd jB$. Now R$\wedge$ yields $j(A \wedge B) \rhd jA \wedge jB$.
2. By reflexivity and monotonicity, $A, B \rhd A$ and $A, B \rhd B$, which by R$\wedge$ yield $A, B \rhd A \wedge B$. Applying first Rj and then Lj gives $jA, jB \rhd j(A \wedge B)$, which by L$\wedge_1$ and L$\wedge_2$ leads to $jA \wedge jB, jA \wedge jB \rhd j(A \wedge B)$, i.e. $jA \wedge jB \rhd j(A \wedge B)$.

Example 5. If j is a nucleus, then R$\to$ is admissible in $\rhd_j$ if and only if every open nucleus $A \to$ [5,11] commutes with j up to $\lhd\rhd$.

By Remark 9.2, if j is a nucleus, then the admissibility in $\rhd_j$ is an issue at most for the rules R$*$ of the five *critical operators* $* \in \{\to, \forall, \bigwedge, \Box, \Diamond\}$.

Corollary 5. *$\rhd_j = \rhd^j$ precisely when j is a nucleus over $\rhd$ and the j-shifts are derivable in $\rhd$ for all the critical operators $* \in \{\to, \forall, \bigwedge, \Box, \Diamond\}$ relevant for $\rhd$.*

Proof. Theorem 1, Theorem 2, Corollary 3, Remark 7, and Remark 9. □

Double Negation. We look back at the cases of conservation we have started from, with $\neg\neg$ as j and $\vdash_m$ or $\vdash_i$ as $\rhd$, for which $\rhd^j$ always is $\vdash_c$. If $\rhd$ is $\vdash_m$ or one of the inductive extensions given before, then conservation amounts to the admissibility for $\rhd_j$ of the critical right rules R$*$ tantamount to DNS$_*$ as follows:

$\Gamma \vdash_c A \Longrightarrow \Gamma, \mathrm{DNS}_\to \vdash_m \neg\neg A$ in propositional logic (Segerberg)

$\Gamma \vdash_c A \Longrightarrow \Gamma, \mathrm{DNS}_\to, \mathrm{DNS}_\forall \vdash_m \neg\neg A$ in first-order logic (Kuroda)

$\Gamma \vdash_c A \Longrightarrow \Gamma, \mathrm{DNS}_\to, \mathrm{DNS}_\bigwedge \vdash_m \neg\neg A$ in infinitary logic (Tesi)

$\Gamma \vdash_c A \Longrightarrow \Gamma, \mathrm{DNS}_\to, \mathrm{DNS}_\Diamond, \mathrm{DNS}_\Box \vdash_m \neg\neg A$ in basic modal logic

Note that one sometimes needs to stack more than one DNS$_*$. If, however, $\rhd$ is $\vdash_i$, then DNS$_\to$ is derivable in $\rhd$ and R$\to$ is admissible in $\rhd_j$; whence Glivenko's original theorem in propositional logic (2): $\Gamma \vdash_c A \Longrightarrow \Gamma \vdash_i \neg\neg A$.

Acknowledgments. The authors wish to thank Thorsten Altenkirch, Benno van den Berg, Ingo Blechschmidt, Pierluigi Graziani, Hugo Herbelin, Hajime Ishihara, Dominik Kirst, Daniel Misselbeck-Wessel, Mirko Tagliaferri, Matteo Tesi, Tarmo Uustalu, Vincenzo Fano, Niccolò Veltri, Cheng-Syuan Wan, Joshua L. Wrigley, and last but not least all three anonymous referees for their constructive critique.

The research we report in the present note was carried out within the projects "A New Dawn of Intuitionism: Mathematical and Philosophical Advances" (ID 60842) funded by the John Templeton Foundation, and "Reducing complexity in algebra, logic, combinatorics - REDCOM" belonging to the programme "Ricerca Scientifica di Eccellenza 2018" of the Fondazione Cariverona. Parts of the writing were done during the Dagstuhl Seminars 24021 "From Proofs to Computation in Geometric Logic and Generalizations", 24341 "Proof Representations: From Theory to Applications", and 26121 "Proof Systems in Actual Practice: Reasoning and Computation". All three authors are members of the Gruppo Nazionale per le Strutture Algebriche, Geometriche e le loro Applicazioni (GNSAGA) of the Istituto Nazionale di Alta Matematica (INdAM). The second author further acknowledges the MIUR Excellence Department Project for the Dipartimento di Matematica, Università di Genova, CUP D33C23001110001, and the project "Infinity and Intensionality: Towards A New Synthesis" of the Norwegian Research Council (NFR), project no. 314435.

References

1. Aczel, P.: The Russell-Prawitz modality. Math. Struct. Comput. Sci. **11**, 541–554 (2001)
2. Aczel, P., Rathjen, M.: Constructive Set Theory (2010). https://michrathjen.github.io/book.pdf
3. Bartoli, G., Fellin, G., Tesi, M.: Infinitary negative translations and Glivenko logic. Submitted (2025)
4. Bellin, G., de Paiva, V., Ritter, E.: Extended Curry-Howard correspondence for a basic constructive modal logic. In: Proceedings of M4M, vol. 2 (2001)
5. van den Berg, B.: A Kuroda-style j-translation. Arch. Math. Logic **58**, 627–634 (2019)
6. van den Berg, B., Briët, M.: Arrow algebras. Ann. Pure Appl. Logic **177**(3), 1–39 (2026)
7. Borsetto, R., Fellin, G., Uustalu, T., Wan, C.-S.: Glivenko's theorem underneath structure. In: Brattka, V., Fernau, H., Galeotti, L. (eds.) Timeless Machines: Computability Across Eras, CiE 2026 (2026)
8. Brouwer, L.E.J.: Intuitionistische Zerlegung mathematischer Grundbegriffe. Jahresber. Dtsch. Math.-Ver. **33**, 251–256 (1925)
9. Cignoli, R., Torrens, A.: Glivenko-like theorems in natural expansions of BCK-logic. MLQ Math. Log. Q. **50**, 111–125 (2004)
10. Dalmonte, T.: Minimal modal logics, constructive modal logics and their relations. Rev. Symb. Log. **18**(2), 463–504 (2025)
11. Escardó, M., Oliva, P.: The Peirce translation. Ann. Pure Appl. Logic **163**, 681–692 (2012)
12. Fellin, G., Schuster, P.: A general Glivenko–Gödel theorem for nuclei. In: Sokolova, A. (ed.) Proceedings of MFPS, Electronic Notes in Theoretical Computer Science **351**, 51–66 (2021)
13. Fellin, G., Schuster, P.: Conservation as translation. Rev. Symb. Log. **18**(1), 316–348 (2025)
14. Fellin, G., Schuster, P., Wessel, D.: The Jacobson radical of a propositional theory. Bull. Symb. Log. **28**, 163–181 (2022)
15. Galatos, N., Jipsen, P., Kowalski, T., Ono, H.: Residuated Lattices: An Algebraic Glimpse at Substructural Logics. Elsevier (2007)
16. Galatos, N., Ono, H.: Glivenko theorems for substructural logics over FL. J. Symb. Log. **71**, 1353–1384 (2006)

17. Ghilardi, S., Lenzi, G.: Unification in lax logic. J. Algebr. Hyperstruct. Log. Algebras **3**, 61–75 (2022)
18. Glivenko, V.: Sur la logique de M. Brouwer. Académie royale de Belgique. Bulletin de la Classe des Sciences **14**, 225–228 (1928)
19. Glivenko, V.: Sur quelques points de la logique de M. Brouwer. Académie royale de Belgique. Bulletin de la Classe des Sciences **15**, 183–188 (1929)
20. Johansson, I.: Der Minimalkalkül, ein reduzierter intuitionistischer Formalismus. Compos. Math. **4**, 119–136 (1937)
21. Johnstone, P.T.: Sketches of an Elephant: A Topos Theory Compendium, vol. 2. Oxford University Press (2002)
22. Johnstone, P.T.: Stone Spaces. Cambridge University Press (1982)
23. Kock, A.: Strong functors and monoidal monads. Arch. Math. **23**, 113–120 (1972)
24. Kolmogoroff, A.N.: Sur le principe de tertium non datur. Mat. Sbornik **32**, 646–667 (1925)
25. Kuroda, S.: Intuitionistische Untersuchungen der formalistischen Logik. Nagoya Math. J. **3**, 35–47 (1951)
26. Lawvere, F.W.: Introduction. In: Toposes, algebraic geometry and logic (Conf., Dalhousie Univ., Halifax, N.S., 1971. Lecture Notes in Mathematics, vol. 274, pp. 1–12 (1972)
27. MacLane, S., Moerdijk, I.: Sheaves in Geometry and Logic: A First Introduction to Topos Theory. Springer (1994)
28. Manes, E.G.: Algebraic Theories. Springer, New York (1976)
29. Marmolejo, F., Wood, R.J.: Monads as extension systems–no iteration is necessary. Theory Appl. Categ. **24**(4), 84–113 (2010)
30. Moggi, E.: Notions of computation and monads. Inform. and Comput. **93**(1), 55–92 (1991)
31. Negri, S., von Plato, J.: Structural Proof Theory. Cambridge University Press (2001)
32. Negri, S., von Plato, J.: Proof Analysis. Cambridge University Press (2011)
33. Odintsov, S.: Logic of classical refutability and class of extensions of minimal logic. Log. Log. Philos. **9**, 91–107 (2002)
34. Ono, H.: Glivenko theorems revisited. Ann. Pure Appl. Logic **161**, 246–250 (2009)
35. Rinaldi, D., Schuster, P., Wessel, D.: Eliminating disjunctions by disjunction elimination. Bull. Symb. Log. **23**, 181–200 (2017)
36. Rinaldi, D., Schuster, P., Wessel, D.: Eliminating disjunctions by disjunction elimination. Indag. Math. (N.S.) **29**, 226–259 (2018)
37. Schuster, P., Wessel, D.: Resolving finite indeterminacy: a definitive constructive universal prime ideal theorem. In: Proceedings of the 35th Annual ACM/IEEE Symposium on Logic in Computer Science, pp. 820–830 (2020)
38. Scott, D.: On engendering an illusion of understanding. J. Phil. **68**, 787–807 (1971)
39. Scott, D.S.: Background to formalization. In: Leblanc, H. (ed.) Truth, Syntax and Modality (Proc. Conf. Alternative Semantics, Temple Univ., Philadelphia, Pa., 1970). North-Holland, Amsterdam, Studies in Logic and the Foundations of Mathematics, vol. 68, pp. 244–273 (1973)
40. Scott, D.S.: Completeness and axiomatizability in many-valued logic. In: Henkin, L. et al. (eds.) Proceedings of the Tarski Symposium, American Mathematical Society, pp. 411–435 (1974)
41. Segerberg, K.: Propositional logics related to Heyting's and Johansson's. Theoria **34**, 26–61 (1968)
42. Simmons, H.: A framework for topology. In: Macintyre, A., Pacholski, L., Paris, J. (eds.) Logic Colloquium '77, North-Holland, pp. 239–251 (1978)

43. Tesi, M.: Gödel–Gentzen. Scuola Normale Superiore di Pisa (2020)
44. Wolter, F., Zakharyaschev, M.: Intuitionistic modal logic In: Cantini, A., Casari, E., Minari, P. (eds.) Logic and Foundations of Mathematics, vol. 280. Springer, Dordrecht, Synthese Library (1999)
45. Wrigley, J.L.: Some properties of internal locale morphisms externalised. Theory Appl. Categ. **41**, 1160–1202 (2024)

On the Σ-Hierarchy of the Logical Principles Over Intuitionistic Predicate Logic

Makoto Fujiwara(✉)

Department of Applied Mathematics, Faculty of Science Division I, Tokyo University of Science, 1-3 Kagurazaka, Shinjuku-ku, Tokyo 162-8601, Japan
makotofujiwara@rs.tus.ac.jp

Abstract. We investigate the hierarchy of the Σ-variants of several logical principles, which are valid in classical propositional logic but unprovable in intuitionistic propositional logic, over intuitionistic predicate logic. This hierarchy can be viewed as a fine-grained version of both the hierarchy of the Σ_1-fragments of these principles over intuitionistic arithmetic and the hierarchy of the intermediate propositional logics obtained from them.

Keywords: Σ-hierarchy · Logical principles · Intermediate logics · Fragments of classical logic

1 Introduction

There are many logical principles that are classically valid but not provable in intuitionistic theories. Typical examples include

$$
\begin{aligned}
\text{PEM (the principle of excluded middle)} &: \varphi \vee \neg\varphi;\\
\text{WPEM (the weak PEM)} &: \neg\varphi \vee \neg\neg\varphi;\\
\text{RPEM (the restricted PEM)} &: (\varphi \leftrightarrow \neg\psi) \to \varphi \vee \neg\varphi;\\
\text{DML (the De Morgan law)} &: \neg(\varphi \wedge \psi) \to \neg\varphi \vee \neg\psi;\\
\text{WDML (the weak DML)} &: \neg(\neg\varphi \wedge \neg\psi) \to \neg\neg\varphi \vee \neg\neg\psi;\\
\text{DNE (the double negation elimination)} &: \neg\neg\varphi \to \varphi.
\end{aligned}
$$

By adding such principles to intuitionistic propositional logic, we obtain intermediate propositional logics lying between intuitionistic propositional logic **IPC** and classical propositional logic **CPC** (cf. [3, Section 4.1]). On the other hand, the arithmetical hierarchy of such logical principles has been extensively studied in the context of intuitionistic arithmetic and analysis over the past several decades (cf. [1,4–8,10,11,17]).

As mentioned in [6, Section 1], there is some connection between the hierarchy of intermediate propositional logics and the hierarchy of semi-classical arithmetic, which are obtained from intuitionistic arithmetic HA by adding fragments

V. Brattka et al. (Eds.): CiE 2026, LNCS 16674, pp. 289–303, 2027.
https://doi.org/10.1007/978-3-032-31348-5_19

of logical principles. An intermediate propositional logic **L** is said to possess the de Jongh property if the propositional logic of the arithmetic obtained by extending HA with **L** for all HA-formulae coincides exactly with **L**. In [16], de Jongh, Verbrugge and Visser showed that any intermediate propositional logic with the finite frame property possesses the de Jongh property.

However, the hierarchy of intermediate propositional logics and the hierarchy of semi-classical arithmetic with respect to the corresponding logical principles are incomparable. This incomparability arises for the following two reasons. First, intermediate propositional logics are closed under substitution by definition, whereas this is not the case for semi-classical arithmetic in general. Second, in the context of arithmetic, one can use arguments about numbers that cannot be simulated in the propositional setting. In fact, the intermediate propositional logic obtained from

$$\mathrm{LIN}_1 : (\varphi \to \psi) \vee (\psi \to \varphi)$$

is strictly stronger than that of

$$\mathrm{LIN}_2 : (\varphi \to \neg\psi) \vee (\neg\psi \to \varphi),$$

while semi-classical arithmetic $\mathsf{HA} + \Sigma_1\text{-}\mathrm{LIN}_1$ is strictly weaker than $\mathsf{HA} + \Sigma_1\text{-}\mathrm{LIN}_2$, where Σ_1-LIN_1 and Σ_1-LIN_2 are restrictions of LIN_1 and LIN_2 to $\varphi, \psi \in \Sigma_1$ respectively (cf. [9, Section 5]). Motivated by this fact, in [6] and [5], the author investigated six variants of the linearity axiom (including the two above) and revealed both the hierarchy of the intermediate propositional logics and the Σ_n-hierarchy of semi-classical arithmetic with respect to PEM, WPEM, DML, WDML, DNE, $\mathrm{LIN}_1, \dots, \mathrm{LIN}_6$. More recently, in [7], in connection with the Kreisel-Putnam axiom, two additional variants LIN_7 and LIN_8 of the linearity axiom were introduced and it was shown that they yield essentially new layers in the Σ_n-hierarchy of semi-classical arithmetic. The results are summarized in Fig. 1.

On the other hand, there is another way of restricting logical principles in the language of predicate logic. For example, the following restriction of DNE (in the language of HA) is well-known as Markov's principle in the context of constructive mathematics (see [20, Section 4.5.1]):

$$\forall x\, (\varphi(x) \vee \neg\varphi(x)) \wedge \neg\neg\exists x \varphi(x) \to \exists x \varphi(x), \tag{1}$$

which may contain free variables as parameters. In fact, (1) is equivalent over HA to the Σ-variant of DNE, namely,

$$\Sigma\text{-DNE} : \begin{array}{l} \forall x_1, \dots, x_k\, (\varphi(x_1, \dots, x_k) \vee \neg\varphi(x_1, \dots, x_k)) \\ \to (\neg\neg\exists x_1, \dots, x_k \varphi(x_1, \dots, x_k) \to \exists x_1, \dots, x_k \varphi(x_1, \dots, x_k)) , \end{array}$$

in the sense of [9, Definition 3.11]. Moreover, these are equivalent to Σ_1-DNE (allowing function parameters) in a finite-type extension of HA with countable choice (see Proposition 1), which is a standard framework to formalize Bishop's constructive mathematics (cf. [2]). In particular, countable choice is accepted in

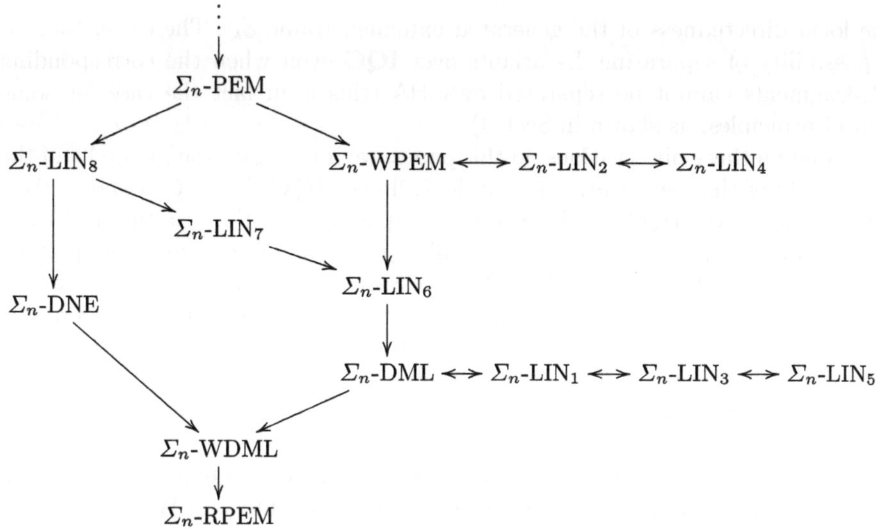

Fig. 1. Hierarchy of the logical axioms over $\mathsf{HA} + \Sigma_{n-1}$-DNE where no implication holds unless it follows by transitivity from the displayed implications (cf. [7])

Bishop's constructive mathematics. These equivalences hold not only for DNE but also for the other logical principles considered in this paper. Note that Σ_1-PEM and Σ_1-DML are well-known as LPO (the limited principle of omniscience) and LLPO (the lesser limited principle of omniscience) respectively in constructive mathematics. Thus Σ-variant is not an artificial formulation but one of the equivalent forms of the restricted logical principles from the perspective of constructive reverse mathematics based on Bishop's constructive mathematics (cf. [14,15]). In contrast to Σ_n-fragments, Σ-variants of logical principles are closed under substitution.

In [9], the author and others established a logical separation theorem for separating Σ_n-fragments of logical principles in the context of HA. In fact, the meta-theorem (cf. [9, Theorem 4.23]) is very useful for this purpose and provides a uniform technique for separating many Σ_n-fragments (cf. [4–7,9]). To apply the logical separation theorem, it suffices to construct a propositional Kripke model $\mathcal{K}$ that refutes one logical principle but satisfies another logical principle (in the sense of extended frames) and to confirm the following two conditions:

1. The Kripke frame $(I_{\mathcal{K}}, \leq_{\mathcal{K}})$ generated by $\mathcal{K}$ is a rooted finite tree;
2. The extended frame $\mathcal{E}_{\mathcal{K}}$ generated by $\mathcal{K}$ is locally directed.

On the other hand, before proving [9, Theorem 4.23], a logical separation theorem in the context of pure intuitionistic predicate logic **IQC** was first presented in [9, Section 3]. This meta-theorem (cf. [9, Theorem 3.15]) provides separation results between Σ-fragments of logical principles over **IQC**. In contrast to the logical separation theorem for HA, the logical separation theorem for **IQC** requires only the local directedness of the generated extended frame $\mathcal{E}_{\mathcal{K}}$. Therefore, there is a possibility of separating Σ-variants over **IQC** even when the corresponding Σ_n-fragments cannot be separated over HA (this is in fact the case for some logical principles, as shown in Sect. 4).

Based on these observations, in this paper, we investigate the hierarchy of the Σ-variants of the logical principles in [6,7,9] over **IQC**. In Sects. 3 and 4, their interrelations over **IQC** are determined completely, and the resulting hierarchy is summarized in Fig. 2. In particular, all separations are obtained by applying the logical separation theorem for **IQC** (see Sect. 4). In addition, we observe that the hierarchy of Σ-variants over **IQC** is a fine-grained version of the hierarchy of Σ_1-fragments of logical principles over HA (see Sect. 5), and simultaneously, that of the hierarchy of intermediate propositional logics obtained from those logical principles (see Sect. 6). Thus the Σ-hierarchy over **IQC** can be viewed as a common framework underlying these two incomparable hierarchies.

We assume some familiarity with the Kripke model techniques developed in [9]. Throughout the paper, n and m denote positive integers. We use classical logic and set theory at the meta-level.

2 Σ-Variants

We use the standard language of intuitionistic propositional logic **IPC** containing the countable set $\mathcal{V}$ of propositional variables, and $\wedge, \vee, \to$ and $\bot$ as primitive logical connectives. We use the abbreviations $\neg\varphi$ and $\varphi \leftrightarrow \psi$ for $\varphi \to \bot$ and $(\varphi \to \psi) \wedge (\psi \to \varphi)$ respectively. In addition, we use the standard language of intuitionistic predicate logic **IQC** containing the propositional connectives and $\forall, \exists$ as primitive logical connectives. See [18] for the description of intuitionistic arithmetic HA.

Notation. For a propositional formula φ with distinct propositional variables $p_1, \dots, p_m$, we write $\varphi[p_1, \dots, p_m]$ to display the distinct propositional variables occurring in φ. For a propositional formula $\varphi[p_1, \dots, p_m]$, $\varphi[\chi_1/p_1, \dots, \chi_m/p_m]$ denotes a formula obtained from φ by substituting propositional variables $p_1, \dots, p_m$ by (propositional or predicate) formulae $\chi_1, \dots, \chi_m$ simultaneously. Let $\boldsymbol{x}$ denote a list of variables $x_1, \dots, x_n$. In the context of arithmetic, $t \neq t'$ denotes $\neg(t = t')$ for (number) terms t and t'.

We first recall the definition of the Σ-variants of propositional formulae.

Definition 1 (cf. [9, Definition 3.11]) *For each propositional formula $\varphi[p_1, \dots, p_m]$, let Σ-φ be the following schema:*

$$\forall \boldsymbol{x}(\psi_1(\boldsymbol{x}) \vee \neg\psi_1(\boldsymbol{x})) \wedge \ldots \wedge \forall \boldsymbol{x}(\psi_m(\boldsymbol{x}) \vee \neg\psi_m(\boldsymbol{x}))$$
$$\to \varphi[\exists \boldsymbol{x}\psi_1(\boldsymbol{x})/p_1, \dots, \exists \boldsymbol{x}\psi_m(\boldsymbol{x})/p_m],$$

where $\psi_1, \dots, \psi_m$ are predicate formulae. Note that $\boldsymbol{x}$ may not list all the free variables in $\psi_1, \dots, \psi_m$.

Proposition 1. *Let* HA^ω *be the finite-type extension of* HA *in [19, Section 1.6.15] and* $\mathrm{AC}^{0,0}$ *denote the following schema of* HA^ω *:*

$$\forall x \exists y \chi(x,y) \to \exists f \forall x \chi(x, fx),$$

where x *and* y *are number variables and* f *is a variable of the function-type. For any propositional formula* $\varphi[p_1,\dots,p_m]$*,* Σ-φ *is equivalent over* $\mathsf{HA}^\omega + \mathrm{AC}^{0,0}$ *to the schema of* Σ^0_1*-substitution fragments of* φ*, namely,*

$$\varphi[\exists x_1 \psi^1_{\mathrm{qf}}/p_1,\dots,\exists x_m \psi^m_{\mathrm{qf}}/p_m] \tag{2}$$

where x_i *is a number variable and* ψ^i_{qf} *is quantifier-free for each* $i \in \{1,\dots,m\}$*.*

Proof. We first show (2) from Σ-φ over HA^ω. Without loss of generality, assume that $x_1,\dots,x_m$ are distinct and x_i does not occur freely in ψ^j_{qf} if $i \neq j$ for all $i,j \in \{1,\dots,m\}$. Since quantifier-free formulae are decidable in HA^ω, for each $i \in \{1,\dots,m\}$, HA^ω proves $\forall x_1,\dots,x_m \left(\psi^i_{\mathrm{qf}} \vee \neg\psi^i_{\mathrm{qf}}\right)$. Therefore, by Σ-φ, we have

$$\varphi[\exists x_1,\dots,x_m \psi^1_{\mathrm{qf}}/p_1,\dots,\exists x_1,\dots,x_m \psi^m_{\mathrm{qf}}/p_m],$$

and hence, $\varphi[\exists x_1 \psi^1_{\mathrm{qf}}/p_1,\dots,\exists x_m \psi^m_{\mathrm{qf}}/p_m]$.

For the converse direction, reasoning in $\mathsf{HA}^\omega + \mathrm{AC}^{0,0}$, assume

$$\forall \boldsymbol{x}(\psi_1(\boldsymbol{x}) \vee \neg\psi_1(\boldsymbol{x})) \wedge \dots \wedge \forall \boldsymbol{x}(\psi_m(\boldsymbol{x}) \vee \neg\psi_m(\boldsymbol{x})).$$

Note that a finite sequence of number terms can be coded as a number term in HA^ω (cf. [19, Section 1.3.9]). For each $i \in \{1,\dots,m\}$, since $\forall \boldsymbol{x}(\psi_i(\boldsymbol{x}) \vee \neg\psi_i(\boldsymbol{x}))$ is equivalent over HA^ω to

$$\forall \boldsymbol{x} \exists k \left((k=0 \to \psi_i(\boldsymbol{x})) \wedge (k \neq 0 \to \neg\psi_i(\boldsymbol{x}))\right),$$

by $\mathrm{AC}^{0,0}$, there exists f_i such that

$$\forall \boldsymbol{x} \left((f_i\boldsymbol{x}=0 \to \psi_i(\boldsymbol{x})) \wedge (f_i\boldsymbol{x} \neq 0 \to \neg\psi_i(\boldsymbol{x}))\right). \tag{3}$$

Then, by (2), we have

$$\varphi[\exists \boldsymbol{x}(f_1\boldsymbol{x}=0)/p_1,\dots,\exists \boldsymbol{x}(f_m\boldsymbol{x}=0)/p_m].$$

Therefore, by (3), we have

$$\varphi[\exists \boldsymbol{x}\psi_1(\boldsymbol{x})/p_1,\dots,\exists \boldsymbol{x}\psi_m(\boldsymbol{x})/p_m].$$

□

Remark 1. In contrast to Proposition 1, Σ-φ is strictly stronger than the schema of Σ_1-substitution fragments of φ over HA. For example, $\mathsf{HA} + \Sigma$-DNE proves Σ_1-DNE while $\mathsf{HA} + \Sigma_1$-DNE does not prove Σ-DNE.

3 Derivability Results

In this paper, as our logical principles, we investigate the following propositional formulae:

$$\begin{aligned}
\mathrm{PEM}[p] &\equiv p \vee \neg p;\\
\mathrm{WPEM}[p] &\equiv \neg p \vee \neg\neg p;\\
\mathrm{RPEM}[p,q] &\equiv (p \leftrightarrow \neg q) \to p \vee \neg p;\\
\mathrm{DML}[p,q] &\equiv \neg(p \wedge q) \to \neg p \vee \neg q;\\
\mathrm{WDML}[p,q] &\equiv \neg(\neg p \wedge \neg q) \to \neg\neg p \vee \neg\neg q;\\
\mathrm{DNE}[p] &\equiv \neg\neg p \to p;\\
\mathrm{LIN}_1[p,q] &\equiv (p \to q) \vee (q \to p);\\
\mathrm{LIN}_2[p,q] &\equiv (p \to \neg q) \vee (\neg q \to p);\\
\mathrm{LIN}_3[p,q] &\equiv (\neg p \to \neg q) \vee (\neg q \to \neg p);\\
\mathrm{LIN}_4[p,q] &\equiv (\neg p \to \neg\neg q) \vee (\neg\neg q \to \neg p);\\
\mathrm{LIN}_5[p,q] &\equiv (\neg\neg p \to \neg\neg q) \vee (\neg\neg q \to \neg\neg p);\\
\mathrm{LIN}_6[p,q] &\equiv (p \to \neg\neg q) \vee (\neg\neg q \to p);\\
\mathrm{LIN}_7[p,q] &\equiv (p \to q) \vee (\neg\neg q \to p);\\
\mathrm{LIN}_8[p,q] &\equiv (\neg\neg p \to q) \vee (\neg\neg q \to p).
\end{aligned}$$

The principles $\mathrm{PEM}, \mathrm{WPEM}, \mathrm{RPEM}, \mathrm{DML}, \mathrm{WDML}, \mathrm{DNE}, \mathrm{LIN}_1, \mathrm{LIN}_2$ were treated already in [9] as examples of logical principles. After that, the hierarchy of those principles with $\mathrm{LIN}_3, \dots, \mathrm{LIN}_6$ was studied systematically in [6] (see also [5] for the erratum). The principles LIN_7 and LIN_8 were considered recently in [7].

Lemma 1. IPC *proves the following:*

1. $\mathrm{PEM}[p] \to \mathrm{LIN}_8[p,q]$;
2. $\mathrm{LIN}_8[p/p, p/q] \to \mathrm{DNE}[p]$;
3. $\mathrm{LIN}_8[p,q] \to \mathrm{LIN}_7[p,q]$;
4. $\mathrm{PEM}[p] \to \mathrm{WPEM}[p]$;
5. $\mathrm{WPEM}[q/p] \to \mathrm{LIN}_2[p,q]$;
6. $\mathrm{LIN}_2[p,q] \to \mathrm{LIN}_4[p,q]$;
7. $\mathrm{LIN}_4[p/p, p/q] \to \mathrm{WPEM}[p]$;
8. $\mathrm{WPEM}[q/p] \to \mathrm{LIN}_6[p,q]$;
9. $\mathrm{LIN}_7[p,q] \to \mathrm{LIN}_6[p,q]$;
10. $\mathrm{LIN}_7[p,q] \to \mathrm{LIN}_1[p,q]$;
11. $\mathrm{LIN}_1[p,q] \to \mathrm{LIN}_3[p,q]$;
12. $\mathrm{LIN}_6[p,q] \to \mathrm{LIN}_3[p,q]$;
13. $\mathrm{LIN}_3[p,q] \to \mathrm{LIN}_5[p,q]$;
14. $\mathrm{LIN}_5[p,q] \to \mathrm{LIN}_3[p,q]$;
15. $\mathrm{LIN}_5[p,q] \to \mathrm{DML}[p,q]$;
16. $\mathrm{LIN}_3[p,q] \to \mathrm{WDML}[p,q]$;

17. DML$[p,q] \to$ RPEM$[p,q]$;
18. WDML$[p,q] \to$ RPEM$[p,q]$.

Proof. The proofs of (1),..., (14) are trivial.
(15): Assume $\neg(p \wedge q)$. By LIN$_5[p,q]$, we have $(\neg\neg p \to \neg\neg q) \vee (\neg\neg q \to \neg\neg p)$. In the former case, we have that p implies $\neg\neg p \wedge \neg\neg q$, which is equivalent to $\neg\neg(p \wedge q)$, which contradicts our assumption. Thus we have $\neg p$. In the latter case, we have $\neg q$ analogously. Then, in any case, we have $\neg p \vee \neg q$.
(16): Assume $\neg(\neg p \wedge \neg q)$. By LIN$_3[p,q]$, we have $(\neg p \to \neg q) \vee (\neg q \to \neg p)$. In the former case, by our assumption, we have $\neg\neg p$ in a similar way to the proof of (15). In the latter case, we have $\neg\neg q$ analogously. Then, in any case, we have $\neg\neg p \vee \neg\neg q$.
(17) Assume $p \leftrightarrow \neg q$. Then we have $\neg(p \wedge q)$. By DML$[p,q]$, we have $\neg p \vee \neg q$. In the latter case, by our assumption, we have p. Then, in any case, we have $p \vee \neg p$.
(18) Assume $p \leftrightarrow \neg q$. Then we have $\neg(\neg p \wedge \neg q)$. By WDML$[p,q]$, we have $\neg\neg p \vee \neg\neg q$. In the former case, by our assumption, we have p. In the latter case, again by our assumption, we have $\neg p$. Then, in any case, we have $p \vee \neg p$.

Remark 2. For any propositional formulae φ and ψ, if ψ is derivable from some propositional substitution instance of φ over **IPC**, then we have that **IQC** + Σ-φ proves Σ-ψ by mimicking the proof. Therefore, by Lemma 1, we have the derivability results between the corresponding Σ-variants of the logical principles over **IQC**.

Lemma 2. *The schema Σ-WDML is provable in* **IQC** + *Σ-DNE.*

Proof. Assume $\forall \boldsymbol{x}(\psi_1(\boldsymbol{x}) \vee \neg\psi_1(\boldsymbol{x}))$, $\forall \boldsymbol{x}(\psi_2(\boldsymbol{x}) \vee \neg\psi_2(\boldsymbol{x}))$, and $\neg(\neg\exists \boldsymbol{x}\psi_1(\boldsymbol{x}) \wedge \neg\exists \boldsymbol{x}\psi_2(\boldsymbol{x}))$. Then, by intuitionistic reasoning, we have $\neg\neg(\exists \boldsymbol{x}\psi_1(\boldsymbol{x}) \vee \exists \boldsymbol{x}\psi_2(\boldsymbol{x}))$, which is equivalent to $\neg\neg\exists \boldsymbol{x}(\psi_1(\boldsymbol{x}) \vee \psi_2(\boldsymbol{x}))$. On the other hand, by our decidability assumptions, we have

$$\forall \boldsymbol{x}\left((\psi_1(\boldsymbol{x}) \vee \psi_2(\boldsymbol{x})) \vee \neg(\psi_1(\boldsymbol{x}) \vee \psi_2(\boldsymbol{x}))\right).$$

Therefore, by Σ-DNE, we have $\exists \boldsymbol{x}(\psi_1(\boldsymbol{x}) \vee \psi_2(\boldsymbol{x}))$, which is equivalent to $\exists \boldsymbol{x}\psi_1(\boldsymbol{x}) \vee \exists \boldsymbol{x}\psi_2(\boldsymbol{x})$. Then we have $\neg\neg\exists \boldsymbol{x}\psi_1(\boldsymbol{x}) \vee \neg\neg\exists \boldsymbol{x}\psi_2(\boldsymbol{x})$. □

As a conclusion, by Lemma 1, Remark 2 and Lemma 2, we have the derivability results between the Σ-variants over **IQC** displayed in Fig. 2.

4 Underivability Results

First, we briefly recall some notions in [9]. A **frame** is a nonempty partially ordered set $(K, \leq)$. For a frame $(K, \leq)$ and $k \in K$, $\uparrow k$ denotes the upward closed subset $\{k' \in K \mid k \leq k'\}$ of K. An **IPC-Kripke model** is a triple $\mathcal{K} = (K, \leq, \Vdash)$, where $(K, \leq)$ is a frame, and $\Vdash$ is a valuation on it. A formula φ is **valid** in $\mathcal{K}$ if $k \Vdash \varphi$ for all $k \in K$ and we then write $\mathcal{K} \Vdash \varphi$. In addition, φ is valid on $(K, \leq)$ if $\mathcal{K} \Vdash \varphi$ for any valuation $\Vdash$ on $(K, \leq)$. See e.g. [20, Section 2.5.2] for more details.

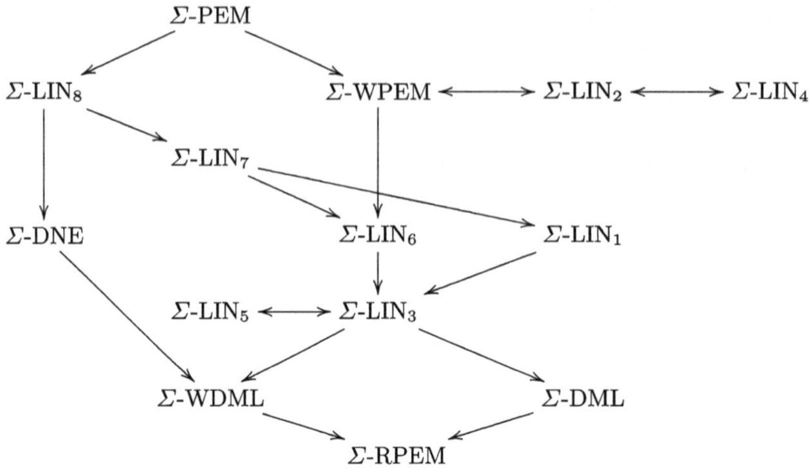

Fig. 2. Derivability relations between the Σ-variants over **IQC**.

Definition 2 (cf. [9, Definitions 2.3 and 2.9, and Remark 2.6]). *An* **extended frame** $\mathcal{E} = ((K, \leq), f, (I, \leq_I))$ *is a triple of frames* $(K, \leq)$ *and* $(I, \leq_I)$*, and a monotone mapping* f *between them, that is,* $k \leq k'$ *implies* $f(k) \leq_I f(k')$ *for each* $k, k' \in K$*. For a frame* $(K, \leq)$*,* $X \subseteq K$ *is* **directed** *if* $l, l' \in X$*, then there exists* $l'' \in X$ *such that* $l'' \leq l$ *and* $l'' \leq l'$*. An extended frame* $\mathcal{E} = ((K, \leq), f, (I, \leq_I))$ *is* **locally directed** *if* $f^{-1}(\uparrow i) \cap \uparrow k$ *is directed for all* $i \in I$ *and* $k \in K$*. Each IPC-Kripke model* $\mathcal{I} = (I, \leq_I, \Vdash)$ *induces an IPC-Kripke model* $\mathcal{K}_{\mathcal{E},\mathcal{I}} = (K, \leq, \Vdash_{\mathcal{E},\mathcal{I}})$ *by defining*

$$k \Vdash_{\mathcal{E},\mathcal{I}} p :\Leftrightarrow f(k) \Vdash p$$

for each $k \in K$ *and propositional variable* p*. A propositional formula* φ *is* **valid** *on* $\mathcal{E}$ *if* $\mathcal{K}_{\mathcal{E},\mathcal{I}} \Vdash_{\mathcal{E},\mathcal{I}} \varphi$ *for each IPC-Kripke model* $\mathcal{I} = (I, \leq_I, \Vdash)$*, that is, for each valuation* $\Vdash$ *on* $(I, \leq_I)$*; we then write* $\mathcal{E} \models \varphi$*. For an extended frame* $\mathcal{E}$*, define* $T(\mathcal{E}) := \{\varphi \mid \mathcal{E} \models \varphi\}$*. Then* Σ*-*$T(\mathcal{E})$ *denotes the schema consisting of* Σ*-*φ *where* $\varphi \in T(\mathcal{E})$*.*

Definition 3 (cf. [9, Example 2.4]). *Let* $\mathcal{K} = (K, \leq, \Vdash)$ *be an IPC-Kripke model, and define a set* $\Phi_{\mathcal{K}}$ *of upward closed subsets of* K *by*

$$\Phi_{\mathcal{K}} := \{\{k \in K \mid k \Vdash p\} \mid p \text{ is a propositional variable}\}.$$

Define binary relations $\preceq_{\mathcal{K}}$ *and* $\sim_{\mathcal{K}}$ *on* K *by*

$$k \preceq_{\mathcal{K}} k' :\Leftrightarrow k \in U \text{ implies } k' \in U \text{ for all } U \in \Phi_{\mathcal{K}},$$
$$k \sim_{\mathcal{K}} k' :\Leftrightarrow k \preceq_{\mathcal{K}} k' \text{ and } k' \preceq_{\mathcal{K}} k.$$

Then $\preceq_{\mathcal{K}}$ *is a preorder and* $\sim_{\mathcal{K}}$ *is an equivalence relation on* K*. Let*

$$I_{\mathcal{K}} := K/\sim_{\mathcal{K}}, \qquad [k]_{\mathcal{K}} \leq_{\mathcal{K}} [k']_{\mathcal{K}} :\Leftrightarrow k \preceq_{\mathcal{K}} k', \qquad f_{\mathcal{K}}(k) := [k]_{\mathcal{K}},$$

where $[k]_{\mathcal{K}}$ *is the equivalence class of* k *with respect to* $\sim_{\mathcal{K}}$. *Then*

$$\mathcal{E}_{\mathcal{K}} := ((K, \leq), f_{\mathcal{K}}, (I_{\mathcal{K}}, \leq_{\mathcal{K}}))$$

is an extended frame, and we call it the extended frame **generated by** *the IPC-Kripke model* $\mathcal{K}$.

The following is the logical separation theorem for **IQC**:

Theorem 1 (cf. [9, Theorem 3.15]). *Let* $\mathcal{K} = (K, \leq, \Vdash)$ *be an finite IPC-Kripke model and* φ *be a propositional formula. If* $\mathcal{K} \nVdash \varphi$ *and* $\mathcal{E}_{\mathcal{K}}$ *is locally directed, then*

$$L(K, \leq)^* + \Sigma\text{-}T(\mathcal{E}_{\mathcal{K}}) \nvdash \Sigma\text{-}\varphi,$$

where $L(K, \leq)^*$ *is the set of schemata of* $\varphi'[\psi_1/p_1, \dots, \psi_m/p_m]$ *for predicate formulae* $\psi_1, \dots, \psi_m$ *and propositional formulae* $\varphi'[p_1, \dots, p_m]$ *which are valid on the frame* $(K, \leq)$.

The following is the logical separation theorem for HA:

Theorem 2 (cf. [9, Theorems 4.17 and 4.23]). *Let* $\mathcal{K} = (K, \leq, \Vdash)$ *be a finite IPC-Kripke model such that* $(I_{\mathcal{K}}, \leq_{\mathcal{K}})$ *is a rooted tree, and let* φ *be a propositional formula. If* $\mathcal{K} \nVdash \varphi$ *and* $\mathcal{E}_{\mathcal{K}}$ *is locally directed, then for each* n,

$$\mathsf{HA} + \Sigma_n\text{-}PEM + L(K, \leq)^* + \Sigma\text{-}T(\mathcal{E}_{\mathcal{K}}) \nvdash \Sigma_{n+1}\text{-}\varphi.$$

For the separation of Σ-φ_1 from Σ-φ_2 by Theorem 1, it suffices to construct a finite IPC-Kripke model $\mathcal{K} = (K, \leq, \Vdash)$ such that

1. $\mathcal{K} \nVdash \varphi_1$;
2. $\varphi_2 \in L(K, \leq) \cup T(\mathcal{E}_{\mathcal{K}})$ where $L(K, \leq)$ is the set of valid formulae in $(K, \leq)$;
3. $\mathcal{E}_{\mathcal{K}}$ is locally directed.

For the application of Theorem 2, it is additionally required that $(I_{\mathcal{K}}, \leq_{\mathcal{K}})$ is a rooted tree. The fourth requirement is necessary for Theorem 2 as shown in [9, Section 5].

Fortunately, all the separations in Fig. 1 are done by constructing finite IPC-Kripke models which satisfies the above four conditions and applying Theorem 2 (cf. [5–7,9]). Therefore, by applying Theorem 1 instead of Theorem 2, we obtain the corresponding separation results between Σ-variants over **IQC**. Furthermore, for the application of Theorem 1, we do not need the fourth condition that $(I_{\mathcal{K}}, \leq_{\mathcal{K}})$ is a rooted tree. According to the relaxation of the assumptions, additional separations between Σ-variants over **IQC** become possible as we show in what follows.

Theorem 3. **IQC** + WPEM *does not prove* Σ-LIN$_1$.

Proof. Let p and q be distinct propositional variables. Consider a finite IPC-Kripke model $\mathcal{K} = (K, \leq, \Vdash)$ given by $K := \{0, 1, 2, 3\}$, $\leq := \{(0,1), (0,2), (1,3), (2,3), (0,3)\}$ and $\Vdash := \{(1,p), (3,p), (2,q), (3,q)\}$. This is exactly the same model as $\mathcal{K}_6 = (K_6, \leq_6, \Vdash_6)$ in [9, Example 5.1]. Then it is straightforward to see that $(p \to q) \vee (q \to p)$ is not valid in the model $\mathcal{K}$ and also that $\mathcal{E}_{\mathcal{K}}$ is locally-directed (which is in fact not required in this case by the proof of Theorem 1). Since $\neg\varphi \vee \neg\neg\varphi$ is valid on the frame $(K, \leq)$, by Theorem 1, we have that $\mathbf{IQC} + \mathrm{WPEM}$ does not prove Σ-LIN_1. □

In contrast to the fact that $\mathsf{HA} + \Sigma$-DML proves Σ-WDML (established essentially in [13, Proposition 1.(2)]), we have the following underivability result:

Theorem 4. $\mathbf{IQC} + \Sigma$*-DML does not prove* Σ*-WDML.*

Proof. Let p, q, r and s be distinct propositional variables. Consider a finite IPC-Kripke model $\mathcal{K} = (K, \leq, \Vdash)$ given by $K := \{0, 1, 2, 3, 4, 5\}$,

$$\leq := \{(0,1), (0,2), (1,3), (1,4), (2,4), (2,5), (0,3), (0,4), (0,5)\}$$

and $\Vdash := \{(1,p), (3,p), (4,p), (2,q), (4,q), (5,q)\}$, which is depicted in Fig. 3.

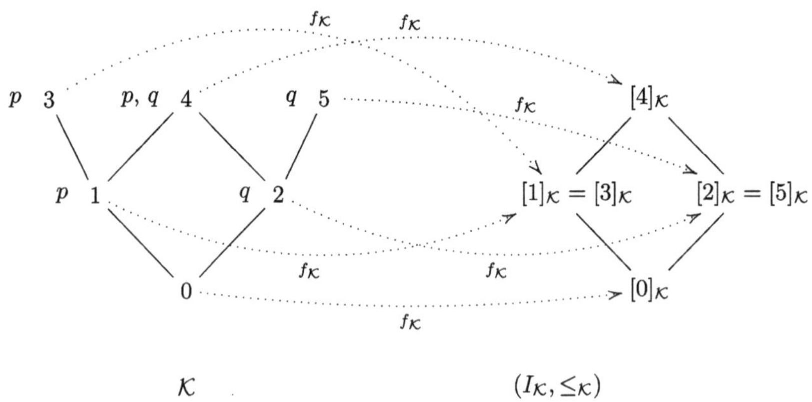

Fig. 3. IPC-Kripke model $\mathcal{K} = (K, \leq, \Vdash)$ which separates Σ-WDML from Σ-DML over **IQC**.

Then $\neg(\neg p \wedge \neg q) \to \neg\neg p \vee \neg\neg q$ is not valid in the model $\mathcal{K}$. In the following, we claim that $T(\mathcal{E}_{\mathcal{K}})$ contains $\neg(r \wedge s) \to \neg r \vee \neg s$. Consider an IPC-Kripke model $\mathcal{I} = (I_{\mathcal{K}}, \leq_{\mathcal{K}}, \Vdash')$. If $[4]_{\mathcal{K}} \Vdash' r$ and $[4]_{\mathcal{K}} \Vdash' s$, for $i \in \{0, 1, 2, 4\}$, we have $i \nVdash_{\mathcal{E}_{\mathcal{K}}, \mathcal{I}} \neg(r \wedge s)$. In addition, for $i \in \{3, 5\}$, if $i \Vdash_{\mathcal{E}_{\mathcal{K}}, \mathcal{I}} \neg(r \wedge s)$, then $i \Vdash_{\mathcal{E}_{\mathcal{K}}, \mathcal{I}} \neg r$ or $i \Vdash_{\mathcal{E}_{\mathcal{K}}, \mathcal{I}} \neg s$. Therefore we have $0 \Vdash_{\mathcal{E}_{\mathcal{K}}, \mathcal{I}} \neg(r \wedge s) \to \neg r \vee \neg s$. Otherwise, $[4]_{\mathcal{K}} \nVdash' r$ or $[4]_{\mathcal{K}} \nVdash' s$. Then we have $0 \Vdash_{\mathcal{E}_{\mathcal{K}}, \mathcal{I}} \neg r$ or $0 \Vdash_{\mathcal{E}_{\mathcal{K}}, \mathcal{I}} \neg s$, and hence, $0 \Vdash_{\mathcal{E}_{\mathcal{K}}, \mathcal{I}} \neg(r \wedge s) \to \neg r \vee \neg s$ in any case. In addition, it is straightforward to see that $\mathcal{E}_{\mathcal{K}}$ is locally directed. Therefore, by Theorem 1, we have that $\mathbf{IQC} + \Sigma$-DML does not prove Σ-WDML. □

Remark 3. For each Kripke model $\mathcal{K}$ in the proofs of Theorems 3 and 4, $(I_{\mathcal{K}}, \leq_{\mathcal{K}})$ is not a rooted tree, and hence, Theorem 2 is not applicable.

By Theorems 3 and 4 together with the remark before Theorem 3, we can conclude that no implication holds unless it follows by transitivity from the displayed implications in Fig. 2.

5 Comparison with the Hierarchy of the Σ_1-Substitution Fragments over HA

As shown in Sect. 4 (see Fig. 2), **IQC** + Σ-DML does not prove Σ-LIN$_1$ (even Σ-WDML). However, the following lemma shows that Σ-DML implies Σ-LIN$_1$ in the presence of the basic axioms of arithmetic:

Lemma 3. HA + *Σ-DML proves Σ-LIN$_1$.*

Proof. The proof is obtained by imitating the proof of the fact that HA+Σ_1-DML proves Σ_1-LIN$_1$ in [6, Lemma 11]. In particular, the proof uses a comparison argument about numbers, which is not available without the basic axioms of arithmetic. □

Therefore, by Lemma 1, Remark 2, Lemmas 2 and 3, we have the derivability results between the Σ-variants over HA displayed in Fig. 4. On the other hand, our separation proofs between Σ_1-substitution fragments of the logical principles in [5–7,9] are based on Theorem 2 (especially [9, Theorem 4.17]). Since Σ-φ implies the Σ_1-substitution fragments of φ over HA, the proofs show the corresponding separation results between Σ-variants over HA. Therefore no implication holds unless it follows by transitivity from the displayed implications in Fig. 4.

Note that Fig. 4 has exactly the same structure as Fig. 1. In this sense, the hierarchy of Σ-variants over **IQC** is a fine-grained version of the hierarchy of Σ_1-fragments over HA.

6 Comparison with the Hierarchy of the Intermediate Propositional Logics

We recall the standard definition of intermediate propositional logic (see [3, Section 4.1] for a comprehensive treatment). A set **L** of propositional formulae is **intermediate propositional logic** if it satisfies the following three conditions:

1. **IPC** $\subseteq$ **L** $\subseteq$ **CPC**, where **IPC** and **CPC** are the sets of theorems of intuitionistic and classical propositional logic respectively;
2. If $\varphi \to \psi$ and φ are in **L**, then ψ is in **L**;
3. If $\varphi[p_1, \dots, p_m]$ is in **L**, then $\varphi[\chi_1/p_1, \dots, \chi_m/p_m]$ is in **L** for any propositional formulae $\chi_1, \dots, \chi_m$.

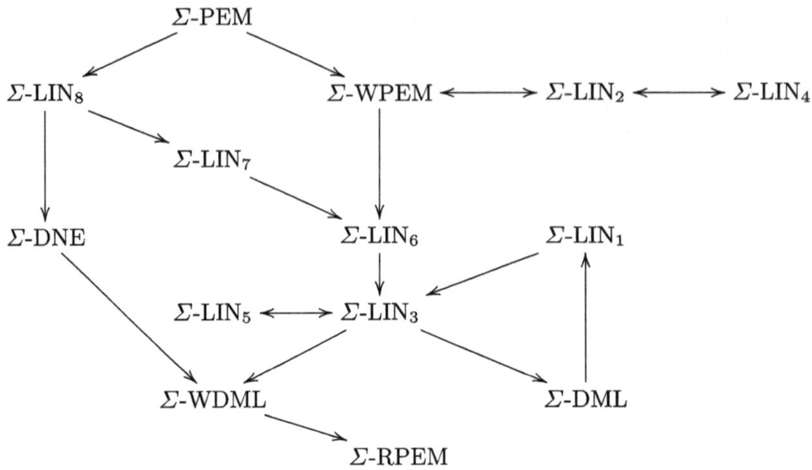

Fig. 4. Derivability relations between the Σ-variants over HA.

For a propositional formula φ, let $\mathbf{L}(\varphi)$ denote the minimum set S of propositional formulae such that $\mathbf{IPC} \cup \{\varphi\} \subseteq S$ and S satisfies the second and third conditions. If φ is a propositional formula which is classically valid, then $\mathbf{L}(\varphi)$ is an intermediate propositional logic. By the third condition, $\mathbf{L}(\varphi)$ does not depend on the choice of the propositional variables used to express φ. Note that for each IPC-Kripke frame $(K, \leq)$, $L(K, \leq)$ forms an intermediate propositional logic.

The hierarchy of the intermediate logics for PEM, WPEM, DML, WDML, DNE, $\mathrm{LIN}_1, \dots, \mathrm{LIN}_6$ has been established already in [6, Section 2]. In what follows, we determine the status of $\mathbf{L}(\mathrm{LIN}_7), \mathbf{L}(\mathrm{LIN}_8)$ and $\mathbf{L}(\mathrm{RELM})$ in the hierarchy, and demonstrate that our Σ-hierarchy in Fig. 2 is a fine-grained version of the hierarchy of the intermediate logics.

Lemma 4. IPC *proves the following*:

1. $\mathrm{DNE}[p \vee \neg p \,/\, p] \to \mathrm{PEM}[p]$;
2. $\mathrm{RPEM}[\neg p/p, \neg\neg p/q] \to \mathrm{WPEM}[p]$.

Proof. (1): Since **IPC** proves $\neg\neg(p \vee \neg p)$, by $\mathrm{DNE}[p \vee \neg p \,/\, p]$, we have $p \vee \neg p$. (2) Since **IPC** proves $\neg p \leftrightarrow \neg\neg\neg p$, by $\mathrm{RPEM}[\neg p/p, \neg\neg p/q]$, we have $\neg p \vee \neg\neg p$. □

By Lemmata 1 and 4, we have the following:

Proposition 2 (Extensions of [6, Propositions 3 and 4]).

1. $\mathbf{CPC} = \mathbf{L}(\mathrm{PEM}) = \mathbf{L}(\mathrm{DNE}) = \mathbf{L}(\mathrm{LIN}_8)$.
2. $\mathbf{L}(\mathrm{WPEM}) = \mathbf{L}(\mathrm{DML}) = \mathbf{L}(\mathrm{WDML}) = \mathbf{L}(\mathrm{LIN}_2) = \mathbf{L}(\mathrm{LIN}_3) = \mathbf{L}(\mathrm{LIN}_4) = \mathbf{L}(\mathrm{LIN}_5) = \mathbf{L}(\mathrm{LIN}_6) = \mathbf{L}(\mathrm{RPEM})$.

Lemma 5. *1.* $\mathbf{L}(\mathrm{LIN}_7)$ *does not contain* $\mathrm{PEM}[p]$;
2. $\mathbf{L}(\mathrm{LIN}_1)$ *does not contain* $\mathrm{LIN}_7[p,q]$.

Proof. (1): Let p be a propositional variable. Consider a finite IPC-Kripke model $\mathcal{K} = (K, \leq, \Vdash)$ given by $K := \{0,1\}$, $\leq := \{(0,1)\}$ and $\Vdash := \{(1,p)\}$. Then it is straightforward to see that $p \vee \neg p$ is not valid in the model $\mathcal{K}$. By the soundness of the Kripke semantics, it suffices to show that $(\varphi \to \psi) \vee (\neg\neg\psi \to \varphi)$ is valid on the frame $(K, \leq)$. Fix a valuation $\Vdash'$ on $(K, \leq)$. Suppose that $0 \nVdash' (\varphi \to \psi) \vee (\neg\neg\psi \to \varphi)$. Since $1 \Vdash' (\varphi \to \psi) \vee (\neg\neg\psi \to \varphi)$, we have $0 \nVdash' \varphi \to \psi$ and $0 \nVdash' \neg\neg\psi \to \varphi$. By $0 \nVdash' \neg\neg\psi \to \varphi$, we have that $1 \Vdash' \psi$ and $0 \nVdash' \varphi$, and hence, $0 \Vdash' \varphi \to \psi$. This is a contradiction. Thus we have shown that $0 \Vdash' (\varphi \to \psi) \vee (\neg\neg\psi \to \varphi)$.
(2): Let p and q be distinct propositional variables. Consider a finite IPC-Kripke model $\mathcal{K} = (K, \leq, \Vdash)$ given by $K := \{0,1,2\}$, $\leq := \{(0,1),(1,2),(0,2)\}$ and $\Vdash := \{(1,p),(2,p),(2,q)\}$. Then $1 \nVdash p \to q$ and $0 \nVdash \neg\neg q \to p$. Therefore, we have that $(p \to q) \vee (\neg\neg q \to p)$ is not valid in the model $\mathcal{K}$. Since $\mathbf{L}(\mathrm{LIN}_1)$ is sound (and even complete) for IPC-Kripke models with finite linear frames and our frame $(K, \leq)$ is linear, we have that $\mathbf{L}(\mathrm{LIN}_1)$ does not contain $\mathrm{LIN}_7[p,q]$. □

By Lemmata 1 and 5, we have the following:

Theorem 5. $\mathbf{L}(\mathrm{PEM}) \supsetneq \mathbf{L}(\mathrm{LIN}_7) \supsetneq \mathbf{L}(\mathrm{LIN}_1)$.

Remark 4 (due to Satoru Niki) Since $\mathbf{L}(\mathrm{LIN}_7)$ contains $\neg q \vee \neg\neg q$ (derived from Lemma 1 and Lemma 4.(2)) and **IPC** proves

$$((p \to q) \vee (\neg\neg q \to p)) \wedge (\neg q \vee \neg\neg q) \to (p \vee (p \to q) \vee \neg q),$$

we have that $\mathbf{L}(\mathrm{LIN}_7)$ contains $p \vee (p \to q) \vee \neg q$. On the other hand, it is straightforward to see that **IPC** proves $(p \vee (p \to q) \vee \neg q) \to (p \to q) \vee (\neg\neg q \to p)$. Therefore we have $\mathbf{L}(\mathrm{LIN}_7) = \mathbf{L}(p \vee (p \to q) \vee \neg q)$. In fact, $\mathbf{L}(p \vee (p \to q) \vee \neg q)$ is an intermediate propositional logic that is sound and complete for the IPC-Kripke models with the frame $(K, \leq)$ in the proof of Lemma 5.(1) (cf. [12, Lemma 3.4]).

By Proposition 2 and Theorem 5 together with [6, Proposition 5], we have the hierarchy of the intermediate propositional logics displayed in Table 1. In fact, the entailment relations are obtained from the Σ-hierarchy in Fig. 2 together with Lemma 4. The separations are given by using the fact that these intermediate propositional logics are valid on certain Kripke frames. The fine-grained Σ-hierarchy in Fig. 2 is obtained by the logical separation theorem for **IQC** (Theorem 1), which is in fact a generalization of the separation arguments by Kripke frames.

Table 1. Hierarchy of the intermediate propositional logics

$$
\begin{array}{c}
\mathrm{PEM} = \mathrm{DNE} = \mathrm{LIN}_8 \\
\cup \\
\mathrm{LIN}_7 \\
\cup \\
\mathrm{LIN}_1 \\
\cup \\
\mathrm{WPEM} = \mathrm{RPEM} = \mathrm{DML} = \mathrm{WDML} = \mathrm{LIN}_2 = \mathrm{LIN}_3 = \mathrm{LIN}_4 = \mathrm{LIN}_5 = \mathrm{LIN}_6
\end{array}
$$

Acknowledgments. The author thanks Satoru Niki for his comments and the observations in Remark 4. The author also thanks Tenyo Takahashi for helpful discussion. This work is supported by JSPS KAKENHI Grant Numbers JP21KK0045 and JP23K03205.

References

1. Akama, Y., Berardi, S., Hayashi, S., Kohlenbach, U.: An arithmetical hierarchy of the law of excluded middle and related principles. In: Ganzinger, H. (ed.) Proceedings of the 19th Annual IEEE Symposium on Logic in Computer Science (LICS), pp. 192–201. IEEE Computer Society Press, Washington (2004)
2. Bishop, E.: Foundations of Constructive Analysis. McGraw-Hill Book Co., New York-Toronto (1967)
3. Chagrov, A., Zakharyaschev, M.: Modal Logic. The Clarendon Press, Oxford University Press, New York (1997)
4. Fujiwara, M.: Δ^0_1 variants of the law of excluded middle and related principles. Arch. Math. Logic **61**(7–8), 1113–1127 (2022)
5. Fujiwara, M.: Correction of the proof of Theorem 16 in [6] (unpublished note) (2026). https://researchmap.jp/makotofujiwara/misc/53544109?lang=en
6. Fujiwara, M.: On the hierarchy of linearity axioms. Arch. Math. Logic **65**(3), 319–331 (2026)
7. Fujiwara, M.: On the Σ^0_1-fragments of the Kreisel-Putnam axiom and two variants of the linearity axiom in intuitionistic arithmetic and analysis. Arch. Math. Logic. 65(5), 675–690, (2026). published online first
8. Fujiwara, M., Ishihara, H., Nemoto, T.: Some principles weaker than Markov's principle. Arch. Math. Logic **54**(7–8), 861–870 (2015)
9. Fujiwara, M., Ishihara, H., Nemoto, T., Suzuki, N.Y., Yokoyama, K.: Extended frames and separations of logical principles. Bull. Symb. Log. **29**(3), 311–353 (2023)
10. Fujiwara, M., Kohlenbach, U.: Interrelation between weak fragments of double negation shift and related principles. J. Symb. Log. **83**(3), 991–1012 (2018)
11. Fujiwara, M., Kurahashi, T.: Refining the arithmetical hierarchy of classical principles. Math. Log. Q. **68**(3), 318–345 (2022)
12. Hosoi, T.: The axiomatization of the intermediate propositional systems S_n of Gödel. J. Fac. Sci., Univ. Tokyo, Sect. I **13**, 183–187 (1966)
13. Ishihara, H.: Markov's principle, Church's thesis and Lindelöf's theorem. Indag. Math. (N.S.) **4**(3), 321–325 (1993)

14. Ishihara, H.: Reverse mathematics in Bishop's constructive mathematics. Philosophia Scientiae CS **6**, 43–59 (2006)
15. Ishihara, H.: An introduction to constructive reverse mathematics. In: Bridges, D., Ishihara, H., Rathjen, M., Schwichtenberg, H. (eds.) Handbook of Constructive Mathematics, pp. 636–660. Cambridge University Press, Cambridge (2023)
16. de Jongh, D., Verbrugge, R., Visser, A.: Intermediate logics and the de Jongh property. Arch. Math. Logic **50**(1–2), 197–213 (2011)
17. Kohlenbach, U.: On the disjunctive Markov principle. Stud. Logica. **103**(6), 1313–1317 (2015)
18. Smoryński, C.A.: Applications of Kripke models. In: Metamathematical Investigation of Intuitionistic Arithmetic and Analysis, pp. 324–391. Springer, Berlin, Heidelberg (1973)
19. Troelstra, A.S. (ed.): Metamathematical Investigation of Intuitionistic Arithmetic and Analysis, Lecture Notes in Mathematics, vol. 344. Springer-Verlag, Berlin (1973)
20. Troelstra, A.S., van Dalen, D.: Constructivism in Mathematics: A An Introduction, Vol. I, Studies in Logic and the Foundations of Mathematics, vol. 121. North-Holland Publishing Co., Amsterdam (1988)

Towards Settling the Complexity of the Lettericity Problem

Mario Grobler[1], Nils Morawietz[2,3], and Silas Cato Sacher[4(✉)]

[1] University of Bremen, Bremen, Germany
grobler@uni-bremen.de
[2] LaBRI, CNRS, University of Bordeaux, Bordeaux INP, Talence, France
nils.morawietz@uni-jena.de
[3] Institute of Computer Science, Friedrich Schiller University Jena, Jena, Germany
[4] Fachbereich IV, Informatikwissenschaften, Universität Trier, Trier, Germany
sacher@uni-trier.de

Abstract. The *lettericity* of a graph $G = (V, E)$ is defined as the smallest size of an alphabet Σ such that there is a word $w_1 \dots w_{|V|} \in \Sigma^*$ and a decoder $\mathcal{D} \subseteq \Sigma^2$ with the property that G is isomorphic to the *letter graph* $G(\mathcal{D}, w)$, that is, the graph with vertex set $\{1, \dots, n\}$ and edge set $\{ij \mid 1 \leq i < j \leq n, w_i w_j \in \mathcal{D}\}$. Note that $G(\mathcal{D}, w)$ can be seen as a graph with inherent coloring $\chi \colon V(G) \to \Sigma$. It is unknown whether the lettericity of a given graph can be computed in polynomial time. The problem to determine the lettericity of a given graph is called the *lettericity problem*. As a step towards answering the complexity of this problem, we investigate the following *retrieval problems*: given a graph G together with two of the three solution-objects (word w, decoder $\mathcal{D}$, and coloring χ), the goal is to compute the third solution-object. We show that word retrieval and decoder retrieval are solvable in polynomial time, while coloring retrieval is equivalent to the graph isomorphism problem. Beyond this, we introduce *symmetric lettericity* which is a restricted version of lettericity where each decoder needs to be symmetrical ($ab \in \mathcal{D}$ if and only if $ba \in \mathcal{D}$). As we show, the symmetric lettericity of a graph always equals the neighborhood diversity of the graph, which in fact can be computed in linear time.

Keywords: Lettericity · Letter Graphs · Neighborhood Diversity

1 Introduction

For a given alphabet Σ, a word $w = w_1 \dots w_n \in \Sigma^*$ and a *decoder* $\mathcal{D} \subseteq \Sigma^2$, we define the *letter graph* $G(\mathcal{D}, w)$ as the graph with the vertex set $\{1, \dots, n\}$ and the edge set $\{ij \mid 1 \leq i < j \leq n, w_i w_j \in \mathcal{D}\}$. The *lettericity* of a graph G is defined as the smallest size of an alphabet Σ such that there is a word $w \in \Sigma^*$ and a decoder $\mathcal{D} \subseteq \Sigma^2$, such that G is isomorphic to $G(\mathcal{D}, w)$. Lettericity was introduced by Petkovšek in 2000 [13]. For instance, for every $m \geq 2$, the star graph S_m with m leaves has lettericity 2. This can be certified by the word

V. Brattka et al. (Eds.): CiE 2026, LNCS 16674, pp. 304–319, 2027.
https://doi.org/10.1007/978-3-032-31348-5_20

$w := ab^m$ and the decoder $\mathcal{D} := \{ab\}$. Note that, S_m does not have lettericity 1 because only complete and edge-less graphs have lettericity 1. The *lettericity problem* is the problem of determining the lettericity of a given graph G.

In their original work, Petkovšek gave bounds or precise values for the lettericity of some graph classes such as threshold graphs, cycles and paths. For paths, the exact value was determined to be within one of two values depending on the length of the path. The problem of determining the lettericity of paths precisely remained open for approximately 20 years and was solved by Ferguson in 2020 [9]. This goes to show that determining the exact lettericity even of relatively simple graph classes is far from trivial. Petkovšek already proved in 2000 that for each constant k, the recognition problem for k-letter graphs, that is, graphs with lettericity at most k, is solvable in polynomial time because the set of k-letter graphs is well-quasi-ordered and has a finite set of minimal forbidden subgraphs [13]. In 2024, Alecu et al. [2] improved upon this by showing that determining if a given graph has lettericity at most k is FPT for the parameter k using MSO-formulas. However the complexity of the lettericity problem remained open and they conjectured it to be NP-complete. For more literature on lettericity, see [1,3–5,10,11]. Furthermore, Feng et al. introduced a generalization of lettericity [8], based on a generalized definition of decoders.

A graph parameter related to lettericity is *thinness*, which was introduced by Mannino and Oriolo [12]. They define the *thinness* of a graph G as the smallest number k such that there exists an ordering of the vertices $v_1 < \cdots < v_{|G(V)|}$ and a partition of $V(G)$ into k classes $V_1, \ldots, V_k$ such that for any triple (r, s, t) with $r < s < t$, if v_r and v_s belong to the same class and $v_t v_r \in E(G)$, then $v_t v_s \in E(G)$. The *thinness problem*, that is the problem of determining the thinness of a given graph, was proven to be NP-complete by Shitov in 2025 [14]. They used a result by Bonomo and de Estrada [6], who showed that the thinness problem with a given partition is NP-complete. Shitov reduced the problem with a given partition to the general thinness problem, which implies that determining the thinness of a graph is NP-hard. In 2026, Feng et al. [7] observed that any k-letter graph is a k-thin graph. In their proof they used the inherent coloring $\chi\colon V(G) \to \Sigma$ as a partition. This motivates the idea that it might be possible to prove a complexity lower bound or upper bound for the lettericity problem with a similar approach, that is, studying the lettericity problem with a given part of the solution already fixed. In this work we call this kind of problem a *retrieval problem*: given a graph G together with two of the three *solution objects*, word w, decoder $\mathcal{D}$, and coloring χ, the goal is to compute the third solution object. Indeed, we show that the retrieval of the coloring χ is GI-complete. Furthermore, we show that retrieving the word or the decoder can be done in polynomial time. Whether there exist reductions between the lettericity problem and these retrieval problems that would clarify the complexity of the lettericity problem remains open. Finally, we also show that a version of the lettericity problem where the decoder has to be symmetric (that is, where $ab \in \mathcal{D}$ if and only if $ba \in \mathcal{D}$) can be solved in polynomial time. Hence, if the lettericity problem would not be solvable in polynomial time, then a respective hardness proof would need to explicitly construct instances for which no decoder can be symmetrical.

Proofs of statements marked with ($\star$) are omitted due to space constraints.

2 Preliminaries

We denote by $\mathbb{N}$ the set of positive integers (excluding 0) and define $[n] = \{1, \ldots, n\}$ for every $n \in \mathbb{N}$. For a set X, we denote by $\binom{X}{2}$ the collection of all size-2 subsets of X.

2.1 Words and Languages

Let Σ be an alphabet, that is, a finite non-empty set and denote by Σ^* the set of all finite words over Σ. A language is a subset $L \subseteq \Sigma^*$. For a word $w \in \Sigma^*$, we denote by $|w|$ the length of w and by w_i the i-th symbol in w. The word of length 0 is called *empty word* and denoted by ε. We say a word $v \in \Sigma^*$ is a *subsequence* of a word $w \in \Sigma^*$, denoted by $v \preccurlyeq w$, if v can be obtained from w by removing symbols at arbitrary positions of w. Formally, $v = v_1 \ldots v_m$ is a subsequence of $w = w_1 \ldots w_n$ if there is a function $f\colon [m] \to [n]$ such that $f(i) < f(j)$ if $i < j$ and $v_i = w_{f(i)}$ for all $i \leq m$. For a word $w = w_1 \ldots w_n$, y is a *factor* of w if $y = w_i \ldots w_j$ for some $1 \leq i \leq j \leq n$. For a word w and a letter c, we call a maximal non-empty factor of w that only contains the letter c a *c-run*. For a set of letters $\Sigma' \subseteq \Sigma$ and a word $w \in \Sigma^*$, we denote by $w[\Sigma']$ the unique maximal subsequence of w that only contains letters of Σ'. For sets of constant size, we may omit the set-brackets in this notation, that is, we may for example write $w[a, b]$ instead of $w[\{a, b\}]$.

2.2 Graph Theory

Throughout this paper mainly we consider undirected, simple, loopless graphs. A graph $G = (V, E)$ consists of its vertex set $V(G) := V$ and edge set $E(G) := E$, where E is a subset of all size-2 subsets of V. For vertex sets A and B, we denote by $E(A, B)$ the set of all edges of E that have one endpoint in A and one endpoint in B. We define the (open) neighborhood of a vertex $v \in V$ by $N(v) = \{u \in V \mid \{u, v\} \in E\}$. Similarly, we define the closed neighborhood of v by $N[v] = N(v) \cup \{v\}$. We call a pair of distinct vertices $u, v \in V$ *(true) twins* if $N[u] = N[v]$ and *false twins* if $N(u) = N(v)$. Moreover, we call two vertices u and v *generalized twins*, if $N(u) \setminus \{v\} = N(v) \setminus \{u\}$. Note that u and v are generalized twins if and only if they are true twins or false twins. Given a subset $U \subseteq V$ of vertices, we denote by $G[U]$ the graph induced by U, that is, the graph with vertex set U and edge set $\{\{u, v\} \in E \mid u, v \in U\}$. A *graph isomorphism* between two graphs G and G' is a bijection $f\colon V(G) \to V(G')$ such that for each $u, v \in V(G)$ we have $\{u, v\} \in E(G)$ if and only if $f(u)f(v) \in E(G')$. An instance of the Graph Isomorphism problem consists of two graphs G_1, G_2, and the goal is to decide whether there exists a graph isomorphism between G_1 and G_2. Recall that GRAPH ISOMORPHISM is GI-complete by definition.

A *permutation* is a sequence of the elements of a finite set X that contains each element from X exactly once. In this sense a permutation of X can be seen as a word over the alphabet X. A directed graph $H = (V, A)$ consists of a set of vertices and a set of arcs $A \subseteq V \times V$. A *topological ordering* of a directed graph $H = (V, A)$ is a permutation of V satisfying $(v, u) \notin A$ if u precedes v. A *directed acyclic graph (DAG)* is a directed graph without cycles. A directed graph is a DAG if and only if it has a topological ordering.

2.3 Letter Graphs

Let Σ be an alphabet of size k. We call a language $\mathcal{D} \subseteq \Sigma^2$ a *decoder*. For a word $w = \Sigma^*$ and a decoder $\mathcal{D}$, we define the *k-letter graph* $G(\mathcal{D}, w)$ to be the graph with vertex set $V(G(\mathcal{D}, w)) = [|w|]$ and edge set $E(G(\mathcal{D}, w)) = \{ij \mid 1 \leq i < j \leq |w|, w_i w_j \in \mathcal{D}\}$. The *lettericity* of a graph G is the smallest value k such that G is isomorphic to a k-letter graph. Note that every letter graph inherently comes with a vertex coloring $\chi_w \colon V(G(\mathcal{D}, w)) \to \Sigma$ defined as $\chi_w(i) = w_i$ for each $i \in [|w|]$. For each graph $G = (V, E)$ isomorphic to a letter graph $G(\mathcal{D}, w)$, we say that the graph isomorphism $f \colon V \to V(G(\mathcal{D}, w))$ *respects the coloring* $\chi \colon V \to \Sigma$ if $\chi(v) = w_{f(v)}$. Note that this is well-defined because $V(G(\mathcal{D}, w)) = [|w|]$.

3 Retrieval Problems

In this section we study *retrieval problems* which are of the following form. Given a graph $G = (V, E)$ together with two of the three objects, word $w \in \Sigma^*$, decoder $\mathcal{D} \subseteq \Sigma^2$, and coloring $\chi \colon V \to \Sigma$, the goal is to compute the third object such that G is isomorphic to $G(\mathcal{D}, w)$ respecting the coloring. We call an instance of any retrieval problem *consistent* if such a third object exists.

3.1 Word Retrieval

The WORD RETRIEVAL problem is defined as follows. Given a graph $G = (V, E)$ with coloring $\chi \colon V \to \Sigma$ and a decoder $\mathcal{D} \subseteq \Sigma^2$, compute a word w such that G is isomorphic to $G(\mathcal{D}, w)$ respecting the coloring or report that no such word exists.

In the following, let $I = (G = (V, E), \Sigma, \chi, \mathcal{D})$ be an instance of WORD RETRIEVAL. We define a directed auxiliary graph $H = (V, A)$ as follows: Initially, the arc set A is empty. For each $(u, v) \in V \times V$ with $u \neq v$, we add the arc (u, v) to A if (i) $\{u, v\} \in E$ and $\chi(v)\chi(u) \notin \mathcal{D}$ or (ii) $\{u, v\} \notin E$ and $\chi(v)\chi(u) \in \mathcal{D}$. An example for this construction is given in Fig. 1.

The arcs of H describe the partial order that the nodes of G must have in relation to each other in all words that are a solution to I.

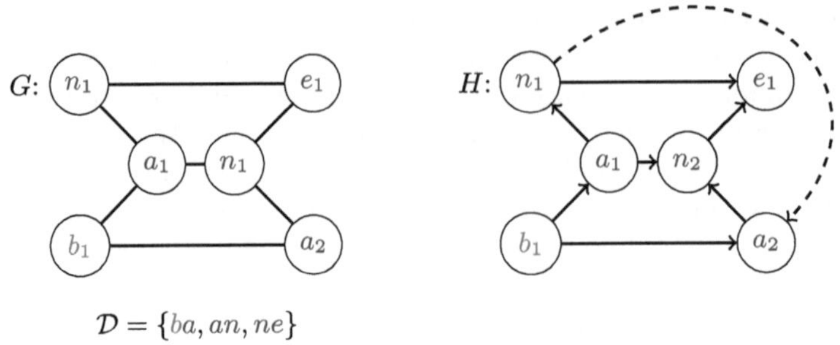

Fig. 1. An example for the transformation of the WORD RETRIEVAL-instance to a directed graph. Here, a vertex c_i implies that this vertex receives color c under the labeling χ. The solid arcs are added due to Condition (i) of the definition of the arc set and the single dashed arc is added due to Condition (ii). The unique topological ordering $b_1a_1n_1a_2n_2e_1$ of this example graph is the unique generalized solution for the WORD RETRIEVAL-instance and implies that the word *banane* is a solution for the WORD RETRIEVAL-instance.

We will show that we are dealing with a yes-instance of WORD RETRIEVAL if and only if H is a DAG. To this end, note that we are dealing with a yes-instance of WORD RETRIEVAL if and only if there is a *generalized solution*, that is, a permutation $\pi(1) \cdot \ldots \cdot \pi(n)$ of the vertices, such that $\chi(\pi(i)) \cdot \ldots \cdot \chi(\pi(n))$ is a solution for our WORD RETRIEVAL-instance.

Lemma 1. *Let π be a permutation of V. Then, π is a topological ordering of H if and only if π is a generalized solution for I.*

Proof. ($\Rightarrow$) We show that for any $1 \leq i < j \leq |V|$, $\{u, v\} \in E$ if and only if $\chi(u)\chi(v) \in \mathcal{D}$, where $u := \pi[i]$ and $v := \pi[j]$.

Assume $\{u, v\} \in E$. If $\chi(u)\chi(v) \notin \mathcal{D}$, then by definition of the arc set A, $(v, u) \in A$. As π is a topological ordering of H, and u precedes v in π, (v, u) is not an arc of A, which implies that $\chi(u)\chi(v) \in \mathcal{D}$.

Assume $\{u, v\} \notin E$. If $\chi(u)\chi(v) \in \mathcal{D}$, then by definition of the arc set A, $(v, u) \in A$. As π is a topological ordering of H, and u precedes v in π, (v, u) is not an arc of A, which implies that $\chi(u)\chi(v) \notin \mathcal{D}$.

Hence, $\{u, v\} \in E$ if and only if $\chi(u)\chi(v) \in \mathcal{D}$ and π is a generalized solution for I.

($\Leftarrow$) We show that for any $1 \leq i < j \leq |V|$, (v, u) is not an arc of A, for $u := \pi[i]$ and $v := \pi[j]$.

Assume towards a contradiction that $(v, u) \in A$. Then, by definition of A, (i) $\{u, v\} \in E$ and $\chi(u)\chi(v) \notin \mathcal{D}$ or (ii) $\{u, v\} \notin E$ and $\chi(u)\chi(v) \in \mathcal{D}$.

As π is a solution for I, $\{u, v\} \in E$ if and only if $\chi(u)\chi(v) \in \mathcal{D}$. This contradicts the assumption that (v, u) is an arc of A. Consequently, π is a topological ordering of H. $\square$

As H can be constructed in polynomial time and deciding whether a topological ordering of a directed graph exists can be done in linear time (and, if one exists, a topological ordering can also be found in that time), we obtain the following.

Theorem 1. *WORD RETRIEVAL can be solved in polynomial time.*

3.2 Decoder Retrieval

An instance I of DECODER RETRIEVAL consists of a graph $G = (V, E)$, an alphabet Σ, a coloring $\chi\colon V \to \Sigma$, and a word $w \in \Sigma^{|V|}$. For each letter $a \in \Sigma$, we denote by V_a the vertices of color a, that is, $V_a := \chi^{-1}(a)$. We further assume that for each letter $a \in \Sigma$, $V_a \neq \varnothing$ and that w contains the letter a exactly $|V_a|$ times. The goal is to decide whether there is some $\mathcal{D} \subseteq \Sigma^2$, such that G can be obtained from χ and w, by using $\mathcal{D}$ as the decoder. We call each such $\mathcal{D} \subseteq \Sigma^2$ a *solution* (for I). If the vertex set of the graph is clear, we may simply define the edge set of the graph in an instance of DECODER RETRIEVAL. That is, we may say that (E, Σ, χ, w) is an instance of DECODER RETRIEVAL, where E is a set of edges.

We show that DECODER RETRIEVAL can be solved in polynomial time. To present the algorithm, we proceed in four steps. In the first step, we show that we can decide whether a given decoder $\mathcal{D}$ is in fact a solution for our given DECODER RETRIEVAL-instance. Afterwards, we show that we can make the choice on whether ab or ba is in a solution greedily for many letter pairs $\{a, b\}$. The only cases where we cannot do this are pairs $\{a, b\}$ where $w[a, b]$ is a palindrome and edges between V_a and V_b are non-trivial, that is, some possible edges between these vertex sets exists and some do not exist. For these remaining letter pairs $\{a, b\}$ and $\{b, c\}$ where the choice cannot be made trivially, a choice for the decoder word of $\{ab, ba\}$ in a solution immediately forces a choice for the decoder word of $\{cb, bc\}$ in the same solution. Based on this, we finally reduce the decision on whether there is a solution for the DECODER RETRIEVAL instance to the satisfiability of a computable 2-SAT formula (after some sanity checks).

Lemma 2. *Let $I = (G, \Sigma, \chi, w)$ be an instance of DECODER RETRIEVAL. Moreover, let $\mathcal{D} \subseteq \Sigma^2$. Then, in polynomial time, we can decide whether D is a solution for I.*

Proof. We show that Algorithm 1 solves this task. Clearly, the running time of the algorithm is polynomial. Intuitively, the algorithm makes a choice for the vertex which will be isomorphic to the first vertex of the letter graph $G(\mathcal{D}, w)$. This is then done recursively. Since without loss of generality the first letter of w is a, the first vertex v needs to be from V_a and its neighborhood N_b towards any vertex of V_b with $b \in \Sigma$ has to be either $V_b \setminus \{v\}$ or $\varnothing$. Here, we need to remove v from V_b, as for $b = a$, v never has an edge towards itself, and for $b \neq a$, $V_b \backslash \{v\} = V_b$. If no such vertex v exists, the algorithm correctly outputs NO. That is, if $\mathcal{D}$ is not a solution for the DECODER RETRIEVAL-instance, the algorithm is correct. It thus remains to show that the algorithm is also correct if $\mathcal{D}$ is a solution. By the above, Line 6 has to hold for some vertex $w \in V_a$. Now assume

Algorithm 1: VerifyDecoder($G, \Sigma, \chi, w, \mathcal{D}$)

Input: Graph $G = (V, E)$, an alphabet Σ, a coloring $\chi \colon V \to \Sigma$, word $w = w_1 \dots w_{|V|} \in \Sigma^*$, and $\mathcal{D} \subseteq \Sigma^2$

Output: YES if and only if $\mathcal{D}$ is a solution for the instance (G, Σ, χ, w) of DECODER RETRIEVAL.

1 **if** $|w| = 0$ **then**

2   **return** YES

3 $a \cdot w' \leftarrow w$

4 **for** $b \in \Sigma$ **do**

5     $N_b \leftarrow \begin{cases} V_b & ab \in \mathcal{D} \\ \varnothing & ab \notin \mathcal{D} \end{cases}$

6 **if** $\exists v \in V_a : N(v) = (\bigcup_{b \in \Sigma} N_b) \setminus \{v\}$ **then**

7   **return** VerifyDecoder($G - v, \Sigma, \chi|_{V \setminus \{v\}}, w', \mathcal{D}$)

8 **else**

9   **return** NO

towards a contradiction that the algorithm chose the wrong vertex $v \in V_a$ at this line. As both v and w fulfill the condition of Line 6, we derive that v and w are generalized twins. That is, v and w are completely indistinguishable. Hence, v cannot be a wrong choice, if w would be a right choice. Consequently, the algorithm correctly answers YES if $\mathcal{D}$ is a solution. □

In the remainder of the section, we assume that we are given an instance I of DECODER RETRIEVAL and that in this instance, each color of Σ is assigned to at least one vertex. We now continue with our sanity checks and greedy choice for some letter pairs. The first observation considers the choice on $\mathcal{D} \cap \{aa\}$ for each letter a.

Observation 1. *Let $a \in \Sigma$. If V_a is a clique in G, then for every solution $\mathcal{D}$, $\mathcal{D} \cup \{aa\}$ is a solution. If V_a is an independent set in G, then for every solution $\mathcal{D}$, $\mathcal{D} \setminus \{aa\}$ is a solution. If V_a is neither a clique nor an independent set, then we are dealing with a no-instance.*

The next observation considers the choice on $\mathcal{D} \cap \{ab, ba\}$ for each letter pairs $\{a, b\}$ where either all or no edges exists between V_a and V_b.

Observation 2. *Let $\{a, b\} \subseteq \Sigma$ with $a \neq b$. If $|E(V_a, V_b)| = |V_a| \cdot |V_b|$, then for every solution $\mathcal{D}$, $\mathcal{D} \cup \{ab, ba\}$ is a solution. If $|E(V_a, V_b)| = 0$, then for every solution $\mathcal{D}$, $\mathcal{D} \setminus \{ab, ba\}$ is a solution.*

We now give a name to letter pairs for which none of the above greedy choices applies.

Definition 1. *We call an unordered letter pair $\{a, b\}$ with $a \neq b$* one-sided, *if $0 < |E(V_a, V_b)| < |V_a| \cdot |V_b|$.*

In the following, let $\mathcal{P}$ denote the set of all one-sided letter pairs.

Lemma 3 ($\star$). *Let $\{a, b\}$ be a one-sided letter pair. Then for each solution $\mathcal{D}$, $|\mathcal{D} \cap \{ab, ba\}| = 1$.*

Next, we show that one-sided letter pairs $\{a, b\}$ for which $w[a, b]$ contains only a single a-run and a single b-run guarantee that we are dealing with a no-instance. Note that $w[a, b]$ contains a single a-run and a single b-run if and only if $w[a, b] \in a^+b^+ \cup b^+a^+$.

Lemma 4 ($\star$). *Let $\{a, b\}$ be a one-sided letter pair, such that $w[a, b]$ contains only one a-run and only one b-run, then I is a no-instance of* Decoder Retrieval.

In the following we thus gain insights for one-sided letter pairs $\{a, b\}$ where $w[a, b]$ contains at least two a-runs or at least two b-runs. Additionally, we observe the following if $w[a, b]$ is not a palindrome.

Lemma 5. *Let $\{a, b\}$ be a one-sided letter pair such that $w[a, b]$ is not a palindrome. We can compute in polynomial time a decoder word $d_{ab} \in \{ab, ba\}$, such that for each solution $\mathcal{D}$, $d_{ab} \in \mathcal{D}$.*

Proof. We describe a constructive proof that shows that there is decoder word $d_{ab} \in \{ab, ba\}$, such that for each solution $\mathcal{D}$, $d_{ab} \in \mathcal{D}$. Let $\mathcal{D}$ be a solution for I. Note that $\mathcal{D}' := \mathcal{D} \cap \{ab, ba\}$ is a solution for $I' := (E(V_a, V_b), \{a, b\}, \chi|_{V_a \cup V_b}, w[a, b])$. We thus analyze $\mathcal{D}'$ and show via induction over the length of $w[a, b]$ that there is no solution for I' where $\mathcal{D}$ contains both, ab and ba. Without loss of generality we assume that $w[a, b]$ starts with an a.

Base Case. For a non-palindrome $w[a, b]$ of length at most 4, the statement trivially holds.

Inductive Step. We consider two cases.

Case 1: $w[a, b]$ ends with letter b. If there is a vertex $v_a \in V_a$ that is adjacent to all vertices of V_b, then the word ab is in every solution for I'. If there is a vertex $v_a \in V_a$ that is adjacent no vertex of V_b, then the word ba is in every solution for I'. If neither of these is the case, there is no solution for I'.

Case 2: $w[a, b]$ ends with letter a. Thus, $w[a, b]$ starts and ends with an a. Let R_f and R_ℓ be the indices of the first and the last a-run in $w[a, b]$, respectively. Moreover, let X_a denote the vertices of V_a that are adjacent to all vertices of V_b. Similarly, let Y_a denote the vertices of V_a that are adjacent to no vertex of V_b. As I' is a yes-instance of Decoder Retrieval and $\{a, b\}$ is a one-sided letter pair, it holds that (1) $|R_f| = |X_a|$ and $|R_\ell| = |Y_a|$ or (2) $|R_f| = |Y_a|$ and $|R_\ell| = |X_a|$.

If $|R_f| = |R_\ell|$, let $V'_a := V_a \setminus (X_a \cup Y_a)$ and let w' be the word obtained from $w[a, b]$ by removing the first and the last a-run. By the induction hypothesis, there is a word $d_{ab} \in \{ab, ba\}$ such that d_{ab} is contained in every solution for the instance $(E(V'_a, V_b), \{a, b\}, \chi|_{V'_a \cup V_b}, w')$ of DECODER RETRIEVAL. This is due to the fact that w' is again not a palindrome.

Otherwise, that is, if $|R_f| \neq |R_\ell|$, let without loss of generality $|R_f| > |R_\ell|$. Moreover, recall that this implies $|X_a| \neq |Y_a|$, that is, there are more vertices of V_a that are universal for the vertices of V_b than vertices of V_a that have no neighbors in V_b, or vice versa. If $|R_f| = |X_a|$, then each solution has to contain ab, as the first a-run has to contain all vertices of V_a that are universal for all vertices of V_b. If $|R_f| = |Y_a|$, then each solution has to contain ba, as the first a-run has to contain all vertices of V_a that have no neighbors in V_b. □

In the next step, we show that the choice of the decoder word for $\{a, b\}$ also determines the choice of the decoder word for $\{b, c\}$ if both $w[a, b]$ and $w[b, c]$ contain at least two b-runs each.

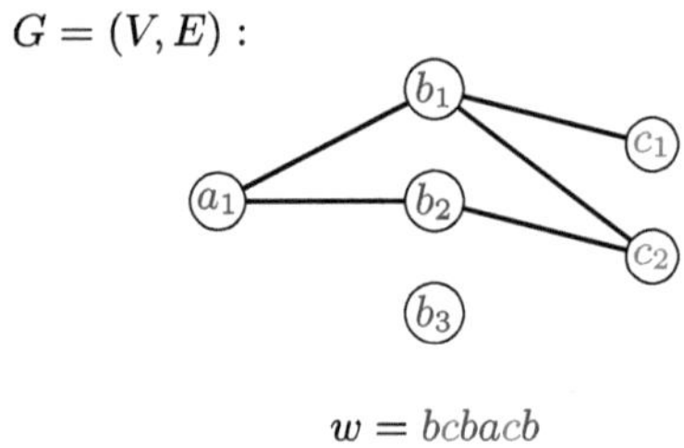

Fig. 2. An example for a word w where $\{a, b\}$ and $\{b, c\}$ are one-sided letter pairs, $w[a, b]$ and $w[b, c]$ contain at least two b-runs each, and where $w[b, c]$ is a palindrome. The choice for the word of $\{bc, cb\}$ for the decoder is determined by the choice for the word of $\{ab, bc\}$ for the decoder. Note that the latter is fixed, as $w[a, b]$ is not a palindrome. Namely, for each decoder $\mathcal{D}$ which is a solution, we need to have $\mathcal{D} \cap \{ab, ba\} = \{ba\}$. This implies that the first two bs of w are associated with the vertices b_1 and b_2. Consequently, solution $\mathcal{D}$ cannot fulfill $\mathcal{D} \cap \{bc, cb\} = \{cb\}$, as otherwise, b_3 (which by necessity would be associated with the last b in w) would have a common edge with c_1 in the respective letter graph $G(\mathcal{D}, w)$.

Lemma 6. *Let $\{a, b\}$ and $\{b, c\}$ be distinct one-sided letter pairs, such that $w[a, b]$ and $w[b, c]$ contain at least two b-runs each, and where $w[b, c]$ is a palindrome. Then for each $d_{ab} \in \{ab, ba\}$, in polynomial time, we can compute $d_{bc} \in \{bc, cb\}$, such that for each solution $\mathcal{D}$ with $d_{ab} \in \mathcal{D}$, $\mathcal{D}$ also contains d_{bc}.*

Proof. An example for this scenario is illustrated in Fig. 2. Let $d_{ab} \in \{ab, ba\}$. First, we show that if there is a solution that contains d_{ab}, then there is a unique $d_{bc} \in \{bc, cb\}$, such that for each solution $\mathcal{D}$ with $d_{ab} \in \mathcal{D}$, $\mathcal{D}$ also contains d_{bc}. Assume towards a contradiction that there are two solutions $\mathcal{D}_1$ and $\mathcal{D}_2$ with $\{d_{ab}, bc\} \subseteq \mathcal{D}_1$ and $\{d_{ab}, cb\} \subseteq \mathcal{D}_2$. For a letter pair $p \in \{\{a, b\}, \{b, c\}\}$, let R^p_f and R^p_ℓ be the indices of the b's in w of the first and the last b-run in $w[p]$, respectively. That is, if the length of the first b run in $w[p]$ is α, then R_f^p contains the indices of the first α many b's in w. Note that $|R_f^{\{b,c\}}| = |R_\ell^{\{b,c\}}| < \frac{|V_b|}{2}$, as $w[b,c]$ is a palindrome. Assume without loss of generality that the length $|R_f^{\{a,b\}}|$ of the first b-run in $w[a,b]$ is at most the length $|R_\ell^{\{a,b\}}|$ of the last b-run in $w[a,b]$. This implies that $|R_f^{\{a,b\}}| < \frac{|V_b|}{2}$. Consequently, $R_f^{\{a,b\}} \cap R_\ell^{\{b,c\}} = \varnothing$.

Under both $\mathcal{D}_1$ and $\mathcal{D}_2$, the vertices of V_b from G associated with the indices of $R_f^{\{a,b\}}$ are the same, as both solutions contain only d_{ab} from $\{ab, ba\}$. Let X_b be this set of vertices of V_b. As $\mathcal{D}_1$ is a solution and $bc \in \mathcal{D}_1$, there is no vertex of X_b that has smallest degree towards the vertices of V_c among all vertices of V_b, as $R_f^{\{a,b\}} \cap R_\ell^{\{b,c\}} = \varnothing$. However, since $\mathcal{D}_2$ is a solution and $cb \in \mathcal{D}_2$, the vertex associated to the first b (which is a vertex of X_b) under $\mathcal{D}_2$ has the smallest degree towards the vertices of V_c among all vertices of V_b. This contradicts the assumption that both $\mathcal{D}_1$ and $\mathcal{D}_2$ are solutions.

It remains to show how to compute the respective word d_{bc} in polynomial time. Consider the instance $I' := (E(V_b, V_a \cup V_c), \{a,b,c\}, \chi|_{V_a \cup V_b \cup V_c}, w[\{a,b,c\}])$ of Decoder Retrieval. By the above arguments, at most one of $\{d_{ab}, bc\}$ and $\{d_{ab}, cb\}$ is a solution for I'. As we can construct I' in polynomial time, we can check in polynomial time (see Lemma 2) which of the two possible candidate solutions is in fact a solution for I'. If one of both is a solution, we output the respective word of $\{bc, cb\}$. Otherwise, that is if neither of both is a solution, then there is no solution that contains d_{ab}, and we can correctly output any word from $\{bc, cb\}$. □

For each letter $a \in \Sigma$, let P_a denote the set of letters b of $\Sigma \setminus \{a\}$ for which $\{a,b\}$ is a one-sided letter pair and $w[a,b]$ has at least two a-runs. Based on this definition, we characterize when a given decoder is a solution.

Lemma 7. *Let $\mathcal{D} \subseteq \Sigma^2$ and assume that for each $\{a,b\} \in \mathcal{P}$, $w[a,b]$ contains at least two a-runs or at least two b-runs. Then $\mathcal{D}$ is a solution for I if and only if*

- *for each letter $a \in \Sigma$, $\mathcal{D} \cap \{aa\}$ is a solution for the Decoder Retrieval-instance $(E(V_a), \{a\}, \chi|_{V_a}, w[a])$,*
- *for each $\{a,b\} \in \binom{\Sigma}{2} \setminus \mathcal{P}$, $\mathcal{D} \cap \{ab, ba\}$ is a solution for the Decoder Retrieval-instance $(E(V_a, V_b), \{a,b\}, \chi|_{V_a \cup V_b}, w[a,b])$, and*
- *for each letter $a \in \Sigma$, $\mathcal{D} \cap \{ab, ba \mid b \in P_a\}$ is a solution for the Decoder Retrieval-instance $(\bigcup_{b \in P_a} E(V_a, V_b), P_a \cup \{a\}, \chi|_{\bigcup_{b \in P_a \cup \{a\}} V_b}, w[a, P_a])$.*

Proof. ($\Rightarrow$) If $\mathcal{D}$ is a solution for I, then all item need to hold, as they describe exactly the edge sets of subgraphs of G.

($\Leftarrow$) Assume that all three items hold. We show that $\mathcal{D}$ is a solution for I. Assume towards a contradiction that this is not the case. Hence, Algorithm 1 outputs NO on input $(G, \chi, w, \mathcal{D})$. Let $(G^* = (V^*, E^*), \Sigma, \chi^*, w^*, \mathcal{D})$ be the input of the recursive call at which the algorithm first outputs NO. Moreover, let a be the first letter of w^* and let $(v_a^1, \dots, v_a^p)$ be the sequence of vertices of $V_a \setminus V^*$ in which they were chosen by previous calls of the algorithm. Since the algorithm outputs NO, there is no vertex $v_a^{p+1} \in V_a \cap V^*$, such that for each $b \in \Sigma$, $N(v_a^{p+1}) \cap V_b \cap V^* = N_b \cap V^* \setminus \{v\}$. To achieve a contradiction, we, however, show that such a vertex exists. To this end, we first observe that for each $b \in \Sigma \setminus P_a$, each vertex of V_a has the exactly same neighborhood towards the vertices of V_b, that is, $N(v) \cap V_b \setminus \{w\} = N(w) \cap V_b \setminus \{v\}$ for each two vertices $v, w \in V_a$. Here, recall that P_a contains all letters b for which $w[a, b]$ contains at least two a-runs.

First, for $b = a$, the statement holds, as Lemma 7 implies that $V_a = V_b$ is a clique or an independent set. For $b \in \Sigma$ with $b \neq a$, such that $\{a, b\}$ is not one-sided, the statement holds as Lemma 7 implies that G contains either all or no edge between V_a and V_b. Finally, consider a letter $b \in \Sigma$, such that $\{a, b\}$ is one-sided but $b \notin P_a$. The latter implies that $w[a, b]$ contains only a single a-run. By the condition that there are no one-sided letter pairs that contain only a single run for each letter, this implies that $w[a, b]$ contains at least/exactly two b-runs. Hence, $a \in P_b$. Thus, Lemma 7 (for reversing the roles of a and b) implies that there is a subset $V_b' \subseteq V_b$, such that for each vertex $v \in V_a$, $N(v) \cap V_b = V_b'$. Consequently, for each letter $b \in \Sigma \setminus P_a$, $N(v) \cap V_b \setminus \{w\} = N(w) \cap V_b \setminus \{v\}$ for each two vertices $v, w \in V_a$.

Intuitively, this implies that the algorithm outputted NO, since there is no vertex $v_a^{p+1} \in V_a \cap V^*$, such that for some $b \in P_a$, $N(v_a^{p+1}) \cap V_b \cap V^* \neq N_b \cap V^*$. However, since Lemma 7 holds, Algorithm 1 outputs YES on the input of Lemma 7. This implies that there is an ordering $(w_a^1, \dots, w_a^{|V_a|})$ in which the algorithm selected the vertices of V_a in the recursive calls. Note that w_a^i and v_a^i for $1 \leq i \leq p$ are guaranteed generalized twins due to the observation that for all $b \in \Sigma \setminus P_a$, all vertices of V_a have the same neighborhood towards V_b. Hence, we can further assume without loss of generality that $w_a^i = v_a^i$ for each $1 \leq i \leq p$, as generalized twins of the same color are indistinguishable. This implies that w_a^{p+1} is a vertex of $V_a \setminus \{v_a^1, \dots, v_a^p\} = V_a \cap V^*$. Hence, $N(w_a^{p+1}) \cap V_b \cap V^* = N_b \cap V^*$ for each $b \in P_a$. The contradicts the assumption that the algorithm outputted NO on input $(G^*, \Sigma, \chi^*, w^*, \mathcal{D})$. Consequently, $\mathcal{D}$ is a solution for I. □

We are now ready to describe our algorithm to solve DECODER RETRIEVAL. To this end, we first make some sanity checks. For each letter $a \in \Sigma$, we check whether $G[V_a]$ is a clique or an independent set. If this is not the case, we deal with a no-instance and output this. Similarly, we check for each one-sided letter pair $\{a, b\}$, whether $w[a, b]$ contains at least two a-runs or at least two b-runs. If this is not the case, we also deal with a no-instance (see Lemma 4) and output this. Note that all these checks can be done in polynomial time and that the latter sanity check implies that for each one-sided letter pair $\{a, b\}$, $a \in P_b$ or $b \in P_a$ holds.

If we have not outputted yet that we are dealing with a no-instance, we proceed as follows: Recall that $\mathcal{P}$ denotes the collection of all one-sided letter pairs. Based on our previous observations, we describe a reduction of our problem to 2-SAT. To this end, we describe a formula F over the variable set $\{ab, ba \mid \{a,b\} \in \mathcal{P}\}$.

1. For each $\{a,b\} \in \mathcal{P}$, we add the clauses $ab \vee ba$ and $\neg ab \vee \neg ba$ that ensure that we chose exactly one of $\{ab, ba\}$ for the decoder.
2. For each $\{a,b\} \in \mathcal{P}$ where $w[a,b]$ is not a palindrome, we add the clause d_{ab}, where d_{ab} is the decoder word that is contained in every solution and that we can compute in polynomial time due to Lemma 5.
3. For each two distinct $\{a,b\} \in \mathcal{P}$ and $\{b,c\} \in \mathcal{P}$ where both $w[a,b]$ and $w[b,c]$ contain at least two b-runs and where $w[b,c]$ is a palindrome, do the following: For each $d_{ab} \in \{ab, ba\}$, in polynomial time compute $d_{bc} \in \{bc, cb\}$, such that for each solution $\mathcal{D}$ with $d_{ab} \in \mathcal{D}$, $\mathcal{D}$ also contains d_{bc} (see Lemma 6) and add the clause $\neg d_{ab} \vee d_{bc}$.
4. For each $a \in \Sigma$, let P_a^p be the subset of P_a of those letters b for which $w[a,b]$ is a palindrome and let $P_a^n := P_a \setminus P_a^p$. We consider two cases in which we are interested in potential solutions for the subinstance $I_a := (\bigcup_{b \in P_a} E(V_a, V_b), P_a \cup \{a\}, \chi|_{\bigcup_{c \in P_a \cup \{a\}} V_c}, w[P_a \cup \{a\}])$ of Decoder Retrieval.
   (a) If $P_a^n \neq \emptyset$, let $b \in P_a^n$. For each $c \in P_a^n$ let d_{ac} be the word of $\{ac, ca\}$ outputted by the algorithm behind Lemma 5. Moreover, for each $c \in P_a^p$, let d_{ac} be the word of $\{ac, ca\}$ for the word d_{ab} outputted by the algorithm behind Lemma 6. If $\{d_{ac} \mid c \in P_a\}$ is not a solution for I_a, we add the clause $\neg d_{ab}$ (which in fact leads to an unsatisfiable formula, as we also have the clause d_{ab} due to Item 2).
   (b) If $P_a^n = \emptyset$, let $b \in P_a^p$. For each $d_{ab} \in \{ab, ba\}$, do the following: For each $c \in P_a^p \setminus \{b\}$, let d_{ac} be the word of $\{ac, ca\}$ for the word d_{ab} outputted by the algorithm behind Lemma 6. If $\{d_{ac} \mid c \in P_a\}$ is not a solution for I_a, we add the clause $\neg d_{ab}$.

To prove our polynomial time algorithm, we now show the following.

Lemma 8. *It holds that I is a yes-instance of* Decoder Retrieval *if and only if the 2-SAT formula F is satisfiable.*

Proof. ($\Rightarrow$) Let $\mathcal{D}$ be a solution for I. We show that setting exactly the variables of $A := \mathcal{D} \cap \{ab, ba \mid \{a,b\} \in \mathcal{P}\}$ to True satisfies the formula F. As $\mathcal{D}$ is a solution and for each $\{a,b\} \in \mathcal{P}$ is a one-sided letter pair, Lemma 3 implies that $|\mathcal{D} \cap \{ab, ba\}| = 1$. Hence, the clauses of Item 1 are satisfied. Further, Lemma 5 implies that for each $\{a,b\} \in \mathcal{P}$ where $w[a,b]$ is not a palindrome, the word d_{ab} is part of $\mathcal{D}$. Hence, the clauses of Item 2 are satisfied. Now, let $\{a,b\} \in \mathcal{P}$ and $\{b,c\} \in \mathcal{P}$ be distinct letter pairs where both $w[a,b]$ and $w[b,c]$ contain at least two b-runs and where $w[b,c]$ is a palindrome. Let d_{ab} be the unique word of $\{ab, ba\}$ that is contained in $\mathcal{D}$. By Lemma 6, there is a unique $d_{bc} \in \{bc, cb\}$, such that $d_{bc} \in \mathcal{D}$. Hence, the clauses of Item 3 are satisfied. Finally, we consider the clauses of Item 4. To this end, let $a \in \Sigma$ and we distinguish whether $P_a^n = \emptyset$.

If $P_a^n \neq \varnothing$, let $b \in P_a^n$ be as chosen in Item 4a. As $\mathcal{D}$ is a solution, Lemma 5 implies that for each $c \in P_a^n$, $d_{ac} \in \mathcal{D}$. In particular, $d_{ab} \in \mathcal{D}$. Thus, Lemma 6 implies that $d_{ac} \in \mathcal{D}$ for each $c \in P_a^p$. This in particular implies that $\mathcal{D} \cap \{ac, ca \mid c \in P_a\} = \{d_{ac} \mid c \in P_a\}$ is a solution for I'. Hence, the clause $\neg d_{ab}$ was not added in Item 4a, which implies that the clauses of Item 4a for letter a are satisfied.

If $P_a^n = \varnothing$, let $b \in P_a^p$ be as chosen in Item 4b. Moreover, let d_{ab} be the unique word of $\{ab, ba\} \cap \mathcal{D}$. As $\mathcal{D}$ is a solution, Lemma 6 implies that $d_{ac} \in \mathcal{D}$ for each $c \in P_a^p$. This in particular implies that $\mathcal{D} \cap \{ac, ca \mid c \in P_a\} = \{d_{ac} \mid c \in P_a\}$ is a solution for I'. Hence, the clause $\neg d_{ab}$ was not added in Item 4b, which implies that the clauses of Item 4b for letter a are satisfied.

Thus, F is satisfied by assigning exactly the variables of A to True.

($\Leftarrow$) Let A be exactly the variables assigned to True by an arbitrary but fixed satisfying assignment of F. We show that there is a solution $\mathcal{D}$ for I with $\mathcal{D} \cap \{ab, ba \mid \{a, b\} \in \mathcal{P}\} = A$. In particular, we show that $\mathcal{D} := A \cup B$ is a solution for I, where $B := \{aa \mid a \in \Sigma, G[V_a] \text{ is a clique}\} \cup \{ab, ba \mid \{a, b\} \in \binom{\Sigma}{2} \setminus \mathcal{P}, |E(V_a, V_b)| = |V_a| \cdot |V_b|\}$. As we only constructed the formula F after our sanity check in which we checked whether for each $a \in \Sigma$, $G[V_a]$ is a clique or an independent set, Observation 1 and Observation 2 imply that the choice of B satisfies the first two items of Lemma 7. To show that $\mathcal{D}$ is a solution, it thus remains to show that also the last point of Lemma 7 holds. This follows immediately by the fact that the clauses of Item 4a and Item 4b are satisfied by setting exactly the variables of A to True. Hence, Lemma 7 implies that $\mathcal{D}$ is a solution for I. □

This thus implies the following, as the formula F can be computed in polynomial time and the previous sanity check also runs in polynomial time.

Theorem 2. *Decoder Retrieval can be solved in polynomial time.*

3.3 Coloring Retrieval

The Coloring Retrieval problem is defined as follows. Given a graph $G = (V, E)$, a word $w \in \Sigma^*$ and a decoder $\mathcal{D} \subseteq \Sigma^2$, decide whether there is a coloring $\chi \colon V \to \Sigma$ that is consistent with the inherent coloring of $G(\mathcal{D}, w)$.

Theorem 3. *Coloring Retrieval is* GI*-complete.*

Proof. The containment in GI holds immediately: An instance $(G, \Sigma, \mathcal{D}, w)$ of Coloring Retrieval is a yes-instance if and only if G and $G(\mathcal{D}, w)$ are isomorphic. As $G(\mathcal{D}, w)$ can be computed in polynomial time, there is a trivial reduction from Coloring Retrieval to Graph Isomorphism.

The GI-hardness can be seen as follows. Let $I := (G = (V, E), H = (U, F))$ be an instance of Graph Isomorphism. We obtain an instance $I' := (G', \Sigma, \mathcal{D}, w)$ of Coloring Retrieval as follows: We set $G' := G$, $\Sigma := U$, $\mathcal{D} := \{uv, vu \mid \{u, v\} \in F\}$, and define w to be an arbitrary permutation of U. Note that $H = G(\mathcal{D}, w)$. This immediately implies that I is a yes-instance if and only if G is isomorphic to $G(\mathcal{D}, w)$. Hence, Coloring Retrieval is also GI-hard. □

4 Symmetric Lettericity

In this section, we consider the lettericity problem, when each allowed decoder needs to be *symmetric*, that is, $ab \in \mathcal{D}$ if and only if $ba \in \mathcal{D}$.

Let $G = (V, E)$ be a graph. We let $\mathrm{sl}(G)$ denote the smallest size of any alphabet Σ, such that there is a symmetric decoder $\mathcal{D} \subseteq \Sigma^2$ and a word $w \in \Sigma^{|V|}$ for which G is isomorphic to $G(\mathcal{D}, w)$. Note that $G(\mathcal{D}, w)$ is isomorphic to $G(\mathcal{D}, w')$ for every permutation w' of w.

Observation 3. *For each graph G we have $l(G) \leq sl(G)$.*

We show that $\mathrm{sl}(G)$ equals the neighborhood diversity of G, a parameter that can be computed in linear time. Here, the neighborhood diversity $\mathrm{nd}(G)$ of G denotes the smallest partition $(V_1, \ldots, V_p)$ of V, such that for each $i \in [1, p]$ and for any two vertices $u, v \in V_i$, u and v are generalized twins, that is $N(u) \setminus \{v\} = N(v) \setminus \{u\}$. Note that this implies that for each $i \in [1, p]$, V_i is a clique or an independent set.

We start by showing that two vertices that are not generalized twins cannot receive the same color under.

Lemma 9 ($\star$). *Let $G = (V, E)$ be a graph and let $u, v \in V$ with $N(u) \setminus \{v\} \neq N(v) \setminus \{u\}$. Then for every symmetric decoder $\mathcal{D}$ and each word w for which there is a graph isomorphism $f\colon V \to V(G(\mathcal{D}, w))$, $w_{f(u)} \neq w_{f(v)}$.*

Note that this immediately implies that for each graph G, $\mathrm{sl}(G) \geq \mathrm{nd}(G)$. We now show that the inequality also holds in the other direction.

Theorem 4 ($\star$). *For every graph G, $\mathrm{sl}(G) = \mathrm{nd}(G)$.*

5 Conclusion

We now briefly discuss another direction to approach the complexity of the lettericity problem. Namely, in contrast to asking whether we can derive/retrieve one of the three objects for a solution of the lettericity problem, we can ask for a single given such object, whether we can derive the other two objects. We call the respective problems *extension problems*. For example, DECODER EXTENSION would be the problem, where we are given a graph G, an alphabet Σ, and a decoder $\mathcal{D} \subseteq \Sigma^2$, and we ask whether there is a word w and a coloring χ, such that $G(\mathcal{D}, w) = G$, while respecting the coloring χ.

For example, DECODER EXTENSION can easily be observed to be GI-hard. This hardness follows from the reduction behind Theorem 3, as the chosen word w for the COLORING RETRIEVAL-instance was chosen arbitrarily among all possible permutations of the alphabet. Moreover, we can ensure that no two vertices in the GRAPH ISOMORPHISM-instance are generalized twins, since subdividing each edge in both G and H ensures this property (assuming there are no connected components of size at most 2) and thus yields an equivalent instance.

For future work, it would be interesting to analyze the other two extension problems, namely Word Extension, where we try to find a decoder and a coloring for a given word, and Coloring Extension, where we try to find a decoder and a word for a given coloring. Our presented algorithms for the retrieval problems do not seem to be adaptable to work for these extension problems, as the algorithms highly relied on having two of the three objects being known.

Other interesting directions for future work could be to look into generalizations of the lettericity problem. For example, what if we allow each vertex to receive an arbitrary set of color instead of just a single one. That is, w would not be a word of Σ^n anymore but rather a sequence of n subsets of Σ. For such a generalization, multiple possible rules for when an edge is added to the letter graph can be considered. For example, let i and j be positions in w with $i < j$ and let C_i and C_j be the set of colors of these two positions respectively. One way to define the edges of the letter graph could be to require that for each $\alpha \in C_i$ and each $\beta \in C_j$, $\alpha\beta$ is a word of the decoder $\mathcal{D}$. Other edge definition rules could require that (i) there is at least one $\alpha \in C_i$ and at least one $\beta \in C_j$, such that $\alpha\beta \in \mathcal{D}$ or (ii) the number of pairs $(\alpha, \beta) \in C_i \times C_j$ with $\alpha\beta \in \mathcal{D}$ is at least a given threshold t (for example $t = \frac{|C_i \times C_j|}{2}$). Such a generalization of letter graphs could allow for an even more compact representation of graphs. Note that for the normal definition of w, that is, where each vertex as only a single color, all three example definitions coincide.

References

1. Alecu, B., Ferguson, R., Kanté, M.M., Lozin, V.V., Vatter, V., Zamaraev, V.: Letter graphs and geometric grid classes of permutations. SIAM J. Discret. Math. **36**(4), 2774–2797 (2022)
2. Alecu, B., Kanté, M.M., Lozin, V.V., Zamaraev, V.: Lettericity of graphs: an FPT algorithm and a bound on the size of obstructions. Algorithmica **88**(1), 2 (2026)
3. Alecu, B., Lozin, V.: Understanding lettericity I: a structural hierarchy. CoRR abs/2106.03267 (2021). https://doi.org/10.48550/ARXIV.2106.03267
4. Atminas, A., Collins, A., Lozin, V.V., Zamaraev, V.: Implicit representations and factorial properties of graphs. Discret. Math. **338**(2), 164–179 (2015)
5. Bera, S., Mahalingam, K.: Structural properties of word representable graphs. Math. Comput. Sci. **10**(2), 209–222 (2016)
6. Bonomo, F., de Estrada, D.: On the thinness and proper thinness of a graph. Discret. Appl. Math. **261**, 78–92 (2019)
7. Feng, Z., Fernau, H., Fleischmann, P., Kindermann, P., Sacher, S.C.: Determining factorial speed fast. CoRR abs/2602.24064 (2026). https://doi.org/10.48550/ARXIV.2602.24064
8. Feng, Z., Fernau, H., Mann, K., Raman, I., Sacher, S.C.: Generalized lettericity of graphs. In: Gaur, D., Mathew, R. (eds.) CALDAM 2025. LNCS, vol 15536. pp. 134–146. Springer, Cham (2025). https://doi.org/10.1007/978-3-031-83438-7_12
9. Ferguson, R.: On the lettericity of paths. Austr. J. Combin. **78**, 348–351 (2020)
10. Ferguson, R., Vatter, V.: Letter graphs and modular decomposition. Discret. Appl. Math. **309**, 215–220 (2022)

11. Mandrick, S., Vatter, V.: Bounds on the lettericity of graphs. Electron. J. Combin. **31**(4) (2024)
12. Mannino, C., Oriolo, G., Ricci, F.: Solving stability problems on a superclass of interval graphs. Universit'a di Roma La Sapienza. Dipartimento di Informatica e Sistemistica **26**(02) (2002)
13. Petkovsek, M.: Letter graphs and well-quasi-order by induced subgraphs. Discret. Math. **244**(1–3), 375–388 (2002)
14. Shitov, Y.: Graph thinness: a lower bound and complexity. Pac. J. Math. **339**(2), 333–343 (2025)

Closure Properties of Left NP Real Numbers and NP Real Functions

Peter Hertling(✉) and Leonard Schulte-Michels

Fakultät für Informatik, Universität der Bundeswehr München, 85579 Neubiberg, Germany
{peter.hertling,leonard.schulte-michel}@unibw.de

Abstract. In the 1980s Ker-I Ko introduced the left NP real numbers and the NP real functions in order to give a complexity-theoretic answer to the question how complex the maximum values of polynomial time computable real functions can be and how difficult it can be to compute them. We prove closure properties of NP real functions, of left NP real numbers, and of differences of left NP real numbers.

We show that if $f_1, \ldots, f_m \colon [0,1]^k \to \mathbb{R}$ are NP real functions with range in $[0,1]$ and $g \colon [0,1]^m \to \mathbb{R}$ is a monotone (that means: nondecreasing with respect to each argument) NP real function then $g \circ (f_1, \ldots, f_m)$ is an NP real function as well. We conclude that, if $\alpha_1, \ldots, \alpha_m$ are left NP real numbers in the interval $[0,1]$ and $g : [0,1]^m \to \mathbb{R}$ is a monotone NP real function, then $g(\alpha_1, \ldots, \alpha_m)$ is a left NP real number as well. This implies that the sum of two left NP real numbers is a left NP real number as well, and so is the product of two nonnegative left NP real numbers. We also look at the set of all differences of left NP real numbers. We show that it is closed under functions that are real analytic and polynomial time computable. It is the smallest subfield of the field of real numbers that contains all left NP real numbers.

Keywords: left NP real numbers · NP real functions · closure properties

1 Introduction

In the 1980s Ker-I Ko [2–5] introduced the left NP real numbers and the NP real functions in order to give a complexity-theoretic answer to the question how complex the maximum values of polynomial time computable real functions can be and how difficult it can be to compute them. For example, he showed that a real number is a left NP real number if, and only if, it is the maximum value of a polynomial time computable function $f \colon [0,1] \to \mathbb{R}$. It is clear that every polynomial time computable number is a left NP real number. It is not known whether the converse is true. Concerning this question, Ko showed the following two implications that relate this question to fundamental open question in structural complexity theory:

V. Brattka et al. (Eds.): CiE 2026, LNCS 16674, pp. 320–334, 2027.
https://doi.org/10.1007/978-3-032-31348-5_21

- If the complexity classes P and NP are identical then every left NP real number is polynomial time computable.
- If every left NP real number is polynomial time computable then the complexity classes EXP and NEXP are identical.

It is not known whether the converse implications are true.

In this article we look at closure properties. The starting point of our investigation was the question whether the sum of two left NP real numbers is a left NP real number as well. We are going to show that this is true. Actually, we are going to show considerably stronger assertions. The first main result is the following closure result for NP real functions. If $f_1, \dots, f_m \colon [0,1]^k \to \mathbb{R}$ are NP real functions with range in $[0,1]$ and $g \colon [0,1]^m \to \mathbb{R}$ is a monotone (that means: nondecreasing with respect to each argument) NP real function then $g \circ (f_1, \dots, f_m)$ is an NP real function as well. As a constant function is an NP real function if, and only if, its value is a left NP real number, we conclude that, if $\alpha_1, \dots, \alpha_m$ are left NP real numbers in the interval $[0,1]$ and $g : [0,1]^m \to \mathbb{R}$ is a monotone NP real function, then $g(\alpha_1, \dots, \alpha_m)$ is a left NP real number as well. Clearly, every polynomial time computable function is an NP real function. Therefore, if $g \colon [0,1]^m \to \mathbb{R}$ is a monotone polynomial time computable function and $\alpha_1, \dots, \alpha_m$ are left NP real numbers in the interval $[0,1]$ then $g(\alpha_1, \dots, \alpha_m)$ is a left NP real number as well. This implies that the sum of two left NP real numbers is a left NP real number as well, and so is the product of two nonnegative left NP real numbers.

We also look at the set of all differences of left NP real numbers and show that it is a subfield of the field of real numbers. It is clear that this set is closed under to function $x \mapsto -x$. And that it is closed under addition and multiplication follows from the fact that the set of nonnegative left NP real numbers is closed under addition and multiplication. But for the assertion that this set is closed under the operation $x \mapsto 1/x$ (of course, defined only for $x \neq 0$), we have to argue differently. In fact, our second main result says that the set of differences of left NP real numbers is closed under univariate functions that are real analytic and polynomial time computable. We finally conclude that the set of differences of left NP real numbers is the smallest subfield of the field $\mathbb{R}$ of real numbers that contains all left NP real numbers.

Finally we would like to point out an analogy. Similarly as the NP sets can be considered as a complexity-theoretic counterpart of the computable enumerable sets, the left NP real numbers can be considered as a complexity-theoretic counterpart of the left-computable real numbers, where a real number is called *left-computable* or *left-c.e.* or *c.e.* if there is an increasing computable sequence of dyadic numbers converging to it. Of course, the sum of two left-computable numbers is left-computable as well and so is the product of two nonnegative left-computable numbers. Similarly, the differences of left NP real numbers can be considered as a complexity-theoretic counterpart of the differences of left-computable real numbers. And our result that the set of these numbers is the smallest field containing all left-NP real numbers is a counterpart of the result by Ambos-Spies, Weihrauch, and Zheng [1, Theorem 3.7] that the set of differences of left-computable real numbers is the smallest field containing all left-computable real numbers.

In the following two sections we introduce the left NP real numbers and the NP real functions, following Ko [2,4,5]. In Sect. 4 we formulate and prove our results concerning closure properties of NP real functions and left NP real numbers. In the final section we state and prove our results concerning the set of differences of left NP real numbers.

2 Left NP Real Numbers

A real number α is called a *dyadic* number if there exist numbers $z \in \mathbb{Z}$ and $k \in \mathbb{N}$ such that $\alpha = z/2^k$. Let λ denote the empty string. For $\ell \in \mathbb{N}$ we define a set $\mathbf{D}_\ell \subseteq \{0, 1, ., -\}^*$ as follows:

$$\mathbf{D}_\ell := \{\lambda, -\} \left(\{\lambda\} \cup \{1\}\{0,1\}^*\right).\{0,1\}^\ell$$

Then the set $\mathbf{D} := \bigcup_{\ell=0}^{\infty} \mathbf{D}_\ell$ is the disjoint union of the sets $\mathbf{D}_\ell$ for $\ell \in \mathbb{N}$. For $d \in \mathbf{D}$ we define

$$\text{prec}(d) := \text{ the uniquely determined } \ell \in \mathbb{N} \text{ with } d \in \mathbf{D}_\ell$$

and call $\text{prec}(d)$ the *precision* of d. We define a notation $\nu_{\mathbb{D}}$ of the dyadic numbers by $\text{dom}(\nu_{\mathbb{D}}) := \mathbf{D}$ and

$$\nu_{\mathbb{D}}(a_k \dots a_0.a_{-1} \dots a_{-\ell}) := \sum_{i=-\ell}^{k} a_i \cdot 2^i$$

as well as

$$\nu_{\mathbb{D}}(-a_k \dots a_0.a_{-1} \dots a_{-\ell}) := -\sum_{i=-\ell}^{k} a_i \cdot 2^i$$

for $k \geq -1$, $\ell \in \mathbb{N}$, $(\forall i \in \{-\ell, \dots, k\})\ a_i \in \{0,1\}$, and if $k \geq 0$ then $a_k = 1$.

Definition 1. 1. A set $A \subseteq \mathbf{D}$ is called a *left cut* of a real number α if it has the following two properties:
   (a) If $b \in A$ and if $a \in \mathbf{D}$ satisfies $\text{prec}(a) = \text{prec}(b)$ and $\nu_{\mathbb{D}}(a) \leq \nu_{\mathbb{D}}(b)$ then $a \in A$ as well.
   (b) For every $\ell \in \mathbb{N}$ the number $q_\ell := \max \nu_{\mathbb{D}}(A \cap \mathbf{D}_\ell)$ exists and satisfies $|\alpha - q_\ell| \leq 2^{-\ell}$.
2. A real number α is called a *left NP real number* if there exists a left cut $A \subseteq \mathbf{D}$ of α with $A \in \mathbf{NP}$.

Remark 2. In view of our wish to approximate real numbers from the left one may wonder whether it would not make sense to consider those real numbers x whose so-called *standard left cut* $L_x := \{d \in \mathbf{D} \mid d \leq x\}$ is in $\mathbf{NP}$. But that seems to be an unnatural condition: it is known that there exists a polynomial time computable real number x whose standard left cut L_x is not in $\mathbf{NP}$; compare [2,

4]. A real number x is called *polynomial time computable* if given a number $\ell \in \mathbb{N}$, one can compute in time polynomial in ℓ a dyadic number $d \in \mathbf{D}_\ell$ with $|x - d| \leq 2^{-\ell}$, or, equivalently, if it possesses a polynomial time decidable left cut.

Remark 3. Concerning the history of the definition of left NP real numbers, we mention that left NP real numbers were also defined in the article [8] by Selman, which appeared one year earlier than [2,5]. But Selman [8] cites preliminary versions (technical reports) from 1980 of [2,3,5] and explicitly gives credit to Ker-I Ko for the definition of left NP real numbers.

We say that a deterministic or nondeterministic Turing machine M with input alphabet Σ *works in polynomial time* if there exists a univariate polynomial p with coefficients in $\mathbb{N}$ such that, for any string $x \in \Sigma^*$, any computation of M on input x stops after at most $p(|x|)$ steps. For a deterministic or nondeterministic Turing machine M with input alphabet Σ let $L(M)$ be the set of all strings $x \in \Sigma^*$ such that there exists an accepting computation path of M on input x. Remember that **NP** is the set of all languages A such that there exists a nondeterministic Turing machine M working in polynomial time and with $L(M) = A$.

Lemma 4. *If α is a left NP real number and $r, s, t \in \mathbb{Z}$ then $2^r \cdot \alpha + s \cdot 2^t$ is a left NP real number as well.*

Proof. Let α be a left NP real number. Note that $2^r \cdot \alpha + s \cdot 2^t = 2^t \cdot (2^{r-t} \cdot \alpha + s)$. Therefore, it is sufficient to prove the following two claims:

1. For $z \in \mathbb{Z}$ the number $2^z \cdot \alpha$ is a left NP real number as well.
2. For $z \in \mathbb{Z}$ the number $\alpha + z$ is a left NP real number as well.

The first assertion is easy as multiplying a dyadic number by 2^z simply corresponds to shifting the binary point and, if necessary, by adding or deleting zeros in a name in $\mathbf{D}$ of the dyadic number. We leave the details to the reader and prove the second assertion.

Let us fix some $z \in \mathbb{Z}$. Let $A \subseteq \mathbf{D}$ be a left cut of α with $A \in \mathbf{NP}$. We claim that there exists a left cut B of $\alpha + z$ with $B \in \mathbf{NP}$. Let M be a nondeterministic Turing machine working in polynomial time with $L(M) = A$. Let M' be a nondeterministic Turing machine that works as follows. Given a string x it first checks whether x is an element of $\mathbf{D}$. If it is not then its stops in a rejecting state. Otherwise, that is, if $x \in \mathbf{D}$, then it computes a string $x' \in \mathbf{D}$ with $\nu_{\mathbb{D}}(x') = \nu_{\mathbb{D}}(x) - z$ and with $\text{prec}(x') = \text{prec}(x)$. This can be done in polynomial time, only the part of x before the binary point needs to be modified. Finally, it simulates a computation of M on input x' and stops in an accepting state iff at the end of the simulation M stops in an accepting state.

This machine M' is a nondeterministic Turing machine working in polynomial time. Hence, the set $B := L(M')$ is in $\mathbf{NP}$. Furthermore, clearly B is a left cut of $\alpha + z$. □

For a real number α let $\lfloor\alpha\rfloor := \max\{z \in \mathbb{Z} \mid z \leq \alpha\}$. Then the number $\alpha - \lfloor\alpha\rfloor$ satisfies $0 \leq \alpha - \lfloor\alpha\rfloor < 1$, and we call it the *fractional part* of α.

Corollary 5. *A real number is a left NP real number if, and only if, its fractional part is a left NP real number.*

3 NP Real Functions

We consider oracle Turing machines that work as follows given an oracle $\varphi : \mathbb{N} \to \Gamma^*$. The machine has a special oracle tape, a special query state, and a special answer state. At any stage of a computation, the machine may write a string onto the oracle tape and enter the query state. Then the oracle responds as follows: if ℓ is the length of the string on the oracle tape then this string is replaced by the string $\varphi(\ell)$ and the read/write head of the oracle tape is placed onto the field to the left of the string $\varphi(\ell)$. Then the answer state is entered and the machine can continue with its computation. This response of the oracle is counted as only one step.

For a nondeterministic oracle Turing machine M with input alphabet Σ and an oracle φ let $L(M^\varphi)$ be the set of all strings $x \in \Sigma^*$ such that there exists an accepting computation path of M^φ (that means: of M with oracle φ) on input x.

Definition 6. 1. Let us consider a real number $\alpha \in [0, 1]$. Any sequence $\varphi : \mathbb{N} \to \{0, 1\}^*$ satisfying

$$|\varphi(n)| = n \quad \text{and} \quad |\alpha - \nu_{\mathbb{D}}(.\varphi(n))| \leq 2^{-n}$$

for all $n \in \mathbb{N}$ is called an *oracle* for α.

2. Let us fix a positive integer k. Let us consider a real vector $\alpha = (\alpha_1, \ldots, \alpha_k) \in [0, 1]^k$. Any sequence $\varphi : \mathbb{N} \to \{0, 1, \#\}^*$ such that there exist oracles φ_i of α_i for $i = 1, \ldots, k$ such that, for all n,

$$\varphi(n) = \varphi_1(n)\#\ldots\#\varphi_k(n),$$

is called an *oracle* for α.

3. Let us fix a positive integer k. A function $f : D \subseteq [0, 1]^k \to \mathbb{R}$ is called an *NP real function on* $D \subseteq [0, 1]^k$ if there exist a nondeterministic oracle Turing machine M and a polynomial p with coefficients in $\mathbb{N}$ such that for any $\alpha \in D$ and any oracle φ for α the set $L(M^\varphi)$ is a left cut of $f(\alpha)$ and for any input string x the machine M^φ (that means: M with oracle φ) stops after at most $p(|x|)$ steps.

Lemma 7. *A constant function* $f : [0, 1]^k \to \mathbb{R}$ *is an NP real function if, and only if, its value is a left NP real number.*

Proof. First, let β be a left NP real number. Let $f\colon [0,1]^k \to \mathbb{R}$ be the constant function with value β. We wish to show that f is an NP real function. Let B be a left cut of β with $B \in \mathbf{NP}$. Let M_B be a nondeterministic Turing machine working in polynomial time and with $L(M_B) = B$. Let M be the nondeterministic oracle Turing machine that works as follows when started with an oracle φ of a vector in $[0,1]^k$and an input string x: it simply works exactly like M_B with input string x, completely ignoring the oracle φ. Clearly, there is a polynomial p such that, for any oracle φ of any vector in $[0,1]^k$, the machine M^φ stops after at most $p(|x|)$ steps, for any input string $x \in \Sigma^*$. For every $\alpha \in [0,1]^k$ and every oracle φ for α we have $L(M^\varphi) = L(M_B) = B$. Hence, $L(M^\varphi)$ is a left cut of $\beta = f(\alpha)$. We have shown that f is an NP real function.

Now, let $f\colon [0,1]^k \to \mathbb{R}$ be a constant NP real function, and let β be its value. We wish to show that β is a left NP real number. Let M_f be a nondeterministic oracle Turing machine as in Definition 6.3. Let φ_{zero} be the oracle for the point $\mathbf{0} := (0,\ldots,0) \in [0,1]^k$ defined by $\varphi_{\text{zero}}(n) := 0^n\#\ldots\#0^n$ (with k copies of 0^n). Let M be the nondeterministic Turing machine that, started with a string x, simulates the oracle Turing machine M_f with oracle φ_{zero}. That means, M simulates M_f and, whenever M_f asks for the value $\varphi(n)$ of an oracle φ, M continues the simulation of M_f with the oracle value $\varphi_{\text{zero}}(n)$. On the one hand, this machine is nondeterministic and works in polynomial time, hence, the language $L(M)$ is in $\mathbf{NP}$. On the other hand, we obtain $L(M) = L(M_f^{\varphi_{\text{zero}}})$, and by the definition of M_f this is a left cut of $f(\mathbf{0}) = \beta$. We have shown that β is a left NP real number. □

Definition 8. Let us fix a positive integer k. A function $f\colon D \subseteq [0,1]^k \to \mathbb{R}$ is called a *polynomial time computable real function* if there exist a deterministic oracle Turing machine M and a polynomial p with coefficients in $\mathbb{N}$ such that for any $\alpha \in D$ and any oracle φ for α and any number $\ell \in \mathbb{N}$, on input 0^ℓ the output of M^φ is a string $F_{M^\varphi}(0^\ell) \in \mathbf{D}_\ell$ satisfying $|f(\alpha) - \nu_{\mathbb{D}}(F_{M^\varphi}(0^\ell))| \leq 2^{-\ell}$ and the machine stops after at most $p(\ell)$ steps.

Lemma 9. *Any polynomial time computable real function $f\colon D \subseteq [0,1]^k \to \mathbb{R}$ is an NP real function.*

Proof. Let $f\colon D \subseteq [0,1]^k \to \mathbb{R}$ be a polynomial time computable function, and let M_f be a deterministic oracle Turing machine as in Definition 8. Let M be an oracle Turing machine that works as follows. Started with an oracle φ and a string x, first it checks whether x is an element of $\mathbf{D}$. If not then its stops in a rejecting state. Otherwise, that is, if $x \in \mathbf{D}$, it computes the length $\ell := |x|$ of the string x, then it tries to compute $F_{M_f^\varphi}(0^\ell)$ (by simulating M_f^φ with input 0^ℓ; this will be successful and will be be done in polynomial time if φ is an oracle for a vector in D), and finally it accepts x iff $\nu_{\mathbb{D}}(x) \leq \nu_{\mathbb{D}}(F_{M^\varphi}(0^\ell))$. This oracle Turing machine works in polynomial time for any oracle φ of a vector in D. Actually, it is even deterministic. For any $\alpha \in D$ and any oracle φ for α the language $L(M^\varphi)$ is a left cut of $f(\alpha)$. □

4 Closure Properties

It is our next goal to show that the sum of two left NP real numbers is a left NP real number as well. Actually, we shall prove stronger results. Let us fix a positive integer m. We define a partial order $\leq$ on $\mathbb{R}^m$ by

$$(\alpha_1,\ldots,\alpha_m) \leq (\beta_1,\ldots,\beta_m) :\iff (\forall i \in \{1,\ldots,m\})\, \alpha_i \leq \beta_i.$$

We call a function $g :\subseteq \mathbb{R}^m \to \mathbb{R}$ with $\mathrm{dom}(g) \subseteq \mathbb{R}^m$ *monotone* if for all $\alpha,\beta \in \mathrm{dom}(g)$: if $\alpha \leq \beta$ then $g(\alpha) \leq g(\beta)$.

Theorem 10. *Let us fix two positive integers k,m. Let $f_1,\ldots,f_m\colon [0,1]^k \to \mathbb{R}$ be NP real functions with range in $[0,1]$. Let $g\colon [0,1]^m \to \mathbb{R}$ be a monotone NP real function. Then the composition $g \circ (f_1,\ldots,f_m)\colon [0,1]^k \to \mathbb{R}$ is an NP real function.*

The following notion, Lemma 13, and Corollary 15 will be useful for the proof of Theorem 10.

Definition 11. We call a finite, nonempty set A of binary strings *consistent* if there exist a real number $\alpha \in [0,1]$ and an oracle $\varphi : \mathbb{N} \to \{0,1\}^*$ for α such that $A \subseteq \mathrm{range}(\varphi)$.

Lemma 12. *Let A be a consistent set. If a real number $\alpha \in [0,1]$ satisfies $|\alpha - \nu_{\mathbb{D}}(.x)| \leq 2^{-|x|}$ for every $x \in A$ then there is an oracle $\varphi : \mathbb{N} \to \{0,1\}^*$ for α such that $A \subseteq \mathrm{range}(\varphi)$.*

Proof. Let $\alpha \in [0,1]$ be a real number satisfying $|\alpha - \nu_{\mathbb{D}}(.x)| \leq 2^{-|x|}$ for every $x \in A$. We define a function $\varphi : \mathbb{N} \to \{0,1\}^*$ by

$$\varphi(n) := \begin{cases} \text{the unique string in } A \text{ of length } n & \text{if there is a string} \\ & \text{in } A \text{ of length } n, \\ \text{the unique binary string } x \in \{0,1\}^n & \\ \text{with } 0 \leq \alpha - \nu_{\mathbb{D}}(.x) < 2^{-n} & \text{otherwise.} \end{cases}$$

Then φ is an oracle for α such that $A \subseteq \mathrm{range}(\varphi)$. □

Lemma 13. *Let $A \subseteq \{0,1\}^*$ be a consistent set such that $\{.\}A \subseteq B$ for a left cut B of a real number $\beta \in [0,1]$. Then there is a real number α with $0 \leq \alpha \leq \beta$ such that there is an oracle φ for α with $A \subseteq \mathrm{range}(\varphi)$.*

Proof. The set $[0,1] \cap \bigcap_{x\in A} \left[\nu_{\mathbb{D}}(.x) - 2^{-|x|}, \nu_{\mathbb{D}}(.x) + 2^{-|x|}\right]$ is nonempty because A is consistent. Furthermore, it is compact. Let α be its minimum. By Lemma 12 there exists an oracle φ for α with $A \subseteq \mathrm{range}(\varphi)$. We claim that $\alpha \leq \beta$. This is clear if $\alpha = 0$. Let us assume that $\alpha > 0$. Then there exists a string $x \in A$ such that $\alpha = \nu_{\mathbb{D}}(.x) - 2^{-|x|}$. The number $q_{|x|} := \max \nu_{\mathbb{D}}(B \cap \mathbf{D}_{|x|})$ satisfies $\nu_{\mathbb{D}}(.x) \leq q_{|x|}$ because $x \in A$ and $\{.\}A \subseteq B$ and B is a left cut. Furthermore, it satisfies $|\beta - q_{|x|}| \leq 2^{-|x|}$. We conclude $\alpha = \nu_{\mathbb{D}}(.x) - 2^{-|x|} \leq q_{|x|} - 2^{-|x|} \leq \beta$. □

Lemma 14. *Let A be a consistent set. Let w be the longest binary string in A. Then at least one $\alpha \in \{\nu_{\mathbb{D}}(.w) - 2^{-|w|}, \nu_{\mathbb{D}}(.w), \nu_{\mathbb{D}}(.w) + 2^{-|w|}\}$ satisfies $|\alpha - \nu_{\mathbb{D}}(.x)| \leq 2^{-|x|}$ for every $x \in A$.*

Proof. Let $\beta \in [0,1]$ be a real number such that there exists an oracle $\varphi : \mathbb{N} \to \{0,1\}^*$ for β such that $A \subseteq \text{range}(\varphi)$. Then for every $x \in A$ we have $|\beta - \nu_{\mathbb{D}}(.x)| \leq 2^{-|x|}$. So, if $\beta \in \{\nu_{\mathbb{D}}(.w) - 2^{-|w|}, \nu_{\mathbb{D}}(.w) + 2^{-|w|}\}$ then we are done. Let us assume that β is not one of the two numbers $\nu_{\mathbb{D}}(.w) - 2^{-|w|}, \nu_{\mathbb{D}}(.w) + 2^{-|w|}$. Then $|\beta - \nu_{\mathbb{D}}(.w)| < 2^{-|w|}$. We claim that in this case the number $\alpha := \nu_{\mathbb{D}}(.w)$ has the desired property (namely: $|\alpha - \nu_{\mathbb{D}}(.x)| \leq 2^{-|x|}$ for all $x \in A$). Let us consider an arbitrary $x \in A$. If $x = w$ then $|\alpha - \nu_{\mathbb{D}}(.x)| = 0 \leq 2^{-|x|}$. Let us assume $x \neq w$. Then $|x| < |w|$. We obtain

$$|\alpha - \nu_{\mathbb{D}}(.x)| \leq |\nu_{\mathbb{D}}(.w) - \beta| + |\beta - \nu_{\mathbb{D}}(.x)| < 2^{-|w|} + 2^{-|x|}.$$

As both α and $\nu_{\mathbb{D}}(.x)$ can be written as integers multiplied by $2^{-|w|}$, this implies $|\alpha - \nu_{\mathbb{D}}(.x)| \leq 2^{-|x|}$. □

Corollary 15. *Given a string of the form*

$$x_1\#x_2\#\ldots\#x_m$$

such that $m \geq 1$ and each x_i for $i = 1, \ldots, m$ is a binary string, one can check in polynomial time whether the set $\{x_i \mid i \in \{1, \ldots, m\}\}$ is consistent.

Proof. Let us assume that a string $x_1\#x_2\#\ldots\#x_m$ of the indicated form is given. First, one checks whether the set $\{x_i \mid i \in \{1, \ldots, m\}\}$ contains two different strings of the same length. If this is the case then this set is not consistent. Otherwise, one computes an index j such that x_j has maximal length among the strings $x_1, \ldots, x_m$. Then one checks for each of the three numbers α in the set $\{\nu_{\mathbb{D}}(.x_j) - 2^{-|x_j|}, \nu_{\mathbb{D}}(.x_j), \nu_{\mathbb{D}}(.x_j) + 2^{-|x_j|}\}$ whether it satisfies $|\alpha - \nu_{\mathbb{D}}(.x_i)| \leq 2^{-|x_i|}$ for every $i \in \{1, \ldots, m\} \setminus \{j\}$. Clearly, if this is true for at least one of these three numbers α then the set $\{x_i \mid i \in \{1, \ldots, m\}\}$ is consistent. By Lemma 14 we can conclude that otherwise it is not. It is clear that this algorithm works in polynomial time. □

Proof (of Theorem 10). We give a detailed proof for the case $k = m = 1$ and in the end explain how this proof can be modified in order to give the desired assertion for arbitrary k and m.

So, let us assume $k = m = 1$. Instead of f_1 we write f. Let M_f and M_g be nondeterministic oracle Turing machines as in Definition 6.3 for the functions f and g, respectively. We wish to construct a nondeterministic oracle Turing machine M that works in polynomial time and such that $L(M^\varphi)$ is a left cut of $g(f(\alpha))$ for any real number $\alpha \in [0,1]$ and any oracle φ for α.

The idea is to simulate the machine M_g where, whenever M_g asks for the value $\psi(m)$ of an oracle ψ (which would ideally be an oracle for $f(\alpha)$) then we use the machine M_f with oracle φ in order to try to guess a suitable oracle value. As $L(M^\varphi)$ is only a left cut of $f(\alpha)$, in general our guessed string will denote a dyadic number that may be smaller than desired. This is the reason why we assume the function g to be monotone. But we can also be lucky and guess a sequence of dyadic numbers that can be part of an oracle for $f(\alpha)$.

We come to the definition of M. Let M be the nondeterministic oracle Turing machine that, started with an oracle φ and an input string x, works as follows. First M^φ checks whether $x \in \mathbf{D}$. If this is not the case then it immediately stops in a rejecting state. Otherwise, that is, if $x \in \mathbf{D}$, then M^φ simulates M_g on input x in the following way. Whenever M_g asks for the value $\psi(m)$ of an oracle ψ (which would ideally be an oracle for $f(\alpha)$ if φ is an oracle for $\alpha \in [0,1]$) then M^φ nondeterministically chooses a binary string y of length m and proceeds as follows.

- It checks deterministically in polynomial time (compare Corollary 15) whether the set consisting of this string y and of all strings that have been obtained in previous steps when M_g asked for the value of an oracle, is consistent. If it is not consistent then the machine M^φ stops in a rejecting state.
- ("Second" step) Otherwise the machine applies M_f^φ to $.y$ once (note that here some nondeterministic computation steps may take place as M_f is a nondeterministic oracle Turing machine). If at the end of this computation of M_f^φ on input $.y$ the string $.y$ is not accepted then the machine M^φ stops in a rejecting state.
- Otherwise the machine M^φ answers the current request of M_g for an oracle value with the string y and continues with the simulation of M_g.

We demand that the machine M^φ stops in an accepting resp. rejecting state if the simulation of M_g stops in an accepting resp. rejecting state. This ends the description of the machine M.

Let us fix a real number $\alpha \in [0,1]$ and an oracle φ for α. Given the oracle φ, the machine M_g works in polynomial time. Indeed, the additional steps carried out as described above whenever during the simulation of M_g the machine M_g asks for an oracle value do not require more than a polynomial (in $|x|$) number of steps. Hence, M^φ will also stop after a polynomial (in $|x|$) number of steps, say, after $p(|x|)$ steps, where p is a polynomial that does not depend on φ. We still have to show that the set $L(M^\varphi)$ is a left cut of $g(f(\alpha))$. In order to show this, we show three assertions:

1. If $x \in L(M^\varphi)$ then $x \in \mathbf{D}$ and $\nu_{\mathbb{D}}(x) \le g(f(\alpha)) + 2^{-|x|}$.
2. If $x \in L(M^\varphi)$ and $x' \in \mathbf{D}_{\text{prec}(x)}$ satisfies $\nu_{\mathbb{D}}(x') \le \nu_{\mathbb{D}}(x)$ then $x' \in L(M^\varphi)$ as well.
3. There exists an oracle ξ of $f(\alpha)$ such that $L(M_g^\xi) \subseteq L(M^\varphi)$.

Note that for an oracle ξ of $f(\alpha)$ the set $L(M_g^\xi)$ is a left cut of $g(f(\alpha))$. Therefore for any $\ell \in \mathbb{N}$ and for the ξ in the third assertion, this set (and then $L(M^\varphi)$ as well) contains a string $x \in \mathbf{D}_\ell$ with $|g(f(\alpha)) - \nu_{\mathbb{D}}(x)| \le 2^{-\ell}$. Thus, the third assertion assures that, for any $\ell \in \mathbb{N}$, the set $L(M^\varphi)$ contains a string with sufficiently large numerical value while the first assertion assures that the strings in the set $L(M^\varphi)$ do not have a too large numerical value. Finally, the second assertion ensures the closure property demanded in part (a) of Definition 1.1. Thus, these three assertions together imply that $L(M^\varphi)$ is a left cut of $g(f(\alpha))$.

First let us consider some string $x \in L(M^\varphi)$. Then x is a string in $\mathbf{D}$ and there exists an accepting computation path of M^φ with input x. The set A of all strings y (all of them are binary strings!) that have been obtained as answers to oracle requests of M_g during the simulation of M_g on this computation path is consistent and the strings $.y$ for $y \in A$ are elements of $L(M_f^\varphi)$. As $L(M_f^\varphi)$ is a left cut of $f(\alpha)$, we conclude by Lemma 13 that there exist a real β with $0 \le \beta \le f(\alpha)$ and an oracle $\psi : \mathbb{N} \to \{0,1\}^*$ for β such that $A \subseteq \text{range}(\psi)$. Hence, the simulation of M_g performed during this computation of M is identical with a computation path of M_g^ψ on input x for some oracle ψ of a number β with $0 \le \beta \le f(\alpha)$. We conclude $x \in L(M_g^\psi)$ and

$$\nu_{\mathbb{D}}(x) \le g(\beta) + 2^{-|x|} \le g(f(\alpha)) + 2^{-|x|}$$

(here we use that g is monotone).

Now let us consider additionally a string $x' \in \mathbf{D}_{\text{prec}(x)}$ with $\nu_{\mathbb{D}}(x') \le \nu_{\mathbb{D}}(x)$. We consider the same real β with $0 \le \beta \le f(\alpha)$ as above and the same oracle ψ for β as above. As $x \in L(M_g^\psi)$ and $\text{prec}(x') = \text{prec}(x)$ and $\nu_{\mathbb{D}}(x') \le \nu_{\mathbb{D}}(x)$ and $L(M_g^\psi)$ is a left cut, we conclude $x' \in L(M_g^\psi)$. Thus, there exists an accepting computation path of M_g^ψ on input x'. When M^φ is started with x' as the input string, then it is possible that exactly this accepting computation path of M_g^ψ on input x' is simulated. Then at the end of such a computation of M the string x' is accepted, hence, we obtain $x' \in L(M^\varphi)$.

Finally, we remember that $L(M_f^\varphi)$ is a left cut of $f(\alpha)$. Let $\xi : \mathbb{N} \to \{0,1\}^*$ be defined by $|\xi(\ell)| = \ell$ and

$$\nu_{\mathbb{D}}(.\xi(\ell)) = \max\{\nu_{\mathbb{D}}(.y) \mid y \in \{0,1\}^\ell \text{ and } .y \in L(M_f^\varphi)\}.$$

We claim that ξ is an oracle for $f(\alpha)$. Indeed, for every $\ell \in \mathbb{N}$ the number $q_\ell := \max(L(M_f^\varphi) \cap \mathbf{D}_\ell)$ satisfies $|f(\alpha) - q_\ell| \le 2^{-\ell}$. We see $\nu_{\mathbb{D}}(.\xi(\ell)) = \min\{q_\ell, 1 - 2^{-\ell}\}$. On the one hand, we obtain $\nu_{\mathbb{D}}(.\xi(\ell)) \le q_\ell \le f(\alpha) + 2^{-\ell}$. On the other hand, using also $f(\alpha) \le 1$ we obtain $f(\alpha) - 2^{-\ell} \le \min\{q_\ell, 1 - 2^{-\ell}\} = \nu_{\mathbb{D}}(.\xi(\ell))$. This implies $|f(\alpha) - \nu_{\mathbb{D}}(.\xi(\ell))| \le 2^{-\ell}$. Thus, ξ is an oracle for $f(\alpha)$. Let $x \in L(M_g^\xi)$. There is an accepting computation path of M_g^ξ on input x. When M^φ is started with x as the input string, then it is possible that exactly this accepting computation path of M_g^ξ on input x is simulated. Then at the end of such a computation of M^φ the string x is accepted, hence, we obtain $x \in L(M^\varphi)$.

We have proved the assertion of the theorem in the case $k = m = 1$. Going from $k = 1$ to arbitrary k makes only a minor difference in notation. Going from $m = 1$ to arbitrary m requires a slightly bigger change: in the discussion of how to treat an oracle call of the simulated machine M_g, instead of working with a binary string y one has to work with a vector of m binary strings of the same length. □

Corollary 16. *Let $\alpha_1, \ldots, \alpha_m$ be m left NP real numbers in the interval $[0,1]$. Let $g : [0,1]^m \to \mathbb{R}$ be a monotone NP real function. Then $g(\alpha_1, \ldots, \alpha_m)$ is a left NP real number as well.*

Proof. For $i = 1, \ldots, m$ let $f_i\colon [0,1] \to \mathbb{R}$ be the constant function with value α_i. By Lemma 7 the functions f_i are NP real functions. And their range is in $[0,1]$. By Theorem 10 the composition $g \circ (f_1, \ldots, f_m)$ is an NP real function. It is a constant function with value $g(\alpha_1, \ldots, \alpha_m)$. Hence, by Lemma 7 this number is a left NP real number. □

Corollary 17. *1. The sum of two left NP real numbers is a left NP real number as well.*
2. The product of two nonnegative left NP real numbers is a left NP real number as well.

Proof. 1. Let α_1, α_2 be two left NP real numbers. By Corollary 5 the fractional parts $\alpha' := \alpha - \lfloor\alpha\rfloor$ and $\beta' := \beta - \lfloor\beta\rfloor$ are left NP real numbers as well. Let $g : [0,1]^2 \to \mathbb{R}$ be the restriction of the addition function to the square $[0,1]^2$. The function g is monotone and polynomial time computable, hence, by Lemma 9 it is an NP real function. By Corollary 16 the number

$$\alpha' + \beta' = \alpha_1 - \lfloor\alpha_1\rfloor + \alpha_2 - \lfloor\alpha_2\rfloor$$

is a left NP real number. It has the same fractional part as the number $\alpha_1 + \alpha_2$. Hence, by Corollary 5 the number $\alpha_1 + \alpha_2$ is a left NP real number as well.

2. The proof starts very similarly. Let α_1, α_2 be two nonnegative left NP real numbers. Then by Corollary 5 their fractional parts $\alpha - \lfloor\alpha\rfloor$ and $\beta - \lfloor\beta\rfloor$ are left NP real numbers as well. Let $g : [0,1]^2 \to \mathbb{R}$ be the restriction of the multiplication function to the square $[0,1]^2$. The function g is monotone and polynomial time computable, hence, by Lemma 9 it is an NP real function. By Corollary 16 the number

$$(\alpha_1 - \lfloor\alpha_1\rfloor) \cdot (\alpha_2 - \lfloor\alpha_2\rfloor) = \alpha_1 \cdot \alpha_2 - \lfloor\alpha_1\rfloor \cdot \alpha_2 - \lfloor\alpha_2\rfloor \cdot \alpha_1 + \lfloor\alpha_1\rfloor \cdot \lfloor\alpha_2\rfloor.$$

is a left NP real number. Note that the numbers $\lfloor\alpha_1\rfloor$ and $\lfloor\alpha_2\rfloor$ are non-negative integers. Hence, by adding the left NP real numbers α_1 and α_2 sufficiently many times and by applying the first assertion we conclude that also the number

$$\alpha_1 \cdot \alpha_2 + \lfloor\alpha_1\rfloor \cdot \lfloor\alpha_2\rfloor$$

is a left NP real number. It has the same fractional part as the number $\alpha_1 \cdot \alpha_2$. Hence, by Corollary 5 the number $\alpha_1 \cdot \alpha_2$ is a left NP real number as well. □

5 Differences of Left NP Real Numbers

Let S be a subset of some set T. Let k be a positive integer, and let $f\colon D \to T$ be a function whose domain of definition D is a subset of T^k. We say that S *is closed under* f if for every $s \in D \cap S^k$ the value $f(s)$ is an element of S as well. A subset $R \subseteq \mathbb{R}$ is called a *subfield* of $\mathbb{R}$ if it contains the numbers 0 and 1 and is closed under the functions addition $(x, y) \mapsto x + y$, additive inverse $x \mapsto -x$, multiplication $(x, y) \mapsto x \cdot y$, and multiplicative inverse $x \mapsto 1/x$ (only defined on $\mathbb{R} \setminus \{0\}$, of course).

In deviation from Definition 8 we now wish to consider functions on more general domains than just $[0, 1]^k$. But note that for functions $f : [0, 1]^k \to \mathbb{R}$ the following definitions lead to the same notion of polynomial time computability as Definition 8. The main difference is that the real numbers considered do not have to be elements of $[0, 1]$ and that now for $\alpha \in \mathbb{R}$ any function $\varphi : \mathbb{N} \to \mathbf{D}$ satisfying

$$\text{prec}(\varphi(n)) = n \quad \text{and} \quad |\alpha - \nu_{\mathbb{D}}(\varphi(n))| \leq 2^{-n},$$

for any $n \in \mathbb{N}$, is called an *oracle* for α. Similarly as in Definition 8, for any $\alpha = (\alpha_1, \ldots, \alpha_k) \in \mathbb{R}^k$ any function $\varphi : \mathbb{N} \to \{0, 1, -, \#\}^*$ such that there exist oracles φ_i of α_i for $i = 1, \ldots, k$ such that, for all n,

$$\varphi(n) = \varphi_1(n)\#\ldots\#\varphi_k(n),$$

is called an *oracle* for α.

Let $D \subseteq \mathbb{R}^k$, $k > 0$ arbitrary. Let $t : \mathbb{N} \to \mathbb{N}$ be a total function. We say that a function $f : D \to \mathbb{R}$ is *computable in time* t if there exists an oracle Turing machine M which, given any oracle φ for any vector $x \in D$ and any unary input string 0^ℓ, produces a string $F_{M^\varphi}(0^\ell) \in \mathbf{D}_\ell$ satisfying $|f(\alpha) - \nu_{\mathbb{D}}(F_{M^\varphi}(0^\ell))| \leq 2^{-\ell}$ and stops after at most $t(\ell)$ steps. We say that a function $f : D \to \mathbb{R}$ is *polynomial time computable* if there exists a polynomial p with coefficients in $\mathbb{N}$ such that f is computable in time p.

Definition 18. We call a real number γ a *dNP number* if it can be written as the difference of two left NP real numbers, that is, if there exist left NP real numbers α and β such that $\gamma = \alpha - \beta$.

Let $f : U \to \mathbb{R}$ be a function whose domain of definition U is an open subset of $\mathbb{R}$. Such a function is called *real analytic in a point* $x_0 \in U$ if there exists a positive number r such that in the interval $(x_0 - r, x_0 + r)$ the function f is given by a convergent power series with center x_0; compare Krantz and Parks [6]. Such a function is called *real analytic* if it is real analytic in every point $x_0 \in U$.

Theorem 19. *If $y \in \mathbb{R}$ is a dNP number, if $U \subseteq \mathbb{R}$ is an open neighbourhood of y, and if $f : U \to \mathbb{R}$ is a polynomial time computable function that is real analytic in y then $f(y)$ is a dNP number as well.*

Proof. Let $y \in U$ be a dNP number. We assume that y is not a dyadic number because otherwise the assertion is trivial (if y is a polynomial time computable number and f is a polynomial time computable function then $f(y)$ is a polynomial time computable number as well and, hence, a dNP number). We can also assume that y is positive (otherwise instead of f we can consider the function $x \mapsto f(-x)$ defined on $-U$). As we assume that f is real analytic in y, there exists a number $m \in \mathbb{N}$ such that for all x in the interval $(y - 2^{-m}, y + 2^{-m})$ the Taylor series $\sum_{i=0}^{\infty} \frac{f^{(i)}(y)}{i!} \cdot (x - y)^i$ converges to $f(x)$. By choosing m larger, we can even assume that this series converges absolutely and uniformly on the closed interval $[y - 2^{-m}, y + 2^{-m}]$ and that $0 < y - 2^{-m}$. There are left NP real numbers $\widetilde{\alpha}$ and $\widetilde{\beta}$ with $y = \widetilde{\alpha} - \widetilde{\beta}$. Let $d \in \mathbf{D}_{m+3}$ be the string such that the dyadic number $\nu_{\mathbb{D}}(d)$ satisfies $\widetilde{\beta} - 2^{-(m+3)} < \nu_{\mathbb{D}}(d) \leq \widetilde{\beta}$. Next, let $e \in \mathbf{D}_{m+3}$ be the string such that the dyadic number $\nu_{\mathbb{D}}(e)$ satisfies $\widetilde{\alpha} - \nu_{\mathbb{D}}(d) - 2^{-(m+3)} < \nu_{\mathbb{D}}(e) \leq \widetilde{\alpha} - \nu_{\mathbb{D}}(d)$. Then by Lemma 4 the numbers $\gamma := \widetilde{\alpha} - \nu_{\mathbb{D}}(d) - \nu_{\mathbb{D}}(e)$ and $\beta := \widetilde{\beta} - \nu_{\mathbb{D}}(d)$ are left NP real numbers as well and satisfy $\gamma - \beta = y - \nu_{\mathbb{D}}(e)$ and $0 \leq \beta < 2^{-(m+3)}$ and $0 \leq \gamma < 2^{-(m+3)}$. Together this implies $|y - \nu_{\mathbb{D}}(e)| = |\gamma - \beta| < 2^{-(m+2)}$. It is well known (compare [6, Prop. 1.2.3]) that for x in the interval $((\nu_{\mathbb{D}}(e) - 3 \cdot 2^{-(m+2)}, \nu_{\mathbb{D}}(e) + 3 \cdot 2^{-(m+2)})$ the Taylor series $\sum_{i=0}^{\infty} \frac{f^{(i)}(\nu_{\mathbb{D}}(e))}{i!} \cdot (x - \nu_{\mathbb{D}}(e))^i$ converges to $f(x)$ as well, and on the closed interval

$$C := \left[\nu_{\mathbb{D}}(e) - 2 \cdot 2^{-(m+2)}, \nu_{\mathbb{D}}(e) + 2 \cdot 2^{-(m+2)}\right]$$

it converges absolutely and uniformly. Let $M \in \mathbb{N}$ be an integer such that $\sum_{i=0}^{\infty} \frac{|f^{(i)}(\nu_{\mathbb{D}}(e))|}{i!} \cdot \left(2 \cdot 2^{-(m+2)}\right)^i \leq 2^M$. Then for all $x \in C$ we have

$$\sum_{i=0}^{\infty} \frac{|f^{(i)}(\nu_{\mathbb{D}}(e))|}{i!} \cdot |x - \nu_{\mathbb{D}}(e)|^i \leq 2^M.$$

We define two formal power series $g(x, y)$ and $h(x, y)$ in two variables x, y by taking for $g(x, y)$ all terms with positive coefficients in the formal series $\sum_{i=0}^{\infty} \frac{f^{(i)}(\nu_{\mathbb{D}}(e))}{i!} \cdot (x - y)^i$ and by taking for $-h(x, y)$ all terms with negative coefficients in this series such that

$$\sum_{i=0}^{\infty} \frac{f^{(i)}(\nu_{\mathbb{D}}(e))}{i!} \cdot (x - y)^i = g(x, y) - h(x, y).$$

To be precise, we define

$$g(x, y) := \sum_{i=0}^{\infty} pterms_i(x, y) \quad \text{and} \quad h(x, y) := \sum_{i=0}^{\infty} nterms_i(x, y)$$

where

$$pterms_i(x, y) := \begin{cases} \frac{f^{(i)}(\nu_{\mathbb{D}}(e))}{i!} \cdot \sum_{j=0,\ldots,i,\, j \text{ even}} \binom{i}{j} \cdot x^{i-j} \cdot y^j & \text{if } f^{(i)}(\nu_{\mathbb{D}}(e)) > 0, \\ 0 & \text{if } f^{(i)}(\nu_{\mathbb{D}}(e)) = 0, \\ \frac{f^{(i)}(\nu_{\mathbb{D}}(e))}{i!} \cdot \sum_{j=0,\ldots,i,\, j \text{ odd}} \binom{i}{j} \cdot x^{i-j} \cdot y^j & \text{if } f^{(i)}(\nu_{\mathbb{D}}(e)) < 0, \end{cases}$$

$$nterms_i(x, y) := \begin{cases} \frac{f^{(i)}(\nu_{\mathbb{D}}(e))}{i!} \cdot \sum_{j=0,\ldots,i,\, j \text{ odd}} \binom{i}{j} \cdot x^{i-j} \cdot y^j & \text{if } f^{(i)}(\nu_{\mathbb{D}}(e)) > 0, \\ 0 & \text{if } f^{(i)}(\nu_{\mathbb{D}}(e)) = 0, \\ \frac{f^{(i)}(\nu_{\mathbb{D}}(e))}{i!} \cdot \sum_{j=0,\ldots,i,\, j \text{ even}} \binom{i}{j} \cdot x^{i-j} \cdot y^j & \text{if } f^{(i)}(\nu_{\mathbb{D}}(e)) < 0. \end{cases}$$

The series defining these functions contain only terms with nonnegative coefficients. We claim that they are converging uniformly for all $(x, y) \in [0, 2^{-(m+3)}]^2$

and we can even give an explicit convergence modulus. Indeed, for both series and for $(x, y) \in [0, 2^{-(m+3)}]^2$ the sum of all terms with $i \geq n$ is bounded from above by

$$\begin{aligned}
&\sum_{i=n}^{\infty} \frac{|f^{(i)}(\nu_{\mathbb{D}}(e))|}{i!} \cdot (x+y)^i \\
&\leq \sum_{i=n}^{\infty} \frac{|f^{(i)}(\nu_{\mathbb{D}}(e))|}{i!} \cdot \left(2^{-(m+3)} + 2^{-(m+3)}\right)^i \\
&\leq \sum_{i=n}^{\infty} \frac{|f^{(i)}(\nu_{\mathbb{D}}(e))|}{i!} \cdot (2^{-(m+2)})^i \\
&\leq \frac{(2^{-(m+2)})^n}{(2 \cdot 2^{-(m+2)})^n} \cdot \sum_{i=n}^{\infty} \frac{|f^{(i)}(\nu_{\mathbb{D}}(e))|}{i!} \cdot \left(2 \cdot 2^{-(m+2)}\right)^i \\
&\leq \frac{1}{2^n} \cdot 2^M,
\end{aligned}$$

Thus, the series $g(x, y)$ and $h(x, y)$ define monotone functions on $[0, 2^{-(m+3)}]^2$; we will denote them by g and h as well. As f is polynomial time computable, by a result by Müller [7] the sequence of values $(f^{(i)}(\nu_{\mathbb{D}}(e))/i!)_i$ is polynomial time computable. That means, there exists a deterministic Turing machine which, given 0^i and 0^ℓ, computes a string $d' \in \mathbf{D}_\ell$ such that $\left|\frac{f^{(i)}(\nu_{\mathbb{D}}(e))}{i!} - \nu_{\mathbb{D}}(d')\right| \leq 2^{-\ell}$. This implies that the functions g and h are polynomial time computable. Note that this is true although in the definition of $g(x, y)$ and $h(x, y)$ we made a distinction between positive terms and negative terms, and it might seem that this requires that one has to decide whether the values $f^{(i)}(\nu_{\mathbb{D}}(e))$ are positive or negative, which is, of course, not possible effectively. The reason why this is not a problem is the fact that these terms appear as factors in terms that are added. And when a term $f^{(i)}(\nu_{\mathbb{D}}(e))$ is close to zero (which would be the problematic case) then the corresponding term that is to be added is small. Therefore, if we wish to compute $g(x, y)$ or $h(x, y)$ up to precision $2^{-\ell}$, then it is sufficient to compute the terms $f^{(i)}(\nu_{\mathbb{D}}(e))/i!$ only up to a certain precision $2^{-q(\ell)}$, where q is some polynomial, and by the estimate above it is sufficient to take these terms into account only for $i = 0, \ldots, \ell + M + 2$. As g and h are polynomial time computable and monotone functions on $[0, 2^{-(m+3)}]^2$ and γ and β are left NP real numbers in $[0, 2^{-(m+3)})$, by Corollary 16 we conclude that $g(\gamma, \beta)$ and $h(\gamma, \beta)$ are left NP real numbers. To be precise, as in Corollary 16 we considered only monotone and polynomial time computable functions defined on a full cube $[0, 1]^k$, here we should apply Corollary 16 to the functions $\widetilde{g}, \widetilde{h} \colon [0, 1]^2 \to \mathbb{R}$ defined by $\widetilde{g}(x, y) := g(x \cdot 2^{-(m+3)}, y \cdot 2^{-(m+3)})$, similarly for $\widetilde{h}$, and to the numbers $\gamma \cdot 2^{m+3}$ and $\gamma \cdot 2^{m+3}$, which are left NP real numbers by Lemma 4. Clearly, these functions $\widetilde{g}$ and $\widetilde{h}$ are monotone and polynomial time computable as well. We conclude that $g(\gamma, \beta) = \widetilde{g}(\gamma \cdot 2^{m+3}, \beta \cdot 2^{m+3})$ and $h(\gamma, \beta) = \widetilde{h}(\gamma \cdot$

$2^{m+3}, \beta \cdot 2^{m+3})$ are left NP real numbers. Finally, we observe that

$$f(y) = f(\gamma - \beta + \nu_{\mathbb{D}}(e)) = \sum_{i=0}^{\infty} \frac{f^{(i)}(\nu_{\mathbb{D}}(e))}{i!} \cdot (\gamma - \beta)^i = g(\gamma, \beta) - h(\gamma, \beta).$$

This shows that $f(y)$ is a dNP number. □

Corollary 20. *1. The set $\mathbb{R}_{\text{dNP}}$ of dNP numbers is a subfield of $\mathbb{R}$.*
2. It is the smallest subfield of $\mathbb{R}$ that contains all left NP real numbers.

Proof. It is clear that 0 and 1 are elements of the set $\mathbb{R}_{\text{dNP}}$. It is clear as well that this set is closed under the additive inverse operation $x \mapsto -x$. Let x_1 and x_2 be dNP numbers, and let $\alpha_1, \beta_1, \alpha_2, \beta_2$ be left NP real numbers with $x_1 = \alpha_1 - \beta_1$ and $x_2 = \alpha_2 - \beta_2$. We can assume without loss of generality that the numbers $\alpha_1, \beta_1, \alpha_2, \beta_2$ are nonnegative because by Lemma 4 we may add a sufficiently large integer to all of them. Clearly this does not change the differences $\alpha_1 - \beta_1$ and $\alpha_2 - \beta_2$. By Corollary 17 the numbers $x_1 + x_2$ and $x_1 \cdot x_2$ are dNP numbers as well:

$$x_1 + x_2 = \alpha_1 - \beta_1 + \alpha_2 - \beta_2 = (\alpha_1 + \alpha_2) - (\beta_1 + \beta_2),$$
$$x_1 \cdot x_2 = (\alpha_1 - \beta_1) \cdot (\alpha_2 - \beta_2) = (\alpha_1 \cdot \alpha_2 + \beta_1 \cdot \beta_2) - (\alpha_1 \cdot \beta_2 + \alpha_2 \cdot \beta_1).$$

Thus, the set $\mathbb{R}_{\text{dNP}}$ is closed under addition and multiplication as well. Finally, the function $x \mapsto 1/x$, defined on $\mathbb{R} \setminus \{0\}$, is real analytic, and for every x in its domain there exists an open neighbourhood U of x such that this function is computable in polynomial time on U. Thus, by Theorem 19 the set $\mathbb{R}_{\text{dNP}}$ is closed under the multiplicative inverse function $x \mapsto 1/x$ as well. We have proved the first assertion. Then the second assertion is clear. □

References

1. Ambos-Spies, K., Weihrauch, K., Zheng, X.: Weakly computable real numbers. J. Complex. **16**(4), 676–690 (2000). https://doi.org/10.1006/jcom.2000.0561
2. Ko, K.I.: The maximum value problem and NP real numbers. J. Comput. Syst. Sci. **24**(1), 15–35 (1982). https://doi.org/10.1016/0022-0000(82)90053-8
3. Ko, K.I.: On the definitions of some complexity classes of real numbers. Math. Syst. Theory **16**, 95–109 (1983). https://doi.org/10.1007/BF01744572
4. Ko, K.I.: Complexity Theory of Real Functions. Progress in Theoretical Computer Science. Birkhäuser, Boston (1991). https://doi.org/10.1007/978-1-4684-6802-1
5. Ko, K.I., Friedman, H.: Computational complexity of real functions. Theoret. Comput. Sci. **20**, 323–352 (1982). https://doi.org/10.1016/S0304-3975(82)80003-0
6. Krantz, S.G., Parks, H.R.: A Primer of Real Analytic Functions, 2nd edn. Birkhäuser Boston, Inc., Boston (2002). https://doi.org/10.1007/978-0-8176-8134-0
7. Müller, N.T.: Uniform computational complexity of Taylor series. In: Ottmann, T. (ed.) ICALP 1987. LNCS, vol. 267, pp. 435–444. Springer, Heidelberg (1987). https://doi.org/10.1007/3-540-18088-5_37
8. Selman, A.L.: Some observations on NP real numbers and P-selective sets. J. Comput. Syst. Sci. **23**(3), 326–332 (1981). https://doi.org/10.1016/0022-0000(81)90068-4

Counting Random Oracles for the Polynomial-Time Hierarchy and Quantum Complexity Classes

John M. Hitchcock[1](✉), Adewale Sekoni[2], and Hadi Shafei[3]

[1] Department of Electrical Engineering and Computer Science, University of Wyoming, Laramie, USA
jhitchco@uwyo.edu

[2] Department of Mathematics, Computer Science and Physics, Roanoke College, Salem, USA
sekoni@roanoke.edu

[3] Department of Computer Science, University of Huddersfield, Huddersfield, UK
H.Shafei@hud.ac.uk

Abstract. We utilize the framework of counting martingales and resource-bounded counting measures introduced by Hitchcock, Sekoni, and Shafei (2025) to study *counting random oracles*. An oracle A is Δ-random if no Δ-computable martingale succeeds on A. In this paper we consider the "sharp" resource bounds #P and #QP and the "gap" resource bounds GapP, GapQP, and GapE. Here "QP" means quasipolynomial.

We show that PH^A is infinite relative to every #QP-random oracle A. We also prove that a #QP-random oracle A separates the counting classes $\oplus\mathsf{P}^A$, PP^A, and $\mathsf{C_=P}^A$ from each other and from PH^A and PSPACE^A. Furthermore, we show that $\mathsf{IP}^A \neq \mathsf{PSPACE}^A$ for every #QP-random A.

In the quantum domain, we strengthen a result of Bennett, Bernstein, Brassard, and Vazirani (1997) by proving that $\mathsf{NTIME}^B(n)$ is not contained in $\mathsf{BQTIME}^B(o(2^{n/2}))$ relative to GapP, GapQP, or GapE random oracles depending on how much oracle use is allowed. We also extend results of Aaronson, Ingram, and Kretschmer (2022) to show that $\Sigma_{k+1}^{\mathsf{P},A}$ is not contained in $\mathsf{BQP}^{\Sigma_k^{\mathsf{P},A}}$ for all $k \geq 1$, relative to #QP-random oracles A.

We apply Stockmeyer's approximate counting of #P (1985) to show that separating PH with deterministic QP-random oracles yields $\mathsf{P} \neq \mathsf{NP}$. We also address a question of Aaronson et al. about whether $\mathsf{P} = \mathsf{NP}$ implies that BQP is a "small" complexity class, relating the question to approximate counting of GapP functions.

This research was supported in part by NSF grant 2431657.

V. Brattka et al. (Eds.): CiE 2026, LNCS 16674, pp. 335–349, 2027.
https://doi.org/10.1007/978-3-032-31348-5_22

1 Introduction

1.1 Overview

Bennett and Gill [7] introduced measure theory into computational complexity when they proved that $\mathsf{P}^A \neq \mathsf{NP}^A$ relative to a random oracle A with probability 1. Formally, they proved the class of oracles A that separate P^A and NP^A has Lebesgue measure 1. Equivalently, the class of oracles A where the collapse $\mathsf{P}^A = \mathsf{NP}^A$ holds has Lebesgue measure 0. Bennett and Gill proved a number of other random oracle results including $\mathsf{P}^A = \mathsf{BPP}^A$ and $\mathsf{NP}^A \neq \mathsf{coNP}^A$ relative to a random oracle A with probability 1.

Subsequent work has focused on individual random oracles rather than almost all oracles. Book, Lutz, Wagner [8] proved that $\mathsf{P}^A \neq \mathsf{NP}^A \neq \mathsf{coNP}^A$ for every Martin-Löf random oracle A [32]. Here an oracle is Martin-Löf random if it avoids all effectively measure-zero subsets of $\{0,1\}^\infty$. In the resource-bounded setting, Allender and Strauss [2] proved that $\mathsf{P}^A = \mathsf{BPP}^A$ relative to every P-random oracle. Informally, an oracle is P-random if it avoids all polynomial-time effective measure-zero sets, or equivalently, if no polynomial-time martingale succeeds on it. There are a number of other random-oracle results that have not been given the individual random oracle treatment, particularly related to counting complexity, the polynomial-time hierarchy, and quantum computation. This paper investigates the interplay between counting complexity classes and resource-bounded random oracles, initiating the study of *counting random oracles*, with applications to the polynomial-time hierarchy and quantum complexity classes.

1.2 Counting Random Oracles

Resource-bounded measure, introduced by Lutz [26–28], provides a way to quantify the relative "size" of complexity classes within classes like EXP and EXPSPACE [27,33]. Hitchcock, Sekoni, and Shafei [19] recently extended resource-bounded measure to counting complexity resource bounds through *counting martingales* defined by #P, SpanP, and GapP counting functions. We extend their framework to include #QP, GapQP, and GapE resource bounds. Here #QP and GapQP are the quasipolynomial-time versions of #P and GapP, respectively, and GapE is the exponential-time version of GapP. An oracle A is said to be Δ-random if no Δ-computable martingale succeeds on it. Formal definitions are given in Sect. 2. The relationships among the randomness notions used in this paper are depicted in Fig. 1.

We make several contributions to the understanding of random oracles in complexity theory. We show that many random oracle separations hold relative to #QP-random oracles.

1. Key counting classes ($\oplus\mathsf{P}$, PP, $\mathsf{C_=P}$) separate from each other and from PH and PSPACE relative to #QP-random oracles, extending results of Cai [9], Babai [5], and Aspnes, Beigel, Furst, and Rudich [4].

2. The polynomial-time hierarchy is infinite, extending the almost-everywhere result of Rossman, Servedio, and Tan [16,37,38] and the QPSPACE-random oracle result of Hitchcock, Sekoni, Shafei [18].
3. Bennett and Gill's Random Oracle Hypothesis [7] asserts that for any acceptable fully relativized statement, the statement holds in the unrelativized setting if and only if it holds with probability 1 relative to a randomly chosen oracle. This hypothesis was proved false by Kurtz [24] and Chang et al. [10]. We show the random oracle hypothesis fails ($\mathsf{IP}^A \neq \mathsf{PSPACE}^A$ and more specifically, $\mathsf{coNP}^A \not\subseteq \mathsf{IP}$) relative to #QP-random oracles.

$$
\begin{array}{ccccc}
A \text{ is Martin-Löf-random} & \Rightarrow & A \text{ is } \mathsf{QPSPACE}\text{-random} & \Rightarrow & A \text{ is } \mathsf{PSPACE}\text{-random} \\
\Downarrow & & \Downarrow & & \Downarrow \\
A \text{ is } \mathsf{GapE}\text{-random} & \Rightarrow & A \text{ is } \mathsf{GapQP}\text{-random} & \Rightarrow & A \text{ is } \mathsf{GapP}\text{-random} \\
 & & \Downarrow & & \Downarrow \\
 & & A \text{ is } \#\mathsf{QP}\text{-random} & \Rightarrow & A \text{ is } \#\mathsf{P}\text{-random} \\
 & & \Downarrow & & \Downarrow \\
 & & A \text{ is } \mathsf{QP}\text{-random} & \Rightarrow & A \text{ is } \mathsf{P}\text{-random}
\end{array}
$$

Fig. 1. Relationships Among Types of Individual Random Oracles A.

In the quantum setting, we prove strong oracle separation results relative to #QP, GapP, GapQP, and GapE random oracles:

1. We improve a random oracle result of Aaronson, Ingram, and Kretschmer [1] to show $\Sigma_{k+1}^{\mathsf{P},A}$ is not contained in $\mathsf{BQP}^{\Sigma_k^{\mathsf{P}},A}$ relative to #QP-random oracles A, extending the separation of PH above.
2. We also strengthen a result of Bennett, Bernstein, Brassard, and Vazirani [6] by proving that $\mathsf{NTIME}^A(n)$ is not contained in $\mathsf{BQTIME}^A(o(2^{n/2}))$ relative to GapE-random oracles A. Analogous results hold relative to GapP and GapQP random oracles when the BQTIME^B machine is limited to linear or polynomial use of the oracle, respectively.

1.3 Our Techniques and Framework

Bennett and Gill's paper [7] established a framework for results on random oracles. This includes a test language L_A and a category of relativized machines, denoted as M_1^A, M_2^A, M_3^A, etc. The oracle A influences both the test language and machine behavior. For relativizable complexity classes $\mathcal{C}$ and $\mathcal{D}$, the test language is designed so L_A belongs to $\mathcal{C}^A$ for any oracle A, while the machines form the set $\mathcal{D}^A = \{L(M_i^A) \mid i \geq 1\}$. The goal is to prove with probability 1 that $\mathcal{C}^A \not\subseteq \mathcal{D}^A$ by showing $L_A \neq L(M_i^A)$ for each $i \geq 1$. Bennett and Gill [7] argue that L_A and machine sequences $M_1^A, M_2^A, M_3^A, \ldots$ meet four key conditions on computability and patchability. Their key lemma states that if there's a positive constant ϵ ensuring each machine's language differs from the test language on a set of oracles with measure at least ϵ, the set of oracles for which L_A is in $\mathcal{D}^A$ has measure zero. For more on random oracles, see [13,15,42,45]. Hitchcock, Sekoni, and Shafei [18] gave a classification theorem for QPSPACE-random oracles. We extend this to counting random oracles in Theorem 3.2.

1.4 Approximate Counting

We apply Stockmeyer's approximate counting of #P [39] to show that separating PH with deterministic random oracles yields $\mathsf{P} \neq \mathsf{NP}$. We also address a question of Aaronson et al. [1] about whether $\mathsf{P} = \mathsf{NP}$ implies that BQP is a "small" complexity class, relating the question to approximate counting of GapP functions.

By investigating the intersection of counting complexity and resource-bounded randomness, this paper advances our understanding of computational power within both classical and quantum frameworks. The findings have implications for fundamental questions such as P vs NP, separations within the polynomial-time hierarchy, and limitations of quantum computation.

1.5 Organization

The remainder of this paper is organized as follows: Sect. 2 reviews preliminaries and introduces the new counting measures. Section 3 presents our framework for counting random oracles. Section 4 explores complexity class consequences for counting random oracles. Section 5 addresses approximate counting. Section 6 concludes with a summary of results and open questions. All of the proofs are given in the appendix.

2 Preliminaries

The set of all finite binary strings is $\{0,1\}^*$. The empty string is denoted by λ. We use the standard enumeration of binary strings $s_0 = \lambda, s_1 = 0, s_2 = 1, s_3 = 00, \ldots$. We use the notation $x \sqsubseteq y$ to say that x is a prefix of y. The length of a string $x \in \{0,1\}^*$ is denoted by $|x|$.

All *languages* (decision problems) in this paper are encoded as subsets of $\{0,1\}^*$. The *Cantor space* of all infinite binary sequences is C. We routinely identify a language $A \subseteq \{0,1\}^*$ with the element of Cantor space that is A's characteristic sequence according to the standard enumeration of binary strings, i.e., the infinite bit sequence whose i^{th} bit indicates whether the ith binary string is in A. In this way, each complexity class is identified with a subset of Cantor space. We write $A \upharpoonright n$ for the n-bit prefix of the characteristic sequence of A, and $A[n]$ for the nth-bit of its characteristic sequence.

Lutz used martingales to define resource-bounded measure [27]. We review the basic definitions. More background is available in the survey papers [3,28,30].

A *martingale* is a function $d : \{0,1\}^* \to [0,\infty)$ such that for all $w \in \{0,1\}^*$,

$$d(w) = \frac{d(w0) + d(w1)}{2}.$$

If we relax the equality to an inequality ($\geq$) in the above equation, we call d a *supermartingale.*

Let d be a supermartingale or martingale. We say d *succeeds on* a sequence $A \in \mathsf{C}$ if $\limsup_{n\to\infty} d(A \upharpoonright n) = \infty$. The *success set* $S^\infty[d]$ is the class of sequences that d succeeds on.

Let Δ be any of the time or space resource bounds including P, QP, PSPACE, $\mathsf{QPSPACE}$ considered by Lutz [27] and their relativizations including $\Delta_2^{\mathsf{P}} = \mathsf{P}^{\mathsf{NP}}$, $\Delta_3^{\mathsf{P}} = \mathsf{P}^{\Sigma_2^{\mathsf{P}}}$ [20,34] and let $X \subseteq \mathsf{C}$.

1. X has *Δ-measure 0*, and we write $\mu_\Delta(X) = 0$, if there is a Δ-computable martingale d with $X \subseteq S^\infty[d]$.
2. X has *Δ-measure 1*, and we write $\mu_\Delta(X) = 1$, if $\mu_\Delta(X^c) = 0$, where X^c is the complement of X within C.

In his seminal paper on counting complexity [41], Valiant introduced $\#\mathsf{P}$ as follows. Let M be a polynomial-time nondeterministic Turing machine that accepts or rejects on each computation path. The $\#\mathsf{P}$ function computed by M is defined as

$$f(x) = \text{number of accepting computation paths of } M \text{ on input } x$$

for all $x \in \{0,1\}^*$.

Theorem 2.1. (Stockmeyer [39]) *Let $f \in \#\mathsf{P}$. Then there is a function $g \in \Delta_3^{\mathsf{P}}$ such that for all n, for all $x \in \{0,1\}^n$, $(1 - 1/n)g(x) \leq f(x) \leq (1 + 1/n)g(x)$.*

The class GapP is defined as the closure of $\#\mathsf{P}$ functions under subtraction [11,25].

Hitchcock, Sekoni, and Shafei [19] defined Counting Martingales as martingales that are the ratio of a counting function from $\#\mathsf{P}$, SpanP [21], or GapP to a polynomial-time function that is a power of 2. These definitions naturally extend to the quasipolynomial ($n^{(\log n)^{O(1)}}$) resource bounds $\#\mathsf{QP}, \mathsf{SpanQP}, \mathsf{GapQP}$ and to exponential ($2^{O(n)}$) resource bounds $\#\mathsf{E}, \mathsf{SpanE}, \mathsf{GapE}$. Let the counting resource bounds be

$$\mathcal{CRB} = \{\#\mathsf{P}, \mathsf{SpanP}, \mathsf{GapP}, \#\mathsf{QP}, \mathsf{SpanQP}, \mathsf{GapQP}, \#\mathsf{E}, \mathsf{SpanE}, \mathsf{GapE}\}.$$

Each counting resource bound has a denominator class associated with it. We define the polynomial denominator classes

$$\mathsf{den}(\#\mathsf{P}) = \mathsf{den}(\mathsf{SpanP}) = \mathsf{den}(\mathsf{GapP}) = \mathsf{FP},$$

quasipolynomial denominator classes

$$\mathsf{den}(\#\mathsf{QP}) = \mathsf{den}(\mathsf{SpanQP}) = \mathsf{den}(\mathsf{GapQP}) = \mathsf{FQP},$$

and exponential denominator classes

$$\mathsf{den}(\#\mathsf{E}) = \mathsf{den}(\mathsf{SpanE}) = \mathsf{den}(\mathsf{GapE}) = \mathsf{FE}.$$

In this paper we will focus on the #QP, GapP, GapQP, and GapE counting resource bounds.

We now define resource-bounded counting supermartingales as follows. Let $\Delta \in \mathcal{CRB}$ be a counting resource bound. An *exact Δ-supermartingale* is a supermartingale $d(w) = \frac{f(w)}{g(w)}$, where $f \in \Delta$, $g \in \mathsf{den}(\Delta)$, and $g(w)$ is a power of 2 for all $w \in \{0,1\}^*$. It is also possible to define approximately computable Δ-supermartingales [19], but all the counting supermartingales we construct in this paper will be exact.

Resource-bounded measures are defined using martingales [27]. We adopt this approach to define counting measures using counting martingales. Let $\Delta \in \mathcal{CRB}$ be a counting resource bound. A class $X \subseteq \mathsf{C}$ has *Δ-measure 0*, written $\mu_\Delta(X) = 0$, if there is a Δ-supermartingale d with $X \subseteq S^\infty[d]$. A sequence $S \in \mathsf{C}$ is *Δ-random* if $\{S\}$ does not have Δ-measure 0. Note that a sequence S is Δ-random if no Δ-supermartingale succeeds on it.

3 Counting Random Oracle Framework

In this section, we extend the general random oracle frameworks of Bennett and Gill [7] and Hitchcock, Sekoni, and Shafei [18].

Construction 3.1. (Zero-One Law Martingale Construction) *Let $L_A \subseteq 0^*$ be a test language depending on oracle A such that for some function $q(n)$, the membership of $0^n \in L_A$ depends only on strings of length at most $q(n)$ and can be decided in $2^{q(n)}$ time. Let M be an oracle* TM, PTM, *or* QTM *that runs in time $t(n)$ and makes queries of length at most $l(n)$. Let $r(n)$ be a function upper bounding the maximum of $q(n)$, $t(n)$, and $l(n)$. Note that we do not require $t(n)$, $l(n)$, or $r(n)$ to be polynomials.*

Let $n_0 < n_1 < n_2 < \cdots$ be a sequence of numbers such that $r(n_i) < n_{i+1}$ for all $i \geq 0$. For example, let $n_0 = 1$ and let $n_{i+1} = 2^{2^{n_i}}$ for all $i \geq 0$.

In stage i, let w have length $|w| = 2^{n_i} - 1$. Consider choosing an extension $B \subseteq \{0,1\}^{\leq r(n)}$ with $w \sqsubseteq B$ uniformly at random and consider the random variables:

$$Acc_B(0^n) = \begin{cases} 1 & \text{if } M^B \text{ accepts } 0^n \\ 0 & \text{if } M^B \text{ rejects } 0^n \end{cases}$$

$$Rej_B(0^n) = \begin{cases} 1 & \text{if } M^B \text{ rejects } 0^n \\ 0 & \text{if } M^B \text{ accepts } 0^n \end{cases}$$

$$X_B(0^n) = \begin{cases} Acc_B(0^n) & \text{if } 0^n \in L_B, \\ Rej_B(0^n) & \text{if } 0^n \notin L_B \end{cases}.$$

Suppose that for all i,

$$E[X_B(0^{n_i}) = 1 \mid w \sqsubseteq B] = \Pr[X_B(0^{n_i}) = 1 \mid w \sqsubseteq B] < 1 - \epsilon$$

for all $w \in \{0,1\}^{2^{n_i}-1}$*, where* $\epsilon = 2^{-k}$ *for some* $k \geq 1$*. Let*

$$\gamma = 1 + 2^{-k} = \frac{2^k + 1}{2^k}.$$

Define

$$d(w) = d(w \restriction 2^{n_i} - 1) \cdot E[X_B(0^{n_i}) \mid w \sqsubseteq B] \cdot \gamma$$

for all w *with* $2^{n_i} \leq |w| < 2^{n_{i+1}}$*. Then for* w *of length* 2^{n_i}*,*

$$E[X_B(0^n) \mid w \sqsubseteq B] \cdot \gamma \leq (1-\epsilon)\gamma = (1-2^{-k})(1+2^{-k}) = 1 - 2^{-2k} < 1,$$

so d *is a supermartingale, its value may slightly decrease when we start betting at a new length.*

Suppose that for all n*,* $M^A(0^n) = L_A(0^n)$*, which means* $X_A(0^n) = 1$*. In particular,*

$$E[X_B(0^{n_i}) \mid A \restriction 2^{n_{i+1}} - 1] = 1$$

for all $i \geq 0$*. Therefore*

$$d(A \restriction 2^{n_{i+1}} - 1) \geq (\gamma)^i$$

for all $i \geq 0$*. Therefore* d *succeeds on* $\{A \mid L(M^A) = L_A\}$*.*

We now present our main tool for classifying random oracles, which follows from Construction 3.1.

Theorem 3.2. (Counting Random Oracle Classification) *In Construction 3.1, let*

$$Z = \{R \mid L_R = L(M^R)\}$$

and suppose there exists $\epsilon \in (0,1)$ *such that, for all* $w \in \{0,1\}^{2^n-1}$*,*

$$\Pr_R[M^R(0^n) - L_R(0^n) \mid w \sqsubseteq R] < 1 - \epsilon$$

for sufficiently large n*.*

1. *If* $E[X_B \mid w \sqsubseteq B]$ *is computable as the ratio of a* #P *function and a polynomial-time function, then* Z *has* #P*-measure 0. In particular, every* #P*-random oracle is not in* Z*.*
2. *If* $E[X_B \mid w \sqsubseteq B]$ *is computable as the ratio of a* #QP *function and a quasipolynomial-time function, then* Z *has* #QP*-measure 0. In particular, every* #QP*-random oracle is not in* Z*.*
3. *If* $E[X_B \mid w \sqsubseteq B]$ *is computable as the ratio of a* GapP *function and a polynomial-time function, then* Z *has* GapP*-measure 0. In particular, every* GapP*-random oracle is not in* Z*.*
4. *If* $E[X_B \mid w \sqsubseteq B]$ *is computable as the ratio of a* GapQP *function and a quasipolynomial-time function, then* Z *has* GapQP*-measure 0. In particular, every* GapQP*-random oracle is not in* Z*.*
5. *If* $E[X_B \mid w \sqsubseteq B]$ *is computable as the ratio of a* GapE *function and a* FE *function, then* Z *has* GapE*-measure 0. In particular, every* GapE*-random oracle is not in* Z*.*

6. *Suppose that $M^R(0^n)$ makes queries of length at most $O(n)$, $L_R(0^n)$ depends on the oracle to length $O(n)$, and $M^R(0^n)$ and $L_R(0^n)$ can both be evaluted in time $2^{O(n)}$. Then Z has #P-measure 0. In particular, every #P-random oracle is not in Z.*
7. *Suppose that $M^R(0^n)$ makes queries of length at most $n^{O(1)}$, $L_R(0^n)$ depends on the oracle to length $n^{O(1)}$, and $M^R(0^n)$ and $L_R(0^n)$ can both be evaluted in time $2^{n^{O(1)}}$. Then Z has #QP-measure 0. In particular, every #QP-random oracle is not in Z.*

Hitchcock [17] showed that most random oracle classes of interest in computational complexity have Hausdorff dimension 1, so Theorem 3.2 does not extend to the resource-bounded dimension [29] setting.

4 Relativization of Counting Random Oracles

In this section, we apply the framework of Sect. 3 to concrete complexity classes, strengthening prior results by establishing them under a stricter notion of randomness. We begin with PP and ⊕P as a basic example, then turn to the polynomial hierarchy, interactive proofs, and quantum complexity classes.

4.1 Random Oracles for PP and ⊕P

Aspnes, Beigel, Furst, and Rudich [4] showed that PP and PSPACE are different relative to a random oracle. This proof easily adapts to the setting of Construction 3.1 and Theorem 3.2. Their proof actually separated the ⊕P complexity class [36] from PP relative to a random oracle. We now extend this to #QP-random oracles.

Theorem 4.1. $\oplus\mathsf{P}^A \not\subseteq \mathsf{PP}^A$ *for every #QP-random oracle A.*

Corollary 4.2. $\mathsf{PP}^A \neq \mathsf{PSPACE}^A$ *for every #QP-random oracle A.*

4.2 Random Oracles for the Polynomial-Time Hierarchy

Babai [5] showed that PH does not equal PSPACE relative to a random oracle. His proof shows that PH is not contained in ⊕P. Rossman et al. [37] showed that PH is infinite relative to a random oracle. In this section we show these separations hold relative to a #QP-random oracle.

Corollary 4.3. $\mathsf{PH}^A \neq \mathsf{PSPACE}^A$ *for every #QP-random oracle A.*

Rossman, Servedio, and Tan [37] defined Sipser^n_d, a family of n-variable read-once monotone depth-d circuits for which the following theorem holds.

Theorem 4.4 (Rossman, Servedio, and Tan [37]). *Let $2 \le d \le \frac{c\sqrt{\log n}}{\log\log n}$, where $c > 0$ is an absolute constant. Then any circuit C of depth at most $d-1$ and size at most $2^{n^{\frac{1}{6(d-1)}}}$ over $\{0,1\}^n$ agrees with Sipser^n_d on at most $2^n(\frac{1}{2}+n^{-\omega(1/d)})$ inputs.*

This theorem implies that the polynomial hierarchy is infinite relative to a random oracle [15,37]. Hitchcock et al. [18] showed that PH^R is infinite relative to every QPSPACE-random oracle R. We now strengthen this result to show that the polynomial hierarchy is infinite relative to any #QP random oracle.

The following lemma is a combination of two lemmas from [15].

Lemma 4.5 (Håstad [15]). *Let M^A be a $\sum_k^{\mathsf{P},A}$ oracle TM which runs in time n^c on input x of length n, and $\mathcal{Q}_n(M)$ be the set of boolean variables that represent answers to queries that can be made by M on input 0^n. Then there is a depth $k+2$ circuit C of size 2^{n^c} with input set $\mathcal{Q}_n(M)$ that computes $M^A(0^n)$ for any oracle A.*

Theorem 4.6. *For any $k \in \mathbb{N}$, $\sum_k^{\mathsf{P},A} \neq \sum_{k+1}^{\mathsf{P},A}$ for every #QP-random oracle A.*

Corollary 4.7. $\oplus\mathsf{P}^A \not\subseteq \mathsf{PH}^A$ *for every #QP-random oracle A.*

Wagner [44] introduced the exact counting class $\mathsf{C_=P}$. Toda and Ogiwara [40] proved that $\mathsf{PH} \subseteq \mathsf{BP}\cdot\mathsf{C_=P}$. Fenner, Green, Homer, and Pruim [12] showed that $\mathsf{NQP} = \mathsf{coC_=P}$ and $\mathsf{coNQP} = \mathsf{C_=P}$. (Here NQP means nondeterministic quantum polynomial time.) We now show these classes are separated by counting random oracles.

Theorem 4.8. $\mathsf{PH}^A \subsetneq \mathsf{C_=P}^A \cap \mathsf{coC_=P}^A \subsetneq \mathsf{C_=P}^A \subsetneq \mathsf{PP}^A$ *for every #QP-random oracle A.*

4.3 Random Oracles for Interactive Proofs

Chang et al. [10] proved that the Random Oracle Hypothesis of Bennett and Gill [7] is false by showing that for almost all oracles A, $\mathsf{IP}^A \neq \mathsf{PSPACE}^A$. We strengthen this result and show that the same separations hold for #QP-random oracles.

Theorem 4.9. *Relative to a #QP-random oracle R, $\oplus\mathsf{P}^R \not\subseteq \mathsf{IP}^R$. In particular, $\mathsf{IP}^R \neq \mathsf{PSPACE}^R$.*

We now strengthen the second conclusion of Theorem 4.9.

Theorem 4.10. *Relative to a #QP-random oracle R, $\mathsf{coNP}^R \not\subseteq \mathsf{IP}^R$.*

4.4 Random Oracles for BQP and the QMA Hierarchy

Bennett et al. [6] showed that for any $T(n) = o(2^{n/2})$, $\mathsf{BQTIME}^R(T(n))$ does not contain NP^R (actually $\mathsf{NTIME}^R(n)$) relative to a random oracle R with probability 1. This is tight because Grover's algorithm [14] gives a square root speedup in solving NP problems, i.e. $\mathsf{NTIME}(n) \subseteq \mathsf{BQTIME}(2^{n/2})$.

We first show that NP separates from BQP relative to #QP-random oracles.

Theorem 4.11. *1. For every #P-random oracle R, $\mathsf{NTIME}^R(n)$ is not contained in $\mathsf{BQTIME}^R(n)$.*

2. *For every* #QP*-random oracle* R, $\mathsf{NTIME}^R(n)$ *is not contained in* BQP^R.

In the following theorem, we present three extensions of this result. Here $\mathsf{BQTIME}^{R\restriction f(n)}(T(n))$ means the QTM runs in time $T(n)$ and makes queries of length at most $f(n)$. For GapP-random oracles, we can handle linear use of the oracle. For GapQP-random oracles, we can handle polynomial use of the oracle. For GapE-random oracles, the use of the oracle is unrestricted.

Theorem 4.12. *1. For any* $c > 0$, $T(n) = o(2^{n/2})$, *and every* GapP*-random oracle* A, $\mathsf{NTIME}^A(n)$ *is not contained in* $\mathsf{BQTIME}^{A\restriction cn}(T(n))$.
2. *For any* $c > 0$, $T(n) = o(2^{n/2})$, *and every* GapQP*-random oracle* A, $\mathsf{NTIME}^A(n)$ *is not contained in* $\mathsf{BQTIME}^{A\restriction n^c}(T(n))$.
3. *For any* $c > 0$, $T(n) = o(2^{n/2})$, *and every* GapE*-random oracle* A, $\mathsf{NTIME}^A(n)$ *is not contained in* $\mathsf{BQTIME}^A(T(n)) = \mathsf{BQTIME}^{A\restriction T(n)}(T(n))$.

Aaronson et al. [1] proved that $\Sigma_{k+1}^{\mathsf{P}} \not\subseteq \mathsf{BQP}^{\Sigma_k^{\mathsf{P}}}$ relative to a random oracle. We now prove a resource-bounded version of this random oracle separation.

Theorem 4.13. *For every* #QP*-random oracle* R *and* $k \geq 1$, $\Sigma_{k+1}^{\mathsf{P},R} \not\subseteq \mathsf{BQP}^{\Sigma_k^{\mathsf{P},R}}$

Aaronson, Ingram, and Kretschmer [1] proved that PP^A is not contained in the "QMAH^A hierarchy," consisting of constant-depth towers of the form QMA^A, $\mathsf{QMA}^{\mathsf{QMA}^A}$, $\mathsf{QMA}^{\mathsf{QMA}^{\mathsf{QMA}^A}}$, ..., with probability 1 relative to a random oracle A. We refer to [1] for the precise definitions of QMA and QMAH. We only need that $\mathsf{QMAH}^A \subseteq \mathsf{EXP}^A$ to extend this to #QP-random oracles.

Theorem 4.14. *Relative to a* #QP*-random oracle* A, $\oplus\mathsf{P}^A \not\subseteq \mathsf{QMAH}^A$.

Corollary 4.15. *Relative to every* #QP*-random oracle* A, $\mathsf{PP}^A \not\subseteq \mathsf{QMAH}^A$.

5 Approximate Counting

5.1 Counting Random Oracles Versus Deterministic Random Oracles

In the previous section, we established several separation results relative to counting random oracles. Using Stockmeyer's approximate counting [39], we can extend some of our previous results to deterministic random oracles. Using approximate counting, #P can be efficiently approximated by a polynomial-time machine with access to a $\sum_2^{\mathsf{P}}$ oracle. This allows us to convert the counting martingales used in our proofs into martingales computable within the polynomial-time hierarchy.

The following theorem extends Theorem 4.6 to the quasipolynomial-hierarchy by using approximate counting.

Theorem 5.1. *For any* $k \in \mathbb{N}$, $\sum_k^{\mathsf{P},A} \neq \sum_{k+1}^{\mathsf{P},A}$ *for every* $\mathsf{QP}^{\Sigma_2^{\mathsf{QP}}}$*-random oracle* A.

Hitchcock, Sekoni, and Shafei [18] showed that determining whether P equals NP relative to a P-random oracle would separate complexity classes:

Theorem 5.2. (Hitchcock, Sekoni, and Shafei [18])

1. *If* $\mathsf{P}^A \neq \mathsf{NP}^A$ *relative to every* P*-random oracle* A*, then* $\mathsf{BPP} \neq \mathsf{EXP}$.
2. *If* $\mathsf{P}^A = \mathsf{NP}^A$ *relative to some* P*-random oracle* A*, then* $\mathsf{P} \neq \mathsf{PSPACE}$.

We can now strengthen the second separation to $\mathsf{P} \neq \mathsf{NP}$ if we consider QP-random oracles.

Corollary 5.3. *If* $\mathsf{P}^A = \mathsf{NP}^A$ *relative to some* QP*-random oracle* A*, then* $\mathsf{P} \neq \mathsf{NP}$.

We note that Corollary 5.3 also holds with the stronger conclusion $\mathsf{QP} \neq \mathsf{NQP}$ (here NQP is nondeterministic quasipolynomial time). It is interesting that Corollary 5.3 is proved using counting measures, yet counting complexity does not appear at all in the statement.

5.2 Quantum Approximate Counting and the Size of BQP

Aaronson, Ingram, Kretschmer [1] studied the limitations of approximate counting BQP and asked the following question, among other questions.

Question 5.4. (Aaronson, Ingram, and Kretschmer [1]) Suppose $\mathsf{P} = \mathsf{NP}$. Does it follow that BQP is "small" (say, not equal to EXP)?

A class $\mathcal{C}$ has measure 0 in EXP, written $\mu(\mathcal{C} \mid \mathsf{EXP}) = 0$, if $\mu_{\mathsf{QP}}(\mathcal{C} \cap \mathsf{EXP}) = 0$. A natural answer to this question would be $\mu(\mathsf{BQP} \mid \mathsf{EXP}) = 0$, which is equivalent to $\mu_{\mathsf{QP}}(\mathsf{BQP}) = 0$ since $\mathsf{BQP} \subseteq \mathsf{EXP}$.

Stockmeyer's approximate counting of $\#\mathsf{P}$ functions [39] allows converting a $\#\mathsf{QP}$-martingale into a $\mathsf{QP}^{\Sigma_2^{\mathsf{QP}}}$-martingale as we saw in Theorem 5.1. It is open whether this can be done for GapP martingales.

The complexity class SBP is the class of languages that can be decided by a bounded-error classical machine with a small gap between the acceptance and rejection probabilities [1]. The complexity class SBQP is the class of languages that can be decided by a bounded-error quantum machine with a small gap between the acceptance and rejection probabilities [23,43]. It is the class that corresponds to quantum approximate counting [1]. Kretschmer [22] constructed an oracle where $\mathsf{SBQP}^A \not\subseteq \mathsf{QMA}^A$. He discussed SBP and SBQP in detail.

Proposition 5.5. *If approximate counting for* GapP *holds with an* A*-oracle, then* BQP *has* P^A*-measure 0 and* $\mu(\mathsf{BQP} \mid \mathsf{EXP}^A) = 0$.

If the hypothesis of Proposition 5.5 holds with a PH oracle, then $\mathsf{P} = \mathsf{NP}$ would imply BQP has P-measure 0 in EXP.

Theorem 5.6. *Suppose approximate counting for* GapP *holds with an oracle* $A \in \mathsf{PH}$*. Then* $\mathsf{P} = \mathsf{NP}$ *implies* $\mu_{\mathsf{P}}(\mathsf{BQP}) = \mu(\mathsf{BQP} \mid \mathsf{EXP}) = 0$ *and* $\mathsf{BQP} \neq \mathsf{EXP}$.

Stockmeyer's result that #P may be approximately counted in PH is equivalent to SBP $\subseteq$ PH. Whether the same holds for SBQP is an open question.

Corollary 5.7. *If* SBQP $\subseteq$ PH, *then* P $=$ NP *implies* $\mu_{\mathsf{P}}(\mathsf{BQP}) = \mu(\mathsf{BQP} \mid \mathsf{EXP}) = 0$ *and* BQP $\neq$ EXP.

Van Melkebeek [35] showed that $\mu_{\mathsf{P}}(\mathsf{BPP}) \neq 0$ if and only if the collapse BPP $=$ EXP occurs. In contrast, the contrapositive of Corollary 5.7 says that $\mu_{\mathsf{P}}(\mathsf{BQP}) \neq 0$ separates complexity classes:

Corollary 5.8. *If* $\mu_{\mathsf{P}}(\mathsf{BQP}) \neq 0$, *then* P $\neq$ NP *or* SBQP $\not\subseteq$ PH.

It also follows from our results that counting random oracles separate SBP and SBQP.

Corollary 5.9. *Relative to a #QP-random oracle A,* $\mathsf{SBP}^A \neq \mathsf{SBQP}^A$.

We can extend Corollary 4.15 as follows to show that relative to a #QP-random oracle SBQP is not even contained in the QMA hierarchy. This is an extension of Aaronson et al. [1] as well.

Corollary 5.10. *Relative to every #QP-random oracle A,* $\mathsf{SBQP}^A \not\subseteq \mathsf{QMAH}^A$.

6 Conclusion

Relative to a #QP-random oracle A, we have shown:

1. PH^A is infinite and more generally, $\Sigma_{k+1}^{\mathsf{P},A} \not\subseteq \mathsf{BQP}^{\Sigma_k^{\mathsf{P},A}}$ for every $k \geq 1$.
2. PH^A is a proper subset of $\mathsf{C_=P}^A$.
3. PP^A is a proper subset of PSPACE^A.
4. $\mathsf{coNP}^A \not\subseteq \mathsf{IP}^A$ and $\mathsf{IP}^A \neq \mathsf{PSPACE}^A$.

In particular, we have

$$\mathsf{P}^A \subsetneq \mathsf{NP}^A \subsetneq \Sigma_2^{\mathsf{P},A} \subsetneq \Sigma_3^{\mathsf{P},A} \subsetneq \cdots \subsetneq \mathsf{PH}^A \subsetneq \mathsf{C_=P}^A \subsetneq \mathsf{PP}^A \subsetneq \mathsf{PSPACE}^A$$

relative to every #QP-random oracle A.

We note that all of the above separations also hold relative to QPSPACE-random and Martin-Löf random oracles.

Question 6.1. For any $T(n) = o(2^{n/2})$, does $\mathsf{BPTIME}^R(T(n))$ not contain NP^R relative to a #QP-random or GapQP-random oracle R?

We showed that Question 6.1 has a positive answer relative to GapE-random oracles, where GapE is the $2^{O(n)}$-time version of GapP, and hence also for ESPACE-random oracles.

Question 6.2. What random oracle results hold relative to "span" counting oracles SpanP, SpanQP, and SpanE that do not hold for "sharp" or "gap" random oracles?

References

1. Aaronson, S., Ingram, D., Kretschmer, W.: The acrobatics of BQP. In: 37th Computational Complexity Conference, pp. 20:1–20:17 (2022). https://doi.org/10.4230/LIPICS.CCC.2022.20
2. Allender, E., Strauss, M.: Measure on small complexity classes with applications for BPP. In: Proceedings of the 35th IEEE Symposium on Foundations of Computer Science, pp. 807–818. IEEE Computer Society (1994). https://doi.org/10.1109/SFCS.1994.365713
3. Ambos-Spies, K., Mayordomo, E.: Resource-bounded measure and randomness. In: Sorbi, A. (ed.) Complexity, Logic and Recursion Theory, pp. 1–47. Lecture Notes in Pure and Applied Mathematics, Marcel Dekker, New York, N.Y. (1997). https://doi.org/10.1201/9780429187490-1
4. Aspnes, J., Beigel, R., Furst, M., Rudich, S.: The expressive power of voting polynomials. In: Proceedings of the 23rd Annual ACM Symposium on Theory of Computing, pp. 402–409. ACM Press (1991). https://doi.org/10.1145/103418.103461
5. Babai, L.: Random oracles separate pspace from the polynomial-time hierarchy. Inf. Process. Lett. **26**, 51–53 (1987). https://doi.org/10.1016/0020-0190(87)90036-6
6. Bennett, C.H., Bernstein, E., Brassard, G., Vazirani, U.V.: Strengths and weaknesses of quantum computing. SIAM J. Comput. **26**(5), 1510–1523 (1997). https://doi.org/10.1137/s0097539796300933
7. Bennett, C.H., Gill, J.: Relative to a random oracle A, $\mathrm{P}^A \neq \mathrm{NP}^A \neq \text{co-NP}^A$ with probability 1. SIAM J. Comput. **10**, 96–113 (1981). https://doi.org/10.1137/0210008
8. Book, R.V., Lutz, J.H., Wagner, K.W.: An observation on probability versus randomness with applications to complexity classes. Math. Syst. Theory **27**, 201–209 (1994). https://doi.org/10.1007/bf01578842
9. Cai, J.: With probability one, a random oracle separates PSPACE from the polynomial-time hierarchy. J. Comput. Syst. Sci. **38**, 68–85 (1989). https://doi.org/10.1016/0022-0000(89)90033-0
10. Chang, R., et al.: The random oracle hypothesis is false. J. Comput. Syst. Sci. **49**(1), 24–39 (1994). https://doi.org/10.1016/s0022-0000(05)80084-4
11. Fenner, S.A., Fortnow, L., Kurtz, S.A.: Gap-definable counting classes. J. Comput. Syst. Sci. **48**(1), 116–148 (1994). https://doi.org/10.1016/s0022-0000(05)80024-8
12. Fenner, S., Green, F., Homer, S., Pruim, R.: Determining acceptance possibility for a quantum computation is hard for the polynomial hierarchy. Proc. R. Soc. Lond. Series A: Math. Phys. Eng. Sci. **455**(1991), 3953–3966 (1999). https://doi.org/10.1098/rspa.1999.0485
13. Furst, M., Saxe, J., Sipser, M.: Parity, circuits, and the polynomial time hierarchy. In: Proceedings of the Twenty-second Annual IEEE Symposium on Foundations of Computer Science, pp. 260–270. (1981). https://doi.org/10.1109/sfcs.1981.35
14. Grover, L.K.: Quantum mechanics helps in searching for a needle in a haystack. Phys. Rev. Lett. **79**, 325–328 (1997). https://doi.org/10.1103/physrevlett.79.325
15. Håstad, J.: Computational Limitations for Small-Depth Circuits. The MIT Press, Cambridge (1986)
16. Håstad, J., Rossman, B., Servedio, R.A., Tan, L.: An average-case depth hierarchy theorem for Boolean circuits. J. ACM **64**(5), 35:1–35:27 (2017). https://doi.org/10.1145/3095799

17. Hitchcock, J.M.: Hausdorff dimension and oracle constructions. Theor. Comput. Sci. **355**(3), 382–388 (2006). https://doi.org/10.1016/j.tcs.2006.01.025
18. Hitchcock, J.M., Sekoni, A., Shafei, H.: Polynomial-time random oracles and separating complexity classes. ACM Trans. Comput. Theory **13**(1) (2021). https://doi.org/10.1145/3434389
19. Hitchcock, J.M., Sekoni, A., Shafei, H.: Counting martingales for measure and dimension in complexity classes. In: Proceedings of the 40th Computational Complexity Conference (CCC 2025). Leibniz International Proceedings in Informatics (LIPIcs), vol. 339, pp. 20:1–20:35 (2025). https://doi.org/10.4230/LIPIcs.CCC.2025.20
20. Hitchcock, J.M., Vinodchandran, N.V.: Dimension, entropy rates, and compression. J. Comput. Syst. Sci. **72**(4), 760–782 (2006). https://doi.org/10.1016/j.jcss.2005.10.002
21. Köbler, J., Schöning, U., Toran, J.: On counting and approximation. Acta Informatica **26**(4), 363–379 (1989). https://doi.org/10.1007/BF00276023
22. Kretschmer, W.: QMA lower bounds for approximate counting. Tech. rep., arXiv (2019)
23. Kuperberg, G.: How hard is it to approximate the Jones polynomial? Theory Comput. **11**(6), 183–219 (2015). https://doi.org/10.4086/toc.2015.v011a006
24. Kurtz, S.: On the random oracle hypothesis. Information Control **57**, 40–47 (1983). https://doi.org/10.1016/s0019-9958(83)80023-0
25. Li, L.: On the counting functions. Ph.D. thesis, University of Chicago (1993). https://www.proquest.com/dissertations-theses/on-counting-functions/docview/304080357/se-2
26. Lutz, J.H.: Category and measure in complexity classes. SIAM J. Comput. **19**(6), 1100–1131 (1990). https://doi.org/10.1137/0219076
27. Lutz, J.H.: Almost everywhere high nonuniform complexity. J. Comput. Syst. Sci. **44**(2), 220–258 (1992). https://doi.org/10.1016/0022-0000(92)90020-j
28. Lutz, J.H.: The quantitative structure of exponential time. In: Hemaspaandra, L.A., Selman, A.L. (eds.) Complexity Theory Retrospective II, pp. 225–254. Springer, Cham (1997). https://doi.org/10.1007/978-1-4612-1872-2_10
29. Lutz, J.H.: Dimension in complexity classes. SIAM J. Comput. **32**(5), 1236–1259 (2003). https://doi.org/10.1137/S0097539701417723
30. Lutz, J.H., Mayordomo, E.: Twelve problems in resource-bounded measure. Bull. Eur. Assoc. Theor. Comput. Sci. **68**, 64–80 (1999). also appears as ch22Lutz:TPRBM01
31. Lutz, J.H., Mayordomo, E.: Twelve problems in resource-bounded measure. In: Păun, G., Rozenberg, G., Salomaa, A. (eds.) Current Trends in Theoretical Computer Science: Entering the 21st Century, pp. 83–101. World Scientific Publishing (2001). https://doi.org/10.1142/9789812810403_0001
32. Martin-Löf, P.: The definition of random sequences. Information Control **9**, 602–619 (1966). https://doi.org/10.1016/s0019-9958(66)80018-9
33. Mayordomo, E.: Almost every set in exponential time is P-bi-immune. Theor. Comput. Sci. **136**(2), 487–506 (1994). https://doi.org/10.1016/0304-3975(94)00023-c
34. Mayordomo, E.: Contributions to the study of resource-bounded measure. Ph.D. thesis, Universitat Politècnica de Catalunya (1994). https://eccc.weizmann.ac.il/static/books/Contributions_to_the_Study_of_Resource_Bounded_Measure/
35. van Melkebeek, D.: The zero-one law holds for BPP. Theor. Comput. Sci. **244**(1–2), 283–288 (2000). https://doi.org/10.1016/s0304-3975(00)00191-2

36. Papadimitriou, C.H., Zachos, S.K.: Two remarks on the power of counting. In: Theoretical Computer Science: 6th Gl-Conference Dortmund, January 5–7, 1983, LNCS, pp. 269–275. Springer, Cham (1982). https://doi.org/10.1007/BFB0009651
37. Rossman, B., Servedio, R.A., Tan, L.Y.: An average-case depth hierarchy theorem for Boolean circuits. In: Foundations of Computer Science (FOCS), 2015 IEEE 56th Annual Symposium on, pp. 1030–1048. IEEE (2015). https://doi.org/10.1109/focs.2015.67
38. Rossman, B., Servedio, R.A., Tan, L.: Complexity theory column 89: the polynomial hierarchy, random oracles, and Boolean circuits. ACM SIGACT News **46**(4), 50–68 (2015). https://doi.org/10.1145/2852040.2852052
39. Stockmeyer, L.J.: On approximation algorithms for #P. SIAM J. Comput. **14**, 849–861 (1985). https://doi.org/10.1137/0214060
40. Toda, S., Ogiwara, M.: Counting classes are at least as hard as the polynomial-time hierarchy. SIAM J. Comput. **21**(2), 316–328 (1992). https://doi.org/10.1137/0221023
41. Valiant, L.G.: The complexity of computing the permanent. Theor. Comput. Sci. **8**(2), 189–201 (1979). https://doi.org/10.1016/0304-3975(79)90044-6
42. Vollmer, H., Wagner, K.W.: Measure one results in computational complexity theory. In: Du, D.Z., Ko, K.I. (eds.) Advances in Algorithms, Languages, and Complexity, LNCS, pp. 285–312. Springer, US, Boston, MA (1997). https://doi.org/10.1007/978-1-4613-3394-4_14
43. Vyalyi, M.N.: QMA = PP implies that PP contains PH. Tech. Rep. 03-021, ECCC (2003). https://eccc.weizmann.ac.il/eccc-reports/2003/TR03-021/index.html
44. Wagner, K.W.: The complexity of combinatorial problems with succinct input representation. Acta Informatica **23**(3), 325–356 (1986). https://doi.org/10.1007/BF00289117
45. Yao, A.: Separating the polynomial-time hierarchy by oracles. In: 26th IEEE Symposium on Foundations of Computer Science, pp. 1–10. IEEE Computer Society Press (1985). https://doi.org/10.1109/sfcs.1985.49

Tadaki Numbers and Speedability

Philip Janicki(✉)

Fakultät für Informatik, Institut für Theoretische Informatik, Mathematik und Operations Research, Universität der Bundeswehr München, 85579 Neubiberg, Germany
philip.janicki@unibw.de

Abstract. One of the most important topics in algorithmic information theory is the study of randomness of binary sequences. Perhaps the most important and central randomness notion in this field is Martin-Löf randomness. Fix some optimal prefix-free machine U, and let K be the prefix-free Kolmogorov complexity with respect to U. A binary sequence x is called *Martin-Löf random* if there exists a constant $c \in \mathbb{N}$ with $K(x \upharpoonright n) \geq n - c$ for all $n \in \mathbb{N}$. Since the binary expansion of a real number can be identified as a binary sequence as well, algorithmic randomness can be defined for real numbers analogously.

A real number is called *left-computable* if there exists a computable increasing sequence of rational numbers converging to it. A very iconic example of a left-computable number which is Martin-Löf random is Chaitin's number $\Omega := \sum_{w \in \mathrm{dom}(U)} 2^{-|w|}$. This concept has been generalized by Tadaki. For any computable number $s \in]0,1]$, one can define $\Omega^s := \sum_{w \in \mathrm{dom}(U)} 2^{-\frac{|w|}{s}}$, which is a left-computable number such that there is a constant $c \in \mathbb{N}$ with $K(\Omega^s \upharpoonright n) \geq s \cdot n - c$ for all $n \in \mathbb{N}$.

In this article we will investigate the speedability of these numbers. A left-computable number x is called *speedable* if there exists a computable increasing sequence of rational numbers $(x_n)_n$ converging to x with $\limsup_{n \to \infty} \frac{x_{n+1} - x_n}{x - x_n} > 0$. If we even have $\limsup_{n \to \infty} \frac{x_{n+1} - x_n}{x - x_n} = 1$, this number is called *superspeedable*. Every superspeedable number is speedable, but the converse is not true. While Ω is non-speedable, we show that Ω^s is speedable but not superspeedable for any computable number $s \in]0,1[$.

Keywords: Martin-Löf randomness · partial randomness · universal prefix-free machines · left-computable numbers · speedable numbers · superspeedable numbers

1 Partial Randomness

One of the most important topics in algorithmic information theory is the study of randomness of binary sequences. A very common approach to describe algorithmic randomness is by the prefix-free Kolmogorov complexity of binary strings. Let $\Sigma := \{0,1\}$ be the binary alphabet. We denote the set of all binary

V. Brattka et al. (Eds.): CiE 2026, LNCS 16674, pp. 350–357, 2027.
https://doi.org/10.1007/978-3-032-31348-5_23

strings by Σ^*, the set of all non-empty binary strings by Σ^+ and the set of all binary sequences by $\Sigma^{\mathbb{N}}$. A computable function $M\colon\subseteq \Sigma^* \to \Sigma^*$ with prefix-free domain will be called a *prefix-free machine.* If M is a prefix-free machine, then for any $\tau \in \Sigma^*$ we define

$$K_M(\tau) := \begin{cases} \min\{|\sigma| \mid \sigma \in \mathrm{dom}(M) \wedge M(\sigma) = \tau\} & \text{if } \tau \in \mathrm{range}(M) \\ \infty & \text{otherwise} \end{cases}$$

the *Kolmogorov complexity* of τ with respect to M. A prefix-free machine U is called *optimal* if for every prefix-free machine M there exists a constant $c \in \mathbb{N}$ with $K_U(\tau) \leq K_M(\tau) + c$ for all $\tau \in \Sigma^*$. It is well-known that there exists an optimal prefix-free machine. For the remainder of this article we fix some optimal prefix-free machine U and write $K(\tau)$ for $K_U(\tau)$.

Since the fractional part of the binary expansion of some real number can be identified as a binary sequence, one can extend the definition above to real numbers as well. Every binary sequence $x \in \Sigma^{\mathbb{N}}$ is a binary expansion of the real number $\sum_{k=0}^{\infty} x(k) \cdot 2^{-(k+1)}$ in the interval $[0, 1]$. If not mentioned otherwise, we restrict ourselves to the real numbers in the unit interval. Quite often we will silently identify a binary sequence with the real number it represents.

Another concept [10] for describing the algorithmic randomness of a real number is by statistical tests. Recall that a *Martin-Löf test* is a computable sequence $(U_{i,j})_{i,j}$ of intervals with rational endpoints satisfying $\sum_{j=0}^{\infty} \mu(U_{i,j}) \leq 2^{-i}$ for all $i \in \mathbb{N}$. A real number x is *covered* by some Martin-Löf test $(U_{i,j})_{i,j}$ if for every $i \in \mathbb{N}$ there is some $j \in \mathbb{N}$ with $x \in U_{i,j}$.

Solovay [13] introduced a similar test concept. A *Solovay test* is a computable sequence $(I_n)_n$ of intervals with rational endpoints such that the series $\sum_{k=0}^{\infty} \mu(I_k)$ converges. A real number x is *covered* by some Solovay test $(I_n)_n$ if we have $x \in I_n$ for infinitely many $n \in \mathbb{N}$.

Theorem 1. *For any real number x the following statements are equivalent:*

(1) x is not covered by any Martin-Löf test.
(2) There is some constant $c \in \mathbb{N}$ with $K(x \restriction n) \geq n - c$ for all $n \in \mathbb{N}$.
(3) x is not covered by any Solovay test.
(4) $\lim_{n\to\infty} (K(x \restriction n) - n) = \infty$.

A real number with these properties is called *Martin-Löf random.* Schnorr showed (see Chaitin [5]) that Property 1 and 2 are equivalent, Solovay [13] showed that Property 1 and 3 are equivalent, and Chaitin showed [6, Theorem R2] that Property 3 and 4 are equivalent.

This characterization of Martin-Löf randomness points out an interesting connection between algorithmic information theory and computable analysis. There is a specific subset of the real numbers which plays a very important role in both areas. From classical analysis we know that for every real number there exists an increasing sequence of rational numbers converging to it. We will call such a sequence a *left-approximation* of this number. A real number is called *left-computable* if it has a computable left-approximation. Every computable number is left-computable but not vice versa. There are also left-computable numbers which are Martin-Löf random. For any optimal prefix-machine M, the limit of the series $\sum_{w \in \mathrm{dom}(M)} 2^{-|w|}$ is a left-computable in the unit interval which is Martin-Löf random. We will call such a left-computable number a *Chaitin number*. In fact, we have the following equivalence.

Theorem 2. *For any left-computable number $x \in [0,1]$ the following statements are equivalent:*

(1) There exists an optimal prefix-free machine M with $\sum_{w \in dom(M)} 2^{-|w|} = x$.
(2) x is Martin-Löf random.

The implication (1) $\Longrightarrow$ (2) was shown [5, Theorem 4.3] by Chaitin, while the other implication (2) $\Longrightarrow$ (1) follows from a result [9, Theorem 2.1] by Kucera and Slaman combined with a result [4, Theorem 6.5] by Calude, Hertling, Khoussainov and Wang. So every left-computable number in the unit interval that is Martin-Löf random is actually a Chaitin number for some optimal prefix-free machine.

However, the definitions of algorithmic randomness suggest not only a qualitative, but also a quantitative classification of randomness. For example, a real number x with $K(x \upharpoonright n) \geq \frac{n}{2} - c$ for all $n \in \mathbb{N}$ and some $c \in \mathbb{N}$ might be classified as being $\frac{1}{2}$-random. Tadaki [14] introduced the following concepts for partial randomness.

Definition 3. ([14]) Let $s \in]0,1]$.

(1) A *Martin-Löf s-test* is a computable sequence $(U_{i,j})_{i,j}$ of intervals with rational endpoints satisfying $\sum_{j=0}^{\infty} \mu(U_{i,j})^s \leq 2^{-i}$ for all $i \in \mathbb{N}$.
(2) A real number x is *covered* by some Martin-Löf s-test $(U_{i,j})_{i,j}$ if for every $i \in \mathbb{N}$ there is some $j \in \mathbb{N}$ with $x \in U_{i,j}$.
(3) A *Solovay s-test* is a computable sequence $(I_n)_n$ of intervals with rational endpoints such that the series $\sum_{k=0}^{\infty} \mu(I_k)^s$ converges.
(4) A real number x is *covered* by some Solovay s-test $(I_n)_n$ if we have $x \in I_n$ for infinitely many $n \in \mathbb{N}$.

Tadaki showed ([14, Theorem 3.1] and [14, Theorem 3.4]) the following two equivalences.

Theorem 4. ([14]) *For any real number x and any computable number $s \in]0,1]$ the following are equivalent:*

(1) x is not covered by any Martin-Löf s-test.
(2) There exists a constant $c \in \mathbb{N}$ with $K(x \upharpoonright n) \geq s \cdot n - c$ for all $n \in \mathbb{N}$.

A real number satisfying these properties is called *Martin-Löf s-random.* Tadaki showed [14, Theorem 3.2] that for every optimal prefix-free machine M and for every computable number $s \in]0,1]$, the series $\sum_{w \in \mathrm{dom}(U)} 2^{-\frac{|w|}{s}}$ converges, and its limit is left-computable and Martin-Löf s-random. We will call such a left-computable number a *Tadaki number.*

Theorem 5. ([14]) *For any real number x and any computable number $s \in]0,1]$ the following are equivalent:*

(1) x is not covered by any Solovay s-test.
(2) $\lim_{n\to\infty}(K(x \upharpoonright n) - s \cdot n) = \infty$.

A real number satisfying these properties is called *Solovay s-random*. It follows right from the definition that Solovay s-randomness implies Martin-Löf s-randomness, and it is well-known that the converse holds for $s = 1$. Tadaki left it as an open problem [14, Remark 3.3] whether Martin-Löf s-randomness and Solovay s-randomness are equivalent for some computable number $s \in]0,1[$. This problem was solved [12, Theorem 4.2] by Reimann and Stephan. They showed that both notions are not equivalent for any computable number $s \in]0,1[$.

2 Speedable Numbers

In this section we will show that every Tadaki number which is not a Chaitin number is speedable.

Proposition 6. ([8]) *For any left-computable number x the following are equivalent:*

(1) x has a computable left-approximation $(x_n)_n$ with $\liminf_{n\to\infty} \frac{x-x_{n+1}}{x-x_n} < 1$.
(2) x has a computable left-approximation $(x_n)_n$ with $\limsup_{n\to\infty} \frac{x_{n+1}-x_n}{x-x_n} > 0$.
(3) x has a computable left-approximation $(x_n)_n$ with $\limsup_{n\to\infty} \frac{x_{n+1}-x_n}{x-x_{n+1}} > 0$.

A left-computable number satisfying these properties is called *speedable*. The notion of speedability was introduced [11] by Merkle and Titov. This concept is based on the following result [2, Theorem 1.7] by Barmpalias and Lewis-Pye.

Theorem 7. ([2]) *Let α and β be left-computable numbers, and let $(a_n)_n$ and $(b_n)_n$ be computable left-approximations of α and β, respectively. If β is Martin-Löf random, then the limit* $\lim_{n\to\infty} \frac{\alpha-a_n}{\beta-b_n}$ *exists and does not depend on the choice of $(a_n)_n$ and $(b_n)_n$.*

In particular, if x is a left-computable number which is Martin-Löf random and $(x_n)_n$ is a computable left-approximation of x, then $\lim_{n\to\infty} \frac{x-x_{n+1}}{x-x_n} = 1$. However, if x is not Martin-Löf random, things are very different. Surely, the sequence $\left(\frac{x-x_{n+1}}{x-x_n}\right)_n$ is bounded between 0 and 1, but it does not have to converge at all.

Merkle and Titov studied how such sequences behave for left-computable numbers in general. They made the interesting observation [11, Theorem 10] that all speedable numbers are not Martin-Löf random, and they asked the question whether the converse holds. Hölzl and Janicki [7, Corollary 62] showed that this is not the case. Their proof is by an infinite injury priority construction. Since such a construction is considered as difficult, one might ask whether there are easier and more natural examples for left-computable numbers which are neither speedable nor Martin-Löf random. One possible class of candidates for this hypothesis might be the Tadaki numbers. Recall that for every computable number $s \in]0,1[$ the number Ω^s is left-computable and not Martin-Löf random. However, contrary to Ω, the binary expansion of Ω^s does not have arbitrarily long blocks of ones, as shown [3, Corollary 14] by Calude, Hay and Stephan. We will show that this property already implies speedability. For every $x \in \Sigma^{\mathbb{N}}$ and every $\tau \in \Sigma^+$, we say that τ *occurs in* x *as a substring* if there is some $n \in \mathbb{N}$ with $x(n)\dots x(n+|\tau|-1) = \tau$.

Theorem 8. ([3]) *Let $s \in]0,1[$ be a computable number. Then, for any $\tau \in \Sigma^+$ there is a natural number $c \geq 1$ such that τ^c does not occur in Ω^s as a substring.*

Theorem 9. *Let x be a left-computable number. If there exists a natural number $c \geq 1$ such that 1^c does not occur as a substring in x, then x is speedable.*

Proof. Fix some natural number $c \geq 1$ satisfying the assumption. We can assume that x is an irrational number, otherwise x is computable and therefore speedable. Since x is irrational, its binary expansion is unique. This also implies that, given an arbitrary computable left-approximation of x, one can always compute a subsequence which avoids 1^c as a substring step-by-step. So we can fix some computable left-approximation $(x_n)_n$ of x with $x_n(ck)\dots x_n(c(k+1)-1) \neq 1^c$ for every $n \in \mathbb{N}$ and every $k \in \{0,\dots,n\}$.

Let $\theta\colon \mathbb{N} \to \mathbb{N}$ be the function defined by

$$\theta(n) := \min\{t^* \in \mathbb{N} \mid \forall t > t^*\colon x_t \upharpoonright cn = x_{t^*} \upharpoonright cn\}$$

for every $n \in \mathbb{N}$. Since $(x_t)_t$ is increasing and converging, θ is well-defined, non-decreasing and tends to infinity. So there exist infinitely many $n \in \mathbb{N}$ with $\theta(n) < \theta(n+1)$.

Now consider some $n \in \mathbb{N}$ with $\theta(n) < \theta(n+1)$, and let $t := \theta(n+1)-1$. We claim that we have $x_{t+1} - x_t \geq 2^{-(c(n+2)-1)}$ on the one hand and $x - x_{t+1} \leq 2^{-(c(n+1)-1)}$ on the other hand.

First we explain why the first inequality holds. By the definition of t, the binary expansion of this left-approximation changes on the first $c(n+1)$ bits for the last time. The smallest possible amount which is necessary to do this is $2^{-(c(n+2)-1)}$. This scenerio occurs when we have $x_t(c(n+1))\dots x_t(c(n+2)-1) = 1^{c-1}0$ and $x_t(k) = 1$ for all $k \geq c(n+2)$. Note that we assumed that 1^c does not occur in x as a substring. By the definition of θ, the binary expansion of this left-approximation does not change anymore on the first $c(n+1)$ bits after stage $\theta(n+1)$. Therefore, it is not possible that we have $x_t(c(n+1))\dots x_t(c(n+2)-1) = 1^c$, so it is not possible to change the first $c(n+1)$ bits with a smaller quantity.

The second inequality holds, since $x_{t+1} = x_{\theta(n+1)}$ and x mutually agree on the first $c(n+1)$ bits. So we obtain

$$\frac{x_{t+1} - x_t}{x - x_{t+1}} \geq \frac{2^{-(c(n+2)-1)}}{2^{-(c(n+1)-1)}} = 2^{-c}$$

witnessing $\limsup_{t\to\infty} \frac{x_{t+1}-x_t}{x-x_{t+1}} > 0$. Therefore, x is speedable.

Corollary 10. *Ω^s is speedable for every computable number $s \in]0,1[$.*

3 Superspeedable Numbers

In the last section we have shown that every Tadaki number which is not a Chaitin number is speedable. In contrast to our last result, we will show in this section that no Tadaki number is superspeedable.

Proposition 11. ([8]) *For every left-computable number x the following are equivalent:*

(1) x has a computable left-approximation $(x_n)_n$ with $\liminf_{n\to\infty} \frac{x-x_{n+1}}{x-x_n} = 0$.
(2) x has a computable left-approximation $(x_n)_n$ with $\limsup_{n\to\infty} \frac{x_{n+1}-x_n}{x-x_n} = 1$.
(3) x has a computable left-approximation $(x_n)_n$ with $\limsup_{n\to\infty} \frac{x_{n+1}-x_n}{x-x_{n+1}} = \infty$.

A left-computable number satisfying these properties is called *superspeedable*. Recall that a left-computable number x is speedable if it has a computable left-approximation $(x_n)_n$ with $\frac{x-x_{n+1}}{x-x_n} \le \rho$ for some constant $\rho \in]0,1[$ and infinitely many $n \in \mathbb{N}$. Merkle and Titov also showed [11, Theorem 6] that the property of being speedable does not depend on the choice of ρ, that is, if some left-computable number is speedable via some ρ witnessed by some left-approximation, it is also speedable via any other $\rho' \in]0,1[$ witnessed by some other left-approximation. However, their proof is highly nonuniform, which led to Barmpalias asking [1] whether every speedable number also has a *single* left-approximation with $\liminf_{n\to\infty} \frac{x-x_{n+1}}{x-x_n} = 0$. This question was answered by Hölzl, Janicki, Merkle and Stephan. They showed [8, Corollary 9] that there exists a left-computable number which is speedable but not superspeedable.

Recall that left-computable numbers which are Martin-Löf random are not speedable. Therefore, every Chaitin number is not superspeedable in particular. However, we can even show that every Tadaki number is not superspeedable.

Definition 12. ([15]) Let $s \in]0,1]$.

- An increasing sequence of real numbers $(x_n)_n$ is called *s-convergent* if the series $\sum_{k=0}^{\infty} (x_{k+1} - x_k)^s$ converges.
- A left-computable number is called *s-convergent* if it has a computable left-approximation which is s-convergent.

Every increasing and converging sequence of real numbers is 1-convergent, and thus every left-computable number is 1-convergent. Furthermore, Ω^s is s-convergent for every computable number $s \in]0,1]$, as mentioned [15, Theorem 7] by Tadaki. He also showed [15, Theorem 11] that a left-computable number which is s-convergent for some $s \in]0,1[$ is not Martin-Löf t-random for any $t \in]s,1]$.

Lemma 13. *Let x be a left-computable number, and let $(x_n)_n$ be a computable left-approximation of x.*

(1) If x is superspeedable, then there exists a computable increasing function $r\colon \mathbb{N} \to \mathbb{N}$ with $\limsup_{n\to\infty} \frac{x_{r(n+1)}-x_{r(n)}}{x-x_{r(n+1)}} = \infty$.

(2) *If x is s-convergent for some $s \in]0,1]$, then there exists a computable increasing function $t\colon \mathbb{N} \to \mathbb{N}$ such that the sequence $(x_{t(n)})_n$ is s-convergent.*

Proof. (1) Fix some computable left-approximation $(y_n)_n$ of x witnessing the superspeedability, namely $\limsup_{n\to\infty} \frac{y_{n+1}-y_n}{x-y_{n+1}} = \infty$. Since $(x_n)_n$ and $(y_n)_n$ are both computable left-approximations of x, there exist computable increasing functions $s,t\colon \mathbb{N} \to \mathbb{N}$ with

$$x_{s(n)} < y_{t(n)} < x_{s(n+1)} < y_{t(n+1)} < x_{s(n+2)}$$

for every $n \in \mathbb{N}$. It is clear that we have $\limsup_{n\to\infty} \frac{y_{t(n+1)}-y_{t(n)}}{x-y_{t(n+1)}} = \infty$ as well. Define the computable functions $r_1, r_2\colon \mathbb{N} \to \mathbb{N}$ by $r_1(n) := s(2n)$ and $r_2(n) := s(2n+1)$ for every $n \in \mathbb{N}$. Then at least one of the choices $r := r_1$ or $r := r_2$ satisfies $\limsup_{n\to\infty} \frac{x_{r(n+1)}-x_{r(n)}}{x-x_{r(n+1)}} = \infty$.

(2) Fix some computable left-approximation $(y_n)_n$ of x such that $(y_n)_n$ is s-convergent. Let $L := \sum_{k=0}^{\infty} (y_{k+1} - y_k)^s$. Since $(x_n)_n$ and $(y_n)_n$ are both computable left-approximations of x, there exist computable increasing functions $t, u\colon \mathbb{N} \to \mathbb{N}$ with

$$y_{u(n)} < x_{t(n)} < y_{u(n+1)} < x_{t(n+1)} < y_{u(n+2)}$$

for every $n \in \mathbb{N}$. Then we obtain

$$\begin{aligned}
\sum_{k=0}^{\infty} \left(x_{t(k+1)} - x_{t(k)}\right)^s &< \sum_{k=0}^{\infty} \left(y_{u(k+2)} - y_{u(k)}\right)^s \\
&\le \sum_{k=0}^{\infty} \left(y_{u(k+2)} - y_{u(k+1)}\right)^s + \sum_{k=0}^{\infty} \left(y_{u(k+1)} - y_{u(k)}\right)^s \\
&\le \sum_{k=u(1)}^{\infty} \left(y_{k+1} - y_k\right)^s + \sum_{k=u(0)}^{\infty} \left(y_{k+1} - y_k\right)^s \\
&< 2L,
\end{aligned}$$

showing that $(x_{t(n)})_n$ is s-convergent.

Theorem 14. *Let x be a left-computable number which is s-convergent for some computable number $s \in]0,1]$. If x is superspeedable, then x is not Martin-Löf s-random.*

Proof. Let $(x_n)_n$ be a computable left-approximation of x witnessing both superspeedability and s-convergence (see Lemma 13). Furthermore, fix some constant $d \in \mathbb{N}$ satisfying $\sum_{k=0}^{\infty} (x_{k+1} - x_k)^s \le 2^d$. We define the sequence $(U_{i,j})_{i,j}$ as follows: For all $i, j \in \mathbb{N}$, we let $U_{i,j} := \left[x_{j+1}, x_{j+1} + 2^{-\frac{i+d}{s}} \cdot (x_{j+1} - x_j)\right]$. Note that we have

$$\sum_{j=0}^{\infty} \mu^s(U_{i,j}) = \sum_{j=0}^{\infty} \left(2^{-\frac{i+d}{s}} \cdot (x_{j+1} - x_j)\right)^s \le 2^{-(i+d)} \cdot 2^d = 2^{-i}$$

for every $i \in \mathbb{N}$. Since this sequence is also computable, $(U_{i,j})_{i,j}$ is a Martin-Löf s-test. We still have to show that x is covered by this test. Let $i \in \mathbb{N}$. Since we have $\limsup_{n\to\infty} \frac{x_{n+1}-x_n}{x-x_{n+1}} = \infty$, there is some $j \in \mathbb{N}$ satisfying $\frac{x_{j+1}-x_j}{x-x_{j+1}} > 2^{\frac{i+d}{s}}$. Then we obtain

$$\begin{aligned} x_{j+1} + 2^{-\frac{i+d}{s}} \cdot (x_{j+1} - x_j) &> x_{j+1} + 2^{-\frac{i+d}{s}} \cdot 2^{\frac{i+d}{s}} \cdot (x - x_{j+1}) \\ &= x \\ &> x_{j+1}, \end{aligned}$$

hence $x \in U_{i,j}$, showing that x is not Martin-Löf s-random.

Corollary 15. *Ω^s is not superspeedable for any computable number $s \in]0,1]$.*

References

1. Barmpalias, G.: Personal communication (2023)
2. Barmpalias, G., Lewis-Pye, A.: Differences of halting probabilities. J. Comput. Syst. Sci. **89**, 349–360 (2017)
3. Calude, C.S., Hay, N.J., Stephan, F.: Representation of left-computable ε-random reals. J. Comput. Syst. Sci. **77**(4), 812–819 (2011). JCSS IEEE AINA 2009
4. Calude, C.S., Hertling, P.H., Khoussainov, B., Wang, Y.: Recursively enumerable reals and Chaitin Ω numbers. Theor. Comput. Sci. **255**, 125–149 (2001)
5. Chaitin, G.J.: A theory of program size formally identical to information theory. J. ACM **22**(3), 329–340 (1975)
6. Chaitin, G.J.: Incompleteness theorems for random reals. Adv. Appl. Math. **8**(2), 119–146 (1987)
7. Hölzl, R., Janicki, P.: Benign approximations and non-speedability, 2023. Available at https://arxiv.org/abs/2303.11986
8. Hölzl, R., Janicki, P., Merkle, W., Stephan, F.: Randomness versus superspeedability. In: Královič, R., Kučera, A. (eds)., 49th International Symposium on Mathematical Foundations of Computer Science (MFCS 2024), volume 306 of Leibniz International Proceedings in Informatics (LIPIcs), pp. 62:1–62:14, Dagstuhl, Germany, 2024. Schloss Dagstuhl – Leibniz-Zentrum für Informatik
9. Kucera, A., Slaman, T.: Randomness and recursive enumerability. SIAM J. Comput. **31**(1), 199–211 (2001)
10. Martin-Löf, P.: The definition of random sequences. Inf. Control **9**(6), 602–619 (1966)
11. Merkle, W., Titov, I.: Speedable Left-c.e. Numbers. In: Fernau, H. (ed.) CSR 2020. LNCS, vol. 12159, pp. 303–313. Springer, Cham (2020). https://doi.org/10.1007/978-3-030-50026-9_22
12. Reimann, J., Stephan, F.: Hierarchies of randomness tests. In: Mathematical Logic in Asia: Proceedings of the Ninth Asian Logic Conference, Novosibirsk, Russia 16–19 August 2005, pp. 215–232, 2006
13. Solovay, R.M.: Draft of a paper (or series of papers) on Chaitin's work (1975). Unpublished notes,
14. Tadaki, K.: A generalization of Chaitin's halting probability Ω and halting self-similar sets. Hokkaido Math. J. **31**(1), 219–253 (2002)
15. Tadaki, K.: Partial randomness and dimension of recursively enumerable reals. In: Královič, R., Niwiński, D. (eds.) Mathematical Foundations of Computer Science 2009. MFCS 2009. LNCS, vol. 5734, pp. 687–699. Springer, Berlin, Heidelberg (2009). https://doi.org/10.1007/978-3-642-03816-7_58

Adversarial Synchronization and Reset Length

Anton E. Lipin[1,2] and Mikhail V. Volkov[2,3(✉)]

[1] Krasovskii Institute of Mathematics and Mechanics, 620108 Yekaterinburg, Russia
tony.lipin@yandex.ru
[2] Ural Federal University, 620062 Yekaterinburg, Russia
[3] 620075 Yekaterinburg, Russia
m.v.volkov@urfu.ru
https://s.csproject.org/volkov

Abstract. We study a variant of the synchronization game on finite deterministic automata. In this game, Alice chooses one input letter of an automaton $\mathscr{A}$ on each of her moves while Bob may respond with an arbitrary finite word over the input alphabet of $\mathscr{A}$; Alice wins if the word obtained by interleaving her letters with Bob's responses resets $\mathscr{A}$. We prove that if Alice has a winning strategy in this game on $\mathscr{A}$, then $\mathscr{A}$ admits a reset word whose length is strictly smaller than the number of states of $\mathscr{A}$. In contrast, for any $k \geq 1$, we exhibit automata with shortest reset-word length quadratic in the number of states, on which Alice nevertheless wins a version of the game in which Bob's responses are restricted to arbitrary words of length at most k.

Keywords: Synchronizing automata · reset threshold · synchronization game

1 Introduction

1.1 Automata

All automata in this paper are complete, deterministic, and finite. Such an automaton (a *DFA*) is a pair (Q, Σ) of two finite non-empty sets equipped with a map $Q \times \Sigma \to Q$. We call Q the *state set* and Σ the *input alphabet*; elements of Q and Σ are referred to as *states* and *letters*, respectively. The map $Q \times \Sigma \to Q$ is called the *action* of letters on states. For a state $q \in Q$ and a letter $a \in \Sigma$, the result of the action of a on q is denoted by $q{\cdot}a$ by default, although we may also use $\circ$ to denote the action, when working with multiple automata in one argument.

Any DFA $\mathscr{A} = (Q, \Sigma)$ can be represented as a labeled digraph with vertex set Q and edges of the form $q \xrightarrow{a} q{\cdot}a$ for all $q \in Q$ and $a \in \Sigma$. We will take the liberty of applying graph-theoretical notions to automata; that is, we speak, for instance, of paths or strongly connected components of $\mathscr{A}$, meaning those of the digraph of $\mathscr{A}$.

A *word* over Σ is a finite sequence of letters. If $w = a_1 \cdots a_\ell$, where $a_1, \ldots a_\ell$ are letters, the number ℓ is called the *length* of the word w and is denoted by $|w|$. We allow the *empty word*, denoted by ε, which is the empty sequence, and set $|\varepsilon| := 0$. The set of all words over Σ is denoted by Σ^*. For $q \in Q$ and $w \in \Sigma^*$, the action of w on q in (Q, Σ) is

V. Brattka et al. (Eds.): CiE 2026, LNCS 16674, pp. 358–373, 2027.
https://doi.org/10.1007/978-3-032-31348-5_24

defined recursively as $q \cdot w := \begin{cases} q & \text{if } w = \varepsilon, \\ (q \cdot w') \cdot a & \text{if } w = w'a \text{ for some } w' \in \Sigma^* \text{ and } a \in \Sigma, \end{cases}$ and the action of w on a subset $P \subseteq Q$ is defined by $P \cdot w := \{p \cdot w : p \in P\}$.

1.2 Synchronizing Automata

A DFA (Q, Σ) is called *synchronizing* if there exists a word $w \in \Sigma^*$ such that applying w to any state in Q results in the same state:

$$q \cdot w = q' \cdot w \ \text{ for all } \ q, q' \in Q.$$

Any word w satisfying this condition is called a *reset* word for the automaton. The minimum length of reset words for a synchronizing automaton $\mathscr{A}$ is called its *reset threshold* and is denoted by $\mathrm{rt}(\mathscr{A})$.

Synchronizing automata are widely studied as transparent and natural models of error-resistant systems, with numerous applications in different areas of computer science, including coding theory, robotics, and the testing of reactive systems. They also reveal fascinating connections with symbolic dynamics, substitution systems, permutation groups, and other areas of mathematics. Per se, synchronizing automata attract considerable attention because of the Černý conjecture, arguably the most longstanding open problem in the combinatorial theory of finite automata. (The conjecture, formulated in the late 1960s, states that the reset threshold of every synchronizing automaton with n states does not exceed $(n-1)^2$.) We refer the reader to chapter [12] of the 'Handbook of Automata Theory' and to the overview article [15] for an introduction to the field of automata synchronization. In connection with the Černý conjecture, the 'living survey' [16] provides a regularly updated reference source.

1.3 Synchronization Games

Fominykh and the second-named author [10] initiated the study of synchronizing automata from a game-theoretic perspective, motivated, on the one hand, by a game-theoretical approach to software testing proposed in [5] and, on the other hand, by a peg-solitaire-like game used in [3,4] to justify new examples of synchronizing automata whose reset thresholds are close to the Černý conjecture bound. In the game introduced in [10], two players, Alice (Synchronizer) and Bob (Desynchronizer), take turns choosing letters from the input alphabet of a DFA $\mathscr{A}$. Alice who wants to synchronize $\mathscr{A}$ wins when the word formed by the letters chosen by her and Bob, taken in the order in which they were selected, resets $\mathscr{A}$. Bob, on the other hand, strives to prevent synchronization or, if synchronization is unavoidable, to delay it for as long as possible. Assuming both players play optimally, the outcome of the game depends solely on the automaton. This raises the problem of classifying synchronizing automata into those for which Alice has a winning strategy and those for which Bob does. Several results in this direction were obtained in [6–10, 14].

1.4 New Synchronization Games and Our Contribution

It was observed in [8, Section 5.4] that many of the results established in [6–10] remain valid for a modified game in which Bob is allowed to choose an arbitrary word in response to each of Alice's moves. Synchronization games under such modified rules appear to better model reliably controlled systems that must perform tasks in the presence of natural or artificial interference. Indeed, nature or an adversary does not necessarily act in the same rhythm as we do, nor are they obliged to respond to each of our actions with exactly one obstacle or counteraction.

In [13], we investigated several aspects of synchronization games with various 'degrees of freedom' for an adversary. Due to the space constraints, here we restrict ourselves to a single aspect, namely, the speed of synchronization. Automata on which Alice can win against an adversary appear to be more amenable to synchronization. Consequently, it is natural to expect that, in the absence of resistance, they should admit short reset words. We show that the reset threshold of a DFA (Q,Σ) on which Alice can win even if Bob may respond with arbitrary word from Σ^* is always less than $|Q|$. As a corollary, we obtain a new, fully combinatorial proof (and a strengthening) of a result by Almeida and Steinberg [2] on the reset threshold of synchronizing automata subject to certain algebraic restrictions. In contrast, for any k, we exhibit automata with a reset threshold quadratic in the number of states on which Alice wins provided that Bob's responses are restricted to words of length less than k. For $k = 2$, this resolves a question left open in [7].

1.5 Prerequisites

A fair effort has been made to keep the paper reasonably self-contained. We use only a few standard notions from graph theory; no game- or automata-theoretic background is presupposed.

2 Variants of Adversarial Synchronization

2.1 Definition and Graphical Representation

$\mathbb{N}$ is the set of all positive integers, and $\overline{\mathbb{N}}$ denotes $\mathbb{N}$ augmented with the symbol ω. We extend the usual order on $\mathbb{N}$ to $\overline{\mathbb{N}}$ by declaring $k < \omega$ for all $k \in \mathbb{N}$. For $k \in \overline{\mathbb{N}}$ and an alphabet Σ, we write $\Sigma^{<k}$ for the set of all words over Σ of length less than k. In particular, $\Sigma^{<1} = \{\varepsilon\}$ and $\Sigma^{<\omega} = \Sigma^*$.

For $k \in \overline{\mathbb{N}}$ and a DFA $\mathscr{A} = (Q,\Sigma)$, we call the following game the *k-synchronization game* on $\mathscr{A}$. Two players, Alice (Synchronizer) and Bob (Desynchronizer), take turns moving. On her turns, Alice chooses letters from Σ. Bob, on his turns, chooses words from $\Sigma^{<k}$ (thus, when $k = \omega$, Bob may choose arbitrary finite words).

The concatenation of all letters chosen by Alice and all words chosen by Bob up to some turn t, taken in the order in which they were selected, is called the *history* of the game up to turn t. If, at some turn, this history is a reset word for the automaton $\mathscr{A}$, then Alice wins the game. The goal of Bob is to continue the game indefinitely, or if synchronization is unavoidable, to prolong the game for as long as possible.

The reader may have noticed that we have not yet specified who starts the game, Alice or Bob. This does not actually matter. If, given a DFA, Alice wins a k-synchronization game in which she makes the first move, then she can also win the game in which Bob moves first, because a reset word remains a reset word when multiplied on the left by an arbitrary word. If, given a DFA, Bob has a winning strategy in a k-synchronization game which Alice starts, then he can also win the game in which he moves first by passing the first turn to Alice, that is, by choosing the empty word as his initial move.

For brevity, the following two conventions will be used throughout this paper. First, we will tacitly assume that DFAs under consideration have at least three states, since Alice trivially wins on each synchronizing automaton with two or less states. Second, since no games other than synchronization games are considered, we will shorten the expression 'k-synchronization game' to 'k-game'.

Using the standard graphical representation, one can visualize the k-game on a DFA $\mathscr{A}$ as a two-player peg-solitaire-like game with (the digraph of) $\mathscr{A}$ as the game board. At the beginning each state of $\mathscr{A}$ holds one token. Tokens are moved and removed as follows. If Alice chooses a letter a, all tokens still present on the board simultaneously slide along the edges labeled a. Similarly, if Bob chooses a word w, all tokens still present on the board simultaneously slide along the paths formed by consecutive edges whose labels form w. Whenever several tokens arrive at the same state after this, all of them except one are removed, so that after the move is completed, each state holds at most one token.

We refer to the set of states carrying tokens after some turn t as the *position* at turn t. Since token movements correspond to actions of letters and words, the position at turn t in the k-game on a DFA (Q,Σ) equals $Q{\cdot}u$, where u is the history of the game up to turn t.

In the graphical representation, Alice wins the game when all but one token have been removed. Bob wins if he can keep at least two tokens on the board indefinitely.

2.2 Hierarchy of A_k-Automata

Provided that both Alice and Bob play optimally, the outcome of the k-game on a DFA $\mathscr{A}$ depends only on $\mathscr{A}$. If Alice has a winning strategy in the k-game on $\mathscr{A}$, we say that $\mathscr{A}$ is an *A_k-automaton.*

The following observations are immediate:

(O1) a DFA is an A_1-automaton if and only if it is synchronizing;
(O2) for each $k \in \mathbb{N}$, every A_{k+1}-automaton is an A_k-automaton;
(O3) a DFA is an A_ω-automaton if and only if it is an A_k-automaton for each $k \in \mathbb{N}$.

For $k \in \overline{\mathbb{N}}$, denote the class of all A_k-automata by $\mathbf{A}_k$. By (O2) and (O3), the classes $\mathbf{A}_k$ form a decreasing hierarchy:

$$\mathbf{A}_1 \supseteq \mathbf{A}_2 \supseteq \cdots \supseteq \mathbf{A}_k \supseteq \ldots \mathbf{A}_\omega. \tag{1}$$

In fact, all inclusions in (1) are strict. For a series of examples witnessing this fact, see [13, Lemma 2.1].

Two questions about the hierarchy (1) arise naturally. The first is how to efficiently determine the level of (1) which a given synchronizing automaton $\mathscr{A}$ reaches, that is, to compute the largest $k \in \overline{\mathbb{N}}$ such that $\mathscr{A} \in \mathbf{A}_k$. The second is how the reset threshold of $\mathscr{A}$ depends on its level. In this paper, we address the second question; for polynomial algorithms deciding the level of a given DFA, see [13, Sect. 4].

3 Synchronization Speed

3.1 Reset Thresholds of A_ω-Automata

We aim to show that the reset threshold of every A_ω-automaton is less than the number of its states. In a rough approximation, the idea of the proof is that we assign to each position in the ω-game on a DFA (Q,Σ) the value of a certain parameter with the following properties: first, Alice always has a move that decreases the current value (in an appropriate sense), while no move of Bob can increase it; and second, the total number of possible values of this parameter is less than $|Q|$. To embody this idea, we need several concepts and pieces of notation.

For a DFA $\mathscr{A} = (Q,\Sigma)$, we define a sequence $\{E_\ell\}_{\ell=0,1,2,\dots}$ of binary relations on the set Q by induction:

- $E_0 := \{(q,q) : q \in Q\}$;
- $E_{\ell+1}$ is the set of all pairs $(p,q) \in Q \times Q$ such that for every word $w \in \Sigma^*$, there exists a letter $x \in \Sigma$ such that $(p{\cdot}wx,\ q{\cdot}wx) \in E_\ell$.

In terms of the ω-game on $\mathscr{A}$, the definition says that $(p,q) \in E_\ell$ if and only if Alice has a strategy that allows her to move tokens from the states p and q to a common state in at most ℓ moves, provided that Bob moves first.

A straightforward induction on ℓ shows that $(p,q) \in E_\ell$ if and only if $(q,p) \in E_\ell$ for all ℓ; that is, the relations E_ℓ are symmetric. We therefore regard the pairs (Q,E_ℓ) as (undirected) graphs and use the corresponding terminology. In this context, we take the liberty of occasionally referring to pairs $(p,q) \in E_\ell$ as *edges*, although, strictly speaking, the edges of (Q,E_ℓ) are the sets $\{p,q\}$ with $(p,q) \in E_\ell$.

We record a few properties of the relations E_ℓ.

Proposition 1. *Let $\mathscr{A} = (Q,\Sigma)$ be a DFA, ℓ a nonnegative integer.*

(Stability) if $(p,q) \in E_\ell$, then $(p{\cdot}w, q{\cdot}w) \in E_\ell$ for every word $w \in \Sigma^*$;
(Containment) $E_\ell \subseteq E_{\ell+1}$;
(Criterion) $\mathscr{A}$ is an A_ω-automaton if and only if $\bigcup_{\ell\geq 0} E_\ell = Q \times Q$.

Proof. Stability. The claim is immediate for $\ell = 0$. Let $\ell > 0$, and denote $p' := p{\cdot}w$ and $q' := q{\cdot}w$. Take an arbitrary word $w' \in \Sigma^*$. Then $p'{\cdot}w' = p{\cdot}ww'$ and $q'{\cdot}w' = q{\cdot}ww'$. Applying the definition of E_ℓ to the pair (p,q) and the word ww' ensures the existence of a letter $x \in \Sigma$ such that $(p{\cdot}ww'x,\ q{\cdot}ww'x) \in E_{\ell-1}$. Thus, $(p'{\cdot}w'x,\ q'{\cdot}w'x) \in E_{\ell-1}$, and this verifies that $(p',q') \in E_\ell$.

Containment. We induct on ℓ. The base $E_0 \subseteq E_1$ is clear. Let $\ell > 0$. By definition, for each $(p,q) \in E_\ell$ and each $w \in \Sigma^*$, there is a letter $x \in \Sigma$ such that $(p{\cdot}wx,\ q{\cdot}wx) \in E_{\ell-1}$. By the inductive assumption, $E_{\ell-1} \subseteq E_\ell$, whence $(p{\cdot}wx,\ q{\cdot}wx) \in E_\ell$. Thus, $(p,q) \in E_{\ell+1}$.

Criterion. The 'only if' part. Since the set Q is finite and $E_\ell \subseteq E_{\ell+1}$ for all $\ell \geq 0$, there exists an integer L such that $\bigcup_{\ell\geq 0} E_\ell = E_L$. Let $H := (Q \times Q) \setminus E_L$ and suppose, towards a contradiction, that $H \neq \varnothing$. Then, for every pair $(p,q) \in H$, we have $p \neq q$, and there exists a word $w_{(p,q)} \in \Sigma^*$ such that $(p \cdot w_{(p,q)} x,\, q \cdot w_{(p,q)} x) \in H$ for every letter $x \in \Sigma$. Indeed, otherwise, for every $w \in \Sigma^*$, one could find a letter $x \in \Sigma$ such that $(p \cdot wx,\, q \cdot wx) \in E_L$ which would mean that $(p,q) \in E_{L+1} = E_L$.

Now, Bob wins the ω-game on $\mathscr{A}$, using the following strategy. He first chooses a pair $(p_0,q_0) \in H$ and plays the word $w_1 := w_{(p_0,q_0)}$. If Alice then plays a letter x_1 and we set $p_1 := p_0 \cdot w_1 x_1$ and $q_1 := q_0 \cdot w_1 x_1$, Bob responds by playing the word $w_2 := w_{(p_1,q_1)}$, and so on. By construction, after each turn the pair of states reached from (p_0,q_0) belongs to H, and hence consists of two distinct states. Therefore, at no turn can the history of the game be a reset word for $\mathscr{A}$. This shows that Bob has a winning strategy on $\mathscr{A}$, contradicting the assumption that $\mathscr{A}$ is an A_ω-automaton.

The 'if' part. First we show that Alice wins from every position with only two tokens. Take any such position, and let p and q be the states holding the tokens. Since $\bigcup_{\ell\geq 0} E_\ell = Q \times Q$, there exists ℓ such that the pair (p,q) belongs to the relation E_ℓ. As observed after the definition of E_ℓ, Alice then has a strategy that allows her to move the tokens from the states p and q to a common state in at most ℓ moves.

Now, Alice may proceed as follows to win from every position P (including the initial position of the game). She selects a pair of tokens, say T and T', and plays as if no other tokens were present, that is, she applies her winning strategy from the position in which T and T' occupy the same states as they do in P and all other tokens are removed. This brings the game to a position in which T or T' is removed. Alice then selects a new pair of tokens and plays as if these were the only tokens, and so on. Since at least one token is removed in each round, Alice eventually wins. □

By Proposition 1, in any A_ω-automaton (Q,Σ), every pair $(p,q) \in Q \times Q$ belongs to some relation E_ℓ. We denote by $d(p,q)$ the smallest integer ℓ with this property. All items in the next proposition follow easily from the definitions and Proposition 1.

Proposition 2. *In any A_ω-automaton (Q,Σ), the following hold for all $p,q \in Q$:*

(D1) $d(p,q) = 0$ *if and only if* $p = q$;
(D2) $d(p,q) = d(q,p)$;
(D3) $d(p \cdot w, q \cdot w) \leq d(p,q)$ *for every word* $w \in \Sigma^*$;
(D4) *if* $d(p,q) > 0$, *then there is a letter* $x \in \Sigma$ *such that* $d(p \cdot x, q \cdot x) < d(p,q)$.

We need some notation related to the connected components of the graphs (Q,E_ℓ), where (Q,Σ) is an A_ω-automaton. For brevity, we refer to the set of all vertices of a connected component of a graph G as a *component* of G.

For every integer $\ell \geq 0$, let $\mathscr{D}_\ell$ denote the set of all components of the graph (Q,E_ℓ). Each $\mathscr{D}_\ell$ is a partition of the set Q, and when viewed as such, $\mathscr{D}_{\ell+1}$ is coarser than $\mathscr{D}_\ell$. Thus, every component in $\mathscr{D}_{\ell+1}$ is a union of components from $\mathscr{D}_\ell$. We therefore obtain a descending chain of partitions with respect to the refinement order. The chain starts with the one-block partition $\{Q\}$ (which is $\mathscr{D}_K$ for some integer K such that $E_K = Q \times Q$; such a K exists since $\bigcup_{\ell\geq 0} E_\ell = Q \times Q$ and the set Q is finite) and ends with $\mathscr{D}_0 := \{\{q\} : q \in Q\}$. We set $\mathscr{D} := \bigcup_{\ell \geq 0} \mathscr{D}_\ell$ and $\mathscr{D}^+ := \mathscr{D} \setminus \mathscr{D}_0$. Notice that any two components in $\mathscr{D}$ are either disjoint or comparable with respect to inclusion.

For any component $C \in \mathscr{D}$, let $h(C)$ denote the smallest integer ℓ such that $C \in \mathscr{D}_\ell$. Notice that $h(C) = 0$ if and only if $C \in \mathscr{D}_0$, that is, if and only if $|C| = 1$. Each $C \in \mathscr{D}^+$ is partitioned in more than one components from $\mathscr{D}_{h(C)-1}$. We denote by $\mathscr{V}(C)$ the set of components of this partition of C. Thus, $\mathscr{V}(C) \subseteq \mathscr{D}_{h(C)-1}$ and $|\mathscr{V}(C)| > 1$.

For every nonempty subset $A \subseteq Q$, there is the least (with respect to inclusion) component in $\mathscr{D}$ that contains A. We denote this component by C_A. Notice that $|C_A| > 1$ if and only if $|A| > 1$. Denoting $h(C_A)$ by ℓ, we introduce a further bundle of notation:

- $\mathscr{T}_A$ is the family of all connected graphs $G = (V, E)$ with $A \subseteq V \subseteq C_A$ and $E \subseteq E_\ell$;
- $\theta_\ell(G)$ for $G = (V, E) \in \mathscr{T}_A$ is the number of edges $e \in E$ such that $d(e) = \ell$;
- $\gamma(A) := \min\limits_{G \in \mathscr{T}_A} \theta_\ell(G)$.

Lemma 1. *For every A_ω-automaton (Q, Σ) and every nonempty subset $A \subseteq Q$,*

$$\gamma(A) \leq \gamma(C_A) = \begin{cases} |\mathscr{V}(C_A)| - 1 & \textit{if } |A| > 1; \\ 0 & \textit{otherwise.} \end{cases}$$

Proof. To lighten notation, we write C for C_A in this proof.

The inequality $\gamma(A) \leq \gamma(C)$ follows from the fact that the minimum in the expression for $\gamma(A)$ is taken over a larger set since $\mathscr{T}_C$ is a subfamily of the family $\mathscr{T}_A$.

To prove the equality

$$\gamma(C) = \begin{cases} |\mathscr{V}(C)| - 1 & \text{if } C \in \mathscr{D}^+; \\ 0 & \text{otherwise,} \end{cases} \tag{2}$$

first observe that $C_C = C$; hence all graphs in the family $\mathscr{T}_C$ have C as their vertex set.

If $|C| = 1$, then the one-vertex graph with no edges occurs in the family $\mathscr{T}_C$, whence $\min\limits_{G \in \mathscr{T}_C} \theta_\ell(G) = 0$, which agrees with (2).

Recall that each set $C \in \mathscr{D}^+$ is partitioned into $|\mathscr{V}(C)|$ components from $\mathscr{D}_{\ell-1}$. For a graph $G = (C, E)$ with $E \subseteq E_\ell$ to be connected, any two such components must be connected by a path involving edges in $E_\ell \setminus E_{\ell-1}$.

For each edge $e \in E_\ell \setminus E_{\ell-1}$, we have $d(e) = \ell$ by the definition of the function d, and at least $|\mathscr{V}(C)| - 1$ edges are required to connect $|\mathscr{V}(C)|$ components. Hence, $\theta_\ell(G) \geq |\mathscr{V}(C)| - 1$ for every graph $G \in \mathscr{T}_C$. On the other hand, $|\mathscr{V}(C)| - 1$ edges suffice. To see this, consider the graph $H := (\mathscr{V}(C), F)$ in which two vertices $V_1, V_2 \in \mathscr{V}(C)$ are connected by an edge f if and only if there exists an edge $e \in E_\ell \setminus E_{\ell-1}$ connecting V_1 and V_2 in the graph $(C, (C \times C) \cap E_\ell)$; we say that e *induces* f. The graph H is connected, and any spanning tree of H has $|\mathscr{V}(C)| - 1$ edges. Take a spanning tree H' of H and consider the graph $G' = (C, ((C \times C) \cap E_{\ell-1}) \cup E')$, where E' consists of $|\mathscr{V}(C)| - 1$ edges in $E_\ell \setminus E_{\ell-1}$ that induce the edges of the tree H'. The graph G' is connected and so it belongs to the family $\mathscr{T}_C$. By construction, $\theta_\ell(G') = |\mathscr{V}(C)| - 1$. Thus, $\min\limits_{G \in \mathscr{T}_C} \theta_\ell(G) = |\mathscr{V}(C)| - 1$, which again agrees with (2). □

For a nonempty subset $A \subseteq Q$, the pair $\chi(A) := (C_A, \gamma(A))$ is called the *characteristic* of A. The characteristics of singletons have the form $(\{q\}, 0)$ for some $q \in Q$; we call such characteristics *trivial*. For every nontrivial characteristic $(C_A, \gamma(A))$, we have $|C_A| > 1$ and $1 \le \gamma(A) \le \gamma(C_A)$. The next two propositions show that nontrivial characteristics are suited to play the role of the 'parameter' from the proof outline at the beginning of this subsection.

Proposition 3. *For every A_ω-automaton (Q, Σ), there are at most $|Q| - 1$ distinct nontrivial characteristics.*

Proof. For every component $C \in \mathscr{D}^+$, there are at most $\gamma(C)$ distinct characteristics with first entry C because the possible values of the second entry belong to the set $\{1, \ldots, \gamma(C)\}$. Summing over all such components yields at most $\sum_{C \in \mathscr{D}^+} \gamma(C)$ distinct nontrivial characteristics. Clearly, $\sum_{C \in \mathscr{D}^+} \gamma(C) = \sum_{C \in \mathscr{D}} \gamma(C)$ as extending the summation to components from $\mathscr{D} \setminus \mathscr{D}^+ = \mathscr{D}^0$ adds only zero summands.

To compute the latter sum, we set, for every component $R \in \mathscr{D}$,

$$\Gamma(R) := \sum_{\substack{C \in \mathscr{D} \\ C \subseteq R}} \gamma(C). \tag{3}$$

We prove that $\Gamma(R) = |R| - 1$ by induction on $|R|$. The equality is immediate if $|R| = 1$. Suppose that $|R| > 1$. Then $|\mathscr{V}(R)| > 1$, whence the inductive assumption applies to every component in $\mathscr{V}(R)$. Clearly, for every $C \in \mathscr{D}$ such that $C \subsetneq R$, there is a unique component $S \in \mathscr{V}(R)$ such that $C \subseteq S$. We then have

$$\begin{aligned}
\Gamma(R) &= \gamma(R) + \sum_{\substack{C \in \mathscr{D} \\ C \subsetneq R}} \gamma(C) = \gamma(R) + \sum_{S \in \mathscr{V}(R)} \sum_{\substack{C \in \mathscr{D} \\ C \subseteq S}} \gamma(C) \\
&= \gamma(R) + \sum_{S \in \mathscr{V}(R)} \Gamma(S) && \text{by (3)} \\
&= \gamma(R) + \sum_{S \in \mathscr{V}(R)} (|S| - 1) && \text{by the inductive assumption} \\
&= |\mathscr{V}(R)| - 1 + \sum_{S \in \mathscr{V}(R)} (|S| - 1) && \text{by Lemma 1} \\
&= |\mathscr{V}(R)| - 1 + \sum_{S \in \mathscr{V}(R)} |S| - |\mathscr{V}(R)| = |R| - 1 && \text{since } |R| = \sum_{S \in \mathscr{V}(R)} |S|.
\end{aligned}$$

In particular, $\Gamma(Q) = \sum_{C \in \mathscr{D}} \gamma(C) = |Q| - 1$. This establishes the claimed bound on the number of different nontrivial characteristics. □

We say that a nonempty subset $A \subseteq Q$ *avoids* a characteristic μ if $\mu \neq \chi(A \cdot w)$ for every word $w \in \Sigma^*$.

Proposition 4. *For every A_ω-automaton (Q, Σ) and every subset $A \subseteq Q$ with $|A| > 1$, there exists a letter $x \in \Sigma$ such that the set $A \cdot x$ avoids the characteristic $\chi(A)$.*

Proof. Denote $\ell = h(C_A)$ and choose a graph $G = (V,E) \in \mathscr{T}_A$ such that $\theta_\ell(G) = \gamma(A)$. Recall that by the definition of the family $\mathscr{T}_A$, the graph G is connected and $E \subseteq E_\ell$. Choose an edge $(p_0, q_0) \in E$ such that $d(p_0, q_0) = \ell$. As registered in Proposition 2, item (D4), there exists a letter $x \in \Sigma$ such that $d(p_0{\cdot}x,\, q_0{\cdot}x) < \ell$. We now show that $A{\cdot}x$ avoids the characteristic $\chi(A)$.

Take an arbitrary word $w \in \Sigma^*$ and set $B := A{\cdot}xw$. We aim to prove that $\chi(B) \neq \chi(A)$. This is immediate if $C_A \neq C_B$, so assume that $C_A = C_B$.

Consider the image G' of the graph G under the action of the word xw, that is, $G' := (V', E')$, where $V' := V{\cdot}xw$ and $E' := \{(p{\cdot}xw,\, q{\cdot}xw) : (p,q) \in E\}$. We aim to prove that the graph G' lies in the family $\mathscr{T}_B$, which amounts to verifying the following:

(i) $E' \subseteq E_{h(C_B)}$;
(ii) G' is connected;
(iii) $B \subseteq V' \subseteq C_B$.

The inclusion (i) readily follows from the equality $h(C_B) = h(C_A) = \ell$ (ensured by the equality $C_A = C_B$), the inclusion $E \subseteq E_\ell$, and the stability of E_ℓ (Proposition 1).

For (ii), it suffices to notice that if a sequence of consecutive edges

$$(p_1, p_2), (p_2, p_3), \ldots, (p_{m-1}, p_m) \in E$$

connects two states $p = p_1$ and $p_m = q$ from A, then, after removing from its image

$$(p_1{\cdot}xw,\, p_2{\cdot}xw), (p_2{\cdot}xw,\, p_3{\cdot}xw), \ldots, (p_{m-1}{\cdot}xw,\, p_m{\cdot}xw)$$

the pairs belonging to E_0, if any, we obtain a sequence of consecutive edges in E' connecting the states $p{\cdot}xw$ and $q{\cdot}xw$ from B.

A similar argument can be used to justify the inclusion $V' \subseteq C_B$ in (iii). Indeed, every state in V' is of the form $q{\cdot}xw$ for some $q \in V$. Since $V \subseteq C_A$ and C_A is the component of the graph (Q, E_ℓ) containing A, the state q is connected by a sequence of consecutive edges in E_ℓ with a state $p \in A$. Then the image of this sequence connects $q{\cdot}xw$ with $p{\cdot}xw \in B$ and each edge in the image belongs to E_ℓ due to the stability of E_ℓ. Hence, $q{\cdot}xw$ lies in the component of (Q, E_ℓ) containing B, that is, $q{\cdot}xw \in C_B$.

The inclusion $B \subseteq V'$ in (iii) is ensured by the inclusion $A \subseteq V$.

Thus, G' is a graph in $\mathscr{T}_B$. By item (D3) of Proposition 2, $d(p{\cdot}xw,\, q{\cdot}xw) \leq \ell$ for every edge $(p,q) \in E_\ell$. By the same property, for the edge (p_0, q_0), this inequality is strict. Hence, $\theta_\ell(G') < \gamma(A)$, so $\gamma(B) < \gamma(A)$. Hence, $\chi(B) \neq \chi(A)$. □

We are ready for the proof of our main result.

Theorem 1. *On every A_ω-automaton $\mathscr{A} = (Q, \Sigma)$, Alice has a strategy that allows her to win in less than $|Q|$ moves. In particular, the reset threshold of $\mathscr{A}$ is less than $|Q|$.*

Proof. We may assume that Bob moves first. Denote by w_i the history up to Bob's i-th turn, and let $A_i := Q{\cdot}w_i$. Alice's strategy is to respond with a letter x_i such that the set $A_i{\cdot}x_i$ avoids the characteristic $\chi(A_i)$. By Proposition 4, such a letter x_i exists as long as $|A_i| > 1$, that is, as long as w_i is not a reset word for $\mathscr{A}$.

The described strategy ensures that $\chi(A_i) \neq \chi(A_j)$ whenever $i \neq j$. Indeed, suppose that $i < j$; then $w_j = w_i x_i u$ for some word $u \in \Sigma^*$, and consequently $A_j = A_i{\cdot}x_i u$.

By construction, $A_i{\cdot}x_i$ avoids $\chi(A_i)$, which by the definition of avoidability means that $\chi(A_i) \neq \chi(A_i{\cdot}xw)$ for every word $w \in \Sigma^*$. Taking $w = u$ yields $\chi(A_i) \neq \chi(A_j)$. Thus, all characteristics $\chi(A_1), \chi(A_2), \ldots$ are distinct.

By Proposition 4, there are at most $|Q|-1$ nontrivial characteristics. Hence, using the described strategy, Alice makes at most $|Q|-1$ moves in order to win.

The history of the game up to its final move is a reset word for $\mathscr{A}$, independent of which words Bob has chosen on his moves. In particular, the final history of the game in which Bob has chosen the empty word on each of his moves is a reset word whose length equals the number of Alice's moves and thus does not exceed $|Q|-1$. □

Remarks. 1. It is easy to see that the bound in Theorem 1 is tight. As a simple witness, for every $n \geq 3$, consider the DFA $(\{0,1,\ldots,n-1\},\{a\})$ in which $0{\cdot}a := 0$ and $m{\cdot}a := m-1$ for $0 < m < n$. This is an A_ω-automaton with n states and reset threshold $n-1$, and Alice needs exactly $n-1$ moves to win on it.

2. The converse of the second conclusion of Theorem 1 does not hold. As a witness, for every $n \geq 4$, consider the DFA $(\{0,1,\ldots,n-1\},\{a,b\})$ in which the letters move the states in the opposite directions: $0{\cdot}a := 0$ and $m{\cdot}a := m-1$ for $0 < m < n$, whereas $(n-1){\cdot}b := n-1$ and $m{\cdot}b := m+1$ for $0 \leq m < n-1$. The DFA is synchronizing with reset threshold $n-1$ since each of a^{n-1} and b^{n-1} is a reset word of minimum length. On the other hand, Bob wins even the 2-game on this DFA. Bob's winning strategy is to play b whenever Alice plays a, and to play a if she plays b. In this way, he can ensure that each of the 'middle' states $1, 2, \ldots, n-2$ holds a token after his responses.

3.2 Reset Threshold of DS-Automata

Theorem 1 confirms the natural expectation that automata on which Alice wins against a powerful adversary should be quickly synchronizable. However, the property of being an A_ω-automaton may appear too strong, and it is far from clear whether Theorem 1 applies to any meaningful class of DFAs not defined in terms of synchronization games. Thanks to an observation made in [8, Section 5.4], we are in a position to exhibit such a class. To introduce it, we recall the notion of the transition monoid of a DFA and a few basic concepts from semigroup theory.

For a DFA $\mathscr{A} = (Q,\Sigma)$, the map $\tau_a\colon Q \to Q$ defined by $q\tau_a := q{\cdot}a$ is a transformation of the set Q. The *transition monoid* of a DFA $\mathscr{A} = (Q,\Sigma)$ is the submonoid of the monoid of all transformations on the set Q generated by the set $\{\tau_a : a \in \Sigma\}$. We denote the transition monoid of a DFA $\mathscr{A}$ by $T(\mathscr{A})$. In $T(\mathscr{A})$, any product of the form $\tau_{a_1}\tau_{a_2}\cdots\tau_{a_n}$ is precisely the transformation $\tau_w\colon Q \to Q$ defined by $q\tau_w := q{\cdot}w$, where $w := a_1a_2\cdots a_n$. Thus, the transition monoid $T(\mathscr{A})$ can equivalently be defined as the monoid of all transformations on the set Q induced by the action of words over Σ.

If $\mathscr{A} = (Q,\Sigma)$ is a synchronizing DFA and w is a reset word for $\mathscr{A}$, then the transformation τ_w is a constant map on Q, that is, $Q\tau_w = \{q\}$ for some $q \in Q$. Thus, the transition monoid of a synchronizing automaton always contains a constant transformation. Conversely, if $\zeta \in T(\mathscr{A})$ is a constant transformation, then any word w with $\tau_w = \zeta$ is a reset word for $\mathscr{A}$, and hence $\mathscr{A}$ is synchronizing. We therefore see that synchronization is actually a property of the transition monoid of an automaton rather than the automaton itself: for DFAs $\mathscr{A} = (Q,\Sigma)$ and $\mathscr{A}' = (Q,\Sigma')$ with the same states but different input alphabets, the equality $T(\mathscr{A}) = T(\mathscr{A}')$ guarantees that $\mathscr{A}'$ is synchronizing if and only if $\mathscr{A}$ is.

Recall the definition of *Green's relations* $\mathfrak{R}$, $\mathfrak{L}$, and $\mathfrak{D}$ on an arbitrary semigroup S:

- $a \,\mathfrak{R}\, b \iff$ either $a = b$ or $a = bs$ and $b = at$ for some $s,t \in S$;
- $a \,\mathfrak{L}\, b \iff$ either $a = b$ or $a = sb$ and $b = ta$ for some $s,t \in S$;
- $a \,\mathfrak{D}\, b \iff a \,\mathfrak{R}\, c$ and $c \,\mathfrak{L}\, b$ for some $c \in S$.

The relations $\mathfrak{R}$ and $\mathfrak{L}$ are clearly equivalences. The definition of $\mathfrak{D}$ means that $\mathfrak{D}$ is the product of $\mathfrak{R}$ and $\mathfrak{L}$ as binary relations. As observed in [11], $\mathfrak{D}$ is also the product of $\mathfrak{L}$ and $\mathfrak{R}$, which implies that $\mathfrak{D}$ is the least equivalence containing both $\mathfrak{R}$ and $\mathfrak{L}$.

A $\mathfrak{D}$-class D is called *regular* if it contains an element a such that $aba = a$ for some $b \in D$. DFAs in whose transition monoids all regular $\mathfrak{D}$-classes are subsemigroups often appear in the literature; see, e.g., [1,2]. In [7], such DFAs were coined *DS-automata*, and we adopt this terminology here. We refer the reader to [7, Section 4.1] for an overview of studies of synchronizing automata in various subclasses of DS-automata.

The main result of [8] states that, on every synchronizing DS-automaton, Alice has a *uniform* winning strategy in the synchronization game in which Bob's responses are restricted to letters. Here, uniformity means that Alice's winning sequence of moves depends only on the underlying automaton and does not depend on Bob's responses. In [8, Section 5.4], this strategy is shown to persist for the ω-game. Hence, synchronizing DS-automata are A_ω-automata, and Theorem 1 yields the following:

Corollary 1. *The reset threshold of every synchronizing DS-automaton is less than the number of its states.*

The result of Corollary 1 is contained in [2, Theorem 2.6], where its proof was based on the classical theory of linear representations of semigroups (the Munn–Ponizovsky theory). Our proof is quite different, as it is purely combinatorial, apart from invoking the results from [8], which themselves use only a little bit of algebra.

Observe that an A_ω-automaton need not be a DS-automaton. The simplest witness here is the automaton $\mathscr{B}_2$ shown in Fig. 1.

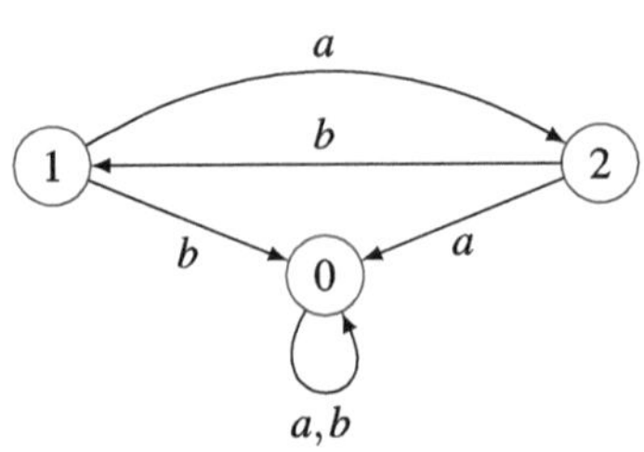

Fig. 1. An A_ω-automaton which is not a DS-automaton

Its transition monoid $T(\mathscr{B}_2)$ is easy to compute. It consists of six transformations of $\{0,1,2\}$: the identity transformation τ_ε, the constant transformation τ_{a^2}, and the transformations τ_a, τ_b, τ_{ab}, and τ_{ba} that form a $\mathfrak{D}$-class in $T(\mathscr{B}_2)$. This $\mathfrak{D}$-class is not a subsemigroup, since, for instance, it does not contain the transformation $\tau_a^2 = \tau_{a^2}$. Thus, $\mathscr{B}_2$

is not a DS-automaton. On the other hand, Alice wins the ω-game on $\mathscr{B}_2$ by playing a and thus leaving tokens on states 0 and 2. After any non-losing response of Bob, tokens either remain on 0 and 2 or they move to states 0 and 1. Alice then wins by playing a in the former case or b in the latter.

Hence, Theorem 1 not only provides an alternative proof of Corollary 1 but also extends its conclusion to the strictly larger class $\mathbf{A}_\omega$.

Notice that $\mathbf{A}_\omega$ cannot be characterized solely in terms of transition monoids. In fact, for every synchronizing DFA $\mathscr{A} = (Q,\Sigma)$, there exists an A_ω-automaton $\mathscr{A}' = (Q,\Sigma')$ such that $T(\mathscr{A}) = T(\mathscr{A}')$. To see this, let $\Sigma' := \Sigma \cup \{c\}$ where the action of the added letter c coincides with the action of a fixed reset word for $\mathscr{A}$. The transformations induced by the letters in Σ' generate the same submonoid of the monoid of all transformations of the set Q as do the transformations induced by the letters in Σ. Still, Alice wins the ω-game on $\mathscr{A}'$ instantly by choosing the letter c on her first move.

That said, the question of which properties of the monoid $T(\mathscr{A})$ ensure that $\mathscr{A}$ is an A_ω-automaton appears to be worth further exploration.

3.3 Iterations of synchronizing automata

What happens if Alice is allowed to play words rather than single letters in an ω-game? Of course, if Alice may play arbitrary words, the game trivializes as Alice can instantly win on any synchronizing automaton by spelling out a reset word. The variant in which Alice's moves are subject to a length restriction is more interesting. To handle it, the following construction is well suited.

For $m \in \mathbb{N}$ and an alphabet Σ, let $\Sigma^{\leq m}$ denote the set of all nonempty words over Σ of length at most m. The *m-th iteration* of a DFA $\mathscr{A} = (Q,\Sigma)$ is the DFA $\mathscr{A}^{(m)} := (Q,\Sigma^{\leq m})$, with the action of the 'letters' in $\Sigma^{\leq m}$ defined in a natural way: $q \circ w := q \cdot w$ for all $q \in Q$ and $w \in \Sigma^{\leq m}$, with the right-hand side denoting the action in $\mathscr{A}$.

By an m/ω-game we mean the variant in which Alice's moves are nonempty words of length at most m, whereas Bob may play arbitrary words. It is then immediate that Alice wins the m/ω-game on a DFA $\mathscr{A}$ if and only if she wins the standard ω-game played on the DFA $\mathscr{A}^{(m)}$. It follows that the reset threshold of any n-state synchronizing automaton on which Alice has a winning strategy in the m/ω-game does not exceed $m(n-1)$. We can show that this upper bound can be slightly improved; see [13, Sect. 3.3] for the proof.

Proposition 5. *The reset threshold of an n-state automaton on which Alice has a winning strategy in the m/ω-game does not exceed $m(n-2)+1$.*

Remark. For $m = 2$, the bound $2(n-2)+1 = 2n-3$ provided by Proposition 5 is tight; see [13, Sect. 3.3] for a series of examples witnessing this fact. We do not know whether the bound $m(n-2)+1$ is tight for $m \geq 3$, and we tend to believe that it is not.

Proposition 5 also leads to a new class of synchronizing automata satisfying the Černý conjecture bound:

Corollary 2. *The reset threshold of an n-state automaton on which Alice has a winning strategy in the n/ω-game does not exceed $n(n-2)+1 = (n-1)^2$.*

Although the observation in Corollary 2 appears to be worth recording, it is fair to say that, in contrast to Theorem 1, we have so far not found any application of this observation to DFAs not defined in terms of synchronization games.

3.4 Reset Thresholds of A_k-Automata with $k \in \mathbb{N}$

Here, for each $k \in \mathbb{N}$, we present a series of examples showing that, in contrast to A_ω-automata, A_k-automata with $k \in \mathbb{N}$ can have reset thresholds vastly exceeding the number of states. More precisely, the reset threshold of the DFAs in our series is bounded from below by a quadratic function of the number of states, with leading coefficient $\frac{1}{2k}$.

For each $k, m \in \mathbb{N}$, let $n = k+m+1$ and consider the automaton

$$\mathscr{L}_m^k := (\{0,1,\ldots,n-1\}, \{a_1,\ldots,a_{k+m}\})$$

in which each letter a_i with $i = 1,\ldots,k$ sends state i to 0 and fixes all other states, whereas each letter a_{k+j} with $j = 1,\ldots,m$ cyclically permutes the states j, $j+1$, …, $k+j-1$, $k+j$ and fixes all other states. Thus, for each $q \in \{0,1,\ldots,n-1\}$, we have

$$q{\cdot}a_i := \begin{cases} 0 & \text{if } q = i, \\ q & \text{otherwise,} \end{cases} \quad i = 1,\ldots,k; \quad q{\cdot}a_{k+j} := \begin{cases} q+1 & \text{if } j \le q < k+j, \\ j & \text{if } q = k+j, \\ q & \text{otherwise,} \end{cases} \quad j = 1,\ldots,m.$$

Notice that, in particular,

$$q{\cdot}a_q = \begin{cases} 0 & \text{if } 1 \le q \le k, \\ q-k & \text{if } k+1 \le q \le k+m, \end{cases} \quad \text{and } p{\cdot}a_q \ge p \text{ whenever } p \ne q. \tag{4}$$

Notice also that 0 is a *sink* of $\mathscr{L}_m^k$, that is, a state fixed by all letters.

Proposition 6. *The DFA $\mathscr{L}_m^k$ is an A_k-automaton.*

Proof. Assume that Bob moves first. Denote by w_s the history up to Bob's s-th turn, and let $A_s := Q{\cdot}w_s$. If $A_s = \{0\}$, Alice has already won. Otherwise, if $A_s \ne \{0\}$ and $q_s := \min(A_s \setminus \{0\})$, Alice plays the letter a_{q_s}. Then $q_s{\cdot}a_{q_s} = 0$ or $q_s{\cdot}a_{q_s} = q_s - k$ by (4). Whichever word v_{s+1} of length less than k Bob plays in response, $q_s{\cdot}a_{q_s}v_{s+1} < q_s$. Hence, either $q_s{\cdot}a_{q_s}v_{s+1} = 0$ (and then $|A_{s+1}| < |A_s|$) or $q_{s+1} < q_s$. Therefore, the strategy guarantees that eventually $A_t = \{0\}$ for some t, and Alice wins. □

Proposition 7. *The reset threshold of the automaton $\mathscr{L}_m^k$ is not less than $\frac{n(n-1)}{2k}$.*

Proof. Suppose $w = x_1 \cdots x_\ell$ is a reset word for $\mathscr{L}_m^k$. For each $s \in \{0,1,\ldots,\ell\}$, denote by u_s the prefix of w of length s; in particular, $u_0 = \varepsilon$.

The reset word w brings all states of $\mathscr{L}_m^k$ to a single state, and since 0 is fixed by all input letters, this state must be 0. Thus, $q{\cdot}w = 0$ for every state $q \in \{0,1,\ldots,n-1\}$. Let $p_t := q{\cdot}u_t$ and consider the sequence of states

$$q = p_0,\ p_1,\ \ldots\ p_{t-1},\ p_t, \ldots\ p_\ell = q{\cdot}w = 0.$$

If $q \neq 0$, then some states in this sequence must be strictly greater than their immediate successors. Thus, for every $q \in \{1,\ldots,k+m\}$, the set

$$E_q := \{s \in \{1,\ldots,\ell\} : p_s < p_{s-1}\}$$

is nonempty. For $s \in E_q$, we have $p_{s-1} \neq 0$. Since $p_s = p_{s-1}\cdot x_s$, the first part of (4) implies that $p_{s-1} - p_s = k$ if $k+1 \leq p_{s-1} \leq k+m$ and $p_{s-1} - p_s \leq k$ if $1 \leq p_{s-1} \leq k$. Thus, $0 < p_{s-1} - p_s \leq k$ for every $s \in E_q$, while $p_{t-1} - p_t \leq 0$ for all $t \notin E_q$ by the inequality in (4). Now we have

$$\begin{aligned} q &= \sum_{t=1}^{\ell}(p_{t-1} - p_t) && \text{(telescopic sum)}\\ &\leq \sum_{s\in E_q}(p_{s-1} - p_s) && \text{(dropping nonpositive summands)}\\ &\leq \sum_{s\in E_q} k = k|E_q| && \text{(using that } p_{s-1} - p_s \leq k \text{ for } s \in E_q). \end{aligned}$$

Hence $|E_q| \geq \frac{q}{k}$.

For any distinct $q, q' \in \{1,\ldots,k+m\}$, the sets E_q and $E_{q'}$ are disjoint. Indeed, suppose to the contrary that $E_q \cap E_{q'} \neq \varnothing$, and let $s \in E_q \cap E_{q'}$. Then $p_s = p_{s-1}\cdot x_s < p_{s-1}$ and, at the same time, $p'_s = p'_{s-1}\cdot x_s < p'_{s-1}$, where $p_t := q\cdot u_t$ and $p'_t := q'\cdot u_t$. By (4), for each letter $x \in \{a_1,\ldots,a_{k+m}\}$, there is exactly one state $p \in \{0,1,\ldots,n-1\}$ such that $p\cdot x < p$. Hence, $p_{s-1} = p'_{s-1}$, that is, $q\cdot u_{s-1} = q'\cdot u_{s-1}$. By the construction of the DFA $\mathscr{L}_m^k$, if two distinct states are mapped to the same state by a word, then the resulting state is 0. Since $p_{s-1} \neq 0$, this is a contradiction.

We thus have

$$|w| \geq \left|\bigcup_{q=1}^{k+m} E_q\right| = \sum_{q=1}^{k+m} |E_q| \geq \sum_{q=1}^{k+m} \frac{q}{k} = \frac{(k+m+1)(k+m)}{2k} = \frac{n(n-1)}{2k}.$$

Hence, $\mathrm{rt}(\mathscr{L}_m^k) \geq \frac{n(n-1)}{2k}$, as claimed. □

Remarks. 1. Question 2 in [7] asked whether every A-automaton with n states has a reset word of length $n-1$. (The A-automata of [7] are precisely DFAs on which Alice has a winning strategy in a variant of the 2-game where Bob is not allowed to play the empty word.) Propositions 6 and 7 imply a negative answer for $n > 4$.

2. Since $|E_q|$ is an integer, the proof of Proposition 7 shows that in fact $|E_q| \geq \lceil \frac{q}{k} \rceil$ for each q. Hence, $\mathrm{rt}(\mathscr{L}_m^k) \geq L := \sum_{q=1}^{k+m} \lceil \frac{q}{k} \rceil$. We present a reset word of length L for $\mathscr{L}_m^k$, thus showing that L is the precise value of $\mathrm{rt}(\mathscr{L}_m^k)$.

Let $v_q := a_q$ for $1 \leq q \leq k$ and $v_q := a_q a_{q-k} a_{q-2k} \cdots a_t$, where $1 \leq t \leq k$, for $k+1 \leq q \leq k+m$. Then $|v_q| = \lceil \frac{q}{k} \rceil$ and $q\cdot v_q = 0$, whereas $r\cdot v_q = r$ for all $r > q$. Hence, the word $v_1 v_2 \cdots v_{k+m}$ resets $\mathscr{L}_m^k$ and has length L.

4 Conclusion and Future Work

We summarize our results and highlight questions that have so far remained open.

If $\mathbf{S}$ is a class of DFAs, its *Černý function* $\mathfrak{C}_{\mathbf{S}}(n)$ is defined as the maximum reset threshold among all n-state synchronizing automata from $\mathbf{S}$. With this notation, Theorem 1, together with the remark following it, establishes the equality $\mathfrak{C}_{\mathbf{A}_\omega}(n) = n-1$, while Proposition 7 yields the inequality $\mathfrak{C}_{\mathbf{A}_k}(n) \geq \frac{n(n-1)}{2k}$ for $k \in \mathbb{N}$. We believe that the lower bound $\frac{n(n-1)}{2k}$ for $\mathfrak{C}_{\mathbf{A}_k}(n)$, with $k \in \mathbb{N}$, can be improved. Recall that the DFAs witnessing this bound are automata with a sink. For several classes of automata with known lower bounds on their Černý functions, the bound for the subclass consisting of automata with a sink is around half of the bound for the entire class. If this empirical observation extends to the class $\mathbf{A}_k$, one might therefore expect a lower bound with leading term $\frac{n^2}{k}$ rather than $\frac{n^2}{2k}$.

As for an upper bound for $\mathfrak{C}_{\mathbf{A}_k}(n)$ with $k \in \mathbb{N}$, the class $\mathbf{A}_k$ is not contained in any class of synchronizing automata, for which a quadratic upper bound on the Černý function has been established so far; see [16] for a catalogue of such classes. Therefore, finding an $O(n^2)$ upper bound on $\mathfrak{C}_{\mathbf{A}_k}(n)$ is of significant interest.

With regard to reset thresholds of A_k-automata, we pose two particular questions.

Question 1. Is the reset threshold of every A_k-automaton with k states less than k?

The answer is affirmative for automata with a sink; see [13].

Question 2. Do there exist real positive constants C and D such that $\mathrm{rt}(\mathscr{A}) \leq Cn$ whenever $\mathscr{A}$ is an A_k-automaton with n states and $k \geq Dn$?

Proposition 5 gives the upper bound $m(n-2)+1$ on the reset thresholds of m/ω-automata with n states. The bound is tight for $m=2$ but is likely not tight for larger m. For the case $m=3$, there is some evidence that the bound can be improved to $3n-6$, and we therefore pose the following question.

Question 3. Does each $3/\omega$-automaton with n states have a reset word of length $3n-6$?

Acknowledgements. The authors thank the anonymous reviewers for their feedback.

References

1. Almeida, J., Margolis, S., Steinberg, B., Volkov, M.: Representation theory of finite semigroups, semigroup radicals and formal language theory. Trans. Amer. Math. Soc. **361**(3), 1429–1461 (2009). https://doi.org/10.1090/S0002-9947-08-04712-0
2. Almeida, J., Steinberg, B.: Matrix mortality and the Černý–Pin conjecture. In: Diekert, V., Nowotka, D., (eds.) Developments in Language Theory (DLT 2009), Lect. Notes Comput. Sci. vol. 5583, pp. 67–80. Springer, Berlin (2009). https://doi.org/10.1007/978-3-642-02737-6_5
3. Ananichev, D.S., Volkov, M.V., Zaks, Y.I.: Synchronizing automata with a letter of deficiency 2. In: Ibarra, O.H., Dang, Z,. (eds.) Developments in Language Theory (DLT 2006), Lect. Notes Comput. Sci. vol. 4036, pp. 433–442. Springer, Berlin (2006). https://doi.org/10.1007/11779148_39

4. Ananichev, D.S., Volkov, M.V., Zaks, Y.I.: Synchronizing automata with a letter of deficiency 2. Theor. Comput. Sci. **376**, 30–41 (2007). https://doi.org/10.1016/j.tcs.2007.01.010
5. Blass, A., Gurevich, Y., Nachmanson, L., Veanes, M.: Play to test. In: Grieskamp, W., Weise, C., (eds.) Formal Approaches to Software Testing (FATES 2005), Lect. Notes Comput. Sci. vol. 3997, pp. 32–46. Springer, Berlin (2006). https://doi.org/10.1007/11759744_3
6. Fernau, H., Haase, C., Hoffmann, S.: The synchronization game on subclasses of automata. In: Fraigniaud, P., Uno, Y., (eds.) Fun with Algorithms (FUN 2022), Leibniz Int. Proc. Inform. vol. 226, pp. 14:1–14:17. Schloss Dagstuhl (2022). https://doi.org/10.4230/LIPIcs.FUN.2022.14
7. Fernau, H., Haase, C., Hoffmann, S., Volkov, M.: Winning strategies for the synchronization game on subclasses of finite automata. In: Manea, F., Pighizzini, G., (eds.) Non-Classical Models of Automata and Applications. Electron. Proc. Theor. Comput. Sci. pp. 77–85 (2024). https://doi.org/10.4204/EPTCS.407.6
8. Fernau, H., Haase, C., Hoffmann, S., Volkov, M.: Uniform winning strategies for the synchronization game on subclasses of finite automata. J. Autom. Lang. Comb. (2025). arXiv:2512.11007. https://doi.org/10.48550/arXiv.2512.11007
9. Fominykh, F.M., Martyugin, P.V., Volkov, M.V.: P(l)aying for synchronization. Int. J. Found. Comput. Sci. **24**(6), 765–780 (2013). https://doi.org/10.1142/S0129054113400170
10. Fominykh, F., Volkov, M.: P(l)aying for synchronization. In: Moreira, N., Reis, R., (eds.) Implementation and Application of Automata (CIAA 2012), Lect. Notes Comput. Sci. vol. 7381, pp. 159–170. Springer, Berlin (2012). https://doi.org/10.1007/978-3-642-31606-7_14
11. Green, J.A.: On the structure of semigroups. Ann. Math. (2) **54**(1), 163–172 (1951). https://doi.org/10.2307/1969317
12. Kari, J., Volkov, M.: Černý's conjecture and the road colouring problem. In: Pin, J.-E., (ed.) Handbook of Automata Theory, vol. I, pp. 525–565. EMS Press, Zürich (2021). https://doi.org/10.4171/AUTOMATA-1/15
13. Lipin, A.E., Volkov, M.V.: Adversarial synchronization (2026). arXiv:2601.18362. https://doi.org/10.48550/arXiv.2601.18362
14. Montoya, J.A., Nolasco, C.: Some remarks on synchronization, games and planar automata. Electron. Notes Theor. Comput. Sci. **339**, 85–97 (2018). https://doi.org/10.1016/j.entcs.2018.06.006
15. Volkov, M.V.: Synchronization of finite automata. Russ. Math. Surv. **77**(5), 819–891 (2022). https://doi.org/10.4213/rm10005e
16. Volkov, M.V.: List of results on the Černý conjecture and reset thresholds for synchronizing automata (2025). arXiv:2508.15655. https://doi.org/10.48550/arXiv.2508.15655

Lines in Every Direction with No ee-Random Points

Neil Lutz(✉), Spencer Park Martin, and Rain White

Swarthmore College, Swarthmore, PA 19081, USA
nlutz1@swarthmore.edu

Abstract. We prove that in every direction in the Euclidean plane, there exists a line containing no double exponential time random (ee-random) points. This means each point on these lines has an algorithmically predictable location, to the extent that a gambler in an environment with fair payouts can, using double exponential time computing resources, amass unbounded capital placing bets on increasingly precise estimates of the point's location. Our proof relies on effectivizing the construction of the lineal extension of a Kakeya set. This resolves an open question of Lutz and Lutz (2015).

Keywords: Algorithmic randomness · Resource-bounded measure · Kakeya sets · Lineal extensions · Geometric measure theory

1 Introduction

The theory of algorithmic information is fruitfully applied in continuous geometric spaces by regarding individual points as infinite data streams and quantifying the predictability or compressibility of those streams according to various notions of algorithmic randomness or dimension. There has been significant recent work using this type of analysis to study points on lines [6,9,10,15–18,20]. In this paper, we show that in every direction in $\mathbb{R}^2$, there is a line on which no point is *double exponential time random* (or ee-*random*). We define this algorithmic randomness notion in Sect. 2.2, but informally it means that every point on the line is at least slightly predictable to an algorithm running in double exponential time. Thus a gambler with double exponential time computing power could, in an environment with fair payouts, accumulate unbounded capital by placing bets on the location of the point.

This paper resolves an open question posed in 2015 by Lutz and Lutz [16], which addressed two related questions by proving the existence of lines in all directions containing no computably random points and line segments in every direction containing no double exponential time random points. The authors asked whether the latter result could be extended to include full lines. Such extensions are often nontrivial in geometric measure theory. Keleti [12] proved in 2016, for example, that given any union of line segments in $\mathbb{R}^2$, taking the *lineal extension*—the union of all lines containing these segments—does not change the

V. Brattka et al. (Eds.): CiE 2026, LNCS 16674, pp. 374–388, 2027.
https://doi.org/10.1007/978-3-032-31348-5_25

Hausdorff dimension; the *line segment extension conjecture* in that work is that the same statement holds in all higher-dimensional Euclidean spaces. Both the results of [16] and our new result are obtained by effectivizing constructions of different *Kakeya sets.*

Algorithmic information theoretic analysis of points on lines is closely tied to Kakeya sets. In this paper, a Kakeya set is a set in a Euclidean space $\mathbb{R}^n$ that contains unit-length line segments in all directions. Besicovitch [2,3] proved in 1919 that such a set can have Lebesgue measure 0. Revisiting the problem decades later, Besicovitch [4] in 1964 gave a simpler construction of a set with Lebesgue measure 0 that contains full lines in all directions, and Davies [8] proved in 1971 that every Kakeya set in $\mathbb{R}^2$ has Hausdorff dimension 2. Since then, there has been a prolific line of research studying the properties of Kakeya sets and their lineal extensions, including a major 2025 breakthrough in which Wang and Zahl [22] proved that every Kakeya set in $\mathbb{R}^3$ has Hausdorff dimension 3. By an equivalence established by Keleti and Máthé [13], the result of Wang and Zahl also proves the line segment extension conjecture in $\mathbb{R}^3$. Our result can be equivalently stated as saying there is a Kakeya set in $\mathbb{R}^2$ whose lineal extension contains no double exponential time random points.

2 Double Exponential Time Measure and Randomness in $\mathbb{R}^2$

We now define double exponential time measure 0 and double exponential time randomness in $\mathbb{R}^2$ using martingales. These definitions generalize straightforwardly to other resource bounds and to higher-dimensional Euclidean spaces [16].

2.1 Dyadic Squares and Martingales

For each $r \in \mathbb{N}$, the *r-dyadic rationals* are the elements of the set $2^{-r}\mathbb{Z}$.

For each $r \in \mathbb{N}$ and $u, v \in \mathbb{Z}$, let

$$Q_r(u,v) = 2^{-r}([u, u+1) \times [v, v+1))$$

be the (half-open, half-closed) *r-dyadic square* at (u, v), and let

$$\mathcal{Q} = \{Q_r(u,v) : r \in \mathbb{N} \text{ and } u, v \in \{0, \ldots, 2^r - 1\}\}$$

be the set of all dyadic squares in $[0,1)^2$.

A *martingale* on $[0,1)^2$ is a function $d : \mathcal{Q} \to [0, \infty)$ satisfying

$$d(Q_r(u,v)) = \frac{1}{4} \sum_{\alpha,\beta \in \{0,1\}} d(Q_{r+1}(2u+\alpha, 2v+\beta)). \tag{1}$$

for all $Q_r(u,v) \in \mathcal{Q}$. This condition places a limit on how quickly the martingale's value can grow with r; it implies, for each r-dyadic square $Q_r(u,v)$ and all $\alpha, \beta \in \{0,1\}$, that

$$d(Q_{r+1}(2u+\alpha, 2v+\beta)) \leq 4d(Q_r(u,v)),$$

so

$$d(Q_r(u,v)) \leq 4^r d(Q_0(0,0)). \tag{2}$$

holds for all $Q_r(u,v) \in \mathcal{Q}$. We call $d(Q_0(0,0))$ the *initial capital* of the martingale d.

Informally, we regard a martingale as a betting strategy for a gambler placing bets on increasingly precise estimates for the location of an unknown point—which may be adversarially chosen from some known set—in an environment with fair payouts. When the point is revealed to be in the dyadic square Q, the gambler has amassed $d(Q)$ capital, which they now allocate among the four immediate dyadic subsquares of Q. The point will be revealed to be in one of those subsquares, and the gambler will receive capital equal to four times the amount they allocated to that subsquare. As there were four subsquares to choose from, this payout is fair. The amount allocated to the other subsquares is forfeited by the gambler, and the process is repeated ad infinitum on ever-smaller squares.

For each martingale d, $r \in \mathbb{N}$, and $(x,y) \in [0,1)^2$, define

$$d^{(r)}(x,y) = d(Q_r(\lfloor 2^r x \rfloor, \lfloor 2^r y \rfloor)),$$

the value of the martingale on the unique r-dyadic square containing the point (x,y). The martingale d *succeeds* at a point $(x,y) \in [0,1)^2$ if

$$\limsup_{r\to\infty} d^{(r)}(x,y) = \infty,$$

and its *success set*

$$S^\infty[d] = \left\{ (x,y) \in [0,1)^2 : \limsup_{r\to\infty} d^{(r)}(x,y) = \infty \right\}$$

is the set of points at which it succeeds.

2.2 Martingales and Measure

Originally working in the Cantor space, Ville [21] proved that a set E has Lebesgue measure $\lambda(E) = 0$ if and only if it is contained in the success set of some martingale. This is also true for $E \subseteq [0,1)^2$, and this correspondence is extended to the entire plane by taking a union of unit square intersections with the set: Given any set $E \subseteq \mathbb{R}^2$, define

$$\#E = [0,1)^2 \cap \bigcup_{i,j\in\mathbb{Z}} E - (i,j). \tag{3}$$

Then $\lambda(E) = 0$ if and only if there is a martingale d such that $\#E \subseteq S^\infty[d]$.

Effective notions of measure 0 are obtained by placing resource bounds on this martingale. A martingale $d : \mathcal{Q} \to [0,\infty)$ is *double exponential time computable* if there is a function

$$\hat{d} : \mathcal{Q} \times \mathbb{N} \to \mathbb{Q}$$

such that, for all $Q_r(u,v) \in \mathcal{Q}$ and all $t \in \mathbb{N}$,

$$\left|d(Q_r(u,v)) - \hat{d}(Q_r(u,v),t)\right| < 2^{-t} \tag{4}$$

and $\hat{d}(Q_r(u,v),t)$ is computable in time $2^{2^{O(r+t)}}$.

A set E has *double exponential time measure 0* (or ee-*measure 0*) if there is some double exponential time computable martingale d such that $\#E \subseteq S^{\infty}[d]$. This type of *resource-bounded measure* was originally used to quantify the size of complexity classes [14] but is also useful for classifying geometric sets of Lebesgue measure 0. A point $(x,y) \in \mathbb{R}^2$ is *double exponential time random* (or ee-*random*) if it does not belong to any set of ee-measure 0.

2.3 Open Set Martingales

We will make essential use of a class of martingales defined in [16] which, for some given set E, for each dyadic square Q, allocate value to the four immediate dyadic subsquares of Q proportional to the measure of each subsquare's intersection with E. Formally, for any set $E \subseteq [0,1)^2$, the martingale $d_E : \mathcal{Q} \to [0,\infty)$ is defined as follows.

- The initial capital is

$$d_E(Q_0(0,0)) = \begin{cases} 0 & \lambda(E) = 0 \\ 1 & \text{otherwise.} \end{cases}$$

- For all $r \in \mathbb{N}$, $u,v \in \{0,\ldots,2^r-1\}$ and $\alpha,\beta \in \{0,1\}$, the value

$$d_E(Q_{r+1}(2u+\alpha, 2v+\beta))$$

  is 0 if $d_E(Q_r(u,v)) = 0$ and is otherwise given by

$$4d_E(Q_r(u,v))\frac{\lambda(E \cap Q_{r+1}(2u+\alpha, 2v+\beta))}{\lambda(E \cap Q_r(u,v))}.$$

When E is an open set, d_E is called the *open set martingale* for E.

Theorem 2.1 (Lutz and Lutz [16]). *For every nonempty set E that is open as a subset of $[0,1)^2$ and every $(x,y) \in E$,*

$$d_E^{(r)}(x,y) = \frac{1}{\lambda(E)}$$

for all sufficiently large r.

3 Constructing the Set

We follow a Kakeya set construction from the textbook of Bishop and Peres [5], who attribute this construction to Bishop while also acknowledging an earlier, related construction by Sawyer [19]. The authors prove that the closure of the set has Lebesgue measure 0 and further that δ-neighborhoods of the set have Lebesgue measure $O\left(\frac{1}{-\log\delta}\right)$.

Here we modify that construction—in a manner loosely suggested by an exercise in [5]—to construct a set that contains full lines in all directions. In Sect. 7 we will show that this set has ee-measure 0.

Define the sequence $\{x_k\}_{k\in\mathbb{N}}$ of dyadic rationals as the concatenation, over all $\ell \in \mathbb{N}$, of the blocks

$$X_\ell = \left((-1)^\ell(i2^{-\ell} - \ell) : i = 0, \dots, 2^\ell(2\ell+1) - 1\right).$$

Thus for even ℓ the block X_ℓ consists of all ℓ-dyadic rationals in the range $[-\ell, \ell+1)$ in ascending order, and for odd ℓ the block X_ℓ consists of all ℓ-dyadic rationals in the range $(-\ell-1, \ell]$ in descending order.

Using this sequence, define the function $b : [0,1] \to \mathbb{R}$ by

$$b(a) = \sum_{n=1}^{\infty}(x_{n-1} - x_n)(a - 2^{-n}\lfloor a2^n \rfloor). \tag{5}$$

For each slope $a \in [0,1]$, our set F will contain the line with slope a and y-intercept $b(a)$: We define

$$F = \{(x, ax + b(a)) : a \in [0,1], x \in \mathbb{R}\}.$$

(See Figs. 1 and 2).

Fig. 1. The set $\{(a, b_8(a)) : a \in (2^{-8}\mathbb{Z}) \cap [0,1)\}$.

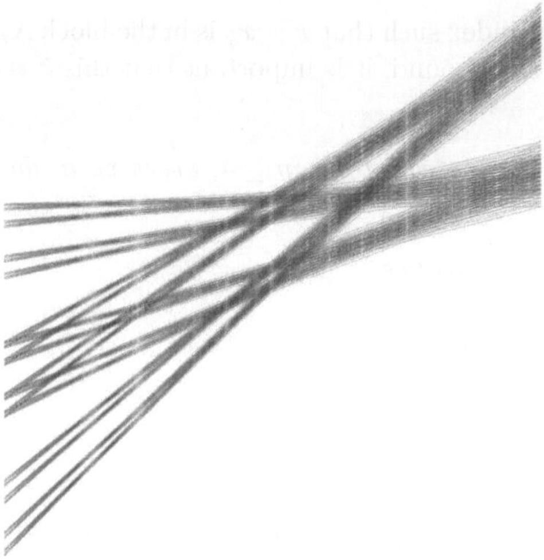

Fig. 2. The set $F_8 \cap ([-1,1] \times \mathbb{R})$.

4 Finitary Precursors to *F*

We now define and analyze finitary precursors to the set F, which will be used to build an effective martingale that succeeds everywhere in $\#F$, the set obtained by applying the $\#$ operator defined in (3) to F.

For each positive integer k, define the function $b_k : [0,1] \to \mathbb{R}$ by

$$b_k(a) = \sum_{n=1}^{k} (x_{n-1} - x_n)(a - 2^{-n}\lfloor 2^n a \rfloor), \tag{6}$$

and define the set

$$F_k = \{(x, ax + b_k(a)) : a \in (2^{-k}\mathbb{Z}) \cap [0,1), x \in \mathbb{R}\}.$$

Thus F_k is the union of 2^k lines, each of which has a slope and intercept that can be explicitly calculated in finite time.

For every set $E \subseteq \mathbb{R}^2$ and every $\delta > 0$, let $E[\delta]$ denote the open δ-neighborhood of E under the ℓ^∞ norm. That is, a point $(x,y) \in \mathbb{R}^2$ belongs to $E[\delta]$ if and only if there is some point $(x',y') \in E$ such that

$$\max\{|x - x'|, |y - y'|\} < \delta.$$

Using the ℓ^∞ norm instead of the ordinary Euclidean norm will slightly simplify our analysis, in part because neighborhoods of polygons under the ℓ^∞ norm are polygons.

For every $m \in \mathbb{N}$ and every m-dyadic rational $\hat{x} \in [-m, m]$, define the set

$$\tilde{F}_{m,\hat{x}} = (F_k \cap ([\hat{x}, \hat{x} + 2^{-m}) \times \mathbb{R}))[2^{1-m-k}],$$

where k is the unique index such that $\hat{x} = x_k$ is in the block X_m. For our eventual double exponential time bound, it is important that this k is only exponentially larger than m.

Observation 4.1. *For every integer $m \geq 4$, every x_k in the block X_m satisfies $k \leq 2^{2m}$.*

Proof. Every x_k in X_m satisfies

$$\begin{aligned} k &\leq \sum_{\ell=1}^{m} |X_\ell| \\ &= \sum_{\ell=1}^{m} (2^\ell(2\ell+1) - 1) \\ &= (2^{m+2} - 1)m - 2(2^m - 1) \\ &\leq 2^{2m}, \end{aligned}$$

where the last inequality holds whenever $m \geq 4$.

The set $\tilde{F}_{m,\hat{x}}$ can be obtained by taking the union of 2^k thin tubes, i.e., neighborhoods of lines, and then taking a thin vertical section of that union. As we now show, this vertical section must have small measure. Furthermore, within the scope of an even thinner vertical section, these 2^k tubes cover the union of the infinitely many lines in F.

Lemma 4.2. *For every $m \in \mathbb{N}$ and every m-dyadic rational $\hat{x} \in [-m, m]$, we have*

$$\lambda(\tilde{F}_{m,\hat{x}}) \leq 2^{2-2m} \tag{7}$$

and

$$F \cap ([\hat{x}, \hat{x} + 2^{-m}) \times \mathbb{R}) \subseteq \tilde{F}_{m,\hat{x}} \cap ([\hat{x}, \hat{x} + 2^{-m}) \times \mathbb{R}). \tag{8}$$

Proof. Fix any $m \in \mathbb{N}$ and $\hat{x} \in (2^{-m}\mathbb{Z}) \cap [-m, m]$. Letting

$$\delta = 2^{1-m-k},$$

observe that $\tilde{F}_{m,\hat{x}}$ can equivalently be written as

$$\bigcup_{a \in (2^{-k}\mathbb{Z}) \cap [0,1)} L_{a,b_k(a)}[\delta] \cap ([\hat{x}, \hat{x} + 2^{-m}) \times \mathbb{R})[\delta],$$

where $L_{a,b_k(a)}$ denotes the line of slope a and y-intercept $b_k(a)$. Each of these lines intersects

$$([\hat{x}, \hat{x} + 2^{-m}) \times \mathbb{R})[\delta]$$

in a segment of horizontal length $2^{-m} + 2\delta$, and each vertical slice of $L_{a,b_k(a)}[\delta]$ has length 2δ. Therefore the area of the intersection for each $L_{a,b_k(a)}$ is at most

$(2^{-m}+2\delta)\cdot 2\delta$. As there are 2^k lines, the area of the union of these intersections is at most

$$2^k \cdot (2^{-m}+2\delta)\cdot 2\delta \leq 2^{2-2m},$$

proving that (7) holds.

For (8), observe first that the function b_k defined in (6) closely approximates the function b defined in (5), in the sense that for all $a \in [0,1]$ we have

$$\begin{aligned} |b(a)-b_k(a)| &\leq \sum_{n=k+1}^{\infty} |x_{n-1}-x_n|(a-2^{-n}\lfloor 2^n a\rfloor) \\ &\leq 2^{-m}\sum_{n=k+1}^{\infty} 2^{-n} \\ &= 2^{-m-k}. \end{aligned}$$

The function $b_k(a)$ is piecewise linear and right-continuous, with slope $-x_k = -\hat{x}$ at all $a \in [0,1]\setminus 2^{-k}\mathbb{Z}$. Therefore for all $x \in [\hat{x}, \hat{x}+2^{-m})$, $ax+b_k(a)$ is also piecewise linear and right-continuous as a function of a, with slope of absolute value

$$\begin{aligned} \left|\frac{d}{da}(ax+b_k(a))\right| &= |x-\hat{x}| \\ &< 2^{-m} \end{aligned}$$

at all $a \in [0,1]\setminus 2^{-k}\mathbb{Z}$.

Let $a \in [0,1]$ and $x \in [\hat{x}, \hat{x}+2^{-m})$ be such that

$$(x, ax+b(a)) \in F,$$

and let $\hat{a} = 2^{-k}\lfloor a2^k\rfloor$. Then

$$(x, \hat{a}x+b_k(\hat{a})) \in F_k,$$

and

$$\begin{aligned} |ax+b(a)-(\hat{a}x+b_k(\hat{a}))| &\leq |b(a)-b_k(a)| + |ax+b_k(a)-(\hat{a}x+b_k(\hat{a}))| \\ &\leq 2^{-m-k}+2^{-m}|a-\hat{a}| \\ &< 2^{1-m-k}. \end{aligned}$$

The vertical slice at x of $\tilde{F}_{m,\hat{x}}$ includes a 2^{1-m-k} open neighborhood of $\hat{a}x+b_k(\hat{a})$. This shows that

$$(x, ax+b(a)) \in \tilde{F}_{m,\hat{x}} \cap ([\hat{x}, \hat{x}+2^{-m})\times\mathbb{R}).$$

As $x \in [\hat{x}, \hat{x}+2^{-m})$ was arbitrary, this completes the proof of (8) and of the lemma.

5 Martingales for Finitary Precursor Sets

5.1 Open Set Martingales for the $\Theta(2^{-m}) \times 1$ Scale

Lemma 4.2 tells us that F is covered by a collection of thin vertical sets, each of which has small measure. We now further divide these vertical sets into pieces of height 1, translate those pieces to the unit square $[0,1)^2$, and define an effective martingale corresponding to each such piece.

For each $m \in \mathbb{N}$, $\hat{x} \in (2^{-m}\mathbb{Z}) \cap [-m, m]$, and $j \in \mathbb{Z}$, let

$$\tilde{F}_{m,\hat{x},j} = [0,1)^2 \cap (\tilde{F}_{m,\hat{x}} - (\lfloor \hat{x} \rfloor, j)),$$

and let $\tilde{d}_{m,\hat{x},j}$ be the open set martingale (as defined in Sect. 2.3) for $\tilde{F}_{m,\hat{x},j}$.

Lemma 5.1. *For every $m \in \mathbb{N}$, $\hat{x} \in 2^{-m}\mathbb{Z} \cap [-m, m]$, $j \in \mathbb{Z}$, and $Q_r(u,v) \in \mathcal{Q}$, the martingale value*

$$\tilde{d}_{m,\hat{x},j}(Q_r(u,v))$$

is computable in $2^{2^{O(m+r)}}$ time.

Proof. Fix any integer $m \geq 4$, $\hat{x} \in 2^{-m}\mathbb{Z} \cap [-m, m]$, and $j \in \mathbb{Z}$. We will prove the lemma by induction on r. It is trivially true for $r = 0$, so fix $r \geq 0$ and assume that $\tilde{d}_{m,\hat{x},j}(Q)$ is computable in $2^{2^{O(m+r)}}$ time for each r-dyadic square $Q \in \mathcal{Q}$.

Let $Q_{r+1}(2u+\alpha, 2v+\beta)$, where $u, v \in \{0, \ldots, 2^r - 1\}$ and $\alpha, \beta \in \{0,1\}$, be any $(r+1)$-dyadic square. Then $\tilde{d}_{m,\hat{x},j}(Q_{r+1}(2u+\alpha, 2v+\beta))$ is given by either 0 or

$$4\tilde{d}_{m,\hat{x},j}(Q_r(u,v)) \frac{\lambda(\tilde{F}_{m,\hat{x},j} \cap Q_{r+1}(2u+\alpha, 2v+\beta))}{\lambda(\tilde{F}_{m,\hat{x},j} \cap Q_r(u,v))},$$

so it suffices to show that $\lambda(\tilde{F}_{m,\hat{x},j} \cap Q_{r+1}(2u+\alpha, 2v+\beta))$ and $\lambda(\tilde{F}_{m,\hat{x},j} \cap Q_r(u,v))$ can be found in $2^{2^{O(m+r)}}$ time.

Let k be the unique index such that $\hat{x} = x_k$ is in the block X_m, and let $\delta = 2^{1-m-k}$, noting that $k \leq 2^{2m}$ for all $m \geq 4$ by Observation 4.1. Then $\tilde{F}_{m,\hat{x},j}$ is the intersection with $[0,1)^2$ of a translated copy of $\tilde{F}_{m,\hat{x}}$, i.e., of

$$\bigcup_{a \in (2^{-k}\mathbb{Z}) \cap [0,1)} L_{a,b_k(a)}[\delta] \cap ([\hat{x}, \hat{x} + 2^{-m}) \times \mathbb{R})[\delta], \tag{9}$$

where $L_{a,b(a)}$ denotes the line with slope a and y-intercept $b(a)$. We can find the set (9) as follows. Initially set $S = \emptyset$. For $a \in (2^{-k}\mathbb{Z}) \cap [0,1)$, we do the following.

First find $b_k(a)$ by directly applying (6), which involves finding the sum of 2^k terms, each of which is an $m + k \leq (2k)$-dyadic number. This runs in poly(2^k) time.

Then set

$$S \leftarrow S \cup \left(L_{a,b_k(a)}[\delta] \cap ([\hat{x}, \hat{x} + 2^{-m}) \times \mathbb{R})[\delta]\right).$$

Each $L_{a,b_k(a)}[\delta] \cap ([\hat{x}, \hat{x}+2^{-m}) \times \mathbb{R})[\delta]$ is a parallelogram, and at all stages, each vertex of S is the intersection of segments on the boundaries of these 2^k parallelograms. Therefore S never has more than $\binom{4\cdot 2^k}{2} < 2^{4+2k}$ vertices. Standard algorithms from computational geometry can find the union of two polygons, each with at most N vertices, in poly(N) time; see Sect. 2.4 of [1]. Thus, each union operation takes poly(2^{4k+2}) = poly(2^k) time. In total, finding the polygon $\tilde{F}_{m,\hat{x}}$ takes 2^k(poly(2^k) + poly(2^k)) = poly(2^k) time.

Translating $\tilde{F}_{m,\hat{x}}$ and finding the area of the translated polygon's intersection with $Q_r(u,v)$ takes poly$(2^k + r)$ time, again using standard algorithms from computational geometry [1]. Therefore the denominator $\lambda(\tilde{F}_{m,\hat{x},j} \cap Q_r(u,v))$ can be found in poly$(2^k + r) \leq 2^{2^{O(m+r)}}$ time. For the same reason, the numerator $\lambda(\tilde{F}_{m,\hat{x},j} \cap Q_{r+1}(2u+\alpha, 2v+\beta))$ can be found in $2^{2^{O(m+r)}}$ time, completing the proof of the lemma.

5.2 Combining Martingales for the 1×1 Scale

We now average the 2^m martingales $\tilde{d}_{m,\hat{x},j}$ corresponding to each integer unit square piece of F to get an effective martingale corresponding to that whole 1×1 square.

For every $m \in \mathbb{N}$ and $i, j \in \mathbb{Z}$, let $d_{m,i,j} : \mathcal{Q} \to [0,\infty)$ be the martingale defined by

$$d_{m,i,j}(Q) = 2^{-m} \sum_{\hat{x}\in(2^{-m}\mathbb{Z})\cap[i,i+1)} \tilde{d}_{m,\hat{x},j}(Q). \tag{10}$$

By Lemma 5.1, $d_{m,i,j}(Q)$ is computable in

$$2^m \cdot 2^{2^{O(m+r)}} \leq 2^{2^{O(m+r)}}$$

time for each r dyadic square $Q \in \mathcal{Q}$.

Lemma 5.2. *Let $i, j \in \mathbb{Z}$ and $m \geq \max\{|i|+1, 4\}$. For every $(x,y) \in F \cap ([i,i+1) \times [j,j+1))$, for all sufficiently large $r \in \mathbb{N}$,*

$$d^{(r)}_{m,i,j}(x-i, y-j) \geq 2^{m-2}.$$

Proof. For each $\hat{x} \in (2^{-m}\mathbb{Z}) \cap [i, i+1)$,

$$\tilde{F}_{m,\hat{x},j} \subseteq \tilde{F}_{m,\hat{x}} - (i,j).$$

As

$$\begin{aligned}\lambda(\tilde{F}_{m,\hat{x}} - (i,j)) &= \lambda(\tilde{F}_{m,\hat{x}})\\ &\leq 2^{2-2m}\end{aligned}$$

by Lemma 4.2, Theorem 2.1 tells us that for all sufficiently large r and all $(x,y) \in F \cap ([\hat{x}, \hat{x}+2^{-m}) \times [j,j+1))$,

$$\tilde{d}^{(r)}_{m,\hat{x},j}(x-i, y-j) \geq 2^{2m-2}.$$

Therefore for all sufficiently large r and all $(x,y) \in F \cap ([i,i+1) \times [j,j+1)$, letting $\hat{x} = 2^{-m}\lfloor 2^m x \rfloor$, we have

$$\begin{aligned} d^{(r)}_{m,i,j}(x-i,y-j) &\geq 2^{-m}\tilde{d}^{(r)}_{m,\hat{x},j}(x-i,y-j) \\ &\geq 2^{-m} \cdot 2^{2m-2} \\ &= 2^{m-2}, \end{aligned}$$

which is the desired bound.

6 A Martingale for #*F*

In Sect. 5.2, we constructed, for each $m \in \mathbb{N}$ and $i,j \in \mathbb{Z}$, a martingale that attains value $\Omega(2^m)$ on the piece of F that lies in the unit square with lower left-hand corner (i,j) (after it has been translated). To attain unbounded value on all of F (again, after being translated), we now sum these martingales over all m, i, and j, with exponential discounting to ensure that the resulting sum is a martingale with finite initial capital.

Let $\pi : \mathbb{Z}^2 \to \mathbb{N}$ be any standard pairing function, and define the function $\hat{d} : \mathcal{Q} \times \mathbb{N} \to \mathbb{Q}$ by

$$\hat{d}(Q_r(u,v),t) = \sum_{m=0}^{2r+t+1} \sum_{\substack{i,j\in\mathbb{Z} \\ \pi(i,j)<m}} 2^{-m-\pi(i,j)} d_{m,i,j}(Q_r(u,v))$$

and the function $d : \mathcal{Q} \to [0,\infty)$ by

$$d(Q_r(u,v)) = \sum_{m\in\mathbb{N}} \sum_{\substack{i,j\in\mathbb{Z} \\ \pi(i,j)<m}} 2^{-m-\pi(i,j)} d_{m,i,j}(Q_r(u,v)).$$

Lemma 6.1. *The function d is a double exponential time computable martingale.*

Proof. The martingale property (1) is preserved by linear combinations, so to show that d is a martingale we only need to verify that its initial capital

$$d(Q_0(0,0))$$

is finite.

Each $\tilde{d}_{m,\hat{x},j}$ is an open set martingale, so each one's initial capital

$$\tilde{d}_{m,\hat{x},j}(Q_0(0,0))$$

is either 0 or 1. It is immediate from (10) that each $d_{m,i,j}(Q_0(0,0))$ is therefore at most 1. As π is a pairing function, we have

$$\begin{aligned}\sum_{\substack{i,j\in\mathbb{Z}\\ \pi(i,j)<m}} 2^{-\pi(i,j)} d_{m,i,j}(Q_0(0,0)) &\leq \sum_{\substack{i,j\in\mathbb{Z}\\ \pi(i,j)<m}} 2^{-\pi(i,j)}\\ &\leq \sum_{i,j\in\mathbb{Z}} 2^{-\pi(i,j)}\\ &= \sum_{p\in\mathbb{N}} 2^{-p}\\ &= 2\end{aligned}$$

for all m, so

$$\begin{aligned}d(Q_0(0,0)) &\leq \sum_{m\in\mathbb{N}} 2^{1-m}\\ &= 4.\end{aligned}$$

To prove that d is double exponential time computable, it suffices to show, for all $Q_r(u,v) \in \mathcal{Q}$ and all $t \in \mathbb{N}$, that

$$\left| d(Q_r(u,v)) - \hat{d}(Q_r(u,v),t) \right| < 2^{-t}$$

and $\hat{d}(Q_r(u,v),t)$ is computable in $2^{2^{O(r+t)}}$ time.

For the former condition, note that $d_{m,i,j}(Q_r(u,v)) \leq 4^r$ by (2). Therefore

$$\begin{aligned}\left| d(Q_r(u,v)) - \hat{d}(Q_r(u,v),t) \right| &\leq 4^r \sum_{m=2r+t+2}^{\infty} 2^{-m} \sum_{\substack{i,j\in\mathbb{Z}\\ \pi(i,j)<m}} 2^{-\pi(i,j)}\\ &= 2^{2r+1} \sum_{m=2r+t+2}^{\infty} 2^{-m}\\ &= 2^{2r+1} 2^{-2r-t-1}\\ &= 2^{-t}.\end{aligned}$$

To see that $\hat{d}(Q_r(u,v),t)$ is computable in $2^{2^{O(r+t)}}$ time, observe that computing each of the inner sums in $\hat{d}(Q_r(u,v),t)$ involves computing the martingale value $d_{m,i,j}(Q_r(u,v))$ for exactly m pairs $(i,j) \in \mathbb{Z}^2$. As noted in Sect. 5.2, it follows from Lemma 5.1 that each of these martingale values is computable in $2^{2^{O(r+t)}}$ time, so the time to compute m of them, scale each one by $2^{-\pi(i,j)}$, sum the results, and repeat this for $2r+t+2$ values of m, is still $2^{2^{O(r+t)}}$.

7 Main Theorem

Theorem 7.1. *There is a plane set of ee-measure 0 that contains lines in all directions.*

Proof. As the union of F with three rotated copies of F contains lines in all directions, it suffices to prove that F has ee-measure 0. By Lemma 6.1 and the definition of ee-measure 0 in Sect. 2.2, it suffices to show that $\#F \subseteq S^\infty[d]$.

To this end, let $(x', y') \in \#F$. Then there are some $i, j \in \mathbb{Z}$ such that

$$(x, y) := (x' + i, y' + j) \in F \cap ([i, i+1) \times [j, j+1)).$$

By Lemma 5.2, for all integers $m \geq |i| + 1$, for all sufficiently large $r \in \mathbb{N}$, we have

$$d^{(r)}_{m,i,j}(x', y') \geq 2^{m-2}.$$

Therefore

$$\begin{aligned}\limsup_{r\to\infty} d^{(r)}(x', y') &\geq \sum_{m=|i|+1}^{\infty} 2^{-m-\pi(i,j)} 2^{m-2} \\ &= \sum_{m=|i|+1}^{\infty} 2^{-\pi(i,j)-2} \\ &= \infty.\end{aligned}$$

As the double exponential time martingale d succeeds on every $(x', y') \in \#F$, we conclude that F has ee-measure 0.

By the definition of ee-randomness, the following corollary follows immediately.

Corollary 7.2. *For every direction $\theta \in [0, 2\pi)$, there is a line in $\mathbb{R}^2$ with direction θ containing no ee-random points.*

8 Conclusion

While we restricted our analysis to $\mathbb{R}^2$, higher-dimensional versions of Theorem 7.1 and Corollary 7.2 also hold. The union of $F \times \mathbb{R}^{n-2}$ with three rotated copies of itself contains lines in all directions in $\mathbb{R}^n$, and using the definitions and techniques of [16], it is routine to show that this product also has ee-measure 0.

It is natural to ask if it is possible to improve on our double exponential time resource bound. Indeed, [16] conjectured there are lines in every direction that miss every polynomial time random point. Our effectivization, like those in [16], relies on open set martingales and is fundamentally brute-force in character; our martingale d is estimated by a function $\hat{d}(Q, t)$ that exhaustively iterates through a very large number of polygons. Known lower bounds on the measure of neighborhoods of Kakeya sets [5,7,11] indicate that any Kakeya set with exponentially small area has a double exponential number of "pieces." This presents a barrier to satisfying (4) with fewer than double exponentially many pieces, suggesting that improvements on the resource bound—such as showing that a Kakeya set's lineal extension can miss every exponential time random point—would require a different approach that implicitly or explicitly handles many pieces of a Kakeya set in aggregate.

Acknowledgments. We thank Zheye Caroline Yao and an anonymous reviewer for helpful corrections.

References

1. de Berg, M., Cheong, O., van Kreveld, M., Overmars, M.: Computational Geometry: Algorithms and Applications, 3rd edn. Springer-Verlag, Santa Clara, CA, USA (2008). https://doi.org/10.5555/1370949
2. Besicovitch, A.S.: Sur deux questions d'intégrabilité des fonctions. Journal de la Société de physique et de mathématique de l'Université de Perm **2**, 105–123 (1919)
3. Besicovitch, A.S.: On Kakeya's problem and a similar one. Math. Z. **27**, 312–320 (1928). https://doi.org/10.1007/BF01171101
4. Besicovitch, A.S.: On fundamental geometric properties of plane line sets. J. Lond. Math. Soc. **39**, 441–448 (1964). https://doi.org/10.1112/jlms/s1-39.1.441
5. Bishop, C.J., Peres, Y.: Fractals in Probability and Analysis. Cambridge University Press, Cambridge (2017). https://doi.org/10.1017/9781316460238
6. Bushling, R.E.G., Fiedler, J.B.: Bounds on the dimension of lineal extensions. J. Fractal Geometry **12**(1/2), 105–133 (2025). https://doi.org/10.4171/JFG/161
7. Cordóba, A.: The fat needle problem. Bull. Lond. Math. Soc. **25**(1), 81–82 (1993). https://doi.org/10.1112/blms/25.1.81
8. Davies, R.O.: Some remarks on the Kakeya problem. Math. Proc. Cambridge Philos. Soc. **69**, 417–421 (1971). https://doi.org/10.1017/S0305004100046867
9. Fiedler, J.B.: The information content of points on lines and k-plane extensions (2025). https://arxiv.org/abs/2510.11645
10. Fiedler, J.B.: On the packing dimension of unions and extensions of k-planes. J. Geom. Anal. **36**(5), 185 (2026). https://doi.org/10.1007/s12220-026-02441-w
11. Keich, U.: On L^p bounds for Kakeya maximal functions and the Minkowski dimension in $\mathbb{R}^2$. Bull. Lond. Math. Soc. **31**(2), 213–221 (1999). https://doi.org/10.1112/S0024609398005372
12. Keleti, T.: Are lines much bigger than line segments? Proc. Amer. Math. Soc. **144**, 1535–1541 (2016). https://doi.org/10.1090/proc/12978
13. Keleti, T., Máthé, A.: Equivalences between different forms of the Kakeya conjecture and duality of Hausdorff and packing dimensions for additive complements (2023). https://arxiv.org/abs/2203.15731
14. Lutz, J.H.: Almost everywhere high nonuniform complexity. J. Comput. Syst. Sci. **44**(2), 220–258 (1992). https://doi.org/10.1016/0022-0000(92)90020-J
15. Lutz, J.H., Lutz, N.: Algorithmic information, plane Kakeya sets, and conditional dimension. ACM Trans. Comput. Theory **10** (2018). https://doi.org/10.1145/3201783
16. Lutz, J.H., Lutz, N.: Lines missing every random point. Computability **4**(2), 85–102 (2015). https://doi.org/10.3233/COM-150038
17. Lutz, N., Stull, D.: Bounding the dimension of points on a line. Inf. Comput. **275** (2020). https://doi.org/10.1016/j.ic.2020.104601
18. Lutz, N., Stull, D.: Dimension spectra of lines. Computability **11**(2), 85–112 (2022). https://doi.org/10.3233/COM-190292
19. Sawyer, E.: Families of plane curves having translates in a set of measure zero. Mathematika **34**(1), 69–76 (1987). https://doi.org/10.1112/S0025579300013292

20. Stull, D.M.: The dimension spectrum conjecture for planar lines. J. Lond. Math. Soc. **111**(6), e70216 (2025). https://doi.org/10.1112/jlms.70216
21. Ville, J.: Étude Critique de la Notion de Collectif. Gauthier-Villars, Paris (1939)
22. Wang, H., Zahl, J.: Volume estimates for unions of convex sets, and the Kakeya set conjecture in three dimensions (2025). https://arxiv.org/abs/2502.17655

Self-assembly of Strings and Languages Revisited: Efficient Membership Algorithms

Katalin Anna Lázár[1], Florin Manea[2], Stefan Siemer[2], and Timo Specht[2](✉)

[1] ELTE Eötvös Loránd University, Budapest 1117, Hungary
lazarkati@inf.elte.hu
[2] University of Göttingen, 37077 Göttingen, Germany
{florin.manea,stefan.siemer}@cs.uni-goettingen.de, timo.specht@uni-goettingen.de

Abstract. The self-assembly operation $\mathcal{S}$, or self-assembly for short, is a bio-inspired formal string operation that takes two strings xy and yz, with a nonempty common overlap, y, as arguments and combines them to obtain the string xyz. The binary string operation $\mathcal{S}$ can be canonically extended to a unary operation applied to languages. For a language L, the set $\mathcal{S}(L)$ is the set of all possible strings that can be obtained by applying the self-assembly operation, $\mathcal{S}$, to pairs of strings from L. Despite being extensively studied from a formal language-theoretic perspective, the algorithmic properties of self-assembly remain largely unexplored. In this paper, we address the membership problem for languages obtained by iteratively applying $\mathcal{S}$ to an initial language L: Given a string w and a language L, determine whether $w \in \mathcal{S}^+(L)$. We present efficient algorithms, along with matching fine-grained lower bounds, for this problem for language classes in the Chomsky hierarchy.

1 Introduction

The string operation called self-assembly (of strings and languages, also known as overlap assembly) was introduced in [13]. As pointed out in the respective seminal work (and paraphrasing it), this string operation offers a framework that allows shorter strings (i.e., simpler objects) to aggregate into longer strings (i.e., larger structures). This replicates in a string-based framework the general principles of self-assembly, one of the important operations for nano-scale engineering. The bioinformatics-driven motivations for introducing and analyzing this formal string operation are thoroughly discussed in [13], as well as in a series of subsequent papers [15,16], to which we refer the interested reader.

The self-assembly operation $\mathcal{S}$, or self-assembly for short, is a bio-inspired formal string operation that takes two strings $u_1 = xy$ and $u_2 = yz$, with a nonempty common overlap, y, as arguments and combines them to obtain the string xyz. If there are multiple overlaps between u_1 and u_2, the operation

V. Brattka et al. (Eds.): CiE 2026, LNCS 16674, pp. 389–404, 2027.
https://doi.org/10.1007/978-3-032-31348-5_26

yields a distinct string for each overlap. The resulting set of strings is denoted by $\mathcal{S}(u_1, u_2)$. Note that the operation is not commutative; i.e., $\mathcal{S}(u_1, u_2)$ and $\mathcal{S}(u_2, u_1)$ are not necessarily equal. The binary string operation $\mathcal{S}$ extends naturally to languages. For a language L, the set $\mathcal{S}(L)$ can be defined as the set of all strings obtained by applying the self-assembly operation, $\mathcal{S}$, to pairs of strings from L. This extension allows us to consider the transitive closure $\mathcal{S}^+$, as originally proposed in [13], and supports the analysis of the behaviour of the operation on languages.

Since the introduction of the self-assembly operation [13], numerous results regarding its language- and complexity-theoretic properties have been established [9,12,15,16]. The seminal paper [13] presents foundational results regarding the generative capacity of self-assembly. Specifically, it addresses the classes of languages that can be obtained through (iterated) self-assembly when starting from regular, context-free, or context-sensitive languages. The language-theoretic properties, such as closure properties, of the language families produced by iterated self-assembly are also investigated, and the sets of strings that can be generated from a given base set are characterized.

Subsequent studies, including [15,16], further examine the closure properties of fundamental language families (including certain complexity classes) under this operation and the complexity of several decision problems of language-theoretic relevance. From an applied standpoint, the potential use of iterated overlap assembly in constructing combinatorial DNA libraries has also been explored. Building on the observation in [13] that the $\mathcal{S}$-operation is regularity preserving, in [9] the state complexity of languages of $\mathcal{S}(L)$, for L regular, is analyzed. Further works, e.g., [12] explore this operation in the context of novel, bio-inspired computational models. Other closely related string operations have also been investigated in the literature, we recall here, among others, the word blending operation [5,17], the hairpin completion [20,24,27,28], hairpin reduction [26], hairpin lengthening [25], superposition [7], outfix/site-directed insertion [10,11]. Experimental gene-shuffling methodologies [29] have employed parallel overlap assembly of DNA strands mediated by DNA polymerase.

One algorithmic problem that has been thoroughly studied for many of the aforementioned string operations, but not yet for self-assembly, is the membership problem. Given a string w and a language L, is $w \in \mathcal{S}^+(L)$? While general results regarding the complexity class to which $\mathcal{S}^+(L)$ belongs based on the class of L can be derived from the literature, we are interested in fine-grained results here. More precisely, we provide optimal algorithms (along with corresponding lower bounds) for the membership problem when the languages considered are finite, regular (whether specified as deterministic or non-deterministic finite automata or regular expressions), and context-free. Interestingly, not only do we provide a complete picture of the membership problem for iterated self-assembly when the language classes belong to the Chomsky hierarchy, but we also deliver results on tight algorithmic upper and lower bounds for a general formal language theoretical problem. Given a language L from the aforementioned classes and a string w, does w contain a suffix that belongs to L?

The paper is organized as follows. We first provide an overview of the basic definitions, introduce the membership problem for self-assembly of languages, and present several preliminary results. Subsequent sections are structured by the class of the input language. First, we address finite languages, followed by regular languages specified via deterministic automata. Next, we cover regular languages given as nondeterministic automata with ε–transitions. Finally, we discuss context-free and context-sensitive languages. A concluding section summarizes our findings.

2 Preliminaries

Let $\mathbb{N} = \{1, 2, \ldots\}$ denote the natural numbers and $\mathbb{N}_0 = \mathbb{N} \cup \{0\}$ as well as $[n] = \{1, \ldots, n\}$ and $[i : n] = \{i, i+1, \ldots, n\}$, for all $i, n \in \mathbb{N}_0$ with $i \leq n$. Let Σ be a finite set of terminal symbols (*letters*). A *string* w is a finite sequence of letters where the total number of letters, the *length* of w, is denoted by $|w|$. The string with no letters is called the *empty string* ε with $|\varepsilon| = 0$. The set of all finite strings over the alphabet Σ, denoted by Σ^*, forms a free monoid $(\Sigma, \cdot, \varepsilon)$ generated by Σ under concatenation, with ε as the neutral element. The set of all nonempty strings over Σ is $\Sigma^+ = \Sigma^* \setminus \{\varepsilon\}$. A subset $L \subseteq \Sigma^*$ is called a *(formal) language*. Let Σ^n denote all strings in Σ^* of length $n \in \mathbb{N}_0$. For $1 \leq i \leq j \leq |w|$ let $w[i]$ denote the ith letter of w, whereas $w[i : j] = w[i] \cdots w[j]$ the *factor* of w starting at position i and ending at position j. If $i = 1$, the factor is a *prefix*; if $j = |w|$, then it is a *suffix* of w.

Definition 1 ([13]). *Let $u, v \in \Sigma^*$. The (self-)assembly operation $\mathcal{S}(u, v)$ is defined as $\mathcal{S}(u, v) = \{u'wv' \in \Sigma^+ \mid u = u'w, v = wv', w \neq \varepsilon\}$. Note that if one of u or v is empty, then $\mathcal{S}(u, v) = \emptyset$.*

For two languages $L_1, L_2 \subseteq \Sigma^*$, let $\mathcal{S}(L_1, L_2) = \bigcup_{w_1 \in L_1, w_2 \in L_2} \mathcal{S}(w_1, w_2)$.

Further, we define $\mathcal{S}^n(L)$ for $L \subseteq \Sigma^+$ and $n \geq 1$ as $\mathcal{S}^n(L) = \mathcal{S}(L, \mathcal{S}^{n-1}(L))$, where $\mathcal{S}^0(L) = L$. The iterated assembly of L is defined as $\mathcal{S}^+(L) = \bigcup_{n \geq 1} \mathcal{S}^n(L)$.

If we denote $\mathcal{S}^*(L) = \bigcup_{n \geq 0} \mathcal{S}^n(L)$, we can observe that $\mathcal{S}^*(L) \setminus \mathcal{S}^+(L) \subseteq \{\varepsilon\}$. Indeed, if $\varepsilon \in L$, then when computing $\mathcal{S}(L)$, the only string of L that is not in $S(L)$ is ε, since it cannot be produced by any self-assembly operation. A language $L \subseteq \Sigma^*$ is closed under iterated assembly if $L = \mathcal{S}^*(L)$ (if $L \subseteq \Sigma^+$, it suffices to require that $L = \mathcal{S}^+(L)$). Since the distinction between $\mathcal{S}^+(L)$ and $\mathcal{S}^*(L)$ is irrelevant for the membership problem we consider in this paper, we will focus, from now on, on $\mathcal{S}^+(L)$ and languages L that do not contain ε. We establish a strengthened form of Lemma 1 from [13].

Lemma 1 ([13]). *For any language $L \subseteq \Sigma^+$, any integer $n \geq 1$, and any string $x \in \mathcal{S}^n(L) \setminus \mathcal{S}^{n-1}(L)$, there are strings $u_i \in L$ and $\alpha_j, \beta_k \in \Sigma^*$, $i \in [0, n]$, $j, k \in [1, n]$ such that:*

1. $x = \beta_1 \alpha_1 \beta_2 \alpha_2 \cdots \beta_n \alpha_n \beta_{n+1}$;
2. $u_i = \alpha_{i-1} \beta_i \alpha_i \in L$, *for all* $i \in [1, n+1]$;

3. $\alpha_0 = \alpha_{n+1} = \varepsilon, \alpha_i \neq \varepsilon$, *for all* $i \in [1, n]$, *and* $\beta_1, \beta_{n+1} \neq \varepsilon$.

Lemma 1 implies that $\mathcal{S}^n(L) = \mathcal{S}(\mathcal{S}^{n-1}(L), L)$, a fact that plays a key role in our algorithms.

We will briefly review several concepts from formal language theory. See [19] for further background.

Automata and Formal Languages. A ε-NFA is a 5-tuple $A = (Q, \Sigma, q_0, F, \delta)$ with a finite set of states Q, an input alphabet Σ, a transition function $\delta : Q \times (\Sigma \cup \{\varepsilon\}) \to 2^Q$, an initial state $q_0 \in Q$, and the set of accepting states $F \subseteq Q$. By $L(A)$ we denote the language accepted by A, and the total size of A, i.e., its number of transitions, by $|A|$. A DFA is a ε-NFA, with the restriction that $\delta : Q \times \Sigma \to Q$, it has a single initial state $q_0 \in Q$, and its set of accepting states is $F \subseteq Q$. Both ε-NFAs and DFAs accept exactly the class of regular languages, and each ε-NFA can be determinized, although this often leads to an exponential increase in size. Regular languages can also be described by regular expressions. A regular expression r can be converted in time $O(|r|)$ into an ε-NFA A accepting the language described by the regular expression r, of total size $|A| = O(|r|)$.

A context-free grammar is a 4-tuple $G = (V, \Sigma, P, I)$ consisting of a finite set V of nonterminal symbols, a finite set Σ of terminal symbols with $V \cap \Sigma = \emptyset$, a nonempty set $P \subseteq V \times (V \cup \Sigma)^*$ of productions, and a start symbol $I \in V$. We assume that every nonterminal and terminal appears in at least one production. The language generated by a context-free grammar G is denoted $L(G)$. A language L is context-free (for short, CFL) if there is a CFG G such that $L(G) = L$. The size of a grammar G, i.e., the total number of symbols occurring in its productions, is denoted by $|G|$. A CFG G is in Chomsky normal form (CNF) if and only if $P \subseteq \big(V \times (V^2 \cup \Sigma)\big) \cup \{I \to \varepsilon\}$. According to [23] and the references therein, any context-free grammar G can be converted into an equivalent G' CNF grammar in time $\mathcal{O}(|G|^2)$, where $|G'| \in \mathcal{O}(|G|^2)$ and $L(G) = L(G')$.

Computational Model. We assume the standard unit-cost word RAM with logarithmic word-size $\mathtt{w}$, where each memory word consists of $\mathtt{w} \geq \log \mathtt{n}$ bits. Standard arithmetical and bitwise operations (and indirect addressing) on such memory-words are assumed to take $\mathcal{O}(1)$ time [18].

Data Structures. The suffix array SA_w for a string w with $|w| = n$ is defined as $SA_w[i] = j$ if and only if $w[j : n]$ is the i-th suffix of w in lexicographic order. We may omit the subscript when the underlying string is clear from context.

Lemma 2 (**[22]**). *For a string w with $|w| = n$, the suffix array SA_w can be constructed in $\mathcal{O}(n)$ time.*

For an array $A[1 : n]$, with elements from a totally ordered set (e.g., integers), a range maximum query (RMQ), denoted $\mathtt{RMaxQ}_A(a, b)$, retrieves the position $j \in [n]$ if and only if $j \in [a : b]$, $A[j] > A[j']$ for all $j' \in [a : j-1]$, and $A[j] \geq A[j'']$ for all $j'' \in [j+1 : b]$. In other words, $A[j]$ is the maximum element in range $A[a : b]$, and j is the smallest position at which this maximum occurs in $A[a : b]$.

Lemma 3 **([6]).** *An array $A[1:n]$ can be preprocessed in $\mathcal{O}(n)$ time to answer range maximum queries in $\mathcal{O}(1)$ time.*

3 The Membership Problem

Let us define the self-assembly membership problem (SAM):

$\mathtt{SAM}(w, L)$
Input: Given a string w of length n over Σ, and a language $L \subseteq \Sigma^+$.
Task: Decide whether $w \in \mathcal{S}^+(L)$.

Remark 1. In this paper we consider $\mathtt{SAM}(\cdot, L)$ when L is finite, regular (given as a DFA or as NFA), context-free, or context-sensitive. Note that in all these cases, SAM is at least as hard as the membership problem for the respective classes of languages. Indeed, assume that \$ and # are two symbols not contained in Σ. Clearly, if L is finite, regular (given as a DFA or as NFA), context-free (given as a CFG), or context-sensitive (given as a context-sensitive grammar or linearly bounded automaton), then $\$L\#$ is from the same class as L and the representation of L can be trivially modified to obtain a representation of $\$L\#$ (without changing the size of the representation asymptotically). In this framework, solving $\mathtt{SAM}(\$w\#, \$L\#)$ is equivalent to testing whether $w \in L$. Indeed, because $\mathcal{S}^+(\$L\#) = \$L\#$, $w \in L$ if and only if $\$w\# \in \$L\#$ if and only if $\$w\# \in \mathcal{S}^+(\$L\#)$. Therefore, any lower bound for the complexity of the membership problem for a certain class of languages is also a lower bound for the complexity of SAM for the respective class.

As established in Lemma 1, $w \in \mathcal{S}^+(L)$ if and only if there exists a sequence of strings $(u_0, u_1, \dots, u_m)$, with $u_i \in L$ for all $i \in [m]$, such that $w = \mathcal{S}(\mathcal{S}(\dots\mathcal{S}(\mathcal{S}(u_0, u_1), u_2)\dots), u_m)$. We call the sequence $(u_0, u_1, \dots, u_m)$ an L-decomposition of w in L. The sequence of operations $w = \mathcal{S}(\mathcal{S}(\dots\mathcal{S}(\mathcal{S}(u_0, u_1), u_2)\ \dots), u_m)$ naturally induces a sequence $(i_0, i_1, \dots, i_m = n)$, where, for $t \in [m]$, i_t is defined such that, after the first $t \in [m]$ assembly steps in this derivation of w, the obtained string is exactly the prefix $w[1 : i_t]$. More precisely, w.r.t. the derivation $w = \mathcal{S}(\mathcal{S}(\dots\mathcal{S}(\mathcal{S}(u_0, u_1), u_2)\dots), u_m)$, we have:

- $u_0 = w[1 : i_0]$, where $i_0 = |u_0|$,
- $\mathcal{S}(\mathcal{S}(\dots\mathcal{S}(\mathcal{S}(u_0, u_1), u_2)\dots), u_t) = w[1 : i_t]$,
- $\mathcal{S}(\mathcal{S}(\dots\mathcal{S}(\mathcal{S}(u_0, u_1), u_2)\dots), u_m) = w[1 : i_m] = w$.

An L-decomposition $(u_0, u_1, \dots, u_m)$ of w is called greedy, if it satisfies the following conditions:

- u_0 is the longest prefix of w in L.

– For $t \in [0, m-1]$, let $w[1 : i_t] = \mathcal{S}(\mathcal{S}(\ldots \mathcal{S}(\mathcal{S}(u_0, u_1), u_2) \ldots), u_t)$ and $w[1 : i_{t+1}] = \mathcal{S}(\mathcal{S}(\ldots \mathcal{S}(\mathcal{S}(u_0, u_1), u_2) \ldots), u_{t+1})$. Then $w[1 : i_{t+1}]$ is the longest prefix of w that can be obtained as the result of applying $\mathcal{S}(w[1 : i_t], x)$, for $x \in L$. Moreover, u_{t+1} is the longest string in L such that $\mathcal{S}(w[1 : i_t], u_{t+1}) = w[1 : i_{t+1}]$.

In other words, every element of a greedy L-decomposition is chosen to maximize the extension of the current prefix. Among all such extensions, the admissible longest string is selected.

Lemma 4. *If an L-decomposition for w in L exists, then a greedy L-decomposition exists.*

By the definition of greedy decomposition, every string $w \in \mathcal{S}^+(L)$ has a unique greedy L-decomposition.

Lemma 5. *For any language L and string w, the greedy L-decomposition is unique.*

The following result shows that, for any string $w \in \mathcal{S}^+(L)$, the greedy L-decomposition is the shortest L-decomposition.

Lemma 6. *For L, w, and the greedy L-decomposition $(u_0, u_1, \ldots, u_m)$ of w, there is no L-decomposition $(v_0, v_1, \ldots, v_k)$ of w such that $k < m$.*

According to Lemma 4, it is sufficient to consider the computation of a greedy L-decomposition. Moreover, by Lemma 1 and Lemma 4, if a string w has a greedy L-decomposition with n elements, then $w \in \mathcal{S}^n(L) \setminus \mathcal{S}^{n-1}(L)$. The data structure introduced below plays a central role in our results.

Definition 2. *The array $MaxExt_{w,L}$ stores for each index i in w, the largest index j such that the factor $w[i : j]$ belongs to L. If no such factor exists, the value is -1. Formally, for $i \in [1, n]$:*

ïż£

$$\text{MaxExt}_{w,L}[i] = \begin{cases} \max\{j \mid i \leq j \leq n, w[i : j] \in L\}, & \text{if such } j \text{ exists,} \\ -1, & \text{otherwise.} \end{cases}$$

Lemma 7. *Given the precomputed array $MaxExt_{w,L}$ for language L and string w with $|w| = n$, then $\texttt{SAM}(w, L)$ can be solved in $\mathcal{O}(n)$ time.*

Proof. For brevity, we omit the subscripts w and L when referring to $\text{MaxExt}_{w,L}$. Using Lemma 3, we construct, in linear time, the range maximum query data structures $\texttt{RMaxQ}_{\text{MaxExt}}$. This allows us to retrieve the answer to $\texttt{RMaxQ}_{\text{MaxExt}}(a, b)$-queries, for $a, b \in [n]$, in constant time.

According to Lemma 4, it is sufficient to decide the existence of a greedy L-decomposition $g_0, g_1, \ldots g_m$ of w. We define $g_0 = w[1 : \text{MaxExt}[1]]$, and set $i = \text{MaxExt}[1]$, if $\text{MaxExt}[1] \neq -1$. We also set $\ell = 1$. Then, while $i \neq -1$ and $i \neq n$, we proceed as follows. We set $j = \texttt{RMaxQ}_{\text{MaxExt}}(1, i)$ and if $\text{MaxExt}[j] > i$, we set $g_\ell = w[j : \text{MaxExt}[j]]$, $i = \text{MaxExt}[j]$, and $\ell = \ell + 1$, and iterate. Otherwise, we set $i = -1$ and iterate.

The idea implemented by this algorithm is to simulate the definition of the greedy L-decomposition of w. More precisely, given the prefix $w[1 : i]$ covered by the elements $g_0, \ldots, g_\ell$ of the greedy decomposition, the next element of the respective decomposition is chosen to extend the covered prefix as much as possible. In other words, we look for the position j between 1 and i, for which MaxExt$[j]$ is maximum. If MaxExt$[j] > i$, then using the string $w[j :$ MaxExt$[j]]$ maximally extends the current prefix $w[1 : i]$. Using `RMaxQ` to retrieve this position allows us to also have the minimum j with these properties, and ensures, in this way, the fact that the the string we choose for this extension is the longest among all that extend $w[1 : i]$ maximally. We define $g_{\ell+1}$ to be $w[j : \text{MaxExt}[j]]$ and iterate. If MaxExt$[j] <= i$, then no extension is possible, and no greedy L-decomposition can be computed.

The time needed to implement this algorithm is clearly $\mathcal{O}(n)$, under the assumption that we are provided the array MaxExt. □

Lemma 7 shows that we can compute a greedy L-decomposition of a string w in linear time, once MaxExt$[\cdot]$ is available. Consequently, it suffices to focus on computing the array $\text{MaxExt}_{w,L}$ efficiently. We discuss how this is done (optimally) for various classes of formal languages.

Finite Languages. In this section, we discuss the problem `SAM`(w, L) when L is a finite language. For the sake of generality, assume that L is given as a finite list of strings $L = \langle w_1, \ldots, w_k \rangle$, which may contain duplicates.

Lemma 8. *Given a finite language $L = \langle w_1, \ldots, w_k \rangle$ and a string w with $|w| = n$, the array $MaxExt_{w,L}$ can be computed in $\mathcal{O}(n + \sum_{i=1}^{k} |w_i|)$ time.*

Proof. To begin with, we construct the suffix array $SA_w[\cdot]$ of w (see [22]).

Firstly, we can process the language $L = \langle w_1, \ldots, w_k \rangle$ in time $\mathcal{O}(\sum_{i=1}^{k} |w_i|)$ to eliminate duplicate strings from this language. By the definition of the suffix array, note that all suffixes of w that start with the string w_i, for every $i \in [k]$, correspond to a range $SA_w[a_{w_i} : b_{w_i}]$ of the respective suffix array. Another property of the suffix array is that w_i is a prefix of w_j, for some $i, j \in [k]$, if and only if $a_{w_i} \leq a_{w_j} \leq b_{w_j} \leq b_{w_i}$; in other words, the range $SA_w[a_{w_i} : b_{w_i}]$ completely includes the range $SA_w[a_{w_j} : b_{w_j}]$, inducing a sort of nesting relationship between these ranges. This is, indeed, true as every suffix of w which starts with w_j also starts with w_i.

Furthermore, for w and $w_1, \ldots, w_k$, we compute in total $\mathcal{O}(n + \sum_{i=1}^{k} |w_i|)$ time the ranges $SA_w[a_{w_i} : b_{w_i}]$ containing, for each $i \in [k]$, exactly the positions j such that $w[j : j + |w_i| - 1] = w_i$.

We construct a list $\mathcal{L} = \{(a_{w_i}, b_{w_i}, |w_i|) \mid i \in k\}$. We use radix sort [2] to sort this list in ascending order by the first component. To break ties, we implement the radix sort procedure in such a way that, if for some i and j we have that $a_{w_i} = a_{w_j}$ and $b_{w_i} \geq b_{w_j}$, then $(a_{w_i}, b_{w_i}, |w_i|)$ comes before $(a_{w_j}, b_{w_j}, |w_j|)$ in the sorted list. Due to the properties of the suffix array stated above, for some i and j such that $a_{w_i} < a_{w_j}$ (meaning that $(a_{w_i}, b_{w_i}, |w_i|)$ occurs before $(a_{w_j}, b_{w_j}, |w_j|)$ in the sorted list) and $b_{w_i} \geq a_{w_j}$, we have $b_{w_i} \geq b_{w_j}$. From now on, we assume that the list $\mathcal{L}$ is sorted according to the above-defined criteria. Since all elements of the triples stored in $\mathcal{L}$ are integers from $[n]$, using radix sort to sort this list takes $\mathcal{O}(n+k) \subseteq \mathcal{O}(n+\sum_{i=1}^{k}|w_i|)$ time.

Recall that we want to compute, for every $i \in [n]$, the value MaxExt$[i]$, which is the end position of the longest string from L that occurs at position i in w. This implies that $\text{MaxExt}[i] = i-1+\max\{|w_j| \mid j \in [k] \text{ and } i \in [a_{w_j} : b_{w_j}]\}$. Thus, for every i, we need to compute the triple $(a_{w_j}, b_{w_j}, |w_j|)$ such that $i \in [a_{w_i} : b_{w_i}]$ and $|w_j|$ is maximum, w.r.t. all other triples $(a_{w_f}, b_{w_f}, |w_f|)$ such that $i \in [a_{w_f} : b_{w_f}]$. A first consequence of this observation is that, we can once more go through $\mathcal{L}$ and if it contains multiple triples $(a, b, \ell_0), \ldots, (a, b, \ell_t)$ (i.e., triples that share the first two components), we only keep from all these the one with the largest third component. This traversal of $\mathcal{L}$ can be implemented in $O(k) \subseteq \mathcal{O}(\sum_{i=1}^{k}|w_i|)$, since the list was sorted. Note that, at this point, if the range $SA_w[a_{w_j} : b_{w_j}]$ is strictly included in the range $SA_w[a_{w_i} : b_{w_i}]$ then $(a_{w_i}, b_{w_i}, |w_i|)$ occurs before $(a_{w_j}, b_{w_j}, |w_j|)$ in the sorted list $\mathcal{L}$.

To compute $\text{MaxExt}[i] = i-1+\max\{|w_j| \mid j \in [k] \text{ and } i \in [a_{w_j} : b_{w_j}]\}$ it is enough to traverse the string w from left to right (i.e., consider i from 1 to n) while keeping track of the ranges that cover position i, via a simultaneous traversal of $\mathcal{L}$. This allows us to define $\text{MaxExt}[i] = i-1+|w_{j_i}|$, where the range $[a_{w_{j_i}} : b_{w_{j_i}}]$ is the smallest range (w.r.t. set inclusion) from our list that includes i. This process can be implemented in $\mathcal{O}(n)$ time. □

From Lemma 8 and 7 we obtain the following:

Theorem 1. *Given a finite language* $L = \langle w_1, \ldots, w_k \rangle$ *and a string* w *with* $|w| = n$, $\texttt{SAM}(w, L)$ *can be solved in* $\mathcal{O}(n+\sum_{i=1}^{k}|w_i|)$ *time.*

Regular Languages Given as DFA.

Lemma 9. *Given a DFA* $A = (\Sigma, Q, \delta, q_0, F)$ *and a string* w *with* $|w| = n$, *the array* $MaxExt_{w,L(A)}$ *can be computed in* $\mathcal{O}(n \cdot |Q|)$ *time.*

Proof. We begin by constructing a directed acyclic graph (DAG, for short) $G = (V, E)$ from w and A (also called the product graph for w and A), where $V = \{(q, i) \mid q \in Q, i \in [n+1]\}$ and $E = \{((q, i), (q', i+1)) \mid q' = \delta(q, w[i]), i \in [n]\}$. Given that G is a DAG and A is a DFA (so (q, i) may only have at most one outgoing edge), we can compute in total time $\mathcal{O}(n|Q|)$, for every node (q, i) the length $d(q, i)$ of the longest path from (q, i) to a node (q', j), such that $q' \in F$ (or $d(q, i) = -\infty$ if there is no such path). Then, in $\mathcal{O}(n)$ time, we can set, for all $i \in [n]$, $\text{MaxExt}[i] = \max\{-1, i+d(q_0, i)-1\}$. □

From Lemma 9 and 7 we obtain the following result:

Theorem 2. *Given a DFA* $A = (\Sigma, Q, \delta, q_0, F)$ *and a string* w *with* $|w| = n$, $\texttt{SAM}(w, L(A))$ *can be solved in* $\mathcal{O}(n \cdot |Q|)$ *time.*

§ *Conditional Lower Bound.* Let us first recall the OV problem [31].

Orthogonal Vectors: OV(U, V)
Input: Two sets $U, V \subseteq \{0,1\}^d$ of n vectors, where $d \in \omega(\log n)$.
Task: Do there exist vectors $u \in U, v \in V$, such that $\sum_{i=1}^{d} u[k] \cdot v[k] = 0$, i.e., for all $1 \leq k \leq d$, $u[k] \cdot v[k] = 0$ holds?

This problem can be solved naively in $\mathcal{O}(n^2 d)$, however, there is the following conditional lower bound.

Theorem 3 ([31]). *There is no $\mathcal{O}(n^{2-\varepsilon} d^c)$ algorithm for any $\varepsilon > 0$ and constant c solving* OV, *unless the Strong Exponential Time Hypothesis (SETH) fails.*

Consequently, the *OV conjecture* states that, for any $\varepsilon > 0$ and constant c, there is no $\mathcal{O}(n^{2-\varepsilon} d^c)$ algorithm solving OV. We can show the following result.

Theorem 4. SAM($w, L(A)$), *for input DFA $A = (\Sigma, Q, \delta, q_0, F)$ and string w with $|w| = n$, cannot be solved in $\mathcal{O}((n \cdot |Q|)^{1-\varepsilon})$ time for any $\varepsilon > 0$, unless the* OV *conjecture fails (which would falsify SETH, as well).*

Proof. We reduce an OV instance $U, V \subseteq \{0,1\}^d$ with n vectors in $\mathcal{O}(nd)$ time to an equivalent instance for SAM($w, L(A)$) where $A = (\Sigma, Q, \delta, q_0, F)$ is a DFA having $\mathcal{O}(nd)$ states with $|\Sigma| = \mathcal{O}(1)$, and $w \in \Sigma^*$ with $|w| = \mathcal{O}(nd)$. This reduction shows that if an $\mathcal{O}((n \cdot |Q|)^{1-\varepsilon})$ algorithm exists for SAM($w, L(A)$), then we solve OV in time $\mathcal{O}(nd + ((nd)(nd))^{1-\varepsilon}) = \mathcal{O}(n^{2-2\varepsilon} poly(d))$, hence refuting the OV conjecture. We show how to construct such an instance by providing gadgets for encoding vector coordinates, entire vectors, and then the two vector sets. The set U is encoded using the string w and the set V is encoded using the DFA A.

§ *Vector coordinate gadgets.* For a vector $u_i \in U$ we construct the following strings for all coordinates $u_i[k]$ with $k \in [d]$.

$$CW(u_i[k]) = \begin{cases} \text{“0”, if } u_i[k] = 0 \\ \text{“1”, if } u_i[k] = 1 \end{cases}$$

For a vector $v_j \in V$ we construct the two DFAs depicted in Fig. 1 for all coordinates $v_j[k]$ with $k \in [d]$.

$$CA(v_j[k]) = (\{0,1\}, \{q_{j,k}, p_{j,k}, e_j\}, \delta, q_{j,k}, \{p_{j,k}\})$$

$$\delta = \begin{cases} \{\delta(q_{j,k}, 1) = p_{j,k}, \delta(q_{j,k}, 0) = p_{j,k}\} & \text{, if } v_j[k] = 0 \\ \{\delta(q_{j,k}, 0) = p_{j,k}, \delta(q_{j,k}, 1) = e_j\} & \text{, if } v_j[k] = 1 \end{cases}$$

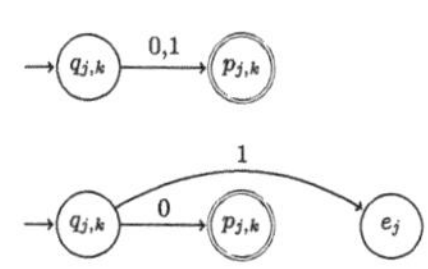

Fig. 1. Vector coordinate gadgets, transition functions (left) and graphs (right; $CA(v_j[k])$ for $v_j[k] = 0$ is pictured above, and $CA(v_j[k])$ for $v_j[k] = 1$ below).

Clearly, $CA(v_j[k])$ accepts both strings 0 and 1, for $v_j[k] = 0$, and $CA(v_j[k])$ accepts only the string 0, for $v_j[k] = 1$. We observe that $CW(u_i[k])$ is accepted by $CA(v_j[k])$ if and only if $u_i[k]v_j[k] = 0$. It is worth noting that all the DFAs $CA(v_j[k])$, for $k \in [d]$, share a special state, e_j. The gadgets constructed in this step (both strings and DFAs) have $\mathcal{O}(1)$ size.

§ *Vector gadgets.* For a vector $u_i \in U$ we construct the following string by concatenating the encodings of its coordinates.

$$VW(u_i) = CW(u_i[1]) \cdot CW(u_i[2]) \cdot \ldots \cdot CW(u_i[d])$$

For a vector $v_j \in V$, we construct the DFA

$$VA(v_j) = CA(v_j[1]) \circ CA(v_j[2]) \circ \ldots \circ CA(v_j[d])$$

depicted in Fig. 2 using a specialized version of DFA concatenation, denoted by $\circ$, outputting DFAs by the fact that there are no transitions leaving the singular final state of any DFA $CA(v_j[k])$. As noted when constructing the vector coordinate gadgets, all DFAs $CA(v_j[1]), CA(v_j[2]), \ldots, CA(v_j[d])$ share the special state e_j.

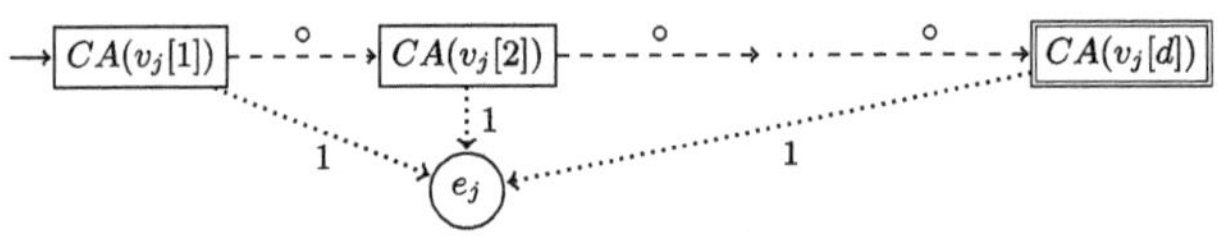

Fig. 2. Vector gadgets. The dotted lines indicate potential transitions, depending on the coordinate gadget.

Clearly, the constructed DFA $VA(v_j)$ accepts the concatenation of the languages $L(CA(v_j[1])), \ldots, L(CA(v_j[d]))$. To this end, we note that $VW(u_i)$ is accepted by $VA(v_j)$ if and only if every two components $u_i[k]$ and $v_j[k]$, for $k \in [d]$, are orthogonal, meaning that u_i and v_j are orthogonal. The run of $VA(v_j)$ on $VW(u_i)$ ends in state e_j if u_i and v_j are not orthogonal. The gadgets constructed in this step are of size $\mathcal{O}(d)$.

§ *Vector set gadget.* For the set of vectors U we construct a string w which starts with a fresh symbol B followed by two copies of $VW(u_i)$ for $i \in [n]$ and ends with a new symbol E. More precisely w looks as follows.

$$w = B(\#VW(u_1)\#VW(u_2)\#\ldots\#VW(u_n))^2\#E$$

For the set of vectors V, we construct the union-DFA $A = A_1 \cup A_2$, where the transition graph of A_1 is given in Fig. 3 and the one of A_2 in Fig. 4. More precisely, the DFA A_1 accepts all strings starting with B, ending with $\#$, and containing only $\{0, 1, \#\}$ in between the first and the last symbol. The DFA A_2 accepts strings starting with $\#$ and ending with E, which contain at least one substring $\#x\#$ with $x \in L(VA(v_j))$, for some $j \in [n]$. The construction of A_2 is such that we have a new initial state s, in which we can read $\#$ (all other transitions will lead to an error state g, not depicted in the figure). With this transition we move to $VA(v_1)$. Then, for $i \in [n]$, from $VA(v_i)$, if we reach the final state of the respective DFA $VA(v_i)$, we move with # to the state o, in which we then read the rest of the input string and move to the final state f by reading the symbol E (this case corresponds to finding in the input string a substring $\#x\#$ with $x \in L(VA(v_i))$). If, in $VA(v_i)$, we reach the state e_i, we read the input string until we read a # symbol, and, with this symbol, we move to the initial state of $VA(v_{i+1})$, and continue. Once A_2 is constructed, s will be its only initial state, and f its only final state.

Next, we create the union of A_1 and A_2, by merging their initial states s_0 and s into a single state s. This is possible as these two states have no common outgoing transitions, and may therefore be merged into a single common initial state s. All yet-undefined transitions are defined to lead to the error state g.

The string w is of length $\mathcal{O}(nd)$ and A consists of $\mathcal{O}(nd)$ states and, because the alphabet $\Sigma = \{0, 1, \#, B, E\}$ is of constant cardinality, of $\mathcal{O}(nd)$ transitions. This concludes our reduction, which can be implemented in $\mathcal{O}(nd)$ time.

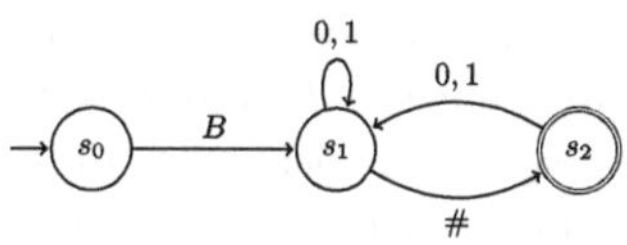

Fig. 3. A_1 accepts any nontrivial prefix of w that starts with B and ends with #. A_1 does not accept strings starting with # and ending with E.

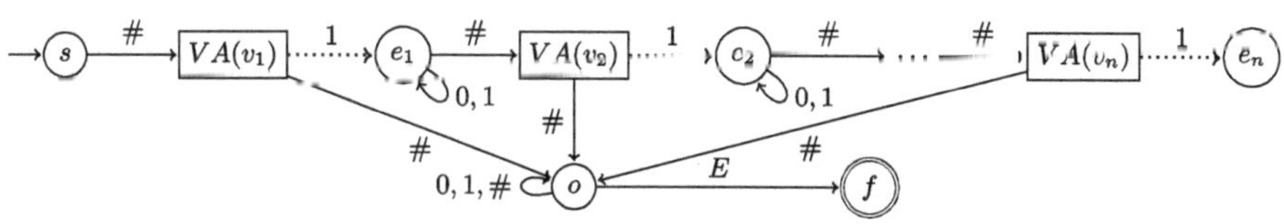

Fig. 4. A_2 accepts nontrivial suffixes of strings from $L(A_1)$ that begin with # and contain any substring $VW(u_i)$ accepted by some $VA(v_j)$ (i.e., v_j is orthogonal to u_i). The strings accepted by A_2 end with E and do not start with B.

We show that the reduction is correct. More precisely, we show that there exist $u_i \in U$ and $v_j \in V$ such that u_i is orthogonal to v_j if and only if $\mathtt{SAM}(w, L(A))$ is decided positively.

($\Rightarrow$) : Let us assume there exists $u_i \in U$ and $v_j \in V$ such that u_i is orthogonal to v_j. We show existence of strings $s_1 \in L(A_1)$ and $s_2 \in L(A_2)$ such that $\mathcal{S}(s_1, s_2) = w$. We distinguish two cases and provide the run of s_2 in A_2 under the string.

In the first case, when $i \leq j$, we have:

$$s_1 = B\#VW(u_1)\#\ldots\#VW(u_{n-(j-i)})\#\ldots\#VW(u_1)\#\ldots\#VW(u_i)\#\ldots\#VW(u_n)\#$$
$$w = B\#VW(u_1)\#\ldots\#VW(u_{n-(j-i)})\#\ldots\#VW(u_1)\#\ldots\#VW(u_i)\#\ldots\#VW(u_n)\#E$$
$$s_2 = \underbrace{\#VW(u_{n-(j-i)})\#\ldots\#VW(u_1)\#\ldots}_{s\ldots e_{j-1}}\underbrace{\#VW(u_i)}_{VA(v_j)}\underbrace{\#\ldots\#VW(u_n)\#E}_{o}$$

In the second case, when $i > j$, we have:

$$s_1 = B\#VW(u_1)\#\ldots\#VW(u_{i-j})\#\ldots\#VW(u_i)\#\ldots\#VW(u_1)\#\ldots\#VW(u_n)\#$$
$$w = B\#VW(u_1)\#\ldots\#VW(u_{i-j})\#\ldots\#VW(u_i)\#\ldots\#VW(u_1)\#\ldots\#VW(u_n)\#E$$
$$s_2 = \underbrace{\#VW(u_{i-j})\#\ldots}_{s\ldots e_{j-1}}\underbrace{\#VW(u_i)}_{VA(v_j)}\underbrace{\#\ldots\#VW(u_1)\#\ldots\#VW(u_n)\#E}_{o}$$

In conclusion, if there exist $u_i \in U$ and $v_j \in V$ such that u_i is orthogonal to v_j then $\texttt{SAM}(w, L(A))$ is decided positively.
$(\Leftarrow)$: Assume $\texttt{SAM}(w, L(A))$ returns a positive answer. It is not hard to show that $\mathcal{S}^+(L) = \mathcal{S}(L)$ (due to the presence of B and E symbols), so there must exist $s_1, s_2 \in L(A)$ such that $\mathcal{S}(s_1, s_2) = w$. Clearly, s_1 must be from $L(A_1)$ because w starts with a B and s_2 must be from $L(A_2)$ because w ends with an E. Further, all strings from $L(A_2)$ must contain a substring $\#VW(u_i)\#$ that is accepted by one of the $VA(v_j)$ gadgets to complete a run through A_2, ending in the final state. Therefore, $w = \mathcal{S}(s_1, s_2)$ must contain such a substring $\#VW(u_i)\#$ as well. Hence, there exists vectors u_i, v_j that are orthogonal. $\square$

Regular Languages Given as NFA.

Lemma 10. *Given an ε-NFA $A = (\Sigma, Q, \delta, q_0, f)$ and a string w with $|w| = n$, the array $MaxExt_{w,L(A)}$ can be computed in $\mathcal{O}(n \cdot |Q|)$ time.*

Proof. We can assume, w.l.o.g., that A has exactly one final state f and we can also modify the ε−NFA A, in $\mathcal{O}(|A|)$ time, and obtain a new automaton A' which does not have any cycles consisting only of ε-transitions, while accepting the same language as A. For that, we simply collapse to single states the strongly connected components of the graph consisting of the ε-transitions of A. Moreover, we topologically sort (see, e.g., [21]) the directed acyclic graph D whose vertices are the states of A and whose directed edges are the ε-transitions of A; this induces an ordering $\pi = (q_1 < \ldots < q_p)$ of the states of A (where $p = |Q|$) that will be used in the following. All these preprocessing steps can be implemented in $\mathcal{O}(|A|)$ time. From now on, we assume that the input ε-NFA A has no cycles consisting only of ε-transitions and its states are ordered w.r.t. π.

In our main algorithm, we construct the product graph G, for the input automaton A and the input string w. The vertices of this graph are tuples (q, i), with $q \in Q$ and $i \in [1, n+1]$; for simplicity, we say, for $i \in [n+1]$, that the nodes $\{(q, i) \mid q \in Q\}$ are on level i of G. To define the edges of G, let us consider an arbitrary vertex (q, i) of G and define the edges leaving this vertex. If $\delta(q, w[i]) \neq \emptyset$, we connect (q, i) with $(q', i+1)$ for every $q' \in \delta(q, w[i])$. Moreover, if $\delta(q, \varepsilon) \neq \emptyset$, we connect (q, i) with (q', i) for every $q' \in \delta(q, \varepsilon)$. The main difference in this construction is that, compared to the deterministic case from Lemma 9, there may be more edges leaving a node both towards nodes on the next level and towards nodes on the same level. Note also that the nodes on level i are canonically ordered according to the ordering π of the states of Q.

We define the function $d : Q \times [n+1] \to \{-\infty\} \cup [n+1]$, where $d(q, i)$ is the length ℓ of the longest prefix of $w[i : n]$ such that there is a path labeled with $w[i : i+\ell-1]$ going from state q to f in the ε-NFA A (or $-\infty$ if there is no such prefix). Note that this path may also include ε-transitions, and $d(q, i)$ counts only the non-ε-transitions along it. The values $d(q, i)$ can be computed by dynamic programming, in $\mathcal{O}(n|A|)$ time, by considering the values i from n to 1, and, for each i, the states q in the order given by the topological sorting π. Once the values of d are computed, we set $\text{MaxExt}[i] = \max\{-1, i + d(q_0, i) - 1\}$, for $i \in [n]$. The whole algorithm (including the preprocessing) runs in $\mathcal{O}(n|A|)$ time. □

Lemma 10 and 7 yield the following theorem.

Theorem 5. *Given an ε-NFA $A = (\Sigma, Q, \delta, q_0, f)$ and a string w with $|w| = n$, $\texttt{SAM}(w, L(A))$ can be solved in $\mathcal{O}(n \cdot |A|)$ time.*

Recall that any truly subquadratic algorithm, running in $\mathcal{O}((|w| \cdot |A|)^{1-\varepsilon})$, for the regular-language membership problem, given as ε-NFA A, with input string w, would refute the strong exponential time hypothesis (SETH) [3,8]. By Remark 1, this lower bounds also applies to $\texttt{SAM}(w, A)$, thus, an algorithm solving this problem in $\mathcal{O}((|w| \cdot |A|)^{1-\varepsilon})$ time would likewise contradict SETH. Hence, the algorithm in the previous theorem is conditionally optimal under SETH.

Context-Free Languages. In the following, let ω be the matrix multiplication exponent, i.e., ω is the smallest number such that computing the product of two $n \times n$ matrices is done in $\mathcal{O}(n^\omega)$ time; it is known that $2.3716 \geq \omega \geq 2$.

Lemma 11. *Given a CFG $G = (V, T, P, I)$ in CNF and a string w with $|w| = n$, the array $MaxExt_{w,L(G)}$ can be computed in $\mathcal{O}(\min\{|w|^\omega \cdot |G|^2, |w|^3 \cdot |G|\})$ time.*

Proof. (Sketch). We use Valiant's algorithm [30], which runs in $\mathcal{O}(|w|^\omega \cdot |G|^2)$ time, or the Cocke-Younger-Kasami algorithm (CYK for short, see [19]), which runs in $\mathcal{O}(|w|^3 \cdot |G|)$ time, to check for all $i, j \in [n]$ and $A \in V$ whether $w[i : j]$ can be generated by A. Then, $\text{MaxExt}_{w,L(G)}[i] = max\{j \mid S \text{ generates } w[i : j]\}$. □

The next result follows from Lemma 11 and 7.

Theorem 6. *Given a CFG $G = (V, T, P, I)$ in CNF and string w, $\texttt{SAM}(w, L(G))$ can be solved in $\mathcal{O}(\min\{|w|^\omega \cdot |G|^2, |w|^3 \cdot |G|\})$ time.*

In a combinatorial sense, the classical Cocke-Younger-Kasami algorithm falls within the framework of [14], where the intermediate computations have natural combinatorial interpretations in terms of the underlying problem. In line with [4], it can also be viewed as an algorithm that avoids oracle calls to fast ring matrix multiplication or other algebraic techniques. Therefore, applying the CYK algorithm in Theorem 6 yields a combinatorial algorithm for SAM. It has been conjectured that no truly subcubic combinatorial algorithm exists for Boolean matrix multiplication [14,32]. Consequently, Valiant's algorithm, whose performance relies on Boolean matrix multiplication, cannot asymptotically outperform CYK when restricted to the combinatorial setting.

Regarding lower bounds, [1] shows that any algorithm for the context-free language membership problem (with grammars in CNF) running in $\mathcal{O}(f(|G|)n^{\omega-\varepsilon})$, for some $\varepsilon > 0$ and a computable function f, would yield a breakthrough algorithm for the k-Clique problem (which asks whether a graph on n vertices contains a clique of size k; see also [31] for the *k-Clique hypothesis*). Likewise, the existence of a combinatorial algorithm with running time $\mathcal{O}(f(|G|)n^{3-\varepsilon})$, would imply a truly faster combinatorial algorithm for the k-Clique problem than the current state-of-the-art, which is conjectured to be optimal. These conditional lower bounds extend to SAM. Consequently, assuming the optimality of the best known (respectively, combinatorial) algorithms for k-Clique, equivalently, assuming the k-Clique hypothesis [31], the SAM for context-free languages, cannot be solved in $\mathcal{O}(|w|^{\omega-\varepsilon} \cdot f(|G|))$ time (respectively, $\mathcal{O}(|w|^{3-\varepsilon} \cdot f(|G|))$ time), for any $\varepsilon > 0$ and computable function f.

Finally, $\mathtt{SAM}(w, L)$ is PSPACE-complete if L is context-sensitive (given either as a grammar or a linearly-bounded Turing machine).

4 Conclusions

In this paper, we have presented efficient algorithms for solving $\mathtt{SAM}(w, L)$, for the cases when L is a finite language, a regular language given by a DFA or NFA (or regular expression), or a context-free language specified by a CNF grammar. By Lemma 6 and 7, our algorithms also canonically compute the minimum ℓ such that $w \in \mathcal{S}^{\ell}(L)$, which naturally serves as a distance measure between the language L and the string w, with respect to the $\mathcal{S}$ operation.

We have also proved that these algorithms are optimal. For finite languages, optimality is immediate. For context-free languages, optimality holds under the k-Clique hypothesis. For regular languages, whether given as DFAs or NFAs, our algorithms are optimal under the Strong Exponential Time Hypothesis (SETH). Notably, for finite languages, NFAs (or regular expressions), and context-free grammars, our running times match the complexity of the standard membership problem for the corresponding language class, and are therefore (conditionally) optimal by Remark 1.

The most striking case arises when the input given as a string w of length n and DFA with m states. In this case, SAM can be solved in $\mathcal{O}(nm)$ time, and—conditional on SETH—no algorithm running in $\mathcal{O}((nm)^{1-\varepsilon})$ time. This contrasts sharply with the classical membership problem for DFAs, which is solvable in linear time $O(n)$. In fact, Theorem 4 demonstrates an even stronger statement: deciding whether $w \in S(L)$ (or equivalently, whether w can be derived in a single $\mathcal{S}$-step) cannot be done in $\mathcal{O}((nm)^{1-\varepsilon})$.

Moreover, the reduction underlying Theorem 4 reveals an even more interesting result: assuming SETH, there is no $\mathcal{O}((nm)^{1-\varepsilon})$ algorithm for deciding whether a string w of length n has a suffix in a regular language L specified by a DFA A with m states. This is particularly remarkable because, for finite languages, NFAs, and context-free grammars, the analogous problem is as hard as the membership problem for the respective language class, yet this correspondence breaks down for DFAs.

Acknowledgements. Florin Manea, Stefan Siemer, and Timo Specht are supported by the German Research Foundation (Deutsche Forschungsgemeinschaft, DFG) through the Heisenberg project 466789228 and research project 562495345.

References

1. Abboud, A., Backurs, A., Vassilevska Williams, V.: If the current clique algorithms are optimal, so is Valiant's parser. SIAM J. Comput. **47**(6), 2527–2555 (2018)
2. Aho, A.V., Hopcroft, J.E., Ullman, J.D.: The Design and Analysis of Computer Algorithms. Addison-Wesley (1974)
3. Backurs, A., Indyk, P.: Which regular expression patterns are hard to match? In: FOCS, pp. 457–466 (2016)
4. Bansal, N., Williams, R.: Regularity lemmas and combinatorial algorithms. Theory Comput. **8**(1), 69–94 (2012)
5. Bellamoli, F., Franco, G., Kari, L., Lampis, S., Ng, T., Wang, Z.: Conjugate word blending: formal model and experimental implementation by XPCR. Nat. Comput. **20**(4), 647–658 (2021)
6. Bender, M.A., Farach-Colton, M.: The LCA problem revisited. In: Gonnet, G.H., Viola, A. (eds.) LATIN 2000. LNCS, vol. 1776, pp. 88–94. Springer, Heidelberg (2000). https://doi.org/10.1007/10719839_9
7. Bottoni, P., Labella, A., Manca, V., Mitrana, V.: Superposition based on Watson-Crick-like complementarity. Theory Comput. Syst. **39**(4), 503–524 (2006)
8. Bringmann, K., Grønlund, A., Künnemann, M., Larsen, K.G.: The NFA acceptance hypothesis: non-combinatorial and dynamic lower bounds. TheoretiCS **3** (2024)
9. Brzozowski, J.A., Kari, L., Li, B., Szykuła, M.: State complexity of overlap assembly. Int. J. Found. Comput. Sci. **31**(8), 1113–1132 (2020)
10. Cho, D.J., Han, Y.S., Ng, T., Salomaa, K.: Outfix-guided insertion. Theor. Comput. Sci. **701**, 70–84 (2017)
11. Cho, D.-J., Han, Y.-S., Salomaa, K., Smith, T.J.: Site-directed insertion: decision problems, maximality and minimality. In: Konstantinidis, S., Pighizzini, G. (eds.) DCFS 2018. LNCS, vol. 10952, pp. 49–61. Springer, Cham (2018). https://doi.org/10.1007/978-3-319-94631-3_5
12. Csuhaj-Varjú, E., Lázár, K.A.: String assembly in networks of evolutionary processors. J. Autom. Lang. Comb. **21**(1–2), 41–54 (2016)
13. Csuhaj-Varjú, E., Petre, I., Vaszil, G.: Self-assembly of strings and languages. Theor. Comput. Sci. **374**(1), 74–81 (2007)

14. Das, D., Koucký, M., Saks, M.E.: Lower bounds for combinatorial algorithms for boolean matrix multiplication. In: STACS. LIPIcs, vol. 96, pp. 23:1–23:14 (2018)
15. Enaganti, S.K., Ibarra, O.H., Kari, L., Kopecki, S.: Further remarks on DNA overlap assembly. Inf. Comput. **253**(1), 143–154 (2017)
16. Enaganti, S.K., Ibarra, O.H., Kari, L., Kopecki, S.: On the overlap assembly of strings and languages. Nat. Comput. 1–11 (2017)
17. Enaganti, S.K., Kari, L., Ng, T., Wang, Z.: Word blending in formal languages. Fundam. Inform. **171**(1–4), 151–173 (2020)
18. Fredman, M.L., Willard, D.E.: BLASTING through the information theoretic barrier with FUSION TREES. In: STOC, pp. 1–7 (1990)
19. Hopcroft, J.E., Ullman, J.D.: Introduction to Automata Theory, Languages and Computation. Addison-Wesley (1979)
20. Ito, M., Leupold, P., Manea, F., Mitrana, V.: Bounded hairpin completion. Inf. Comput. **209**(1–2), 471–485 (2011)
21. Kahn, A.B.: Topological sorting of large networks. Commun. ACM **5**(11), 558–562 (1962)
22. Kärkkäinen, J., Sanders, P., Burkhardt, S.: Linear work suffix array construction. J. ACM **53**(6), 918–936 (2006)
23. Lange, M., Leiß, H.: To CNF or not to CNF? An efficient yet presentable version of the CYK algorithm. Informatica Didact **8** (2009)
24. Manea, F., Martín-Vide, C., Mitrana, V.: On some algorithmic problems regarding the hairpin completion. Discrete Appl. Math. **157**(9), 2143–2152 (2009)
25. Manea, F., Martín-Vide, C., Mitrana, V.: Hairpin lengthening: language theoretic and algorithmic results. J. Log. Comput. **25**(4), 987–1009 (2015)
26. Manea, F., Mitrana, V., Yokomori, T.: Two complementary operations inspired by the DNA hairpin formation: completion and reduction. Theor. Comput. Sci. **410**(4–5), 417–425 (2009)
27. Manea, F., Mitrana, V., Yokomori, T.: Some remarks on the hairpin completion. Int. J. Found. Comput. Sci. **21**(5), 859–872 (2010)
28. Păun, G., Rozenberg, G., Yokomori, T.: Hairpin languages. Int. J. Found. Comput. Sci. **12**, 837–847 (2001)
29. Stemmer, W.P.: DNA shuffling by random fragmentation and reassembly: in vitro recombination for molecular evolution. Proc. Natl. Acad. Sci. **91**(22), 10747–10751 (1994)
30. Valiant, L.G.: General context-free recognition in less than cubic time. J. Comput. Syst. Sci. **10**(2), 308–315 (1975)
31. Vassilevska Williams, V.: On some fine-grained questions in algorithms and complexity. In: ICM, pp. 3447–3487. World Scientific (2018)
32. Vassilevska Williams, V., Williams, R.: Subcubic equivalences between path, matrix and triangle problems. In: FOCS, pp. 645–654 (2010)

Efficiently Finding All Shortest Absent Subsequences in a String

Florin Manea(✉), Tina Ringleb, Stefan Siemer, and Maximilian Winkler

Department of Computer Science, University of Göttingen, Göttingen, Germany
{florin.manea,tina.ringleb,stefan.siemer, maximilian.winkler}@cs.uni-goettingen.de

Abstract. For a string w, another string v is said to be a subsequence of w if v can be obtained from w by removing some of its letters; on the other hand, v is called an absent subsequence of w if v is not a subsequence of w. The existing literature on absent subsequences focused on understanding, for a string w, the set of its shortest absent subsequences (i.e., the shortest strings which are absent subsequences of w) and that of its minimal absent subsequences (i.e., those strings which are absent subsequences of w but whose every proper subsequence occurs in w). In this paper, we present the first optimal algorithm (with linear time preprocessing and constant delay) for the enumeration of the shortest absent subsequences, based on a novel, succinct graph representation of these objects.

1 Introduction

A string v of length ℓ is a *subsequence* of a string w if there exist (possibly empty) strings $x_1, \ldots, x_{\ell+1}$ and letters $v_1, \ldots, v_\ell$ such that $v = v_1 \cdots v_\ell$ and $w = x_1 v_1 \cdots x_\ell v_\ell x_{\ell+1}$. That is, v is obtained from w by removing some of its letters. Any other string u, which is not a subsequence of w, is an *absent subsequence* of w.

The concept of subsequence (also called scattered factor, or subword) appears and plays important roles in many different areas of computer science, such as combinatorics on strings [36,38], formal languages [40,41,24,2,14] and automata theory [22,45]. Moreover, subsequences are important objects of study in the area of algorithm-design. On the one hand, they appear in string-algorithmic problems, such as computing the longest common subsequences or shortest common supersequences (see the survey [8] and the references therein) or testing the Simon's congruence of strings (see [42,20] and the references therein). On the other hand, the combinatorial and algorithmic study of subsequences (or patterns) occurring in permutations (such as, e.g., longest increasing subsequences [16]) is an important and classical topic of research (see, e.g., [25,37]). For a more detailed discussion regarding combinatorial pattern matching problems related to subsequences see the survey [31]. Moreover, algorithmic problems on subsequences were revisited in the context of fine-grained complexity [11,1], and

V. Brattka et al. (Eds.): CiE 2026, LNCS 16674, pp. 405–420, 2027.
https://doi.org/10.1007/978-3-032-31348-5_27

lower bounds for their complexity were shown. Subsequences also appear prominently in applied settings: for modeling concurrency [35,39], in database theory, in settings related to event stream processing [21,46], in data mining [32], or in problems related to bioinformatics [10].

This paper is centered around the notion of *absent subsequence*, introduced in [29]. The study of absent subsequences has several motivations. We first overview the theoretical motivations, rooted in combinatorics. The notion of absent subsequences is related to the congruence $\sim_k$ (called Simon's Congruence, see, e.g., [40,34]) between strings having exactly the same set of subsequences of length at most k. The problems of deciding whether two given strings are $\sim_k$-equivalent, for a given k, and to find the largest k such that two given strings are $\sim_k$-equivalent (and their applications) were heavily investigated in the literature, with optimal solutions for both these problems being obtained recently [20]. One of the central ingredients in this context was the notion of a distinguishing string, i.e., the shortest subsequence present in one string and *absent* from the other. Further, one of the most important notions, from a combinatorial perspective, in the study of subsequences occurring in a string is that of subsequence-universality. A string w is said to be k-universal, for some positive integer k, if all strings of length k are subsequences of w, and the universality index of w is the largest k for which w is k-universal. Computing the universality index k of a string is equivalent to computing the minimum number $k+1$ for which there is a string of length $k+1$ which is an absent subsequence of w. To this end, the study of m-nearly k-universal strings, where one is considering strings which contain as subsequences all strings of length k except exactly m such strings (which are, therefore, absent), was pursued in [15,3]. On a related note, absent subsequences also naturally occur in the study of patterns avoided by permutations [25], with the important difference that a permutation is essentially a string whose letters are pairwise distinct. From a more applied perspective, subsequences are used to model sequences of events occurring in a string or a stream, and, as such, are relevant to the area of complex event processing and recognition (see, e.g., [26,27,18] and the references therein); consequently, absent subsequences correspond, similarly, to sequences of events avoided by the respective string or stream. On a related note, if a string or stream formalizes the trace of some computation [35,39], absent subsequences correspond to sequences of events avoided by the respective computation. Lastly, we should recall the study of *missing substrings* in strings, where the focus is on the set of strings which do not occur contiguously in w. These missing factors have practical applications (see [12] and the references therein) but are also theoretically interesting (see, e.g., [9,6,19)].

Two classes of absent subsequences were investigated in the seminal paper [29] and subsequent works [30,43,3]. On the one hand, the class of *shortest absent subsequences* (SAS, for short) of a string w: the absent subsequences of minimal length of w. On the other hand, the class of *minimal absent subsequences* (MAS, for short) of a string w: the absent subsequences whose every proper subsequence is a subsequence of w. In [29,30,3], combinatorial characterisations of the sets of SAS and MAS of a string, as well as compact representations of these sets, and tight bounds on their cardinality, were shown. In [29,30,43], optimal algorithms testing if a string is a shortest or minimal absent subsequence in a string were given, alongside efficient (yet not optimal) algorithms for counting and enumerating them. Understanding the *set of shortest absent subsequences of some string* w (denoted $\mathrm{SAS}(w)$) seems to play an important role in almost all contexts where absent subsequences appear. For instance, their length allows us to compute the universality index of a given string w, and the set $\mathrm{SAS}(w)$ contains exactly the strings certifying why w has the respective universality index and not a larger one; to this end, they are central in the works related to m-nearly k-universal strings: the size of the set $\mathrm{SAS}(w)$ defines the parameter m for that string w. From an applied perspective, the SAS of a string or stream are the shortest subsequences of events avoided in that string/stream, and, thus, any string extending them is also avoided.

§ *Our Results.* In this context, we focus in this paper on efficiently enumerating the set $\mathrm{SAS}(w)$, for a given string w of length n, over an alphabet of size σ. We improve significantly a series of algorithmic results reported in the literature and give an optimal algorithm for the considered enumeration problem, having an $O(n)$ time preprocessing and constant delay; the fastest known algorithms solving this problem required $O(n\sigma)$ time preprocessing with output-linear delay (see [43], where the results of [30] were already improved). The main ingredient of our algorithm is a novel and efficient graph-theoretical implicit representation of the set of SAS, suitable both for efficient enumeration as well as counting; as such, one important result presented here is an efficient algorithm for enumerating all source-sink paths in directed acyclic graphs from a certain class, defined by a series of structural constraints. Coming back to the enumeration of SAS, the previous works [43, 30] constructed explicit representations of $\mathrm{SAS}(w)$ as directed acyclic graphs (or finite automata), and explored these graphs in a straightforward manner; as said, unlike those works, we propose an implicit, more succinct representation of $\mathrm{SAS}(w)$ and data-structures allowing us to efficiently construct this representation and enumerate the elements of the respective set. As a direct consequence of our result regarding the efficient enumeration of $\mathrm{SAS}(w)$, we can also retrieve efficiently only a fragment of the set $\mathrm{SAS}(w)$; indeed, as our enumeration has constant delay, if we only want to check if $\mathrm{SAS}(w)$ has at least m elements, and retrieve these elements, then we can stop the enumeration after $O(m)$ time. Secondly, the enumeration of the objects is done in a natural order: w.r.t. the currently enumerated absent subsequence s, the next absent subsequence is the remaining one that shares the longest common prefix with s, or one of them if there are more with this property. Finally, the used representation of $\mathrm{SAS}(w)$ allows us to also efficiently count the number of elements of $\mathrm{SAS}(w)$.

2 Preliminaries

Let $\mathbb{N}$ be the set of natural numbers. For $m, n \in \mathbb{N}$, we let $[m : n] = \{m, m + 1, \ldots, n\}$ and $[n] = [1 : n]$. An alphabet Σ is a non-empty finite set of symbols called *letters*. A *string* is a finite sequence of letters from Σ; for the rest of the paper, we assume that all strings we consider are over an alphabet $\Sigma = \{1, 2, \ldots, \sigma\}$. The set of all strings over Σ is denoted by Σ^*, and let $\Sigma^+ = \Sigma^* \setminus \{\varepsilon\}$, where ε is the empty string. The *length* of a string $w \in \Sigma^*$ is denoted by $|w|$; let $\Sigma^n = \{w \in \Sigma^* \mid |w| = n\}$. The ith letter of $w \in \Sigma^*$ is denoted by $w[i]$, for $i \in [|w|]$. For $m, n \in \mathbb{N}$, we let $w[m : n] = w[m]w[m+1] \cdots w[n]$ and $|w|_a = |\{i \in [|w|] \mid w[i] = a\}|$. A string $u = w[m : n]$ is a *factor* of w, and we have $w = xuy$ for some $x, y \in \Sigma^*$. If $x = \varepsilon$ (resp., $y = \varepsilon$), u is called a *prefix* (resp., *suffix*) of w. Let $\mathrm{alph}(w) = \{x \in \Sigma \mid |w|_x > 0\}$ be the smallest subset $S \subseteq \Sigma$ such that $w \in S^*$. For $i = 0$, we define $w^0 = \varepsilon$ and, for $i > 0$, we define $w^i = ww^{i-1}$. We can now recall the classical notion of subsequence.

Definition 1. *A string v is a* subsequence *of length $|v| = k$ of string w, where $|w| = n$, if there exist positions $1 \leq i_1 < \ldots < i_k \leq n$, such that $v = w[i_1] \cdots w[i_k]$. For $j \in [k]$, we say that $v[j]$ is* embedded *on position i_j of w.*

If v is a subsequence of w, with $|v| = k$ and $|w| = n$, the *canonical embedding* of v in w is defined as the sequence of positions $1 \leq i_1 < i_2 < \ldots < i_k \leq n$ such that i_1 is the leftmost occurrence of $v[1]$ in w, and, for $j \in [2 : k]$, i_j is the leftmost occurrence of $v[j]$ in $w[i_{j-1} + 1 : n]$. We recall the notion of k-universality of a string as introduced in [7]: a string w is *k-universal* if every string v of length $|v| \leq k$ over $\mathrm{alph}(w)$ appears as a subsequence of w. For a string w, its *universality index* $\iota(w)$ is the largest integer k such that w is k-universal. Clearly, if $\iota(w) = k$, then w is ℓ-universal for all $\ell \leq k$. Note that the universality index of a string w is always defined w.r.t. $\mathrm{alph}(w)$, the alphabet of the string w, see [7,13].

We also recall the *arch factorisation* [23]. For $w \in \Sigma^*$, with $\Sigma = \mathrm{alph}(w)$, the *arch factorisation* of w is defined as $w = \mathrm{ar}_w(1) \cdots \mathrm{ar}_w(k)\, \mathrm{r}(w)$ where for all $\ell \in [k]$ the last letter of $\mathrm{ar}_w(\ell)$ occurs exactly once in $\mathrm{ar}_w(\ell)$, each string $\mathrm{ar}_w(\ell)$ is 1-universal over Σ, and $alph(\mathrm{r}(w)) \subsetneq \Sigma$. The strings $\mathrm{ar}_w(\ell)$ are called *arches* of w, $\mathrm{r}(w)$ is called the *rest*. Note that every string has a unique arch factorisation and by definition each arch $\mathrm{ar}_w(\ell)$ from a string w is 1-universal; clearly, the number of arches in the arch factorisation of a string w is exactly $\iota(w)$. By an abuse of notation, we can write $i \in \mathrm{ar}_w(\ell)$ if i is a natural number such that $|\mathrm{ar}_w(1) \cdots \mathrm{ar}_w(\ell - 1)| < i \leq |\mathrm{ar}_w(1) \cdots \mathrm{ar}_w(\ell)|$, i.e., i is a position of w contained in the ℓth arch of w.

The main concepts discussed in this paper were introduced in [30]:

Definition 2 ([30]). *A string v is an* absent subsequence *of w if v is not a subsequence of w. An absent subsequence v of w is a* shortest absent subsequence *(for short,* SAS*) of w if $|v| \leq |v'|$ for any other absent subsequence v' of w. We denote the set of all* SASs *of w by* SAS(w).

As an extension of the previous definition, an absent subsequence v of w is a *minimal absent subsequence* (for short, MAS) of w if every proper subsequence of v is a subsequence of w. Any SAS of w has length $\iota(w) + 1$ (as w contains all strings of length $\iota(w)$ as subsequences, and at least one string of length $\iota(w)+1$ is not a subsequence of w). For a string w, with $|w| = n$ and $\iota(w) = k$, and v a SAS of w, let $i_1 < i_2 < \ldots < i_k$ be the canonical embedding of $v[1:k]$ in w. Then $i_\ell \in \mathrm{ar}_w(\ell)$, for all $\ell \in [k]$, and $v[k+1] \notin \mathrm{alph}(w[i_k+1:n])$, see [30]. Interestingly, every SAS is a MAS, but not vice versa. Moreover, the length of the shortest MAS of a string w is $\iota(w)+1$ (as the shortest MASs are the SASs), but it is not obvious which is the length of a longest MAS of w (except that it is at most $|w|+1$).

A series of data structures were defined in [30] for the efficient processing of the absent subsequences of a string. We recall some of them here. The array $\mathrm{dist}[\cdot]$, with n elements, is defined as $\mathrm{dist}[i] = \min\{|u| \mid u$ is an absent subsequence of $w[i:n]$, starting with $w[i]\}$, for $i \in [n]$. This array is computed in $O(n)$ time, see [30, Lemma 4.16]. The main property of this array, shown in [30], is the following: if $\iota(w) = k$, then, for $i \in \mathrm{ar}_w(\ell)$, $\ell \in [k]$, there exists a SAS v of w such that i is part of the canonical embedding of $v[1:k]$ in w if and only if $\mathrm{dist}[i] = k+1-(\ell-1) = k-\ell+2$ and i is the first occurrence of $w[i]$ in $\mathrm{ar}_w(\ell)$. Indeed, if we have a string of length ℓ which is canonically embedded in w, such that this embedding ends on position i, we can extend it and obtain a SAS (which has length $k+1$) if and only if there exists an absent subsequence of $w[i:n]$, starting with $w[i]$, of length $k-\ell+2$, i.e., $\mathrm{dist}[i] = k-\ell+2$. Going more into details, when constructing such a SAS, the first $(\ell-1)$ positions of the embedding correspond to (and are found in) the first $(\ell-1)$ arches of w and, in order to use i as the next position of the respective embedding, corresponding to arch ℓ, we need to still be able to find suitable positions for the remaining $k-\ell$ arches, as well as one last letter, that cannot be embedded in w, after the last position of the embedding; $\mathrm{dist}[i]$ allows us to efficiently test if this is the case for position i. We also use two $k \times \sigma$ matrices $\mathrm{firstInArch}[\cdot,\cdot]$ and $\mathrm{lastInArch}[\cdot,\cdot]$. For $\ell \in [k]$ and $a \in \Sigma$, $\mathrm{firstInArch}[\ell,a]$ is the leftmost position of $\mathrm{ar}_w(\ell)$ where a occurs and $\mathrm{lastInArch}[\ell,a]$ is the rightmost position of $\mathrm{ar}_w(\ell)$ where a occurs. To avoid confusions, for all ℓ and a, the values stored in $\mathrm{firstInArch}[\ell,a]$ and $\mathrm{lastInArch}[\ell,a]$ are positions of w (so, between 1 and $|w|$). These two arrays are computed in $O(n)$ time, see [30, Lemma 4.7].

Our algorithms work on the standard unit-cost word RAM with logarithmic word-size $\omega \geq \log n$ (where n is the size of the input), and assume an *integer input alphabet*, see the extended version of this paper [33].

Finally, we can show a preliminary lemma, used later in Sect. 3. Let T be a rooted tree, with n nodes and root r, with the edges oriented away from the root and with some of its nodes being coloured. The depth of a node y in tree T, denoted $\mathrm{depth}_T(y)$, is the length (number of edges) of the path from r to y.

Lemma 1. *We can preprocess T in $O(n)$ time such that* $\texttt{colour}(x,y)$ *queries can be answered in $O(1)$ time: for x, y nodes of T, such that x is on the path from r to y, return the coloured node $u \neq x$ of minimum depth on the path from x to y (or $\bot$, if no such node exists).*

3 Enumeration Algorithms for Paths in DAGs

We begin with a series of definitions related to directed acyclic graphs, which are important in our algorithms. A directed graph is a structure $G = (V, E)$, where V is the set of nodes (or vertices) and $E \subseteq V \times V$ is the set of edges. The direction of the edge $(x, y) \in E$ is said to be from x to y, and x is referred to as the origin of (x, y), while y is the target; we also say that y is a child of x. A (non-trivial) walk of length ℓ in the directed graph G is a sequence of $\ell - 1$ edges $(x_1, x_2), (x_2, x_3), \ldots, (x_{\ell-1}, x_\ell)$, where $\ell \geq 2$; $x_1, \ldots, x_\ell$ are the nodes on that walk, and the walk is called a path if the respective walk contains only pairwise distinct nodes. An $x - y$-path in G is a path connecting nodes x and y. A directed acyclic graph (DAG, for short) G is a directed graph that has no cycles (i.e., if there is an $x - y$-path, then there is no $y - x$-path in G); note that, for DAGs, the notions of path and walk coincide and there is only a finite set of paths between two given nodes.

§ *Enumeration Framework.* While enumeration problems and enumeration algorithms can be defined in a more general setting (see, e.g., [44]), here we focus on enumerating paths between two given nodes in a given DAG, to keep the presentation simple. Note that we show a bijection between the set SAS(w) and the source-to-sink paths in certain DAGs in Theorem 3, and explain how the SAS can be retrieved efficiently from the respective paths.

Let $G = (V, E)$ be a DAG with two nodes $x, y \in V$. We aim to design an algorithm which takes G, x, and y as input, and reports all $x - y$-paths in G, without repetitions. As typical in enumeration (see [28]), our algorithm maintains a data structure $\mathcal{S}$, storing a (succinct) representation of the currently enumerated path (as concatenation of default subpaths, to be explained later), allowing us to directly retrieve the actual path in time proportional to its length. Then, during the execution of the algorithm a procedure `report` is called, each time $\mathcal{S}$ contains the representation of a path which has not been enumerated yet; we say that, with this call of `report`, the current path is reported. Intuitively, after each path is computed, it is given as input (via `report`) to another process; this can either output it, or process it in some other way. The main motivation of this setting is that we do not want to mix the computation of the enumerated paths with the processing done to them after their reporting. Indeed, in a further processing step we might want to completely print these paths (which would inherently take time linear in the output), we might only want to output the changes applied on the previously enumerated path to get this one, instead of the full path descriptions (which would take as much time as is needed to write these changes), or we might assign each path a score and only output explicitly the ones with the highest score. Using the `report` function we can then pipeline the enumeration and further processing; the runtime of the later processing steps is irrespective of the time needed to compute and enumerate (that is, `report`) all $x - y$-paths.

The *enumeration problem* considered in our setting is therefore the following: Given a DAG $G = (V, E)$, and two nodes $x, y \in V$ of G, enumerate all $x - y$-

paths of G. An *enumeration algorithm* solving such an enumeration problem is an algorithm $\mathcal{A}$ that `reports` every $x-y$-path of G exactly once, and stops once the last path was `reported`.

To analyse the complexity of such an algorithm $\mathcal{A}$, we distinguish between two phases. The *preprocessing phase* consists of the computation done by the algorithm $\mathcal{A}$ before the first call of `report`; the *preprocessing time* is the time needed, in the worst case, by the respective computation, as a function in $|V|+|E|$ (i.e., the size of the DAG $G=(V,E)$). The rest of the computation is the actual *enumeration phase*; the *delay* is the time elapsed, in the worst case, expressed as a function in $|V|+|E|$, between two consecutive calls of `report` (as a side note, our algorithms spend at most $O(1)$ time between the last call of `report` and the end of the algorithm). The algorithm $\mathcal{A}$ is said to have *output-linear delay* if the delay between the calls to `report` corresponding to path s_{i-1} and path s_i is $O(|s_i|)$ (i.e., linear in the length of path s_i; note that this does not include the time needed to output that path, just the time needed to construct its representation and store it in $\mathcal{S}$). In our setting, the best time complexity we can hope for is linear preprocessing time and constant delay. Examples of enumeration algorithms developed in similar settings can be found in [44, 4, 5, 3].

§ *Paths in DAGs.* We can now formalise the representation of paths, used to maintain them in our algorithm. Let $G=(V,E)$ be a DAG. We assume that we have a function $de : V \to E \cup \{\perp\}$ which associates to every node $x \in V$ an edge $de(x)=(x,y)$, called the *default edge* of x, or sets $de(x)=\perp$, meaning that there is no default edge leaving node x. Thus, we have at most $|V|$ default edges in a DAG. The path from x to y consisting of only default edges is called *default path*, denoted by $\text{defPath}(x,y)$; clearly, there exists at most one default path from node x to node y. With respect to a function de, defining the default edges of G, an arbitrary $x-y$-path α in G can be *encoded*, recursively, as follows:

- if α is the path of length 0, its encoding is $\Phi(\alpha)=\varepsilon$,
- if α is the (unique) default $x-y$-path, its encoding is $\Phi(\alpha)=\text{defPath}(x,y)$,
- if α is the sequence $\beta_1,(z,z'),\beta_2$, where (z,z') is not the default edge $de(z)$, β_1 is an $x-z$-path, and β_2 is a $z'-y$-path, then its encoding is $\Phi(\alpha)=\Phi(\beta_1)(z,z')\Phi(\beta_2)$.

Let us now explain how the solution data structure $\mathcal{S}$ is maintained during the enumeration of the $x-y$-paths in the DAG G. We define $\mathcal{S}$ as a stack, and for $\mathcal{S}=(s_1,\ldots,s_r)$ (with $top(\mathcal{S})=s_r$), every element s_i, for $i\in[r]$, is the encoding $\Phi(p_i)$ of some path p_i connecting two nodes x_i and x_{i+1} of V, with $x_1=x$ and $x_{r+1}=y$. The stack $\mathcal{S}$ then stores an encoded representation of the path $p=p_1p_2\cdots p_r$ obtained by concatenating the paths $p_1,\ldots,p_r$. Clearly, given stack $\mathcal{S}$ and DAG G, we can directly retrieve the path p in $O(|p|)$ time (i.e., linear in the length of p): follow the edges of $p_1,\ldots,p_r$ in order. Note that, if p only consists of non-default edges, then $|p|\le|V|$ yields a naive upper bound for the height of $\mathcal{S}$.

Our enumeration algorithms work on input DAGs which can be specified in a succinct, implicit way. We thus introduce a notion, important for our approach.

Definition 3. *A directed acyclic graph (for short, DAG) $G = (V, E)$ is an m-*skeleton DAG *if the following conditions hold:*

- *There is a surjective function level : $V \to [0 : m]$ such that $(x, y) \in E \Rightarrow$ level$(x) \leq$ level(y).*
- *G has two distinguished nodes, the source s and sink f, with level$(s) = 0$ and level$(f) = m$. Node s (resp., f) is the unique node with level$(s) = 0$ (resp., with level$(f) = m$).*
- *For $\ell \in [m-1]$, there is a total ordering $v_1, \ldots, v_r$ of $V_\ell = \{v \in V \mid$ level$(v) = \ell\}$ such that $(v_i, v_{i+1}) \in E$, for all $i \in [r-1]$, and no other edges (u, v) with level(u) = level(v) exist in E. We define the relation $v \to_\ell u$, called the* sibling *relation, such that $v \to_\ell u$ if and only if $(v, u) \in E$, and the relation $\to$ as the union of $\to_\ell$, for all $\ell \in [m-1]$.*
- *For all $v \in V \setminus \{s, f\}$ there is exactly one edge $(v, v') \in E$, $v' \in V$, such that level$(v) <$ level(v'). The source s can have multiple outgoing edges, but for any two edges $(s, v), (s, v')$ of G we have level$(v) \neq$ level(v').*

Clearly, for an m-skeleton DAG $G = (V, E)$ we have that $|E| \in O(|V|)$. We now define the directed acyclic graph encoded by an m-skeleton DAG. See Fig. 1.

Definition 4. *An m-skeleton DAG $G = (V, E)$ encodes the DAG $D(G) = (V, E_1 \cup E_2)$, where $E_1 = \{(v, v') \in E \mid$ level$(v) <$ level$(v')\}$ and $E_2 = \{(v, v'') \mid (v, v') \in E_1, v' \to^* v''\}$, (where $\to^*$ is the transitive and reflexive closure of the sibling relation $\to$).*

Note that an m-skeleton DAG $G = (V, E)$ is a succinct representation of $D(G)$, as $D(G)$ might have significantly more edges than G (e.g., when all the nodes of each level of $D(G)$ are connected to all the nodes on the next level, we only have two outgoing edges for each node, other than the source, in the skeleton G). However, not all DAGs can be encoded by skeletons (because, e.g., in a skeleton-encoded DAG, there can be no edge between two children of a node $v \in V \setminus \{s\}$), but all trees can be encoded by skeletons, and the representation of trees by skeletons is very similar to the child-sibling representation of trees [17]. In the case of a DAG $D(G)$ encoded by an m-skeleton DAG $G = (V, E)$, with source s and sink f, we assume that the default edge $de(x)$ of each node $x \in V \setminus \{s, f\}$ in $D(G)$ is the single edge (x, y) of G (which is also an edge of $D(G)$) with level$(y) >$ level(x), and $de(s) = de(f) = \perp$ (i.e., there are no default edges leaving s and f). The enumeration of $s - f$-paths in $D(G)$ is done w.r.t. this definition of default paths.

We can show the following result, see the full proof in [33].

Theorem 1. *Given m-skeleton DAG $G = (V, E)$, with $|V| = n$, after an $O(n)$ time preprocessing, we can enumerate the $s - f$-paths in $D(G)$ with $O(1)$ delay.*

Proof (Sketch). § *Setup.* For each node v of G (except s and f) we have exactly one default outgoing edge: the edge leaving v and going to a different level. Consequently, for each node $v \in V \setminus \{s\}$ there is exactly one default path (consisting

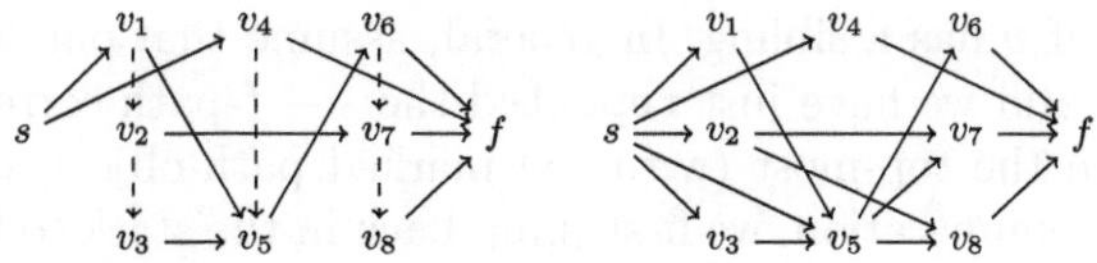

Fig. 1. 4-Skeleton DAG G (a) and its encoded DAG $D(G)$ (b); levels are drawn vertically, and dashed arrows represent the sibling relation.

only of default edges) leading to f and the union of these $v-f$-default paths, for $v \in V \setminus \{s\}$, forms a rooted tree with root f (with edges oriented towards the root). Let T be the tree obtained from the respective rooted tree by changing the orientation of the edges. The nodes of $D(G)$ which also have non-default outgoing edges (edges of $D(G)$ which do not occur in G) are called *branching nodes*; we can process G in $O(n)$ time and identify all branching nodes, and colour them in T. Using Lemma 1, we can preprocess T in linear time so that we can answer in $O(1)$ queries $\texttt{colour}(u, v)$ requesting the coloured node $t \neq u$ of minimum depth (i.e., closest to f) on the path from u to v in T (or $\perp$ if no such node exists). It is worth noting that constructing T from $D(G)$, identifying the branching (coloured) nodes, and then constructing the data structures of Lemma 1 is already the first non-trivial step in our algorithm, and is crucial in having a relatively simple, efficient enumeration.

As each $s-f$-path in $D(G)$ can be represented as an alternating sequence of non-default edges and default paths, the enumerated paths are stored and maintained using a stack $\mathcal{S} = (s_1, \ldots, s_r)$ (with $top(\mathcal{S}) = s_r$) where, for $i \in [r]$, s_i is either the encoding of some default path p_i connecting two nodes x_i and x_{i+1} of V or the non-default edge (x_i, x_{i+1}) of $D(G)$, with $x_1 = s$ and $x_{r+1} = f$. We say that $\mathcal{S}$ stores path-objects (i.e., edges or default paths), and some of these objects are marked. If a marked object is a default path from node x to node y, the marking means that the respective path contains at least one branching node, and, for the branching node u closest to y on that $x-y$-default path, we did not yet explore the paths sharing the same prefix from s to u with the current path, but then continuing with a non-default edge. If a marked object is a non-default edge (u, u'), the marking means that we did not yet explore the paths sharing the same prefix from s to u with the current path, but then continuing with the non-default edge (u, u''), where u'' is a sibling of u'. Again, as a side note, we mention that the marking of path-objects, keeping track of which objects are marked in $\mathcal{S}$, and being able to move efficiently from one such marked object to another is another crucial aspect in the efficient enumeration algorithm, that we want to achieve.

§ *Enumeration Algorithm.* For each edge (s, v) leaving s in G, we enumerate the paths starting with this edge. The first $\texttt{report}$ed path consists of (s, v), followed by the $v-f$-default path; these two path-objects are stored in $\mathcal{S}$ (as encoding, in the case of the default path). The $v-f$-default path is marked if there is any branching node on this path (checked using Lemma 1), while

(s, v) is marked if v has a sibling. In general, assume that our solution-stack is $\mathcal{S} = (s_1, \ldots, s_r)$ and we have just `reported` the $s - f$-path corresponding to $\mathcal{S}$. Now, we retrieve the top-most (w.r.t. $\mathcal{S}$) marked path-object s_i. To reach the next path in our enumeration, we first jump back in the stack to the path-object s_{i-1} (we implement $\mathcal{S}$ statically, so we can do this in $O(1)$ time, by simply using a different stack to keep track of the marked objects of $\mathcal{S}$, and updating the top of $\mathcal{S}$ to point to the object below the top-most marked object, i.e., s_{i-1}); in this way, we only keep the prefix $(s_1, \ldots, s_{i-1})$ in the stack (the other objects are not deleted explicitly, but simply disregarded, as they are above the top of the stack, and will be overwritten later). If the marked path-object s_i is a non-default edge (u, u'), we push to $\mathcal{S}$ the non-default edge (u, u'') (marked, if u'' still has a sibling) and then the $u'' - f$-default path (marked, if this default path contains branching nodes, tested with Lemma 1). If the marked path-object s_i is the default path connecting nodes x and y, we use Lemma 1 to retrieve the branching node u of minimum depth in T (i.e., closest to f, and, thus, to y as well) on the default path from x to y. Then, if (u, q) is the default edge leaving u, we proceed as follows. If $x \neq u$, we push to $\mathcal{S}$ the $x - u$-default path (marked if it contains a branching node, other than u). Both for $x = u$ and $x \neq u$, we then push the non-default edge (u, u'), where u' is the sibling of q (marked if u' also has a sibling). This leads to a partial solution, defined by the prefix $(s_1, \ldots, s_{i-1}, \text{defPath}(x, u), (u, u'))$, that did not appear before in any solution. Finally, we push the $u' - f$-default path (marked, if it contains branching nodes). In all cases, we now `report` the current path, and repeat the process.

§ *Conclusion.* Our algorithm corresponds to a backtracking (depth-first) exploration of the $s - f$-paths in $D(G)$. The crucial point is that, after `reporting` a solution, we can find in $O(1)$ time the node where we need to backtrack (lowest unexplored branching node on the currently enumerated path), in order to start building the next solution, using Lemma 1 and our marking policy; once this node is found, default paths are used to quickly complete a partial solution, and the correct marking of the path-objects stored on the stack is maintained efficiently using, again, Lemma 1. Moving from one path to another takes $O(1)$ time, so constructing the new path requires only a constant number of changes w.r.t. the current one: drop a suffix of the current path, add a new edge, add a new default path, `report`. In conclusion, after a linear time preprocessing, we can `report` the first path in $O(1)$ time, and enumerate all the others with constant delay. □

It is not hard to see that counting the $s - f-$paths in $D(G)$ can also be done efficiently, using a bottom-up dynamic programming algorithm on the skeleton representation of this DAG. However, such DAGs can have exponentially many paths (in the size of G); this follows, e.g., from the results of the next section, where a size-preserving skeleton-encoding for the SAS of a string will be presented, and from the fact that a string of length n can have exponentially many SAS. Therefore, in defining the complexity of our counting algorithm, we will explicitly state the number of arithmetical operations used to compute the number of paths, and note that these operations are additions which may involve large numbers, with up to n bits (where n is the number of nodes of G).

Theorem 2. *Given an m-skeleton DAG $G = (V, E)$, with $|V| = n$, we can count the number of $s - f$-paths of $D(G)$ using $O(n)$ arithmetical operations.*

4 Skeleton-Based Enumeration of SAS

It has previously been shown by Kosche et al. [30, Theorem 4.19] that, given a k-universal string $w \in \Sigma^*$ of length n with $|\Sigma| = \sigma$, we can enumerate SAS(w) with $O(n)$ preprocessing time and $O(k\sigma)$ delay. The main idea of their proof is the following: a string $v \in \Sigma^*$ of length $k+1$ is a SAS of w if and only if $i_\ell \in \mathrm{ar}_w(\ell)$, for all $\ell \in [k]$, and $v[k+1] \notin \mathrm{alph}(w[i_k+1 : n])$, where $0 < i_1 < i_2 < \ldots < i_k \leq n$ is the canonical embedding of $v[1 : k]$ in w. This means that, when greedily embedding v in w, every arch of w contains exactly one letter of v and $v[k+1]$ cannot be embedded. Further, if there is a SAS-prefix u (i.e., u can be extended to a SAS) with canonical embedding $0 < i_1 < i_2 < \ldots < i_{|u|}$ and $i_{|u|} \in \mathrm{ar}_w(|u|)$ (and, consequently, $i_\ell \in \mathrm{ar}_w(\ell)$ for all $\ell \in [|u|]$), then ua, for $a \in \Sigma$, is a SAS-prefix (resp., a SAS) if and only if the first occurrence of a after $i_{|u|}$ is in arch $|u|+1$ (or a does not occur in $w[i_{|u|}+1 : n]$, if $|u| = k$). Using a series of data structures (precomputable in $O(n)$ time), Kosche et al. [30] were able to compute a tree $\mathcal{T}_w$ of SAS-prefixes, such that the root-to-leaf paths of $\mathcal{T}_w$ bijectively correspond to SAS(w). Since $|\,\mathrm{SAS}(w)|$ can be exponential in n [30], explicitly storing $\mathcal{T}_w$ is also not feasible. Instead, the authors enumerated SAS(w) using a depth-first traversal of $\mathcal{T}_w$, computing the children of the currently inspected node on-the-fly. An alternative idea was given in [43], where the author defined a deterministic finite automaton accepting the absent subsequences of a string, and then enumerated the shortest strings accepted by this automaton. The preprocessing time was, in that case, $O(n\sigma)$ and the delay $O(k)$.

We improve here significantly upon the ideas from [30]. Based on the same basic data structures, we show that, instead of building $\mathcal{T}_w$, one can construct a DAG K with $O(k\sigma) \subseteq O(n)$ nodes and $O(k\sigma^2)$ edges. Moreover, we show that we can build a skeleton DAG G with $O(n)$ nodes and edges such that $K = D(G)$, which allows us to apply the efficient enumeration framework defined in the previous section. We obtain the following result, whose full proof is given in [33].

Theorem 3. *Given string $w \in \Sigma^n$, with $\iota(w) = k$, we can construct in $O(n)$ time a $(k+2)$-skeleton DAG G, with $O(n)$ nodes, source s, sink f, such that* SAS(w) *can be bijectively mapped to the $s - f$-paths of $D(G)$. Hence, after an $O(n)$ time preprocessing,* SAS(w) *can be enumerated with $O(1)$ delay.*

Proof (Sketch). We use the data structures introduced in the preliminaries: the array dist$[1 : n]$, as well as the matrices firstInArch$[1 : k, 1 : \sigma]$ and lastInArch$[1 : k, 1 : \sigma]$. Let $F_\ell = \{i \in \mathrm{ar}_w(\ell) \mid i = \mathrm{firstInArch}[\ell, w[i]] \text{ and } \mathrm{dist}[i] = k-\ell+2\}$ and recall that, for $\ell \in [k]$, if $i \in \mathrm{ar}_w(\ell)$ then there exists a SAS v, with $|v| = k+1$, of w such that i is part of the canonical embedding of $v[1:k]$ in w if and only if $i \in F_\ell$ (i.e., F_ℓ contains all positions that can be embedded as the ℓ-th letter of a SAS). We observe that $|F_\ell| \leq \sigma$, and, moreover, all the sets F_ℓ, represented as increasingly sorted lists, can be constructed in $O(n)$ time.

We first define a DAG K, with a source node s, a sink node f, and $k+3$ levels (numbered 0 to $k+2$), such that there is a bijection between SAS(w) and the set of $s-f$-paths in K. Then we construct a $(k+2)$-skeleton DAG G such that $K = D(G)$. In constructing this graph K, the main observation is that, given a SAS-prefix u with canonical embedding $0 < i_1 < \ldots < i_{|u|}$, the choice of possible letters (resp., positions) extending the SAS-prefix (resp., embedding) only depends on the position $i_{|u|}$ of the $|u|$-th letter; the choice of letters $u[1], \ldots, u[|u|-1]$ does not matter. We therefore do not need to store information regarding earlier positions in the nodes of the graph. Instead, we can define the nodes of DAG $K = (V, E)$ as the positions in w. Additionally, as remarked above, for $\ell \in [k]$, a position i_ℓ can embed the ℓ-th letter of a SAS only if i_ℓ is the first occurrence of letter $w[i_\ell]$ in arch $\mathrm{ar}_w(\ell)$. Thus, there are at most σ possible positions on which letters of a SAS can be embedded in each arch.

Basically, K will have on level 0 just the node s, on level $k+2$ the sink f, and on level ℓ, for $\ell \in [k]$, the elements of the set F_ℓ as nodes. On level $k+1$, we have σ nodes, denoted by $n+a$, for every $a \in \Sigma$: these are the nodes corresponding to the last letter of a SAS. Node s is connected to all nodes on level 1 (since every position $i \in F_1$ is contained in the canonical embedding of some SAS of w), and node $i \in F_\ell$ on level $\ell \in [k-1]$ is connected to the nodes $C(i) = \{j \in F_{\ell+1} \mid \mathrm{lastInArch}[\ell, w[j]] \leq i\}$; intuitively, these are the positions j such that there is some SAS-prefix v_j of length $\ell+1$ with the canonical embedding $0 < i_1 < i_2 < \cdots < i_\ell < i_{\ell+1} \leq n$, such that $i_\ell = i$ and $i_{\ell+1} = j$. There is such a SAS-prefix v_j for each $j \in C(i)$, since the next occurrence of $w[j]$ in $w[i+1:n]$ is in the $(\ell+1)$-th arch. Finally, node i on level k is connected to the nodes $C(i) = \{n+a \mid a \in \Sigma, |w[i+1:n]|_a = 0\}$, which correspond to the last letter of a SAS whose prefix of length k has a canonical embedding ending on i. An important observation is that for $i, i' \in \mathrm{ar}_w(\ell)$, $C(i) \subseteq C(i')$ if and only if $i < i'$. One can show that the DAG K has the desired property: there is a bijection between SAS(w) and the set of $s-f$-paths in K.

The crucial part in constructing the $(k+2)$-skeleton G with $D(G) = K$ is to define the order $\rightarrow_\ell$ of the nodes on each level ℓ and the edges of the skeleton DAG which connect nodes on different levels. Intuitively, we want to order the nodes $j, j' \in F_\ell$ such that $j \rightarrow_\ell^* j'$ if and only if the last occurrence of $w[j']$ in $\mathrm{ar}_w(\ell-1)$ is before the last occurrence of $w[j]$ in $\mathrm{ar}_w(\ell-1)$, i.e., j' is contained in more sets $C(i)$, for $i \in F_{\ell-1}$, than j. For this, we define, for $\ell \in [k]$, the increasingly sorted list H_ℓ of the positions $\mathrm{lastInArch}[\ell, a]$, for $a \in \Sigma$ (as a side note, these sorted lists can be constructed in overall time $O(n)$). Then, the order $\rightarrow_1$ of the elements of F_1 is the inverse natural order on integers (i.e., $i \rightarrow_1^* j$ if and only if $i > j$), and the elements of level $\ell+1$, for $\ell \in [k]$, are ordered such that $i \rightarrow_{\ell+1}^* j$ (meaning that there is a path on that level of G from i to j) if and only if $\mathrm{lastInArch}[\ell, w[j]] < \mathrm{lastInArch}[\ell, w[i]]$ (where, when $\ell+1 = k+1$, we use the notation $w[n+a] = a$ for $a \in \Sigma$). That is, the order of the nodes on level $\ell+1$ is induced by the inverse-order of the elements of H_ℓ (when considered w.r.t. $<$, the natural order on integers). This means that the elements of $C(i)$, for some node i on level ℓ, are the last $|C(i)|$ nodes w.r.t. the total order $\rightarrow_{\ell+1}$. Now, for node i on level $\ell \in [k]$ we define an edge connecting it to the first element of $C(i)$ w.r.t. $\rightarrow_{\ell+1}$, on level $\ell+1$ (note that this corresponds to the default edge of i and all other elements of $C(i)$ are the targets of non-default edges from i in the encoded DAG $D(G)$). The first element of $C(i)$ w.r.t. $\rightarrow_{\ell+1}$, on level $\ell+1$, can be identified using H_ℓ: we retrieve the largest element $j' \leq i$ of H_ℓ, and then the first element of $C(i)$ is $\text{firstInArch}[\ell+1, w[j']]$ (if $\ell+1 \leq k$) or $n + w[j']$ (if $\ell+1 = k+1$). Moreover, all nodes on level $k+1$ are connected to f, and there is an edge connecting s to the smallest node $\min\{v \in F_1\}$ w.r.t. $\rightarrow_1$ on level 1. One can show that $D(G) = K$.

Finally, G can be constructed in time $O(n)$, the main challenge being here the part where the edges going from nodes on level ℓ to level $\ell+1$ are constructed. However, these edges can be determined by an algorithm reminiscent of merging the sorted sets (i.e., lists) F_ℓ and H_ℓ. Our claim now follows from Theorem 1.□

5 Conclusion

In this paper we have shown two main results. In the first one, Theorem 1, we show that one can enumerate, after a linear time preprocessing and with constant delay, all the $s-f$–paths in a DAG which has a skeleton-representation. This result, although regarding a relatively constrained class of DAGs, is then applied directly to efficiently enumerate and count all the shortest absent subsequences in a string. The most important aspect in getting this efficient enumeration algorithm (presented in our second main result, Theorem 3) for the set of SAS of a string is a novel representation of the respective set as DAG which can be encoded using skeletons.

But, there are more applications where skeletons and Theorem 1 may play a relevant role (albeit, still related to the study of subsequences). In [33] (the complete version of this extended abstract) we show that the set of minimal absent subsequences of a string w (of length n, over an alphabet of size σ) can also be represented as a DAG, which, at its turn, can be encoded by a skeleton of size $O(n\sigma)$. This immediately leads to an enumeration algorithm for the set $\text{MAS}(w)$, with $O(n\sigma)$ preprocessing and constant delay, and to a counting algorithm which computes the number of MAS of string w with $O(n\sigma)$ arithmetical operations (matching the complexity of the algorithm reported in [43]). While not optimal, this approach improves the state-of-the-art for the respective problem ([43] reported an enumeration algorithm for $\text{MAS}(w)$ with the same preprocessing time, but output-linear delay). Moreover, while optimal enumeration algorithms for $\text{MAS}(w)$ can actually be derived (see [33]), there is no obvious way to adapt the respective algorithms to counting the elements of $\text{MAS}(w)$ without becoming more complex, and the skeleton-based algorithm for counting the elements of this set and the one reported in [43] remain the best we know.

As potential directions for future study, it would be interesting to extend the efficient enumeration algorithm of Theorem 1 to other classes of graphs, and find more applications where such algorithms can be used.

Acknowledgements. The authors of this paper are partly supported by the German Research Foundation (DFG) through the Heisenberg project 466789228 and research project 562495345.

References

1. Abboud, A., Backurs, A., Williams, V.V.: Tight hardness results for LCS and other sequence similarity measures. In: FOCS 2015, pp. 59–78 (2015)
2. Adamson, D., Fleischmann, P., Huch, A., Koß, T., Manea, F., Nowotka, D.: k-universality of regular languages. In: ISAAC 2023. LIPIcs, vol. 283, pp. 4:1–4:21 (2023)
3. Adamson, D., Fleischmann, P., Huch, A., Manea, F., Sarnighausen-Cahn, P., Wiedenhöft, M.: Tight bounds for the number of absent subsequences. In: FCT 2025. LNCS, vol. 16106, pp. 15–29 (2025)
4. Adamson, D., Gawrychowski, P., Manea, F.: Enumerating m-length walks in directed graphs with constant delay. In: LATIN 2024. LNCS, vol. 14578, pp. 35–50 (2024)
5. Amarilli, A., Monet, M.: Enumerating regular languages with bounded delay. In: STACS 2023. LIPIcs, vol. 254, pp. 8:1–8:18 (2023)
6. Badkobeh, G., Charalampopoulos, P., Pissis, S.P.: Internal shortest absent word queries. In: CPM 2021. LIPIcs, vol. 191, pp. 6:1–6:18 (2021)
7. Barker, L., Fleischmann, P., Harwardt, K., Manea, F., Nowotka, D.: Scattered factor-universality of words. In: DLT 2020. pp. 14–28. LNCS, Cham (2020)
8. Bergroth, L., Hakonen, H., Raita, T.: A survey of longest common subsequence algorithms. In: SPIRE 2000, pp. 39–48 (2000)
9. Bernardini, G., Marchetti-Spaccamela, A., Pissis, S.P., Stougie, L., Sweering, M.: Constructing strings avoiding forbidden substrings. In: CPM 2021. LIPIcs, vol. 191, pp. 9:1–9:18 (2021)
10. Bille, P., Gørtz, I.L., Vildhøj, H.W., Wind, D.K.: String matching with variable length gaps. Theoretical Comput. Sci. **443**, 25–34 (2012)
11. Bringmann, K., Künnemann, M.: Multivariate fine-grained complexity of longest common subsequence. In: SODA 2018, pp. 1216–1235 (2018)
12. Charalampopoulos, P., Crochemore, M., Fici, G., Mercaş, R., Pissis, S.P.: Alignment-free sequence comparison using absent words. Inf. Comput. **262**, 57–68 (2018)
13. Day, J.D., Fleischmann, P., Kosche, M., Koß, T., Manea, F., Siemer, S.: The edit distance to k-subsequence universality. In: STACS 2021. LIPIcs, vol. 187, pp. 25:1–25:19 (2021)
14. Fazekas, S.Z., Koß, T., Manea, F., Mercas, R., Specht, T.: Subsequence matching and analysis problems for formal languages. In: ISAAC 2024. LIPIcs, vol. 322, pp. 28:1–28:23 (2024)
15. Fleischmann, P., Haschke, L., Höfer, J., Huch, A., Mayrock, A., Nowotka, D.: Nearly k-universal words - investigating a part of Simon's congruence. Theor. Comput. Sci. **974**, 114113 (2023)

16. Fredman, M.L.: On computing the length of longest increasing subsequences. Discrete Math. **11**(1), 29–35 (1975)
17. Fredman, M.L., Sedgewick, R., Sleator, D.D., Tarjan, R.E.: The pairing heap: a new form of self-adjusting heap. Algorithmica **1**(1), 111–129 (1986)
18. Frochaux, A., Kleest-Meißner, S.: Puzzling over subsequence-query extensions: Disjunction and generalised gaps. In: AMW 2023. CEUR Workshop Proceedings, vol. 3409 (2023)
19. Fujishige, Y., Tsujimaru, Y., Inenaga, S., Bannai, H., Takeda, M.: Computing DAWGs and minimal absent words in linear time for integer alphabets. In: MFCS 2016. LIPIcs, vol. 58. Schloss Dagstuhl-Leibniz-Zentrum fuer Informatik (2016)
20. Gawrychowski, P., Kosche, M., Koß, T., Manea, F., Siemer, S.: Efficiently testing Simon's congruence. In: STACS 2021. LIPIcs, vol. 187, pp. 34:1–34:18 (2021)
21. Giatrakos, N., Alevizos, E., Artikis, A., Deligiannakis, A., Garofalakis, M.N.: Complex event recognition in the big data era: a survey. VLDB J. **29**(1), 313–352 (2020)
22. Halfon, S., Schnoebelen, P., Zetzsche, G.: Decidability, complexity, and expressiveness of first-order logic over the subword ordering. In: LICS 2017, pp. 1–12 (2017)
23. Hébrard, J.J.: An algorithm for distinguishing efficiently bit-strings by their subsequences. Theoretical Comput. Sci. **82**(1), 35–49 (1991)
24. Karandikar, P., Schnoebelen, P.: The height of piecewise-testable languages and the complexity of the logic of subwords. Log. Methods Comput. Sci. **15**(2) (2019)
25. Kitaev, S.: Patterns in Permutations and Words. Springer (2011)
26. Kleest-Meißner, S., Sattler, R., Schmid, M.L., Schweikardt, N., Weidlich, M.: Discovering event queries from traces: laying foundations for subsequence-queries with wildcards and gap-size constraints. In: ICDT 2022. LIPIcs, vol. 220, pp. 18:1–18:21 (2022)
27. Kleest-Meißner, S., Sattler, R., Schmid, M.L., Schweikardt, N., Weidlich, M.: Discovering multi-dimensional subsequence queries from traces - from theory to practice. In: BTW 2023. LNI, vol. P-331, pp. 511–533 (2023)
28. Knuth, D.E.: The Art of Computer Programming, vol. 4A: Combinatorial Algorithms, Part 1. Addison-Wesley (2014)
29. Kosche, M., Koß, T., Manea, F., Siemer, S.: Absent subsequences in words. In: RP 2021. LNCS, vol. 13035, pp. 115–131 (2021)
30. Kosche, M., Koß, T., Manea, F., Siemer, S.: Absent subsequences in words. Fundam. Informaticae **189**(3–4), 199–240 (2022)
31. Kosche, M., Koß, T., Manea, F., Siemer, S.: Combinatorial algorithms for subsequence matching: a survey. In: NCMA 2022. EPTCS, vol. 367, pp. 11–27 (2022)
32. Li, C., Yang, Q., Wang, J., Li, M.: Efficient mining of gap-constrained subsequences and its various applications. ACM TKDD **6**(1), 2:1–2:39 (2012)
33. Manea, F., Ringleb, T., Siemer, S., Winkler, M.: Efficiently finding all minimal and shortest absent subsequences in a string. CoRR abs/2504.21471 (2025)
34. Pin, J.: The influence of Imre Simon's work in the theory of automata, languages and semigroups. Semigroup Forum **98**, 1–8 (2019)
35. Riddle, W.E.: An approach to software system modelling and analysis. Comput. Lang. **4**(1), 49–66 (1979)
36. Rigo, M., Salimov, P.: Another generalization of abelian equivalence: binomial complexity of infinite words. Theoret. Comput. Sci. **601**, 47–57 (2015)
37. Romik, D.: The Surprising Mathematics of Longest Increasing Subsequences. Cambridge (2015)
38. Salomaa, A.: Connections between subwords and certain matrix mappings. Theoret. Comput. Sci. **340**(2), 188–203 (2005)

39. Shaw, A.C.: Software descriptions with flow expressions. IEEE Trans. Softw. Eng. **4**(3), 242–254 (1978)
40. Simon, I.: Hierarchies of events with dot-depth one - Ph.D. thesis (1972)
41. Simon, I.: Piecewise testable events. In: Automata Theory and Formal Languages. pp. 214–222. LNCS, Berlin, Heidelberg (1975)
42. Simon, I.: Words distinguished by their subwords (extended abstract). In: WORDS 2003. TUCS General Publication, vol. 27, pp. 6–13 (2003)
43. Tronícek, Z.: On problems related to absent subsequences. In: COCOA 2023. LNCS, vol. 14462, pp. 351–363 (2023)
44. Wasa, K.: Enumeration of enumeration algorithms. CoRR abs/1605.05102 (2016)
45. Zetzsche, G.: The complexity of downward closure comparisons. In: ICALP 2016. LIPIcs, vol. 55, pp. 123:1–123:14 (2016)
46. Zhang, H., Diao, Y., Immerman, N.: On complexity and optimization of expensive queries in complex event processing. In: SIGMOD 2014, pp. 217–228 (2014)

Comparing the Effective Content of Subshifts

A. Nakid Cordero(✉) and I. Scott(✉)

University of Wisconsin–Madison, Madison, WI 53705, USA
{nakidcordero,iscott6}@wisc.edu

Abstract. Motivated by Ziegler's computability-theoretic characterisation of finite absolute presentability between groups, we prove an analogous theorem in symbolic dynamics. We introduce the notion of one subshift being *finitely determined over* another and show that this relation is characterised by *Ziegler reducibility* between their co-languages. We further investigate a notion of *existential closure* for subshifts.

Keywords: Subshifts · Ziegler reducibility · Existential closure

1 Introduction

The study of multidimensional subshifts—colourings of $\mathbb{Z}^d$ that avoid a set of forbidden finite patterns—has greatly benefited from the tools of computability theory. A signature example is the simulation theorem by Hochman [4] and later refinements by Aubrun–Sablik [1] and Durand–Romashchenko–Shen [3], which state that every effectively closed shift is simulated by a subshift of finite type (SFT) of higher dimension.

In this vein, we characterise in computability-theoretic terms when a subshift is "completely determined" by another subshift using a finite amount of extra information. Namely, in Theorem 2 we prove that this relationship coincides with *Ziegler reducibility*, denoted $\leq^*$, (see Definition 8) between their co-languages.

The definition of $\leq^*$ dates back to Ziegler's analysis of the finitely generated subgroups of existentially closed groups [10]. This fits into a growing "dictionary" between notions and results in combinatorial group theory and in symbolic dynamics, as developed in [6,7]. We also raise an issue with extending the dictionary further, to capture existentially closed groups. We show that the most natural definition and construction of "existentially closed subshifts" (following the theory for groups) does not satisfy Ziegler's Theorem [10, Folgerungen III.1.8].

2 Preliminaries

In what follows, we will be working in subspaces of $\Sigma^{\mathbb{Z}^d}$ where Σ is a finite set and $d < \omega$. Recall that this space comes endowed with a natural topology generated

V. Brattka et al. (Eds.): CiE 2026, LNCS 16674, pp. 421–436, 2027.
https://doi.org/10.1007/978-3-032-31348-5_28

by the basic clopen sets $[w] = \left\{x \in \Sigma^{\mathbb{Z}^d} \mid x \restriction_{\mathrm{dom}(w)} = w\right\}$, where $w : D \to \Sigma$ is a map from a *rectangle* $[m_0, n_0] \times \cdots \times [m_{d-1}, n_{d-1}] \subseteq \mathbb{Z}^d$. We write Σ^{*d} for the set of such maps and call them *(d-dimensional) words*, when there is no risk of ambiguity, we simply write Σ^*. For $k < d$, we identify $\mathbb{Z}^k \subseteq \mathbb{Z}^d$ via the map $(v_0, \ldots, v_{k-1}) \mapsto (v_0, \ldots, v_{k-1}, 0, \ldots, 0)$, and hence, with this convention $\Sigma^{*k} \subseteq \Sigma^{*d}$. It is a standard fact that $\Sigma^{\mathbb{Z}^d}$ with this topology is a compact space.

The *shift map* $\mathrm{sh} : \mathbb{Z}^d \times \Sigma^{\mathbb{Z}^d} \to \Sigma^{\mathbb{Z}^d}$ is given by $\mathrm{sh}(v, x)(u) = x(u + v)$.

Definition 1. *A* subshift S *is a topologically closed, shift-invariant subset of* $\Sigma^{\mathbb{Z}^d}$*, for some set* Σ*, called the* alphabet of S *and* $d < \omega$ *called the* dimension of S. *Unless otherwise specified, we will assume that* Σ *is finite. An element* $x \in S$ *is called a* point *or* configuration *of* S.

We write Σ_S *and* d_S *for the alphabet and dimension of* S*, respectively.*

Because subshifts are closed and shift-invariant, we may identify S with the set of words that induce open sets disjoint from S.

Definition 2. *The* co-language *of a subshift* S *is*

$$\mathcal{L}^c(S) = \{w : D \to \Sigma_S \mid D \text{ is a rectangle in } \mathbb{Z}^{d_S} \text{ and } \forall x \in S \ (x \restriction_{\mathrm{dom}(w)} \neq w)\}.$$

The language *of* S *is* $\mathcal{L}(S) = \Sigma^{*d} \setminus \mathcal{L}^c(S)$. *Equivalently, the language of* S *is the set of words that appear in some* $x \in S$. *For* $x \in S$*, we define* $\mathcal{L}(x)$ *and* $\mathcal{L}^c(x)$ *accordingly.*

As part of the philosophy of this paper of translating between notions in group theory and symbolic dynamics, and following the lead of [7], we write $\langle \Sigma_S \mid F\rangle^{d_S}$ to denote the largest (under inclusion) subshift S such that $F \subseteq \mathcal{L}^c(S)$. Further, we denote by

$$S \models L(G) \wedge L^c(B)$$

that $G \subseteq \mathcal{L}(S)$ and $B \subseteq \mathcal{L}^c(S)$. If $G = \{v\}$ and $B = \{v\}$ then we write $S \models L(v) \wedge L^c(w)$.

If S can be expressed as $\langle \Sigma_S \mid F\rangle^{d_S}$ for F finite, we call S a *subshift of finite type (SFT)*. Similarly, if F can be chosen computable (equivalently, c.e.), we say that S is *effectively closed*.

Example 1. The *full shift* on alphabet Σ and dimension d is the shift $\langle \Sigma \mid \emptyset\rangle^d$. Notice that every shift is a subset of the full shift in the same alphabet and dimension.

On the other end of this spectrum, a *minimal shift* is a shift S such that if $T \subseteq S$ is a subshift, then T is empty or equal to S. Equivalently, S is minimal if for every $x, y \in S$, we have $\mathcal{L}(x) = \mathcal{L}(y)$.

Given a subshift S, we have several ways to form a new subshift from S.

Definition 3. *Let S be a subshift. The* slice of S to alphabet $\Delta \subseteq \Sigma_S$ *is the subshift* $S\restriction_\Delta = S \cap \Delta^{\mathbb{Z}^{d_S}}$. *The* slice of S to dimension $d < d_S$ *is the subshift*

$$S\restriction_d = \{x \in \Sigma_S^{\mathbb{Z}^d} \mid \exists y \in S\ \forall v \in \mathbb{Z}^d\ (x(v) = y(v, 0, \dots, 0)\}.$$

In the other direction, we form a $d_S + k$-dimensional subshift from S as

$$S^{(k)} = \{x \in \Sigma_S^{\mathbb{Z}^{d_S+k}} \mid \exists y \in S\ (x(u_1, \dots, u_{d_S+k})) = y(u_1, \dots, u_{d_S})\}.$$

Notice that the two notions of slice do not commute; in general $(S\restriction_\Sigma)\restriction_d \subseteq (S\restriction_d)\restriction_\Sigma$.

Given a subshift S and a set of words F, we can also build a new subshift by enlarging the co-language. We write S/F for the subshift $\langle \Sigma_S \mid \mathcal{L}^c(S) \cup F\rangle^{d_S}$ and note that this is a *subset* of S.

There are a number of ways in the literature of recognising a subshift as a "subsubshift" of another. We mention two that will be relevant to this manuscript.

Definition 4. *If T and S are subshifts of dimension d, then $f : T \to S$ is a* factor map *if f is surjective and commutes with the shift action.*

If, in addition, f is injective, f is called a conjugacy *or an* isomorphism.

By the Curtis–Hedlund–Lyndon theorem, a factor map $f : T \to S$ is given by a *block code* $\hat{f} : \Sigma_T^D \to \Sigma_S$ such that $f(x)(u) = \hat{f}(\mathrm{sh}(u, x)\restriction_D)$. By passing to an isomorphic copy of T, we can always assume that $\hat{f} : \Sigma_T \to \Sigma_S$. For simplicity of notation, when it makes sense, we identify $\hat{f}$ with its extension to a map from Σ_T^* to Σ_S^*. In particular, for $w \in \Sigma_S^*$, we always have that $\hat{f}^{-1}(w)$ is well-defined and finite.

We also borrow a notion from [7], better adapted to our approach of identifying a shift with its co-language.

Definition 5. *Let S and T be subshifts. Then S is a* full restriction of T, *written* $S \sqsubseteq T$, *if* $\mathcal{L}(T) \cap \Sigma^{*d} = \mathcal{L}(S)$. *In particular, if* $S \sqsubseteq T$, *we must have* $\Sigma_S \subseteq \Sigma_T$ *and* $d_S \leq d_T$.

To allay this dependence on the alphabet, we write $S \mathrel{\underset{\sim}{\sqsubseteq}} T$ if there is an injective map $\hat{f} : \Sigma_S \hookrightarrow \Sigma_T$ such that $f(S) \sqsubseteq T$.

Observe that $\sqsubseteq$ is transitive and that $S \sqsubseteq T$ iff $T \models L(\mathcal{L}(S)) \wedge L^c(\mathcal{L}^c(S))$.

3 A Characterisation of Computable Subshifts

The key definition is that of "finitely determined over" (Definition 7 below). The definition should capture the intuition that from T, together with a finite amount of extra information, one can determine S. So, what does it mean to have access to T? How will S be presented? And what kind of finite information are we allowed?

We identify T and S with their co-languages (and also implicitly with their alphabets and dimensions). Thus, having access to T should correspond to "coding" the co-language of T (assuming that we know the alphabet and dimension of T). This is captured by Jeandel–Vanier's notion of full restriction (Definition 5). Similarly, "determining S" will correspond to the resulting subshift having S as a full restriction. Thus, "S is finitely determined over T" should have the following flavour:

$$\text{If } T \sqsubseteq R \text{ and } R \text{ satisfies } (*), \text{ then } S \sqsubseteq R,$$

where (*) is some finite set of conditions, not dependent on T or S.

Here the analogy with groups provides some insight. The relevant condition in the group case is that of a group G being $\exists_1$-*isolated* over another group H ([10, Bem. III.1.6.2]).

Definition 6. *A group G is* $\exists_1$-isolated[1] *over H if there is an existential formula $\varphi(\overline{x}, \overline{h})$ such that any $F \models \varphi(\overline{a}, \overline{h})$ with $\langle \overline{h} \rangle \cong H$ has $\langle \overline{a} \rangle \cong G$.*

Thus we should require that (*), in the framework above, enforces finitary constraints on the language and co-language of R.

The next condition on R arises from the observation that the notion of "substructure" in group theory (and indeed more generally in algebra or model theory) bifurcates for subshifts: the first notion (that T is a substructure of R if the atomic diagram of T is a subset of the atomic diagram of R) is captured by $\sqsubseteq$, and this is what we have used so far. However, there is a second notion; namely that there is a "structure-preserving" injection $f : T \to R$. In subshift theory, the analogue to this notion is that of a factor map; thus T is a "substructure" of R if there is a surjective, shift-commuting map $f : R \to T$ (see Definition 4). Indeed, this is borne out by the Simulation Theorems—shift-theoretic analogues of Higman's Embedding Theorem in group theory—proved by Durand–Romashchenko–Shen (Theorem 1 of [3]) and Aubrun–Sablik (Theorem 3.1 of [1]). We shall need both notions of substructure.

Finally, for this to all be well-defined, we need to fix the dimension of R.

Definition 7. *Let S and T be subshifts. S is* finitely determined over T *if there are:*

- *A finite alphabet Σ*
- *A dimension d*
- *A pair of finite sets of words (G, B) in Σ*

[1] Ziegler called this relation *finitely absolutely presented over H* (see [10, Bem. III.1.6.2]) Unfortunately this conflicts with the standard notation in group theory that a presentation of a group corresponds to the unique group on the generators which is "maximally free" subject to the relations (and irrelations). Indeed, in general even if $F \leq G$, one will need to take a larger group F in order to "present" G over H. Thus, in the absence of better notation, we choose the hopefully-explanatory "$\exists_1$-isolated".

– *A block code* $\hat{f} : \Sigma^D \to \Delta$, *for some* $\Delta \supseteq \Sigma_S \cup \Sigma_T$

such that

1. *There is some subshift* R *in dimension* d *and alphabet* Σ *such that* $G \subseteq \mathcal{L}(R)$, $B \subseteq \mathcal{L}^c(R)$, *and* $T \sqsubseteq f(R)$.
2. *Any such subshift* R *further satisfies that* $S \sqsubseteq f(R)$.

In this case, say that S *is* finitely determined over T by $(\Sigma, d, G, B, \hat{f})$.

If S *is finitely determined over the empty subshift, we say that* S *is* finitely determined.

Proposition 1. *Every minimal SFT is finitely determined.*

Proof. Let $S = \langle \Sigma_S \mid F \rangle^{d_S}$ be minimal. We claim that S is finitely determined by $(\Sigma_S, d_S, \{w\}, F, \mathrm{id})$, where w is any word in $\mathcal{L}(S)$. Indeed, if R is any subshift satisfying Definition 7 for this tuple, then R is a nontrivial subset of S and hence by minimality is equal to S. Thus $\mathrm{id}(R) \sqsupseteq S$. □

In contrast to Proposition 1, not every SFT is finitely determined over any subshift. We show in Theorem 2 that S is finitely determined over T if and only if $\mathcal{L}^c(S)$ is Ziegler-reducible to $\mathcal{L}^c(T)$. In this section, we warm up by showing that $\mathcal{L}^c(S)$ is computable if and only if S is finitely determined over $\emptyset$. This result can be seen as an analogue to Rips' theorem that a group has computable word problem if and only if there is a $\exists$-formula φ which defines it [9].

Theorem 1. *Let* S *be a subshift in a finite alphabet. Then* $\mathcal{L}^c(S)$ *is computable if and only if* S *is finitely determined over* $\emptyset$.

Proof. Firstly, let S be a d_S-dimensional computable subshift. By [7, Corollary 27], there is a minimal SFT T in dimension $d_S + 2$ with $S \sqsubseteq T$. Then M is finitely determined by Proposition 1, and hence S is also finitely determined.

For the other direction, suppose that S is finitely determined over $\emptyset$ by $(\Sigma, k, G, B, \hat{f})$. Then for any subshift $R \subseteq \Sigma^{\mathbb{Z}^k}$ with $R \models L(G) \wedge L^c(B)$, we have that $S \sqsubseteq f(R)$; in particular $S \sqsubseteq f(\langle \Sigma \mid B \rangle^k)$. Thus, $w \in \mathcal{L}^c(S)$ if and only if $f^{-1}(w) \subseteq \mathcal{L}^c(\langle \Sigma \mid B \rangle^k)$. But $\mathcal{L}^c(\langle \Sigma \mid B \rangle^k)$ is c.e. since B is finite, so it follows that $\mathcal{L}^c(S)$ is c.e.

On the other hand, $w \in \mathcal{L}(S)$ if and only if $f(\langle \Sigma \mid B \rangle^d)/\{w\} \not\models L(G)$. Since G and B are finite, this is also c.e. □

4 Relativising the Characterisation

In this section, we show that Theorem 1 relativises in the Ziegler-degrees, defined below. The definition is unusual at first pass, so we invite the reader to consult Example 2 for a discussion.

Definition 8. *Let $(D_u)_{u<\omega}$ be an effective enumeration of the finite subsets of ω and $(D_v^1)_{v<\omega}$ an effective enumeration of the subsets of ω of cardinality at most 1 (e.g., $D_v^1 = \{v-1\} \cap \omega$).*

Let $X, Y \subseteq \omega$. Then X is enumeration-reducible to *Y, denoted $X \leq_e Y$, if there is a c.e. set W such that*

$$n \in X \iff \exists u\ (\langle n, u\rangle \in W \wedge D_u \subseteq Y).$$

X is 1-enumeration-reducible to *Y, $X \leq_e^1 Y$, if there is a c.e. set W such that*

$$n \in X \iff \exists u, v\ (\langle n, u, v\rangle \in W \wedge D_u \subseteq Y \wedge D_v^1 \subseteq Y^c).$$

Finally, X is Ziegler-reducible to *Y, $X \leq^* Y$, if $X \leq_e Y$ and $X^c \leq_e^1 Y$.*

Example 2. Since we can give an effective list $\{w_i\}_{i\in\omega}$ of all words on alphabet Σ, we can identify a word w with its index. Now, for every $B \subseteq \Sigma^*$, we have $\mathcal{L}^c(\langle\Sigma \mid B\rangle^d) \leq_e B$. The reduction is given by the c.e. set

$$W = \{\langle w, u\rangle \mid \exists n \text{ (every extension } v \text{ of } w \text{ of length } n \text{ contains a word from } D_u)\}$$

Intuitively, $X \leq^* Y$ if and only if positive information about X (i.e., statements of the form $n \in X$) can be determined by asking finitary positive questions about Y, while negative information can be determined by asking a series of questions involving at most one negative query and finitely many positive queries. The notion was introduced by Martin Ziegler in [10], in which he showed that it is closely related to parameter-definability in groups.

Theorem 2. *Let S and T be subshifts in finite alphabets. Then S is finitely determined over T if and only if $\mathcal{L}^c(S) \leq^* \mathcal{L}^c(T)$.*

We prove the two directions independently.

4.1 Finitely Determined Implies Ziegler Reducible

Lemma 1. *If S is finitely determined over T, then $\mathcal{L}^c(S) \leq^* \mathcal{L}^c(T)$.*

Proof. Let S be finitely determined over T via $(\Sigma, d, G, B, \hat{f})$, and let R be a d-dimensional subshift over Σ with $B \subseteq \mathcal{L}^c(R)$, $G \subseteq \mathcal{L}(R)$, and $T \sqsubseteq f(R)$.

Let $U = \hat{f}^{-1}(\mathcal{L}^c(T))$ and set $Q = \langle\Sigma \mid B \cup U\rangle^d$. Then, since $T \sqsubseteq f(R)$, we have $U \cap \mathcal{L}(R) = \emptyset$ and $R \subseteq Q$, so $f(R) \subseteq f(Q)$. Hence $Q \models L(G) \wedge L^c(B)$, and $T \sqsubseteq f(Q)$. Since this determines S over T, we have $S \sqsubseteq f(Q)$. Furthermore, since B is finite and $\hat{f}$ is a finite block code, $\mathcal{L}^c(Q) \leq_e U \leq_e \mathcal{L}^c(T)$.

Because $\hat{f}$ is a block code, $\mathcal{L}(f(Q)) \leq_e \mathcal{L}(Q)$. Putting this together, we get

$$\mathcal{L}^c(T) \geq_e \mathcal{L}^c(Q) \geq_e \mathcal{L}^c(f(Q)) \geq_e \mathcal{L}^c(S).$$

To show that $\mathcal{L}(S) \leq_e^1 \mathcal{L}^c(T)$, let $w \in \Sigma_S^{*ds}$ and $U_w = \hat{f}^{-1}(w)$. (Since $\hat{f}$ is finite-to-one, this will be a finite set of words in Σ^{*d}.) Thus, $w \in \mathcal{L}^c(S)$ if and only if $U_w \subseteq \mathcal{L}^c(Q)$.

If $w \in \mathcal{L}^c(S)$, then $Q/U_w = Q$ and so $\mathcal{L}^c(f(Q/U_w)) \cap (\mathcal{L}(T) \cup G) = \emptyset$. On the other hand, if $w \in \mathcal{L}(S)$, then $S \not\sqsubseteq f(Q/U_w)$, so if $G \subseteq \mathcal{L}(Q/U_w)$, we must have $T \not\sqsubseteq f(Q/U_w)$. Since $f(Q/U_w) \subseteq f(Q)$, this must be because some $u \in \mathcal{L}(T)$ is in the co-language of $f(Q/U_w)$. Hence, if $w \in \mathcal{L}(S)$, then $\mathcal{L}^c(f(Q/U_w)) \cap (\mathcal{L}(T) \cup G) \neq \emptyset$.

Observe that $\mathcal{L}^c(f(Q/U_w)) \leq_e \mathcal{L}^c(T)$, since U_w is finite and $\mathcal{L}^c(f(Q)) \leq_e \mathcal{L}^c(T)$, as argued above. Thus $\mathcal{L}(S) \leq_e^1 \mathcal{L}^c(T)$ via the following algorithm for each w: Using $\mathcal{L}^c(T)$, enumerate $\mathcal{L}^c(f(Q/U_w))$. For each element, check if it is in G (since G is finite, this can be built into the 1-enumeration operator) or $\mathcal{L}(T)$. The latter check requires information about a single bit of $\mathcal{L}(T)$ at a time. Thus, $\mathcal{L}(S) \leq_e^1 \mathcal{L}^c(T)$, and so $\mathcal{L}^c(S) \leq^* \mathcal{L}^c(T)$. □

4.2 Ziegler Reducible Implies Finitely Determined

We now complete the proof of Theorem 2.

Lemma 2. *Let S and T be subshifts with $\mathcal{L}^c(S) \leq^* \mathcal{L}^c(T)$. Then S is finitely determined over T.*

The proof relies heavily on the construction of a subshift which acts as an oracle for the language of a given subshift. This *oracle shift* was introduced in the proof of Theorem 23 in [7], their analogue of the relativised Higman Embedding Theorem. The other key ingredient is the Simulation Theorem in [1,3], which shows that every effectively closed shift of dimension d is the d-dimensional slice of a factor of a $d+1$-dimensional SFT.

The Oracle Shift. Since it was sketched in [7], we provide details of the construction here. It may be useful to also reference the proof of Theorem 23 in [7]. The idea of the oracle shift is that, given a d-dimensional shift S in a finite alphabet, it effectively produces a $d+1$-dimensional shift O_S that displays the language of S in a uniform way. The forbidden words of O_S can be chosen as $F \cup \mathcal{L}^c(S)$, where the set F is c.e. Further, $\mathcal{L}(S) \leq_1 O_S$—and in fact, a stronger relation holds: $S \sqsubseteq O_S$.

Let S be a d-dimensional subshift in the alphabet Σ and let $\Sigma_{O_S} = \Sigma \cup \{\#\}$.

The subshift O_S can be thought of as a "stack" of d-dimensional shifts, which we, borrowing the terminology of the $d=1$ case, call "rows". Each d-dimensional row in this stack will be associated with a number $i \in \omega \cup \{\infty\}$ called the *type* of the row. Rows of type i consist of words $w \in \mathcal{L}(S)$ which are on hypercubes of length $2^i - 1$, surrounded by a single layer of #s on each side. These rows will be periodic, with period $2^i \cdot |\Sigma|^{2^i}$ in each of their d dimensions.

Further, the rows interact. Each of the rows of type i are identical, and appear every 2^i rows. Thus every other row is a row of type 1, and every other of the remaining rows is a row of type 2, etc. Moreover, each connected word in Σ on a row of type i appears as a subword of some word in the rows of type $i+1$.

The construction of O_S is then essentially the following: first $O_{\langle\Sigma|\rangle}$ is constructed, which has c.e. set of forbidden words. O_S is then obtained by further forbidding $\mathcal{L}^c(S)$.

Say that a word d-dimensional word w on alphabet Σ_{O_S} is a *valid pattern of type* i if its domain is a hypercube of length $2^i + 1$ where the hypercube of length $2^i - 1$ in the interior only has letters from Σ and it is wrapped by a layer of #, which we call the *boundary* of the valid pattern. Each row of type i will consist of only valid pattern of type i with shared boundaries. Let F be the set of forbidden $d+1$-dimensional words described below.

Building a grid: For each i, forbid the words on hypercubes of length $2^i + 1$ such that none of its rows is consistent with a valid pattern of type i.

Fixing the type: For each i, forbid the words that contain a valid pattern of type i and something else inconsistent with a valid pattern of type i in the same row.

Ensuring periodicity: For each i, forbid words with length-$2^{i-1}-1$-hypercubes of letters from Σ, surrounded by #s, which are not periodic in one of the first d-dimensions with period length-$2^{i-1} \cdot |\Sigma|^{2^{d(i-1)}}$.

Recurrence of slices: For each i, forbid words that contain a valid pattern of type i and anything other than the same valid pattern 2^i rows above it.

Subword-closed: For each i, forbid words which contain a valid pattern of type i, which contains a subword of length $2^{i-1} - 1$ not in a row of type $i - 1$.

This c.e. set of forbidden words provides the scaffolding for the oracle shift. Then $O_S = \langle \Sigma, \# \mid F \cup \mathcal{L}^c(S) \rangle^{d_S+1}$.

Proof of Lemma 2. Without loss of generality, suppose $\Sigma_S \cap \Sigma_T = \emptyset$. Further, by considering $T^{(n)}$ or $S^{(n)}$ as appropriate (cf. Definition 3), we may assume that S and T have the same dimension d. Further, by Theorem 1, we may assume $T \neq \emptyset$.

Let W_i witness that $\mathcal{L}^c(S) \leq_e \mathcal{L}^c(T)$ and W_j witness that $\mathcal{L}(S) \leq_e^1 \mathcal{L}^c(T)$. Then, for any R with $T \sqsubseteq R$, we have $S \sqsubseteq R$ if and only if the following c.e. sets of implications hold:

$$I = \left\{ \bigwedge_{v \in D_n} v \in \mathcal{L}^c(R) \to w \in \mathcal{L}^c(R) \mid \langle w, n \rangle \in W_i \right\}$$

$$J = \left\{ \bigwedge_{v \in D_n} v \in \mathcal{L}^c(R) \wedge D^1_m \subseteq L(R) \to w \in \mathcal{L}(R) \mid \langle w, n, m \rangle \in W_j \right\}.$$

We show that $I \cup J$ can be coded into a set of forbidden words in dimension $d+1$. The intuition is to modify the construction of the oracle shift described above by starting with a c.e. set of forbidden words that ensure the shift consists of interleaved d-dimensional rows from $O_{\langle \Sigma_S | \rangle}$ and $O_{\langle \Sigma_T | \rangle}$ (see Fig. 1). We can then add a further c.e. set of forbidden words that code the implications $I \cup J$ (see Fig. 2). We assume that $O_{\langle \Sigma_S | \rangle}$ and $O_{\langle \Sigma_T | \rangle}$ have distinct safe symbols. That it, their alphabets are $\Sigma_S \cup \{\#\}$ and $\Sigma_T \cup \{\%\}$, respectively. Further, recalling that our convention that the domains of words be rectangles, write $|w|$ for the longest side of w.

Let $\mathcal{F}$ be the set of forbidden words on alphabet $\Sigma_S \cup \Sigma_T \cup \{\#, \%\}$ as follows.

Alternation of S-rows and T-rows: Forbid words that do not consist of a word in Σ_{O_S} directly below (in the last dimension) a word in Σ_{O_T}.

Synch the S-rows and the T-rows: Forbid words to ensure that S-rows of type i sit directly above T-rows of type i.

Oracle for T: The forbidden words that ensure the T-rows are configurations of O_T (see Sect. 4.2).

Oracle for S: Forbidden words to ensure that the S-rows are of the form of configurations of an oracle shift in the alphabet Σ_{O_S} (see Sect. 4.2).

Coding I: For each $\langle w, u\rangle \in I$, forbid all configurations containing w on an S-row for words of type $\geq \log_2(|w|+1)$ but no element of D_u in the corresponding places on the T-rows (see Fig. 2).

Coding J: Forbid all configurations that have no w on an S-row in the appropriate type, no $v \in D_n$ in the corresponding places in the oracle rows and which do have D^1_m in the corresponding place in the oracle rows (see Fig. 2).

Note that, while in the coding of I and J we are forbidding configurations (instead of words), thus can be accomplished by finite patterns because of the horizontal periodicity of O_T and O_S, and the regularity of the rows of type i. In fact, for each implication $\langle w, u\rangle \in W_i$ or $\langle w, n, m\rangle \in W_j$ it is more than enough to look at hypercubes of length $2^l \cdot |\Sigma|^{2^{d(l)}}$, where the longest word in the implication has length l. Moreover, observe that $\mathcal{F}$ is c.e.

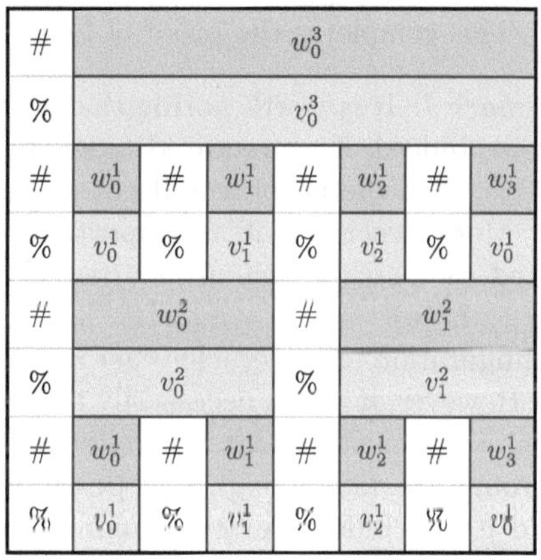

Fig. 1. A window of the combined oracle shift. The green words are from $\mathcal{L}(T)$ while the blue are from $\mathcal{L}(S)$. The superscript on each word represents the *type* of the row. By subword closure, all the subwords of each word of length 2^i-1 appear in the rows of type $i-1$.

Let $Q = \langle \Sigma_S \cup \Sigma_T \cup \{\#, \%\} \mid \mathcal{F}\rangle^{d+1}$. Then Q is effectively closed, so by Theorem 1 of [3] or Theorem 3 of [1], there is a SFT R in $d+2$ dimensions and a factor map $f : \Sigma_R^{\mathbb{Z}^{d+2}} \to \Sigma_Q^{\mathbb{Z}^{d+2}}$ such that $Q = f(R) \restriction_{d+1}$. Let B be the finite set of forbidden words of R and $\hat{f}$ a finite block code for f. We claim that S is finitely determined by $(\Sigma_R, d+2, \emptyset, B, \hat{f})$ over T.

Indeed, suppose that P is a subshift in dimension $d+2$ and alphabet Σ_R with $B \subseteq \mathcal{L}^c(P)$. Let $P_0 = f(P) \restriction_{d+1}$ and assume $T \sqsubseteq P_0$. Then, P_0 has more forbidden words than $f(R)$. In particular, it contains the forbidden words above, which force P_0 to be a subset of the oracle shift. Let $w \in \mathcal{L}(S)$. Then there is an axiom $\langle w, n, m\rangle \in W_j$ such that $D_n \subseteq \mathcal{L}^c(T)$ and $D^1_m \subseteq \mathcal{L}(T)$. Let $D^1_m = \{v\}$. Since $T \sqsubseteq f(P)$, take a configuration x containing v, which must not contain any word in D_n. Thus, by forbidden words coding J, w must appear in x.

For the other direction, let $w \in \mathcal{L}^c(S)$. This is witnessed by an enumeration axiom of the form $\langle w, u\rangle \in W_i$, where $D_u \subseteq \mathcal{L}^c(T)$. Then, *on every configuration*

w_0^2			#	w_1^2			#	w_2^2		
v_0^2			%	v_1^2			%	v_2^2		
w_0^1	#	w_1^1	#	$\boldsymbol{w}$	#	w_3^1	#	w_4^1	#	w_0^1
v_0^1	%	v_1^1	%	v_2^1	%	v_0^1	%	v_1^1	%	v_2^1

w_0^2			#	w_1^2			#	w_2^2		
v_0^2			%	v_1^2			%	v_2^2		
w_0^1	#	w_1^1	#	w_2^1	#	w_3^1	#	w_4^1	#	w_0^1
v_0^1	%	$\boldsymbol{t}$	%	v_2^1	%	v_0^1	%	$\boldsymbol{t}$	%	v_2^1

Fig. 2. Example of tiles coding the axioms of I (left) and J (right). The I-tile represents an axiom $\langle w, u \rangle$ where D_u only has words of type 2 and are not any v_i^2 (red row). The J-tile represents an axiom $\langle w, u, v \rangle$ where w and all the words in D_u have type 2 and do not appear in the corresponding row (red) and $D_v^1 = \{t\}(Color figure online)$ with t of type 1.

of P_0, $D_u \subseteq \mathcal{L}^c(x)$. Since, since P_0 is a subset of the oracle shift, and forbids the words coding I, it must be that w does not appear on any configuration, and hence $w \in \mathcal{L}^c(P_0)$.

This completes the proof of Lemma 2.

Remark 1. It is worth noting that where Ziegler reduction features in this argument. Indeed, if we required more than the single piece of negative information in $\leq_e^1$, then the proof would not go through.

More precisely, it is important that the Ziegler axioms can be formulated as *positive, quantifier-free Horn formulas*; i.e., formulas of the form $\bigwedge_{u \in D} L^c(u) \implies L^c(w)$. The key use of this fact is in the coding of J: each configuration in the oracle is set up to contain a subset of the forbidden words of T. However, it is not necessarily the case that if u and v are both words in $\mathcal{L}(T)$, that they must appear on the *same* configuration. For example, after passing through the factor map, it is possible that in P_0, every word containing both u and v is forbidden, even though $u, v \in \mathcal{L}(P_0)$. In this case, if we needed to see *both* words in order to verify that $w \in \mathcal{L}(S)$, it could be that $w \notin \mathcal{L}(P_0)$.

5 Existential Closure

Ziegler originally introduced $\leq^*$ in the context of *existentially closed groups.* Say that a group M is *existentially closed* (ec) if for any quantifier-free $\varphi(\overline{x}, \overline{m})$, where $\overline{m} \subseteq M$, if there is $N \geq M$ with $N \models \exists \overline{x} \varphi$, then $M \models \exists \overline{x} \varphi$.

Then Ziegler showed in [10] that for finitely generated groups G and H, $W(G) \leq^* W(H)$ iff G embeds into every ec group that H embeds into iff G is $\exists_1$-isolated over H (Definition 6; see also [8, Chapter 4] for more detail on the background). Thus there is an equivalence between (1) relative effective content, (2) relative embeddability in an algebraic structure, and (3) relative $\exists_1$-definability for groups.

Notably, Belegradek [2] gave conditions for Ziegler's theorem to hold in a general algebraic context. In Theorem 2, we translate the equivalence between (1) and (3) into the theory of subshifts. In this section, we introduce a notion of existential closure for subshifts, namely being *existentially closed for consistent systems*, in parallel to the definition for groups (Definition 10).

Subshifts which are existentially closed for consistent systems have some commonality with existentially closed groups (for example, both classes are amenable to model-theoretic forcing arguments), but the class nonetheless fails to satisfy an analogue of Ziegler's Theorem—indeed, there are no subshifts S such that $S \precsim M$ for every M which is existentially closed for consistent systems.

Definition 9. *Let U be a set of words. We write $U(\overline{a})$ to denote that the range of each word in U is contained in $\overline{a}$. Further, we write $U(\overline{x}, \overline{a})$ to highlight that $\overline{a}$ comes from some given alphabet Σ and $\overline{x}$ comes from an alphabet disjoint from Σ, interpreted as "variables".*

Let T be a subshift. Then $\exists \overline{x}\ (L(G(\overline{x}, \overline{a})) \wedge L^c(B(\overline{x}, \overline{a})))$ is consistent with *T if $\overline{a} \subseteq \Sigma_T$ and there is $S \sqsupseteq T$ with $\overline{b} \subseteq \Sigma_S$ such that $S \models L(G(\overline{b}, \overline{a})) \wedge L^c(B(\overline{b}, \overline{a}))$. In this case we say that $L(G(\overline{x}, \overline{a})) \wedge L^c(B(\overline{x}, \overline{a}))$* is satisfied *in S.*

If there is T with alphabet Σ which satisfies $L(G(\overline{x}, \overline{a})) \wedge L^c(B(\overline{x}, \overline{a}))$, then say that $(G(\overline{x}, \overline{a}), B(\overline{x}, \overline{a}))$ is a consistent pair for Σ.

Remark 2. The alphabet Σ is important in this last condition: $(\{a\}, \{aa\})$ is not a consistent pair for $\Sigma = \{a\}$ but is for $\Sigma = \{a, \#\}$, as witnessed by the set of strings in $\{a, \#\}^{\mathbb{Z}}$ with no repeated as.

The following proposition shows that it is straightforward to check if $L(G(\overline{x}, \overline{a})) \wedge L^c(B(\overline{x}, \overline{a}))$ is consistent with T.

Proposition 2. *Let T be a subshift. Then $\exists \overline{x}\ (L(G(\overline{x}, \overline{a})) \wedge L^c(B(\overline{x}, \overline{a})))$ is consistent with T iff $G \cap \mathcal{L}^c(T) = \emptyset$ and for every $w \in G$ and $v \in B$, w does not contain v as a subword. Further, if $\exists \overline{x}\ (L(G(\overline{x}, \overline{a})) \wedge L^c(B(\overline{x}, \overline{a})))$ is consistent with T, then it can be satisfied in a subshift in the same dimension as T.*

Moreover, $(G(\overline{x}, \overline{a}), B(\overline{x}, \overline{a}))$ is a consistent pair for Σ with $|\Sigma \setminus \overline{a}| > 1$ if and only if for every $w \in G$ and every $v \in B$, w does not contain v as a subword.

Proof. Suppose that G and B satisfy the hypothesis of the proposition. Let $\Sigma = \Sigma_T \cup \{\#\} \cup \{\overline{b}\}$, where $|\overline{b}| = |\overline{x}|$. Then consider the configuration y consisting of, for each $w(\overline{x}, \overline{a}) \in G$, the word $w(\overline{b}, \overline{a})$ surrounded by $\#$. For example, in one dimension, if $G = \{a_0 x, a_1 x a_0\}$, then y is the configuration

$$\cdots \#\#\# a_0 b \# a_1 b a_0 \#\#\# \cdots$$

Let S be the least subshift in alphabet Σ which is a superset of $T \cup \{y\}$. Notice S has dimension d_T, and it is evident from the construction that $T \sqsubseteq S$.

The same argument works for the second part of the claim.

Definition 10. *A subshift M in alphabet Σ is* existentially closed for consistent systems (in dimension d), *or* ecfcs, *if for every pair of finite sets $U(\overline{x}, \overline{a})$ and $V(\overline{x}, \overline{a})$ of words in $(\Sigma \cup X)^{*d}$, if $\exists \overline{x}\ (L(U(\overline{x}, \overline{a})) \wedge L^c(V(\overline{x}, \overline{a})))$ is consistent with M, then $L(U(\overline{x}, \overline{a})) \wedge L^c(V(\overline{x}, \overline{a}))$ is satisfied in M.*

Remark 3. We will show below that if M is existentially closed for consistent systems, then Σ_M must be infinite. From this we lose the compactness of the space $\Sigma^{\mathbb{Z}^d}$, which means that many classical results, like the Curtis–Hedlund–Lyndon Theorem, are not available in this setting.

We start by verifying that existentially closed subshifts for dimension d exist.

Theorem 3. *For every d, there is a subshift that is ecfs in dimension d.*

The proof is a simple forcing argument. Because of its simplicity and flexibility, we expect that this construction may have other applications. As examples, we present some properties of subshifts which are ecfcs and which are obtained by straightforward modifications of the following proof.

Proof. Let $\Sigma = \{a_i \mid i < \omega\}$ be an infinite alphabet and let $\{F_k \mid k < \omega\}$ be an effective enumeration of the finite subsets of $(\Sigma \cup X)^{*d}$, where X is an infinite subset, disjoint from Σ, thought of as "variable letters". We build by finite approximation sets G, B of words in Σ such that $G \subseteq \mathcal{L}(\langle \Sigma \mid B\rangle)$. $\langle \Sigma \mid B\rangle$ will be the required subshift.

Stage 0: $G_0 = \varnothing$; $B_0 = \varnothing$.

Stage $s+1$: Let $s = \langle j, k\rangle$ and consider the pair (F_j, F_k). If there is T with $\Sigma_T \supseteq \Sigma$ that satisfies $L(G_s(\overline{a}) \cup F_j(\overline{x}, \overline{a})) \wedge L^c(B_s(\overline{a}) \cup F_k(\overline{x}, \overline{a}))$, let $\overline{b} \subseteq \Sigma \setminus \overline{a}$ and set $G_{0,s+1} = G_s(\overline{a}) \cup F_j(\overline{b}, \overline{a})$ and $B_{s+1} = B_s(\overline{a}) \cup F_k(\overline{b}, \overline{a})$. Then, let G_{s+1} be obtained from $G_{0,s+1}$ by adding, for each $w \in G_{0,s+1}$ a word w^+ in Σ that contains w as a subword such that the taxicab metric from any letter in w to the edge of the word is $\geq s$ and such that w^+ does not contain as a subword any word from B_{s+1}. We check below that, for every s, $G_{0,s+1} \subseteq \langle \Sigma \mid B_{s+1}\rangle^d$, so, looking at a point of $\langle \Sigma \mid B_{s+1}\rangle^d$, we may find words with arbitrarily large domains containing w and not containing any element of B_{s+1}; hence we will always be able to find such w^+.

This ends the construction. Let $G = \bigcup_{s<\omega} G_s$ and $B = \bigcup_{s<\omega} B_s$, and let $M = \langle \Sigma \mid B\rangle^d$.

We now verify that $G = \mathcal{L}(M)$ and that M is ecfcs.

We start by showing that for each s, $G_s \subseteq \mathcal{L}(\langle \Sigma \mid B_s\rangle^d)$. This is evidently true for $s = 0$. Suppose that $G_s \subseteq \langle \Sigma \mid B_s\rangle^d$. If there is no T with $\Sigma_T \supseteq \Sigma$ that satisfies $(L(G_s(\overline{a}) \cup F_j(\overline{x}, \overline{a}) \wedge L^c(B_s(\overline{a}) \cup F_k(\overline{x}, \overline{a}))$, then $B_{s+1} = B_s$ and so $G_s \subseteq \langle \Sigma \mid B_{s+1}\rangle$ and we may pick w^+ so that $G_{s+1} \subseteq \mathcal{L}(\langle \Sigma \mid B_{s+1}\rangle)$. If there is such T, then by Proposition 2, $(G_{0,s+1}, B_{s+1})$ is a consistent pair for Σ, so $G_{0,s+1} \subseteq \mathcal{L}(\langle \Sigma \mid B_{s+1}\rangle)$. Thus, we may ensure that $G_{s+1} \subseteq \mathcal{L}(\langle \Sigma \mid B_{s+1}\rangle)$.

Now we verify that $G = \mathcal{L}(M)$. Let $w \in G$. Then $w \in G_s$ for some $s < \omega$. Thus, by the second part of each stage, we obtain a sequence of words such that w is a subword of w_r is a subword of w_t for every $s \leq r \leq t$. Then, since each w_t does not contain a subword from B_t, the limit word (obtained by fixing some element of w at the origin) is in $\langle \Sigma \mid B\rangle$. On the other hand, let $w \in \mathcal{L}(M)$ and $s = \langle j, k\rangle$ where $F_j = \{w\}$ and $F_k = \emptyset$. Then, M witnesses that $L(G_s \cup F_j) \wedge L^c(B_s)$ is consistent, so $w \in G_{s+1} \subseteq G$.

Finally, we check that M is ecfcs. Let $U(\overline{x}, \overline{a})$ and $V(\overline{x}, \overline{a})$ be finite sets of words with $\overline{a} \subseteq \Sigma$ and such that $\exists \overline{x}\ (L(U) \wedge L^c(V))$ is consistent with M, as witnessed by $T \sqsupseteq M$. Let j, k be such that $F_j = U$ and $F_k = V$. Then T satisfies $(L(G_s(\overline{a}) \cup F_j(\overline{x}, \overline{a}) \wedge L^c(B_s(\overline{a}) \cup F_k(\overline{x}, \overline{a}))$, so $G_{s+1} \supseteq G_s(\overline{a}) \cup F_j(\overline{b}, \overline{a})$ and $B_{s+1} = B_s(\overline{a}) \cup F_k(\overline{b}, \overline{a})$. Since $G_{s+1} \subseteq G \subseteq \mathcal{L}(M)$, we have that M satisfies $L(U) \wedge L^c(V)$, as required. □

The next proposition states that any subshift is $\sqsubseteq$-embedded in one that is existentially closed for consistent systems.

Proposition 3. *For any subshift S, there is a subshift $M \sqsupseteq S$ with M ecfcs.*

Proof sketch. Run the argument as in the proof of Theorem 3, but start with $G_0 = L(S)$ and $B_0 = L^c(S)$. □

Proposition 4. *Let M be existentially closed for consistent systems in dimension d. Then Σ_M is infinite.*

Proof. Suppose that $\Sigma_M = \{a_0, \ldots, a_{n-1}\}$ and, without loss of generality, that $a_i \in \mathcal{L}(M)$ for all $i < n$. Then, by Proposition 2, $\exists x\ (L(a_0 x) \wedge L^c(xa_0, \ldots, xa_{n-1}))$ is consistent with M, and hence satisfied in M, say by b.

However, b cannot be a_i for any i: for every i, because $a_i \in \mathcal{L}(M)$, a_i satisfies one of $\{xa_0, \ldots, xa_{n-1}\}$. □

A similar argument implies that existentially closed shifts for consistent systems are transitive.

Proposition 5. *Let M be existentially closed for consistent systems in dimension d. Then for any $v, w \in L(M)$, there is $x \in M$ with $v, w \in L(x)$.*

We note that the following argument fundamentally relies on the convention that words are connected.

Proof. (We work in dimension 1 for simplicity, but the argument evidently generalises.) Let $v, w \in \mathcal{L}(M)$. Then, by Proposition 2, $\exists x\ L(wxv)$ is consistent with M and hence satisfied in M. □

However, this notion of existential closure does not satisfy an analogue of Ziegler's Theorem. We use $\precsim$ rather than $\sqsubseteq$ in this theorem, otherwise the statement is trivially true, by choosing M and N ecfcs's but with $\Sigma_M \cap \Sigma_N = \emptyset$.

Theorem 4. *There is no subshift $S \neq \emptyset$ such that for every subshift M that is existentially closed for consistent systems in dimension d, $S \precsim M$.*

Proof. We implicitly fix throughout a dimension d. We start by showing that any S which satisfies the condition of the theorem is "strongly finitely determined".

Suppose that $S \precsim M$ for every ecfcs M and let $\Sigma_S = \{a_0, \ldots, a_{n-1}\}$ be an ordering of Σ_S. Let $\Sigma = \{c_i \mid i < \omega\}$ be an infinite alphabet and let $\{F_k \mid k < \omega\}$ again be an effective enumeration of the finite subsets of $(\Sigma \cup X)^{*d}$, where X is an infinite subset, disjoint from Σ. If $|\overline{c}| = |\Sigma_S|$, we write $S_{\overline{c}}$ for the subshift on alphabet $\overline{c}$ obtained by replacing a_i with c_i.

Claim. There is an alphabet Δ, a pair (G, B) of finite sets of words in Δ, and some $\bar{c} \subseteq \Sigma$ such that for every $\Delta' \supseteq \Delta$, and every T with alphabet Δ' if $T \models L(G) \wedge L^c(B)$, then $S_{\bar{c}} \sqsubseteq T$.

Proof of Claim. Suppose S does not satisfy the conclusion of the claim. Then we will build M by modifying the proof of Theorem 3 with $S \not\sqsubseteq M$. Let $(\bar{c}_e)_{e<\omega}$ be an enumeration of the tuples of length $|\Sigma_S|$ from Σ.

At even stages s, let (G_s, B_s) be the stage s-approximation to M. As in the previous proof, add the $\frac{s}{2}$th pair (F_j, F_k) if it is consistent to do so; and do nothing otherwise. Further add, for every $w \in G$, a word w^+ as before. This ensures that M is existentially closed for consistent systems.

At odd stages s, we ensure that the $(s-1)/2$th tuple $\bar{c} \subseteq \Sigma$ does not give rise to a full restriction of M equal to $S_{\bar{c}_{(s-1)/2}}$. Let (G_s, B_s) be the approximation to M at stage s and let Σ_s be the symbols used so far. By assumption, there is some $\Delta \supseteq \Sigma_s$ and some T with alphabet Δ such that $T \models L(G_s) \wedge L^c(B_s)$ but $S_{\bar{c}_{(s-1)/2}} \not\sqsubseteq T$. Thus, we may take a word $w \in \Delta^{*d}$ such that $(G_s \cup \{w\}, B_s)$ is a consistent pair for Δ and $w \in \mathcal{L}^c(S_{\bar{c}_{(s-1)/2}})$ or $(G_s, B_s \cup \{w\})$ is a consistent pair for Δ but $w \in \mathcal{L}(S_{\bar{c}_{(s-1)/2}})$. This ensures that $S \not\sqsubseteq M$ via $\bar{c}_{(s-1)/2}$.

We finish the proof by showing that no $S \neq \emptyset$ satisfying the claim exists. Indeed, let S be a subshift in alphabet Σ_S and let (G, B) be a consistent pair for some alphabet Λ.

Fix $\Delta \supseteq \Lambda \cup \{\#\}$. For each of the finitely many $\bar{c} \subseteq \Lambda$ with $|\bar{c}| = |\Sigma_S|$, choose $w_{\bar{c}} \in \mathcal{L}(S_{\bar{c}})$ such that, for every $v \in G$, $w_{\bar{c}}$ is not a subword of v. Here we use that $S \neq \emptyset$, so $\mathcal{L}(S_{\bar{c}})$ is infinite. Let

$$B^+ = B \cup \{w_{\bar{c}} \mid \bar{c} \subseteq \Lambda \wedge |\bar{c}| = |\Sigma_S|\}.$$

Then by Proposition 2, (G, B^+) is a consistent pair for Δ. Further, the construction of B^+ evidently guarantees that for every T with $T \models L(G) \wedge L^c(B^+)$ and every $\bar{c} \subseteq \Lambda$, $S_{\bar{c}} \not\sqsubseteq T$. □

Remark 4. One important concept in group theory that has no natural analogue in subshift theory is that of a finitely generated substructure. It would be natural, for example, to look at the alphabet slices $M \restriction_\Delta$ for $\Delta \subseteq_{fin} \Sigma_M$. However, in general, ecfcs subshifts may not have any nontrivial alphabet slices:

In fact, a construction like the one in Theorem 3 *cannot* guarantee that there are nontrivial slices. For instance, one could, at odd stages $s = 2t + 1$, add a word v^+ to B for each $v \in \mathcal{L}(\langle \bar{c} \mid B_{s+1} \rangle)$ such that v^+ contains v as a subword and $B^+ = B \cup \{v^+ \mid v \in \mathcal{L}(\langle \bar{c} \mid B_{s+1} \rangle)\}$ forms a consistent pair with G_s, this guarantees that $M \restriction_{\bar{c}} = \emptyset$.

This underscores a key tension in subshifts between any notions of existential closure and finitary definability. On the one hand, the notions of definability, as exemplified in the previous proof (via exemplified in the previous proof (via Proposition 2), are very sensitive to the ambient alphabet. However, forcing constructions tend to produce subshifts with various mixing properties; in particular, with very little "local structure".

As a final application, we observe that the technique of the proof of Theorem 4 together with the construction of Hodges in Theorem 4.1.5 in [5] allows one to show the following.

Proposition 6. *There are $2^{\aleph_0}$ ecfcs subshifts $(M_\alpha)_{\alpha<2^\omega}$ on a countably infinite alphabet Σ such that if $\alpha \neq \beta$, then $M_\alpha \not\preceq M_\beta$.*

Notice that this is optimal—any subshift in Σ is determined by a subset of Σ^{*d}, of which there are continuum many.

Proof sketch. The idea of this argument is to run the by-now standard construction of existentially closed subshifts for consistent systems, but, at each stage s, ensure that there at least 2^s different subshifts $(M_\sigma)_{\sigma\in 2^s}$. This will be done by running the construction on a binary tree. One further ensures that stage s that for every pair of tuples $\overline{c}$ and $\overline{d}$ from Σ and every pair of approximations M_σ and M_τ, there is a word $w_{\overline{c},\overline{d}}(\overline{x})$ such that $w_{\overline{c},\overline{d}}(\overline{c}) \in \mathcal{L}^c(M_\sigma)$ and $w_{\overline{c},\overline{d}}(\overline{d}) \in \mathcal{L}(M_\tau)$, or vice versa. This ensures that, for any $\alpha \succeq \sigma$ and $\beta \succeq \tau$, we have $M_\alpha \not\preceq M_\beta$ via a map extending $\overline{c} \mapsto \overline{d}$. □

Acknowledgments. The first author was supported by the Secretaría de Ciencia, Humanidades, Tecnología e Innovación de México through scholarship number 836694. Both authors would like to thank Mariya Soskova for many helpful conversations, and the anonymous referees for their detailed feedback. The authors used AI tools to aid in the construction of their figures.

Disclosure of Interests. The authors have no competing interests to declare that are relevant to the content of this article.

References

1. Aubrun, N., Sablik, M.: Simulation of Effective Subshifts by Two-dimensional Subshifts of Finite Type. Acta Appl. Math. **126**(1), 35–63 (2013)
2. Belegradek, O.V.: Higman's Embedding Theorem in a General Setting and Its Application to Existentially Closed Algebras. Notre Dame J. Formal Logic **37**(4) (1996)
3. Durand, B., Romashchenko, A., Shen, A.: Effective Closed Subshifts in 1D Can Be Implemented in 2D. In: Blass, A., Dershowitz, N., Reisig, W. (eds.) Fields of Logic and Computation. LNCS, vol. 6300, pp. 208–226. Springer, Heidelberg (2010). https://doi.org/10.1007/978-3-642-15025-8_12
4. Hochman, M.: On the dynamics and recursive properties of multidimensional symbolic systems. Invent. Math. **176**(1), 131–167 (2009)
5. Hodges, W.: Building Models by Games. London Mathematical Society Student Texts, Cambridge University Press (1985)
6. Jeandel, E.: Enumeration reducibility in closure spaces with applications to logic and algebra. In: LICS (2017)
7. Jeandel, E., Vanier, P.: A Characterization of Subshifts with Computable Language. In: 36th International Symposium on Theoretical Aspects of Computer Science. LIPIcs, vol. 126, pp. 40:1–40:16. Schloss Dagstuhl, Dagstuhl, Germany (2019)

8. Nakid Cordero, A.: The Computability of Asymmetric Information. Ph.D. thesis, University of Wisconsin–Madison (2026)
9. Rips, E.: Another characterization of finitely generated groups with a solvable word problem. Bull. Lond. Math. Soc. **14**(1), 43–44 (1982)
10. Ziegler, M.: Algebraisch Abgeschlossene Gruppen. In: Word Problems II, Studies in Logic and the Foundations of Mathematics, vol. 95. Elsevier (1980)

Termination of Real Linear Loops

Eike Neumann[(✉)] and Margret Tembo

Swansea University, Swansea, UK
{e.f.neumann,m.k.tembo}@swansea.ac.uk

Abstract. We study the problem of deciding universal termination of linear and affine loops over the reals in the bit-model of real computation. We show that both problems are as close to decidable as one can expect them to be: there exist sound partial algorithms that halt on all problem instances whose answer is robust under all sufficiently small perturbations. We further show that in each case the set of non-robust problem instances has Lebesgue measure zero.

Keywords: Computable analysis · Automated verification · Dynamical systems

1 Introduction

Discrete-time linear systems arise naturally in a wide range of application areas across systems and control theory and verification. A fundamental liveness verification task is to decide whether every system state in a given set eventually escapes the set under the dynamics. Here we consider the following instantiation of this problem: does every point in a given polyhedron P escape P under the iteration of a given matrix A? Tiwari [15] has shown that this problem is decidable when A and P are specified by exact rational data.

In applications, the system under consideration may not always be specifiable by exact rational data – for example, it may not be possible to assume that the transition probabilities of a Markov chain be exactly known. This leads to the question whether the problem remains decidable when the matrix or polyhedron are known only approximately, but with guaranteed error bounds. Following [7], we model this situation by studying the problem for matrices and polyhedra specified by real data in the sense of computable analysis [3,5,16]. This line of work also connects to recent developments in exact real computation and program verification, such as the Hoare-logic framework of Park et al. [9], which provides a natural setting for verification problems over real data. Further, Tiwari's method in general requires exact computation with algebraic numbers, in particular symbolic computation of the Jordan normal form, which is computationally very expensive (cf. the discussion in [15, p.80]). We will solve the real-number version of the problem using validated numerical methods, which are potentially much more efficient.

V. Brattka et al. (Eds.): CiE 2026, LNCS 16674, pp. 437–452, 2027.
https://doi.org/10.1007/978-3-032-31348-5_29

Let us define the decision problem we consider more precisely. For real vectors $v, w \in \mathbb{R}^m$ of the same dimension, we write $v \rhd w$ if $v_j > w_j$ for all indexes $j \in \{1, \ldots, m\}$. The relations $v \unrhd w$, $v \lhd w$, and $v \unlhd$ are defined analogously. For a matrix $B \in \mathbb{R}^{m \times n}$, we let $P(B) = \{x \in \mathbb{R}^n \mid Bx \rhd 0\}$ denote the open polyhedron defined by B. We let $\overline{P}(B) = \{x \in \mathbb{R}^n \mid Bx \unrhd 0\}$ denote the closed polyhedron defined by B. Observe that, since we do not impose any constraints on B, the set $\overline{P}(B)$ is not necessarily equal to the closure of $P(B)$. The Linear Universal Escape Problem is the following decision problem: given two matrices $A \in \mathbb{R}^{n \times n}$ and $B \in \mathbb{R}^{m \times n}$, decide whether for all $x \in \mathbb{R}^n$ there exists $k \in \mathbb{N}$ with $A^k x \notin P(B)$.

Since computability implies continuity, no non-trivial subset of Euclidean space has a computable characteristic function. Thus, we cannot hope to decide any interesting problem over real data in the naïve sense of computing its characteristic function. However, we can ask for the next best thing: A subset S of a represented space X is called *maximally partially decidable* [7] if there exists an algorithm which takes as input an element $x \in X$, and either runs forever or halts in finite time. If it halts, it must report correctly whether x belongs to S. Moreover, it is required to halt on all inputs $x \notin \partial S$. Observe that if X is discrete in the sense that equality for elements of X is semidecidable, then a subset of X is decidable in the usual sense if and only if it is maximally partially decidable. Thus, maximal partial decidability generalises decidability over discrete spaces – generally in a more meaningful way than the naïve definition. See [7] for further discussion and motivation of this concept. When we think of $S \subseteq X$ as a decision problem, we also refer to points $x \in X$ as *instances of* S. A point $x \in \partial S$ is called a *boundary instance.* A point $x \in X \setminus \partial S$ is called a *robust instance.*

Our first main result is the following:

Theorem 1. *The Linear Universal Escape Problem is maximally partially decidable.*

Tiwari's decision method [15] for the universal escape problem for rational inputs relies on the computation of the Jordan normal form, which is well-known to be uncomputable for real matrices for continuity reasons (cf. [17]). Hence, we pursue an entirely different approach: We characterise the robust problem instances via first-order formulas, where all universal quantification ranges over compact sets, and all existential quantification ranges over overt sets. This immediately yields a maximal partial algorithm.

To state our characterisation results, let us introduce some notation. For a matrix $A \in \mathbb{R}^{n \times n}$, we let $\sigma(A) \subseteq \mathbb{C}$ denote its spectrum. For a real number r, we let $\sigma_{\geq r}(A) = \sigma(A) \cap [r, +\infty)$ and $\sigma_{>r}(A) = \sigma(A) \cap (r, +\infty)$. We let $\sigma^{\text{odd}}_{>r}(A)$ denote the eigenvalues in $\sigma_{>r}(A)$ with odd algebraic multiplicity. We let A_j denote the jth row of A. Let $S^{n-1} \subseteq \mathbb{R}^n$ denote the (Euclidean) unit sphere. We call an instance (A, B) *trapped* if some $x \in P(B)$ satisfies $A^k x \in P(B)$ for all $k \geq 0$ and *escaping* otherwise. Robust escaping and robust trapped instances are those lying in the interiors of the escaping and trapped sets, respectively.

Theorem 2. *Let $(A, B) \in \mathbb{R}^{n\times n} \times \mathbb{R}^{m\times n}$ be an instance of the Linear Universal Escape Problem. Then:*

1. *(A, B) is a robust escaping instance if and only if it satisfies the formula*

$$\forall \lambda \in \sigma_{\geq 0}(A). \forall v \in S^{n-1}. (Av = \lambda v \implies \exists j \in \{1, \ldots, m\}. B_j v < 0). \quad (1)$$

2. *(A, B) is a robust trapped instance if and only if it satisfies the formula*

$$\exists \lambda \in \sigma_{>0}^{odd}(A). \forall v \in S^{n-1}. (Av = \lambda v \implies (Bv \rhd 0 \vee Bv \lhd 0)). \quad (2)$$

So far, we have focused on (homogeneous) linear systems. It is natural to consider the same problem for affine systems. The Affine Universal Escape Problem is the following: Given $A \in \mathbb{R}^{n\times n}$, $b \in \mathbb{R}^n$, $B \in \mathbb{R}^{m\times n}$, $\eta \in \mathbb{R}^m$, decide whether every point escapes the affine polyhedron $P(B, \eta) = \{x \in \mathbb{R}^n \mid Bx \rhd \eta\}$ under iteration of the map $f(x) = Ax + b$ in the sense that for all $x \in \mathbb{R}^n$ there exists $k \in \mathbb{N}$ with $f^k(x) \notin P(B, \eta)$. Over discrete data, the Linear Universal Escape Problem is a special case of the Affine Universal Escape Problem. Further, the Affine Universal Escape Problem reduces straightforwardly to its homogeneous linear counterpart via homogenisation. Explicitly, given an instance (A, B, b, η) of the Affine Universal Escape Problem, we can effectively compute the homogenised instance $\left(\widehat{A}, \widehat{B}\right) \in \mathbb{R}^{(n+1)\times(n+1)} \times \mathbb{R}^{(m+1)\times(n+1)}$, given as

$$\widehat{A} = \begin{pmatrix} A & b \\ 0 & 1 \end{pmatrix}, \; \widehat{B} = \begin{pmatrix} B & -\eta \\ 0 & 1 \end{pmatrix}.$$

It is easy to see that (A, B, b, η) escapes if and only if $\left(\widehat{A}, \widehat{B}\right)$ escapes. However, in order for this mapping to be a reduction in the sense of maximal partial decidability, it needs to map robust instances of the Affine Universal Escape Problem to robust instances of the Linear Universal Escape Problem. This turns out not to be the case, even for one-dimensional instances[1]. Similarly, it is easy to see that there exist robust instances (A, B) of the Linear Universal Escape Problem such that the instance $(A, 0, B, 0)$ is not a robust instance of the Affine Universal Escape Problem[2]. Thus, while the two problems are closely related, there is no obvious reduction in either direction when it comes to maximal partial decidability. Nonetheless, we obtain:

Theorem 3. *The Affine Universal Escape Problem is maximally partially decidable.*

Theorem 3 is based on a characterisation of the robust instances, similar to Theorem 2:

[1] Take for example $A = 1/2$, $b = -1$, $B = 1$, and $\eta = 0$.
[2] Take for example $A = 1/2$ and $B = 1$.

Theorem 4. *Let $(A, B, b, \eta) \in \mathbb{R}^{n \times n} \times \mathbb{R}^{m \times n} \times \mathbb{R}^n \times \mathbb{R}^m$ be an instance of the Affine Universal Escape Problem. Then:*

1. *(A, B, b, η) is a robust escaping instance if and only if it satisfies the formula*

$$\forall \lambda \in \sigma_{\geq 1}\left(\widehat{A}\right).\forall \widehat{v} \in S^n.\left(\widehat{A}\widehat{v} = \lambda \widehat{v} \implies \exists j \in \{1, \ldots, m+1\}.\widehat{B}_j \widehat{v} < 0\right). \quad (3)$$

2. *(A, B, b, η) is a robust trapped instance if and only if it satisfies the formula*

$$\begin{aligned} &(1 \notin \sigma(A) \wedge B \cdot (A - I)^{-1} b \lhd -\eta) \\ &\vee \exists \lambda \in \sigma^{odd}_{>1}(A).\forall v \in S^{n-1}.(Av = \lambda v \implies (Bv \rhd 0 \vee Bv \lhd 0)). \end{aligned} \quad (4)$$

Observe that (4) contains a slight abuse of notation: the matrix $(A - I)^{-1}$ is well defined if and only if $1 \notin \sigma(A)$. Thus, the second conjunct in the first disjunctive clause cannot be evaluated independently of the first conjunct.

Since our decision methods cannot halt on every input, it is natural to investigate the size of their halting sets. We observe that our algorithms halt on almost every input:

Theorem 5.

1. *The set of boundary instances of the Linear Universal Escape Problem has Lebesgue measure zero.*
2. *The set of boundary instances of the Affine Universal Escape Problem has Lebesgue measure zero.*

The rest of the paper is structured as follows: In Sect. 2 we summarise required results and definitions from computable analysis and linear systems. In Sect. 3 we show that the formulas in Theorems 2 and 4 are semidecidable. This proves Theorems 1 and 3 relative to Theorems 2 and 4. The rest of the paper is dedicated to the proofs of Theorems 2, 4, and 5. In Sect. 4 we prove a few technical lemmas that are required for the proofs of Theorems 2 and 4. The first part of Theorem 2 is proved in Sect. 5. The second part is proved in Sect. 6. Theorem 4 is proved in Sects. 7 and 8. Theorem 5 is proved in Sect. 9. The full version of this paper is available as [8].

2 Preliminaries

2.1 Represented Spaces

We assume some familiarity with the basic results and definitions of computable analysis and the theory of represented spaces. For details we refer to [10] or [5]. A *represented space* is a set X together with a partial surjective function $\delta_X \colon \subseteq \mathbb{N}^{\mathbb{N}} \to X$. Examples of represented spaces include computable metric spaces with the Cauchy representation [4,6,16]. A map $f \colon X \to Y$ is called *computable* if there exists a computable function $F \colon \operatorname{dom} \delta_X \to \operatorname{dom} \delta_Y$ satisfying

$\delta_Y \circ F = f \circ \delta_X$. It is called *continuous* if there exists a continuous such function. Clearly, computability implies continuity. Continuity in this sense is closely linked to topological continuity with respect to the final topology induced by the representation. See [10,12,13] for details.

Represented spaces enjoy excellent closure properties. In particular, they admit countable products and exponentials. This is the basis for important hyperspace constructions. Let $\Sigma = \{\top, \bot\}$ denote Sierpinski space, with the representation $\delta_\Sigma \colon \mathbb{N}^\mathbb{N} \to \Sigma$ where $\delta_\Sigma(0^\omega) = \bot$ and $\delta_\Sigma(x) = \top$ for all $x \in \mathbb{N}^\mathbb{N} \setminus \{0^\omega\}$. Then, for a represented space X, the space $\mathcal{O}(X)$ of open subsets of X is given by identification of an open set with its characteristic function as an element of the space Σ^X of all continuous functions from X to Σ. The space $\mathcal{A}(X)$ of closed subsets of X is given by identifying a closed set A with its complement as an element of $\mathcal{O}(X)$.

A subset $K \subseteq X$ is *compact* if and only if the set $\Box K = \{U \in \mathcal{O}(X) \mid K \subseteq U\}$ is an open subset of $\mathcal{O}(X)$. The set K is called *computably compact* if $\Box K$ is a computable open subset of $\mathcal{O}(X)$. Dually a subset $A \subseteq X$ is *overt* if and only if the set $\Diamond A = \{U \in \mathcal{O}(X) \mid A \cap U \neq \emptyset\}$ is an open subset of $\mathcal{O}(X)$. The set A is called *computably overt* if $\Diamond A$ is a computable open subset of $\mathcal{O}(X)$. It is easy to see that *every* set is overt, so that only computable overtness yields a (classically) meaningful notion.

For a represented space X the *space* $\mathcal{K}(X)$ *of compact subsets of* X is the space whose underlying set is the set of saturated compact subsets of X and whose representation is obtained by identifying a saturated compact set K with the set $\Box K \in \mathcal{O}(\mathcal{O}(X))$. The restriction to *saturated* compact sets is necessary, since the set $\Box K$ determines K only up to saturation. Dually, the *space* $\mathcal{V}(X)$ *of overt subsets of* X is the space whose underlying set is the set of closed subsets of X and whose representation is obtained by identifying a closed set A with the set $\Diamond A \in \mathcal{O}(\mathcal{O}(X))$. The restriction to *closed* sets is necessary, since the set $\Diamond A$ determines A only up to closure. See [5,10] for effective closure properties of compact and overt sets.

The following important proposition follows immediately from the definition:

Proposition 1. *Let X be a represented space. Given compact sets $K_1, \dots, K_n \in \mathcal{K}(X)$, overt sets $A_1, \dots, A_n \in \mathcal{V}(X)$, and an open set $U \in \mathcal{O}(X^{2n})$, we can semidecide if the sentence*

$$\forall x_1 \in K_1. \exists y_1 \in V_1. \dots \forall x_n \in K_n. \exists y_n \in V_n. (x_1, y_1, \dots, x_n, y_n) \in U$$

holds true.

2.2 Linear Dynamical Systems

We require two results on the long-term behaviour of linear dynamical systems. We will decompose the systems under consideration into systems with only positive real eigenvalues and systems without any positive real eigenvalues. For our purpose, the following theorem characterises the behaviour of the latter:

Theorem 6 (Bell and Gerhold, [2]). *Let $f : \mathbb{N} \to \mathbb{R}$ be a linear recurrence sequence without positive real eigenvalues. Then f is either eventually zero or f assumes infinitely many (strictly) negative values.*

Corollary 1. *Let E be a finite-dimensional real vector space. Let $\mathcal{L} : E \to E$ be a linear map without positive real eigenvalues. Let $\varphi : E \to \mathbb{R}$ be a linear functional. Then for all $x \in E$ there exist infinitely many $k \geq 0$ with $\varphi \circ \mathcal{L}^k x \leq 0$.*

Proof. It follows from the Cayley-Hamilton theorem that the sequence of real numbers $\varphi \circ \mathcal{L}^k x$ satisfies a linear recurrence whose characteristic polynomial is that of the matrix $\mathcal{L}$. The result now immediately follows from Theorem 6.

The long-term behaviour of linear dynamical systems with only positive real eigenvalues is captured by the next lemma:

Lemma 1. *Let E be a finite-dimensional real vector space. Let $\mathcal{L} : E \to E$ be a linear map with $\sigma(\mathcal{L}) \subseteq (0, +\infty)$. Let $\rho_1 < \cdots < \rho_N$ denote the eigenvalues of $\mathcal{L}$. Let E_j denote the space of all generalised eigenvectors for the eigenvalue ρ_j. Let $x \in E \setminus \{0\}$. Write $x = x_1 + \cdots + x_N$ with $x_j \in E_j$. Let $1 \leq M \leq N$ be maximal with the property that $x_M \neq 0$. Then:*

1. *There exists a non-negative integer d such that the sequence $\binom{k}{d}^{-1} \rho_M^{-k} \mathcal{L}^k x$ converges to an eigenvector for the eigenvalue ρ_M as $k \to \infty$.*
2. *For all $\rho > \rho_M$, the sequence $\rho^{-k} \mathcal{L}^k x$ converges to zero as $k \to \infty$.*

Proof. See [8, Lemma 9].

3 Recognising Robust Instances

Proposition 2. *The formulas* (1), (2), (3), *and* (4) *in Theorems 2 and 4 are semidecidable.*

Proof. Polynomial root finding is computable [14]: given a vector of complex numbers $(a_1, \ldots, a_d) \in \mathbb{C}^d$ we can compute complex numbers $(z_1, \ldots, z_d) \in \mathbb{C}^d$ such that $z_1, \ldots, z_d$ contains all roots of the polynomial $X^d + a_1 X^{d-1} + \cdots + a_d$. It follows that given a real matrix $A \in \mathbb{R}^{n \times n}$ we can compute the spectrum $\sigma(A)$ as an overt and compact subset of $\mathbb{C}$. To semidecide the formula (1), observe that the set $[0, +\infty)$ is a computable closed subset of $\mathbb{C}$. It follows that $\sigma_{\geq 0}(A) = \sigma(A) \cap [0, +\infty)$ is a computably compact subset of $\mathbb{C}$, uniformly in A. The set S^{n-1} is a computably compact subset of $\mathbb{R}^n$, uniformly in n. The set

$$\{(\lambda, v) \in \mathbb{C} \times \mathbb{R}^n \mid Av \neq \lambda v \vee \exists j \in \{1, \ldots, m\} : B_j v < 0\}$$

is easily seen to be a computably open subset of $\mathbb{C} \times \mathbb{R}^n$. Semidecidability of (1) follows. Semidecidability of (3) is proved analogously.

To show that (2) is semidecidable, we first observe that $\sigma_{>0}^{\mathrm{odd}}(A)$ is computable as an overt subset of $\mathbb{R}$, uniformly in A. Indeed, it follows from the fundamental theorem of algebra that a rational interval (a, b) contains a real root of a polynomial f of odd multiplicity if and only if there exist rational numbers $a < a' < b' < b$ with $f(a') \cdot f(b') < 0$. This is clearly semidecidable, yielding the claim of overtness. Now, the result follows easily from Proposition 1.

To show that (4) is semidecidable, we treat both disjunctive clauses separately. The second disjunct is semidecided analogously to (2). For the first disjunct, we use that we can compute $\sigma(A)$ as a closed subset of $\mathbb{C}$, so that we can semidecide if $1 \notin \sigma(A)$. If the we find this to be true, $(A - I)^{-1}$ is well-defined and uniformly computable in A via Gaussian elimination (cf. [17]). It is then obvious that we can semidecide the condition $B \cdot (A - I)^{-1}b \lhd -\eta$.

4 Perturbation Lemmas

Our proofs of Theorems 2 and 4 require the existence of certain matrix perturbations with prescribed properties. We summarise them in this section. Let X be a metric space and let $x \in X$. We say that *there exist arbitrarily small perturbations of x with the property P* if for all $\varepsilon > 0$ there exists $\widetilde{x} \in X$ with $d(x, \widetilde{x}) < \varepsilon$ such that $\widetilde{x}$ has the property P.

Lemma 2. *Let $A \in \mathbb{R}^{n\times n}$. Let $r \geq 0$. Let $\sigma^{\text{odd}}_{>r}(A) = \{\lambda_1, \dots, \lambda_N\}$ where $N \geq 0$. For $j = 1, \dots, N$, let v_j be an eigenvector of A for the eigenvalue λ_j. Then there exist arbitrarily small perturbations $\widetilde{A}$ of A such that $\sigma_{>r}\left(\widetilde{A}\right) = \{\lambda_1, \dots, \lambda_N\}$, all eigenvalues of $\widetilde{A}$ in $(r, +\infty)$ are simple, and $\widetilde{A}$'s eigenspace for λ_j is spanned by v_j.*

Proof. Extend the (possibly empty) set $\{v_1, \dots, v_N\}$ to a Jordan basis of $\mathbb{C}^n$. Let Q be the corresponding base-change matrix, whose first N columns are $v_1, \dots, v_N$. Then we have $A = QJQ^{-1}$ where J is of the form $J = \begin{pmatrix} D & B \\ 0 & C \end{pmatrix}$ with $D \in \mathbb{R}^{N\times N}$ being a diagonal matrix with diagonal entries $\lambda_1, \dots, \lambda_N$, $B \in \mathbb{R}^{N\times N}$, and a matrix in Jordan normal form $C \in \mathbb{R}^{(n-N)\times(N-n)}$, all of whose real eigenvalues above r have even multiplicity.

It remains to show that there exist arbitrarily small perturbations of C such that the perturbed matrices have no real eigenvalues greater than r. Up to permutation we may assume that C is a block diagonal matrix with diagonal entries $C_1, \dots, C_M$, where each C_i is a matrix of the form

$$C_i = \begin{pmatrix} \alpha_i & x_{i,1} & & \\ & \alpha_i & \ddots & \\ & & \ddots & x_{i,K_i} \\ & & & \alpha_i \end{pmatrix}$$

with α_i being an eigenvalue of C and $x_{i,j} \in \{0, 1\}$ for $j = 1, \dots, K_i$. Further $\alpha_i \neq \alpha_j$ for $i \neq j$.

Now, if α_i is a real eigenvalue of C with $\alpha_i > r$, then the multiplicity of α_i as an eigenvalue of C is even, so that by grouping the diagonal entries into blocks of two, the matrix C_i can be written in the form

$$C_i = \begin{pmatrix} C_{i,1} & Y_{i,1} & & \\ & C_{i,2} & \ddots & \\ & & \ddots & Y_{i,(K_i-1)/2} \\ & & & C_{i,(K_i+1)/2} \end{pmatrix}$$

where $Y_{i,j} \in \{0,1\}^{2\times 2}$ and the $C_{i,j}$s are 2×2-matrices of the form

$$\begin{pmatrix} \alpha_i & \\ & \alpha_i \end{pmatrix} \text{ or } \begin{pmatrix} \alpha_i & 1 \\ & \alpha_i \end{pmatrix}.$$

Now, we have $\det\left(C_i - tI_{(K_i+1)\times(K_i+1)}\right) = \prod_{i=1}^{(K_i+1)/2} \det\left(C_{i,j} - tI_{2\times 2}\right)$, so that it suffices to show that the matrices of the above form admit arbitrarily small perturbations without real eigenvalues. Now, for any $\varepsilon > 0$ the above matrices can respectively be perturbed into the matrices

$$\begin{pmatrix} \alpha_i & -\varepsilon \\ \varepsilon & \alpha_i \end{pmatrix} \text{ and } \begin{pmatrix} \alpha_i & 1 \\ -\varepsilon & \alpha_i \end{pmatrix}$$

with respective characteristic polynomials $\chi_1(z) = (\alpha_i - z)^2 + \varepsilon^2$ and $\chi_2(z) = (\alpha_i - z)^2 + \varepsilon$. Clearly, neither matrix has any real eigenvalues. This proves the claim.

Lemma 3. *Let $v \in \mathbb{R}^n$. Let $B_1, B_2 \in \mathbb{R}^n$ be linearly independent. Assume that $B_1 \cdot v \geq 0$ and $B_2 \cdot v \leq 0$. Then there exist arbitrarily small perturbations $\widetilde{v}$ of v with $B_1 \cdot \widetilde{v} > 0$ and $B_2 \cdot \widetilde{v} < 0$.*

Proof. See [8, Lemma 12].

Lemma 4. *Let $A \in \mathbb{R}^{n\times n}$ be a real matrix. Let $v_1, \ldots, v_N \in \mathbb{R}^n$ be linearly independent eigenvectors for real eigenvalues $\rho_1, \ldots, \rho_N$ of A. Then for all $\varepsilon > 0$ there exists $\delta > 0$ such that if $w_1, \ldots, w_N \in \mathbb{R}^n$ are vectors with $||v_j - w_j|| < \delta$ for all j, then there exists a matrix $\widetilde{A}$ with $\left|\left|A - \widetilde{A}\right|\right| < \varepsilon$ such that $\widetilde{A}$ has the same eigenvalues as A, counted with multiplicity, and for all $j \in \{1, \ldots, N\}$, we have $\widetilde{A}w_j = \rho_j w_j$.*

Proof. See [8, Lemma 13].

Lemma 5. *Let $A \in \mathbb{R}^{n\times n}$, $B \in \mathbb{R}^{m\times n}$, $b \in \mathbb{R}^n$, $\eta \in \mathbb{R}^m$. Assume that $1 \in \sigma(A)$ and that $B \neq 0$. Then there exist arbitrarily small perturbations $\widetilde{A}$ of A and $\widetilde{\eta}$ of η satisfying the following properties:*

1. $1 \notin \sigma\left(\widetilde{A}\right)$.

2. *The unique fixed point of the map $x \mapsto \widetilde{A}x + b$ is not contained in the closed polyhedron $\overline{P}(B, \widetilde{\eta}) = \{x \in \mathbb{R}^n \mid Bx \trianglerighteq \eta\}$.*
3. *A complex number $\lambda \in \mathbb{C}$ with $|\lambda| > 1$ is an eigenvalue of A with multiplicity μ if and only if it is an eigenvalue of $\widetilde{A}$ with multiplicity μ. Moreover, a vector $v \in \mathbb{C}^n$ is a generalised eigenvector of rank r of A for the eigenvalue λ if and only if it is a generalised eigenvector of rank r of $\widetilde{A}$ for the eigenvalue λ.*

Proof. See [8, Lemma 14].

5 Robust Escaping Instances of the Linear Problem

We now prove the first item of Theorem 2. Assume that (A, B) satisfies (1). Let $x \in \mathbb{R}^n$ be a point with $Bx \triangleright 0$. By standard linear algebra, the matrix A admits a real Jordan decomposition. Thus, we can write $\mathbb{R}^n = E_0 \oplus E_{>0} \oplus E_{\text{rest}}$, where $E_0, E_{>0}, E_{\text{rest}} \subseteq \mathbb{R}^n$ are linear subspaces of $\mathbb{R}^n$, potentially equal to zero, where E_0 is spanned by the generalised eigenvectors for the eigenvalue 0 of A, $E_{>0}$ is spanned by the generalised eigenvectors for the positive real eigenvalues of A, E_{rest} is invariant under A, and the restriction of A to E_{rest} has no non-negative real eigenvalues.

In particular, we can uniquely write $x = x_0 + x_{>0} + x_{\text{rest}}$ for vectors $x_0 \in E_0$, $x_{>0} \in E_{>0}$, $x_{\text{rest}} \in E_{\text{rest}}$. By Corollary 1, the sequence $\left(B_j A^k (x_0 + x_{\text{rest}})\right)$ assumes infinitely many non-positive values for all j. It hence suffices to show that at least one of the sequences $(B_j A^k x_{>0})_k$ for $j \in \{1, \dots, m\}$ is eventually non-positive.

If $x_{>0} = 0$, then the sequence $(B_j A^k x_{>0})_k$ is zero and hence non-positive for all j. If $x_{>0} \neq 0$, then by Lemma 1 there exists a positive real eigenvalue ρ of A and a non-negative integer such that the sequence $\binom{k}{d}^{-1} \rho^k A^k x_{>0}$ converges to an eigenvector v of A for the eigenvalue ρ.

Since the instance satisfies (1), there must exist an index $j \in \{1, \dots, m\}$ with $B_j v < 0$. Now, since multiplication with B_j is continuous and $\binom{k}{d}^{-1} \rho^k$ is positive for all k, we must have $B_j A^k x_{>0} < 0$ for all sufficiently large k. This shows that the instance is escaping. It must be robustly escaping, since the formula (1) is semidecidable by Proposition 2, so that the set of all instances that satisfy the formula is open.

Now, assume that the formula (1) is violated for (A, B). Then, by definition,

$$\exists \lambda \in \sigma_{\geq 0}(A). \exists v \in S^{n-1}. Av = \lambda v \wedge \forall j \in \{1, \dots, m\}. B_j v \geq 0.$$

Thus, there exists a non-negative eigenvalue $\lambda \geq 0$ of A with eigenvector v such that $B_j v \geq 0$ for all j. It is easy to see that there exists an arbitrarily small perturbation $\widetilde{A}$ of A such that $\widetilde{A}$ has a positive real eigenvalue $\widetilde{\lambda} > 0$ with $\widetilde{A}v = \widetilde{\lambda}v$. Now, we can replace each row B_j of B by $\widetilde{B}_j = B_j + \varepsilon v^T$ for an arbitrarily small $\varepsilon > 0$ to obtain $\widetilde{B}_j v = \widetilde{B}_j v + \varepsilon \,||v|| > 0$ for all j. Then the point $x = v$ is trapped in the polyhedron $P\left(\widetilde{B}\right)$ under $\widetilde{A}$, for we have $\widetilde{B}\widetilde{A}^k x = \widetilde{\lambda}^k \widetilde{B} x \triangleright 0$ for all k.

6 Robust Trapped Instances of the Linear Problem

We now prove the second item of Theorem 2. First, assume that (2) holds true. Then A has a positive real eigenvalue $\lambda > 0$ with odd multiplicity. Pick a unit eigenvector v for the eigenvalue λ. Then we either have $Bv \rhd 0$ or $B(-v) \rhd 0$. In the first case, the point v is trapped in $P(B)$, in the second case, the point $-v$ is trapped. This shows that the instance is trapped. It must be robustly trapped, since the formula (2) is semidecidable by Proposition 2, so that the set of all instances that satisfy the formula is open.

Now, assume that (2) is false. We will construct arbitrarily small perturbations of the instance that are escaping. By assumption we have:

$$\forall \lambda \in \sigma_{>0}^{\text{odd}}(A).\exists v \in S^{n-1}.Av = \lambda v \wedge \exists j.B_j v \leq 0 \wedge \exists j.B_j v \geq 0. \tag{5}$$

Let $0 < \rho_1 < \cdots < \rho_N$, for $N \geq 0$, denote the positive real eigenvalues of A with odd algebraic multiplicity. For each $\ell \in \{1, \ldots, N\}$, pick a witness v_ℓ of (5) for ρ_ℓ, *i.e.*, an eigenvector for the eigenvalue ρ_ℓ such that $B_{j_1} v_\ell \leq 0$ and $B_{j_2} v_\ell \geq 0$ for some indexes $j_1, j_2 \in \{1, \ldots, m\}$.

By Lemma 2, there exists an arbitrarily small perturbation $\widetilde{A}_1$ of A such that the positive real eigenvalues of $\widetilde{A}_1$ are $\rho_1, \ldots, \rho_N$, each ρ_ℓ is a simple eigenvalue of $\widetilde{A}_1$, and the corresponding eigenspace is spanned by v_ℓ.

Assume that at least two of the rows $B_1, \ldots, B_m$ of B are linearly independent. We will return to the (easy) case where all of these vectors are pairwise linearly dependent at the end of the proof. We will show that there exist arbitrarily small perturbations $w_1, \ldots, w_m$ of the vectors $v_1, \ldots, v_m$ such that for all ℓ there exist j_1 and j_2 such that $B_{j_1} w_\ell < 0$ and $B_{j_2} w_\ell > 0$. Let $j \in \{1, \ldots, m\}$. Then by construction there exist j_1 and j_2 such that $B_{j_1} v_\ell \leq 0$ and $B_{j_2} v_\ell \geq 0$. If B_{j_1} and B_{j_2} are linearly independent, then we can apply Lemma 3 to find an arbitrarily small perturbation w_ℓ of v_ℓ with the desired property. If they are linearly dependent, then $B_{j_1} = cB_{j_2}$ for some constant c. If $c < 0$ then for all $x \in \mathbb{R}^n$, the inequality $B_{j_1} x > 0$ implies $B_{j_2} x < 0$, so that the polyhedron is empty and the instance is escaping for trivial reasons. Hence, let us assume that $c \geq 0$. Then we must have $B_{j_1} v_\ell = B_{j_2} v_\ell = 0$. Now, since not all of the vectors $B_1, \ldots, B_m$ are pairwise linearly dependent, there must exist j_3 such that B_{j_3} is linearly independent of B_{j_1}. Since $B_{j_1} v_\ell \geq 0$ and $B_{j_1} v_\ell \leq 0$, we can apply Lemma 3 to B_{j_1} and B_{j_3} to obtain the desired perturbation. This proves the claim.

By Lemma 4 we can further perturb the matrix $\widetilde{A}_1$ into a matrix $\widetilde{A}_2$ such that the positive real eigenvalues of $\widetilde{A}_2$ are $\rho_1, \ldots, \rho_N$, all of these eigenvalues are simple, and their eigenspaces are spanned by the vectors $w_1, \ldots, w_N$ respectively. We can write $\mathbb{R}^n$ as a direct sum $\mathbb{R}^n = \mathbb{R}w_1 \oplus \cdots \oplus \mathbb{R}w_N + R$ where R is a linear subspace of $\mathbb{R}^n$, invariant under $\widetilde{A}_2$, and the restriction of $\widetilde{A}_2$ to R has no positive real eigenvalues. Let $x \in \mathbb{R}^n$. Then we can write $x = \alpha_1 w_1 + \cdots + \alpha_N w_N +$ r for unique real numbers $\alpha_1, \ldots, \alpha_N$ and a vector $r \in R$.

We have by definition for all j and k:

$$B_j \widetilde{A}_2^k x = (\alpha_1 B_j w_1)\, \rho_1^k + \cdots + (\alpha_N B_j w_N)\, \rho_N^k + B_j \widetilde{A}_2^k r.$$

By Corollary 1, the sequence $B_j \widetilde{A}_2^k r$ assumes non-positive values infinitely often for all j.

It hence suffices to show that the sequence $(\alpha_1 B_j w_1)\, \rho_1^k + \cdots + (\alpha_N B_j w_N)\, \rho_N^k$ assumes non-positive values infinitely often. If $\alpha_1 = \cdots = \alpha_N = 0$, then this is obvious. If not all α_ℓs are zero, let L be the largest integer such that $\alpha_L \neq 0$.

Now, by construction of w_L there exists j such that $\alpha_L B_j w_L < 0$. Since $\rho_L > \rho_j$ for all $j < L$ and $\alpha_j = 0$ for $j > L$, the sequence $(\alpha_1 B_j w_1)\, \rho_1^k + \cdots + (\alpha_N B_j w_N)\, \rho_N^k$ is eventually negative. Since $B_j \widetilde{A}_2^k r$ is non-positive infinitely often, the sequence $(B_j \widetilde{A}_2^k x)_k$ assumes a negative value infinitely often, so that x escapes the polyhedron $P(B)$ under iteration of $\widetilde{A}_2$.

It remains to consider the possibility that $B_1, \ldots, B_m$ are all linearly dependent. It follows as before that either the polyhedron is empty or we must have $B_j v_\ell = 0$ for all j and all ℓ. In the first case, the instance is escaping, so it remains to consider the second case. Any $x \in \mathbb{R}^n$ can be decomposed as above: $x = \alpha_1 v_1 + \cdots + \alpha_N v_N + r$. We have by definition for all j, using that $B_j v_\ell$ vanishes for all ℓ: $B_j \widetilde{A}_2^k x = B_j \widetilde{A}_2^k r$. It follows as above that $B_j \widetilde{A}_2^k x$ assumes non-positive values infinitely often.

7 Robust Escaping Instances of the Affine Problem

We now prove the first item of Theorem 4. Assume that the instance satisfies the formula. We show that every $x \in P(\widehat{B})$ escapes $P(\widehat{B})$ under iteration of $\widehat{A}$. We have $\mathbb{R}^{n+1} = E_{\geq 1} \oplus E_{<1} \oplus E_{\text{rest}}$ where $E_{\geq 1}$ is the space spanned by the generalised eigenvectors of $\widehat{A}$ for eigenvalues in $(1, +\infty)$, $E_{<1}$ is the space spanned by the generalised eigenvectors of $\widehat{A}$ for eigenvalues in $(0, 1)$, and E_{rest} is invariant under A, and the restriction of A to E_{rest} has no positive real eigenvalues.

Let $x \in P(\widehat{B})$. Then we can write $x = x_{\geq 1} + x_{<1} + x_{\text{rest}}$ for unique $x_{\geq 1} \in E_{\geq 1}$, $x_{<1} \in E_{<1}$, and $x_{\text{rest}} \in E_{\text{rest}}$. Since the $(n+1)^{\text{st}}$ component of x is strictly greater than 0, we must have $x_{\geq 1} \neq 0$. Let $1 \leq \rho_1 < \cdots < \rho_N$ denote the real eigenvalues of A above 1. We can further write $E_{\geq 1} = E_1 \oplus \cdots \oplus E_N$ where E_ℓ is the space of generalised eigenvectors for the eigenvalue ρ_ℓ. Then we have $x_{\geq 1} = x_1 + \cdots + x_N$ for unique vectors $x_\ell \in E_\ell$. Let M be the largest integer such that $x_M \neq 0$. Then by Lemma 1, there exists $d \geq 0$ such that the sequence $\binom{k}{d}^{-1} \rho_M^{-k} \widehat{A}^k (x_{\geq 1} + x_{<1})$ converges to an eigenvector v for the eigenvalue ρ_M. By assumption, there exists j such that $\widehat{B}_j v < 0$. It follows that the sequence $\widehat{B}_j \widehat{A}^k (x_{\geq 1} + x_{<1})$ is eventually negative. Now, by Corollary 1, the sequence $\widehat{B}_j \widehat{A}^k x_{\text{rest}}$ assumes a non-positive value infinitely often. Overall, it follows that the sequence $\widehat{B}_j \widehat{A}^k x$ assumes infinitely many negative values. In particular, the point escapes the polyhedron under iteration of $\widehat{A}$. It follows that the point robustly escapes the polyhedron, since the formula is semidecidable by Proposition 2, so that the set of all instances that satisfy the formula is open.

Now assume that the instance does not satisfy the formula. Then we have:

$$\exists \lambda \in \sigma_{\geq 1}\left(\widehat{A}\right). \exists \widehat{v} \in S^n. \left(\widehat{A}\widehat{v} = \lambda \widehat{v} \wedge \forall j. \widehat{B}_j \widehat{v} \geq 0.\right)$$

We show that there exists a trapped point under arbitrarily small perturbations of A and B. Let λ and $\widehat{v}$ witness the above. Let z be the $(n+1)^{\mathrm{st}}$ entry of $\widehat{v}$. We have $z = \widehat{B}_{m+1}\widehat{v} \geq 0$. If $z > 0$, then we must have $\lambda = 1$. It follows that the point $x^* = (\widehat{v}_1/z, \ldots, \widehat{v}_n/z)$ is a fixed point of the map $x \mapsto Ax + b$. Further, we have for all $j \in \{1, \ldots, m\}$: $\widehat{B}_j\widehat{v} = z\,(B_j x^* - \eta_j) \geq 0$. Since $z > 0$ we obtain $B_j x^* - \eta_j \geq 0$. Up to arbitrarily small perturbation of B, these inequalities are strict. Then the map $x \mapsto Ax + b$ has a fixed point in the polyhedron, which in particular is trapped.

If $z = 0$, then $v = (\widehat{v}_1, \ldots, \widehat{v}_n)$ is an eigenvector of A for the eigenvalue λ. It is easy to see that there exists an arbitrarily small perturbation $\widetilde{A}$ of A such that $1 \notin \sigma\left(\widetilde{A}\right)$ and v is an eigenvector of $\widetilde{A}$ for an eigenvalue $\widetilde{\lambda} > 1$. Then the map $f(x) = \widetilde{A}x + b$ has a unique fixed point x^*. Consider the point $x = x^* + v$. Then we have $f^k(x) = x^* + \widetilde{\lambda}^k v$. Now, letting $\widetilde{B}$ be the matrix with columns $\widetilde{B}_j = B_j + \delta v^T$ for arbitrarily small $\delta > 0$, we have $\widetilde{B}v \rhd 0$, so that $\widetilde{B}f^k(x) \rhd \eta$ for all sufficiently large k.

8 Robust Trapped Instances of the Affine Problem

We now prove the second item of Theorem 4. Assume that the instance satisfies the formula. If the instance satisfies the clause $1 \notin \sigma(A) \wedge B \cdot (A - I)^{-1}b \lhd -\eta$ then the map $x \mapsto Ax + b$ has a unique fixed point, given by $x^* = -(A - I)^{-1}b$. We have $Bx^* = -B \cdot (A - I)^{-1}b \rhd \eta$. Thus, x^* is trapped in the polyhedron under $x \mapsto Ax + b$.

If the instance does not satisfy the first clause, then it must satisfy

$$\exists \lambda \in \sigma^{\mathrm{odd}}_{>1}(A). \forall v \in S^{n-1}. \left(Av = \lambda v \implies (\forall j. B_j v > 0 \vee \forall j. B_j v < 0)\right).$$

Let $\lambda > 1$ witness the above. Then there exists an eigenvector v for λ satisfying $Bv \rhd 0$. Let E_1 be the space of generalised eigenvectors of A for the eigenvalue 1. Let E_2 be the space spanned by the generalised eigenvectors of A for all other eigenvalues. Then E_1 and E_2 are invariant under A and we have $\mathbb{R}^n = E_1 \oplus E_2$. In particular, there exist unique linear projections $\pi_j \colon \mathbb{R}^n \to E_j$ such that we have $x = \pi_1(x) + \pi_2(x)$ for all $x \in \mathbb{R}^n$. Let $f(x) = Ax + b$. Since the restriction of A to the space E_2 does not have the eigenvalue 1, there exists a unique $x^* \in E_2$ with $f(x^*) = x^* + x_1$, where $x_1 \in E_1$. Consider the point $x = x^* + v$. Then we have $f^k(x) = A^k v + f^k(x^*) = A^k v + x^* + A^k x_1 = \lambda^k v + x^* + A^k x_1$. Since $x_1 \in E_1$, the norm of the vector $A^k x_1$ is a polynomial in k. It follows that $\lambda^{-k} f^k(x) \to$ v as $k \to \infty$. Hence, $\lambda^{-k} B f^k(x) \to Bv \rhd 0$ as $k \to \infty$. It follows that $Bf^k(x) \rhd \eta$ for all sufficiently large k. Hence, the instance is trapped. It must be robustly trapped, since the formula is semidecidable by Proposition 2, so that the set of all instances that satisfy the formula is open.

Now, assume that the formula does not hold true. Then we have:

$$\begin{aligned}
&\left(1 \in \sigma(A) \vee B \cdot (A - I)^{-1}b \not\lhd -\eta\right) \\
&\wedge \forall \lambda \in \sigma^{\mathrm{odd}}_{>1}(A). \exists v \in S^{n-1}. \left(Av = \lambda v \wedge \exists j. B_j v \leq 0 \wedge \exists j. B_j v \geq 0\right).
\end{aligned}$$

If $B = 0$, then it is easy to see that $P(B, \eta)$ is empty up to an arbitrarily small perturbation of B and η, so that the instance escapes for trivial reasons. We may hence assume that $B \neq 0$. If $1 \notin \sigma(A)$, then by the above formula, the unique fixed point x^* of the map $x \mapsto Ax + b$ is not contained in $P(B, \eta)$. Thus, there exists an arbitrarily small perturbation $\widetilde{\eta}$ of η such that x^* is not contained in $\overline{P}(B, \widetilde{\eta})$. If $1 \in \sigma(A)$, then by Lemma 5, there exist arbitrarily small perturbations $\widetilde{A}_1$ of A and $\widetilde{\eta}$ of η such that $1 \notin \sigma\left(\widetilde{A}_1\right)$, the unique fixed point of the map $x \mapsto \widetilde{A}_1 x + b$ is not contained in the closed polyhedron $\overline{P}(B, \widetilde{\eta})$, a complex number $\lambda \in \mathbb{C}$ with $|\lambda| > 1$ is an eigenvalue of A with multiplicity μ if and only if it is an eigenvalue of $\widetilde{A}_1$ with multiplicity μ, and a vector $v \in \mathbb{C}^n$ is a generalised eigenvector of rank r of A for the eigenvalue λ if and only if it is a generalised eigenvector of rank r of $\widetilde{A}_1$ for the eigenvalue λ. In either case, there exist arbitrarily small perturbations $\widetilde{A}_1$ of A and $\widetilde{\eta}$ of η with the above properties.

Let $\sigma_{>1}^{\text{odd}}(\widetilde{A}_1) = \{\rho_1, \ldots, \rho_N\}$ where $\rho_1 < \cdots < \rho_N$. Then for all $\ell \in \{1, \ldots, N\}$ there exists an eigenvector v_ℓ of $\widetilde{A}_1$ for ρ_ℓ and a $j \in \{1, \ldots, m\}$ such that $B_j v_\ell \leq 0$. By Lemma 2 there exists an arbitrarily small perturbation $\widetilde{A}_2$ of $\widetilde{A}_1$ such that $\sigma\left(\widetilde{A}_2\right)_{>1} = \{\rho_1, \ldots, \rho_N\}$, and the eigenspace for ρ_ℓ is spanned by v_ℓ for all ℓ. Since the unique fixed point of $x \mapsto \widetilde{A}_1 x + b$ depends continuously on $\widetilde{A}_1$ and the polyhedron $\overline{P}(B, \widetilde{\eta})$ is closed, we can further ensure that the unique fixed point of the map $x \mapsto \widetilde{A}_2 x + b$ is not contained in the closed polyhedron $\overline{P}(B, \widetilde{\eta})$.

Assume that B has at least two linearly independent rows. Then, arguing as in the proof of Theorem 2(2.), we can (without loss of generality) use Lemmas 3 and 4 to find an arbitrarily small perturbation $\widetilde{A}_3$ of $\widetilde{A}_2$ such that $\widetilde{A}_3$ has the same eigenvalues as $\widetilde{A}_2$, counted with multiplicities and for all $\ell \in \{1, \ldots, N\}$ the eigenspace of $\widetilde{A}_3$ for ρ_ℓ is spanned by a vector w_ℓ with $B_j w_\ell < 0$ for some j. As above, we can further ensure that the unique fixed point of the map $x \mapsto \widetilde{A}_3 x + b$ is not contained in the closed polyhedron $\overline{P}(B, \widetilde{\eta})$.

Now, let $x \in \mathbb{R}^n$. Let x^* be the unique fixed point of the map that sends x to $\widetilde{A}_3 x + b$. Let E_0 denote the space spanned by the generalised eigenvectors of $\widetilde{A}_3$ for eigenvalues in $\mathbb{C} \setminus (0, +\infty)$. Let E_1 denote the space spanned by the generalised eigenvectors of $\widetilde{A}_3$ for eigenvalues in $(0, 1)$. Since 1 is not an eigenvalue of $\widetilde{A}_3$, the spectrum of $\widetilde{A}_3$ is covered by the sets $(0, 1)$, $(1, +\infty)$, and $\mathbb{C} \setminus (0, +\infty)$. It follows that there exist unique real numbers $c_1, \ldots, c_N$ and vectors $r_0 \in E_0$, $r_1 \in E_1$ such that $x = x^* + c_1 w_1 + \cdots + c_N w_N + r_0 + r_1$. Let $f(x) = \widetilde{A}_3 x + b$. Then we have: $f^k(x) = x^* + \widetilde{A}_3^k r_1 + c_1 \rho_1^k w_1 + \cdots + c_N \rho_N^k w_N + \widetilde{A}_3^k r_0$. Hence: $Bf^k(x) = B\left(x^* + \widetilde{A}_3^k r_1\right) + B\left(c_1 \rho_1^k w_1 + \cdots + c_N \rho_N^k w_N\right) + B\widetilde{A}_3^k r_0$. Now, the sequence $x^* + \widetilde{A}_3^k r_1$ converges to x^* with $Bx^* \lhd \widetilde{\eta}$. By Corollary 1, the sequence $B_j \widetilde{A}_3^k r_0$ assumes a non-positive value infinitely often for all j. It is easy to see that the sequence $B_j\left(c_1\rho_1^k w_1 + \cdots + c_N\rho_N^k w_N\right)$ is either zero or eventually negative for all j. It follows that $B_j f^k(x) \lhd \tilde{\eta}$ for all sufficiently large k.

The case where all rows of B are linearly dependent is treated as in the proof of Theorem 2(2.).

9 On the Measure of the Boundary Instances

We now prove Theorem 5 for the Linear Universal Escape Problem. The proof for the Affine Universal Escape Problem is analogous. We may assume without loss of generality that A is non-singular, that B has full rank, and that all eigenvalues of A are simple, for the set of remaining problem instances already has measure zero. By Theorem 2(1.) an escaping instance is a boundary instance if and only if it satisfies the formula $\exists\lambda \in \sigma_{\geq 0}(A).\exists v \in S^{n-1}.Av = \lambda v \wedge Bv \unrhd 0$. The set of all boundary escaping instances is hence contained in the projection of the set

$$\left\{(A, B, \lambda, v) \in \mathbb{R}^{n\times n} \times \mathbb{R}^{m\times n} \times \mathbb{R} \times \mathbb{R}^n \mid ||v||_2^2 = 1, \lambda \geq 0, Av = \lambda v, Bv \unrhd 0\right\}.$$

If a problem instance further satisfies $\lambda > 0$ and $Bv \rhd 0$, then it is trapped. Hence, the set of all boundary escaping instances is contained in the projection of the set

$$\left\{(A, B, \lambda, v) \in \mathbb{R}^{n\times n} \times \mathbb{R}^{m\times n} \times \mathbb{R} \times \mathbb{R}^n \mid ||v||_2^2 = 1, \lambda = 0, Av = \lambda v\right\}$$
$$\cup \bigcup_{j=1}^{m} \left\{(A, B, \lambda, v) \in \mathbb{R}^{n\times n} \times \mathbb{R}^{m\times n} \times \mathbb{R} \times \mathbb{R}^n \mid ||v||_2^2 = 1, Av = \lambda v, B_j v = 0\right\}.$$

Using that A and B have full rank, we find that the above is a real algebraic set of dimension $n^2 + mn - 1$. Its projection to $\mathbb{R}^{n\times n} \times \mathbb{R}^{m\times n}$ is hence a semi-algebraic set of codimension at most one [1, Lemma 5.30]. It follows that the set has zero measure.

Now, assume that (A, B) is a boundary trapped instance. Then by Theorem 2(2.), the instance satisfies the formula

$$\forall\lambda \in \sigma_{>0}^{\text{odd}}(A).\exists v \in S^{n-1}.Av = \lambda v \wedge \exists j.B_j v \leq 0 \wedge \exists j.B_j v \geq 0.$$

Since we assume that A has only simple eigenvalues, $\sigma_{>0}^{\text{odd}}(A) = \sigma_{>0}(A)$. Since A and B have full rank, A must have a positive real eigenvalue, for otherwise the instance is escaping by Corollary 1. The instance hence satisfies the formula

$$\exists\lambda \in \sigma_{>0}(A).\exists v \in S^{n-1}.Av = \lambda v \wedge \exists j.B_j v \leq 0 \wedge \exists j.B_j v \geq 0.$$

It is hence contained in the projection of the set

$$\left\{(A, B, \lambda, v) \in \mathbb{R}^{n\times n} \times \mathbb{R}^{m\times n} \times \mathbb{R} \times \mathbb{R}^n \mid ||v||_2^2 = 1, Av = \lambda v, \exists j.B_j v \leq 0\right\}.$$

Now, if for all $\lambda \in \sigma_{>0}(A)$ and all v with $Av = \lambda v$ there exists j such that $B_j v < 0$, then the instance is robustly escaping by the proof of Theorem 2(1.). It follows that the instance is contained in the projection of the set

$$\left\{(A, B, \lambda, v) \in \mathbb{R}^{n\times n} \times \mathbb{R}^{m\times n} \times \mathbb{R} \times \mathbb{R}^n \mid ||v||_2^2 = 1, Av = \lambda v, \exists j. B_j v = 0\right\}.$$

This is again a real algebraic set of dimension $n^2 + mn - 1$. Its projection is hence a semi-algebraic set of dimension at most $n^2 + mn - 1$. In particular, it has zero measure.

Acknowledgements. The authors grant to Swansea University a non-exclusive, irrevocable, sub-licensable, worldwide license to make the accepted manuscript available on its institutional repository.

References

1. Basu, S.: Pollack, Richard, and Roy. Algorithms in Real Algebraic Geometry, Springer-Verlag, Berlin, Heidelberg, Marie-Françoise (2006)
2. Bell, J.P., Gerhold, S.: On the positivity set of a linear recurrence. Israel J. Math. **157**, 333–345 (2007). https://doi.org/10.1007/s11856-006-0015-1
3. Vasco, B., Peter, H., (eds.), Handbook of Computability and Complexity in Analysis, Theory and Applications of Computability, Cham, Springer (2021) https://doi.org/10.1007/978-3-030-59234-9
4. Brattka, V., Presser, G.: Computability on subsets of metric spaces. Theoret. Comput. Sci. **305**(1–3), 43–76 (2003)
5. Collins, P.: Computable analysis with applications to dynamic systems. Math. Struct. Comput. Sci. **30**(2), 173–233 (2020)
6. Ker-I, K.: Complexity theory of real functions, Progress in Theoretical Computer Science, Birkhäuser Boston, Inc., Boston, MA, MR 1137517 (93i:03057) (1991). https://doi.org/10.1007/978-1-4684-6802-1
7. Eike, N.: Decision problems for linear recurrences involving arbitrary real numbers, Logical Methods in Computer Science Volume 17, Issue 3, (3:16) (2021). https://doi.org/10.46298/lmcs-17
8. Eike, N., Margret, T.: Termination of Real Linear Loops (2026). https://doi.org/10.48550/arXiv.2605.00947
9. Sewon, P., et al.: Semantics, Specification Logic, and Hoare Logic of Exact Real Computation, Logical Methods Comput. Sci. **20**(2)(2:17) (2024) https://doi.org/10.46298/LMCS-20
10. Pauly, A.: On the topological aspects of the theory of represented spaces. Computability **5**(2), 159–180 (2016). https://doi.org/10.3233/COM-150049
11. Pour-El, Marian, B., Ian, R.J.: Computability in Analysis and Physics, Springer (1989)
12. Matthias, S.: Admissible Representations for Continuous Computations, Ph.D. thesis, FernUniversität Hagen (2002)
13. Extended admissibility, Theoretical Computer Science 284, 519–538 (2002)
14. Ernst, S.: The fundamental theorem of algebra in recursive analysis, Constructive Aspects of the Fundamental Theorem of Algebra (Dejon, B. and Henrici, P., eds.), Wiley-Interscience, London, vol. 329 pp. 321–15 (1969)

15. Tiwari, A.: Termination of linear programs. In: Alur, R., Peled, D.A. (eds.) CAV 2004. LNCS, vol. 3114, pp. 70–82. Springer, Heidelberg (2004). https://doi.org/10.1007/978-3-540-27813-9_6
16. Computable Analysis. TTCSAES, Springer, Heidelberg (2000). https://doi.org/10.1007/978-3-642-56999-9_9
17. Ziegler, M., Brattka, V.: Computability in linear algebra. Theoret. Comput. Sci. **326**(1–3), 187–211 (2004)

Safety and Reachability in k-Control Games on Integer Vector Addition Systems with States

Reino Niskanen[1(✉)], Igor Potapov[2], and James Topley[1]

[1] Liverpool John Moores University, Liverpool, UK
{r.niskanen,csmjtopl}@ljmu.ac.uk
[2] University of Liverpool, Liverpool, UK
potapov@liverpool.ac.uk

Abstract. Integer vector addition systems with states ($\mathbb{Z}$-VASSs) are vector addition systems with states, where the configurations are from $\mathbb{Z}^d$. Two-player games on $\mathbb{Z}$-VASS are known to be undecidable when played on two-dimensional $\mathbb{Z}$-VAS(S)s. Here, we study a restricted variant of the game, where one of the players, Adam, can play k non-zero moves from his move set. Adam's (safety) goal is to ensure that the other player, Eve, cannot reach a particular point in the space. We show that, for any dimension, deciding whether Eve has a winning strategy is in NEXPTIME and is NP-hard if k is part of the input. Additionally, we extend the safety objective from a single point to a hyperplane and prove that the complexity does not increase. We also study the dual variant (reachability), where Eve's resources are limited and show that deciding whether Eve has a winning strategy is also in EXPTIME. On the other hand, if k is fixed, safety in k-control games is NP-complete, while reachability is in PTIME.

Keywords: vector addition game · decidability · winning strategy

1 Introduction

Vector addition systems with states (VASSs) are a well-known computational model that can be viewed as an equivalent formalism to Petri nets [31]. The stateless variant was introduced in [21] and with states in [19]. A vector addition system with states consists of a directed graph labelled with d-dimensional integer vectors and d counters over natural numbers. Every time a transition of a graph is taken, the label of the transition is added to the counters. As the counters are over natural numbers, only the transitions respecting this can be taken. One of the central decision problems for VASSs is whether for two given configurations, i.e., a vertex and counter values, there exists a path in the directed graph. This decision problem is known as the *reachability problem* and it was proved to be decidable in [28]. The result was simplified later on in [22,23]. Unfortunately, the techniques used in the proofs seemed to be beyond

V. Brattka et al. (Eds.): CiE 2026, LNCS 16674, pp. 453–468, 2027.
https://doi.org/10.1007/978-3-032-31348-5_30

complexity analysis and it wasn't until a decade ago that Leroux and Schmitz derived the first upper bound for the reachability problem [26]. Since then, the tight bounds have been obtained; it is an Ackermann-complete problem [9].

Fixing the dimension of a VASSs does not make the problem significantly easier as only one and two-dimensional VASSs have tight complexity bounds. Namely, when the numbers in transition vectors are encoded in binary, the reachability problem for one-dimensional VASSs is NP-complete [18] and for two-dimensional VASSs is PSPACE-complete [3]. If the numbers are encoded in unary, the problem is NL-complete in both cases [12,36]. There has been substantial research done on related problems (e.g., coverability [27,32], boundedness [32,35]), and extensions or generalisations of VASSs (e.g., branching VASS [13,24], reset Petri nets [11], integer VASS [5], polynomial register machines [15]). In this paper, we consider games on integer vector addition systems with states ($\mathbb{Z}^d$-VASSs). The $\mathbb{Z}^d$-VASS, where the counter values are no longer restricted to natural numbers, is a natural extension of VASS [5]. This makes the system more amenable to analysis. The reachability problem for $\mathbb{Z}^d$-VASSs is NP-complete already when numbers are encoded in unary.

Compared to the reachability problem, there is a wider variety of complexity results for two-player games on VASS or $\mathbb{Z}^d$-VASS. In a game, the states of VASS are partitioned into two sets, one for each player, and an edge is picked by the owner of the state. The decision problem associated with these games is the question of existence of a winning strategy. That is, whether one of the players has a strategy that ensures that their objectives are met. The undecidability of two-dimensional VASS games follows in rather straightforward fashion from the undecidability of the halting problem for two-counter Minsky machines [1,4]. On the other hand, one-dimensional VASS games were proved to be EXPSPACE-complete [20,34]. The same complexity results hold for $\mathbb{Z}^d$-VASS games [20,34].

In this paper, we investigate a middle ground between the reachability problem and a game. Indeed, the reachability problem is an existential problem, namely whether there exists a sequence of transitions from the initial configuration to the target configuration, while the question of the existence of a winning strategy in a game is an infinite alternation of existential and universal problems. A simple extension of a reachability problem by allowing a single move of the universal player transforms the reachability problem into a problem where, at any given point of a computation, a configuration can be transformed into a set of configurations from which the reachability problem has to be considered. We consider a scenario, where one of the players can play their move fixed number of times, that is, there is a finite number of alternations of quantifiers in a computation.

This study is motivated by results of [1,4,30,34], where the considered games are either undecidable or lie in high complexity classes such as EXPTIME or EXPSPACE. In this paper, we investigate modifications to $\mathbb{Z}^d$-VASS games that could yield lower complexity bounds whilst remaining nontrivial. We observe that the gadgets present in [1,4,30,34] that ensure a faithful simulation, rely on an infinite punishing behaviour by Adam. For example, in [1], if Eve makes an incorrect simulating move, then Adam has a simple winning strategy, where he enters a gadget that ensures that he wins. But if we limit how many times Adam can play, then Eve can exhaust Adam's moves and then reach $(0,0)$ with an unfaithful simulation. Motivated by this behaviour, in this paper, we consider two variants of $\mathbb{Z}^d$-VASS games where players' resources are limited.

- k-control $\mathbb{Z}^d$-VASS games with a safety objective, where Adam can apply his moves only k times during any play and has to skip otherwise.
- k-control $\mathbb{Z}^d$-VASS games with a reachability objective, where Eve can apply her moves only k times during any play and has to skip otherwise.

Table 1. Summary of results on k-control games. Note that the games with safety and reachability objectives are the same for $k = \infty$.

Game	$k = 0$	k	$k = \infty$	
	any dimension		$d = 1$	$d \geq 2$
$\mathbb{Z}^d$-VASS games (safety)	NP-c [5,33]	NP-h [5], NEXPTIME (Thm. 7)	EXPSPACE-c [30]	undecid. [33]
$\mathbb{Z}^d$-VASS games (reachability)	LIN	EXPTIME (Cor. 14)		

The results are highlighted in Table 1. The main result of the paper is that checking whether Eve has a winning strategy to ensure safety of a given configuration is NP-hard and in NEXPTIME. Observe that the complexity of checking whether Adam has a winning strategy is in the respective complement class. The main observation is that, if Adam has a winning strategy in a d-dimensional k-control $\mathbb{Z}^d$-VASS game, then he has a winning strategy where his moves are played "close" to the target configuration. The intuition is the following: if the configuration is "far" from the target, then Adam does not have to use one of his moves to influence the play. On the other hand, if the configuration is "close" to the target, then Adam has to play a move in order to avoid the target. The area "close" to the target is exponential in size of input, and to check whether Eve can reach it is in NP.

The second variant of the k-control $\mathbb{Z}^d$-VASS game, where Eve has limited number of moves, relies on different observations. We notice that if Eve skips her move, then Adam has a winning strategy as he can always choose a vector such that the zero vector is not reachable. On the other hand, as Eve can play only k times, eventually, she will have to only skip. That is, only a finite duration of the game needs to be considered. If k is fixed and is not a part of the input, then in both variants the aforementioned areas are polynomial in size of the input and hence the problems are NP-complete and in PTIME respectively.

This paper is based on a chapter in the first author's PhD thesis [29].

2 Preliminaries

We denote the sets of integers and non-negative integers (that is, natural numbers) by $\mathbb{Z}$ and $\mathbb{Z}^+$ respectively. Unless stated otherwise, all numbers are encoded in binary. By $\mathbf{0}_d$ we denote the d-dimensional zero vector. If dimension d is clear from the context, we omit the subindex.

A *game on integer vector addition system with states* ($\mathbb{Z}^d$-VASS game) is a two-player d-dimensional game. The game consists of two graphs, one for Eve and one for Adam, labelled with vectors in $\mathbb{Z}^d$, and the initial configuration. That is, the game is given as (A, E, c_0), where $A = (Q_A, T_A)$ and $T_A \subseteq Q_A \times \mathbb{Z}^d \times Q_A$, and $E = (Q_E, T_E)$ and $T_E \subseteq Q_E \times \mathbb{Z}^d \times Q_E$, and $c_0 \in Q_E \times Q_A \times \mathbb{Z}^d$ is the initial configuration. The graph A consists of moves that Adam can apply during his turn and E consists of moves that Eve can apply during her turn. Note that A and E are $\mathbb{Z}^d$-VASSs. We abuse notation and frequently equate A and E with T_A and T_E as state sets are usually clear from the context. We denote $|A| = |T_A| = \ell$ and $|E| = |T_A| = m$. A *configuration* is a triple $[s, t, \mathbf{v}]$ consisting of Eve's control state s, Adam's control state t and the *counter vector* $\mathbf{v} \in \mathbb{Z}^d$. Eve updates her control state when she makes a move: in the configuration $[s, t, \mathbf{v}]$, for any vector $\mathbf{v}$, only moves of the form $(s, \mathbf{x}, s')$ are enabled, and with one such move the new configuration is $[s', t, \mathbf{v} + \mathbf{x}]$. Adam updates his control state when he makes a move in a similar fashion. Eve *wins* if, and only if, after her turn, the configuration is $[s, t, \mathbf{0}_d]$ for any $s \in Q_E$ and $t \in Q_A$. Note that in [1,4,34], where VASS games are considered, the game is played on a directed graph with vertices partitioned between the two players. This definition is not convenient for the k-control games we define later on. In [30], it was shown that these two definitions are equivalent for one-dimensional games. The result extends in a natural way to d-dimensional games. In order to indicate whose turn it is in the configuration $[s, t, \mathbf{v}]$, we put a dot above s if it is Eve's turn, or above t if it is Adam's turn. That is, the respective configurations are $[\dot{s}, t, \mathbf{v}]$ and $[s, \dot{t}, \mathbf{v}]$. In figures, Adam's moves are in green and Eve's moves are in blue.

A d-dimensional *game on integer vector addition system* ($\mathbb{Z}^d$-VAS game), introduced as robot game in [10], is a special case of the $\mathbb{Z}^d$-VASS games, where $|Q_A| = |Q_E| = 1$ and the moves are self-loops. That is, a $\mathbb{Z}^d$-VAS game consists of two players, Eve and Adam, having a set of vectors E, A over $\mathbb{Z}^d$, respectively, and an *initial vector* $\mathbf{x}_0$. Starting from $\mathbf{x}_0$ players add a vector from their respective sets to the current configuration of the game in turns. As in $\mathbb{Z}^d$-VASS games, Eve tries to reach the origin while Adam tries to keep Eve from reaching the origin. The simpler formulation is convenient for our examples.

We define plays and strategies in terms of $\mathbb{Z}^d$-VASS games. A *play* is an infinite sequence of configurations of a game, $\pi = c_0 c_1 c_2 \cdots$, where if $c_i = [\dot{s}, t, \mathbf{x}]$, then $c_{i+1} = [s', \dot{t}, \mathbf{x}+\mathbf{e}]$ if $(s, \mathbf{e}, s') \in E$, or if $c_i = [s, \dot{t}, \mathbf{x}]$, then $c_{i+1} = [\dot{s}, t', \mathbf{x}+\mathbf{a}]$ if $(t, \mathbf{a}, t') \in A$. A play is *winning* for Eve, if the target configuration appears in it. On the other hand, if the target configuration never appears in the play, then the play is winning for Adam. A (positional) *strategy* of Eve is a function $\sigma_E : Q_E \times Q_A \times \mathbb{Z}^d \to E$ that tells Eve which move to use in the current configuration. We define Adam's strategy σ_A symmetrically. We say that a play π is *consistent* with a strategy σ_E if for configurations $c_i = [\dot{s}, t, \mathbf{x}]$ and $c_{i+1} = [s', \dot{t}, \mathbf{x}']$, it holds $\sigma_E([s, t, \mathbf{x}]) = (s, \mathbf{x}' - \mathbf{x}, s')$. We define consistency with Adam's strategy σ_A symmetrically. A strategy σ_E (resp., σ_A) is a *winning strategy*, if all consistent plays are winning for Eve (resp., Adam). Due to the positional determinacy of this setting [14], Adam has a winning strategy if and only if Eve does not have one. See [14,17] for more details on games.

In the decision problem associated with $\mathbb{Z}$-VAS(S) games, we are asked whether Eve has a winning strategy from the given initial configuration.

A k-control $\mathbb{Z}^d$-VASS game is a variant of $\mathbb{Z}^d$-VASS games where one of the players can *skip* their turn and he or she can play at most k non-skipping moves during a play. We say that the k-control $\mathbb{Z}^d$-VASS game has a *safety* objective if Adam has k moves and *reachability* objective if Eve has k moves. That is, in safety of k-control $\mathbb{Z}^d$-VASS game, the game now consists of a $\mathbb{Z}^d$-VASS game (A, E, c_0) together with a counter $i \geq 0$ that starts at k. If SKIP is played by Adam in a configuration $[s, \dot{t}, \mathbf{x}]$, the next configuration is $[\dot{s}, t, \mathbf{x}]$. A strategy $\sigma_A : Q_E \times Q_A \times \mathbb{Z}^d \to A \cup \{\text{SKIP}\}$ of Adam takes i into account and returns SKIP if $i = 0$, otherwise it either returns a move in A and decrements i or returns SKIP and does not decrement i. The reachability variant is defined analogously.

Example 1. Let (A, E, x_0) be a $\mathbb{Z}$-VAS game, where $A = \{-2\}$, $E = \{1\}$ and $x_0 < 0$. In the $\mathbb{Z}$-VAS game, the unique strategy of Adam is winning as, after the players play their moves, the resulting configuration $x = x_0 - 2 + 1$ is less than the initial configuration x_0. If we consider it as a k-control $\mathbb{Z}$-VAS game with a safety objective for any $k \in \mathbb{Z}^+$, then Eve has a winning strategy. Indeed, with at most $2k + |x_0|$ moves, Eve can reach the origin as Adam can apply his -2 move at most k times. Two plays are depicted in Fig. 1.

Fig. 1. A play of the $\mathbb{Z}$-VAS game of Example 1 (left) and a play in the 1-control $\mathbb{Z}$-VAS game, where Adam can play his move only once (right).

Let us consider the one-player version of the game, that is, where $A = \{\{q\}, \{(q, 0, q)\}\}$. This is now the reachability problem in $\mathbb{Z}^d$-VASS. For the sake of readability, we omit Adam's component and denote a configuration by $(q, \mathbf{x})$, where $q \in Q_E$. Let $(q, \mathbf{x})$ and $(q', \mathbf{x}')$ be two configurations. We denote $(q, \mathbf{x}) \to (q', \mathbf{x}')$ if there is a move $(q, \mathbf{x}' - \mathbf{x}, q') \in E$. We use notation $\to^i$ and $\to^*$ to denote a sequence of i configurations and a sequence of any number of configurations, respectively.

3 Safety of k-control $\mathbb{Z}^d$-VASS games

In this section, we consider k-control $\mathbb{Z}^d$-VASS games with a safety objective. We have an immediate lower bound from the reachability problem for $\mathbb{Z}^d$-VASS as discussed in the previous paragraph by assuming that Adam has no influence on the game. In [5], it was proved that the reachability problem is NP-complete. Thus, k-control $\mathbb{Z}^d$-VASS games are NP-hard.

Let us then study 1-control $\mathbb{Z}^d$-VASS games to illustrate the idea of the proof better. We denote $E^1_{\text{SKIP}} = \{(s, -\mathbf{e}) \mid (s, \mathbf{e}, s') \in E\}$, where E is the move set of Eve. The set E^1_{SKIP} consists of pairs in $Q_E \times \mathbb{Z}^d$ from which she can win with one move. In the next lemma, we show that if Adam has a winning strategy, he also has a winning strategy, where the move is played only if the configuration is $[s, \dot{t}, \mathbf{x}]$ for $(s, \mathbf{x}) \in E^1_{\text{SKIP}}$. In the proof, we consider a path reaching $(s, \mathbf{x}) \in E^1_{\text{SKIP}}$ using only moves of Eve. Because Adam has a winning strategy, then in the game following this path, he plays a move $(t, \mathbf{a}, t')$ making $\mathbf{0}$ unreachable for Eve. If Adam plays $(t, \mathbf{a}, t')$ in $[s, \dot{t}, \mathbf{x}]$, then Eve cannot reach the origin from the resulting configuration $[\dot{s}, t', \mathbf{x} + \mathbf{a}]$. The idea is depicted in Fig. 2.

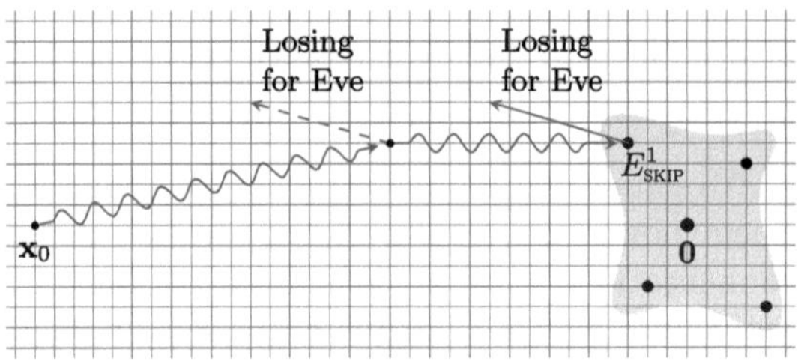

Fig. 2. An illustration of Lemma 2 in a $\mathbb{Z}^2$-VAS game.

Lemma 2. *Let σ_A be a winning strategy of Adam in a 1-control $\mathbb{Z}^d$-VASS game with a safety objective $(A, E, [s_0, t_0, \mathbf{x}_0])$. Then there exists a winning strategy of Adam, σ'_A, such that*

$$\sigma'_A([s, t, \mathbf{x}]) = \begin{cases} \text{SKIP} & \text{if } (s, \mathbf{x}) \notin E^1_{\text{SKIP}} \\ \sigma_A([s', t, \mathbf{y}]) & \text{if } (s, \mathbf{x}) \in E^1_{\text{SKIP}},\ \text{for some } [s', t, \mathbf{y}]. \end{cases}$$

Proof. Let σ_A be a winning strategy of Adam. Consider $(s, \mathbf{x}) \in E^1_{\text{SKIP}}$ that is reachable from $[s_0, t_0, \mathbf{x}_0]$ using Eve's moves only. Let p be this path. Consider then a play in which Eve follows p. That is,

$$p = [s_0, t_0, \mathbf{x}_0][s_{i_j}, t_0, \mathbf{x}_1] \cdots [s', t_0, \mathbf{y}] \cdots [s, t_0, \mathbf{x}].$$

Since σ_A is a winning strategy, there is a configuration $[s', t_0, \mathbf{y}]$ for which the winning strategy $\sigma_A([s', t_0, \mathbf{y}]) = (t_0, \mathbf{a}, t)$, for some $(t_0, \mathbf{a}, t) \in A$, making $\mathbf{0}$ unreachable using Eve's moves. If $(s', \mathbf{y}) = (s, \mathbf{x})$, then we are done. If $(s', \mathbf{y}) \neq (s, \mathbf{x})$, then consider a strategy of Adam, where he skips unless $(s, \mathbf{x})$ is reached and then plays $(t_0, \mathbf{a}, t)$. This play is winning for Adam. Indeed, assume that on the contrary Eve can reach $\mathbf{0}$ from $[s, t, \mathbf{x} + \mathbf{a}]$ by a path p'. Then Eve can also reach $\mathbf{0}$ from $[s', t, \mathbf{y} + \mathbf{a}]$ by playing the same moves as in the suffix of p followed by p', which is not possible as σ_A is a winning strategy.

By repeating this procedure for every $(s, \mathbf{x}) \in E^1_{\text{SKIP}}$, we find a move of Adam that, according to σ_A, needs to be played in $[s, t_0, \mathbf{x}]$. We construct σ'_A such that $\sigma'_A([s, t_0, \mathbf{x}]) = \textsc{Skip}$ if $(s, \mathbf{x}) \notin E^1_{\text{SKIP}}$ and $\sigma'_A([s, t_0, \mathbf{x}]) = \sigma_A([s', t_0, \mathbf{y}])$ if $(s, \mathbf{x}) \in E^1_{\text{SKIP}}$. By the previous argument, σ'_A is a winning strategy of Adam. □

Now, by the previous lemma, it is enough to only consider strategies where Adam always skips or where an element of E^1_{SKIP} is reached after which Adam plays his move.

Let us consider the game with $k = 0$ in more details, that is, reachability in $\mathbb{Z}^d$-VASS. In [5], it was shown that deciding whether $(q_f, \mathbf{0})$ is reachable from $(q_0, \mathbf{x}_0)$ is decidable in NP. We call this algorithm ZVASS$(E, (q_0, \mathbf{x}_0), (q_f, \mathbf{0}))$. In $\mathbb{Z}^d$-VASS games, the target is just the counter value $\mathbf{0}$. This does not increase the complexity as Eve can guess the state q_f, where the counter becomes $\mathbf{0}$, and call ZVASS$(E, (q_0, \mathbf{x}_0), (q_f, \mathbf{0}))$.

input : A 1-control $\mathbb{Z}^d$-VASS game $(A, E, [s_0, t_0, \mathbf{x}_0])$.
output: The player that has a winning strategy in the safety game.

1 **if** ZVASS$(E, (s_0, \mathbf{x}_0), (s, \mathbf{0}))$ = Adam **forall** $s \in Q_E$ **then**
2 | **return** Adam;
3 **else**
4 | **foreach** *element* $(s, \mathbf{e}) \in E^1_{\text{SKIP}}$ **do**
5 | | **if** ZVASS$(E, (s_0, \mathbf{x}_0), (s, \mathbf{e}))$ = Eve **and forall** $(t_0, \mathbf{a}, t) \in A$, ZVASS$(E, (s, \mathbf{e} + \mathbf{a}), (s', \mathbf{0}))$ = Eve **then return** Eve;
6 | **end**
7 | **return** Adam;
8 **end**

Algorithm 1: Solving a 1-control $\mathbb{Z}^d$-VASS game with a safety objective.

Theorem 3. *It is decidable in* NP *whether Eve has a winning strategy in a 1-control* $\mathbb{Z}^d$-VASS *game with a safety objective for Adam.*

Proof. We claim that Algorithm 1 solves the problem in non-deterministic polynomial time. First, we prove the correctness of the algorithm. Assume that Adam has a winning strategy. There are two types of strategies, one where Adam only skips and another where he plays one of his moves. In the first case, Adam has a winning strategy if and only if Eve cannot reach $\mathbf{0}$ from the initial configuration $[s_0, t_0, \mathbf{x}_0]$. That is, $\mathbf{0}$ is not reachable in the underlying $\mathbb{Z}^d$-VASS. This is verified on line 1. In the second case, by Lemma 2, it is enough to consider values in E^1_{SKIP}. As Adam has a winning strategy, for any value $(s, \mathbf{e}) \in E^1_{\text{SKIP}}$, that is reachable using only the moves of Eve, Adam has a move $(t, \mathbf{a}, t')$ such that the origin is not reachable from $(s, \mathbf{e} + \mathbf{a})$. This is verified in the loop on line 4. It both cases the correct output is produced.

It is clear that the algorithm terminates. In total, there are at most $m\ell + m$ calls to solve the reachability in $\mathbb{Z}^d$-VASS, where $|E| = m$ and $|A| = \ell$. The reachability in $\mathbb{Z}^d$-VASS can be solved in non-deterministic polynomial time, and thus Algorithm 1 is in NP. □

Before considering the general case of k-control $\mathbb{Z}^d$-VASS game, let us consider 2-control $\mathbb{Z}^d$-VASS games and the extended definition of E^1_{SKIP}. When Adam has one move in a 1-control game, he needs to react when he is one step from losing. In a 2-control games, he has two moves and hence he needs to react when he is two steps from losing. That is, we need to consider configurations from which Adam loses if he skips or moves only once. Then, we prove an analogous result to Lemma 2 showing that only these finite number of configurations close to the origin need to be considered. To this end, let us define

$$E^2_{\text{SKIP}} = \{(s, \mathbf{x}) \mid (s, \mathbf{x}) \to^2 (s', \mathbf{0}) \text{ for some } s' \in Q_E\}$$

and

$$E^2_{\text{PLAY},t} = \{(s, \mathbf{x}) \mid [\dot{s}, t, \mathbf{x}][s', \dot{t}, \mathbf{x}+\mathbf{e}][\dot{s'}, t', \mathbf{x}+\mathbf{e}+\mathbf{a}][s'', \dot{t'}, \mathbf{0}] \text{ is a valid play}\}$$

for each $t \in Q_A$. We also define $E^2_{\text{PLAY}} = \bigcup_{t \in Q_A} E^2_{\text{PLAY},t}$. The first set, E^2_{SKIP}, contains all state-counter pairs from which Eve can reach the origin in two moves, and the second set, $E^2_{\text{PLAY},t}$, contains all state-counter pairs from which Eve can reach the origin in two turns if Adam plays his move after Eve's first move.

We illustrate the need to look beyond E^1_{SKIP} when $k > 1$ in the next example.

Example 4. Consider a k-control $\mathbb{Z}^2$-VAS game $(A, E, \mathbf{x}_0)$, where Adam's move set is $A = \{(1, 1)\}$, Eve's move set is $E = \{(-1, 0), (0, 1), (2, 1)\}$ and $\mathbf{x}_0 = (-2n, -n-1)$, for some $n > 1$, is the initial configuration. If $k = 1$, then Eve has a winning strategy. Indeed, if Adam skips and the current configuration is not $(0, -1)$, Eve plays $(2, 1)$. Then, at $(0, -1)$, Eve plays $(0, 1)$ and wins the play. Consider then a play where Adam plays his move. After his move, the configuration is $(-2n', -n'-1) + (1, 1) = (-2n'+1, -n')$ for some $0 \le n' \le n$. After this Eve plays $(-1, 0)$, followed by $(2, 1)$ n' times and wins the play.

In contrast, if $k > 1$, then Adam has a winning strategy. His strategy is to play $(1, 1)$ if the configuration is $(-2n', -n'-1)$ or $(-2n', -n')$ for some $n' > 0$. Note, that Adam has to play for the first time before the configuration is $(0, -1)$, as in that case Eve has a winning strategy. Two plays are depicted in Fig. 3.

This example illustrates that this approach of $k = 1$ is not sufficient for larger k. Namely, Adam might not have a winning strategy if he only moves when the configuration is in E^1_{SKIP}.

Lemma 5. *Let σ_A be a winning strategy of Adam in a 2-control $\mathbb{Z}^d$-VASS game with a safety objective $(A, E, [s_0, t_0, \mathbf{x}_0])$. There exists a winning strategy of Adam, σ'_A, such that*

$$\sigma'_A([s, t, \mathbf{x}]) = \begin{cases} \text{SKIP} & \text{if } (s, \mathbf{x}) \notin E^1_{\text{SKIP}} \cup E^2_{\text{SKIP}} \cup E^2_{\text{PLAY}} \\ \sigma_A([s', t, \mathbf{y}]) & \text{if } (s, \mathbf{x}) \in E^1_{\text{SKIP}} \cup E^2_{\text{SKIP}} \cup E^2_{\text{PLAY}} \text{ for some } [s', t, \mathbf{y}]. \end{cases}$$

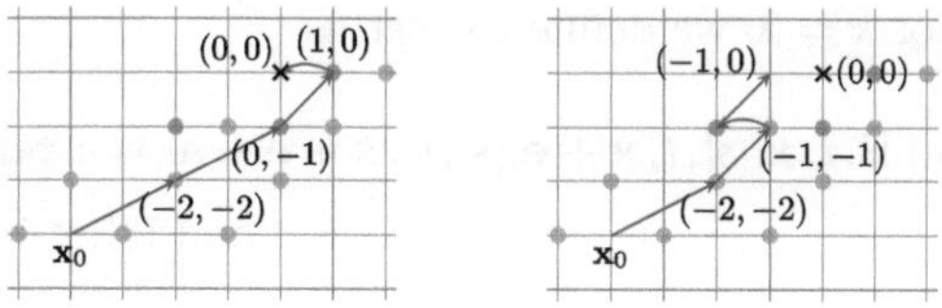

Fig. 3. Plays of two k-control $\mathbb{Z}^d$-VAS games of Example 4, with $k = 1$ (left) and $k > 1$ (right). Red points are elements of E^1_{SKIP} and orange are elements of $E^2_{\text{SKIP}} \cup E^2_{\text{PLAY}}$.

Proof. Let σ_A be a winning strategy of Adam. Consider $(s, \mathbf{x}) \in E^2_{\text{SKIP}} \cup E^2_{\text{PLAY}} \cup E^1_{\text{SKIP}}$ that is reachable from $[s_0, t_0, \mathbf{x}_0]$ using Eve's moves only. Let p be this path. Consider then a play in which Eve follows p. That is,

$$p = [s_0, t_0, \mathbf{x}_0] \cdots [s', t_0, \mathbf{y}] \cdots [s, t_0, \mathbf{x}].$$

Since σ_A is a winning strategy, there are two types of plays that follow p. Either Adam plays once making $\mathbf{0}$ unreachable using Eve moves only, or Adam plays both of his moves. In the former case, Adam can play his move in E^1_{SKIP} and still win by the argument of Lemma 2. In the latter case, consider the first move $(t_0, \mathbf{a}, t_1)$ of Adam that he plays in $[s', t_0, \mathbf{y}]$. If $(s', \mathbf{y}) \in E^2_{\text{PLAY}}$, we are done. Otherwise consider the next move in the play. Eve can play any move $(s', \mathbf{e}, s'')$ and Adam will still win since σ_A was a winning strategy. Now let Adam skip instead. This leads to Eve playing a move of the form $(s', \mathbf{e}, s'')$, and this followed by $(t_0, \mathbf{a}, t_1)$ results in the same configuration. By iteratively applying this reasoning, Adam can skip until a configuration in E^2_{PLAY} is reached whilst maintaining the winning strategy.

By repeating this procedure for every $(s, \mathbf{x}) \in E^2_{\text{SKIP}} \cup E^2_{\text{PLAY}} \cup E^1_{\text{SKIP}}$, we find a move of Adam that, according to σ_A, Adam needs to play in $[s, t, \mathbf{x}]$. We construct σ'_A such that $\sigma'_A([s, t, \mathbf{x}]) = \text{SKIP}$ if $(s, \mathbf{x}) \notin\in E^2_{\text{SKIP}} \cup E^2_{\text{PLAY}} \cup E^1_{\text{SKIP}}$ and $\sigma'_A([s, t, \mathbf{x}]) = \sigma_A([s', t, \mathbf{y}])$ if $(s, \mathbf{x}) \in E^2_{\text{SKIP}} \cup E^2_{\text{PLAY}} \cup E^1_{\text{SKIP}}$. By the previous argument, σ'_A is a winning strategy of Adam. □

We are ready to consider the general case of k-control $\mathbb{Z}^d$-VASS game $(A, E, [s_0, t_0, \mathbf{x}_0])$. Lemmata 2 and 5 generalise in a natural way. We again denote by E^k_{SKIP} the set of all pairs in $Q_E \times \mathbb{Z}^d$ from which there exists a path of length k reaching an element of $Q_E \times \{\mathbf{0}\}$. That is,

$$E^k_{\text{SKIP}} = \{(s, \mathbf{x}) \mid (s, \mathbf{x}) \rightarrow^k (s', \mathbf{0}) \text{ for some } s' \in Q_E\}.$$

We recursively define $E^k_{\text{PLAY},t}$, which is the set of all pairs in $Q_E \times \mathbb{Z}^d$ from which there exists a winning play for Eve, where Eve plays k times and Adam plays $k - 1$ times. That is, for $k \geq 3$

$$E^k_{\text{PLAY},t} = \{(s, \mathbf{x}) \mid [\dot{s}, t, \mathbf{x}][s', \dot{t}, \mathbf{x} + \mathbf{e}][\dot{s'}, t', \mathbf{x} + \mathbf{e} + \mathbf{a}] \text{ is a valid play}$$
$$\text{and } (s', \mathbf{x} + \mathbf{e} + \mathbf{a}) \in E^{k-1}_{\text{PLAY},t'}\}$$

for each $t \in Q_A$. For $k = 2$, we define the set as

$$E^2_{\text{PLAY},t} = \{(s, \mathbf{x}) \mid [\dot{s}, t, \mathbf{x}][s', \dot{t}, \mathbf{x} + \mathbf{e}][\dot{s}', t', \mathbf{x} + \mathbf{e} + \mathbf{a}] \text{ is a valid play} \\ \text{and } (s', \mathbf{x} + \mathbf{e} + \mathbf{a}) \in E^1_{\text{SKIP}}\}$$

for each $t \in Q_A$. Note that this definition is consisting with the definition of $E^2_{\text{PLAY},t}$ given previously. As before we define $E^k_{\text{PLAY}} = \bigcup_{t \in Q_A} E^k_{\text{PLAY},t}$.

Let us denote $\overline{E^k} = E^k_{\text{SKIP}} \cup E^k_{\text{PLAY}}$. This is the set of configurations from which Eve wins in k moves if Adam either skips or plays $k-1$ moves. The size of each $\overline{E^k}$ is at most $m^k \ell^{k-1} + km$, which is exponential in the size of the input.

We are ready to prove the main lemma that generalises Lemmata 2 and 5.

Lemma 6. *Let σ_A be a winning strategy of Adam in a k-control $\mathbb{Z}^d$-VASS game with a safety objective $(A, E, [s_0, t_0, \mathbf{x}_0])$. Then there exists a winning strategy of Adam, σ'_A, such that*

$$\sigma'_A([s, t, \mathbf{x}]) = \begin{cases} \text{SKIP} & \text{if } (s, \mathbf{x}) \notin \overline{E^1} \cup \overline{E^2} \cup \ldots \cup \overline{E^k} \\ \sigma_A([s', t', \mathbf{y}]) & \text{if } (s, \mathbf{x}) \in \overline{E^1} \cup \overline{E^2} \cup \ldots \cup \overline{E^k} \text{ for some } [s', t', \mathbf{y}]. \end{cases}$$

Proof. Let σ_A be a winning strategy of Adam. Consider $(s, \mathbf{x}) \in \overline{E^k}$ that is reachable from $[s_0, t_0, \mathbf{x}_0]$ using Eve's moves only. Because σ_A is a winning strategy, there are $k' \le k$ moves of Adam application of which leads to Eve not being able to reach $Q_E \times \{\mathbf{0}\}$. Using the same reasoning as in Lemmata 2 and 5, it is clear that Adam can play these sequence of moves starting from $\overline{E^{k'}}$ and still win.

By repeating the procedure for every element of $\overline{E^1} \cup \overline{E^2} \cup \ldots \cup \overline{E^k}$, we obtain a winning strategy σ'_A, where Adam skips outside $\overline{E^1} \cup \overline{E^2} \cup \ldots \cup \overline{E^k}$. □

Now, we are ready to prove the main result of the section.

Theorem 7. *It is decidable in* NEXPTIME *whether Eve has a winning strategy in a k-control $\mathbb{Z}^d$-VASS game with a safety objective for Adam.*

Proof. To prove the claim, we construct a finite game graph and use the attractor construction to find the winner. Thanks to Lemma 6, we do not have to consider the infinite number of configurations in $Q_E \times Q_A \times \mathbb{Z}^d$, but instead can focus on configurations in $\overline{E^k}$.

In the game graph, we have nodes of Eve corresponding to the initial configuration, to configurations in $\overline{E^k}$, to configurations that can be reached from a configuration in $\overline{E^k}$, and two sink states—one winning and one losing. Let us define the nodes in more details. Let $\mathcal{Q}_E$ be the nodes of Eve. Now

- if $(s, \mathbf{x}) \in \overline{E^k}$ for some k, then $(\dot{s}, t, \mathbf{x}, k) \in \mathcal{Q}_E$;
- if there exists $(s', \mathbf{x}') \in \overline{E^k}$ such that $[\dot{s}', t', \mathbf{x}'][s, \dot{t}', \mathbf{x}''][\dot{s}, t, \mathbf{x}]$ is a valid play, then $(\dot{s}, t, \mathbf{x}, k-1) \in \mathcal{Q}_E$;
- if there exists $(s', \mathbf{x}') \in \overline{E^k}$ such that $[\dot{s}', t, \mathbf{x}'][s, \dot{t}, \mathbf{x}]$ is a valid play, then $(\dot{s}, t, \mathbf{x}, k) \in \mathcal{Q}_E$;

- $(\dot{s}_0, t_0, \mathbf{x}_0) \in \mathcal{Q}_E$, where $[\dot{s}_0, t_0, \mathbf{x}_0]$ is the initial configuration;
- $\bot, \top \in \mathcal{Q}_E$.

Let us then consider nodes $\mathcal{Q}_A$ of Adam. If there exists $(s', \mathbf{x}') \in \overline{E^k}$ such that $[\dot{s}', t', \mathbf{x}'][s, \dot{t}, \mathbf{x}]$ is a valid play, then $(s, \dot{t}, \mathbf{x}, k) \in \mathcal{Q}_A$.

Let us finally consider the edges in the game graph. The intuition is straightforward: we add edges from Eve's nodes to Adam's nodes without changing the last component, k, and we add edges from Adam's nodes to Eve's nodes either without changing anything (corresponding to Adam skipping) or changing the values and decreasing k by one (corresponding to Adam playing a move). Additionally, we add edges from every state of Eve to $\bot$, which is the losing sink, and we add edges from every state with counter vector $\mathbf{0}$ to $\top$, which is the winning sink. That is,

- if $\texttt{ZVASS}(E, (s_0, \mathbf{x}_0), (s, \mathbf{x}))=\text{TRUE}$, then $((\dot{s}_0, t_0, \mathbf{x}_0), (\dot{s}, t_0, \mathbf{x}, k)) \in \mathcal{E}$;
- if $(s, \mathbf{e}, s') \in E$, then $((\dot{s}, t, \mathbf{x}, k), (s', \dot{t}, \mathbf{x} + \mathbf{e}, k)) \in \mathcal{E}$;
- if $(t, \mathbf{a}, t') \in A$ and $k \geq 1$, then $((s, \dot{t}, \mathbf{x}, k), (\dot{s}, t', \mathbf{x} + \mathbf{a}, k - 1)) \in \mathcal{E}$;
- $((s, \dot{t}, \mathbf{x}, k), (\dot{s}, t, \mathbf{x}, k)) \in \mathcal{E}$;
- for every $q \in \mathcal{Q}_E$, $(q, \bot) \in \mathcal{E}$; and $((\dot{s}, t, \mathbf{0}, k), \top) \in \mathcal{E}$.

The starting position is the node $(\dot{s}_0, t_0, \mathbf{x}_0, k)$ and the objective of Eve is to reach node $\top$. Plays, strategies and winning strategies are defined analogously to the $\mathbb{Z}^d$-VASS games in Sect. 2. The main difference is that the game arena is finite and we can apply the standard attractor construction to the graph to obtain the winner [14,17] in polynomial time in the size of the game. The game graph $(\mathcal{Q}_E \cup \mathcal{Q}_A, \mathcal{E})$ can be constructed in NEXPTIME since the number of nodes is exponential and an NP algorithm is needed to create some of the edges. □

Corollary 8. *Deciding whether Eve has a winning strategy in a k-control $\mathbb{Z}^d$-VASS games with a safety objective is* NP*-complete if k is fixed.*

Proof. The lower bound follows from the observation in the beginning of the section by considering 0-control $\mathbb{Z}^d$-VASS games. The upper bound follows from the observation that the size of $\overline{E^k}$ is polynomial in the size of the input when k is fixed. Thus, the size of the finite game graph constructed in Theorem 7 is in NP when k is not part of the input. □

Next we briefly consider safety in k-control $\mathbb{Z}^d$-VAS games. It is worth noting that one-dimensional $\mathbb{Z}^d$-VAS are easier than one-dimensional $\mathbb{Z}^d$-VASS games. Namely, the deciding the winner in a $\mathbb{Z}^d$-VAS is in EXPTIME rather than EXPSPACE [2]. It is clear that the upper bound from Theorem 7 holds for stateless games as well. The lower bound also holds as it has been proven in [5] that the reachability for $\mathbb{Z}^d$-VAS is NP-complete.

Corollary 9. *Deciding whether Eve has a winning strategy in a k-control $\mathbb{Z}^d$-VAS games with a safety objective is* NP*-hard and in* NEXPTIME.

Corollary 10. *For any fixed k, deciding whether Eve has a winning strategy in a k-control $\mathbb{Z}^d$-VAS games with a safety objective is* NP*-complete.*

We conclude this section, with a generalisation of the objective of k-control $\mathbb{Z}^d$-VASS games with a safety objective. We extend the target from a single point in $\mathbb{Z}^d$ to a hyperplane, and prove that the complexity does not increase. This extension is similar to [6,7], where extensions of the orbit problem were studied. Recall, that in the orbit problem, we are given a single rational matrix and a vector and iterate the vector with the matrix. The authors considered whether a polyhedron can be reached by this iteration [6] or a general vector space [7].

In our extension, Eve's target is given by a system of linear equations with equalities and inequalities. This system of linear equations can be added into the construction as additional equations that need to be satisfied. For example, the construction of Theorem 7 can be modified by removing edges $(\dot{s}, t, \mathbf{0}, k, \top)$ and adding edges $(\dot{s}, t, \mathbf{x}, k, \top)$ iff $\mathbf{x}$ is a solution to the given system of linear equations. Clearly, the complexity does not change as a system of linear equations can be solved in non-deterministic polynomial time.

Theorem 11. *Consider a system of linear equations with equalities and inequalities $\mathcal{D}$. Deciding which player has a winning strategy in a k-control $\mathbb{Z}^d$-VASS game with safety objectives, where the targets satisfy $\mathcal{D}$, is* NP*-hard and in* NEXPTIME.

If the target set is instead defined by a semi-linear set, the complexity grows significantly. The checking whether $\mathbf{x}$ is in a given semi-linear set is in double exponential time [16], which increases the complexity of the whole algorithm.

Note that deciding whether Adam has a winning strategy in the settings of this section is in the respective complementary class, e.g., coNEXPTIME for Theorem 7.

4 Reachability of k-control $\mathbb{Z}^d$-VASS games

In this section, we consider the dual setting, where Eve can play her moves k times and Adam has unlimited resources. First, we present a lemma showing that Adam has a winning strategy in a $\mathbb{Z}^d$-VASS game where he has more moves.

Lemma 12. *Let $(A, E, [s_0, t_0, \mathbf{x}_0])$ be a $\mathbb{Z}^d$-VASS game. Assume that for each $s \in Q_E$ and $t \in Q_A$, $\deg(s) < \deg(t)$ and $\deg(t) > 1$, where $\deg(q)$ is the outdegree of q. Then Adam has a winning strategy.*

Proof. Let $[s, \dot{t}, \mathbf{x}]$ be a configuration of the game. Since $\deg(t) > 1$ and $\deg(t) > \deg(s)$, there exists at least one move $(t, \mathbf{a}, t')$ such that from $[\dot{s}, t', \mathbf{x}+\mathbf{a}]$, no move of Eve results in $[s', t', \mathbf{0}]$. This is obvious as otherwise, there would exist a move of Eve $(s, \mathbf{e}, s') \in Q_E$ such that for two moves of Adam $(t, \mathbf{a}_1, t_1), (t, \mathbf{a}_2, t_2) \in Q_A$, where $\mathbf{a}_1 \neq \mathbf{a}_2$, such that $\mathbf{x} + \mathbf{a}_1 + \mathbf{e} = 0 = \mathbf{x} + \mathbf{a}_2 + \mathbf{e}$. Now, Adam's strategy of playing $(t, \mathbf{a}, t')$ in a configuration $[s, \dot{t}, \mathbf{x}] =$ is a winning strategy. □

We are ready for the main result of the section. We prove the claim by showing that Eve can reach only finite number of points in $\mathbb{Z}^d$, that is, the arena is finite.

Theorem 13. *Deciding which player has a winning strategy in k-control $\mathbb{Z}^d$-VASS games with a reachability objective is in* EXPTIME.

Proof. We observe that only a finite area needs to be considered. By a similar argument as in Lemma 12, Adam can avoid the origin if Eve played SKIP, unless the game is trivial. Indeed, Adam will not choose a move that updates the counter to $\mathbf{0}$ if he has a choice. That is, if the configuration is vector is $-\ell\mathbf{a}$ and Adam can update the counter by $\mathbf{a}$ only. We further observe that after Eve has played all k moves, Adam has a trivial winning strategy.

If we only consider effect of Eve's moves during a play, then there are at most $\binom{m+1+k-1}{k}$ vectors she can reach. Indeed, it is the number of ways to select k moves out of m possible moves with the possibility of selecting the same move repeatedly. We denote this number by q. It is clear that Adam wins as soon as he can reach more than q vectors. Due to the commutativity of vector addition, the arena is finite and exponential in size of the input. We can compute whether the initial vector is winning for Eve in polynomial time using the standard attractor construction; see, e.g., [14,17], completing the proof. □

Corollary 14. *Deciding which player has a winning strategy in a k-control $\mathbb{Z}^d$-VASS games with a reachability objective is in* PTIME *if k is fixed.*

5 Conclusion

In this paper, we considered $\mathbb{Z}^d$-VASS games with limited resources. If we limit the available resources of Adam by allowing him to only play at most k times, then the problem of deciding the winner is NP-hard and in NEXPTIME. If we instead limit number of times Eve is allowed to play, then the problem is in EXPTIME. This is in contrast to $\mathbb{Z}^2$-VASS games, where the problem is undecidable.

The k-control VAS(S) games remain open for future research. The reachability problem for VAS is a hard problem that has been known to be decidable [28], but only recently it was shown to be Ackermann-complete [9]. The corresponding k-control game should be challenging. Indeed, if one of the players only skips, then the underlying VAS is simulated. It would be more fruitful to consider fixed dimensions as it was proven in [18] that the reachability problem, when the inputs are encoded in binary, is NP-complete for one-dimensional VASS and in [3] that it is PSPACE-complete for two-dimensional VASS. Leveraging these results should give us that the k-control game has a similar complexity. There is a considerable complexity gap for the reachability problem starting from three-dimensional VASS; see [8,25].

Another interesting direction would be to consider infinite state games played on finite graphs, where vertices are partitioned between the two players. It would require a different definition of skipping to address the fact that the game could be in a state of a player who has no moves left. This would simulate a situation, where both players are under an asymmetric restriction on moves available.

Acknowledgements. The authors would like to thank the anonymous reviewers for their helpful comments and suggestions. Igor Potapov was supported by the APEX Awards 2024 - APX\ R1\241186.

References

1. Abdulla, P.A., Bouajjani, A., d'Orso, J.: Deciding monotonic games. In: Baaz, M., Makowsky, J.A. (eds.) Computer Science Logic. CSL 2003. LNCS, vol. 2803, pp. 1–14. Springer, Berlin, Heidelberg (2003). https://doi.org/10.1007/978-3-540-45220-1_1
2. Arul, A., Reichert, J.: The complexity of robot games on the integer line. In: Proceedings of QAPL 2013, volume 117 of EPTCS, pp. 132–148, 2013. https://doi.org/10.4204/EPTCS.117.9
3. Blondin, M., Finkel, A., Göller, S., Haase, C., McKenzie, P.: Reachability in two-dimensional vector addition systems with states is PSPACE-complete. In: Proceedings of LICS 2015, pp. 32–43, 2015. https://doi.org/10.1109/LICS.2015.14
4. Brázdil, T., Jančar, P., Kučera, A.: Reachability games on extended vector addition systems with states. In: Abramsky, S., Gavoille, C., Kirchner, C., Meyer auf der Heide, F., Spirakis, P.G. (eds.) Automata, Languages and Programming. ICALP 2010. LNCS, vol. 6199, pp. 478–489. Springer, Berlin, Heidelberg (2010). https://doi.org/10.1007/978-3-642-14162-1_40
5. Chistikov, D., Haase, C., Halfon, S.: Context-free commutative grammars with integer counters and resets. Theor. Comput. Sci. **735**, 147–161 (2018). https://doi.org/10.1016/J.TCS.2016.06.017
6. Chonev, V., Ouaknine, J., Worrell, J.: The polyhedron-hitting problem. In: Proceedings of SODA 2015, pp. 940–956. SIAM, 2015. https://doi.org/10.1137/1.9781611973730.64
7. Chonev, V., Ouaknine, J., Worrell, J.: On the complexity of the orbit problem. J. ACM **63**(3), 23:1–23:18 (2016). https://doi.org/10.1145/2857050
8. Czerwiński, W., Jecker, I., Lasota, S., Leroux, J., Orlikowski, L.: New lower bounds for reachability in vector addition systems. In: Proceedings of FSTTCS 2023, volume 284 of LIPIcs, pp. 35:1–35:22. Schloss Dagstuhl–Leibniz-Zentrum fuer Informatik, 2023. https://doi.org/10.4230/LIPIcs.FSTTCS.2023.35
9. Czerwinski, W., Orlikowski, L.: Reachability in vector addition systems is ackermann-complete. In: Proceedings of FOCS 2022, pp. 1229–1240. IEEE, 2021. https://doi.org/10.1109/FOCS52979.2021.00120
10. Doyen, L., Rabinovich, A.: Robot games. Technical Report LSV-13-02, LSV, ENS Cachan, 2013. http://www.lsv.fr/Publis/RAPPORTS_LSV/PDF/rr-lsv-2013-02.pdf
11. Dufourd, C., Finkel, A., Schnoebelen, P.: Reset nets between decidability and undecidability. In: Proceedings of ICALP 1998, volume 1443 of LNCS, pp. 103–115. Springer, Cham (1998). https://doi.org/10.1007/bfb0055044
12. Englert, M., Lazić, R., Totzke, P.: Reachability in two-dimensional unary vector addition systems with states is NL-complete. In: Proceedings of LICS, pp. 477–484. ACM, 2016. https://doi.org/10.1145/2933575.2933577
13. Figueira, D., Lazić, R., Leroux, J.F., Sutre, G.: Polynomial-space completeness of reachability for succinct branching VASS in dimension one. In: Proceedings of ICALP 2017, vol. 80, pp. 119:1–119:14. Schloss Dagstuhl–Leibniz-Zentrum fuer Informatik, 2017. https://doi.org/10.4230/lipics.icalp.2017.119

14. Fijalkow, N. (ed.): Games on Graphs: From Logic and Automata to Algorithms, 2025. arXiv:2305.10546
15. Finkel, A., Göller, S., Haase, C.: Reachability in register machines with polynomial updates. In: Chatterjee, K., Sgall, J. (eds.) Mathematical Foundations of Computer Science 2013. MFCS 2013. LNCS, vol. 8087, pp. 409–420. Springer, Berlin, Heidelberg (2013). https://doi.org/10.1007/978-3-642-40313-2_37
16. Fischer, M.J., Rabin, M.O.: Super-exponential complexity of Presburger arithmetic. In: Complexity of Computation, SIAM-AMS Proceedings, vol. 7, pp. 27–41. SIAM, 1974. https://doi.org/10.1007/978-3-7091-9459-1_5
17. Grädel, E., Thomas, W., Wilke, T. (eds.): Automata, Logics, and Infinite Games: A Guide to Current Research, LNCS, vol. 2500. Springer, Cham (2002). https://doi.org/10.1007/3-540-36387-4
18. Haase, C., Kreutzer, S., Ouaknine, J., Worrell, J.: Reachability in succinct and parametric one-counter automata. In: Bravetti, M., Zavattaro, G. (eds.) CONCUR 2009 - Concurrency Theory. CONCUR 2009. LNCS, vol. 5710, pp. 369–383. Springer, Berlin, Heidelberg (2009). https://doi.org/10.1007/978-3-642-04081-8_25
19. Hopcroft, J.E., Pansiot, J.-J.: On the reachability problem for 5-dimensional vector addition systems. Theor. Comput. Sci. **8**, 135–159 (1979). https://doi.org/10.1016/0304-3975(79)90041-0
20. Hunter, P.: Reachability in succinct one-counter games. In: Bojanczyk, M., Lasota, S., Potapov, I. (eds.) Reachability Problems. RP 2015. LNCS, vol. 9328, pp. 37–49. Springer, Cham (2015). https://doi.org/10.1007/978-3-319-24537-9_5
21. Karp, R., Miller, R.: Parallel program schemata. J. Comput. Syst. Sci. **3**(2), 147–195 (1969). https://doi.org/10.1016/S0022-0000(69)80011-5
22. Kosaraju, S.R.: Decidability of reachability in vector addition systems (preliminary version). In: Proceedings of STOC, pp. 267–281. ACM, 1982. https://doi.org/10.1145/800070.802201
23. Lambert, J.L.: A structure to decide reachability in Petri nets. Theor. Comput. Sci. **99**(1), 79–104 (1992). https://doi.org/10.1016/0304-3975(92)90173-d
24. Lazić, R.: The reachability problem for branching vector addition systems requires doubly-exponential space. Inf. Process. Lett. **110**(17), 740–745 (2010). https://doi.org/10.1016/j.ipl.2010.06.008
25. Leroux, J.: The reachability problem for petri nets is not primitive recursive. In: Proceedings of FOCS, pp. 1241–1252, 2022. https://doi.org/10.1109/FOCS52979.2021.00121
26. Leroux, J., Schmitz, S.: Demystifying reachability in vector addition systems. In: Proceedings of LICS 2015, pp. 56–67. IEEE, 2015. https://doi.org/10.1109/LICS.2015.16
27. Lipton, R.: The reachability problem requires exponential space. Technical report, Department of Computer Science, Yale University, 1976
28. Mayr, E.: An algorithm for the general Petri net reachability problem. In: Proceedings of STOC 1981, pp. 238–246. ACM, 1981. https://doi.org/10.1145/800076.802477
29. Niskanen, R.: Reachability games and related matrix and word problems. PhD thesis, University of Liverpool, UK, 2018. https://livrepository.liverpool.ac.uk/3018108/
30. Niskanen, R., Potapov, I., Reichert, J.: On decidability and complexity of low-dimensional robot games. J. Comput. Syst. Sci. **107**, 124–141 (2020). https://doi.org/10.1016/J.JCSS.2019.08.003

31. Petri, C.: Fundamentals of a theory of asynchronous information flow. In: IFIP Congress 1962, pp. 386–390, 1962
32. Rackoff, C.: The covering and boundedness problems for vector addition systems. Theor. Comput. Sci. **6**(2), 223–231 (1978). https://doi.org/10.1016/0304-3975(78)90036-1
33. Reichert, J.: Reachability games with counters: decidability and algorithms. Doctoral thesis, LSV, ENS Cachan, France, 2015
34. Reichert, J.: On the complexity of counter reachability games. Fundam. Inform. **143**(3-4), 415–436 (2016). https://doi.org/10.3233/FI-2016-1320
35. Rosier, L., Yen, H.-S.: A multiparameter analysis of the boundedness problem for vector addition systems. J. Comput. Syst. Sci. **32**(1), 105–135 (1986). https://doi.org/10.1016/0022-0000(86)90006-1
36. Valiant, L.G., Paterson, M.: Deterministic one-counter automata. J. Comput. Syst. Sci. **10**(3), 340–350 (1975). https://doi.org/10.1016/S0022-0000(75)80005-5

Mergeable Represented Spaces

Arno Pauly(✉)

Computational Foundary, Swansea University, Swansea, UK
arno.m.pauly@gmail.com

Abstract. We introduce the notion of a *mergeable* represented space, which formalizes the concept that two parallel computations which either will yield the same answer, or have one computation explicitly fail, can be combined into a single computation. It is a very general concept, but we also provide some examples of spaces which are not mergeable. Mergeability can be characterized in a variety of different approaches.

1 Introduction

Consider a function $f : \mathbb{R} \to \mathbb{R}$ defined piecewise through continuous functions $f_1, f_2 : \mathbb{R} \to \mathbb{R}$ via:

$$f(x) = \begin{cases} f_1(x) & \text{if } x < a \\ f_2(x) & \text{if } x \geq a \end{cases}$$

It is well-known that f is continuous itself if and only if $f_1(a) = f_2(a)$. The same holds if we replace continuous by computable, provided that a is a computable real number. We can express this phenomenon in a slightly different way. If f_1, f_2, a are all computable and

$$f(x) = \begin{cases} f_1(x) & \text{if } x \leq a \\ f_2(x) & \text{if } x \geq a \end{cases}$$

actually defines a (singlevalued) function f, then f is computable. In general f defined thus is going to be multivalued, and in that case, not computable.

The general phenomenon here is that we have a non-computable multivalued test (is $x \leq a$ or $x \geq a$) in an otherwise computable procedure, with the added promise that if both branches of the test are correct, the respective computations yield the same answer. If the result of such a computation is a real number, we can actually compute this real number. What we study here is for what other spaces than $\mathbb{R}$ the same holds true. This leads us to define *weakly mergeable spaces* and *mergeable spaces*.

We assume some basic familiarity with computability and represented spaces, and freely refer to notions from [17]. There will be some references to Weihrauch reducibility, although the majority of the paper should be accessible without knowledge on this. For an introduction to Weihrauch reducibility, we direct the reader to [5].

V. Brattka et al. (Eds.): CiE 2026, LNCS 16674, pp. 469–481, 2027.
https://doi.org/10.1007/978-3-032-31348-5_31

To define weakly mergeable spaces, we refer to the space $\mathbf{T}$ of Kleenean truth values. The elements of $\mathbf{T}$ are $0, 1, \bot$ represented via $\delta_{\mathbf{T}}(0^{2n}1p) = 0$, $\delta_{\mathbf{T}}(0^{2n+1}1p) = 1$ and $\delta_{\mathbf{T}}(0^\omega) = \bot$. This space has the property that given $0 \in \mathbf{T}$ or $1 \in \mathbf{T}$ we can be sure which element we are looking at. However, $\bot$ is undefined in the sense that the name for $\bot$ can arbitrarily late change into a name for 0 or 1. The map $\mathrm{Comp}_a : \mathbb{R} \to \mathbf{T}$ defined via

$$\mathrm{Comp}_a(x) = \begin{cases} 0 & \text{if } x < a \\ \bot & \text{if } x = a \\ 1 & \text{if } x > a \end{cases}$$

is computable for computable $a \in \mathbb{R}$.

Definition 1. *Let* $\mathbf{X}$ *be a represented space. The operation* $\mathrm{wMerge}_{\mathbf{X}} :\subseteq \mathbf{T} \times \mathbf{X} \times \mathbf{X} \to \mathbf{X}$ *is defined on all inputs except* $(\bot, x, y)$ *for* $x \neq y$, *and satisfies* $\mathrm{wMerge}_{\mathbf{X}}(0, x, y) = x$, $\mathrm{wMerge}_{\mathbf{X}}(1, x, y) = y$ *and* $\mathrm{wMerge}_{\mathbf{X}}(\bot, x, y) = x = y$. *We call* $\mathbf{X}$ weakly mergeable *if* $\mathrm{wMerge}_{\mathbf{X}}$ *is computable.*

The reason we call this **weak** mergeability is the requirement that both x and y actually have to be elements of $\mathbf{X}$. This means that this only covers the case where the functions used in the individual cases are total. To also be able to consider the case where the functions used in the cases need to be defined only if the case is actually applicable, we need the notion of a completion of a represented space.

We follow the approach taken in [19], but point out that the key observations are due to [3,4,8,11].

Definition 2. *A represented space* $\mathbf{X}$ *is* multiretraceable, *if every computable partial function* $f :\subseteq \{0,1\}^{\mathbb{N}} \to \mathbf{X}$ *has a total computable extension* $F : \{0,1\}^{\mathbb{N}} \to \mathbf{X}$.

A *computable embedding* is a computable function $\iota : \mathbf{X} \to \mathbf{Y}$ with a computable partial left-inverse.

Definition 3. *A* completion *of a represented space* $\mathbf{X}$ *consists of a computable embedding* $\iota : \mathbf{X} \hookrightarrow \mathbf{Y}$ *into a multiretraceable space* $\mathbf{Y}$.

Proposition 1. *Every represented space has a completion.*

Proposition 2. *If* $\iota_1 : \mathbf{X} \hookrightarrow \mathbf{Y}_1$ *and* $\iota_2 : \mathbf{X} \hookrightarrow \mathbf{Y}_2$ *are completions of* $\mathbf{X}$, *then there is a computable multivalued function* $T : \mathbf{Y}_1 \rightrightarrows \mathbf{Y}_2$ *such that* $\iota_2 = T \circ \iota_1$.

Unfortunately, we are unaware of a canonic construction for the completion. Beyond aesthetics, this does not influence our results though. We write $\overline{\mathbf{X}}$ to denote some completion of $\mathbf{X}$ where the embedding is assumed to be the identity. For our uses of the completion, the concrete choice of $\overline{\mathbf{X}}$ does not matter.

Definition 4. *The operation* $\mathrm{Merge}_{\mathbf{X}} :\subseteq \mathbf{T} \times \overline{\mathbf{X}} \times \overline{\mathbf{X}} \to \mathbf{X}$ *satisfies* $\mathrm{Merge}_{\mathbf{X}}(0,x,y) = x$ *for* $x \in \mathbf{X}$, $\mathrm{Merge}_{\mathbf{X}}(1,x,y) = y$ *for* $y \in \mathbf{X}$ *and* $\mathrm{Merge}_{\mathbf{X}}(\bot,x,y) = x = y$ *for* $x = y \in \mathbf{X}$, *and is undefined elsewhere. We call* $\mathbf{X}$ mergeable *if* $\mathrm{Merge}_{\mathbf{X}}$ *is computable.*

Lemma 1. *The computability of* $\mathrm{Merge}_{\mathbf{X}}$ *does not depend on the choice of completion* $\overline{\mathbf{X}}$.

Proof. Let $\mathrm{Merge}^1_{\mathbf{X}}$ and $\mathrm{Merge}^2_{\mathbf{X}}$ be defined using two completions $\iota_1 : \mathbf{X} \hookrightarrow \mathbf{Y}_1$ and $\iota_2 : \mathbf{X} \hookrightarrow \mathbf{Y}_2$ of $\mathbf{X}$, and assume that $\mathrm{Merge}^2_{\mathbf{X}}$ is computable. Let T be the computable multivalued function from Proposition 2. Then $(b,x,y) \mapsto \mathrm{Merge}^2_{\mathbf{X}}(b,T(x),T(y))$ is a computable multivalued map extending $\mathrm{Merge}^1_{\mathbf{X}}$, so $\mathrm{Merge}^1_{\mathbf{X}}$ is computable. This shows that whether $\mathrm{Merge}_{\mathbf{X}}$ is computable does not depend on the choice of completion $\mathbf{X}$.

Clearly, every mergeable space is weakly mergeable.

2 Examples of (weakly) Mergeable Spaces

The conventional wisdom in computable analysis is that one ought to work with admissibly represented spaces whenever possible. We start our investigation of mergeability by showing that it is an even more general notion than admissibility. Recall that admissible represented spaces are those characterized by their topology. A represented space is computably admissible if its elements can be effectively recovered from their open neighbourhood filters.

Proposition 3. *If* $\mathbf{X}$ *is computably admissible, then it is mergeable.*

Proof. A computably admissible space $\mathbf{X}$ is characterized by the fact that a point x can be computably recovered from its open neighborhood filter $\{U \in \mathcal{O}(\mathbf{X}) \mid x \in U\} \subseteq \mathcal{O}(\mathcal{O}(\mathbf{X}))$. Given $(b,x,y) \in \mathrm{dom}(\mathrm{Merge}_{\mathbf{X}})$ we can compute

$$\mathfrak{U}_{b,x,y} := \{U \in \mathcal{O}(\mathbf{X}) \mid (x \in U \wedge b = 0) \vee (y \in U \wedge b = 1) \vee (x \in U \wedge y \in U)\}$$

The possibility that $x \notin \mathbf{X}$ or $y \notin \mathbf{X}$ is unproblematic for triples actually in the domain of $\mathrm{Merge}_{\mathbf{X}}$, as these have no impact on the truth values of the three cases for determining that $U \in \mathfrak{U}_{b,x,y}$. By construction $\mathfrak{U}_{b,x,y}$ is the open neighborhood filter of $\mathrm{Merge}_{\mathbf{X}}(b,x,y)$, so the computable admissibility of $\mathbf{X}$ lets us compute $\mathrm{Merge}_{\mathbf{X}}(b,x,y)$.

A well-known example of a non-admissible space are the reals with the decimal expansion, which we shall denote by $\mathbb{R}_{10}$.

Proposition 4. $\mathbb{R}_{10}$ *is not weakly mergeable.*

Proof. We show that $\mathrm{wMerge}_{\mathbb{R}_{10}}$ is not only non-computable, but Weihrauch-equivalent to C_2 (aka LLPO). The map $d : \mathbf{T} \rightrightarrows \{0,1\}$ mapping 0 to 0, 1 to 1 and $\bot$ non-deterministically to either 0 or 1 is easily seen to be equivalent to C_2. It is immediate that $\mathrm{wMerge}_{\mathbf{X}} \leq_{\mathrm{W}} d$ for any represented space $\mathbf{X}$.

Given some $b \in \mathbf{T}$ we can non-deterministically compute decimal reals x, y such that if $b = \bot$, then $x = y = 1$ if $b = 0$ then $x < 1$ and $y = 1$ and if $b = 1$ then $x = 1$ and $y > 1$. Now we find that $\mathrm{wMerge}_{\mathbb{R}_{10}}(b, x, y) \leq 1$ implies that $b \neq 1$ and $\mathrm{wMerge}_{\mathbb{R}_{10}}(b, x, y) \geq 1$ implies that $b \neq 0$. As we can select a true case between $\mathrm{wMerge}_{\mathbb{R}_{10}}(b, x, y) \leq 1$ and $\mathrm{wMerge}_{\mathbb{R}_{10}}(b, x, y) \geq 1$ given a decimal name for $\mathrm{wMerge}_{\mathbb{R}_{10}}(b, x, y)$, this yields a Weihrauch reduction $d \leq_{\mathrm{W}} \mathrm{wMerge}_{\mathbb{R}_{10}}$.

A separation between weakly mergeable spaces and mergeable spaces can be found by investigating computably discrete spaces. Recall that $\mathbf{X}$ is computably discrete if the map $\neq : \mathbf{X} \times \mathbf{X} \to \mathbb{S}$ is computable. While it may seem at first glance that the computably discrete spaces are rather trivial, they can be a very useful class of examples [14].

Proposition 5. *If* $\mathbf{X}$ *is computably discrete, then it is weakly mergeable.*

Proof. If $\mathbf{X}$ is computably discrete, the cases $x = y$, $b = 0$ and $b = 1$ are all recognizable. At least one of them must be true for (b, x, y) in the domain of wMerge, so we can wait until we have identified a true case and then output an appropriate answer.

Corollary 1. *If* $|\mathbf{X}| = 1$ *then* $\mathbf{X}$ *is weakly mergeable.*

Proof. Any one-element represented space is trivially computably discrete, as we know that all elements are equal.

Example 1. There is a one-point represented space which is not mergeable.

Proof. A one-point space $\mathbf{1}_A$ is characterized by the domain $A \subseteq \{0,1\}^{\mathbb{N}}$ of its representation $\delta_1 : A \to \{1\}$. Let $q, r \in \{0,1\}^{\mathbb{N}}$ be Turing-incomparable. Let $A = \{0p \mid p \equiv_{\mathrm{T}} q\} \cup \{1p \mid p \equiv_{\mathrm{T}} r\}$. We find that $(0^\omega, 0q, 1r)$ consists of names for a valid input to $\mathrm{Merge}_{\mathbf{1}_A}$. If $\mathrm{Merge}_{\mathbf{1}_A}$ had a computable realizer, this would have to output the first digit of an element of A on this input. Wlog, assume it outputs 0 after having read some prefix $(0^n, 0q_{\leq n}, 1r_{\leq n})$. Then it would also output 0 as the first digit if the input is $(0^{2n+1}1^\omega), 0q_{\leq n}0^\omega, 1r)$. However, to correctly continue, the realizer would need to output something in the Turing degree of q, but it can only output names Turing-reducible to r. Outputting 1 as the first digit leads to a mistake in the same fashion.

Corollary 2. *There is a represented space which is weakly mergeable but not mergeable.*

For our next class of represented spaces which are mergeable we recall the finite mindchange endofunctor $^\nabla$. Let the map $M :\subseteq \{0,1\}^\omega \to \{0,1\}^\omega$ be defined by $M(w11\langle p, 0^\omega\rangle) = p$ (i.e., M is defined on those sequences containing a last consecutive pair of 1s, and discards everything up to this last pair and every other 0 afterwards). For a represented space $\mathbf{X} = (X, \delta_\mathbf{X})$, let $\mathbf{X}^\nabla = (X, \delta_\mathbf{X} \circ M)$. For map $f : \mathbf{X} \to \mathbf{Y}$ between represented spaces, let f^∇ be the same set-theoretic function from $\mathbf{X}^\nabla$ to $\mathbf{Y}^\nabla$. This makes $^\nabla$ a monad on the category of represented spaces. The computable maps from $\mathbf{X}$ to $\mathbf{Y}^\nabla$ are exactly the maps from $\mathbf{X}$ to $\mathbf{Y}$ which are computable with finitely many mindchanges. For more detail, see [18].

Proposition 6. *For any represented space* $\mathbf{X}$ *we find that* $\mathbf{X}^\nabla$ *is mergeable.*

Proof. It suffices to show that for any represented space $\mathbf{Y}$, $\text{Merge}_\mathbf{Y}$ is computable with finitely many mindchanges. In fact, a single mindchange suffices: Given an input (b, x, y) for $\text{Merge}_\mathbf{Y}$ we output x. If we ever learn that $b = 1$, we change our mind and output y instead.

Examples for spaces closed under $^\nabla$ include the Turing degrees, $\mathfrak{T}$, any space $\mathbf{Y}'$ obtained by taking the jump, as well as most hyperspaces in descriptive set theory such as the space of Σ^0_n-sets or of all Borel sets. Note that $\mathbf{X}^\nabla$ is never T_0 as long as $\mathbf{X}$ contains two distinct points, thus $\mathbf{X}^\nabla$ cannot be admissible unless $\mathbf{X}$ is empty or a singleton. This actually works computably:

Proposition 7. *If* $\mathbf{X}^\nabla$ *is computably admissible, then either* $\mathbf{X}^\nabla \cong \mathbf{X} \cong \emptyset$ *or* $\mathbf{X}^\nabla \cong \mathbf{X} \cong \mathbf{1}$.

Proof. An admissible space always carries a T_0 topology, while $\mathbf{X}^\nabla$ always carries the indiscrete topology. This shows that if $\mathbf{X}^\nabla$ is admissible, it contains at most one point. The empty represented space is unique anyway. Any space $\mathbf{X}^\nabla$ is computably overt, as if a realizer of an open set in $\mathbf{X}^\nabla$ accepts any element of $\{0,1\}^\mathbb{N}$, it also accepts a name of an element of $\mathbf{X}^\nabla$, which means we can recognize the non-empty open sets in $\mathcal{O}(\mathbf{X}^\nabla)$. If $\mathbf{X}^\nabla$ contains a single point x, the collection of all non-empty open sets is the neighborhoodfilter of x. As the former is computable, computable admissibility of $\mathbf{X}^\nabla$ then shows that the point x itself is computable, which yields that $\mathbf{X}^\nabla \cong \mathbf{X} \cong \mathbf{1}$.

There is one "natural" example of a represented space which is neither admissible nor closed under $^\nabla$. This is the space COrd of countable ordinal, where countable ordinals are represented as the ranks of well-founded countably-branching trees [16].

Proposition 8. COrd *is mergeable.*

Proof. Given two countably-branching trees, at least one of which is well-founded, we can compute a well-founded tree whose rank is the smaller of the ranks of the two initial trees.

3 Closure Properties of the (weakly) Mergeable Spaces

Both the weakly mergeable spaces and the mergeable spaces enjoy very strong closure properties, as we will detail in this section. Each result can either be read with *weakly* in all places where *(weakly)* appears or in none.

Proposition 9. *If* $\mathbf{X}, \mathbf{Y}$ *are (weakly) mergeable spaces, then their disjoint union* $\mathbf{X} + \mathbf{Y}$ *is also (weakly) mergeable.*

Proof. Given an input (b, x, y) for $\text{wMerge}_{\mathbf{X}+\mathbf{Y}}$ we can determine for both x and y whether they come from $\mathbf{X}$ or $\mathbf{Y}$. If they come from the same component, we can run the corresponding $\text{wMerge}_{\mathbf{X}}$ or $\text{wMerge}_{\mathbf{Y}}$ on (b, x, y). If they come from different components, then $x \neq y$ hence $b \neq \bot$, and we can just compute the answer.

Given an input (b, x, y) for $\text{Merge}_{\mathbf{X}+\mathbf{Y}}$ we have to account for the possibility that either x or y is merely an element of the completion of $\mathbf{X} + \mathbf{Y}$. However, at least one of the following must be true: Either $b \neq \bot$, or we can compute for both x and y to which of $\mathbf{X}$ or $\mathbf{Y}$ they would belong if they do belong to $\mathbf{X}+\mathbf{Y}$. Moreover, in the latter case we can demand that both would belong to the same component, because otherwise again $b \neq \bot$ must hold. We then proceed as in the preceding paragraph.

Proposition 9 shows that the classes of spaces shown in Sect. 2 to be (weakly) mergeable do not constitute a characterization of the (weakly) mergeable spaces. For example, if $\mathbf{X}$ is computably admissible but not invariant under ∇ and $\mathbf{Y}^\nabla$ is not computably admissible, then $\mathbf{X} + \mathbf{Y}^\nabla$ is mergeable, yet neither computably admissible nor invariant under ∇. A concrete such example is $\mathbf{2} + \mathbf{2}^\nabla$.

Proposition 10. *If* $\mathbf{X}$ *is a subspace of* $\mathbf{Y}$ *and* $\mathbf{Y}$ *is (weakly) mergeable, then so is* $\mathbf{X}$.

Proof. We find that $\text{wMerge}_{\mathbf{X}}$ is just a restriction of $\text{wMerge}_{\mathbf{Y}}$ here, so it inherits computability from the latter. The same holds for $\text{Merge}_{\mathbf{X}}$ and $\text{Merge}_{\mathbf{Y}}$ if we take into account that every completion of $\mathbf{Y}$ is also a completion of $\mathbf{X}$ and invoke Lemma 1.

Proposition 11. *If* $\mathbf{Y}$ *is (weakly) mergeable, then* $\mathcal{C}(\mathbf{X}, \mathbf{Y})$ *is (weakly) mergeable for every represented spaces* $\mathbf{X}$.

Proof. Given an input (b, f, g) to $\text{wMerge}_{\mathcal{C}(\mathbf{X},\mathbf{Y})}$ we can compute the map $x \mapsto \text{wMerge}_{\mathbf{Y}}(b, f(x), g(x))$ and find that it is indeed $\text{wMerge}_{\mathcal{C}(\mathbf{X},\mathbf{Y})}(b, f, g)$. The same argument works for Merge in place of wMerge.

Corollary 3. *If* $\mathbf{X}, \mathbf{Y}$ *are (weakly) mergeable spaces, then their product* $\mathbf{X} \times \mathbf{Y}$ *is also (weakly) mergeable.*

Proof. By combining Propositions 9 10, 11 and noting that $\mathbf{X} \times \mathbf{Y}$ embeds as a subspace into $\mathcal{C}(\mathbf{2}, \mathbf{X} + \mathbf{Y})$.

4 Characterizing the Mergeable Quotients of $\mathbb{N}$

A general characterization of the mergeable (or the weakly mergeable) spaces eludes us. However, if we restrict our attention to represented spaces arising as a quotient of $\mathbb{N}$ by some equivalence relation, we recover a well-studied class. The quotients of $\mathbb{N}$ in general are those represented spaces whose representations we can conceive of as total numberings in the sense of Ershov [9].

Proposition 12. *The following are equivalent for an equivalence relation* $\equiv \subseteq \mathbb{N} \times \mathbb{N}$*:*

1. $\mathbb{N}/\equiv$ *is computably discrete.*
2. $\mathbb{N}/\equiv$ *is computably admissible.*
3. $\mathbb{N}/\equiv$ *is mergeable.*
4. $\mathbb{N}/\equiv$ *is weakly mergeable.*

Proof. $1 \Rightarrow 2$ [14, Proposition 6]

$2 \Rightarrow 3$ Proposition 3.

$3 \Rightarrow 4$ Trivial.

$4 \Rightarrow 1$ Given $n, m \in \mathbb{N}$ we run $\mathrm{Merge}_{\mathbb{N}/\equiv}(\bot, n, m)$. If $n \equiv m$, then the procedure needs to output something eventually. If $n \not\equiv m$ and $\mathrm{Merge}_{\mathbb{N}/\equiv}(\bot, n, m)$ outputs something, we can subsequently specify $\bot$ to be either 0 or 1 in a way that makes the output wrong. Thus, $\mathrm{Merge}_{\mathbb{N}/\equiv}(\bot, n, m)$ provides an output iff $n \equiv m$, which allows us to detect equality in $\mathbb{N}/\equiv$.

The class of represented spaces covered by Proposition 12 has further characterizations, as was shown in [14]. For example, these are also exactly the computably discrete computably Quasi-Polish spaces. Maybe most importantly, the properties in Proposition 12 are equivalent to $\equiv$ being computably enumerable. The study of computably enumerable equivalence relations (ceers) on $\mathbb{N}$ has received a lot of attention.

5 Generalized Register Machines

Tavana and Weihrauch introduced the notion of a generalized register machine [21]. Each register is associated with a represented space, and can store arbitrary elements thereof. The program of the register machine makes use of a finite collection of (multivalued) functions between represented spaces, and can assign to a register the result of applying such operations to the values in other registers, provided the types match. The program can branch dependend on the outcome of (potentially non-deterministic) tests, which are just (multivaled) functions into **2**. The machine can halt and return the value of the first register. Such a machine computes a multivalued function defined on all those initial assignments where it is guaranteed that the machine will never apply an operation to inputs outside its domain and will halt no matter how the non-determinism is resolved. The valid outputs are the values the machine could return.

We characterize both weakly mergeable and mergeable spaces via the property that certain generalized register machines can only compute computable singlevalued functions with that codomain. The difference between the two settings is that in Theorem 1 the operations used by the generalized register machine need to be total, whereas in Theorem 2 the machines can use partial computable functions, albeit only if it is guaranteed that they will only be invoked on inputs they are defined on.

Theorem 1. *The following are equivalent for a represented space* $\mathbf{X}$*:*

1. $\mathbf{X}$ *is weakly mergeable.*
2. *Every singlevalued function with codomain* $\mathbf{X}$ *which can be computed by a generalized register machine using* **total** *computable operations and the test whether* $x \leq 0$ *or* $x \geq 0$ *for* $x \in \mathbb{R}$ *is already computable.*

Proof. To see that 2 implies 1, we observe that the multivalued map $t : \mathbf{T} \rightrightarrows \mathbb{R}$ mapping $\bot$ to 0, 0 to any $x < 0$ and 1 to any $x > 0$ is computable. This shows that wMerge is computable by a generalized register machine using total computable operations and the weak sign test on $\mathbb{R}$.

For the other direction, we assume that $\mathbf{X}$ is weakly mergeable and that we have a generalized register machine of the given type which computes some function $f : \mathbf{Y} \to \mathbf{X}$ together with some input $y \in \mathbf{Y}$ for $\mathbf{X}$. We can follow the computation of the generalized register machine. We can actually execute any function application. Whenever we encounter a test for $x \leq 0$ or $x \geq 0$, we follow both branches at first. If we ever learn that $x < 0$ or $x > 0$, we retroactively discard the wrong branch. As all the computable operations involved are total, we do not encounter any problems while temporarily exploring the wrong branches.

As we started with an input in the domain of f, we know that every valid branch will halt after finitely many steps. As the tree we explore is finitely branching, weak Köning's lemma implies that the tree of valid branches is finite. This in turn ensures that eventually we will arrive at a configuration where all not yet discarded branches have halted. Of course, it might still be the case that some of the halting branches are actually wrong. As f is a function, we also know that all correct branches yield the same answer. We can apply wMerge pairwise to the results of the branches and together with $b \in \mathbf{T}$ indicating which of the branch is not found to be wrong first.

Theorem 2. *The following are equivalent for a represented space* $\mathbf{X}$*:*

1. $\mathbf{X}$ *is mergeable.*
2. *Every singlevalued function with codomain* $\mathbf{X}$ *which can be computed by a generalized register machine using computable operations and the test whether* $x \leq 0$ *or* $x \geq 0$ *for* $x \in \mathbb{R}$ *is already computable.*

Proof. The proof is very similar to that of Theorem 1, and we only outline the required adjustments here. To see that 2 implies 1 we need in addition the particular computable inverses of the embeddings of $\mathbf{X}$ into its completion $\overline{\mathbf{X}}$.

For the implication from 1 to 2, we can first replace the codomains of all the computable operations used by the generalized register machine by their completions and the computable operations by some total computable continuation. We can also computably extend $\text{Merge}_{\mathbf{X}}$ to all of $\mathbf{T} \times \overline{\mathbf{X}} \times \overline{\mathbf{X}}$ if we change the codomain to $\mathbf{X}$. We can apply this extension pairwise to all the results of terminating branches (including some wrong branches). The ultimate result of that is the value f takes, which we can computably translate back from $\overline{\mathbf{X}}$ to $\mathbf{X}$.

A more powerful machine model than the generalized register machines using computable operations and $x \leq 0 \vee x \geq 0$ tests are the nondeterministic Type-2 Turing machines studied in [2,7,22]. These machines can non-deterministically guess some $z \in \{0,1\}^{\mathbb{N}}$, and then either compute a correct solution or reject the guess, with the requirement that at least one guess must work. If the output is in a computably admissible space (rather than just a mergeable one), it still holds that every singlevalued function computable by such a machine is computable. It seems, however, that concrete implementations constructed via a detour through non-deterministic Type-2 Turing machines will be significantly less efficient than concrete implementations constructed via generalized register machines using non-deterministic order tests. Whether this impression can be turned into a formal theorem is left as an open question.

There are two more related studies to be mentioned: Brattka and Hertling proposed the Feasible Real RAM model [6]. Here the available test chooses a true cases between $x < \varepsilon$ and $x \geq -\varepsilon$ for an adjustable $\varepsilon > 0$. In addition, the Feasible Real RAMs uses the arithmetic operations on the reals and a controlled approximation operation. They can compute exactly the computable functions between powers of $\mathbb{R}$. The Feasible Real RAM model were a response to the Blum-Shub-Smale (BSS) model [1], which allows exact tests on $\mathbb{R}$ and thereby can compute non-computable functions. For the BSS model (but not the Feasible Real RAMs) it is a crucial feature that only arithmetic operations are available. The more general consequences of allowing exact tests on the reals were studied in [13].

6 Weihrauch Degrees Realizable on Certain Types

Mergeability of a space $\mathbf{Y}$ is equivalent to a certain interval in the Weihrauch lattice not containing any functions with codomain $\mathbf{Y}$. The top of this interval is the Weihrauch degree $\mathrm{C}_2^\diamond$, which corresponds to the principle LLPO. The degree $\mathrm{C}_2^\diamond$ provides the ability to pick a valid choice out of finitely many options, some of which may later be revealed to be false. Alternatively, it provides the ability to pick a valid choice out of two options, at most one of which may be revealed false, finitely many times, either in parallel or sequentially.

Corollary 4. *A represented space $\mathbf{Y}$ is mergeable iff every function $f : \mathbf{X} \to \mathbf{Y}$ with $f \leq_W \mathrm{C}_2^\diamond$ is already computable.*

Proof. With the definition of the $^\diamond$-operator from [13], this is just a restatement of Theorem 2.

There are already some similar results known in the literature, which we shall list below. We point out that $C_2^\diamond <_W C_{\{0,1\}^\mathbb{N}}$ while $C_2^\diamond$ and LPO are incomparable w.r.t. Weihrauch reducibility.

Theorem 3 **([2, Theorem 5.1]).**[1] *If* $\mathbf{Y}$ *is computably admissible, then every function* $f : \mathbf{X} \to \mathbf{Y}$ *with* $f \leq_W C_{\{0,1\}^\mathbb{N}}$ *is already computable.*

Theorem 4 **([19, Proposition 9]).** *The following are equivalent for a represented space* $\mathbf{X}$*:*

1. $\mathbf{X}$ *is admissible.*
2. *Every function* $f : (\omega + 1) \to \mathbf{X}$ *satisfies* $(f \equiv_W^* 1) \vee (f \equiv_W^* LPO)$.

The following is a rather trivial observation matching the same pattern:

Observation 5 *The following are equivalent for a represented space* $\mathbf{X}$*:*

1. $\mathbf{X} \cong \mathbf{X}^\nabla$
2. *Whenever* $f : \mathbf{Y} \to \mathbf{X}$ *satisfies* $f \leq_W C_\mathbb{N}$*, then already* $f \leq_W \mathbf{1}$.
3. *Whenever* $f : \mathbf{Y} \rightrightarrows \mathbf{X}$ *satisfies* $f \leq_W C_\mathbb{N}$*, then already* $f \leq_W \mathbf{1}$.

Observation 5 is not specific to the ∇-endofunctor or the Weihrauch degree $C_\mathbb{N}$. Any Weihrauch degree g gives rise to an endofunctor $\mathfrak{G}$ preserving the underlying set of represented spaces, such that $f : \mathbf{X} \to \mathbf{Y}$ satisfies $f \leq_W g$ iff $f : \mathbf{X} \to \mathfrak{G}\mathbf{Y}$ is computable (cf. [15]). If $g \equiv_W g^\diamond$, then $\mathfrak{G}$ is even a monad; and a space $\mathbf{X}$ being invariant under $\mathfrak{G}$ is equivalent to there being no Weihrauch degrees of maps with codomain $\mathbf{X}$ between $\mathbf{1}$ and g.

7 Functions Defined on Unions of Spaces

An additional motivation for the present paper was a question raised by King on MathOverflow [12]. He asked, working in a general constructive setting, when a map $f : [a, b] \cup [b, c] \to S$ can be extended to some $g : [a, c] \to S$, where $a, b, c \in \mathbb{R}$. To make sense of this in the context of computable analysis, we consider what representations we could use on $A \cup B$ where $A, B \subseteq \mathbf{X}$ for a given represented space $\mathbf{X}$: First, we could just use the subspace representation the set $A \cup B$ inherits from $\mathbf{X}$. We will denote this by $A \cup B$. We could use the disjoint union $A \sqcup B$, where each element is tagged as belonging to A or belonging to B, together with its name as an element of $\mathbf{X}$. This in particular means that any $x \in A \cap B$ is duplicated in $A \sqcup B$. Finally, we could use the space $A \uplus B$ we obtain as a quotient of $A \sqcup B$ by identifying both copies of points in $A \cap B$. This then let's us phrase King's question in a computable analysis setting as:

Question 1. For which represented spaces $\mathbf{X}$ can we extend any computable $f : [a, b] \uplus [b, c] \to \mathbf{X}$ to a computable $g : [a, c] \to \mathbf{X}$?

[1] As pointed out on [20, Page 6], the statement of the theorem in [2] is weaker then the present version, but the proof carries over.

From Theorem 1 we immediately conclude that all weakly mergeable spaces have this extension property. King already observed that spaces $\mathbf{X}$ admitting a computable group structure also have the extension property [12]. This leads us to ask:

Open Question 6 *Is there a represented space $\mathbf{X}$ with a computable group structure (i.e. a computable point $0 \in \mathbf{X}$, a computable map $- : \mathbf{X} \to \mathbf{X}$ and a computable map $+ : \mathbf{X} \times \mathbf{X} \to \mathbf{X}$ satisfying the group axioms) which is not weakly mergeable?*

We can adapt Example 1 to show that the answer would be *yes* if we dropped the word *weakly*: For any one-point space $\mathbf{1}_A$ we can equip $\mathbf{1} \sqcup \mathbf{1}_A$ with a computable group structure; and if $\mathbf{1}_A$ is not mergeable, then $\mathbf{1} \sqcup \mathbf{1}_A$ is not mergeable either.

One ingredient for King's use of a group structure on $\mathbf{X}$ to prove the extension property for functions defined on $[a, b] \uplus [b, c]$ is that the metric projections $\pi_1 : [a, c] \to [a, b]$ and $\pi_2 : [a, c] \to [b, c]$ to the closest point are computable. This is a very restrictive demand (cf. [10]), however, the weaker observation that $[a, b]$ and $[b, c]$ are retracts of $[a, c]$ suffices for the argument.

If we move beyond the specific setting of functions defined on the union of adjacent real intervals, we arrive at a clear characterization. It uses the space $\mathcal{A}(\mathbf{X})$ of closed subsets of $\mathbf{X}$, represented by negative information.

Theorem 7. *The following are equivalent for a represented space $\mathbf{X}$:*

1. *$\mathbf{X}$ is mergeable.*
2. *Whenever $A, B \in \mathcal{A}(\mathbf{Y})$ are computable closed subspaces of some space $\mathbf{Y}$ and $f : A \uplus B \to \mathbf{X}$ is a computable function, then f has a computable extension $g : A \cup B \to \mathbf{X}$.*

Proof. For the implication $1 \Rightarrow 2$, we observe that A, B being computable closed subspaces of $\mathbf{Y}$ implies that the map $t : A \cup B \to \mathbf{T}$ mapping $y \in A \cap B$ to $\bot$, $y \in A \setminus B$ to 0 and $y \in B \setminus A$ to 1 is computable. We use $t(y)$ together with the results of attempting to compute $f|_A(y)$ and $f|_B(y)$ as inputs to $\text{Merge}_{\mathbf{X}}$ to compute $g(y)$.

For the other direction, note that $A = (\{0\} \times \mathbf{X} \times \overline{\mathbf{X}}) \cup (\{\bot\} \times \{(x, x) \mid x \in \mathbf{X}\})$ and $B = (\{1\} \times \overline{\mathbf{X}} \times \mathbf{X}) \cup (\{\bot\} \times \{(x, x) \mid x \in \mathbf{X}\})$ are computable closed subspaces of the domain of $\text{Merge}_{\mathbf{X}}$ and their union covers that set. The restrictions of $\text{Merge}_{\mathbf{X}}$ to either A or B are trivially computable, and this tells us that $\text{Merge}_{\mathbf{X}} : A \uplus B \to \mathbf{X}$ is computable. By assumption it has a computable extension, which shows that $\text{Merge}_{\mathbf{X}}$ is computable on its actual domain.

Acknowledgments. This article benefited from the careful attention of the referees. The author is grateful for discussions to the members of the Computability Group at Swansea University.

Disclosure of Interests. The author has no competing interests to declare that are relevant to the content of this article.

References

1. Blum, L., Shub, M., Smale, S.: On a theory of computation and complexity over the real numbers: NP-completeness, recursive functions and universal machines. Bull. Amer. Math. Soc. **21**(1), 1–46 (1989)
2. Brattka, V., de Brecht, M., Pauly, A.: Closed choice and a uniform low basis theorem. Ann. Pure Appl. Log. **163**(8), 986–1008 (2012). https://doi.org/10.1016/J.APAL.2011.12.020
3. Brattka, V., Gherardi, G.: Weihrauch goes Brouwerian. J. Symb. Log. **85**(4), 1614–1653 (2020). https://doi.org/10.1017/JSL.2020.76
4. Brattka, V., Gherardi, G.: Completion of choice. Ann. Pure Appl. Log. **172**(3), 102914 (2021). https://doi.org/10.1016/J.APAL.2020.102914
5. Brattka, V., Gherardi, G., Pauly, A.: Weihrauch complexity in computable analysis. In: Brattka, V., Hertling, P. (eds.) Handbook of Computability and Complexity in Analysis, LNCS, pp. 367–417. Springer, Cham (2021). https://doi.org/10.1007/978-3-030-59234-9_11
6. Brattka, V., Hertling, P.: Feasible real random acess machines. J. Complex. **14**, 490–526 (1998). https://doi.org/10.1006/jcom.1998.0488
7. Brattka, V., Pauly, A.: Computation with advice. Electron. Proc. Theor. Comput. Sci. **24** (2010). http://arxiv.org/html/1006.0551, cCA 2010
8. Dzhafarov, D.D.: Joins in the strong Weihrauch degrees. Math. Res. Lett. **26**(3), 749–767 (2019). https://doi.org/10.4310/MRL.2019.v26.n3.a5, https://dx.doi.org/10.4310/MRL.2019.v26.n3.a5
9. Ershov, Y.L.: Theory of numberings. In: Handbook of Computability Theory, pp. 473–506. Elsevier (1999). https://doi.org/10.1016/S0049-237X(99)80030-5
10. Gherardi, G., Marcone, A., Pauly, A.: Projection operators in the Weihrauch lattice. Computability **8**(3–4), 281–304 (2019). https://doi.org/10.3233/COM-180207
11. Hoyrup, M.: The fixed-point property for represented spaces. Ann. Pure Appl. Log. **173**(5) (2022)
12. (https://mathoverflow.net/users/65915/christopher king), C.K.: When can a function defined on $[a, b] \cup [b, c]$ be constructively extended to a function defined on $[a, c]$? MathOverflow, https://mathoverflow.net/q/464084, (version: 2024-04-12)
13. Neumann, E., Pauly, A.: A topological view on algebraic computations models. J. Complex. **44** (2018). https://doi.org/10.1016/j.jco.2017.08.003, http://arxiv.org/abs/1602.08004
14. Neumann, E., Pauly, A., Pradic, C., Valenti, M.: Computably discrete represented spaces. In: Beckmann, A., Oitavem, I., Manea, F. (eds.) Crossroads of Computability and Logic: Insights, Inspirations, and Innovations - 21st Conference on Computability in Europe, CiE 2025, Lisbon, Portugal, July 14-18, 2025, Proceedings. LNCS, vol. 15764, pp. 349–364. Springer, Cham (2025). https://doi.org/10.1007/978-3-031-95908-0_25
15. Nobrega, H., Pauly, A.: Game characterizations and lower cones in the Weihrauch degrees. Log. Methods Comput. Sci. **15**(3), 11.1–11.29 (2019). https://doi.org/10.23638/LMCS-15(3:11)2019
16. Pauly, A.: Computability on the countable ordinals and the Hausdorff-Kuratowski theorem. In: MFCS 2015 (2015). http://arxiv.org/abs/1501.00386
17. Pauly, A.: On the topological aspects of the theory of represented spaces. Computability **5**(2), 159–180 (2016). https://doi.org/10.3233/COM-150049, http://arxiv.org/abs/1204.3763

18. Pauly, A., de Brecht, M.: Descriptive set theory in the category of represented spaces. In: 30th Annual ACM/IEEE Symposium on Logic in Computer Science, LICS 2015, Kyoto, Japan, 6–10 July 2015, pp. 438–449. IEEE Computer Society (2015). https://doi.org/10.1109/LICS.2015.48
19. Pauly, A., Soldà, G.: Sequential discontinuity and first-order problems. In: Patey, L.L., Pimentel, E., Galeotti, L., Manea, F. (eds.) Twenty Years of Theoretical and Practical Synergies - 20th Conference on Computability in Europe, CiE 2024, Amsterdam, The Netherlands, 8–12 July 2024, Proceedings. LNCS, vol. 14773, pp. 351–365. Springer, Cham (2024). https://doi.org/10.1007/978-3-031-64309-5_28
20. Roux, S.L., Pauly, A.: Finite choice, convex choice and finding roots. Log. Methods Comput. Sci. **11**(4) (2015). https://doi.org/10.2168/LMCS-11(4:6)2015
21. Tavana, N., Weihrauch, K.: Turing machines on represented sets, a model of computation for analysis. Log. Methods Comput. Sci. **7**, 1–21 (2011). https://doi.org/10.2168/LMCS-7(2:19)2011
22. Ziegler, M.: Computability and continuity on the real arithmetic hierarchy and the power of type-2 nondeterminism. In: Cooper, S.B., Löwe, B., Torenvliet, L. (eds.) New Computational Paradigms. CiE 2005. LNCS, vol. 3526, pp. 562–571. Springer, Berlin, Heidelberg (2005). https://doi.org/10.1007/11494645_68

On Normality and Equidistribution for Separator Enumerators

Subin Pulari(✉)

National Research University Higher School of Economics Moscow, Moscow, Russia
spulari@hse.ru

Abstract. A separator is a countable dense subset of $[0, 1)$, and a separator enumerator is a naming scheme that assigns a real number in $[0, 1)$ to each finite word so that the set of all named values is a separator. Mayordomo introduced separator enumerators to define f-normality and a relativized finite-state dimension $\dim_{\mathrm{FS}}^{f}(x)$, where finite-state dimension measures the asymptotic lower rate of finite-state information needed to approximate x through its f-names. This framework extends classical base-k normality, and Mayordomo showed that it supports a point-to-set principle for finite-state dimension. This representation-based viewpoint has since been developed further in follow-up work, including by Calvert et al., yielding strengthened randomness notions such as supernormal and highly normal numbers.

Mayordomo posed the following open question: can f-normality be characterized via equidistribution properties of the sequence $\left(|\Sigma|^n a_n^f(x)\right)_{n=0}^{\infty}$, where $a_n^f(x)$ is the sequence of best approximations to x from below induced by f? We give a strong negative answer: we construct computable separator enumerators f_0, f_1 and a point x such that $a_n^{f_0}(x) = a_n^{f_1}(x)$ for all n, yet $\dim_{\mathrm{FS}}^{f_0}(x) = 0$ while $\dim_{\mathrm{FS}}^{f_1}(x) = 1$. Consequently, no criterion depending only on the sequence $\left(|\Sigma|^n a_n^f(x)\right)_{n=0}^{\infty}$ - in particular, no equidistribution property of this sequence - can characterize f-normality uniformly over all separator enumerators. On the other hand, for a natural finite-state coherent class of separator enumerators we recover a complete equidistribution characterization of f-normality. We also show that beyond finite-state coherence, this characterization can fail even for a separator enumerator computable in nearly linear time.

Keywords: Finite-state dimension · Normality · Separator Enumerators · Equidistribution

1 Introduction

Finite-state dimension is a quantitative notion of the rate of randomness in an individual infinite sequence, as measured by finite automata. It was introduced by Dai, Lathrop, Lutz, and Mayordomo [3] as a finite-state analogue of effective Hausdorff dimension [6,7]. It admits several equivalent characterizations,

V. Brattka et al. (Eds.): CiE 2026, LNCS 16674, pp. 482–497, 2027.
https://doi.org/10.1007/978-3-032-31348-5_32

including formulations via finite-state gambling, finite-state compression, and block entropy rates [1,3,4]. It also has an important connection with the theory of normal numbers: a real number is normal to base k if and only if its base-k digit sequence has finite-state dimension equal to 1 [1].

An important connection between effective dimension and classical fractal dimension is provided by point-to-set principles. Lutz and Lutz [8] proved that the Hausdorff dimension in Euclidean spaces of a set can be obtained by minimizing, over oracles, the supremum of the relativized effective dimensions of its points. This reduces many lower-bound questions in geometric measure theory to analyzing the information density of carefully chosen points, and it has led to several new results and new proofs across classical fractal geometry; see [9]. In a recent work, Mayordomo [11] established an analogous point-to-set principle for finite-state dimension. A key feature of this development is that it is representation-based: instead of fixing a base-k expansion, it uses *separator enumerators*, i.e. naming schemes that assign reals in $[0, 1)$ to finite words so that the range forms a countable dense subset. With such an enumerator $f : \Sigma^* \to [0, 1)$, one can measure how efficiently a finite-state transducer can output an f-name approximating a real x to a given precision, leading to a relativized finite-state dimension $\dim_{\mathrm{FS}}^f(x)$ and the induced notion of *f-normality*, defined by the condition $\dim_{\mathrm{FS}}^f(x) = 1$ [11]. Calvert et al. [2] develop this framework further, introducing strengthened notions such as supernormal and highly normal numbers under broad classes of representations.

In this paper we study the relationships between f-normality and equidistribution. In the classical base-k setting, normality has a sharp equidistribution characterization: x is normal to base k if and only if the sequence $(k^n x)_{n\geq 1}$ is uniformly distributed modulo 1 [5]. Motivated by the classical equivalence between normality and equidistribution, Mayordomo [11] asked whether an analogous equidistribution criterion holds in the setting of f-normality. Let $(a_n^f(x))_{n\geq 0}$ denote the best-approximation-from-below sequence associated with f and x, i.e. $a_n^f(x) = \max\{f(w) : |w| \leq n,\ f(w) \leq x\}$. Mayordomo posed the following open question:

Can f-normality be characterized via equidistribution properties of the sequence $\left(|\Sigma|^n\, a_n^f(x)\right)_{n\geq 0}$?

We show that the answer is negative in a strong way. We construct two total computable rational-valued separator enumerators f_0, f_1 and a point $x \in [0, 1)$ such that $a_n^{f_0}(x) = a_n^{f_1}(x)$ for all n, yet $\dim_{\mathrm{FS}}^{f_0}(x) = 0$ while $\dim_{\mathrm{FS}}^{f_1}(x) = 1$. Consequently, no criterion depending only on the single numeric sequence $\left(|\Sigma|^n a_n^f(x)\right)_{n\geq 0}$—in particular, no equidistribution property of that sequence—can characterize f-normality uniformly over all separator enumerators.

At the same time, an equidistribution characterization does hold under a natural structural restriction on the representation. We identify a class of *finite-state coherent* enumerators obtained from the standard base-k grid by an invertible synchronous Mealy-machine relabeling, and we prove that in this regime f-normality is equivalent to a k-adic equidistribution property of the integer sequence $\left(k^n a_n^f(x)\right)_{n\geq 1}$ (uniform distribution of residues modulo k^m for every fixed m). Beyond finite-state coherence, we show that this equivalence can already break for a separator enumerator f computable in $O(n \log n)$ time: there is a point x whose scaled best-from-below approximations are k-adically equidistributed, but x is not f-normal.

Section 2 defines separator enumerators, finite-state transducers, $\dim_{\mathrm{FS}}^f$, and f-normality. Section 3 proves the negative result via a pair of computable enumerators with identical best-from-below chains but sharply different relativized finite-state dimension. Section 4 establishes the k-adic equidistribution characterization for finite-state coherent enumerators, and we then show that this correspondence can already fail for a separator enumerator computable in nearly linear time.

2 Preliminaries

Let Σ be a finite alphabet with $k = |\Sigma| \geq 2$ and write Σ^* for the set of finite strings over Σ. For $w \in \Sigma^*$, $|w|$ denotes its length, and for an infinite sequence $X \in \Sigma^\infty$ we write $X{\upharpoonright}n$ for its length-n prefix. We use $\lfloor\cdot\rfloor$ for the floor function and interpret congruences $b_n \equiv r \pmod{k^m}$ in the usual sense. Throughout this section we fix an identification $\Sigma = \{0, 1, \ldots, k-1\}$. We view finite words over Σ as base-k numerals and use the associated k-adic grid in $[0, 1)$. For a word $u = u_1 u_2 \cdots u_n \in \Sigma^n$, define its base-$k$ value

$$\mathrm{val}(u) = \sum_{i=1}^{n} u_i\, k^{n-i} \in \{0, 1, \ldots, k^n - 1\}, \qquad \text{and} \qquad \mathrm{grid}(u) = \frac{\mathrm{val}(u)}{k^n} \in [0, 1).$$

Thus $\mathrm{grid}(\Sigma^n) = \{j/k^n : 0 \leq j < k^n\}$. Let $\mathrm{seq}_k(x) \in \Sigma^\infty$ denote the (canonical) base-k expansion of $x \in [0, 1)$ chosen so as *not* to end in $(k-1)^\infty$.

Finite-state transducers model one-pass, constant-memory transformations on words and are used to measure finite-state description length.

Definition 1 (Finite-state transducer (FST)). *A Σ-finite-state transducer (briefly, Σ-FST) is a tuple $T = (Q, \delta, \nu, q_0)$ where Q is a finite nonempty set of states, $\delta : Q \times \Sigma \to Q$ is a transition function, $\nu : Q \times \Sigma \to \Sigma^*$ is an output function, and $q_0 \in Q$ is the start state.*

The transition function extends to words by $\delta(q, \lambda) = q$ and $\delta(q, wa) = \delta(\delta(q, w), a)$. For $q \in Q$ and $w \in \Sigma^*$, define the output $\nu(q, w) \in \Sigma^*$ by $\nu(q, \lambda) = \lambda$ and $\nu(q, wa) = \nu(q, w)\, \nu(\delta(q, w), a)$ for $a \in \Sigma$. The overall output of T on input w is $T(w) = \nu(q_0, w)$.

We measure how concisely a transducer can generate a given target string.

Definition 2 (T-information content [11]). *Let T be a Σ-FST and $w \in \Sigma^*$. Define*

$$K^T(w) = \min\{|\pi| : \ \pi \in \Sigma^* \text{ and } T(\pi) = w\},$$

with the convention $K^T(w) = \infty$ if w is not in the range of T.

A separator enumerator is a naming scheme for a countable dense subset of $[0,1)$, assigning a real in $[0,1)$ to each finite word.

Definition 3 (Separator, separator enumerator [11]). *A set $S \subseteq [0,1)$ is a* separator *if it is countable and dense in $[0,1)$. A function $f : \Sigma^* \to [0,1)$ is a* separator enumerator *(SE) if $\mathrm{Im}(f)$ is a separator.*

We call an SE $f : \Sigma^* \to [0,1)$ *total computable* if there is an algorithm that, on input (w,t) with $w \in \Sigma^*$ and $t \in \mathbb{N}$, outputs a rational q such that $|q - f(w)| \leq 2^{-t}$, and halts on every input. Given an enumerator f, we quantify how many input symbols a transducer needs in order to produce some f-name that δ-approximates a target point.

Definition 4 (Relativized approximation complexity [11]). *Let f be an SE, T a Σ-FST, $\delta > 0$, and $x \in [0,1)$. Define*

$$K_\delta^{T,f}(x) = \min\{K^T(w) : \ w \in \Sigma^* \ \textit{and} \ |f(w) - x| < \delta\}.$$

The induced finite-state dimension is the optimal asymptotic approximation rate achievable by finite-state transducers.

Definition 5 (Relativized finite-state dimension and f-normality [11]). *Let f be an SE and $x \in [0,1)$. Define*

$$\dim_{\mathrm{FS}}^f(x) = \inf_{T \ \Sigma\text{-}FST} \ \liminf_{\delta \to 0^+} \frac{K_\delta^{T,f}(x)}{\log_k(1/\delta)}.$$

*We say x is f-*normal *if $\dim_{\mathrm{FS}}^f(x) = 1$.*

The standard base-k naming map will serve as the baseline enumerator throughout. Let $f_{\mathrm{std}} : \Sigma^* \to [0,1)$ be the standard base-k enumerator

$$f_{\mathrm{std}}(u) = \mathrm{grid}(u) = \frac{\mathrm{val}(u)}{k^{|u|}} \qquad (u \neq \lambda),$$

with $f_{\mathrm{std}}(\lambda) = 0$. We now define the induced *best approximation from below* chain associated with a separator enumerator.

Definition 6 (Best approximation from below [11]). *Let f be an SE and $x \in [0,1)$. For each $n \in \mathbb{N}$, define $a_n^f(x)$ to be any value $f(w)$ with $|w| \leq n$ such that $f(w) \leq x$ and $x - f(w)$ is minimum among all u with $|u| \leq n$ and $f(u) \leq x$. Equivalently,*

$$a_n^f(x) = \max\{f(w) : \ |w| \leq n, \ f(w) \leq x\}.$$

3 Equidistribution Cannot Characterize Normality for Separator Enumerators

We now address Mayordomo's question on equidistribution criteria for f-normality. Fix a separator enumerator $f : \Sigma^* \to [0,1)$ and $x \in [0,1)$, and let $(a_n^f(x))_{n\geq 0}$ be the best approximation from below sequence corresponding to f and x (see Definition 6).

Question 1 [11]. Can f-normality be characterized via equidistribution properties of the sequence $(|\Sigma|^n \, a_n^f(x))_{n\geq 0}$?

We answer this question in the negative by constructing two total computable rational-valued separator enumerators f_0, f_1 and a point $x \in [0,1)$ such that $a_n^{f_0}(x) = a_n^{f_1}(x)$ for all n, yet $\dim_{\mathrm{FS}}^{f_0}(x) = 0$ while $\dim_{\mathrm{FS}}^{f_1}(x) = 1$.

3.1 Preliminary Lemmas

From the finite-state transducer characterization of finite-state dimension due to Doty and Moser (see Theorem 3.11 from [4]), we obtain the following corollary.

Lemma 1. *There exists a computable infinite sequence $Z \in \Sigma^\infty$ such that for every Σ-FST T, $\liminf_{n\to\infty} \frac{K^T(z_n)}{n} \geq 1$, where $z_n := Zn$.*

We also need a second sequence whose prefixes are distinct from the first but have the same property. This is achieved by a fixed symbol permutation.

Lemma 2. *Let $\pi : \Sigma \to \Sigma$ be a bijection and extend it letterwise to $\pi : \Sigma^* \to \Sigma^*$. If $Z \in \Sigma^\infty$ satisfies Lemma 1, then so does $\pi(Z)$, i.e. the prefixes $t_n := \pi(Z)n = \pi(z_n)$ satisfy $\liminf_n K^T(t_n)/n \geq 1$ for every Σ-FST T.*

Fix a bijection $\pi : \Sigma \to \Sigma$ with no fixed points. Choose Z as in Lemma 1 so that its first symbol is not $\pi^{-1}(0)$; this is possible since prescribing finitely many initial symbols does not affect normality. Let $Y = \pi(Z)$, and write $z_n := Zn$ and $y_n := Yn = \pi(z_n)$. Then $z_n \neq y_n$ for every $n \geq 1$ because π has no fixed points. Also $y_n \neq 0^n$ for every $n \geq 1$, since the first symbol of Y is not 0.

3.2 Construction of the Two Separator Enumerators

Fix $x := \frac{1}{2} \in (0,1)$. Let $k = |\Sigma|$. For each $n \in \mathbb{N}$, define $r_n := k^{-(n+2)}$ and

$$A_n := \big(x - r_n,\ x - r_{n+1}\big) \ \cup\ \big(x + r_{n+1},\ x + r_n\big).$$

Then each A_n is a nonempty open set, the sets A_n are pairwise disjoint, and their union is

$$\big((x - r_0, x + r_0) \setminus \{x\}\big) \setminus \{x \pm r_n : n \geq 1\}.$$

The omitted endpoints form a countable set and play no role in the density argument below. Also, $\log_k(1/r_n) = n + 2$. Define the target *best-from-below* values

$$c_n := x - \frac{r_n + r_{n+1}}{2} \in (x - r_n, x - r_{n+1}) \subseteq A_n.$$

These values will later be enforced as the canonical length-n approximants to x for both enumerators, i.e., we will arrange $a_n^{f_i}(x) = c_n$ for each n. Observe that $c_n < x$ and $c_n \uparrow x$ as $n \to \infty$.

For each n, fix a computable sequence $(d_{n,t})_{t\in\mathbb{N}}$ of distinct rational points in $A_n \setminus \{c_n\}$ that is dense in A_n, and set $D_n := \{d_{n,t} : t \in \mathbb{N}\}$. Also fix a computable sequence $(q_t)_{t\in\mathbb{N}}$ of distinct rational points in $[0, 1) \setminus (x - r_0, x + r_0)$ that is dense in this set, and set $D_{\text{far}} := \{q_t : t \in \mathbb{N}\}$. We now define two functions $f_0, f_1 : \Sigma^* \to [0, 1)$.

Step 1 (fixing the best-from-below chain for lengths $\le n$). For each $n \ge 0$, define

$$f_0(0^n) := c_n, \qquad f_1(z_n) := c_n.$$

(Here $0 \in \Sigma$ is a fixed symbol, and 0^n is the all-0 word of length n.)

Step 2 (ensuring density near x using incompressible prefixes). Partition the positive integers into infinite decidable sets $(L_n)_{n\ge0}$ such that $m \in L_n$ implies $m \ge n + 1$. For concreteness, one may take $L_n = \{2^n(2t + 1) : t \in \mathbb{N}\}$; then $m \in L_n$ implies $m \ge 2^n \ge n + 1$ for $n \ge 0$. Fix computable bijections $\varphi_n : L_n \to D_n$ by setting, for $t \in \mathbb{N}$,

$$\varphi_n\big(2^n(2t + 1)\big) := d_{n,t}.$$

Now set, for each $n \ge 0$ and each $m \in L_n$,

$$f_0(y_m) = f_1(y_m) := \varphi_n(m) \subset A_n.$$

Note that there is no conflict with Step 1, since $y_m \ne 0^m$ and $y_m \ne z_m$ for all m. Thus, for each n, the set $\{f_i(y_m) : m \in L_n\}$ is dense in A_n (for both $i = 0, 1$).

Step 3 (defining all remaining values far from x while maintaining density). For $i = 0, 1$, let W_i be the set of strings not yet assigned a value for f_i after Steps 1–2:

$$W_0 := \Sigma^* \setminus \Big(\{0^n : n \in \mathbb{N}\} \,\cup\, \{y_n : n \in \mathbb{N}\}\Big),$$

and

$$W_1 := \Sigma^* \setminus \Big(\{z_n : n \in \mathbb{N}\} \,\cup\, \{y_n : n \in \mathbb{N}\}\Big).$$

Enumerate Σ^* in length-lexicographic order, and for each $i \in \{0, 1\}$ let $(w_j^i)_{j\in\mathbb{N}}$ be the subsequence consisting of the words in W_i. Also enumerate $D_{\text{far}} = \{q_0, q_1, q_2, \dots\}$ according to the fixed computable listing above. Define

$$f_i(w_j^i) := q_j$$

for every $i \in \{0, 1\}$ and every $j \in \mathbb{N}$.

Lemma 3. *The functions f_0, f_1 defined above are total computable rational-valued separator enumerators: each* $\mathrm{Im}(f_i)$ *is countable and dense in* $[0,1)$.

Proof. Countability is immediate. For computability, note that all objects used in the construction are computable: (c_n), (z_n), (y_n), the sets L_n, the maps φ_n, and the enumeration (q_j). Given a word w of length ℓ, the cases from Steps 1–3 are decidable: for f_0 the special Step 1 case is $w = 0^\ell$, for f_1 it is $w = z_\ell$, and for both functions the Step 2 case is $w = y_\ell$ when $\ell \in L_n$ for some n. The remaining case is handled by computing the index of w in the length-lexicographic enumeration of W_i and outputting the corresponding q_j. Thus each f_i is total computable.

Density outside $(x - r_0, x + r_0)$ follows from D_{far}. Inside $(x - r_0, x + r_0)$, every nonempty open interval meets some A_n, up to omitted endpoints which do not affect open intervals, and $\{f_i(y_m) : m \in L_n\} = \varphi_n(L_n)$ is dense in A_n. Hence $\mathrm{Im}(f_i)$ is dense in $[0,1)$.

Since f_0 and f_1 are rational-valued and the above procedure computes the defining case and the corresponding index effectively, it in fact yields exact computation: there is a Turing machine that, on input $w \in \Sigma^*$, halts and outputs the rational value $f_i(w)$ itself (rather than merely producing 2^{-t}-approximations). The next lemma shows that f_0 and f_1 induce exactly the same sequence of length-bounded approximations of x from below (and hence the same associated numeric scaling sequence).

Lemma 4. *For every $n \in \mathbb{N}$, $a_n^{f_0}(x) = a_n^{f_1}(x) = c_n$. Consequently, the numeric sequences $(k^n a_n^{f_0}(x))$ and $(k^n a_n^{f_1}(x))$ are identical.*

Proof. Fix n. We first show $a_n^{f_0}(x) = c_n$. Since $f_0(0^n) = c_n \le x$, we have $a_n^{f_0}(x) \ge c_n$. Now consider any string w with $|w| \le n$ and $f_0(w) \le x$. If $w = 0^m$ for some $m \le n$, then $f_0(w) = c_m \le c_n$, since (c_m) is increasing. If $w = y_m$, then $f_0(w) \in A_j$ for some j with $m \in L_j$. Since $f_0(w) \le x$, we have $f_0(w) < x - r_{j+1}$. Also $m \in L_j$ implies $m \ge j+1$, so $j+1 \le m \le n$ and hence $f_0(w) < x - r_{j+1} \le x - r_n < c_n$. Finally, if $w \in W_0$, then $f_0(w) \in D_{\text{far}}$. Since $f_0(w) \le x$, this gives $f_0(w) \le x - r_0 \le x - r_n < c_n$. Thus no word of length at most n has f_0-value in $(c_n, x]$, while 0^n attains c_n. Hence $a_n^{f_0}(x) = c_n$.

The proof for f_1 is identical, replacing 0^m by z_m and W_0 by W_1. Thus $a_n^{f_1}(x) = c_n$.

We now complete the construction by showing that, for the fixed point $x = \frac{1}{2}$, the two separator enumerators f_0 and f_1 constructed above induce different relativized finite-state dimensions: $\dim_{\mathrm{FS}}^{f_0}(x) = 0$ while $\dim_{\mathrm{FS}}^{f_1}(x) = 1$. Together with Lemma 4, this yields a negative answer to Mayordomo's open question [11].

Theorem 1. *There exist total computable rational-valued separator enumerators $f_0, f_1 : \Sigma^* \to [0,1)$ and a point $x \in [0,1)$ such that*

$$a_n^{f_0}(x) = a_n^{f_1}(x) \quad \textit{for all } n,$$

yet

$$\dim_{\mathrm{FS}}^{f_0}(x) = 0 \qquad \textit{and} \qquad \dim_{\mathrm{FS}}^{f_1}(x) = 1.$$

In particular, x is not f_0-normal, while x is f_1-normal.

Proof. We use the constructions above with the fixed $x = \frac{1}{2}$. By Lemma 4, the sequences $a_n^{f_0}(x)$ and $a_n^{f_1}(x)$ coincide (both equal c_n), so $(k^n a_n(x))$ is identical for f_0 and f_1. It remains to compute the relativized finite-state dimensions.

Part 1: We show $\dim_{\mathrm{FS}}^{f_0}(x) = 0$. For each integer $L \geq 1$, let T_L be the 1-state transducer that outputs 0^L on every input symbol. Then $T_L(\pi) = 0^{L|\pi|}$ for all π, hence

$$K^{T_L}(0^n) \leq \left\lceil \frac{n}{L} \right\rceil.$$

Fix n and consider $\delta_n := r_n/2$ (so $\log_k(1/\delta_n) = n+2+\log_k 2$). Since $f_0(0^n) = c_n$ and $|x - c_n| = |x - (x - (r_n + r_{n+1})/2)| = (r_n + r_{n+1})/2 < r_n = 2\delta_n$, we have $|f_0(0^n) - x| < 2\delta_n$.Therefore

$$K_{2\delta_n}^{T_L, f_0}(x) \leq K^{T_L}(0^n) \leq \left\lceil \frac{n}{L} \right\rceil.$$

Hence

$$\liminf_{n\to\infty} \frac{K_{2\delta_n}^{T_L, f_0}(x)}{\log_k(1/(2\delta_n))} \leq \liminf_{n\to\infty} \frac{\lceil n/L \rceil}{(n+2) + O(1)} = \frac{1}{L}.$$

To justify that this controls the full $\liminf_{\delta\to 0^+}$ (and not only the subsequence $2\delta_n$), note that $K_\delta^{T_L, f_0}(x)$ is monotone non-increasing in δ. Hence for any $\delta \in (2\delta_{n+1}, 2\delta_n]$,

$$K_\delta^{T_L, f_0}(x) \;\leq\; K_{2\delta_{n+1}}^{T_L, f_0}(x).$$

Moreover, for such δ we have $\log_k(1/\delta) \geq \log_k(1/(2\delta_n))$. Therefore,

$$\frac{K_\delta^{T_L, f_0}(x)}{\log_k(1/\delta)} \;\leq\; \frac{K_{2\delta_{n+1}}^{T_L, f_0}(x)}{\log_k(1/(2\delta_n))}.$$

Taking lim inf over all $\delta \to 0^+$ and using that $n \to \infty$ along the corresponding intervals yields

$$\begin{aligned}
\liminf_{\delta\to 0^+} \frac{K_\delta^{T_L, f_0}(x)}{\log_k(1/\delta)} &\leq \liminf_{n\to\infty} \frac{K_{2\delta_{n+1}}^{T_L, f_0}(x)}{\log_k(1/(2\delta_{n+1}))} \cdot \frac{\log_k(1/(2\delta_{n+1}))}{\log_k(1/(2\delta_n))} \\
&\leq \liminf_{n\to\infty} \frac{K_{2\delta_{n+1}}^{T_L, f_0}(x)}{\log_k(1/(2\delta_{n+1}))} \cdot \lim_{n\to\infty} \frac{n+3+O(1)}{n+2+O(1)} \\
&\leq \frac{1}{L}.
\end{aligned}$$

Since the above holds for all T_L, taking the infimum over all transducers T and then letting $L \to \infty$ yields $\dim_{\mathrm{FS}}^{f_0}(x) = 0$.

Part 2: We show $\dim_{\mathrm{FS}}^{f_1}(x) = 1$. Fix an arbitrary Σ-FST T. We prove that $\liminf_{\delta\to 0^+} \frac{K_\delta^{T, f_1}(x)}{\log_k(1/\delta)} \geq 1$, which implies $\dim_{\mathrm{FS}}^{f_1}(x) = 1$ after taking the infimum over T.

Consider the scale $\delta_n := r_n$. Any value $f_1(w)$ within distance $\delta_n = r_n$ of x must lie in $(x - r_n,\ x + r_n)$, hence, by construction, must correspond either to (i) one of the special words z_j with $j \geq n$ (since $f_1(z_j) = c_j \in A_j$), or (ii) one of the special words y_m with $f_1(y_m) \in A_j$ for some $j \geq n$ (these are the values placed densely in the annuli), because by construction all other words are mapped into D_{far} outside $(x - r_0, x + r_0)$. Therefore,

$$K^{T,f_1}_{r_n}(x) \geq \min\Big(\min_{j \geq n} K^T(z_j), \min_{m:\ f_1(y_m) \in \bigcup_{j \geq n} A_j} K^T(y_m)\Big).$$

We now lower bound each term asymptotically by n. By Lemma 1, we have $\liminf_{j\to\infty} K^T(z_j)/j \geq 1$. Hence for every $\varepsilon > 0$ there exists J such that for all $j \geq J$, $K^T(z_j) \geq (1-\varepsilon)j$; consequently, for all $n \geq J$, $\min_{j \geq n} K^T(z_j) \geq (1-\varepsilon)n$. Similarly, by Lemma 2 applied to Y, $\liminf_{m\to\infty} K^T(y_m)/m \geq 1$, so for every $\varepsilon > 0$ there exists M such that for all $m \geq M$, $K^T(y_m) \geq (1 - \varepsilon)m$. In our construction, if $f_1(y_m) \in A_j$, then $m \in L_j$, and by design $m \geq j + 1$; therefore, whenever $f_1(y_m) \in \bigcup_{j \geq n} A_j$, we have $m \geq n + 1$. For all $n \geq M$,

$$\min_{m:\ f_1(y_m) \in \bigcup_{j \geq n} A_j} K^T(y_m) \geq (1 - \varepsilon)(n + 1) \geq (1 - \varepsilon)n.$$

Putting the two bounds together, for all sufficiently large n, $K^{T,f_1}_{r_n}(x) \geq (1-\varepsilon)n$. Since $\log_k(1/r_n) = n + 2$, we obtain

$$\liminf_{n\to\infty} \frac{K^{T,f_1}_{r_n}(x)}{\log_k(1/r_n)} \geq \liminf_{n\to\infty} \frac{(1 - \varepsilon)n}{n + 2} = 1 - \varepsilon.$$

As $\varepsilon > 0$ was arbitrary, this yields $\liminf_{n\to\infty} \frac{K^{T,f_1}_{r_n}(x)}{\log_k(1/r_n)} \geq 1$. By monotonicity of $K^{T,f_1}_{\delta}(x)$ in δ and the fact that any $\delta \in (r_{n+1}, r_n]$ satisfies $\log_k(1/\delta) \in [n + 2, n + 3)$, it follows that the full $\liminf_{\delta\to 0^+}$ is also at least 1. Thus

$$\liminf_{\delta\to 0^+} \frac{K^{T,f_1}_{\delta}(x)}{\log_k(1/\delta)} \geq 1$$

for every T, which implies that $\dim^{f_1}_{\text{FS}}(x) = 1$.

As an immediate consequence, f-normality cannot be characterized by any property of the single numerical sequence $(k^n a^f_n(x))$.

Corollary 1. *There is no property P of the numeric sequence $(k^n a^f_n(x))_{n\in\mathbb{N}}$ (in particular, no equidistribution property of this sequence) such that for all separator enumerators f and all $x \in [0, 1)$,*

$$x \text{ is } f\text{-normal} \iff (k^n a^f_n(x)) \text{ has property } P.$$

Proof. Take f_0, f_1, x from Theorem 1. By Lemma 4, the sequences $(k^n a^{f_0}_n(x))$ and $(k^n a^{f_1}_n(x))$ are identical, so they either both satisfy P or both fail P. But by Theorem 1, x is f_1-normal and not f_0-normal, so no such P can exist.

4 Finite-State Coherent Enumerators and an Equidistribution Characterization of f-Normality

From the previous section, we know that no equidistribution (or other distributional) criterion depending only on the sequence $(k^n a_n^f(x))_{n\geq 1}$ can characterize f-normality uniformly over all separator enumerators. The goal of this section is to isolate a natural structural regime in which such a characterization *does* hold. We do this by restricting to *finite-state coherent enumerators*, i.e. naming schemes obtained from the standard base-k grid by an invertible synchronous Mealy-machine relabeling.

To state an equidistribution characterization in this setting, we must use a notion that remains meaningful for the standard enumerator. For the standard base-k naming map $f_{\text{std}}(w) = \text{val}(w)/k^{|w|}$, the scaled approximation sequence $b_n(x) := k^n a_n^{f_{\text{std}}}(x)$ is integer-valued (indeed $b_n(x) = \lfloor k^n x \rfloor$), so the usual notion of equidistribution modulo 1 becomes trivial. The natural replacement is to ask for uniform distribution of the residues of $b_n(x)$ at every finite base-k resolution, i.e. modulo k^m for each fixed m. This leads to the following notion of k-adic equidistribution, which we adopt in the rest of the paper.

Definition 7 (k-adic equidistribution). *A sequence $(b_n)_{n\geq 1}$ of integers is* k-adically equidistributed *if for every $m \geq 1$ and every residue $r \in \{0, 1, \ldots, k^m - 1\}$,*

$$\lim_{N\to\infty} \frac{1}{N} \#\{1 \leq n \leq N : \ b_n \equiv r \pmod{k^m}\} = \frac{1}{k^m}.$$

For finite-state coherent enumerators, the best-from-below approximants admit an explicit closed form, and the induced f-dimension is invariant under finite-state coherent relabelings. These two facts combine to yield a clean equidistribution characterization of f-normality in terms of k-adic equidistribution of the integer sequence $(k^n a_n^f(x))_{n\geq 1}$.

4.1 Finite-State Coherent Enumerators

We begin by defining invertible synchronous Mealy machines. A Mealy machine is a finite-state device that processes a word one symbol at a time, updating its state and producing an output symbol at each step.

Definition 8 (Invertible synchronous Mealy machine). *An* invertible synchronous Mealy machine *is a tuple $M = (Q, \delta, \tau, q_0)$ where Q is a finite nonempty set of states, $\delta : Q \times \Sigma \to Q$ is a transition function, and $\tau : Q \times \Sigma \to \Sigma$ is an output function such that, for every state $q \in Q$, the map $a \mapsto \tau(q, a)$ is a permutation of Σ.*

The induced map $M : \Sigma^* \to \Sigma^*$ is defined by reading left to right: set $M(\epsilon) = \epsilon$, and for $w = w_1 \cdots w_n$ define $u = M(w) = u_1 \cdots u_n$ by the recursion

$$q_i = \delta(q_{i-1}, w_i), \qquad u_i = \tau(q_{i-1}, w_i) \quad (i = 1, \ldots, n),$$

with q_0 as the initial state. In particular, $|M(w)| = |w|$ for all w.

We state the basic properties of invertible synchronous Mealy machines.

Lemma 5. *Let M be an invertible synchronous Mealy machine.*

1. *For every n, the restriction $M : \Sigma^n \to \Sigma^n$ is a bijection.*
2. *There exists an invertible synchronous Mealy machine M^{-1} such that for all $w \in \Sigma^*$, $M^{-1}(M(w)) = w$.*
3. *M extends letter-by-letter (via the same recursion as in the definition of M) to a bijection $M : \Sigma^\infty \to \Sigma^\infty$ with inverse M^{-1}.*

Now we formalize the finite-state coherence condition, which captures length-preserving, bounded-memory relabelings of the standard base-k grid.

Definition 9 (Finite-state coherent enumerator). *A function $f : \Sigma^* \to [0,1)$ is* finite-state coherent *if there exists an invertible synchronous Mealy machine M such that for every nonempty $w \in \Sigma^n$, $f(w) = \mathrm{grid}(M(w)) = \frac{\mathrm{val}(M(w))}{k^n}$. (For definiteness, set $f(\epsilon) = 0$.)*

Finite-state coherence immediately forces f to enumerate the entire standard k-adic grid at each length.

Lemma 6. *If f is finite-state coherent, then $\mathrm{Im}(f) = \bigcup_{n \geq 1} \{j/k^n : 0 \leq j < k^n\}$, hence f is a separator enumerator.*

For finite-state coherent enumerators, the best-from-below approximation sequence coincides with the usual base-k truncations.

Lemma 7. *Let f be finite-state coherent and $x \in [0,1)$. Then for every $n \geq 1$, $a_n^f(x) = \frac{\lfloor k^n x \rfloor}{k^n}$. In particular, the scaled sequence is integer-valued: $k^n a_n^f(x) = \lfloor k^n x \rfloor \in \{0, 1, \ldots, k^n - 1\}$.*

Proof. By Lemma 6, for each $m \leq n$ the set $f(\Sigma^m)$ equals the full grid $\{j/k^m : 0 \leq j < k^m\}$. Hence the set of all values $f(w)$ with $|w| \leq n$ is exactly $\bigcup_{m=1}^{n} \{j/k^m : 0 \leq j < k^m\}$. Among the level-$n$ grid points $\{j/k^n\}$, the largest one $\leq x$ is $\lfloor k^n x \rfloor / k^n$. It remains to check that no coarser grid point (denominator k^m with $m < n$) can exceed this value while staying $\leq x$. But for each $m < n$, $\frac{\lfloor k^m x \rfloor}{k^m} \leq \frac{\lfloor k^n x \rfloor}{k^n}$, because multiplying both sides by k^n gives $k^{n-m} \lfloor k^m x \rfloor \leq \lfloor k^n x \rfloor$, which holds since $k^{n-m} \lfloor k^m x \rfloor \leq k^n x$ and the left-hand side is an integer.

Next we relate approximation complexity under a finite-state coherent f to approximation complexity under the standard base-k enumerator. Recall that $f_{\mathrm{std}} : \Sigma^* \to [0,1)$ is the standard base-k enumerator.

Lemma 8. *Let f be finite-state coherent via a Mealy machine M, i.e. $f(w) = \mathrm{grid}(M(w))$ for all w. Then for every Σ-FST T, every $x \in [0,1)$, and every $\delta > 0$, $K_\delta^{T,f}(x) = K_\delta^{M \circ T,\ f_{\mathrm{std}}}(x)$, where $M \circ T$ denotes the output-composition transducer $\pi \mapsto M(T(\pi))$.* [1]

[1] If no approximating word is produced, both sides are understood to be ∞.

The next proposition formalizes the key robustness property of finite-state coherence: composing the naming map with an invertible synchronous Mealy relabeling does not change the relativized finite-state approximation complexity, and hence does not change the induced f-dimension.

Proposition 1. *If f is finite-state coherent, then for every $x \in [0,1)$, $\dim_{\mathrm{FS}}^{f}(x) = \dim_{\mathrm{FS}}^{f_{\mathrm{std}}}(x)$. In particular, x is f-normal if and only if x is f_{std}-normal.*

Proof. Let f be finite-state coherent via M. By Lemma 8, for every T and every δ, $K_\delta^{T,f}(x) = K_\delta^{M\circ T, f_{\mathrm{std}}}(x)$, hence the corresponding lim inf ratios coincide. Taking $\inf_T$ over all FSTs on the left equals taking $\inf_S$ over all FSTs on the right, because $T \mapsto M \circ T$ is a bijection on FSTs (with inverse $S \mapsto M^{-1} \circ S$). Therefore the two infima coincide.

The next lemma states that, for the *standard* base-k naming map f_{std}, the paper's notion of f-normality (i.e. $\dim_{\mathrm{FS}}^{f}(x) = 1$) coincides exactly with the classical notion of base-k normality of x.

Lemma 9. *For $x \in [0,1)$, one has $\dim_{\mathrm{FS}}^{f_{\mathrm{std}}}(x) = 1$ if and only if x is base-k normal.*

It is straightforward to verify that k-adic equidistribution of the sequence $(\lfloor k^n x \rfloor)_{n\geq 1}$ coincides with base-k normality [5].

Theorem 2. *Let $x \in [0,1)$ and set $b_n = \lfloor k^n x \rfloor$. Then x is base-k normal if and only if $(b_n)_{n\geq 1}$ is k-adically equidistributed.*

Finally we combine finite-state coherent invariance with the explicit form of $a_n^f(x)$ to obtain the equidistribution characterization in terms of the scaled approximation sequence.

Theorem 3. *Let f be a finite-state coherent separator enumerator over Σ and let $x \in [0,1)$. Then x is f-normal if and only if the integer sequence $(k^n a_n^f(x))_{n\geq 1}$ is k-adically equidistributed.*

Proof. Let $b_n(x) := k^n a_n^f(x)$. Since f is finite-state coherent, Proposition 1 gives x is f-normal if and only if x is f_{std}-normal. By Lemma 9, f_{std}-normality is equivalent to base-k normality of x. On the other hand, Theorem 2 states that base-k normality of x is equivalent to k-adic equidistribution of the integer sequence $\big(\lfloor k^n x \rfloor\big)_{n\geq 1}$. Finally, Lemma 7 identifies the scaled approximation sequence for finite-state coherent f with this canonical sequence, namely $b_n(x) = \lfloor k^n x \rfloor$ for all $n \geq 1$. Substituting this identity into the previous equivalence yields x is f-normal if and only if $(b_n(x))_{n\geq 1}$ is k-adically equidistributed, which is exactly the claim.

4.2 Beyond Finite-State Coherence: a Nearly Linear Time Computable Counterexample

This subsection shows that once one moves beyond finite-state coherent enumerators, a k-*adic equidistribution*-based criterion for f-normality can fail even under very low resource bounds. Concretely, we construct a nearly linear time computable separator enumerator f and a point $x \in [0,1)$ such that $b_n(x) := k^n a_n^f(x)$ is k-adically equidistributed, yet $\dim_{\mathrm{FS}}^f(x) = 0$.

Towards defining the enumerator, fix $\Sigma = \{0,1,\ldots,k-1\}$ with $k = |\Sigma| \geq 2$ and set $x := 1/k$. For $n \geq 1$ define $J_n := k^{n-1} - n$ and $c_n := J_n/k^n = 1/k - n/k^n$. Then $0 \leq c_n < x$ for all n and (c_n) is nondecreasing since $c_{n+1} - c_n = ((k-1)n-1)/k^{n+1} \geq 0$.

Definition of the Enumerator. Define $f : \Sigma^* \to [0,1)$ as follows, where $n = |w|$.

$$f(w) := \begin{cases} c_n, & \text{if } w = 0^n, \\ \frac{\mathrm{val}(w)}{k^n}, & \text{if } w \text{ starts with } 0,\ w \neq 0^n,\ \mathrm{val}(w) \leq J_n, \\ 0, & \text{if } w \text{ starts with } 0,\ w \neq 0^n,\ \mathrm{val}(w) > J_n, \\ x + \frac{1}{k^{n+1}}, & \text{if } w = 1\,0^{n-1}, \\ \frac{\mathrm{val}(w)}{k^n}, & \text{otherwise.} \end{cases}$$

In particular, clauses with leading digit 0 satisfy $f(w) \leq c_n < x$, while clauses with leading digit $\neq 0$ satisfy $f(w) > x$.

Lemma 10. *The function f is a rational-valued separator enumerator, and f is $O(n \log n)$ time computable.*

Proof (Proof sketch). For density in $[0,x)$, note that for each $n \geq 1$ clause (ii) realizes all rationals j/k^n with $1 \leq j \leq J_n$, while clause (iii) realizes 0; since $c_n = J_n/k^n \uparrow x$, these values are dense in $[0,x)$. For density in $(x,1)$, clause (v) realizes the standard base-k grid points $\mathrm{val}(w)/k^n$ with first digit nonzero, except that the single point $x = \mathrm{val}(1\,0^{n-1})/k^n$ is shifted to $x + 1/k^{n+1}$ by clause (iv); removing/perturbing one point per level does not affect density, so $\mathrm{Im}(f)$ is dense in $[0,1)$.

For the time bound, we work in the log-cost RAM model of [10] and output $f(w)$ as a pair (p,m) representing p/k^m with p written in base k. Given w, a single pass determines $|w|$ and the relevant case, and in the only nontrivial case ($w = 0u$) we compute the base-k digit string of $J_{|w|}$ and compare it lexicographically with u. All steps use $O(|w|)$ digit operations with $O(\log |w|)$ overhead from log-cost addressing, giving time $O(|w| \log |w|)$.

We next identify the induced best-from-below approximation chain at the special point $x = 1/k$.

Lemma 11. *For every $n \geq 1$, one has $a_n^f(x) = c_n$. Consequently, $b_n(x) = k^n a_n^f(x) = J_n = k^{n-1} - n$.*

We now verify that the sequence $(b_n(x))_{n\geq 1}$ is k-adically equidistributed.

Lemma 12. *The integer sequence $(b_n(x))_{n\geq 1}$ is k-adically equidistributed.*

Proof. Fix $m \geq 1$. For all $n \geq m+1$ one has $k^m \mid k^{n-1}$, hence $b_n(x) = k^{n-1} - n \equiv -n \pmod{k^m}$ for all $n \geq m+1$. As n ranges over $\{1, 2, \ldots, N\}$, the residues of $-n$ mod k^m are asymptotically uniform, and omitting the finite initial segment $n \leq m$ does not affect limiting frequencies. Therefore, for every $r \in \{0, 1, \ldots, k^m - 1\}$, $\lim_{N\to\infty} \frac{1}{N}\#\{1 \leq n \leq N : \ b_n(x) \equiv r (\text{mod } k^m)\} = \frac{1}{k^m}$.

Finally, despite this equidistribution property, the point x has vanishing relativized finite-state dimension with respect to f.

Proposition 2. *For the above f and $x = 1/k$, one has $\dim^f_{\mathrm{FS}}(x) = 0$.*

Proof. For each $L \geq 1$ let T_L be the 1-state Σ-FST that outputs 0^L on every input symbol. Then $T_L(\pi) = 0^{L|\pi|}$, so $K^{T_L}(0^n) \leq \lceil n/L \rceil$. Let $\delta_n := 2n/k^n$. Since $f(0^n) = c_n$ and $|x - c_n| = \frac{n}{k^n} < \delta_n$ we have $K^{T_L, f}_{\delta_n}(x) \leq K^{T_L}(0^n) \leq \lceil n/L \rceil$. Since $\log_k(1/\delta_n) = \log_k(k^n/(2n)) = n - \log_k(2n)$, we obtain

$$\liminf_{n\to\infty} \frac{K^{T_L,f}_{\delta_n}(x)}{\log_k(1/\delta_n)} \leq \liminf_{n\to\infty} \frac{\lceil n/L \rceil}{n - \log_k(2n)} = \frac{1}{L}.$$

Taking the infimum over transducers (in particular over T_L) and letting $L \to \infty$ yields $\dim^f_{\mathrm{FS}}(x) = 0$.

Taking $x = 1/k$ and f as defined above we obtain the following.

Theorem 4. *There exist a rational-valued $O(n \log n)$-time computable separator enumerator $f : \Sigma^* \to [0, 1)$ and a point $x \in [0, 1)$ such that the integer sequence $(k^n a^f_n(x))_{n\geq 1}$ is k-adically equidistributed, while $\dim^f_{\mathrm{FS}}(x) = 0$.*

Proof. Take $x = 1/k$ and f as defined above. Lemma 11 identifies $k^n a^f_n(x) = b_n(x)$, Lemma 12 shows $(b_n(x))$ is k-adically equidistributed, and Proposition 2 gives $\dim^f_{\mathrm{FS}}(x) = 0$.

5 Discussion and Open Questions

We show that, in general, no distributional property of the scaled best-from-below approximation sequence $(k^n a^f_n(x))_{n\geq 1}$ can characterize f-normality uniformly over all separator enumerators (indeed, we construct computable enumerators f_0, f_1 for which the associated scaled approximation sequences coincide while the corresponding f-normality behavior diverges). At the same time, we identify a structured regime—finite-state coherent enumerators—in which f-normality is equivalent to k-adic equidistribution of $(k^n a^f_n(x))_{n\geq 1}$. We also show that this correspondence can already fail for an efficiently computable separator enumerator, in fact for an $O(n \log n)$-time computable f. This raises the natural question of how far this correspondence persists beyond finite-state coherent relabelings. One concrete setting is when the naming map is computable by a deterministic pushdown transducer. In particular, does there exist such a separator enumerator f and a point $x \in [0, 1)$ for which the integer sequence $(k^n a_n^f(x))_{n \geq 1}$ is k-adically equidistributed while $\dim_{\mathrm{FS}}^f(x) < 1$? It would also be interesting to identify the widest natural family of separator enumerators for which an equidistribution characterization of f-normality remains valid.

A related direction concerns the role of invertibility in the defining Mealy-machine relabeling: invertibility is used to transport approximation complexity via the substitution $u = M(w)$ and to ensure that $T \mapsto M \circ T$ is a bijection on finite-state transducers, yielding invariance of the induced f-dimension. It would be interesting to determine whether some weaker condition (e.g. level-wise surjectivity or bounded-to-one behavior on each Σ^n) suffices for the same equidistribution characterization, or whether non-invertible finite-state relabelings can already break it.

Acknowledgments.. Subin Pulari was supported by the Basic Research Program at HSE University and this paper was prepared in the framework of this program.

Disclosure of Interests.. The author has no competing interests to declare that are relevant to the content of this article.

References

1. Bourke, C., Hitchcock, J.M., Vinodchandran, N.: Entropy rates and finite-state dimension. Theor. Comput. Sci. **349**(3), 392–406 (2005)
2. Calvert, W., Gruner, E., Mayordomo, E., Turetsky, D., Villano, J.D.: Normality, relativization, and randomness. Theory Comput. Syst. **69**(3), 26 (2025)
3. Dai, J.J., Lathrop, J.I., Lutz, J.H., Mayordomo, E.: Finite-state dimension. Theor. Comput. Sci. **310**(1–3), 1–33 (2004)
4. Doty, D., Moser, P.: Finite-state dimension and lossy decompressors (2006). https://arxiv.org/abs/cs/0609096
5. Kuipers, L., Niederreiter, H.: Uniform distribution of sequences. Pure and Applied Mathematics, Wiley-Interscience [John Wiley & Sons], New York-London-Sydney (1974)
6. Lutz, J.H.: Dimension in complexity classes. SIAM J. Comput. **32**(5), 1236–1259 (2003). https://doi.org/10.1137/S0097539701417723
7. Lutz, J.H.: The dimensions of individual strings and sequences. Inf. Comput. **187**(1), 49–79 (2003). https://doi.org/10.1016/S0890-5401(03)00187-1. https://www.sciencedirect.com/science/article/pii/S0890540103001871
8. Lutz, J.H., Lutz, N.: Algorithmic information, plane kakeya sets, and conditional dimension. ACM Trans. Comput. Theory **10**(2) (2018). https://doi.org/10.1145/3201783
9. Lutz, J.H., Lutz, N.: Who asked *us?* How the theory of computing answers questions about analysis. In: Du, D.-Z., Wang, J. (eds.) Complexity and Approximation. LNCS, vol. 12000, pp. 48–56. Springer, Cham (2020). https://doi.org/10.1007/978-3-030-41672-0_4

10. Lutz, J.H., Mayordomo, E.: Computing absolutely normal numbers in nearly linear time. Inf. Comput. **281**, 104746 (2021). https://doi.org/10.1016/j.ic.2021.104746. https://www.sciencedirect.com/science/article/pii/S0890540121000614
11. Mayordomo, E.: A point to set principle for finite-state dimension. In: Beckmann, A., Oitavem, I., Manea, F. (eds.) Crossroads of Computability and Logic: Insights, Inspirations, and Innovations, pp. 299–304. Springer, Cham (2025). https://doi.org/10.1007/978-3-031-95908-0_21

On the Reverse Mathematics of Darboux's Supremum Principle

Sam Sanders(✉)

Department of Philosophy II, RUB Bochum, Bochum, Germany
sasander@me.com

Abstract. Around 1875, Darboux established the supremum principle for arbitrary bounded functions from reals to reals. We study the properties of Darboux's supremum principle for various function classes in higher-order Reverse Mathematics. We obtain a number of equivalences involving historical classes like Riemann integrable and pointwise discontinuous functions on one hand, and Feferman's *projection principle* on the other hand. The latter is among the strongest principles studied in higher-order Reverse Mathematics. Baire's semi-continuous functions constitute an exception, for which the supremum principle implies the weaker coding principle that *every open set has a second-order code.* We also consider equivalences for the latter, including the Tietze extension theorem and Urysohn lemma.

Keywords: Reverse Mathematics · supremum principle · higher-order arithmetic

1 Introduction

1.1 Aim and Motivation

In his 1875 *Mémoire sur les fonctions discontinues*, Darboux defines continuity, presents some examples of *dis*continuous functions, and then establishes the existence of the supremum for *arbitrary* bounded functions on bounded intervals ([8, p. 61, Theorem I]). In this paper, we study Darboux's supremum principle from the point of view of the *Reverse Mathematics* (abbreviated RM) program, Kohlenbach's *higher-order* approach in particular (see Sect. 1.2).

In particular, we shall obtain equivalences involving the supremum principle for various historical classes, including the Riemann integrable, pointwise discontinuous, and measurable functions (Sect. 2). These are equivalent to Feferman's *projection principle* from [11, §5], which is one of the strongest axioms studied in higher-order Reverse Mathematics and introduced in Sect. 1.2. Slight variations will be seen to be provable from arithmetical comprehension.

As will become clear, Baire's notion of *semi-continuity* is exceptional in that the associated supremum principle is equivalent to the weaker coding principle open, which was first introduced in [31] and which expresses that an open set has a second-order code. We obtain equivalences for open in Sect. 3, also involving basic properties of continuous functions on closed sets.

V. Brattka et al. (Eds.): CiE 2026, LNCS 16674, pp. 498–513, 2027.
https://doi.org/10.1007/978-3-032-31348-5_33

1.2 On Higher-Order Arithmetic

We assume basic familiarity with Kohlenbach's higher-order RM, the base theory RCA_0^ω in particular. The original text is [22] while more recent introductions are in [30,36,38]. For the rest of this section, we introduce axioms needed below.

First of all, note that real numbers are defined in RCA_0^ω in the same way as in second-order RM, i.e. as fast-converging Cauchy sequences. Thus, '$x \in \mathbb{R}$' and '$x =_{\mathbb{R}} y$' have their usual meaning. A function from reals to reals, denoted $F : \mathbb{R} \to \mathbb{R}$, is then given by $\Phi : \mathbb{N}^{\mathbb{N}} \to \mathbb{N}^{\mathbb{N}}$ such that

$$(\forall x, y \in \mathbb{R})(x =_{\mathbb{R}} y \to \Phi(x) =_{\mathbb{R}} \Phi(y)).$$

Secondly, we consider the following essential axiom where the functional E is also called *Kleene's quantifier* $\exists^2$ and is discontinuous on Cantor space.

$$(\exists E : \mathbb{N}^{\mathbb{N}} \to \{0,1\})(\forall f \in \mathbb{N}^{\mathbb{N}})(E(f) = 0 \leftrightarrow (\exists n \in \mathbb{N})(f(n) = 0)). \qquad (\exists^2)$$

We write $\mathsf{ACA}_0^\omega \equiv \mathsf{RCA}_0^\omega + (\exists^2)$ and observe that the latter proves the same second-order sentences as ACA_0 (see [20]). We shall mostly work in ACA_0^ω, which is convenient as set of reals are then given via characteristic functions, as follows.

Definition 1 (Sets).

a. *A set of reals A is given by a function $F_A : \mathbb{R} \to \{0,1\}$; we write $x \in A$ for $F_A(x) = 1$, for all $x \in \mathbb{R}$.*
b. *We write '$A \subseteq B$' if we have $(\forall x \in \mathbb{R})(x \in A \to x \in B)$.*
c. *A set $O \subseteq \mathbb{R}$ is* open *in case $x \in O$ implies $(\exists k \in \mathbb{N})(B(x, \frac{1}{2^k}) \subseteq O)$.*
d. *The complement O^c of an open set $O \subset \mathbb{R}$ is called closed.*
e. *A set $O \subseteq \mathbb{R}$ is* RM-open *if there are sequences $(a_n)_{n\in\mathbb{N}}, (b_n)_{n\in\mathbb{N}}$ of reals such that $x \in O$ if and only if $x \in \cup_{n\in\mathbb{N}}(a_n, b_n)$ for all $x \in \mathbb{R}$.*
f. *The complement O^c of an RM-open set $O \subset \mathbb{R}$ is called RM-closed.*
g. *A set $A \subset \mathbb{R}$ has* measure zero *if for any $\varepsilon > 0$ there is a sequence of open intervals $(I_n)_{n\in\mathbb{N}}$ such that $\cup_{n\in\mathbb{N}} I_n$ covers A and $\varepsilon > \sum_{n=0}^{\infty} |I_n|$.*

With this convention in place, we can adopt the usual definitions of nowhere dense set, $\mathbf{F}_\sigma$-set, the Baire property, et cetera from the literature.

Thirdly, the functional S^2 is often called *the Suslin functional* [2,22]:

$$(\exists \mathsf{S} : \mathbb{N}^{\mathbb{N}} \to \{0,1\})(\forall f \in \mathbb{N}^{\mathbb{N}})\big[(\exists g \in \mathbb{N}^{\mathbb{N}})(\forall n \in \mathbb{N})(f(\overline{g}n) = 0) \leftrightarrow \mathsf{S}(f) = 0\big]. \qquad (\mathsf{S}^2)$$

By definition, the Suslin functional S^2 can decide whether a Σ_1^1-formula in normal form, i.e. as in the left-hand side of (S^2), is true or false. The system $\Pi_1^1\text{-}\mathsf{CA}_0^\omega \equiv \mathsf{RCA}_0^\omega + (\mathsf{S}^2)$ proves the same Π_3^1-sentences as $\Pi_1^1\text{-}\mathsf{CA}_0$ [2].

Similarly, we define the functional S_k^2 which decides the truth or falsity of Σ_k^1-formulas in normal form; we define $\Pi_k^1\text{-}\mathsf{CA}_0^\omega$ as $\mathsf{RCA}_0^\omega + (\mathsf{S}_k^2)$, where (S_k^2) expresses that S_k^2 exists. We define Z_2^ω as $\cup_k \Pi_k^1\text{-}\mathsf{CA}_0^\omega$, i.e. higher-order Z_2. Fourth, we introduce Kleene's quantifier $\exists^3$ as follows:

$$(\exists E : (\mathbb{N}^\mathbb{N} \to \mathbb{N}) \to \mathbb{N})(\forall Y : \mathbb{N}^\mathbb{N} \to \mathbb{N})\big[(\exists f \in \mathbb{N}^\mathbb{N})(Y(f) = 0) \leftrightarrow E(Y) = 0\big]. \tag{$\exists^3$}$$

Both systems $\mathsf{Z}_2^\Omega \equiv \mathsf{RCA}_0^\omega + (\exists^3)$ and $\mathsf{Z}_2^\omega \equiv \cup_k \Pi_k^1\text{-}\mathsf{CA}_0^\omega$ are conservative over Z_2 as shown in [20]. Hilbert-Bernays' operator ν from [19, p. 479] is essentially Kleene's $\exists^3$, modulo a non-trivial fragment of the Axiom of Choice.

Fifth, Sol Feferman was a leading Stanford logician with a life-long interest in foundational matters, especially predicativist mathematics following Russell and Weyl. As part of the latter, Feferman introduced the 'projection principle' Proj_1 in [11] as a third-order version of $(\exists^3)$. We study the principle BOOT, which is Feferman's Proj_1 in higher-order arithmetic.

Principle 2 (BOOT). *For Y^2, there is $X \subset \mathbb{N}$ such that*

$$(\forall n \in \mathbb{N})\big[n \in X \leftrightarrow (\exists f \in \mathbb{N}^\mathbb{N})(Y(f, n) = 0)\big].$$

The name of this principle derives from the verb 'to bootstrap'. Indeed, BOOT is third-order and weaker than $(\exists^3)$, but we still have that $\Pi_k^1\text{-}\mathsf{CA}_0^\omega + \mathsf{BOOT}$ proves $\Pi_{k+1}^1\text{-}\mathsf{CA}_0$. In other words, BOOT bootstraps itself along the second-order comprehension hierarchy so central to mathematical logic (see [40]). Both BOOT and $(\exists^3)$ are equivalent to (intimately related) basic statements about continuous functions on metric spaces [37].

Sixth, certain equivalences in second-order RM require extra induction beyond what is available in RCA_0 (see [28]). We sometimes use extra induction in higher-order RM, including the following axiom.

Principle 3 (Σ-IND). *The induction axiom for formulas $\varphi(n)$ defined as $(\exists f \in \mathbb{N}^\mathbb{N})(Y(f, n) = 0)$ for any Y^2.*

Finally, like its second-order counterpart, the development of higher-order RM sometimes takes place over an *extension* of the base theory. The following fragment of countable choice, not provable in ZF, plays an important role.

Principle 4 ($\mathsf{QF\text{-}AC}^{0,1}$). *Let φ be quantifier-free with $(\forall n \in \mathbb{N})(\exists f \in \mathbb{N}^\mathbb{N})\varphi(f, n)$, then there exists a sequence $(f_n)_{n\in\mathbb{N}}$ in $\mathbb{N}^\mathbb{N}$ with $(\forall n \in \mathbb{N})\varphi(f_n, n)$.*

The local equivalence between sequential and 'epsilon-delta' continuity over the reals cannot be proved in ZF [18], but can be established in $\mathsf{RCA}_0^\omega + \mathsf{QF\text{-}AC}^{0,1}$, as first shown by Kohlenbach in [22]. In this light, it should not be a surprise that $\mathsf{RCA}_0^\omega + \mathsf{QF\text{-}AC}^{0,1}$ often occurs as a base theory.

1.3 Some Function Classes

We introduce some function classes from the mathematical mainstream.

First of all, we use the standard definition of the Riemann integral, where a modulus is well-known in RM and related areas.

Definition 5 (Riemann integral). *Let $a <_{\mathbb{R}} b$ be given.*

- *The* mesh $\|P\|$ *of a partition P of $[a,b]$ given by the finite sequence $(x_0 = a, t_0, x_1, t_1, \dots, x_{k-1}, t_{k-1}, x_k = b)$ is $\max_{1\leq i\leq k}(x_i - x_{i-1})$.*
- *For a partition P as in the previous item and $f : [a,b] \to \mathbb{R}$, the* Riemann sum *is $S(f, P, a, b) := \sum_{i=1}^{k} f(t_{i-1})(x_i - x_{i-1})$.*
- *A function $f : [a,b] \to \mathbb{R}$ is* Riemann integrable *on $[a,b]$ if for any $k \in \mathbb{N}$ there is $N \in \mathbb{N}$ such that for all partitions P, Q of $[a,b]$ with $\|P\|, \|Q\| \leq \frac{1}{2^N}$, we have $|S(f,P,a,b) - S(f,Q,a,b)| < \frac{1}{2^k}$.*

Secondly, the Vitali-Lebesgue theorem provides necessary and sufficient conditions for Riemann integrability based on continuity almost everywhere. Now, a function $f : \mathbb{R} \to \mathbb{R}$ is usually called continuous almost everywhere (ae) if the set of discontinuity points D_f has measure zero. However, the set D_f need not exist in weak logical systems as its definition involves quantifiers over $\mathbb{R}$. The following definition avoids introducing the set D_f.

Definition 6. *A function $f : \mathbb{R} \to \mathbb{R}$ is* continuous ae *in case for any $\varepsilon > 0$, there is a sequence of intervals $(I_n)_{n\in\mathbb{N}}$ of total length at most ε such that if f is discontinuous at $x \in \mathbb{R}$, we have $x \in \cup_{n\in\mathbb{N}} I_n$.*

One direction of the Vitali-Lebesgue theorem can be proved in weak systems while the other one gives rise to a new 'Big' system [34] based on Tao's *pigeon hole principle* for measure spaces [41].

Finally, we list the remaining definitions of the function classes to be studied.

Definition 7. *For $f : \mathbb{R} \to \mathbb{R}$, we have the following definitions:*

- *f is* upper semi-continuous *(usco) at $x_0 \in \mathbb{R}$ if for any $k \in \mathbb{N}$, there is $N \in \mathbb{N}$ such that $(\forall y \in B(x_0, \frac{1}{2^N}))(f(y) < f(x_0) + \frac{1}{2^k})$,*
- *f is* lower semi-continuous *(lsco) at $x_0 \in \mathbb{R}$ if for any $k \in \mathbb{N}$, there is $N \in \mathbb{N}$ such that $(\forall y \in B(x_0, \frac{1}{2^N}))(f(y) > f(x_0) - \frac{1}{2^k})$,*
- *f is* quasi-continuous *at $x_0 \in \mathbb{R}$ if for any $k, N \in \mathbb{N}$ there is an open interval $(a,b) \subset B(x_0, \frac{1}{2^N})$ with $(\forall x \in (a,b))(|f(x_0) - f(x)| < \frac{1}{2^k})$,*
- *f is* cliquish *at $x_0 \in \mathbb{R}$ if for any $k, N \in \mathbb{N}$ there is an open interval $(a,b) \subset B(x_0, \frac{1}{2^N})$ with $(\forall x, y \in (a,b))(|f(x) - f(y)| < \frac{1}{2^k})$,*
- *f is* pointwise discontinuous *if for any $x \in \mathbb{R}$ and $k \in \mathbb{N}$, there is $y \in \mathbb{R}$ such that f is continuous at y and $|x - y| < \frac{1}{2^k}$* [16],
- *f is* sub-continuous *at $x_0 \in \mathbb{R}$ if for any $(y_n)_{n\in\mathbb{N}}$ with limit x_0, the sequence of reals $(f(y_n))_{n\in\mathbb{N}}$ has a convergent sub-sequence,*
- *f is* measurable *if for any open $V \subset \mathbb{R}$, $f^{-1}(V)$ is the union of an $\mathbf{F}_\sigma$ and measure zero set see e.g.* [41, Ex. 1.2.19 and Lem. 1.3.9],

- f *is* Baire measurable [25] *if for every open* $V \subset \mathbb{R}$, *the set* $f^{-1}(V)$ *has the Baire property,*
- f *is* simply continuous (*sico*) *if for any open* $V \subset \mathbb{R}$, *the set* $f^{-1}(V)$ *is the union of an open and a nowhere dense set* [4].

Moreover, if $f : \mathbb{R} \to \mathbb{R}$ has a certain property from Definition 7 at all reals $x_0 \in \mathbb{R}$, or on its intended domain, we say that the function has that property.

Many more function classes are studied in [38] and we would be interested in other classes that behave in the same way as the Riemann integrable functions.

2 Equivalences for Feferman's Projection Principle

We establish a number of equivalences for BOOT from Sect. 1.2 and the supremum principle restricted to the function classes from Sect. 1.3. Similar results are in [35], pertaining to weak continuity notions stemming from topology.

First of all, we establish Theorem 8 to be contrasted with Theorem 9.

Theorem 8. ($\mathsf{ACA}_0^\omega + \Sigma$-$\mathsf{IND}$). *The following are equivalent to* BOOT.

a. For any function $f : [0,1] \to [0,1]$, *the supremum* $\sup_{x\in[0,1]} f(x)$ *exists.*
b. For any function $f : [0,1] \to [0,1]$, $\sup_{x\in[p,q]} f(x)$ *exists for* $p, q \in \mathbb{Q} \cap [0,1]$.
c. For continuous ae $f : [0,1] \to [0,1]$, *the supremum* $\sup_{x\in[0,1]} f(x)$ *exists.*
d. For Riemann integrable $f : [0,1] \to \mathbb{R}$, *the supremum* $\sup_{x\in[0,1]} f(x)$ *exists.*
e. For cliquish $f : [0,1] \to [0,1]$, *the supremum* $\sup_{x\in[0,1]} f(x)$ *exists.*
f. For pointwise discontinuous $f : [0,1] \to [0,1]$, $\sup_{x\in[0,1]} f(x)$ *exists.*
g. For simply continuous $f : [0,1] \to [0,1]$, $\sup_{x\in[0,1]} f(x)$ *exists.*
h. For (Baire) measurable $f : [0,1] \to [0,1]$, $\sup_{x\in[0,1]} f(x)$ *exists.*
i. For sub-continuous $f : [0,1] \to [0,1]$, $\sup_{x\in[0,1]} f(x)$ *exists.*

Proof. First of all, we recall that $(\exists^2)$ can convert a real $x \in [0,1]$ to a unique binary representation, choosing a tail of zeros when relevant. Similarly, an element of Baire space $f \in \mathbb{N}^\mathbb{N}$ can be identified with its graph as a subset of $\mathbb{N} \times \mathbb{N}$, which can be represented as an element of Cantor space. Over ACA_0^ω, BOOT is thus equivalent to the statement that for any function $f : (\mathbb{R} \to \mathbb{N}) \to \mathbb{N}$, there is a set $X \subset \mathbb{N}$ such that for all $n \in \mathbb{N}$:

$$(\exists x \in [0,1])(f(x,n) = 0) \leftrightarrow n \in X. \tag{1}$$

Secondly, to show that $\mathsf{ACA}_0^\omega + \mathsf{BOOT}$ proves the supremum principle for any function $f : [0,1] \to [0,1]$, BOOT in the form (1) yields $X \subset \mathbb{Q}^3$ with

$$(\forall p, q, r \in \mathbb{Q})\big[(\exists x \in [p,q])(f(x) > r) \leftrightarrow (p,q,r) \in X\big].$$

Now use $\exists^2$ and the usual interval-halving method to define supremum $\sup_{x\in[p,q]} f(x)$.

For the other implications, fix $Y : (2^\mathbb{N} \times \mathbb{N}) \to \{0,1\}$ and recall that the Cantor set $\mathcal{C}$ comes with a recursive homeomorphism $H : 2^\mathbb{N} \to [0,1]$ defined as

$H(f) := \sum_{n=0}^{\infty} \frac{2f(n)}{3^{n+1}}$. General properties of this specific set are readily verified in RCA_0 (see [39, p. 129]). Now define $g_Y : [0,1] \to [0,1]$ as follows

$$g_Y(x) := \begin{cases} 0 & x \notin \mathcal{C} \text{ or } I(x) = 11\dots \\ Z(f_0, \dots, f_k) & I(x) = \underbrace{11\dots11}_{k \text{ times}} * \langle 0 \rangle * f_0 \oplus \dots \oplus f_k \end{cases}, \tag{2}$$

where $I(x)$ is the unique $h \in 2^{\mathbb{N}}$ such that $H(h) = x$ in case $x \in \mathcal{C}$, and $00\dots$ otherwise and where $Z(f_0, \dots, f_k)$ is the sum of all $\frac{1}{2^i}$ for $i \leq k$ such that there is $j \leq k$ with $Y(f_j, i) = 0$.

Let $\eta : [0,1] \to 2^{\mathbb{N}}$ be the function that converts reals in the unit interval to their binary representation, choosing a tail of zeros when applicable. Using Σ-IND, we have the following equivalence for any $n \in \mathbb{N}$:

$$(\exists f \in 2^{\mathbb{N}})(Y(f,n) = 0) \leftrightarrow \eta(\sup_{x \in [0,1]} g_Y(x))(n) = 1,$$

and hence BOOT readily follows. Since the complement of the Cantor set $\mathcal{C}$ is dense and has measure one, the function g_Y is continuous ae, pointwise discontinuous, and cliquish by definition. To verify Riemann integrability, use the 'epsilon-delta' definition to show that $\int_0^1 g_Y(x)dx = 0$. To show that g_Y is simply continuous, fix an open set $O \subset \mathbb{R}$ and consider the inverse image $g_Y^{-1}(O)$. In case $0 \in O$, we have $g_Y^{-1}(O) = \mathcal{C}^c \cup \{x \in \mathcal{C} : g_Y(x) \in O\}$, which is the union of an open set and a nowhere dense set. In case $0 \notin O$, we have $g_Y^{-1}(O) = \emptyset \cup \{x \in \mathcal{C} : g_Y(x) \in O\}$, which is also the union of an open set and a nowhere dense set. In this way, the function g_Y is indeed simply continuous. The Baire measurability of g_Y is established in much the same way, which is left to the reader. The measurability of g_Y is immediate by observing that RM-open sets are $\mathbf{F}_\sigma$ and that the Cantor set $\mathcal{C}$ has measure zero, working over ACA_0^ω (and also weaker systems). Sub-continuity for g_Y is immediate as it is bounded. □

By contrast, we have the following theorem where we note that quasi-continuous and cliquish functions are considered close in light of Corollary 10.

Theorem 9 (ACA_0^ω). *For quasi-continuous $f : [0,1] \to \mathbb{R}$, $\sup_{x\in[0,1]} f(x)$ exists.*

Proof. This is proved in [30, §2]. By way of a sketch, $(\exists x \in [0,1])(f(x) > q)$ is equivalent to $(\exists r \in [0,1] \cap \mathbb{Q})(f(r) > q)$ by the definition of quasi-continuity. This observation readily yields $\sup_{x\in[0,1]} f(x)$ using $(\exists^2)$. □

Finally, basic facts about cliquish functions also imply BOOT as follows. Proofs of these items may be found in [5,14,24,33].

Corollary 10 ($\mathsf{ACA}_0^\omega + \Sigma$-$\mathsf{IND}$). *The axiom* BOOT *follows from the following.*

- *A cliquish function on $[0,1]$ is the sum of two quasi-continuous functions.*
- *A bounded cliquish function on the reals is the pointwise limit of a sequence of bounded quasi-continuous functions,*

- *For any (cliquish) $f : [0,1] \to \mathbb{R}$ and nowhere dense $A \subset [0,1]$, there is quasi-continuous $g : [0,1] \to \mathbb{R}$ such that f coincides with g on A and such that for any $c, d \in \mathbb{R}$, if $c \le f \le d$ then $c \le g \le d$.*

Proof. For the second item, the proof of Theorem 9 provides a sequence of suprema for the sequence of bounded quasi-continuous functions. From the latter, one readily obtains the supremum of the cliquish function. For the third item, the Cantor set $\mathcal{C}$ is nowhere dense while g_Y from (2) is cliquish. □

In conclusion, Darboux's supremum principle restricted to numerous function classes is equivalent to strong axioms like BOOT. Slight variations are provable in much weaker systems by Theorem 9.

3 Equivalences for the Coding Principle of Open Sets

In this section, we establish equivalences for the following coding principle.

Principle 11 (open). *For any open set $O \subset \mathbb{R}$, there exist sequences $(a_n)_{n\in\mathbb{N}}$, $(b_n)_{n\in\mathbb{N}}$ of reals such that $x \in O \leftrightarrow x \in \cup_{n\in\mathbb{N}}(a_n, b_n)$ for all $x \in \mathbb{R}$.*

It is straightforward to show that open follows from its restriction to $[0,1]$.

Clearly, open expresses that an open set has a code as in second-order RM. This principle is equivalent to the following.

- Basic properties of usco functions on the unit interval, including the supremum principle (Sect. 3.1).
- Basic properties of continuous functions on closed sets, including the supremum principle (Sect. 3.2).
- The Tietze theorem and Urysohn lemma (Sect. 3.3).
- Various special cases of the Lindelöf lemma for closed sets (Sect. 3.4).

On a historical note, the concept of open set dates back to Baire's 1899 doctoral thesis [3] as a small side-note in the study of semi-continuity. Of course, Dedekind had considered open sets twenty years earlier but the associated paper was only published much later [9,26].

On a technical note, one readily shows that $\mathsf{BOOT} \to \mathsf{open}$ over ACA_0^ω; the implication is strict as $\mathsf{ACA}_0^\omega + \mathsf{open}$ is a conservative[1] extension of ATR_0 while $\mathsf{ACA}_0^\omega + \mathsf{BOOT}$ readily implies $\Pi_1^1\text{-}\mathsf{CA}_0$.

3.1 Properties of Semi-continuous Functions

We show that basic properties of usco functions are equivalent to open.

Theorem 12 ($\mathsf{ACA}_0^\omega + \mathsf{QF\text{-}AC}^{0,1}$). *The following are equivalent.*

a. *The coding principle* open.

[1] This can be proved by combining [32] and Hunter's conservation results in [20].

b. The supremum[2] *principle for usco functions on the unit interval.*
c. The maximum (see footnote 2) *principle for usco functions on on* $[0,1]$.
d. A usco function on the unit interval is Baire 1.

We only use $\mathsf{QF\text{-}AC}^{0,1}$ *to obtain item (c) from* open.

Proof. First of all, assume open and fix some usco $f:[0,1]\to\mathbb{R}$. Essentially by definition, the set $E_q := \{x\in[0,1] : f(x)\geq q\}$ is closed. Now, for a code $\cup_{n\in\mathbb{N}}(a_n,b_n)$ of the complement of E_q, we have

$$E_q=\emptyset \leftrightarrow (\exists m\in\mathbb{N})\big([0,1]\subseteq \cup_{n\leq m}(a_n,b_n)\big), \tag{3}$$

where the right-hand side of (3) is (equivalent to) arithmetical. Let $(q_n)_{n\in\mathbb{N}}$ be an enumeration of $\mathbb{Q}$ without repetitions. We place a copy of E_{q_n} in $[n,n+1]$ and observe that the corresponding union is closed. Then open provides a code for the complement of the union and by (3), $(\exists^2)$ can decide whether E_{q_n} is empty. Since f is bounded by [30, §2], the usual interval-halving technique readily yields $\sup_{x\in[0,1]} f(x)$. The previous construction is readily adapted to finding $\sup_{x\in[p,r]} f(x)$ for $p,r\in\mathbb{Q}\cap[0,1]$, i.e. item (b) follows. By definition, we have

$$(\forall n\in\mathbb{N})(\exists x\in[0,1])(f(x)>\sup_{z\in[0,1]} f(z)-\tfrac{1}{2^n}).$$

Apply $\mathsf{QF\text{-}AC}^{0,1}$ and note that the resulting sequence has a convergent subsequence, say with limit $y\in[0,1]$. Since f is usco, y is a maximum, i.e. $(\forall x\in[0,1])(f(x)\leq f(y))$, as required for item (c). To show that f is Baire 1, define the continuous function $f_n:[0,1]\to\mathbb{R}$ as follows:

$$f_n(x) := \sup_{y\in[0,1]}(f(y)-n\cdot|x-y|),$$

where the supremum is defined by first restricting x to the rationals and then taking limits. Clearly, f is the pointwise limit of $(f_n)_{n\in\mathbb{N}}$ as required for item (d).

To obtain item (b) from item (d), we note that by [30, §2], bounded Baire 1 functions have suprema in ACA_0^ω. To obtain open from item (d), if $C\subset[0,1]$ is closed, then $\mathbb{1}_C$ is usco (by definition) and we have for $O:=[0,1]\setminus C$ that

$$B(q,r)\subset O \leftrightarrow [0=\sup_{x\in B(q,r)}\mathbb{1}_C(x)].$$

Hence, we can enumerate $B(q,r)$ with rational $q\in[0,1]$ and $r>0$ that are subsets of O, yielding the representation required for open. □

The reader is invited to show that open is equivalent to the statement that usco functions have codes as in e.g. [12]. The latter codes essentially hard-code a Baire 1 representation in a bit of a roundabout way.

Finally, a function $f:[0,1]\to\mathbb{R}$ is called *effectively usco* if there is a sequence $(O_n)_{n\in\mathbb{N}}$ of RM-codes for open sets such that $\{x\in[0,1] : f(x)<q_n\}=O_n$ for all $n\in\mathbb{N}$; here, $(q_n)_{n\in\mathbb{N}}$ is a fixed enumeration of the rationals $\mathbb{Q}$.

[2] To be clear, our supremum principle states the existence of a sequence of reals $(y_{p,q})_{p,q\in\mathbb{Q}\cap[0,1]}$ with $(\forall x\in[p,q])(f(x)\leq y_{p,q})$. Similarly, our maximum principle states the existence of a sequence $(x_{p,q})_{p,q\in\mathbb{Q}\cap[0,1]}$ with $(\forall x\in[p,q])(f(x)\leq f(x_{p,q}))$..

Theorem 13 (RCA_0^ω+$\mathsf{QF\text{-}AC}^{0,1}$). *The following is equivalent to* WKL_0*:*

for any effectively usco $f : [0,1] \to \mathbb{R}$, $\sup_{x\in[0,1]} f(x)$ *exists.*

Proof. The supremum principle for continuous functions on the unit interval is equivalent to WKL_0 over RCA_0^ω ([30, §2]). What remains to show is that for discontinuous and effectively usco $f : [0,1] \to \mathbb{R}$, the supremum exists. By [22, Prop. 3.14], we may use $(\exists^2)$. Let $\{x \in [0,1] : f(x) < q_n\} = O_n = \cup_{m\in\mathbb{N}}(a_{n,m}, b_{n,m})$ be as per our assumptions. Now consider the equivalence

$$O_n = [0,1] \leftrightarrow (\exists m_0 \in \mathbb{N})\big([0,1] \subset \cup_{m\leq m_0}(a_{m,n}, b_{m,n})\big), \tag{4}$$

which follows from the countable Heine-Borel theorem; the latter is provable in WKL_0 by [39, IV.1.1]. The right-hand side of (4) is decidable using $(\exists^2)$, which readily yields $\sup_{x\in[0,1]} f(x)$ again using $(\exists^2)$. □

3.2 Continuous Functions on Closed Sets

We establish equivalences between open and basic properties of continuous functions with domain a closed set, including the supremum principle. On a historical note, Baire already studies functions on open (and general) domains in [3].

First of all, we study the following supremum principle. The latter is provable in ACA_0 for codes of continuous functions on RM-closed sets by [39, IV.2.11].

Principle 14. *Let* $C \subset \mathbb{R}$ *be closed and* $f : C \to \mathbb{R}$ *be continuous. There is a sequence* $(y_n)_{n\in\mathbb{N}}$ *such that for all* $p, q \in \mathbb{Q}$ *with* $C \cap [p,q] \neq \emptyset$, *there is* $n \in \mathbb{N}$ *such that* $y_n = \sup_{x\in C\cap[p,q]} f(x)$.

To be absolutely clear, we explicitly spell out the difference between '$f : \mathbb{R} \to \mathbb{R}$ is continuous on C' and '$f : C \to \mathbb{R}$ is continuous': the latter means that

$$(\forall x \in C, k \in \mathbb{N})(\exists N \in \mathbb{N})\underline{(\forall y \in C)}(|x-y| < \tfrac{1}{2^N} \to |f(x) - f(y)| < \tfrac{1}{2^k}),$$

where the underlined quantifier is the crucial part. Indeed, C is the domain of $f : C \to \mathbb{R}$, as is perfectly clear from the notation.

Secondly, continuous functions on closed sets readily yield usco functions, explaining the following equivalence.

Theorem 15 (ACA_0^ω). *The following are equivalent.*

- *The principle* open.
- *The supremum principle as in Principle 14.*

Proof. Zeroth of all, observe that both items follow from their restrictions to the unit interval, which is readily proved by mapping unit intervals to intervals of the form $[\frac{1}{2^{n+1}}, \frac{1}{2^n}]$. First observe that open proves[3] CIC as the latter restricted to

[3] The choice axiom CIC states that for a sequence $(C_n)_{n\in\mathbb{N}}$ of non-empty and closed sets of reals, there is $(x_n)_{n\in\mathbb{N}}$ with $x_n \in C_n$ for all $n \in \mathbb{N}$.

RM-codes follows from WKL_0 by [39, IV.1.8]. Now, for the downward implication, let $C \subset [0,1]$ be closed and let $f : C \to \mathbb{R}$ be continuous. To prove that f is bounded on C, suppose $(\forall N \in \mathbb{N})(\exists x \in C)(|f(x)| \geq N)$ and apply CIC to obtain a sequence $(x_N)_{N\in\mathbb{N}}$ in C with $f(x_N) \geq N$. By sequential completeness (of the unit interval), there is a sub-sequence with a limit $y \in C$, as the latter set is closed. Clearly, f is not continuous at y, a contradiction, and f is bounded, say of the form $f : [0,1] \to [0,1]$. Use $(\exists^2)$ to define $h : [0,1] \to \mathbb{R}$ as $f(x)$ if $x \in C$ and 0 otherwise. Essentially by definition, h is usco and Theorem 12 guarantees that the supremum $\sup_{x\in[p,q]} h(x)$ exists as a sequence for rational $p, q \in \mathbb{Q} \cap [0,1]$. Clearly, if $[p,q]\cap C \neq \emptyset$, then $\sup_{x\in[p,q]} h(x) = \sup_{x\in C\cap[p,q]} f(x)$, as required. The generalisation from $[0,1]$ to $\mathbb{R}$ is straightforward. As an alternative proof, open provides an RM-code for $C \subset [0,1]$, which in turn yields a dense sub-sequence $(x_n)_{n\in\mathbb{N}}$ of C by [7, Theorem 3.3] or [10, 10.5.7]. Clearly, the supremum of f on C is readily defined in terms of this dense sequence, as we have

$$(\exists x \in C)(f(x) > q) \leftrightarrow (\exists n \in \mathbb{N})(f(x_n) > q)$$

for any $q \in \mathbb{Q}$, essentially by definition.

Next assume the second item and fix closed $C \subset [0,1]$. Let C_n be the set obtained by translating C to $[2n, 2n+1]$ for $n \geq 1$. Define D as the union $\cup_{n\geq 1} C_n$ extended by the points $2n + \frac{i}{2^{n+2}}$ for $i \leq 2^{n+2}$ for each $n \in \mathbb{N}$. Then define $f : D \to \mathbb{R}$ as the following function:

$$f(x) := \begin{cases} \frac{1}{2^n} & x \in C_n \\ \frac{1}{2^{n+1}} & x \in (D \setminus C_n) \cap [2n, 2n+1] \end{cases}.$$

Essentially by definition, $f : D \to \mathbb{R}$ is continuous. Now define $I_{i,n}$ as the interval $[2n + \frac{i}{2^{n+1}}, 2n + \frac{i+1}{2^{n+1}}]$ and observe that:

$$\tfrac{1}{2^n} =_{\mathbb{R}} \sup_{x\in D\cap I_{i,n}} f(x) \leftrightarrow (\exists x \in I_{i,n})(x \in C_n), \tag{5}$$

for $i < 2^{n+1}$. We note that $D\cap[2n+\frac{i}{2^{n+1}}, 2n+\frac{i+1}{2^{n+1}}]$ is non-empty by construction for $i < 2^{n+1}$. Let $i_0, \ldots, i_{k_n}$ be those $i \leq 2^{n+1}$ for which $I_{i,n} \cap C_n$ is non-empty. These numbers can be defined in terms of $(\exists^2)$ as the latter decides the left-hand side of (5). Moreover, $\cup_{i\leq k_n} I_{i,n}$ covers C_n and consists of intervals of length $\frac{1}{2^{n+1}}$. The complement of the closure of $\cup_{i\leq k_n} I_{i,n}$ is then a subset of $O_n := [2n, 2n+1] \setminus C_n$. As for $\mathsf{LIN}(\mathbb{R})$ in the proof of Theorem 21, the union of these complements yields an RM-code for $O := [0,1] \setminus C$, as required. □

Assuming $\mathsf{QF}\text{-}\mathsf{AC}^{0,1}$, one can show that open is equivalent to the maximum principle associated to Principle 14.

Finally, the previous results provide evidence for the following claim.

*Continuous functions are only 'nice' or 'tame' in case a dense sub***sequence** *of the domain is given.*

Indeed, the domain in Principle 14 is a closed set which does not come with the well-known representations from second-order RM, which are RM-codes or dense subsequences [6,7].

3.3 Tietze Theorem and Urysohn Lemma

We show that open is equivalent to well-known results due to Tietze and Urysohn. The results in this section extend and refine those in [29, §5].

First of all, we need the following theorem from second-order RM.

Theorem 16 ([39, II.7.1]). *The following is provable in* RCA_0.

- *Given (a code for) an open set* $U \subset \mathbb{R}$, *we can effectively find (a code for) a continuous* $h_U : \mathbb{R} \to [0, 1]$ *such that* $(\forall x \in \mathbb{R})(x \in U \leftrightarrow h_U(x) > 0)$.
- *Conversely, given (a code for) a continuous* $f : \mathbb{R} \to \mathbb{R}$, *we can effectively find (a code for) an open set* $V \subset R$ *with* $(\forall x \in \mathbb{R})(x \in V \leftrightarrow f(x) > 0)$.

Now, the Tietze extension theorem and Urysohn lemma are well-known in second-order RM ([39, I.8.3]) as part of the Big Five. Apparently, the third-order versions behave quite differently. We note that both versions of the Urysohn lemma in Theorem 17 occur in the literature.

Theorem 17 (ACA_0^ω+$\mathsf{QF\text{-}AC}^{0,1}$). *The following are equivalent to* open.

a. *(Urysohn) For closed disjoint* $C_0, C_1 \subseteq \mathbb{R}$, *there is continuous* $g : \mathbb{R} \to [0, 1]$ *such that* $x \in C_i \leftrightarrow g(x) = i$ *for any* $x \in \mathbb{R}$ *and* $i \in \{0, 1\}$.
b. *The combination of the following.*
   - *(Weak Urysohn) Item a restricted to '*$x \in C_i \to g(x) = i$*'.*
   - *Every open set of reals is* $\mathbf{F}_\sigma$.
c. *The combination of the following.*
   - *(Tietze) For* $f : \mathbb{R} \to [0, 1]$ *continuous on a closed set* $C \subseteq \mathbb{R}$, *there is continuous* $g : \mathbb{R} \to [0, 1]$ *such that* $f(x) =_{\mathbb{R}} g(x)$ *for* $x \in C$.
   - *Every open set of reals is also* $\mathbf{F}_\sigma$.

We only need $\mathsf{QF\text{-}AC}^{0,1}$ *for the second item.*

Proof. Our proof is based on [30, §2] where the following items are established.

(α) The system $\mathsf{RCA}_0^\omega + \mathsf{WKL}_0$ proves that a (third-order) continuous function $f : [0, 1] \to \mathbb{R}$ has a second-order code as in [39, II.6.1].
(β) The system RCA_0 proves that for any second-order code Φ (of a continuous function on $[0, 1]$), there is a continuous $f : [0, 1] \to \mathbb{R}$ such that $f(x)$ equals the value of Φ at any $x \in [0, 1]$.

In this light, we can freely go from continuous functions to codes and vice versa.

First of all, to obtain open from item a, fix an open set $U \subset \mathbb{R}$ and define disjoint closed sets $C_0 = U^c$ and $C_1 = \emptyset$. For $g : \mathbb{R} \to \mathbb{R}$ provided by the Urysohn lemma, we have $x \in U \leftrightarrow g(x) >_{\mathbb{R}} 0$ for all $x \in \mathbb{R}$. By item (α), g has an RM-code and hence represents an open set, namely U, by Theorem 16. Observe that this proof does not go through for item (b).

Secondly, to establish the first item in ACA_0^ω + open, let $C_0, C_1 \subseteq \mathbb{R}$ be disjoint and closed. Using $(\exists^2)$, the usual interval-halving technique shows that the distance function $d(x, C)$ exists (as a continuous third-order function) in case C is given by an RM-code. Hence, we can define $g(x) := \frac{d(x,C_0)}{d(x,C_0)+d(x,C_1)}$ as in the textbook proof of the Urysohn lemma. Alternatively, use the second-order proof ([39, II.7]) for the RM-codes and use items (α) and (β).

Thirdly, to show that item c implies item b, let C_i be as in item b for $i = 0, 1$ and define f on $C_2 := C_0 \cup C_1$ as follows: $f(x) = 0$ if $x \in C_0$ and 1 otherwise. The obvious proof-by-contradiction shows that f is continuous on C_2, i.e. we may apply item c. The resulting function is as required for item b.

To show that item b implies open restricted to $[0, 1]$, let $F \subset [0, 1]$ be closed and suppose the complement O of F satisfies $O = \cup_{n\in\mathbb{N}} F_n$ where each F_n is closed. Define $f_n(x)$ as 0 if $x \in F$ and $\frac{1}{2^n}$ if $x \in F_n$. As in the previous paragraph, one shows that $f_n : [0, 1] \to \mathbb{R}$ is continuous on $F \cup F_n$. Let g_n be the continuous extension[4] of f_n to $[0, 1]$. Rescale the former if necessary and use the M-test of Weierstrass (provable for codes in RCA_0 by [39, II.6.5]) to conclude that $g(x) := \sum_{n=0}^{\infty} g_n(x)$ exists and is continuous. If $x \in F$, then $g_n(x) = 0$ and hence $g(x) = 0$. In case $x \notin F$, then $x \in F_{n_0} \subset O$ for some $n_0 \in \mathbb{N}$. Hence, $g_{n_0}(x) > 0$ and $g(x) > 0$. As a result, $x \in F \leftrightarrow g(x) = 0$ and item (α) together with Theorem 16 yields an RM-code for F.

Finally, for item (c), assume open and fix closed $C \subset [0, 1]$ and $f : [0, 1] \to [0, 1]$ continuous on C. Consider the following definition due to Hausdorff [17]:

$$g(x) := \begin{cases} f(x) & x \in C \\ \inf_{y\in C} \left(f(y) + \frac{|x-y|}{d(x,C)} - 1\right) & x \notin C \end{cases}. \tag{6}$$

As above, $d(x, C)$ exists for RM-closed $C \subset \mathbb{R}$ while Theorem 15 guarantees that the infimum in (6) is well-defined. Indeed, $\lambda y.\left(f(y) + \frac{|x-y|}{d(x,C)} - 1\right)$ is continuous on C as f is, and g is readily seen to be continuous on $[0, 1]$. Technically, we need to restrict the variable x in the former function to rationals and take limits. □

We note that the Urysohn lemma in item (a) in the theorem expresses *perfect normality* of the underlying space, while the *weak* Urysohn lemma in item (b) only expresses normality ([27, p. 211]).

We discuss some possible additional results in the following remark.

Remark 18. First of all, one readily shows that Urysohn's lemma from Theorem 17 is implied by the Hahn-Katětov-Tong theorem [15,21,42] as follows.

For usco $f : [0, 1] \to \mathbb{R}$ and lsco $g : [0, 1] \to \mathbb{R}$ with $(\forall x \in [0, 1])(f(x) \leq g(x))$, there is continuous $h : [0, 1] \to \mathbb{R}$ with $(\forall x \in [0, 1])(f(x) \leq h(x) \leq g(x))$.

A harder-but-feasible exercise is to verify that open implies this 'insertion' theorem via the proof in [13].

Secondly, one readily shows that ACA_0^ω proves the Urysohn lemma if $g : \mathbb{R} \to \mathbb{R}$ only needs to be usco or cliquish. By contrast, open still follows from the Urysohn lemma even if $g : \mathbb{R} \to \mathbb{R}$ only needs to be Baire 1 or quasi-continuous.

[4] Technically, one needs to shift f_n and $F \cup F_n$ to $[n + 1, n + 2]$ to observe that the Tietze theorem for $\mathbb{R}$ implies a sequential version of the Tietze theorem on $[0, 1]$.

3.4 The Lindelöf Lemma for the Real Line

We connect open to the Lindelöf and Cousin lemmas. Here, HBU is a version of the Heine-Borel theorem for uncountable coverings, hence the acronym.

Principle 19 (LIN($\mathbb{R}$)). *For closed $C \subset \mathbb{R}$ and $\Psi : \mathbb{R} \to \mathbb{R}^+$, there is a sequence $(x_n)_{n\in\mathbb{N}}$ in C such that $\cup_{n\in\mathbb{N}} B(x_n, \Psi(x_n))$ covers C.*

Principle 20 (HBU). *For $\Psi : [0,1] \to \mathbb{R}^+$, there are $x_0, \dots, x_k \in [0,1]$ such that the finite union $\cup_{n\leq k} B(x_n, \Psi(x_n))$ covers $[0,1]$.*

We note that Lindelöf himself studied the coverings of *arbitrary* sets in [23].

Theorem 21 (ACA_0^ω+QF-AC^{0,1}). *We have* LIN($\mathbb{R}$) $\to$ open.

Proof. Fix a closed set $C \subset [0,1]$ and define a new closed set $D \subset \mathbb{R}$ by placing a copy of C in each $[2n, 2n+1]$ for $n \in \mathbb{N}$. Let $\Psi : \mathbb{R} \to \mathbb{R}^+$ be constant $\frac{1}{2^n}$ on the latter intervals and the obvious linear extension on the remaining intervals. For the below, note that Ψ is continuous. By LIN($\mathbb{R}$), there is a sequence $(x_n)_{n\in\mathbb{N}}$ in D such that $\cup_{n\in\mathbb{N}} B(x_n, \Psi(x_n))$ covers D.

We have the Heine-Borel theorem for countable open coverings; this is proved in [29] via the obvious proof-by-contradiction involving QF-AC^{0,1}. Hence, for each $n \in \mathbb{N}$, there are $i_0, \dots, i_{k_n} \in \mathbb{N}$ such that the finite union $\cup_{j\leq k_n} B(x_{i_j}, \Psi(x_{i_j}))$ covers $D \cap [2n, 2n+1]$. Then $\Psi(x_{i_j}) = \frac{1}{2^n}$ by definition and the complement of $\cup_{j\leq k_n} B(x_{i_j}, \frac{1}{2^n})$ can be viewed as an (inner) approximation of a code for the complement of C. A code for $O := [0,1] \setminus C$ is then obtained by the union of the complements of the aforementioned finite unions. □

The previous theorem has some interesting (robust) corollaries. Here, *BV* abbreviates 'bounded variation' from Fourier analysis [1].

Corollary 22 (ACA_0^ω+QF-AC^{0,1}). *The following are equivalent.*

- *The principle* open.
- *The principle* LIN($\mathbb{R}$) *restricted to continuous $\Psi : \mathbb{R} \to \mathbb{R}^+$ in BV.*
- *The principle* LIN($\mathbb{R}$) *restricted to continuous $\Psi : \mathbb{R} \to \mathbb{R}^+$.*
- *The principle* LIN($\mathbb{R}$) *restricted to usco $\Psi : \mathbb{R} \to \mathbb{R}^+$.*

Proof. We only need to prove the final item from open as the function Ψ from the proof of the theorem is even continuous and in *BV*. To this end, fix closed $C \subset \mathbb{R}$ and usco $\Psi : \mathbb{R} \to \mathbb{R}^+$. Define $\Phi : \mathbb{R} \to \mathbb{R}^+$ as $\Psi(x)$ in case $x \in C$ and $\min(\Psi(x), d(x,C)/2)$ otherwise. Since Ψ is usco, so is Φ and hence Baire 1 by Theorem 12. By [30, §2], Cousin's lemma for Baire 1 Ψ is provable in $\mathsf{RCA}_0^\omega + \mathsf{WKL}_0$. Hence, we have that

$$(\forall n \in \mathbb{N})(\exists x_0, \dots, x_k \in [-n,n])\big[[-n,n] \subset \cup_{i\leq k} B(x_i, \Phi(x_i))\big].$$

The formula in square brackets is decidable using $(\exists^2)$, i.e. QF-AC^{0,1} applies. Remove all elements not in C from the resulting sequence, and we are done. □

Corollary 23 (ACA_0^ω+QF-$\mathsf{AC}^{0,1}$). *The following are equivalent.*

- *The Lindelöf lemma as in* $\mathsf{LIN}(\mathbb{R})$.
- *The principle* $\mathsf{LIN}(\mathbb{R})$ *restricted to continuous* $\Psi : \mathbb{R} \to \mathbb{R}^+$, *plus* HBU.
- *The principles* HBU *and* open.

We only need QF-$\mathsf{AC}^{0,1}$ *to derive the first two items from the third.*

Proof. Assume the third item and fix closed $C \subset \mathbb{R}$ and any $\Psi : \mathbb{R} \to \mathbb{R}^+$. Define $\Phi : \mathbb{R} \to \mathbb{R}^+$ as $\Psi(x)$ if $x \in C$ and $d(x, C)/2$ in case $x \notin C$. Then HBU implies

$$(\forall n \in \mathbb{N})(\exists x_0, \ldots, x_k \in [-n, n])\big[[-n, n] \subset \cup_{i \leq k} B(x_i, \Phi(x_i))\big]$$

The formula in square brackets is now decidable using $(\exists^2)$, i.e. QF-$\mathsf{AC}^{0,1}$ applies. Remove all elements not in C from the resulting sequence, and we are done. □

Acknowledgments. We thank Anil Nerode for his valuable advice, especially the suggestion of studying nsc for the Riemann integral, and discussions related to Baire classes. We thank Dave L. Renfro for his efforts in providing a most encyclopaedic summary of analysis, to be found online. Our research was supported by the Klaus Tschira Boost Fund via the grant Projekt KT 43. We express our gratitude towards the latter institution.

Disclosure of Interests. To the best of my knowledge, I have no competing interests to declare that are relevant to the content of this article.

References

1. Appell, J., Banaś, J., Merentes, N.: Bounded Variation and Around, vol. 17. De Gruyter, Berlin (2014)
2. Avigad, J., Feferman, S.: Gödel's functional ("Dialectica") interpretation. Handbook of proof theory. Stud. Logic Found. Math. **137**, 337–405 (1998)
3. Baire, R.: Sur les fonctions de variables réelles. Annali di Mat. **3**(3), 1–123 (1899)
4. Borsík, J.: Limit of simply continuous function. Real Anal. Exchange **18**(1), 270–275 (1992/93)
5. Borsík, J.: Sums of quasicontinuous functions defined on pseudometrizable spaces. Real Anal. Exchange **22**(1), 328–337 (1996/97)
6. Brown, D.K.: Functional analysis in weak subsystems of second-order arithmetic. Ph.D. thesis, The Pennsylvania State University, ProQuest LLC (1987)
7. Brown, D.K.: Notions of closed subsets of a complete separable metric space in weak subsystems of second-order arithmetic. Contemp. Math. **106**, 39–50 (1990)
8. Darboux, G.: Mémoire sur les fonctions discontinues. Annales scientifiques del'École Normale Supérieure 2e série **4**, 57–112 (1875)
9. Dedekind, R.: Gesammelte mathematische Werke. Bände II. Chelsea Publishing Co., New York (1968)
10. Dzhafarov, D.D., Mummert, C.: Reverse Mathematics: Problems, Reductions, and Proofs. Springer, Cham (2022)
11. Feferman, S.: How a Little Bit goes a Long Way: Predicative Foundations of Analysis (2013). Unpublished notes from 1977-1981 with updated introduction, https://math.stanford.edu/~feferman/papers/pfa(1).pdf

12. Fernández-Duque, D., Shafer, P., Yokoyama, K.: Ekeland's variational principle in weak and strong systems of arithmetic. Selecta Math. **26**, 68 (2020)
13. Good, C., Stares, I.: New proofs of classical insertion theorems. Comment. Math. Univ. Carolin. **41**(1) (2000)
14. Grande, Z.: On the prolongation of restrictions of Baire 1 functions to functions which are quasicontinuous and approximately continuous. Colloq. Math. **114**(2), 237–243 (2009)
15. Hahn, H.: Über halbstetige und unstetige Funktionen. Sitzungsberichte Akad. Wiss. Wien Abt. IIa **126**, 91–110 (1917)
16. Hankel, H.: Untersuchungen über die unendlich oft oscillirenden und unstetigen Functionen. L. F. Fues, Memoir presented at the University of Tübingen on 6 March 1870
17. Hausdorff, F.: Grundzüge der Mengenlehre. Veit and Company, Leipzig (1914)
18. Herrlich, H.: Axiom of Choice. Lecture Notes in Mathematics, vol. 1876. Springer (2006)
19. Hilbert, D., Bernays, P.: Grundlagen der Mathematik. II. Zweite Auflage. Die Grundlehren der mathematischen Wissenschaften, Band 50. Springer (1970)
20. Hunter, J.: Higher-order reverse topology. ProQuest LLC, Ann Arbor, MI (2008). Thesis (Ph.D.)–The University of Wisconsin - Madison
21. Katětov, M.: On real-valued functions in topological spaces. Fund. Math. **38**, 85–91 (1951)
22. Kohlenbach, U.: Higher order reverse mathematics. Reverse mathematics 2001, Lect. Notes Log., vol. 21, pp. 281–295. ASL (2005)
23. Lindelöf, E.: Sur Quelques Points De La Théorie Des Ensembles. Comptes Rendus 697–700 (1903)
24. Maliszewski, A.: On the products of bounded Darboux Baire one functions. J. Appl. Anal. **5**(2), 171–185 (1999)
25. Moschovakis, Y.N.: Descriptive set theory, 2nd ed. Mathematical Surveys and Monographs, vol. 155. American Mathematical Society, Providence, RI (2009)
26. Moore, G.H.: The emergence of open sets, closed sets, and limit points in analysis and topology. Historia Math. **35**(3), 220–241 (2008)
27. Munkres, J.R.: Topology, 2nd edn. Prentice-Hall (2000)
28. Neeman, I.: Necessary use of Σ_1^1 induction in a reversal. J. Symbolic Logic **76**(2), 561–574 (2011)
29. Normann, D., Sanders, S.: Open sets in reverse mathematics and computability theory. J. Logic Comput. **30**(8), 40 (2020)
30. Normann, D., Sanders, S.: The biggest five of reverse mathematics. J. Math. Logic 56 (2023). https://doi.org/10.1142/S0219061324500077
31. Normann, D., Sanders, S.: The Vitali covering theorem in reverse mathematics and computability theory. Ann. Pure Appl. Logic. https://doi.org/10.1016/j.apal.2024.103505
32. pastebee (MathOverflow user): The strength of representing open sets. MathOverflow. https://mathoverflow.net/q/504647. Accessed 04 Dec 2025
33. Richter, C.: Representing cliquish functions as quasiuniform limits of quasicontinuous functions. Real Anal. Exchange **27**(1), 209–221 (2001)
34. Sanders, S.: Big in Reverse Mathematics: measure and category. J. Symb. Logic 44 (2023). https://doi.org/10.1017/jsl.2023.65
35. Sanders, S.: Sometimes tame, sometimes wild: weak continuity notions. Bull. London Math. Soc. 15 (2024). https://doi.org/10.1112/blms.70019
36. Sanders, S.: A note on continuous functions on metric spaces. Bull. Symb. Logic **30**, 398–420 (2024). https://doi.org/10.1017/bsl.2024.30

37. Sanders, S.: Coding is hard. J. Symb. Logic 25 (2025). https://doi.org/10.1017/jsl.2025.10
38. Sanders, S.: Reverse Mathematics: there and back again. Monograph under review, p. 468 (2026)
39. Simpson, S.G.: Subsystems of second order arithmetic, 2nd edn. Perspectives in Logic, CUP (2009)
40. Simpson, S.G.: The Gödel hierarchy and reverse mathematics. Kurt Gödel. Essays for his centennial, pp. 109–127 (2010)
41. Tao, T.: An introduction to measure theory. Graduate Studies in Mathematics, vol. 126. American Mathematical Society, Providence (2011)
42. Tong, H.: Some characterizations of normal and perfectly normal spaces. Duke Math. J. **19**, 289–292 (1952)
43. Weyl, H.: Das Kontinuum. von Veit & Comp, Leipzig (1918)

On the Computational Properties of Ambivalent Sets and Functions

Dag Normann[1(✉)] and Sam Sanders[2]

[1] Department of Mathematics, University of Oslo, Oslo, Norway
dnormann@math.uio.no
[2] Department of Philosophy II, RUB Bochum, Bochum, Germany
sasander@me.com

Abstract. Examples of discontinuous functions already appear in the work of Euler, Abel, Dirichlet, Fourier, and Bolzano. A ground-breaking discovery due to Baire was that many discontinuous functions are *well-behaved* in that they are the pointwise limit of a sequence of continuous functions; the latter form a class nowadays simply called 'Baire 1'. We shall study a class strictly between the semi-continuous and Baire 1 functions, called the *ambivalent* functions. In particular, we investigate the computational properties of the class of ambivalent functions and sets, denoted $\mathbf{\Delta}$, working with Kleene's S1-S9 schemes. Computational equivalences for various standard operations (supremum, Baire 1 representation, ...) on $\mathbf{\Delta}$ are established, including the *structure functional* $\Omega_{\mathbf{\Delta}}$ that decides if a given ambivalent set is non-empty. A selector is shown to be computable relative to $\Omega_{\mathbf{\Delta}}$ and Kleene's quantifier $\exists^2$.

Keywords: Kleene S1–S9 · ambivalent sets and functions · structure functionals · real analysis

1 Introduction

1.1 Aim and Motivation

First of all, *computability theory* is a discipline in the intersection of theoretical computer science and mathematical logic where the fundamental question is:

given two objects X and Y, is X computable from Y in principle?

Turing's famous 'machine' model is the standard reference [36] in case X, Y are real numbers. Our computational study of third-order (and beyond) objects is based on Kleene's S1-S9 in an equivalent formulation [24,26] using fixed point operators going back to Platek [29]. An introduction is in Sect. 1.2.1.

Secondly, the authors have already studied the computational properties of regulated [24] and semi-continuous functions [26]. Now, the Baire 1 functions constitute a well-known super-class of the latter with pleasing conceptual properties. Indeed, the Baire 1 class encompasses a large swath of all discontinuous

V. Brattka et al. (Eds.): CiE 2026, LNCS 16674, pp. 514–527, 2027.
https://doi.org/10.1007/978-3-032-31348-5_34

functions, as already observed by Baire himself [2,3]. Moreover, a function is Baire 1 if it is the pointwise limit of continuous functions, i.e. an elementary description not alien to undergraduates. Motivated by these observations, we study the *ambivalent* functions, a class strictly between the semi-continuous and Baire 1 functions. The definition of the former (see Definition 5) can be viewed as a 'hybrid' of the definition of Baire 1 and semi-continuity.

Thirdly, the aforementioned computational study of regulated and semi-continuous functions revolves around so-called *structure functionals*. The latter take as input a set X from a given class Γ and decide whether it is empty or not, as in the following specification:

$$\Omega_\Gamma(X) = \begin{cases} 1 & \text{if } X \in \Gamma \wedge X \neq \emptyset \\ 0 & \text{if } X \in \Gamma \wedge X = \emptyset \\ \text{undefined} & \text{otherwise} \end{cases} . \qquad (\Omega_\Gamma)$$

Now, central to the study of regulated (resp. semi-continuous) functions is Ω_b (resp. Ω_C), which is Ω_Γ where Γ is the class of sets of reals that have at most one element (resp. that are closed) [24,26]. For instance, computing the continuous function from the *Urysohn lemma* is exactly as hard[1] as Ω_C while computing the monotone functions from the *Jordan decomposition theorem* is exactly as hard as Ω_b; we repeat that we interpret 'computation' via Kleene's S1-S9.

Finally, the class $\boldsymbol{\Delta} = \boldsymbol{F}_\sigma \cap \boldsymbol{G}_\delta$ of ambivalent sets (see Definition 2) generalises the open and closed sets. A function is ambivalent if its level sets are ambivalent (see Definition 5). We shall establish the following computational properties of ambivalent sets and functions.

- We obtain preliminary results for $\boldsymbol{\Delta}$ that are effective versions of theorems from the literature (Sect. 2.1)
- We obtain a cluster theorem for the structure functional $\Omega_{\boldsymbol{\Delta}}$, i.e. a number of computational equivalences involving ambivalent functions (Sect. 2.2).
- We show that $\Omega_{\boldsymbol{\Delta}}$ computes a *selector* for $\boldsymbol{\Delta}$-sets, i.e. the Axiom of Choice for the latter is effective (Sect. 3).

All required preliminaries may be found in Sect. 1.2. We add that the notion of ambivalent set goes back to de la Vallée-Poussin circa 1916 [37], i.e. the ambivalent sets have a certain history.

1.2 Preliminaries and Definitions

We briefly introduce Kleene's *higher-order computability theory* in Sect. 1.2.1. We also introduce some essential axioms (Sect. 1.2.2) and definitions (Sect. 1.2.3). A full introduction may be found in e.g. [22, §2] or [15].

Since Kleene's computability theory borrows heavily from type theory, we use common notations from the latter; for instance, the natural numbers are type 0 objects, denoted n^0 or $n \in \mathbb{N}$. Similarly, elements of Baire space are type 1 objects, denoted $f \in \mathbb{N}^\mathbb{N}$ or f^1. Mappings from Baire space $\mathbb{N}^\mathbb{N}$ to $\mathbb{N}$ are denoted $Y : \mathbb{N}^\mathbb{N} \to \mathbb{N}$ or Y^2. An overview of such notations can be found in e.g. [15,24].

[1] We always assume that Kleene's quantifier $\exists^2$ from Sect. 1.2 is given.

1.2.1 Kleene's Computability Theory Our main results are in computability theory and we make our notion of 'computability' precise as follows.

(I) We adopt ZFC, i.e. Zermelo-Fraenkel set theory with the Axiom of Choice, as the official metatheory for all results, unless explicitly stated otherwise.
(II) We adopt Kleene's notion of *higher-order computation* as given by his nine clauses S1-S9 (see [15, Ch. 5] or [11]) as our official notion of 'computable'. We use the formulation based on fixed point operators from [24,26].

We refer to [15,24] for a thorough overview of higher-order computability theory. We do mention the distinction between 'normal' and 'non-normal' functionals based on the following definition from [15, §5.4]. We only make use of $\exists^n$ for $n = 2, 3$, as defined in Sect. 1.2.2.

Definition 1. *For* $n \geq 2$*, a functional of type* n *is called* normal *if it computes Kleene's quantifier* $\exists^n$ *following S1-S9, and* non-normal *otherwise.*

It is a historical fact that higher-order computability theory, based on Kleene's S1-S9, has focused primarily on the world of *normal* functionals; this observation can be found in [15, §5.4] and can be explained by the (then) relative scarcity of interesting non-normal functionals, like the *fan functional*, originally due to Kreisel (see [27] for historical details) and the Superjump due to Gandy. The authors have identified interesting *non-normal* functionals, i.e. those that compute the objects claimed to exist by:

- covering theorems due to Heine-Borel, Vitali, and Lindelöf [19],
- the Baire category theorem [18,32],
- local-global principles like *Pincherle's theorem* [19],
- weak fragments of the Axiom of (countable) Choice [20],
- the Jordan decomposition theorem and related results [23,24],
- the uncountability of $\mathbb{R}$ [21,22].

This paper continues the study of non-normal functionals that originate from basic properties of Baire 1 functions and sub-classes.

Finally, the first example of a non-computable non-normal functional, Kreisel's (aka Tait's) fan functional (see [27]), is rather tame: it is computable in $\exists^2$. By contrast, the functionals based on the previous list are computable in $\exists^3$ but not computable in any S^2_k, where the latter decides Π^1_k-formulas (see Sect. 1.2.2).

1.2.2 Some Comprehension Functionals In Turing-style computability theory, computational hardness is measured in terms of where the oracle set fits in the well-known comprehension hierarchy. For this reason, we introduce some functionals related to *higher-order comprehension* in this section. We are mostly dealing with *conventional* comprehension here, i.e. only parameters over $\mathbb{N}$ and $\mathbb{N}^\mathbb{N}$ are allowed in formula classes like Π^1_k and Σ^1_k.
First of all, *Kleene's quantifier* $\exists^2 : \mathbb{N}^\mathbb{N} \to \{0, 1\}$ is the functional satisfying:

$$(\forall f^1)\big[(\exists n^0)(f(n)=0) \leftrightarrow \exists^2(f)=0\big]. \tag{$\exists^2$}$$

Clearly, $\exists^2$ is discontinuous at $f = 11\dots$ in the usual epsilon-delta sense. In fact, given a discontinuous function on $\mathbb{N}^\mathbb{N}$ or $\mathbb{R}$, *Grilliot's trick* computes $\exists^2$ from the former, via a rather low-level term of Gödel's T (see [12, §3]). Moreover, $\exists^2$ computes *Feferman's* μ^2 defined for any f^1 as follows:

$$\mu(f) := \begin{cases} n & \text{if } n^0 \text{ is the least natural number such that } f(n)=0 \\ 0 & \text{if there are no } m^0 \text{ such that } f(m)=0 \end{cases}. \tag{μ^2}$$

Hilbert and Bernays formalise considerable swaths of mathematics using only μ^2 (with that name) in [10, Supplement IV] .
Secondly, the *Suslin functional* $\mathsf{S}^2 : \mathbb{N}^\mathbb{N} \to \{0,1\}$ (see [1,12]) is the unique functional satisfying the following:

$$(\exists g^1)(\forall n^0)(f(\overline{g}n)=0) \leftrightarrow \mathsf{S}(f)=0. \tag{S^2}$$

By definition, the Suslin functional S^2 can decide whether a Σ^1_1-formula as in the left-hand side of (S^2) is true or false. We similarly define the functional S^2_k which decides the truth or falsity of Σ^1_k-formulas, given in their Kleene normal form (see e.g. [35, IV.1.4]).
Thirdly, *Kleene's quantifier* $\exists^3 : (\mathbb{N}^\mathbb{N} \to \mathbb{N}) \to \{0,1\}$ is the functional satisfying:

$$(\forall Y^2)\big[(\exists f^1)(Y(f)=0) \leftrightarrow \exists^3(Y)=0\big]. \tag{$\exists^3$}$$

Hilbert and Bernays introduce functionals in e.g. [10, Supplement IV, p. 479] that readily compute $\exists^3$.

1.2.3 Some Definitions We introduce some definitions needed in the below, mostly stemming from mainstream mathematics. We note that subsets of $\mathbb{R}$ are given by their characteristic functions (Definition 2), where the latter are common in measure and probability theory.
First of all, we make use of the usual definition of (open) set, where $B(x,r)$ is the open ball with radius $r>0$ centred at $x \in \mathbb{R}$. Note that 'RM' stands for 'Reverse Mathematics', as the associated representations are used there.

Definition 2 (Set).

- *Subsets A of $\mathbb{R}$ are given by their characteristic function $F_A : \mathbb{R} \to \{0,1\}$, i.e. we write $x \in A$ for $F_A(x)=1$ for all $x \in \mathbb{R}$.*
- *We write '$A \subset B$' if we have $F_A(x) \le F_B(x)$ for all $x \in \mathbb{R}$.*
- *A subset $O \subset \mathbb{R}$ is* open *in case $x \in O$ implies that there is $k \in \mathbb{N}$ such that $B(x, \frac{1}{2^k}) \subset O$.*

- *A subset* $F \subset \mathbb{R}$ *is* closed *if the complement* $\mathbb{R} \setminus F$ *is open.*
- *Any* $D \subset \mathbb{R}$ *is* $\mathbf{F}_\sigma$ *if it equals* $\cup_{n\in\mathbb{N}} F_n$ *for a sequence* $(F_n)_{n\in\mathbb{N}}$ *of closed sets.*
- *Any* $D \subset \mathbb{R}$ *is* $\mathbf{G}_\delta$ *if it equals* $\cap_{n\in\mathbb{N}} O_n$ *for a sequence* $(O_n)_{n\in\mathbb{N}}$ *of open sets.*
- *We define the class* $\mathbf{\Delta}$ *as* $\mathbf{F}_\sigma \cap \mathbf{G}_\delta$. *A set* $X \in \mathbf{\Delta}$ *is called ambivalent* [28,30], *bivalent* [6], or *ambiguous* [7,13].
- A subset $O \subset \mathbb{R}$ is *RM-open* if there are sequences $(a_n)_{n\in\mathbb{N}}, (b_n)_{n\in\mathbb{N}}$ of reals such that $O = \cup_{n\in\mathbb{N}}(a_n, b_n)$.
- A subset $C \subset \mathbb{R}$ is *RM-closed* if the complement $\mathbb{R} \setminus C$ is RM-open.
- An RM-code or representation for $X \in \mathbf{\Delta}$ is a sequence of RM-codes of open and closed sets $(O_n, F_n)_{n\in\mathbb{N}}$ with $X = \cap_{n\in\mathbb{N}} O_n = \cup_{n\in\mathbb{N}} F_n$.

We remark that the class of RM-codes for $\mathbf{\Delta}$-sets is complete Π^1_1, as is the canonical class of RM-codes for Baire 1 functions.

No computational data/additional representation is assumed in our definition of open set. As established in [23,24,33], one readily comes across closed sets in basic real analysis (Fourier series) that come with no additional representation.

Secondly, the following sets are often crucial in proofs in real analysis.

Definition 3. *The sets* C_f *and* D_f *respectively gather the points where* $f : \mathbb{R} \to \mathbb{R}$ *is continuous and discontinuous.*

One problem with C_f, D_f is that '$x \in C_f$' involves quantifiers over $\mathbb{R}$. In general, deciding whether a given $\mathbb{R} \to \mathbb{R}$-function is continuous at a given real, is as hard as $\exists^3$ from Sect. 1.2.2. For these reasons, the sets C_f, D_f do exist in general, but are not computable in e.g. $\exists^2$. For quasi-continuous and semi-continuous functions, these sets *are* definable in $\exists^2$ by [24, §2] or [34, Theorem 2.4].

Thirdly, we shall study the following notions, many of which are well-known and hark back to Baire, Darboux, Hankel, and Volterra [2,3,5,8,9,38].

Definition 4. *For* $f : \mathbb{R} \to \mathbb{R}$, *we have the following definitions:*

- f *is* upper semi-continuous *at* $x_0 \in \mathbb{R}$ *if for any* $k \in \mathbb{N}$, *there is* $N \in \mathbb{N}$ *such that* $(\forall y \in B(x_0, \frac{1}{2^N}))(f(y) < f(x_0) + \frac{1}{2^k})$,
- f *is* lower semi-continuous *at* $x_0 \in \mathbb{R}$ *if for any* $k \in \mathbb{N}$, *there is* $N \in \mathbb{N}$ *such that* $(\forall y \in B(x_0, \frac{1}{2^N}))(f(y) > f(x_0) - \frac{1}{2^k})$,
- f *is* regulated *if for every* x_0 *in the domain, the 'left' and 'right' limit* $f(x_0-) = \lim_{x\to x_0-} f(x)$ *and* $f(x_0+) = \lim_{x\to x_0+} f(x)$ *exist.*
- f *is* Baire 0 *if it is a continuous function.*
- f *is* Baire $n+1$ *if it is the pointwise limit of a sequence of Baire* n *functions.*

As to notations, a common abbreviation is 'usco' and 'lsco' for the first two items. Moreover, if a function has a certain weak continuity property at all reals in $\mathbb{R}$ (or its intended domain), we say that the function has that property. We shall generally study $\mathbb{R} \to \mathbb{R}$-functions but note that 'usco' and 'lsco' are also well-defined for $f : [0,1] \to \overline{\mathbb{R}}$ where $\overline{\mathbb{R}} = \mathbb{R} \cup \{+\infty, -\infty\}$ involves two special symbols that satisfy $(\forall x \in \mathbb{R})(-\infty <_{\mathbb{R}} x <_{\mathbb{R}} +\infty)$ by fiat.

Next, we list the following equivalent definitions for usco and Baire 1 functions, which motivate the definition of ambivalent function.

Definition 5. *For $f : \mathbb{R} \to \mathbb{R}$, we have the following definitions:*

- $f : \mathbb{R} \to \mathbb{R}$ *is* usco *in case* $\{x \in \mathbb{R} : f(x) \geq a\}$ *is closed for any* $a \in \mathbb{R}$,
- $f : \mathbb{R} \to \mathbb{R}$ *is* Baire 1 *in case* $f^{-1}(V)$ *is* $\mathbf{F}_\sigma$ *for any open* $V \subset \mathbb{R}$ [16,31],
- $f : \mathbb{R} \to \mathbb{R}$ is *ambivalent* in case $\{x \in \mathbb{R} : f(x) < a\}$ and $\{x \in \mathbb{R} : f(x) > a\}$ are $\mathbf{\Delta}$ for any $a \in \mathbb{R}$ [28,30].

2 A Cluster Theorem

We study the structure functional for ambivalent sets, called $\Omega_{\mathbf{\Delta}}$, and establish a cluster theorem (Sect. 2.2). The required preliminaries are in Sect. 2.1.

2.1 Preliminaries

We obtain some preliminary results needed for Sect. 2.2, which are often effective versions of known results from the literature.

First of all, the following is a partial effective version of [31, Lemma 11.6].

Lemma 6. *Let $X \in \mathbf{\Delta}$ be given. Then $\exists^2$ computes a Baire 1 representation for $\mathbb{1}_X$ from an RM-code of X.*

Proof. First of all, we recall that the (continuous) distance function $d(x, C) = \inf_{y \in C} d(x, y)$ exists in case C is RM-closed; the usual interval-halving technique goes through using $\exists^2$. Now, suppose $X = \cap_{n\in\mathbb{N}} G_n = \cup_{m\in\mathbb{N}} F_m$ where the G_n (resp. F_m) are open (resp. closed) sets given via RM codes and where $G_{n+1} \subseteq G_n$ and $F_m \subseteq F_{m+1}$. Use $\exists^2$ to define the continuous function $f_n(x) := \frac{d(x,\mathbb{R}\setminus G_n)}{d(x,F_n)+d(x,\mathbb{R}\setminus G_n)}$ from these codes. Now observe that $\lim_{n\to\infty} f_n(x) = \mathbb{1}_X(x)$ for all $x \in \mathbb{R}$, as required. □

Secondly, the following lemma is an effective version of [31, Lemma 11.7].

Lemma 7. *Let $(F_{n,m})_{n,m\in\mathbb{N}}$ be a double sequence of continuous functions, let f_n be $\lim_{m\to\infty} F_{n,m}$, and suppose $(f_n)_{n\in\mathbb{N}}$ uniformly converges to f. Then $\exists^2$ computes a Baire 1 representation for f in terms of the double sequence.*

Proof. That f is Baire 1 follows from [31, Lemma 11.7]. The Baire 1 representation of f is defined as follows: since $(f_n)_{n\in\mathbb{N}}$ converges uniformly to f, there is a sub-sequence $(g_n)_{n\in\mathbb{N}}$ with

$$(\forall n \in \mathbb{N}, x \in \mathbb{R})(|f(x) - g_n(x)| < \tfrac{1}{2^n}).$$

This subsequence can be computed using $\exists^2$ by restricting to $\mathbb{Q}$ and using the usual '$\varepsilon/3$-trick'. Similarly, let $(G_{n,m})_{n,m\in\mathbb{N}}$ be the associated sub-sequence of $(F_{n,m})_{n,m\in\mathbb{N}}$. Define $h_n = g_{n+1} - g_n$ and observe that $f = g_1 + \sum_{n=1}^{\infty} h_n$ and $|h_n(x)| < \frac{1}{2^n}$ for all $x \in \mathbb{R}$ and $n \in \mathbb{N}$. Clearly $H_{n,m} := G_{n+1,m} - G_{n,m}$ converges to h_n for $m \to \infty$. If necessary, modify $H_{n,m}$ to guarantee that, similar to h_n, we have $H_{n,m}(x) \leq \frac{1}{2^n}$ for all $m, n \in \mathbb{N}$ and $x \in \mathbb{R}$. Then $\lambda m.\lambda x.\sum_{n=1}^{\infty} H_{n,m}(x)$ is a sequence of continuous functions by Weierstrass' M-test. Moreover, this sequence converges to $\sum_{n=1}^{\infty} h_n$, as required, since $f = g_1 + \sum_{n=1}^{\infty} h_n$. □

Thirdly, we obtain the following version of the Baire characterisation theorem.

Lemma 8. *Let $(f_n)_{n\in\mathbb{N}}$ be a sequence of continuous functions with pointwise limit f and C a closed set. Then $\exists^2$ computes a point of continuity $x \in C$ for $f_{\restriction C}$ in terms of the sequence and an RM-code of C.*

Proof. Given a Baire 1 representation of f, $\exists^2$ computes $\sup_{x\in[p,q]} f(x)$ as a sequence over $p, q \in \mathbb{Q}$ ([25, §2]). Hence, one readily obtains the oscillation function $\mathsf{osc}_f(x)$, defined as follows:

$$\mathsf{osc}_f([a,b]) := \sup_{x\in[a,b]} f(x) - \inf_{x\in[a,b]} f(x) \text{ and } \mathsf{osc}_f(x) := \lim_{k\to\infty} \mathsf{osc}_f(B(x, \tfrac{1}{2^k})).$$

We now have $D_f = \cup_{n\in\mathbb{N}} D_n$ where $D_n := \{x \in \mathbb{R} : \mathsf{osc}_f(x) \geq \frac{1}{2^n}\}$ is closed and nowhere dense. The Baire category theorem guarantees that $C_f = \cap_{n\in\mathbb{N}} C_n \neq \emptyset$ where C_n is the complement of D_n. Moreover, given the Baire 1 representation of f, $\exists^2$ computes the point claimed to exist by the Baire category theorem ([18, §6]). The same works relative to any RM-closed C, and we are done. □

We believe that many results from [31] can be similarly made effective, yielding interesting results regarding $\Omega_{\mathbf{F}_\sigma}$.

2.2 Some Equivalences

We identify some operations on ambivalent sets and functions that are exactly as hard as $\Omega_{\mathbf{\Delta}}$. To make this precise, we introduce the $\Omega_{\mathbf{\Delta}}$-cluster (Definition 9).

First of all, as in [24,26], we now introduce the following equivalence class.

Definition 9. *We say that* the functional Φ^3 belongs to the $\Omega_{\mathbf{\Delta}}$-cluster *in case*

- *the combination $\Phi + \exists^2$ computes $\Omega_{\mathbf{\Delta}}$, and*
- *the combination $\Omega_{\mathbf{\Delta}} + \exists^2$ computes Φ.*

We say that the functionals Φ and $\Omega_{\mathbf{\Delta}}$ are computationally equivalent given $\exists^2$.

Secondly, to avoid complicated definitions and domain restrictions, we will sometimes abuse notation and make statements of the form

any functional Ψ satisfying a given specification (A) *belongs to the $\Omega_{\mathbf{\Delta}}$-cluster.*

The centred statement means that for **any** functional Ψ_0 satisfying the given specification (A), the combination $\Psi_0+\exists^2$ computes $\Omega_{\mathbf{\Delta}}$, while $\Omega_{\mathbf{\Delta}}+\exists^2$ computes **some** functional Ψ_1 satisfying the specification (A).

Thirdly, we now have the following 'cluster theorem' for $\Omega_{\mathbf{\Delta}}$. We note that the third item generalises the Urysohn lemma while the fifth item is based on the pre-image characterisation of ambivalent functions.

Theorem 10. *The following are part of the $\Omega_{\mathbf{\Delta}}$-cluster.*

- *Any functional Φ_0 such that for $X \in \mathbf{\Delta}$, $\Phi_0(X)$ is an RM-code for $X \subset [0,1]$.*
- *Any functional Φ_1 such that for $X \in \mathbf{\Delta}$, the sequence $\Phi_1(X) = (f_n)_{n\in\mathbb{N}}$ is a Baire 1 representation of $\mathbb{1}_X$* [31, Lemma 11.6].
- (Urysohn) Any functional Φ_{1b} such that for disjoint $X, Y \in \mathbf{\Delta}$, $\Phi_{1b}(X,Y)$ is a Baire 1 function plus representation which is 0 on X and 1 on Y [30].
- Any functional Φ_2 such that for ambivalent and bounded $f : [0,1] \to \mathbb{R}$, the real $\Phi_2(f,p,q)$ equals $\sup_{x\in[p,q]} f(x)$ for any $p, q \in \mathbb{Q}$.
- Any functional Φ_{2b} such that for ambivalent $f : \mathbb{R} \to \mathbb{R}$ and open $V \subset \mathbb{R}$, $\Phi_{2b}(f,V)$ equals an RM-code for the ambivalent set $f^{-1}(V)$.
- Any functional Φ_3 such that for ambivalent and bounded $f : [0,1] \to \mathbb{R}$, the sequence $\Phi_3(f)$ equals a Baire 1 representation for f.
- Any functional Φ_4 such that for $X \in \mathbf{\Delta}$, the real $\Phi_4(X)$ equals $\sup X$.
- Any functional Φ_5 such that for $X \in \mathbf{\Delta}$ and continuous $f : [0,1] \to \mathbb{R}$, the real $\Phi_5(X,f)$ equals $\sup_{x\in X} f(x)$.

Proof. First of all, that $\Omega_{\mathbf{\Delta}} + \exists^2$ computes Φ_0 is proved in Theorem 21.

Secondly, assume a functional Φ_0 is given and fix $X \in \mathbf{\Delta}$. Use the former to write $X = \cup_{n\in\mathbb{N}} F_n = \cap_{n\in\mathbb{N}} G_n$ where F_n is closed, G_n is open, and all are given by RM-codes. By Lemma 6, a Baire representation of $\mathbb{1}_X$ can be computed, and a functional Φ_1 is thus computable from $\Phi_0 + \exists^2$. Now, $\exists^2$ computes the supremum of a Baire 1 function assuming a given Baire 1 representation by [25, Theorem 2.6]. Hence, Φ_1 is seen to compute $\Omega_{\mathbf{\Delta}}$ as $(\exists x \in \mathbb{R})(x \in X) \leftrightarrow (\exists n \in \mathbb{N})[\sup_{y\in[-n,n]} \mathbb{1}_X(y) - 1]$. Clearly, for $X \in \mathbf{\Delta}$, X^c is disjoint from the former and also ambivalent. Hence, Φ_{1b} computes Φ_1, and vice versa, as required.

Thirdly, to compute Φ_2 from $\Omega_{\mathbf{\Delta}}$, let $(q_n)_{n\in\mathbb{N}}$ be an enumeration of $\mathbb{Q}$, fix ambivalent $f : \mathbb{R} \to \mathbb{R}$, and use $\Omega_{\mathbf{\Delta}}$ to obtain sequences of reals $(a_{n,m,k})_{n,m,k\in\mathbb{N}}$ and $(b_{n,m,k})_{n,m,k\in\mathbb{N}}$ such that $\{x \in [0,1] : f(x) < q_k\} = \cap_{n\in\mathbb{N}} G_{n,k}$ with $G_{n,k} = \cup_{m\in\mathbb{N}}(a_{n,m,k}, b_{n,m,k})$. Clearly, we have

$$\begin{aligned} &(\forall x \in [0,1])(f(x) < q_k) \\ &\leftrightarrow (\forall n \in \mathbb{N})([0,1] \subset G_{n,k}) \\ &\leftrightarrow (\forall n \in \mathbb{N})(\exists m_0 \in \mathbb{N})([0,1] \subset \cup_{m\le m_0}(a_{n,m,k}, b_{n,m,k})), \end{aligned} \tag{1}$$

where the last step follows by (countable) Heine-Borel compactness. The final formula (1) is (equivalent to) arithmetical, i.e. we can now find the supremum of f using the usual interval-halving method (using $\exists^2$). To compute $\Omega_{\mathbf{\Delta}}$ from Φ_2, one readily verifies that $\mathbb{1}_X$ is ambivalent for $X \in \mathbf{\Delta}$ (by a simple case distinction). The supremum as in the latter functional then readily yields the former functional. For Φ_{2b}, the latter is clearly computable from Φ_0 while RM-codes for $f^{-1}((q,+\infty))$ readily yield $\sup_{x\in[0,1]} f(x)$ using a formula similar to (1); hence Φ_2 follows, as required.

Fourth, to compute Φ_3 from Φ_0, we proceed as follows. Let $f : [0,1] \to \mathbb{R}$ be ambivalent with upper bound $m - 1 \in \mathbb{N}$ and lower bound 0, and consider for $i < m$ the sets

$$A^m_{2i} := \{x \in [0,1] : i/m < f(x) < (i+1)/m\}, A^m_{2i+1} := \{x \in [0,1] : f(x) = i/m\}.$$

Each A_j^m is ambivalent because f is and these sets are pairwise disjoint. Now define $g_m : \mathbb{R} \to [0,1]$ as follows: $g_m(x) := \sum_{i=0}^{m} \mathbb{1}_{A_i^m}(x)\frac{i}{m}$. By known results ([31, Lemma 11.6-7]), each function g_m is Baire 1 and the uniform limit is a Baire 1 function, namely the original f. We can thus use Lemmas 6 and 7 to obtain a Baire 1 representation of f, as required. To compute Φ_2 from Φ_3, we recall that $\exists^2$ computes the supremum of a Baire 1 function assuming a given Baire 1 representation by [25, Theorem 2.6].

To compute Φ_4 from Φ_0, fix $X \in \mathbf{\Delta}$ and let $\cup_{n\in\mathbb{N}} C_n = X$ where each C_n is an RM-code for a closed set. Then $\exists^2$ computes $\sup C_n$ and we put $\Phi_4(X) := \sup_{n\in\mathbb{N}} \sup C_n$. To compute $\Omega_{\mathbf{\Delta}}$ from Φ_4, consider $X \in \mathbf{\Delta}$ and check if $0 \in X$, if so, then $\Omega_{\mathbf{\Delta}}(X) = 1$. If not, then consider the ambivalent set $\{0\} \cup X$. Then X is non-empty if and only if the supremum of the latter set is non-zero. For Φ_5, the latter computes Φ_4 in case f is the identity function. To compute the former, proceed as for Φ_4: fix $X \in \mathbf{\Delta}$ and let $\cup_{n\in\mathbb{N}} C_n = X$ where each C_n is an RM-code for a closed set. Then $\exists^2$ computes $\sup_{x\in C_n} f(x)$ and we put $\Phi_5(X, f) := \sup_{n\in\mathbb{N}} \sup_{x\in C_n} f(x)$. □

In conclusion, various basic operations on ambivalent functions and sets are computationally equivalent. We believe there to be many more inhabitants of the $\Omega_{\mathbf{\Delta}}$-cluster.

2.3 Gauges and Baire 1 Functions

We introduce the equivalent 'gauge' definition of Baire 1 functions (Definition 11) and connect it to our structure functionals (Theorem 14).

First of all, we consider the following equivalent definition of Baire 1 function pioneered in [14]. We refer to the function δ as an *ε-gauge* of a Baire 1 function.

Definition 11. *A function $f : \mathbb{R} \to \mathbb{R}$ is Baire 1 if for any $\varepsilon > 0$, there is $\delta : \mathbb{R} \to \mathbb{R}^+$ such that for all $x, y \in \mathbb{R}$, if $|x - y| < \min(\delta(x), \delta(y))$ then $|f(x) - f(y)| < \varepsilon$.*

Secondly, obtaining gauges is easy from the computational viewpoint, as follows.

Theorem 12. *Let $(f_n)_{n\in\mathbb{N}}$ be a sequence of continuous functions with pointwise limit f. Then $\exists^2$ computes a gauge representation for f in terms of the sequence.*

Proof. Part of the proof of [14, Theorem 1] is effective, as follows. Let $(f_n)_{n\in\mathbb{N}}$ be a sequence of continuous functions with pointwise limit f. Then $\exists^2$ readily computes a modulus of continuity $\delta_n(x, \varepsilon)$, i.e. we have

$$(\forall \varepsilon > 0, \forall n \in \mathbb{N})(\forall x, y \in \mathbb{R})(|x - y| < \delta_n(x, \varepsilon) \to |f_n(x) - f_n(y)| < \varepsilon).$$

Similarly, let $\eta(x, \varepsilon)$ be a modulus for the convergence of f_n to f, i.e.

$$(\forall \varepsilon > 0, \forall x \in \mathbb{R})(\forall n \geq \eta(x, \varepsilon))(|f_n(x) - f(x)| < \varepsilon).$$

Define $\delta_\varepsilon(x) := \max_{1\leq\eta(x,\varepsilon)} \delta_n(x, \varepsilon)$ and verify it is an ε-gauge for f. □

Thirdly, we connect the gauge definition to structure functionals in Theorem 14 below. To this end, we need the following definition from [17] where the authors pioneered the computational study of Cousin's lemma [4].

Definition 13. *Any* $\Theta : (\mathbb{R} \to \mathbb{R}) \to \mathbb{R}$ *is called a* realiser for Cousin's lemma *if for any* $\Psi : [0,1] \to \mathbb{R}^+$, $\Theta(\Psi) = (x_0, \dots, x_k)$ *is such that* $[0,1] \subset \cup_{i \leq k} B(x_i, \Psi(x_i))$.

Let $\Phi_\Gamma : ((\mathbb{R} \to \mathbb{R}) \times \mathbb{N}) \to (\mathbb{R} \to \mathbb{R})$ be such that $\Phi_\Gamma(f,k)$ is a $\frac{1}{2^k}$-gauge for f in the class[2] Γ. The following theorem shows that computing ε-gauges amounts to computing structure functionals, modulo a realiser for Cousin's lemma.

Theorem 14. *Let* Θ *be any realiser for the Cousin lemma.*

- *The combination* $\Phi_{\mathsf{regulated}} + \Theta + \exists^2$ *computes* Ω_b.
- *The combination* $\Phi_{\mathsf{usco}} + \Theta + \exists^2$ *computes* Ω_C.
- *The combination* $\Phi_{\mathbf{\Delta}} + \Theta + \exists^2$ *computes* $\Omega_{\mathbf{\Delta}}$.

Proof. For the first part, let $X \subset \mathbb{R}$ be any set with at most one element. Then $\mathbb{1}_X$ is regulated and consider the 1/2-gauge $\delta(x) := \Phi_{\mathsf{regulated}}(f,2)(x)$. If $x_0 \in X$, then $x_0 \notin B(y, \delta(y))$ for any $y \in B(x_0, \delta(x_0)) \setminus \{x_0\}$, by the definition of gauge function. Hence, if there is $x_0 \in X$, then $\Theta(\lambda x.\delta(x), \overline{B}(x_0, \delta(x_0)/2))$ must include the former real to provide a covering of $\overline{B}(x_0, \delta(x_0)/2)$. In particular, we have

$$X \neq \emptyset \leftrightarrow (\exists p, q \in [0,1] \cap \mathbb{Q})(\exists i \in \mathbb{N})(\Theta(\lambda x.\delta(x), [p,q])(i) \in X).$$

As a result, the combination $\Theta + \Phi_{\mathsf{regulated}} + \exists^2$ computes Ω_b.

For the second part, let $C \subset [0,1]$ be closed and consider the usco function $\mathbb{1}_C$. Define the 1/2-gauge $\delta(x) := \Phi_{\mathsf{usco}}(\mathbb{1}_C, 2)(x)$ and the set $O := [0,1] \setminus C$. If $x_0 \in O$, then Feferman's μ suffices to find $N_0 \in \mathbb{N}$ such that $(B(x_0, \frac{1}{2^{N_0}}) \cap \mathbb{Q}) \subset O$. However, the ball $B(x_0, \frac{1}{2^{N_0}})$ can still contain points of C. We now show how to decide if there are such points (using the gauge and any Θ-functional). Suppose $x_1 \notin O$ but $x_1 \in B(x_0, \frac{1}{2^{N_0}})$. Then $x_1 \notin B(y, \delta(y))$ for any $y \in [O \cap B(x_1, \delta(x_1))]$, by the definition of gauge function. Hence, the finite sequence $\Theta(\lambda x.\delta(x), \overline{B}(x_1, \delta(x_1)/2))$ must include $x_2 \notin O$ to provide a covering of $\overline{B}(x_1, \delta(x_1)/2)$. In this way, we obtain the equivalence between $(\exists x \in B(x_0, \frac{1}{2^{N_0}}))(x \notin O)$ and

$$(\exists p, q \in \mathbb{Q} \cap [0,1])(\exists i \in \mathbb{N})\big[[p,q] \subset B(x_0, \tfrac{1}{2^{N_0}}) \wedge \Theta(\lambda x.\delta(x), [p,q])(i) \notin O\big].$$

where the centred formula is decidable using $\exists^2$. Hence, we can 'shrink' N_0 (if necessary) to N_1 to guarantee $B(x_0, \frac{1}{2^{N_1}}) \subset O$. Restricting to the rationals in O, we obtain an RM-code for O, i.e. Ω_C is obtained.

For the third part, fix $X \in \mathbf{\Delta}$ and consider $\mathbb{1}_X$. The latter is Baire 1 by [31, Lemma 11.6] and ambivalent 'by definition'. Now consider the 1/2-gauge $\delta(x) := \Phi_{\mathbf{\Delta}}(\mathbb{1}_X, 2)(x)$. As in the previous, $(\exists x \in [0,1])(x \in X)$ is equivalent to

$$(\exists p, q \in \mathbb{Q} \cap [0,1])(\exists i \in \mathbb{N})\big[\Theta(\lambda x.\delta(x), [p,q])(i) \in X\big],$$

which immediately yields $\Omega_{\mathbf{\Delta}}$. □

[2] We use '$\mathsf{regulated}$' and 'usco' to denote the classes of regulated and usco functions.

We observe that given a realiser for Cousin's lemma, Ω_Γ and Φ_Γ line up nicely.

Finally, as explored in [39], gauges can be taken to be usco, but the construction is far from elementary, say compared to Theorem 12.

3 Recursion and Selectors

In this section, we show that the structure functional $\Omega_{\mathbf{\Delta}}$ can compute RM-codes for $\mathbf{\Delta}$-sets. As a corollary, the former also computes a *selector* for $\mathbf{\Delta}$-sets, i.e. a functional that outputs an element from a non-empty $\mathbf{\Delta}$-set.

We start with some observations and lemmas. We use 'P' as a variable ranging over the sets in $\mathbf{\Delta}$ without always explicitly mentioning this.

Observation 15. *Let Φ be a functional that outputs an RM-code for P given a set $P \in \mathbf{\Delta}$ as input. Then $\Omega_{\mathbf{\Delta}}$ is computable in Φ and $\exists^2$.*

The following lemma goes back to Baire, and is proved by an application of the Baire category theorem for closed subsets of $\mathbb{R}$.

Lemma 16. *Let X be closed and nonempty and let $P \in \mathbf{\Delta}$. Then there is an open set O such that $O \cap X \neq \emptyset$ and either $O \cap X \subset P$ or $O \cap X \subset P^c$.*

We now let $(B_i)_{i\in\mathbb{N}}$ be a computable enumeration of all open intervals with rational endpoints, including the endpoints $+\infty$ and $-\infty$.

Definition 17. *By recursion on the ordinal α, we define $A_\alpha \subseteq \mathbb{N}$ and the closed set $X_\alpha = \mathbb{R} \setminus \bigcup_{i\in A_\alpha} B_i$ as follows.*

- *We define $A_0 = \emptyset$ and for limit ordinals γ we let $A_\gamma = \bigcup_{\alpha<\gamma} A_\alpha$.*
- *We put $i \in A_{\alpha+1}$ if $B_i \cap X_\alpha = \emptyset$ or if $B_i \cap X_\alpha \neq \emptyset$ and the latter set is contained either in P or in P^c.*

Due to Lemma 16, the transfinite recursion in Definition 17 will terminate with empty X_α and, by construction, with $A_\alpha = \mathbb{N}$. Clearly each recursion step is computable in P, P^c and $\Omega_{\mathbf{\Delta}}$. Furthermore, the recursion will induce a pre-well-ordering (see Definition 18) of the final set $A_\alpha = \mathbb{N}$, and from this pre-well-ordering we can obtain the codes for P and P^c, as described in detail in the proof of Theorem 21.

It remains to introduce the aforementioned pre-well-ordering and show it is computable in $\Omega_{\mathbf{\Delta}}$ and $\exists^2$. For this, we introduce a few concepts.

Definition 18 (Pre-well-orderings)

1. *A* pre-ordering *of a set $A \subseteq \mathbb{N}$ is a relation $\preceq$ on A that is transitive and reflexive.*
2. *We write $a \simeq b$ if $a \preceq b$ and $b \preceq a$, and we write $a \prec b$ if $a \preceq b$ while it is not the case that $a \simeq b$. If $a \prec b$ we will say that a is below b.*
3. *A* descending chain *is a sequence $\{a_n\}_{n\in\mathbb{N}}$ such that $a_{n+1} \prec a_n$ for all n.*

4. A pre-ordering without descending chains is a pre-well-ordering.

All pre-orderings will have a maximal pre-well-ordered initial segment, consisting of those points below which there are no descending sequences.

Now, we again consider the inductive definition as in Definition 17.

Definition 19. *A pre-ordering* $(\mathbb{N}, \preceq)$ *is* agreeable *if the following is satisfied.*

For each $a \in \mathbb{N}$*, let* $X_a = \mathbb{R} \setminus \bigcup_{b \prec a} B_b$*. Then either* $B_a \cap X_a = \emptyset$ *or* $B_a \cap X_a \neq \emptyset$ *and* $B_a \cap X_a$ *is either contained in* P *or in* P^c*.*

This reflects the recursion step in the recursion described above.

By transfinite induction, we also see that given an agreeable pre-ordering, the pre-well-ordered initial segment will coincide with an initial segment of the one obtained from the recursion. Finally, we see that $\exists^2$ and $\Omega_{\mathbf{\Delta}}$ can decide if a pair $(A, \preceq)$ is an agreeable pre-ordering in our sense.

Lemma 20. *If* $(\mathbb{N}, \preceq)$ *is an agreeable pre-ordering, it is actually the one and only agreeable pre-well-ordering on* $\mathbb{N}$*.*

Proof. Assume not, and let A^w be the domain of the pre-well-ordered initial segment. Let $X^w = \mathbb{R} \setminus \bigcup_{a \in A^w} B_a$. We must have that $X^w \neq \emptyset$ since otherwise the induction would have stopped at each point in the non-pre-well-ordered part. Choose n such that $B_n \cap X^w \neq \emptyset$ and is contained in either P or P^c. Now, let $c \in \mathbb{N} \setminus A^w$. Then either $B_n \cap X_c = \emptyset$ or $n \simeq c$ (since they both will satisfy the requirements of the recursion step). Since $c \in \mathbb{N} \setminus A^w$ was arbitrary, it follows that $B_n \cap X_c = \emptyset$ for all $c \in \mathbb{N} \setminus A^w$, contradicting the choice of n. □

The following result is the non-trivial direction in Theorem 10. Here Ω_1 is the functional selecting the one element in a singleton set. The functional Ω_1 is computable in all the structure functionals Ω_Γ we have considered so far.

Theorem 21. *There is a functional* Φ *which is computable in* $\Omega_{\mathbf{\Delta}} + \exists^2$ *and such that* $\Phi(P)$ *is an RM-code for any* $P \in \mathbf{\Delta}$*.*

Proof. By the previous, the class of agreeable pre-orderings is computable in the combination $\Omega_{\mathbf{\Delta}} + \exists^2$. By Lemma 20 there is exactly one agreeable pre-ordering, and that is the one obtained by the transfinite recursion in Definition 17. If we apply Ω_1 to this class, we obtain this pre-(well)-ordering, and from that one we can read off the codes for P and P^c.

In more detail, both P and P^c will be disjoint unions of segments of the form $A_n = B_n \setminus \bigcup_{m \prec n} B_m$, and we can use $\Omega_{\mathbf{\Delta}}$ to tell us which new segments will go into P and which will go into P^c. Since each B_n is $\mathbf{F}_\sigma$ with a code computable uniformly in n we also have that each segment A_n will be an $\mathbf{F}_\sigma$-set, and we obtain a code for A_n arithmetically from n and $\{m : m \preceq n\}$. Then it is trivial to find codes for P and P^c. □

Corollary 22. *We have the following.*

a. *There is a selector operator for* $\mathbf{\Delta}$*-sets computable in* $\Omega_{\mathbf{\Delta}}$.
b. *The functional* $\Omega_{\text{countable}}$*, and thus* $\Omega_{\mathbf{F}_\sigma}$*, is not computable in* $\Omega_{\mathbf{\Delta}} + \exists^2$.

Proof. Given $P \in \mathbf{\Delta}$, $\Omega_{\mathbf{\Delta}}$ provides an $\mathbf{F}_\sigma$-code for an increasing sequence $(P_k)_{k\in\mathbb{N}}$ of closed sets with P as the union. When P is non-empty, we extract the least element in the first non-empty $P_k \cap [-k, k]$. This proves item a. from the corollary. Item b also follows because whenever a functional Ω_Γ provides a selector for Γ, then Ω_Γ cannot compute $\Omega_{\text{countable}}$. □

Acknowledgments. The research of the second author was supported by the Klaus Tschira Boost Fund via the grant Projekt KT 43. We express our gratitude towards the latter institution.

Disclosure of Interests. To the best of our knowledge, we have no competing interests to declare that are relevant to the content of this article.

References

1. Avigad, J., Feferman, S.: Gödel's functional ("Dialectica") interpretation, Handbook of proof theory. Stud. Logic Found. Math. **137**, 337–405 (1998)
2. Baire, R.: Sur les fonctions de variables réelles. Annali di Matematica Pura ed Applicata (1898-1922) **3**(1), 1–123 (1899). https://doi.org/10.1007/BF02419243
3. Baire, R.: Leçons sur les fonctions discontinues, Les Grands Classiques GauthierVillars, Éditions Jacques Gabay, Sceaux, (French) (1995). Reprint of the 1905 original
4. Cousin, P.: Sur les fonctions de n variables complexes. Acta Math. **19**, 1–61 (1895)
5. Darboux, G.: Mémoire sur les fonctions discontinues. In: Annales scientifiques de l'École Normale Supérieure 2e série, vol. 4, pp. 57–112 (1875)
6. Dugundji, J.: Topology. Allyn and Bacon Inc., Boston (1966)
7. Elekes, M.: On a converse to Banach's fixed point theorem. Proc. Amer. Math. Soc. **137**(9), 3139–3146 (2009)
8. Hankel, H.: Untersuchungen über die unendlich oft oscillirenden und unstetigen Functionen., L. F. Fues, Memoir presented at the University of Tübingen on 6 (1870)
9. Hankel, H.: Untersuchungen über die unendlich oft oscillirenden und unstetigen Functionen., Math. Ann. **20** (1882)
10. Hilbert, D., Bernays, P.: Grundlagen der Mathematik. II, Zweite Auflage. Die Grundlehren der mathematischen Wissenschaften, Band 50. Springer, Heidelberg (1970)
11. Kleene, S.C.: Recursive functionals and quantifiers of finite types. I. Trans. Amer. Math. Soc. **91**, 1–52 (1959)
12. Kohlenbach, U.: Higher order reverse mathematics, Reverse mathematics. Lect. Notes Log., **21**., 281–295, ASL **2005** (2001)
13. Kuratowski, K.: Topology, vol. I. Academic Press (1966)
14. Lee, P.-Y., Tang, W.-K., Zhao, D.: An equivalent definition of functions of the first Baire class. Proc. Amer. Math. Soc. **129**(8), 2273–2275 (2001)
15. Longley, J., Normann, D.: Higher-order Computability, Theory and Applications of Computability. Springer, Heidelberg (2015)

16. Myerson, G.I.: First-class functions. Amer. Math. Monthly **98**(3), 237–240 (1991)
17. Normann, D., Sanders, S.: Nonstandard analysis, computability theory, and their connections. J. Symb. Logic **84**(4), 1422–1465 (2019)
18. Normann, D., Sanders, S.: Open sets in reverse mathematics and computability theory. J. Logic Comput. **30**(8), 40 (2020)
19. Normann, D., Sanders, S.: Pincherle's theorem in reverse mathematics and computability theory. Ann. Pure Appl. Logic 171(5), 102788, 41 (2020)
20. Normann, D., Sanders, S.: The axiom of choice in computability theory and reverse mathematics. J. Log. Comput. **31**(1), 297–325 (2021)
21. Normann, D., Sanders, S.: On robust theorems due to Bolzano, Jordan, Weierstrass, and cantor in reverse mathematics. J. Symb. Logic **51** (2022). https://doi.org/10.1017/jsl.2022.71
22. Normann, D., Sanders, S.: On the uncountability of R. J. Symb. Logic 43 (2022). https://doi.org/10.1017/jsl.2022.27
23. Normann, D., Sanders, S.: Betwixt turing and kleene. In: LNCS, vol. 13137, proceedings of LFCS22, p. 18 (2022)
24. Normann, D., Sanders, S.: On the computational properties of basic mathematical notions. J. Logic Comput. 44 (2022). https://doi.org/10.1093/logcom/exac075
25. Normann, D., Sanders, S.: The biggest five of reverse mathematics. J. Math. Logic 56 (2023). https://doi.org/10.1142/S0219061324500077
26. Normann, D., Sanders, S.: On the computational properties of open sets. J. Log. Comput. **26** (2024). https://doi.org/10.1093/logcom/exae048
27. Normann, D., Tait, W.: On the computability of the fan functional. In: Jäger, G., Sieg, W. (eds.) Feferman on Foundations. OCL, vol. 13, pp. 57–69. Springer, Cham (2017). https://doi.org/10.1007/978-3-319-63334-3_3
28. O'Malley, R.J.: Approximately differentiable functions: the r topology. Pac. J. Math. **72**(1), 207–222 (1977)
29. Platek, R.A.: Foundations Of Recursion Theory, ProQuest LLC. Thesis (Ph.D.)–Stanford University (1966)
30. Prus-Wiśniowski, F.: Separation by ambivalent sets. Real Anal. Exchange **30**(1), 289–293 (2004/05)
31. van Rooij, A.C.M., Schikhof, W.H.: A Second Course on Real Functions. Cambridge University Press, Cambridge (1982)
32. Sanders, S.: On the computational properties of the Baire category theorem. Computability (2022). arxiv: arxiv.org/abs/2210.05251
33. Sanders, S.: Big in reverse mathematics: the uncountability of the real numbers. J. Symb. Logic 26 (2023). https://doi.org/10.1017/jsl.2023.42
34. Sanders, S.: Big in reverse mathematics: measure and category. J. Symb. Logic 44 (2023). https://doi.org/10.1017/jsl.2023.65
35. Stephen, G.: Simpson, Subsystems of second order arithmetic. Perspectives in Logic, 2nd edn. CUP (2009)
36. Turing, A.: On computable numbers, with an application to the Entscheidungsproblem. Proc. Lond. Math. Soc. **42**, 230–265 (1936)
37. de la Vallée-Poussin, C.J.: Intégrales de Lebesgue. Gauthier-Villars (1916)
38. Volterra, V.: Alcune osservazioni sulle funzioni punteggiate discontinue. In: Giornale di matematiche XIX, pp. 76–86 (1881)
39. Atok, Z., Tang, W.-K., Zhao, D.: Gauges of Baire class one functions. J. Math. Anal. Appl. **343**(2), 866–870 (2008)

A Polynomial-Time Algorithm for the Automatic Baire Property

Ludwig Staiger(✉)

Martin-Luther-Universität Halle-Wittenberg, Institut für Informatik,
von-Seckendorff-Platz 1, 06099 Halle (Saale), Germany
staiger@informatik.uni-halle.de
https://www.informatik.uni-halle.de/arbeitsgruppen/theorie/mitarbeiter/staiger/

Abstract. A subset of a topological space possesses the Baire property if it can be covered by an open set up to a meagre set. For the Cantor space of infinite words Finkel introduced the automatic Baire category where both sets, the open and the meagre, can be chosen to be definable by finite automata. Here we show that, given a Muller automaton defining the original set, automata for the resulting open and meagre sets can be constructed in polynomial time. The construction is based on the strongly connected subsets and the condensation graph of the automaton graph.

Since the constructed sets are of simple topological structure, it is possible to construct not only Muller automata defining them but also the simpler Büchi automata. To this end we give, for a restricted class of Muller automata, a conversion to equivalent Büchi automata with an at most quadratic increase of the size.

Keywords: ω-automata · Cantor space · Baire property · strongly connected components

1 Introduction

A subset of the Cantor space possesses the Baire property if it differs from an open set only by a small set, that is, by a meagre set. In the papers [5,6] the automatic Baire property as a special case of the ordinary Baire property was introduced, and it was shown that every automaton definable language of infinite words possesses not only the Baire property but also the automatic Baire property. To this end Finkel constructed from a given Muller automaton two Muller automata accepting the involved open and meagre set, respectively. Due to the simpler topological structure of the open and meagre sets in Finkel's definition, we can characterise them also by Büchi automata.

The aim of this paper is to present a polynomial time algorithm for the automatic Baire property. For any given Muller automaton $\mathcal{A}$ accepting a regular ω-language we construct in polynomial time two Muller and two Büchi automata for the required ω-languages.

V. Brattka et al. (Eds.): CiE 2026, LNCS 16674, pp. 528–536, 2027.
https://doi.org/10.1007/978-3-032-31348-5_35

Although the conversion, if possible, from Muller to Büchi automata may result in an exponential blow-up (cf. [2]), we show that in our case it can be performed in polynomial time. To this end we prove that the resulting automata can be kept small, thus allowing for an application of standard polynomial-time graph algorithms.

1.1 Notation

In this section we introduce the notation used throughout the paper. By $\mathbb{N} = \{0, 1, 2, \ldots\}$ we denote the set of natural numbers. Its elements will be usually denoted by letters $i, \ldots, n$. Let X be a finite alphabet of cardinality $|X| = r \geq 2$. Then X^* is the set of finite words on X, including the *empty word* e, and X^ω is the set of infinite strings (ω-words) over X. Subsets of X^* will be referred to as *languages* and subsets of X^ω as *ω-languages.*

For $w \in X^*$ and $\eta \in X^* \cup X^\omega$ let $w \cdot \eta$ be their *concatenation.* This concatenation product extends in an obvious way to subsets $W \subseteq X^*$ and $B \subseteq X^* \cup X^\omega$. For a language W let $W^* := \bigcup_{i\in\mathbb{N}} W^i$, and $W^\omega := \{w_1 \cdots w_i \cdots : w_i \in W \setminus \{e\}\}$ be the set of infinite strings formed by concatenating non-empty words in W. Furthermore, $|w|$ is the *length* of the word $w \in X^*$ and $\mathbf{pref}(B)$ is the set of all finite prefixes of strings in $B \subseteq X^* \cup X^\omega$. We shall abbreviate $w \in \mathbf{pref}(\{\eta\})$ ($\eta \in X^* \cup X^\omega$) by $w \sqsubseteq \eta$.

We assume the reader to be familiar with the basic facts of the theory of regular languages and finite automata. We postpone the definition of regularity for ω-languages to Sect. 2.1. For more details on ω-languages and regular ω-languages see the book [10] or the survey papers [12,16].

As usual, for a set M, we denote by $2^M := \{M' : M' \subseteq M\}$ its *powerset.*

1.2 The Cantor space

We consider X^ω as a topological space (Cantor space). The *closure* (smallest closed set containing F) of a subset $F \subseteq X^\omega, \mathcal{C}(F)$, is described as $\mathcal{C}(F) := \{\xi : \mathbf{pref}(\{\xi\}) \subseteq \mathbf{pref}(F)\}$. The *open sets* in Cantor space are the ω-languages of the form $W \cdot X^\omega$. Countable unions of closed sets are referred to as *Σ_2-sets,* their complements as *Π_2-sets.*

Next we recall some topological notions. As usual, an ω-language $F \subseteq X^\omega$ is *dense in* X^ω if $\mathcal{C}(F) = X^\omega$. This is equivalent to $\mathbf{pref}(F) = X^*$. An ω-language $F \subseteq X^\omega$ is *nowhere dense in* X^ω if its closure $\mathcal{C}(F)$ does not contain a non-empty open subset. This property is equivalent to the fact that for all $w \in \mathbf{pref}(F)$ there is a $v \in X^*$ such that $w \cdot v \notin \mathbf{pref}(F)$. Moreover, a subset $F \subseteq X^\omega$ is *meagre* or of *first Baire category* if it is a countable union of nowhere dense sets.

2 Automata and Regularity

2.1 Regular ω-Languages and Finite Automata

As usual we call a language $W \subseteq X^*$ *regular* if there is a finite (deterministic) automaton $A = (X; S; s_0; \delta)$, where S is the finite set of states, $s_0 \in S$ is the initial state and $\delta : S \times X \to S$ is the transition function[1], such that $W = \{w : \delta(s_0; w) \in S'\}$ for some fixed set $S' \subseteq S$.

An ω-language $F \subseteq X^\omega$ is called *regular* provided there are a finite (deterministic) automaton $A = (X; S; s_0; \delta)$ and a table $\mathcal{T} \subseteq \{Z : Z \subseteq S\}$ such that for $\xi \in X^\omega$ it holds $\xi \in F$ if and only if $\mathrm{Inf}(A;\xi) \in \mathcal{T}$ where $\mathrm{Inf}(A;\xi) := \bigcap_{w \sqsubset \xi}\{\delta(s_0, v) : w \sqsubset v \sqsubset \xi\}$ is the set of all states $s \in S$ through which the automaton A runs infinitely often when reading the input ξ. Observe that $Z = \mathrm{Inf}(A;\xi)$ holds for a subset $Z \subseteq S$ and some $\xi \in X^\omega$ if and only if

1. there is a word $u \in X^*$ such that $\delta(s_0; u) \in Z$, and
2. for all $s, s' \in Z$ there are non-empty words $w, v \in X^*$ such that $\delta(s, w) = s'$ and $\delta(s', v) = s$.

Such sets were referred to as *essential sets* [17] or *loops* [11], [12, Section 5.1], and $\mathrm{LOOP}_{\mathcal{A}} = \{\mathrm{Inf}(\mathcal{A};\xi) : \xi \in X^\omega\}$ is the set of all loops of an automaton $\mathcal{A}$. The ω-language $L(\mathcal{A}, \mathcal{T}) = \{\xi : \mathrm{Inf}(\mathcal{A};\xi) \in \mathcal{T}\}$ is the union of all sets $\{\xi : \mathrm{Inf}(A;\xi) = Z\}$ where $Z \in \mathcal{T}$. The pair $(\mathcal{A}, \mathcal{T})$ is usually called a Muller automaton.

In particular, $\{\xi : \mathrm{Inf}(A;\xi) = Z\}$ and $\{\xi : \mathrm{Inf}(A;\xi) = Z'\}$ are disjoint for $Z \neq Z'$. Thus the following holds.

Lemma 1. *Let $\mathcal{A} = (X; S; s_0; \delta)$ be a deterministic automaton and $\mathcal{T}, \mathcal{T}' \subseteq 2^S$ be tables, and let* **op** *be a Boolean set operation. Then $L(\mathcal{A}, \mathcal{T})$ **op** $L(\mathcal{A}, \mathcal{T}') = L(\mathcal{A}, \mathcal{T}\,\mathbf{op}\,\mathcal{T}')$. Moreover, for $\mathcal{T}, \mathcal{T}' \subseteq 2^S$ we have $L(\mathcal{A}, \mathcal{T}) \subseteq L(\mathcal{A}, \mathcal{T}')$ if and only if $\mathcal{T} \cap \mathrm{LOOP}_{\mathcal{A}} \subseteq \mathcal{T}' \cap \mathrm{LOOP}_{\mathcal{A}}$.*

We are going to represent $F = L(\mathcal{A}, \mathcal{T})$ by languages derived from the automaton $\mathcal{A}$. As in [13,14] refer to a word $v \in X^*, v \neq e$, as $(s; Z)$*-loop completing* if and only if

1. $\delta(s, v) = s$ and $\{\delta(s, v') : v' \sqsubseteq v\} = Z$, and
2. $\{\delta(s, v') : v' \sqsubseteq v''\} \neq Z$ for all proper prefixes $v'' \sqsubset v$ with $\delta(s, v'') = s$.

Denote by $V_{(s;Z)}$ the set of all $(s; Z)$-loop completing words, and by $U_s := \{w : \delta(s_0, w) = s\}$ the set of all words leading from the initial state s_0 to the state $s \in S$. Both languages are regular and constructible from the finite automaton $\mathcal{A} = (X; S; s_0; \delta)$. Moreover, $V_{(s;Z)}$ is prefix-free and $\mathbf{pref}(V^\omega_{(s,Z)}) = \{w : \forall w'(w' \sqsubseteq w \to \delta(s, w') \in Z)\}$.

We obtain the following (cf. with [13, Lemma 3]).

Lemma 2. *Let $\mathcal{A} = (X; S; s_0; \delta)$ be a finite automaton, and let $\mathcal{T}$ be a table. Then*

$$L(\mathcal{A}, \mathcal{T}) = \bigcup\nolimits_{Z \in \mathcal{T}} \bigcup\nolimits_{s \in Z} U_s \cdot V^\omega_{(s;Z)}. \tag{1}$$

Thus every regular ω-language has the form $\bigcup_{j=1}^{\ell} W_j \cdot V_j^\omega$ where W_j, V_j are regular languages (see [3,10,12] or [16]). The converse is also true, that is, if $W \subseteq X^*$ and $F, E \subseteq X^\omega$ are regular then also $W^\omega, W \cdot E$ and $E \cup F$ are regular

[1] We use the same symbol δ to denote the usual extension of the function δ to $S \times X^*$.

ω-languages. Note, however, that the representation of Eq. (1) is finer, since it splits a regular ω-language $F = \bigcup_{j=1}^{\ell} W_j \cdot V_j^\omega$ into parts $U_s \cdot V_{(s;Z)}^\omega$, $i \in \{1, \ldots, n\}$, where, additionally, the languages $V_{(s;Z)}$ are prefix-free.

2.2 Loops, Strongly Connected Components and Density

We consider the loop structure of an automaton. For $Z_1, Z_2 \subseteq S$ we write $Z_1 \mapsto Z_2$ if $Z_1 \neq Z_2$ and there exists an $s \in Z_1$ and a $w \in X^*$ such that $\delta(s, w) \in Z_2$.

The maximal w.r.t. " $\subseteq$ " loops are the *strongly connected components* $\mathrm{SCC}_{\mathcal{A}} := \{Z : Z \in \mathrm{LOOP}_{\mathcal{A}} \wedge \forall Z'(Z' \in \mathrm{LOOP}_{\mathcal{A}} \to Z' \subseteq Z \vee Z' \cap Z = \emptyset)\}$ of the automaton (multi-)graph of $\mathcal{A}$. They are the vertices of the condensation graph of $\mathcal{A}$. For vertices $Z, Z' \in \mathrm{SCC}_{\mathcal{A}}$ the relation $\mapsto$ is anti-symmetric and transitive, thus a partial order. Its maximal w.r.t. "$\mapsto$" elements are the *terminal strongly connected components* $\mathrm{SCC}_{\mathcal{A}}^t := \{Z : \forall Z'(Z' \in \mathrm{SCC}_{\mathcal{A}} \wedge Z \neq Z' \to Z \not\mapsto Z')\}$. The following properties are immediate.

Property 1. Let $\mathcal{A} = (X; S; s_0; \delta)$ be a finite automaton and $Z \in \mathrm{SCC}_{\mathcal{A}}$.

1. If $z, \delta(z, w) \in Z$ then $\delta(z, w') \in Z$ for all $w' \sqsubseteq w$.
2. If $z \in Z$ and $Z \in \mathrm{SCC}_{\mathcal{A}}^t$ then $\delta(z, w) \in Z$ for all $w \in X^*$.

We have the following relation to the density of the ω-languages $V_{s,Z}^\omega$.

Theorem 1. *Let $\mathcal{A} = (X; S; s_0; \delta)$ be an automaton and $Z \in \mathrm{LOOP}_{\mathcal{A}}$.*

1. *If $Z \notin \mathrm{SCC}_{\mathcal{A}}^t$ then $V_{s,Z}^\omega$ is nowhere dense.*
2. *If $Z \in \mathrm{SCC}_{\mathcal{A}}^t$ then $V_{s,Z}^\omega$ is dense in X^ω, that is, $\mathcal{C}(V_{s,Z}^\omega) = X^\omega$.*

Proof. First, observe that $\mathbf{pref}(V_{s,Z}^\omega) = \{w : \delta(s, w) \in Z\}$.

1. Let $w \in \mathbf{pref}(V_{s,Z}^\omega)$ and let $\delta(s', u) \notin Z$ for some $s' \in Z$ and $u \in X^*$. Consider $v \in X^*$ such that $\delta(s, wv) = s'$. Since $\delta(s, wvu) \notin Z$ we have $wvu \notin \mathbf{pref}(V_{s,Z}^\omega)$.

2. If $Z \in \mathrm{SCC}_{\mathcal{A}}^t$ then $\delta(s, w) \in Z$ for every $w \in X^*$. Thus $\mathbf{pref}(V_{s,Z}^\omega) = X^*$, that is, $\mathcal{C}(V_{s,Z}^\omega) = X^\omega$.

As a consequence of Theorem 1 we have the following two lemmata.

Lemma 3. *If $T \cap \mathrm{SCC}_{\mathcal{A}}^t = \emptyset$ then $L(\mathcal{A}, T)$ is of first Baire category.*

Proof. If $F \subseteq X^\omega$ is nowhere dense then also $w \cdot F$ is nowhere dense. Now the assertion follows with Lemma 2 and Theorem 1.

Lemma 4. *If $Z \in \mathrm{SCC}_{\mathcal{A}}^t$ then $L(\mathcal{A}, 2^Z) = \{\xi : \mathrm{Inf}(\mathcal{A}; \xi) \cap Z \neq \emptyset\} = \{w : w \in X^* \wedge \delta(s_0, w) \in Z\} \cdot X^\omega$ is open in X^ω.*

Proof. In view of Property 1.2, $z \in Z$ and $Z \in \mathrm{SCC}_{\mathcal{A}}^t$ imply $\delta(z, w) \in Z$. Thus $Z \in \mathrm{SCC}_{\mathcal{A}}^t$ and $\mathrm{Inf}(\mathcal{A}; \xi) \cap Z \neq \emptyset$ yield $\mathrm{Inf}(\mathcal{A}; \xi) \subseteq Z$. Consequently, $\mathrm{Inf}(\mathcal{A}, \xi) \subseteq Z$ when $\{\delta(s_0, w) : w \sqsubset \xi\} \cap Z \neq \emptyset$.

In particular, for a subset $\mathcal{U} \subseteq \mathrm{SCC}^t_{\mathcal{A}}$ and its downward closure $\mathcal{T}_{\mathcal{U}} := \{Z' : \exists Z(Z \in \mathcal{U} \wedge Z' \subseteq Z)\}$, we can simplify an automaton $\mathcal{A} = (X, S, s_0, \delta)$ by merging the elements of the terminal strongly connected components $Z \in \mathrm{SCC}^t_{\mathcal{A}}$ into a single state.

Corollary 1. *Let* $\mathcal{A} := (X, S, s_0, \delta)$ *and* $\mathcal{U} \subseteq \mathrm{SCC}^t_{\mathcal{A}}$. *Then* $L(\mathcal{A}, \mathcal{T}_{\mathcal{U}}) = L(\mathcal{A}', \mathcal{T}')$ *where* $\mathcal{A}' := (X, S', s'_0, \delta')$ *with*

$$S' := (S \setminus \bigcup\{Z : Z \in \mathrm{SCC}^t_{\mathcal{A}}\}) \cup \{s_0\} \cup \{s_Z : Z \in \mathrm{SCC}^t_{\mathcal{A}}\},$$

$$s'_0 := s_0,$$

$$\delta'(s, x) := \begin{cases} \delta(s, x), & \text{if } \delta(s, x) \in S \setminus \bigcup\{Z : Z \in \mathrm{SCC}^t_{\mathcal{A}}\}, \text{and} \\ s_Z & , \text{if } Z \in \mathrm{SCC}^t_{\mathcal{A}} \text{ and } \delta(s, x) \in Z \text{ or } s = s_Z. \end{cases}$$

and $\mathcal{T}' := \{\{s_Z\} : Z \in \mathcal{T}_{\mathcal{U}} \cap \mathrm{SCC}^t_{\mathcal{A}}\}$.

Proof. By the Lemmata 4 and 1 $L(\mathcal{A}, \mathcal{T}_{\mathcal{U}}) = \bigcup_{Z \in \mathcal{U}} \{w : \delta(s_0, w) \in Z\} \cdot X^\omega$. The construction of $\mathcal{A}'$ shows, that for $w \neq e$ we have $\delta(s_0, w) \in Z$ if and only if $\delta'(s_0, w) = s_Z$. Now apply Lemma 4 to $\mathcal{A}'$.

2.3 The Automatic Baire Property

In this section we are going to prove an automatic version of the result stating that every Borel (and even every analytic) set has the Baire property.

Definition 1. *A subset* $F \subseteq X^\omega$ *has the* Baire property *if there is an open set* $E \subseteq X^\omega$ *such that their symmetric difference* $F \mathbin{\Delta} E$ *is of first Baire category.*

An important result of descriptive set theory is the following result, see [7,9].

Theorem 2. *Every Borel set of the Cantor space has the Baire property.*

In [5,6] an automatic version of the above theorem is proved. We first give the definition.

Definition 2 (Automatic Baire property). *A subset* $F \subseteq X^\omega$ *fulfils the* Automatic Baire property *if there are a regular* ω*-languages* E, F' *such that* $F \mathbin{\Delta} E \subseteq F'$, *where* E *is open and* F' *is a* Σ_2*-set of first Baire category.*

Then the following holds.

Theorem 3 ([5,6]). *Every regular* ω*-language fulfils the Automatic Baire property.*

For the purposes of our paper we give a proof. In [15, Corollary 6] it is shown that the requirement on F' to be a Σ_2-set in Definition 1 can dropped.

Proof. For an automaton $\mathcal{A} := (X, S, s_0, \delta)$ and a table $\mathcal{T} \subseteq 2^S$ we have

$$\begin{aligned} \mathcal{T} &= (\mathcal{T} \cap \mathrm{SCC}^t_{\mathcal{A}}) \cup (\mathcal{T} \setminus \mathrm{SCC}^t_{\mathcal{A}}) \subseteq (\mathcal{T} \cap \mathrm{SCC}^t_{\mathcal{A}}) \cup (2^S \setminus \mathrm{SCC}^t_{\mathcal{A}}) \\ &\subseteq \{Z' : \exists Z(Z \in \mathcal{T} \cap \mathrm{SCC}^t_{\mathcal{A}} \wedge Z' \subseteq Z)\} \cup (2^S \setminus \mathrm{SCC}^t_{\mathcal{A}}). \end{aligned} \tag{2}$$

Since $\mathcal{T} \cap \mathrm{SCC}^t_{\mathcal{A}} = \{Z' : \exists Z(Z \in \mathcal{T} \cap \mathrm{SCC}^t_{\mathcal{A}} \wedge Z' \subseteq Z)\} \cap \mathrm{SCC}^t_{\mathcal{A}}$, we have $\mathcal{T} \,\Delta\, \{Z' : \exists Z(Z \in \mathcal{T} \cap \mathrm{SCC}^t_{\mathcal{A}} \wedge Z' \subseteq Z)\} \subseteq 2^S \setminus \mathrm{SCC}^t_{\mathcal{A}}$. Then Lemma 1 yields $L(\mathcal{A},\mathcal{T}) \,\Delta\, L(\mathcal{A}, \{Z' : \exists Z(Z \in \mathcal{T} \cap \mathrm{SCC}^t_{\mathcal{A}} \wedge Z' \subseteq Z)\}) \subseteq L(\mathcal{A}, 2^S \setminus \mathrm{SCC}^t_{\mathcal{A}})$. According to Lemmata 4 and 3 the ω-languages $L(\mathcal{A}, \{Z' : \exists Z(Z \in \mathrm{SCC}^t_{\mathcal{A}} \wedge Z' \subseteq Z)\})$ and $L(\mathcal{A}, 2^S \setminus \mathrm{SCC}^t_{\mathcal{A}}) = X^\omega \setminus L(\mathcal{A}, \mathrm{SCC}^t_{\mathcal{A}})$ are open and of first Baire category, respectively.

3 Transformation of Muller to Büchi Automata

In this section we consider the *Büchi acceptance* of ω-automata. An ω-language $F \subseteq X^\omega$ is Büchi accepted by an automaton $\mathcal{A} = (X; S; s_0; \delta)$ and a set of states $T \subseteq S$ if $F = \{\xi : \mathrm{Inf}(\mathcal{A}, \xi) \cap T \neq \emptyset\}$.

It is well known [8, Theorem 4.2] (cf. also [10,12,16] or [17]) that an ω-language $F \subseteq X^\omega$ is accepted by a finite deterministic Büchi automaton if and only if for every deterministic finite automaton $\mathcal{A}$ and table $\mathcal{T}$ such that $F = L(\mathcal{A}, \mathcal{T})$ the table $\mathcal{T}$ is *upwards closed* with respect to $\mathcal{A}$, that is, for $Z \in \mathcal{T} \cap \mathrm{LOOP}_{\mathcal{A}}$ and $Z' \in \mathrm{LOOP}_{\mathcal{A}}$ the inclusion $Z \subseteq Z'$ implies $Z' \in \mathcal{T}$. This is equivalent to the fact that the ω-language $F \subseteq X^\omega$ is the complement of a regular Σ_2-set. In particular, the tables $\{Z' : \exists Z(Z \in \mathrm{SCC}^t_{\mathcal{A}} \wedge Z' \subseteq Z)\}$ and $\mathrm{SCC}^t_{\mathcal{A}}$ in the proof of Theorem 3 are upwards closed.

Whether a translation from a Muller automaton $(\mathcal{A}, \mathcal{T})$ to a Büchi automaton $(\mathcal{A}', T)$ is polynomial, is unknown (cf. [2]). In case when the table $\mathcal{T}$ satisfies a certain condition we obtain a quadratic increase in the number of states. This implies that the obtained Büchi automaton has only a polynomial increase in size.

Theorem 4. *Let $\mathcal{A} = (X; S; s_0; \delta)$ be an automaton and $\mathcal{T} \subseteq 2^S$ be a table such that $\mathcal{T} \cap \mathrm{LOOP}_{\mathcal{A}} \subseteq \mathrm{SCC}_{\mathcal{A}}$. Then there is a deterministic Büchi automaton $(\mathcal{A}'; T)$ with $O(|S|^2)$ states such that $L(\mathcal{A}, \mathcal{T}) = L(\mathcal{A}'; T)$.*

Proof. Since all loops in $\mathcal{T}$ are strongly connected components of the automaton graph, they are pairwise disjoint. Let $\mathcal{T} \cap \mathrm{LOOP}_{\mathcal{A}} = \{Z_1, \ldots, Z_n\}$. For the purposes of the proof we assume that $z^{(i)}_1, z^{(i)}_2, \ldots, z^{(i)}_{\kappa_i}, \kappa_i = |Z_i|$, be a fixed ordering of Z_i.

Define $\hat{\mathcal{A}} = (X; \hat{S}; \hat{s}_0; \hat{\delta})$ as follows.

$$\hat{S} := (S \times \{0\}) \cup \textstyle\bigcup_{i=1}^n (Z_i \times \{1, \ldots, \kappa_i\}), \tag{3}$$

$$\hat{s}_0 := (s_0, 0), \tag{4}$$

$$T = \{(z^{(i)}_{\kappa_i}, \kappa_i) : i \in \{1, \ldots, n\} \wedge z \in Z_i\}, \tag{5}$$

for $s \in S$, and $x \in X$

$$\hat{\delta}((s,0),x) := \begin{cases} (\delta(s,x), 0), \text{ if } \delta(s,x) \notin \{z^{(i)}_1 : i \in \{1, \ldots, n\}\}, & (6.a) \\ (z^{(i)}_1, 1) \quad, \text{ if } \delta(s,x) = z^{(i)}_1, i \in \{1, \ldots, n\}, & (6.b) \end{cases} \tag{6}$$

and for $z \in Z_i, x \in X$ and $j \geq 1$

$$\hat{\delta}((z,j),x) := \begin{cases} (\delta(z,x),0) & \text{, if } \delta(z,x) \notin Z_i \vee j = \kappa_i, \quad (7.a) \\ (z_{j+1}^{(i)}, j+1), & \text{if } j < \kappa_i \wedge \delta(z,x) = z_{j+1}^{(i)}, \text{ and, } \quad (7.b) \\ (\delta(z,x),j) & \text{, otherwise., } \quad (7.c) \end{cases} \tag{7}$$

Then $\hat{\delta}((z,j),x) = (\delta(z,x),j')$ with $j' \in \{0,j,j+1\}$, and, consequently, $\hat{\delta}((z,j),w) = (\delta(z,w),j'')$ for $w \in X^*$ and some j''.

Moreover the following items hold.

1. if $z, \delta(z,x) \in Z_i$ and $j < \kappa_i$ then $\hat{\delta}((z,j),x) = (\delta(z,x),j')$ where $j' \in \{j, j+1\}$, and
2. if $\hat{\delta}((s,j),x) = (s', j+1)$ then $s' = \delta(s,x) = z_{j+1}^{(i)}$ for some $i \in \{1,\ldots,n\}$. This implies
3. If $j \geq 1$, $\hat{\delta}((s,0),w) = (\delta(s,w),j)$ and $\delta(s,w) \in Z_i$ then $\{\delta(s,w') : w' \sqsubseteq w\} \supseteq \{z_1^{(i)},\ldots,z_j^{(i)}\}$.

If $\mathrm{Inf}(\hat{\mathcal{A}};\xi) \cap T \neq \emptyset$ then there are infinitely many prefixes $u \sqsubset \xi$ such that $\hat{\delta}((s_0,0),u) = (z_{\kappa_i}^{(i)},\kappa_i)$ for some i. From Item 3 above and $\hat{\delta}((z_{\kappa_i}^{(i)},\kappa_i),x) = (\delta(z_{\kappa_i},x),0)$ we obtain infinitely many $u_\ell, w_\ell \in X^*$ and $x_\ell \in X$ such that $u_\ell \cdot x_\ell \cdot w_\ell \sqsubset u_{\ell+1} \sqsubset \xi$ where $\hat{\delta}((s_0,0),u_\ell) = (z_{\kappa_i}^{(i)},\kappa_i)$, $\hat{\delta}((s_0,0),u_\ell \cdot x_\ell) = (\delta(z_{\kappa_i}^{(i)},x_\ell),0)$ and $\{\delta(z_{\kappa_i}^{(i)},x_\ell \cdot w) : w \sqsubseteq w_\ell\} \supseteq Z_i$.

Since Z_i is a strongly connected component, $\mathrm{Inf}(\mathcal{A};\xi) = Z_i$ and $\xi \in L(\mathcal{A},\mathcal{T})$ follow.

Let now $\mathrm{Inf}(\hat{\mathcal{A}};\xi) \cap T = \emptyset$, and assume $\mathrm{Inf}(\mathcal{A};\xi) = Z_i$ for some i. Since $\mathrm{Inf}(\hat{\mathcal{A}};\xi) \cap T = \emptyset$, there is a $u \sqsubset \xi$ such that $\{\delta(s_0,w) : u \sqsubseteq w \sqsubset \xi\} \subseteq Z_i$ and $\{\hat{\delta}((s_0,0),w) : u \sqsubseteq w \sqsubset \xi\} \cap T = \emptyset$. Then for all $\hat{\delta}((s_0,0),w) = (\delta(s_0,w),j_w), u \sqsubseteq w \sqsubset \xi$, we have $j_w < \kappa_i$. Let $\hat{j} := \max\{j_w : u \sqsubseteq w \sqsubset \xi\}$. Fix $\hat{w}, u \sqsubseteq \hat{w} \sqsubset \xi$ such that $\hat{\delta}((s_0,0),\hat{w}) = (\delta(s_0,\hat{w}),\hat{j})$. Then Item 1 implies $\hat{\delta}((s_0,0),w) = (\delta(s_0,w),\hat{j})$ for all $w, \hat{w} \sqsubseteq w \sqsubset \xi$. Thus, in view of Item 2, $z_{\hat{j}+1}^{(i)} \notin \{\delta(s_0,w) : \hat{w} \sqsubseteq w \sqsubset \xi\}$ and, consequently, $z_{\hat{j}+1}^{(i)} \notin Z_i$ which contradicts $\mathrm{Inf}(\mathcal{A};\xi) = Z_i$.

4 Algorithms

After the prerequisites derived in the preceding sections we present the polynomial constructions. To this end we refer to classical graph algorithms (cf. [1,4]) and to the results of [11]. It is well known that the strongly connected components of a graph can be estimated in polynomial time of the size of the graph. Similarly, its condensation graph, that is, the graph having as vertices the strongly connected components and as edges the connections induced by the underlying graph, can be also constructed in polynomial time. This allows also to find the terminal strongly connected components $\mathrm{SCC}_{\mathcal{A}}^t$ in polynomial time.

Moreover we have the following.

Lemma 5. [11] *Let $\mathcal{A} = (X,S,s_0,\delta)$ be deterministic automaton. Then, for $Z, Z' \subseteq S$, the predicates $Z \subseteq Z'$ and $Z \in \mathrm{LOOP}_{\mathcal{A}}$ are decidable in nondeterministic logarithmic space.*

4.1 Muller Automata for the Automatic Baire Property

We show that the equation

$$L(\mathcal{A},\mathcal{T}) \mathbin{\Delta} L(\mathcal{A}_1,\mathcal{T}_1) \subseteq X^\omega \setminus L(\mathcal{A}_2,\mathcal{T}_2) \tag{8}$$

is satisfied for automata $\mathcal{A}_1, \mathcal{A}_2$ and tables $\mathcal{T}_1, \mathcal{T}_2$ which can be constructed in polynomial time from $\mathcal{A}$ and $\mathcal{T}$. We have here to construct the complement of the first Baire category set $L(\mathcal{A}, 2^S \setminus \mathrm{SCC}^t_{\mathcal{A}})$ in order to avoid a possible exponential blow-up (cf. [2]) in the size of the table $(2^S \setminus \mathrm{SCC}^t_{\mathcal{A}}) \cap \mathrm{LOOP}_{\mathcal{A}}$.

According to Eq. (2) and Lemma 1 we define $\mathcal{A}_2 := \mathcal{A}$ and $\mathcal{T}_2 := \mathrm{SCC}^t_{\mathcal{A}}$.

The definition of $\mathcal{A}_1 := \mathcal{A}'$ and $\mathcal{T}_1 := \{s_Z : Z \in \mathcal{T} \cap \mathrm{SCC}^t_{\mathcal{A}}\}$ is given in Corollary 1.

Then an algorithm constructing $\mathcal{A}_1, \mathcal{A}_2$ and $\mathcal{T}_1, \mathcal{T}_2$ from $\mathcal{A} = (X, S, s_0, \delta)$ and $\mathcal{T} \subseteq 2^S$ proceeds in the following way.

1. Detect the strongly connected components $\mathrm{SCC}_{\mathcal{A}}$ of the automaton $\mathcal{A}$.
2. Construct the condensation graph $(\mathrm{SCC}_{\mathcal{A}}, \mapsto)$.
3. Identify, in $(\mathrm{SCC}_{\mathcal{A}}, \mapsto)$, the terminal strongly connected components $\mathrm{SCC}^t_{\mathcal{A}}$.
4. Compare $\mathrm{SCC}^t_{\mathcal{A}}$ with $\mathcal{T}$ to obtain $\mathrm{SCC}^t_{\mathcal{A}} \cap \mathcal{T}$.

Items 1,2 and 3 are well-known graph algorithms polynomial in size($\mathcal{A}$). For the comparison of $\mathrm{SCC}^t_{\mathcal{A}}$ with $\mathcal{T}$ we use Lemma 5 to check whether $Z \subseteq Z'$ and $|Z| = |Z'|$ for $Z \in \mathrm{SCC}^t_{\mathcal{A}}$ and $Z' \in \mathcal{T}$. This can be done in polynomial time in size($\mathcal{A}$) + size($\mathcal{T}$).

4.2 Büchi automata for the Automatic Baire Property

In order to obtain the Büchi automata $(\mathcal{B}_1, T_1)$ and $(\mathcal{B}_2, T_2)$ we use the Muller automata $\mathcal{A}_1, \mathcal{A}_2$ and tables $\mathcal{T}_1, \mathcal{T}_2$ satisfying Eq. (8). For $(\mathcal{A}_1, \mathcal{T}_1)$ the transformation to $(\mathcal{B}_1, T_1)$ is straightforward. Set $\mathcal{B}_1 := \mathcal{A}_1$ and $T_1 := \mathcal{T} \cap \mathrm{SCC}^t_{\mathcal{A}_1}$. This results in a so-called weak Büchi automaton $(\mathcal{B}, T)$ where either $R \subseteq T$ or $R \cap T = \emptyset$ for $R \in \mathrm{LOOP}_{\mathcal{B}_1}$.

The table $\mathcal{T}_2 = \mathrm{SCC}^t_{\mathcal{A}_2}$ of the Muller automaton $(\mathcal{A}_2, \mathcal{T}_2)$ satisfies the hypothesis of Theorem 4. Thus we can construct the Büchi automaton $(\mathcal{B}_2, T_2)$ according to the construction in the proof. Since the size of $(\mathcal{B}_2, T_2)$ is only quadratic in the number of states of $\mathcal{A}_2$, this can be done in polynomial time.

References

1. Baase, S., Gelder, A.V.: Computer Algorithms - Introduction to Design and Analysis. Addison - Wesley, Reading (2000). http://vig.pearsoned.com/store/product/1,1207,store-12521_isbn-0201612445,00.html
2. Boker, U.: Inherent size blowup in ω-automata. In: Hofman, P., Skrzypczak, M. (eds.) Developments in Language Theory, Lecture Notes in Computer Science, vol. 11647, pp. 3–17. Springer, Cham (2019). https://doi.org/10.1007/978-3-030-24886-4_1

3. Büchi, J.R.: On a decision method in restricted second order arithmetic. In: Logic, Methodology and Philosophy of Science (Proceedings of the 1960 International Congress), pp. 1–11. Stanford Univ. Press, Stanford (1962)
4. Cormen, T.H., Leiserson, C.E., Rivest, R.L., Stein, C.: Introduction to Algorithms. MIT Press, Cambridge (2009)
5. Finkel, O.: The automatic Baire property and an effective property of ω-rational functions. In: Leporati, A., Martín-Vide, C., Shapira, D., Zandron, C. (eds.) LATA 2020. LNCS, vol. 12038, pp. 303–314. Springer, Cham (2020). https://doi.org/10.1007/978-3-030-40608-0_21
6. Finkel, O.: Two effective properties of ω-rational functions. Int. J. Found. Comput. Sci. **32**(7), 901–920 (2021). https://doi.org/10.1142/S0129054121500283
7. Kuratowski, K.: Topology, vol. I. Państwowe Wydawnictwo Naukowe, Warsaw (1966)
8. Landweber, L.H.: Decision problems for ω-automata. Math. Systems Theory **3**, 376–384 (1969)
9. Oxtoby, J.C.: Measure and Category, Graduate Texts in Mathematics, vol. 2. Springer-Verlag, New York-Berlin (1980)
10. Perrin, D., Pin, J.E.: Infinite Words. Automata, Semigroups, Logic and Games. Elsevier/Academic Press, Amsterdam (2004)
11. Selivanov, V.L., Wagner, K.W.: Complexity of topological properties of regular ω-languages. Fundam. Inform. **83**(1–2), 197–217 (2008)
12. Staiger, L.: ω-languages. In: Rozenberg, G., Salomaa, A. (eds.) Handbook of Formal Languages, vol. 3, pp. 339–387. Springer-Verlag, Berlin (1997)
13. Staiger, L.: The Hausdorff measure of regular ω-languages is computable. Bull. Eur. Assoc. Theor. Comput. Sci. EATCS **66**, 178–182 (1998)
14. Staiger, L.: On the Hausdorff measure of regular ω-languages in Cantor space. Discr. Math. Theor. Comput. Sci. **17**(1), 357–368 (2015). https://dmtcs.episciences.org/2112
15. Staiger, L.: A note on automatic Baire property. CoRR abs/2505.18626 (2025). https://doi.org/10.48550/arXiv.2505.18626. to appear in International Journal of Foundations of Computer Science
16. Thomas, W.: Automata on infinite objects. In: van Leeuwen, J. (ed.) Handbook of Theoretical Computer Science, vol. B, pp. 133–191. Elsevier, Amsterdam (1990)
17. Wagner, K.: On ω-regular sets. Inform. Control **43**(2), 123–177 (1979). https://doi.org/10.1016/S0019-9958(79)90653-3

The More Gödels, the Better

Patrick Uftring(✉)

Department of Computer Science, University of the Bundeswehr Munich,
Werner-Heisenberg-Weg 39, 85579 Neubiberg, Germany
patrick.uftring@unibw.de
patrick.uftring@unibw.de

Abstract. Let G denote the *Gödel problem*, which maps computable sequences to their codes. We answer the following question by Vasco Brattka: "Does G have the same Weihrauch degree as the compositional product $\mathsf{G} * \mathsf{G}$?" We show that this is false: G is strictly weaker than $\mathsf{G} * \mathsf{G}$. In fact, we even derive the more general strict reduction $\mathsf{G}^{[n]} <_{\mathsf{W}} \mathsf{G}^{[n+1]}$ for all $n \in \mathbb{N}$ by showing that G satisfies a right-cancellative property.

Keywords: Weihrauch reducibility · compositional product · Gödel problem

1 Introduction

Weihrauch reducibility compares partial multi-valued functions $f, g :\subseteq \mathbb{N}^{\mathbb{N}} \rightrightarrows \mathbb{N}^{\mathbb{N}}$, which we call *problems*.[1] Elements x in $\mathsf{dom}(f) := \{x \in \mathbb{N}^{\mathbb{N}} \mid f(x) \neq \emptyset\}$, the domain of f, are called *instances* and elements in $f(x)$ are their *solutions*. We write $f \leq_{\mathsf{W}} g$ (and say that f (Weihrauch-)*reduces* to g) if a single application of g together with computable pre- and post-processing is enough to solve any instance of f (see Sect. 2 for a precise definition). We write $f \equiv_{\mathsf{W}} g$ if additionally $g \leq_{\mathsf{W}} f$ holds. The ensuing partial order of *Weihrauch degrees* enjoys a rich structure that has been studied intensely in the last years (see [4] for a comprehensive survey).

Using the compositional product (cf. [3, Section 4] and [5, Corollary 3.7]), we can express applications of multiple problems, one after another. For arbitrary $f, g, h :\subseteq \mathbb{N}^{\mathbb{N}} \rightrightarrows \mathbb{N}^{\mathbb{N}}$, the reduction $f \leq_{\mathsf{W}} g * h$ means that any instance of f can be solved by computable transformations together with a single application of h followed by a single application of g (again, see Sect. 2 for a precise definition). For a given problem f, we can ask if $f <_{\mathsf{W}} f * f$ holds, i.e. if multiple applications of f lead to a properly stronger problem. This is true for some problems (e.g. LPO, see [5, Proposition 3.11]) but false for others (e.g. $\mathsf{C}_{\mathbb{N}}$, see [5, Theorem 3.12]). In the latter case, we derive $f \equiv_{\mathsf{W}} f * f$.

In this article, we consider the following Weihrauch problem:

---

1 Problems can also be defined on more general *represented spaces* (cf. [4, Section 11.2]).

V. Brattka et al. (Eds.): CiE 2026, LNCS 16674, pp. 537–552, 2027.
https://doi.org/10.1007/978-3-032-31348-5_36

Definition 1. *The* Gödel problem $\mathsf{G} :\subseteq \mathbb{N}^{\mathbb{N}} \rightrightarrows \mathbb{N}$ *is a partial multi-valued function defined on all computable sequences. We set*

$$\mathsf{G}(x) := \{e \in \mathbb{N} \mid x_n = \varphi_e(n) \text{ for all } n \in \mathbb{N}\}$$

for any such sequence $x \in \mathsf{dom}(\mathsf{G})$.

Hence, $\mathsf{G}(x)$ maps any computable sequence $x \in \mathbb{N}^{\mathbb{N}}$ to its codes. Intuitively, the Gödel problem translates from the type-2 world back to the type-1 world and acts as (the weakest) right inverse to φ.

This problem has been introduced and studied by Vasco Brattka in [1]. He could show that with respect to the continuous Weihrauch degrees (where instead of only computable transformations, we allow *continuous* ones), the Gödel problem is equivalent to the choice principle $\mathsf{C}_{\mathbb{N}}$. From this property, it inherits the equivalence $\mathsf{G} \equiv^*_{\mathsf{W}} \mathsf{G} * \mathsf{G}$: With respect to the *continuous* Weihrauch degrees, a single application of G is as good as two (or any larger finite amount).

However, the employed arguments rely on *non-computable* transformations. With respect to (regular) Weihrauch reducibility, he asked if the same result could be shown and formulated the following question (cf. [1, Problem 26]):

Problem 1. Is $\mathsf{G} \equiv_{\mathsf{W}} \mathsf{G} * \mathsf{G}$?

In [1, Theorem 25], he showed that this can be true, interestingly enough, if the right hand side is restricted to only computable instances. In Theorem 1, however, we will see that the equivalence is *wrong* without such a restriction.

In fact, Theorem 3 shows the following surprising right-cancellative property: For any two problems $f, g :\subseteq \mathbb{N}^{\mathbb{N}} \rightrightarrows \mathbb{N}^{\mathbb{N}}$ with

$$f * \mathsf{G} \leq_{\mathsf{W}} g * \mathsf{G},$$

we already derive

$$f \leq_{\mathsf{W}} g$$

if the domain of f is *overt* (see e.g. [7] and Definition 6). In Theorem 2, this allows us to show our main result $\mathsf{G}^{[n]} <_{\mathsf{W}} \mathsf{G}^{[n+1]}$ inductively, i.e. that n-many applications of G are still strictly weaker than $(n+1)$-many ones, for any $n \in \mathbb{N}$.

Although it is not important for our results, we tried but were not able to lift the overtness-requirement on the domain of f. We invite the reader to solve this open problem, which we formulate as Question 1 in the last section.

In Sect. 2, we fix our notation and introduce all required definitions and results on Weihrauch reducibility with focus on those concerning the compositional product. In Sect. 3, we show that $\mathsf{G} <_{\mathsf{W}} \mathsf{G} * \mathsf{G}$ holds. In Sect. 4, we extend on these ideas to show the mentioned right-cancellative property of G and close with our main result $\mathsf{G}^{[n]} <_{\mathsf{W}} \mathsf{G}^{[n+1]}$ for all $n \in \mathbb{N}$.

2 Preliminaries and Notation

We call partial multi-valued functions $f :\subseteq \mathbb{N}^\mathbb{N} \rightrightarrows \mathbb{N}^\mathbb{N}$ *problems*. We write $\mathsf{dom}(f)$ for the set of all $x \in \mathbb{N}^\mathbb{N}$ with $f(x)$ non-empty. To simplify our notation, we identify numbers in $\mathbb{N}$ with constant sequences in $\mathbb{N}^\mathbb{N}$. Consequently, if $\mathbb{N}$ appears as domain or codomain of a (partial multi-valued) function, then this domain or codomain is actually given by $\mathbb{N}^\mathbb{N}$ but restricted to constant sequences. One of the simplest problems is given by $\mathsf{id} : \mathbb{N}^\mathbb{N} \to \mathbb{N}^\mathbb{N}$ with $\mathsf{id}(x) := x$ for all $x \in \mathbb{N}^\mathbb{N}$.

For any X, the set of all finite sequences with members in X is denoted by X^*. We write $\sigma * \tau$ and $\sigma * x$ for the concatenation of $\sigma \in X^*$ with finite $\tau \in X^*$ and even infinite $x \in X^\mathbb{N}$.

We write $\langle \cdot, \cdot \rangle : \mathbb{N} \times \mathbb{N} \to \mathbb{N}$ for Cantor's pairing function. We overload this notation and use computable (see below) functions $\langle \cdot, \cdot \rangle : \mathbb{N}^\mathbb{N} \times \mathbb{N}^\mathbb{N} \to \mathbb{N}^\mathbb{N}$ as well as $\pi^l, \pi^r : \mathbb{N}^\mathbb{N} \to \mathbb{N}^\mathbb{N}$ with $\langle \pi^l(x), \pi^r(x) \rangle = x$, $\pi^l(\langle x, y \rangle) = x$, and $\pi^r(\langle x, y \rangle) = y$, for all $x, y \in \mathbb{N}^\mathbb{N}$.[2] Using these maps, we can identify $\mathbb{N}^\mathbb{N} \times \mathbb{N}^\mathbb{N}$ with $\mathbb{N}^\mathbb{N}$. Moreover, interpreting natural numbers $\mathbb{N}$ as constant sequences in $\mathbb{N}^\mathbb{N}$ allows us to define mixed tuples $\langle n, x \rangle$ of numbers $n \in \mathbb{N}$ and sequences $x \in \mathbb{N}^\mathbb{N}$. We overload our notation for a third time (cf. [4, p. 347]): Given $f, g :\subseteq \mathbb{N}^\mathbb{N} \rightrightarrows \mathbb{N}^\mathbb{N}$, we write $\langle f, g \rangle :\subseteq \mathbb{N}^\mathbb{N} \rightrightarrows \mathbb{N}^\mathbb{N}$ for the problem with $\mathsf{dom}(\langle f, g \rangle) := \mathsf{dom}(f) \cap \mathsf{dom}(g)$ and $\langle f, g \rangle(x) := \langle f(x), g(x) \rangle$ for all common $x \in \mathsf{dom}(\langle f, g \rangle)$.

Also (cf. [4, Definition 11.1.2]), we set $(f \times g) :\subseteq \mathbb{N}^\mathbb{N} \rightrightarrows \mathbb{N}^\mathbb{N}$ to be the problem with $\mathsf{dom}(f \times g) := \mathsf{dom}(f) \times \mathsf{dom}(g)$ and $(f \times g)(\langle x, y \rangle) := f(x) \times f(y)$ for all $x \in \mathsf{dom}(f)$ and $y \in \mathsf{dom}(g)$. (Here, we make use of our previously mentioned identification of $\mathbb{N}^\mathbb{N} \times \mathbb{N}^\mathbb{N}$ with $\mathbb{N}^\mathbb{N}$.) Finally, we write $(f \circ g) :\subseteq \mathbb{N}^\mathbb{N} \rightrightarrows \mathbb{N}^\mathbb{N}$ for the problem where $\mathsf{dom}(f \circ g)$ consists of all $x \in \mathsf{dom}(g)$ with $g(x) \subseteq \mathsf{dom}(f)$ and it is defined by $(f \circ g)(x) := \{z \in f(y) \mid y \in g(x)\}$ for all $x \in \mathsf{dom}(f \circ g)$.[3]

As usual, we code infinite families of sequences into one. In order to extract them, we define $\pi_n : \mathbb{N}^\mathbb{N} \to \mathbb{N}^\mathbb{N}$ with $\pi_n(x)_i := x_{\langle n, i \rangle}$ for all $n, i \in \mathbb{N}$ and $x \in \mathbb{N}^\mathbb{N}$. In order to make the index dependent on our input, we also define $\pi :\subseteq \mathbb{N}^\mathbb{N} \to \mathbb{N}^\mathbb{N}$ with $\pi(\langle n, x \rangle) := \pi_n(x)$ for all $n \in \mathbb{N}$ and $x \in \mathbb{N}^\mathbb{N}$. Finally, we write $\zeta :\subseteq \mathbb{N}^\mathbb{N} \to \mathbb{N}^\mathbb{N}$ for the function that is defined on all $x \in \mathbb{N}^\mathbb{N}$ with infinitely many positive values and set both $\zeta(\langle 0 \rangle * x) := \zeta(x)$ as well as $\zeta(\langle n+1 \rangle * x) := \langle n \rangle * \zeta(x)$ for all $n \in \mathbb{N}$ and $x \in \mathsf{dom}(\zeta)$. Intuitively, ζ removes all members equal to zero and decrements everything that remains by 1.

The functions π_n for $n \in \mathbb{N}$ as well as ζ will also be applied to finite sequences: Given $\sigma \in \mathbb{N}^*$, the sequence $\pi_n(\sigma)$ has maximal length such that $\langle n, i \rangle < |\sigma|$ holds for all $i < |\pi_n(\sigma)|$. It is defined like above, i.e. we have $\pi_n(\sigma)_i := \sigma_{\langle n, i \rangle}$ for any such i. Similarly, the length of $\zeta(\sigma)$ is given by the amount of positive members in σ. Apart from the additional case $\zeta(\langle\rangle) := \langle\rangle$, it is recursively defined as above.

With $\varphi_e^x :\subseteq \mathbb{N} \to \mathbb{N}$, we denote the partial function computed by the machine with code $e \in \mathbb{N}$ and oracle $x \in \mathbb{N}^\mathbb{N}$. If no oracle is used, we simply write φ_e without superscript. Given $\sigma \in \mathbb{N}^*$, we write φ_e^σ for the longest finite sequence

[2] Usually, $\langle x, y \rangle$ is implemented by mapping x to all even and y to all odd indices.

[3] It should be noted that this last definition distinguishes the category of partial multi-valued functions from the category of relations.

of length below $|\sigma|$ that emerges from the machine with code e after at most $|\sigma|$-many time steps that only access information from our oracle(-prefix) σ on indices below $|\sigma|$. The sequence φ_e^σ is a prefix of φ_e^τ whenever σ is one of τ, for $\sigma, \tau \in \mathbb{N}^*$. Moreover, if φ_e^x is total for $x \in \mathbb{N}^\mathbb{N}$ and $e \in \mathbb{N}$, then $(\varphi_e^{x \restriction n})_{n \in \mathbb{N}} \subseteq \mathbb{N}^*$ converges (without mind changes) to φ_e^x if we interpret the latter as infinite sequence. We call $f :\subseteq \mathbb{N} \to \mathbb{N}$ *computable* if there is $e \in \mathbb{N}$ with $f \subseteq \varphi_e$. We call $g :\subseteq \mathbb{N}^\mathbb{N} \to \mathbb{N}^\mathbb{N}$ *computable* if there is $e \in \mathbb{N}$ with $g(x)_n = \varphi_e^x(n)$ for all elements $x \in \mathsf{dom}(g)$ and $n \in \mathbb{N}$.

Given two problems $f, g :\subseteq \mathbb{N}^\mathbb{N} \rightrightarrows \mathbb{N}^\mathbb{N}$, we say that f *Weihrauch reduces* to g (write $f \leq_\mathsf{W} g$) if and only if there are computable $h :\subseteq \mathbb{N}^\mathbb{N} \to \mathbb{N}^\mathbb{N}$ (for solutions) and $k :\subseteq \mathbb{N}^\mathbb{N} \to \mathbb{N}^\mathbb{N}$ (for instances) such that any arbitrary instance $x \in \mathsf{dom}(f)$ yields an instance $k(x) \in \mathsf{dom}(g)$, and any solution $y \in (g \circ k)(x)$ yields a solution $h(\langle x, y \rangle) \in f(x)$.

If, additionally, $g \leq_\mathsf{W} f$ holds, we write $f \equiv_\mathsf{W} g$. Otherwise, we write $f <_\mathsf{W} g$. Finally, we define the compositional product: Given $f, g :\subseteq \mathbb{N}^\mathbb{N} \rightrightarrows \mathbb{N}^\mathbb{N}$, we define the degree

$$f * g := \max{}_{\leq_\mathsf{W}} \{f' \circ g' \mid f' \leq_\mathsf{W} f \text{ and } g' \leq_\mathsf{W} g\}.$$

This was first defined in [3, Section 4] and proven to exist for all problems in [5, Corollary 3.7]. It is associative and id is its neutral element (cf. [5, Proposition 4.2 and Observation 4.3]). Given a problem $f :\subseteq \mathbb{N}^\mathbb{N} \rightrightarrows \mathbb{N}^\mathbb{N}$, we define $f^{[0]} := \mathsf{id}$ and $f^{[n+1]} := f^{[n]} * f$ for all $n \in \mathbb{N}$. It is easy to see that $f \leq_\mathsf{W} f * f$ holds for any such problem f (cf. [3, Lemma 4.3]).

3 Two Gödels

In this section, we begin with the following result:

Theorem 1. $\mathsf{G} <_\mathsf{W} \mathsf{G} * \mathsf{G}$

Although it will be subsumed by Theorem 2 in the next section, we still present the argument for this special case as it is significantly more elegant and introduces us to the more complicated ideas required by the following section in a simplified setting.

Since compositional products are somewhat cumbersome to work with, our first step is to substitute the right hand side with a more concrete problem:

Definition 2. *The* helper problem $\mathsf{H} :\subseteq \mathbb{N}^\mathbb{N} \rightrightarrows \mathbb{N}$ *is a partial multi-valued function with domain*

$$\mathsf{dom}(\mathsf{H}) := \{x \in 2^\mathbb{N} \mid \pi_n(x) \text{ is computable for each } n \in \mathbb{N}\}.$$

We set

$$\mathsf{H}(x) := \{e_1 \in \mathbb{N} \mid e_1 \text{ is a code for } \pi_{e_0}(x) \text{ where } e_0 \text{ is a code for } \pi_0(x)\}$$

for any $x \in \mathsf{dom}(\mathsf{H})$.

With the next result, we see that Theorem 1 already follows from the more concrete statement $\mathsf{G} <_W \mathsf{H}$:

Lemma 1. *$H \leq_W G * G$*

Proof. Let $x \in \mathsf{dom}(\mathsf{H})$ be arbitrary. First, we are interested in $e_0 \in (\mathsf{G} \circ \pi_0)(x)$. Then, we use e_0 as an index and consider the sequence $\pi_{e_0}(x) = \pi(\langle e_0, x\rangle)$. The second application of the Gödel problem yields our solution $e_1 \in (\mathsf{G} \circ \pi)(\langle e_0, x\rangle)$. Combining everything, we arrive at the reduction

$$\mathsf{H} \leq_W \mathsf{G} \circ \pi \circ \langle \mathsf{G} \circ \pi_0, \mathsf{id}\rangle \leq_W \mathsf{G} * \mathsf{G}.$$

In the second inequality, we used that

$$\langle \mathsf{G} \circ \pi_0, \mathsf{id}\rangle \leq_W (\mathsf{G} \times \mathsf{id}) \circ \langle \pi_0, \mathsf{id}\rangle$$

holds combined with the fact that the computable functions π and $\langle \pi_0, \mathsf{id}\rangle$ reduce to id, which is neutral with respect to the compositional product. Moreover, we employed the property that any problem $f :\subseteq \mathbb{N}^{\mathbb{N}} \rightrightarrows \mathbb{N}^{\mathbb{N}}$ is equivalent to its cylindrification $f \times \mathsf{id}$ (cf. [2, Definition 3.4 and Proposition 3.7]).

Now that our task has been simplified, we prepare our main proof method. We begin with the notion of *nice* sequences, which will approximate elements in $\mathsf{dom}(\mathsf{H})$.

Definition 3. *Let $\# : \mathbb{N}^* \cup \mathbb{N}^{\mathbb{N}} \to \mathbb{N} \cup \{\infty\}$ be the map that counts the amount of positive members in a (finite or infinite) sequence. A finite sequence $\sigma \in 2^*$ is* nice *if and only if $(\# \circ \pi_n)(\sigma) \leq 1$ holds for any $n > 0$.*

The heart of our proof lies in the following definition, which is based on an idea that goes back as far as Spector (cf. [8]). Depending on whether the instance-generating part of our claimed reduction $\mathsf{H} \leq_W \mathsf{G}$ satisfies the following condition, we show that it is either (somewhat) constant or produces non-computable results. Each case will lead to a contradiction in Lemmas 2 and 3.

Definition 4. *A code $e \in \mathbb{N}$ satisfies the* splitting property *if and only if for any nice $\sigma \in 2^*$, there are $\tau_0, \tau_1 \in 2^*$ satisfying the following:*

- *τ_0 and τ_1 are nice extensions of σ,*
- *$\pi_0(\tau_0) = \pi_0(\tau_1)$, and*
- *$\varphi_e^{\tau_0}$ and $\varphi_e^{\tau_1}$ are incomparable (i.e. both sequences differ at a common index)*

For the purpose of the following two lemmas, let us fix two computable functions $h :\subseteq \mathbb{N}^{\mathbb{N}} \to \mathbb{N}^{\mathbb{N}}$ and $k :\subseteq \mathbb{N}^{\mathbb{N}} \to \mathbb{N}^{\mathbb{N}}$. Moreover, let $e \in \mathbb{N}$ be a code for k. We investigate the following statement:

$$\mathsf{H} \leq_W \mathsf{G} \quad \text{is realized by } h \text{ (for solutions) and } k \text{ (for instances)} \qquad (*)$$

Lemma 2. *If e violates the splitting property, then $(*)$ cannot be satisfied.*

Proof. If the splitting property does not hold for e, then there must be some nice $\sigma \in 2^*$ such that $\varphi_e^{\tau_0}$ and $\varphi_e^{\tau_1}$ are comparable for any two nice extensions $\tau_0, \tau_1 \in 2^*$ of σ with $\pi_0(\tau_0) = \pi_0(\tau_1)$.

Let $(x_n)_{n<|\sigma|} \subseteq 2^{\mathbb{N}}$ be a collection of $|\sigma|$-many computable sequences with the following two properties: First, x_0 is an infinite extension of $\pi_0(\sigma)$ such that the smallest code for x_0 lies above $|\sigma|$. (This can be achieved by applying the pigeonhole principle to the set of all computable extensions of σ.) Secondly, x_n is simply the result of extending $\pi_n(\sigma)$ with infinitely many zeros, i.e. we set $x_n := \pi_n(\sigma) * \mathbf{0}$, where $\mathbf{0}$ denotes the sequence that constantly maps to 0, for all remaining n with $0 < n < |\sigma|$.

Then, we define $f :\subseteq 2^{\mathbb{N}} \to \mathsf{dom}(\mathsf{H})$ that maps any $z \in 2^{\mathbb{N}}$ with $\#(z) \leq 1$ to the unique $f(z)$ satisfying $(\pi_n \circ f)(z) = x_n$ for all $n < |\sigma|$ and $(\pi_n \circ f)(z) = z$ for all $n \geq |\sigma|$. Since both z and each x_n for $n < |\sigma|$ are computable, it is guaranteed that $f(z)$ lies in the domain of H. As any code of $(\pi_0 \circ f)(z)$ lies above $|\sigma|$ and $(\pi_n \circ f)(z) = z$ holds for all such $n \geq |\sigma|$, we have $(\mathsf{H} \circ f)(z) = \mathsf{G}(z)$, for all $z \in \mathsf{dom}(f)$.

Assume that $(*)$ holds. We claim that this makes $\mathsf{H} \circ f$ continuous (and even computable). First, we note that k is constant on the range of f: Let $z \in \mathsf{dom}(f)$ be arbitrary. Then, $f(z)$ is an extension of σ: By definition, $x_n = (\pi_n \circ f)(z)$ extends $\pi_n(\sigma)$ for each $n < |\sigma|$. Moreover, $\pi_n(\sigma)$ is empty for $n \geq |\sigma|$, which makes $(\pi_n \circ f)(z)$ a trivial extension. Hence, we conclude that $f(z)$ is an extension of σ. Next, we claim that any initial segment of $f(z)$ is nice. Let $n > 0$ be arbitrary. If $n < |\sigma|$ holds, then $(\pi_n \circ f)(z)$ extends $\pi_n(\sigma)$ with zeros. Since σ is nice, the latter contains at most a single 1, which transfers to the former. If $n \geq |\sigma|$ holds, then $(\pi_n \circ f)(z) = z$ holds, which contains at most a single 1 by assumption on z. We conclude $(\# \circ \pi_n \circ f)(z) \leq 1$ for all $n > 0$ and, hence, the property $(\# \circ \pi_n)(\tau) \leq 1$ for all initial segments τ of $f(z)$, and $n > 0$. This makes any such segment nice by definition.

Consider two sequences $z, w \in \mathsf{dom}(f)$. By the previous paragraph, we know that any two initial segments τ_0 and τ_1 of $f(z)$ and $f(w)$, respectively, are nice. If $|\tau_0|, |\tau_1| \geq |\sigma|$ holds, then they are also extensions of σ. Additionally, if τ_0 and τ_1 have the same length, then $\pi_0(\tau_0) = \pi_0(\tau_1)$ holds as both are initial segments of x_0 with the same length. We conclude that $\varphi_e^{\tau_0}$ and $\varphi_e^{\tau_1}$ are comparable. With $f(z), f(w) \in \mathsf{rng}(f) \subseteq \mathsf{dom}(\mathsf{H})$, we know that k is defined on both $f(z)$ and $f(w)$. Moreover, since e computes k, both yield the same value. In conclusion, $k \circ f$ is constant. Let $i \in \mathbb{N}$ be such that $\varphi_i = (k \circ f)(z)$ holds for all $z \in \mathsf{dom}(f)$. Hence, from $i \in (\mathsf{G} \circ k \circ f)(z)$, we derive

$$h(\langle f(z), i\rangle) \in (h \circ \langle \mathsf{id}, \mathsf{G} \circ k\rangle \circ f)(z) \subseteq (\mathsf{H} \circ f)(z) = \mathsf{G}(z)$$

for all $z \in \mathsf{dom}(f)$. In particular, $z \mapsto h(\langle f(z), i\rangle)$ witnesses the continuity (even computability) of G on $\mathsf{dom}(f)$. But this leads to a contradiction: $h(\langle f(\mathbf{0}), i\rangle)$ computes a code for the constant zero sequence $\mathbf{0}$. By continuity, there is $z \in 2^{\mathbb{N}}$ with $\#(z) = 1$ and $h(\langle f(\mathbf{0}), i\rangle) = h(\langle f(z), i\rangle)$. However, since $\mathbf{0}$ and z are distinct, they clearly do not share codes.

Lemma 3. *If e satisfies the splitting property, then* $(*)$ *cannot hold.*

Proof. If e satisfies the splitting property, we construct a map $T\colon 2^* \to 2^*$ by defining approximations $T_n :\subseteq 2^* \to 2^*$ in stages. At each stage $n \in \mathbb{N}$, the current construction T_n is supposed to satisfy the following invariants: $T_n(\tau)$ properly extends $T_n(\sigma)$ if τ properly extends σ, for $\sigma, \tau \in \mathsf{dom}(T_n)$. The range of T_n only consists of nice sequences. The range of $\pi_0 \circ T_n$ only consists of pairwise comparable sequences. The domain of T_n is a finite tree and whenever it is defined on one of $\sigma * \langle 0\rangle$ or $\sigma * \langle 1\rangle$, then it must already be defined on both. Moreover, $\varphi_e^{T_n(\sigma*\langle 0\rangle)}$ and $\varphi_e^{T_n(\sigma*\langle 1\rangle)}$ must be incomparable, for any such $\sigma \in \mathsf{dom}(T_n)$.

We begin with T_0. This function is only defined on the empty sequence $\langle\rangle$, which it maps to itself. Now, assume that T_n has already been defined. Consider the smallest sequence (i.e. of smallest code) $\sigma \in \mathsf{dom}(T_n)$ such that neither $\sigma * \langle 0\rangle$ nor $\sigma * \langle 1\rangle$ lies in the domain of T_n. By our invariant, we know that $T_n(\sigma)$ is nice. Let ρ be a nice extension of $T_n(\sigma)$ such that $\pi_0(\rho)$ extends any sequence in the range of $\pi_0 \circ T_n$: For this, we use the assumption that $\pi_0 \circ T_n$ only produces sequences that are pairwise comparable. Then, ρ can be chosen such that $\pi_0(\rho)$ is the longest of these sequences while $\pi_n(\rho)$ simply extends $(\pi_n \circ T_n)(\sigma)$ with zeros, for $n > 0$. The latter ensures that ρ stays nice. Now, let $\tau_0, \tau_1 \in 2^*$ be nice extensions of ρ with $\pi_0(\tau_0) = \pi_0(\tau_1)$ such that $\varphi_e^{\tau_0}$ and $\varphi_e^{\tau_1}$ are incomparable. Then, we set $T_{n+1} := T_n \cup \{(\sigma * \langle 0\rangle, \tau_0), (\sigma * \langle 1\rangle, \tau_1)\}$. It is easy to see that T_{n+1} still satisfies our invariants.

After the sequence $(T_n)_{n\in\mathbb{N}}$ has been defined, we set $T := \bigcup_{n\in\mathbb{N}} T_n$. This function now still satisfies all of our invariants except for the finite domain. In fact, since in each step the domain of T_n is extended in a fair manner, T is total on 2^*. Finally, T allows us to define the continuous function $\bar{T}\colon 2^{\mathbb{N}} \to 2^{\mathbb{N}}$ with $\bar{T}(x) := \lim_{n\in\mathbb{N}} T(x \restriction n)$ for all $x \in 2^{\mathbb{N}}$. (It should be noted that this limit does not involve mind changes: $\bar{T}(x)$ extends $T(x \restriction n)$ for any $x \in 2^{\mathbb{N}}$ and $n \in \mathbb{N}$.)

We claim that $\mathsf{rng}(\bar{T}) \subseteq \mathsf{dom}(\mathsf{H})$ holds and that $k \circ \bar{T}$ is injective. Let us begin with the former. First, we return to the construction of $(T_n)_{n\in\mathbb{N}}$ and notice that it is effective: All constructions were explicit and the search for extensions τ_0 and τ_1 of ρ is unproblematic since all involved properties are decidable. Hence, T and $\bar{T}$ are computable functions. Therefore, also $(\pi_0 \circ \bar{T})(\mathbf{0})$ is computable, where $\mathbf{0}$ denotes the constant zero function. Since $(\pi_0 \circ T)(\sigma)$ and $(\pi_0 \circ T)(\tau)$ are comparable for any two $\sigma, \tau \in 2^*$, we conclude that also $(\pi_0 \circ \bar{T})(x) = (\pi_0 \circ \bar{T})(\mathbf{0})$ is computable for any $x \in 2^{\mathbb{N}}$. Moreover, $(\# \circ \pi_n \circ \bar{T})(x) \leq 1$ for $n > 0$ follows from the fact that the range of T only consists of nice sequences. In particular, $(\pi_n \circ \bar{T})(x)$ is trivially computable for any $n > 0$. In conclusion, we have $\bar{T}(x) \in \mathsf{dom}(\mathsf{H})$ for any $x \in 2^{\mathbb{N}}$.

Next, we show that $k \circ \bar{T}$ is injective. Consider two distinct $x, y \in 2^{\mathbb{N}}$. There is $\sigma \in 2^*$ such that, w.l.o.g., $\sigma * \langle 0\rangle$ and $\sigma * \langle 1\rangle$ are initial segments of x and y, respectively. By definition, we know that $\varphi_e^{\tau_0}$ and $\varphi_e^{\tau_1}$ for $\tau_0 := T(\sigma * \langle 0\rangle)$ as well as $\tau_1 := T(\sigma * \langle 1\rangle)$ are incomparable. Hence, $(k \circ \bar{T})(x)$ and $(k \circ \bar{T})(y)$ are distinct.

With $\mathsf{rng}(\bar{T}) \subseteq \mathsf{dom}(\mathsf{H}) \subseteq \mathsf{dom}(k)$ and the injectiveness of $k \circ \bar{T}$, we conclude that $(k \circ \bar{T})(2^{\mathbb{N}}) \subseteq \mathsf{dom}(\mathsf{G})$ must be uncountable. However, the problem G only accepts computable sequences, of which there are only countably many.

Finally, we gather all of our results:

Proof (Theorem 1). Any Weihrauch problem $f :\subseteq \mathbb{N}^{\mathbb{N}} \rightrightarrows \mathbb{N}^{\mathbb{N}}$ satisfies $f \leq_{\mathsf{W}} f * f$ (as already stated at the very end of Sect. 2, see also [3, Lemma 4.3]). Hence, we only have to show that $\mathsf{G} * \mathsf{G}$ does not reduce to G. Assume, for contradiction, that it does. Then, we also have $\mathsf{H} \leq_{\mathsf{W}} \mathsf{G}$ by Lemma 1. This reduction is realized by computable $h :\subseteq \mathbb{N}^{\mathbb{N}} \to \mathbb{N}^{\mathbb{N}}$ (for solutions) and $k :\subseteq \mathbb{N}^{\mathbb{N}} \to \mathbb{N}^{\mathbb{N}}$ (for instances). Let $e \in \mathbb{N}$ be a code of k. Then by Lemmas 2 and 3, we see that it does not matter if e satisfies the splitting condition or not. In either case, h and k cannot realize a reduction $\mathsf{H} \leq_{\mathsf{W}} \mathsf{G}$. We conclude that our assumption from the beginning, i.e. $\mathsf{G} * \mathsf{G} \leq_{\mathsf{W}} \mathsf{G}$, must have been wrong and we arrive at the strict reduction $\mathsf{G} <_{\mathsf{W}} \mathsf{G} * \mathsf{G}$.

4 More Gödels

In this section, we prove the following generalization of Theorem 1:

Theorem 2. *For any $n \in \mathbb{N}$, we have $\mathsf{G}^{[n]} <_{\mathsf{W}} \mathsf{G}^{[n+1]}$.*

The general proof idea is the same as in the previous section. However, the presence of compositional products on the left hand side increases the complexity significantly. Similar to before, we begin with a helper problem. However, this time, we define H_f for any problem $f :\subseteq \mathbb{N}^{\mathbb{N}} \rightrightarrows \mathbb{N}^{\mathbb{N}}$. In the previous section, we showed $\mathsf{H} \leq_{\mathsf{W}} \mathsf{G} * \mathsf{G}$. Now, we even aim for an equivalence $\mathsf{H}_f \equiv_{\mathsf{W}} f * \mathsf{G}$ (see the next Lemma 4).

Definition 5. *Let $f :\subseteq \mathbb{N}^{\mathbb{N}} \rightrightarrows \mathbb{N}^{\mathbb{N}}$ be an arbitrary problem. Then, the helper problem (with respect to f) $\mathsf{H}_f :\subseteq \mathbb{N}^{\mathbb{N}} \rightrightarrows \mathbb{N}^{\mathbb{N}}$ is a partial multi-valued function with domain*

$$\mathsf{dom}(\mathsf{H}_f) := \{x \in \mathbb{N}^{\mathbb{N}} \mid \pi_0(x) \text{ is computable and } (\zeta \circ \pi_{e+1})(x) \in \mathsf{dom}(f) \text{ for any code } e \text{ of } \pi_0(x)\}$$

and we set

$$\mathsf{H}_f(x) := \{\langle e, y\rangle \in \mathbb{N}^{\mathbb{N}} \mid e \text{ is a code of } \pi_0(x) \text{ and } y \text{ is a solution of } (f \circ \zeta \circ \pi_{e+1})(x)\}$$

for any $x \in \mathsf{dom}(\mathsf{H}_f)$.

The problem H from the previous section (roughly) corresponds to H_{G}. The addition of ζ in our new definition allows us to try computing instances of f for any possible result of G since any such computation might stall or, in the worst case, not produce an infinite sequence at all. This is crucial for deriving the following result:

Lemma 4. *For any problem $f :\subseteq \mathbb{N}^{\mathbb{N}} \rightrightarrows \mathbb{N}^{\mathbb{N}}$, we have $\mathsf{H}_f \equiv_{\mathrm{W}} f * \mathsf{G}$.*

Proof. Let $x \in \mathsf{dom}(\mathsf{H}_f)$ be arbitrary. First, we determine a code $e \in (\mathsf{G} \circ \pi_0)(x)$. Then, consider $\pi_{e+1}(x) = \pi(\langle e+1, x\rangle)$ and find $y \in (f \circ \zeta \circ \pi)(\langle e+1, x\rangle)$. Finally, we have to output the pair $\langle e, y\rangle$. Since e can be expressed as $\pi^l(\langle e, x\rangle)$, we are just looking for a solution of $\langle \pi^l, f \circ \zeta \circ \pi \circ (i \times \mathsf{id})\rangle(\langle e, x\rangle)$. Here, $i\colon \mathbb{N} \to \mathbb{N}$ increments e to $e+1$, i.e. we set $i(n) := n+1$ for all $n \in \mathbb{N}$. Hence, we arrive at the reduction

$$\mathsf{H}_f \leq_{\mathrm{W}} \langle \pi^l, f \circ \zeta \circ \pi \circ (i \times \mathsf{id})\rangle \circ \langle \mathsf{G} \circ \pi_0, \mathsf{id}\rangle \leq_{\mathrm{W}} f * \mathsf{G}.$$

For the other direction, consider problems $f', g' :\subseteq \mathbb{N}^{\mathbb{N}} \rightrightarrows \mathbb{N}^{\mathbb{N}}$ with $f' \leq_{\mathrm{W}} f$ as well as $g' \leq_{\mathrm{W}} \mathsf{G}$ such that $f' \circ g' \equiv_{\mathrm{W}} f * \mathsf{G}$ holds. From the assumed reductions, we find computable functions $h_1, h_2, k_1, k_2 :\subseteq \mathbb{N}^{\mathbb{N}} \to \mathbb{N}^{\mathbb{N}}$ with

$$f * \mathsf{G} \leq_{\mathrm{W}} f' \circ g' \leq_{\mathrm{W}} h_1 \circ \langle \mathsf{id}, f \circ k_1\rangle \circ h_2 \circ \langle \mathsf{id}, \mathsf{G} \circ k_2\rangle.$$

Let $e \in \mathbb{N}$ be code of a machine realizing $k_1 \circ h_2$. Let $x \in \mathbb{N}^{\mathbb{N}}$ be an instance of the problem on the right hand side. We describe how to construct an instance $y \in \mathsf{dom}(\mathsf{H}_f)$ such that any solution of $\mathsf{H}_f(y)$ solves the right hand side with respect to our original instance x.

First, we set $\pi_0(y) := k_2(x)$. Moreover, for each $n \in \mathbb{N}$, we study the behavior of the machine of code e with respect to the oracle $\langle x, n\rangle$. The sequence $\pi_{n+1}(y)$ is then defined in parallel to the steps that our machine takes: Whenever it produces a value k, we add $k+1$ to our current prefix of $\pi_{n+1}(y)$. Whenever it stalls and does not produce anything new, we add 0 instead. Clearly, y is computable from x. By assumption, $\pi_0(y) = k_2(x)$ must be computable. Let $n \in \mathbb{N}$ be a code of this sequence. Then $n \in (\mathsf{G} \circ k_2)(x)$ holds and $k_1 \circ h_2$ must be defined on $\langle x, n\rangle$. Hence, φ_e produces infinitely many values using our oracle resulting in $(\zeta \circ \pi_{n+1})(y) = (k_1 \circ h_2)(\langle x, n\rangle) \in \mathsf{dom}(f)$. Thus, y lies in the domain of H_f. Moreover, any solution $\langle n, z\rangle \in \mathsf{H}_f(y)$ satisfies $n \in (\mathsf{G} \circ k_2)(x)$ as well as $z \in (f \circ k_1 \circ h_2)(\langle x, n\rangle)$ and, therefore, allows us to compute a solution of x for our original problem using $h_1(\langle h_2(\langle x, n\rangle), z\rangle)$. By transitivity, $f * \mathsf{G}$ reduces to H_f.

For a reason, which will become apparent later (during the proof of Lemma 8), some restrictions on the domains of our problems will be necessary:

Definition 6. *We call a subset $X \subseteq \mathbb{N}^{\mathbb{N}}$* overt *if and only if the set*

$$\mathit{pre}(X) := \{\sigma \in \mathbb{N}^* \mid \sigma \text{ is prefix of some } x \in X\}$$

is computably enumerable.

Lemma 5. *The domain of (some problem within the degree) $\mathsf{G}^{[n]}$ is overt, for any $n \in \mathbb{N}$.*

Proof. Let us begin by proving that $\mathsf{dom}(\mathsf{H}_f)$ is overt for any $f :\subseteq \mathbb{N}^\mathbb{N} \rightrightarrows \mathbb{N}^\mathbb{N}$. There are two cases: If f has an empty domain, then the same holds for H_f and we are done. Let $\sigma \in \mathbb{N}^*$ and $x \in \mathsf{dom}(f)$ be arbitrary, otherwise. Then, we define $y \in \mathbb{N}^\mathbb{N}$ to be an extension of σ with the following properties: First, $\pi_0(y)$ is computable and all of its codes are greater than $|\sigma|$. Secondly, $(\zeta \circ \pi_n)(y) = x$ holds for all $n \geq |\sigma|$. Since σ was chosen arbitrarily and its extension y lies in the domain of H_f, we conclude that the canonical enumeration of $\mathbb{N}^*$ witnesses that the domain of H_f is overt.

Now, we show our claim: For $n = 0$, the problem $\mathsf{id} \equiv_\mathsf{W} \mathsf{G}^{[0]}$ is total and its domain is, therefore, trivially overt. For positive $n \in \mathbb{N}$, let $f :\subseteq \mathbb{N}^\mathbb{N} \rightrightarrows \mathbb{N}^\mathbb{N}$ be a problem with $f \equiv_\mathsf{W} \mathsf{G}^{[n-1]}$. Hence, the equivalence $\mathsf{H}_f \equiv_\mathsf{W} f * \mathsf{G} \equiv_\mathsf{W} \mathsf{G}^{[n]}$ follows from Lemma 4. Here, we apply the result from the previous paragraph.

As a last preparation, we show that domains can be restricted to binary sequences. This will become important since, as in the previous section, we will be working with binary trees.

Lemma 6. *For any problem $f :\subseteq \mathbb{N}^\mathbb{N} \rightrightarrows \mathbb{N}^\mathbb{N}$, there is a problem $g :\subseteq 2^\mathbb{N} \rightrightarrows \mathbb{N}^\mathbb{N}$ with $f \equiv_W g$. Moreover, if f has an overt domain, then g can be chosen to have an overt domain, as well.*

Proof. Consider the function $h\colon \mathbb{N}^\mathbb{N} \to 2^\mathbb{N}$ defined by

$$h(\langle n\rangle * x) := 0^n * \langle 1\rangle * h(x)$$

for any $n \in \mathbb{N}$ and $x \in \mathbb{N}^\mathbb{N}$, where $0^n \in 2^*$ denotes the unique sequence of length n that only consists of zeros. It is easy to see that there exists a computable function $h^{-1} :\subseteq 2^\mathbb{N} \to \mathbb{N}^\mathbb{N}$, defined on precisely those sequences in $2^\mathbb{N}$ with infinitely many 1s, that is an inverse to h in the following sense: Both $h^{-1} \circ h$ and $h \circ h^{-1}$ are the identity, respectively, on $\mathbb{N}^\mathbb{N}$ and $\mathsf{dom}(h^{-1})$. If we define $g :\subseteq 2^\mathbb{N} \rightrightarrows \mathbb{N}^\mathbb{N}$ with $g := f \circ h^{-1}$, we derive the claimed equivalence as follows:

$$g = f \circ h^{-1} \leq_\mathsf{W} f = f \circ h^{-1} \circ h \leq_\mathsf{W} f \circ h^{-1} = g.$$

Now, let us additionally assume that the domain of f is overt. Consider the function $\bar{h}\colon \mathbb{N}^* \to 2^*$ defined by

$$\bar{h}(\langle\rangle) := \langle\rangle \quad \text{and} \quad \bar{h}(\langle n\rangle * \tau) := 0^n * \langle 1\rangle * \bar{h}(\tau)$$

for any $n \in \mathbb{N}$ and $\tau \in \mathbb{N}^*$, where $0^n \in 2^*$ denotes (just like above) the unique sequence of length n that only consists of zeros. Now, for any $\tau \in \mathbb{N}^*$ that is enumerated in $(\mathsf{pre} \circ \mathsf{dom})(f)$, we enumerate $\bar{h}(\tau)$ and all of its initial segments. We show that this yields an enumeration of $(\mathsf{pre} \circ \mathsf{dom})(g)$:

Let $x \in \mathsf{dom}(g)$. Then, $h^{-1}(x) \in \mathsf{dom}(f)$ must hold. Consider an arbitrary prefix $\sigma \in 2^*$ of x. Then, there is a prefix $\tau \in \mathbb{N}^*$ of $h^{-1}(x)$ such that $\bar{h}(\tau)$ extends σ. Hence, σ is part of our enumeration. Conversely, let $\sigma \in 2^*$ be part of our enumeration. Then, there must be a prefix $\tau \in \mathbb{N}^*$ of some $x \in \mathsf{dom}(f)$ such that $\bar{h}(\tau)$ extends σ. We conclude that σ is a prefix of $h(x) \in \mathsf{dom}(g)$.

The rest of this section is concerned with proving the following crucial result:

Proposition 1. *Let* $f :\subseteq 2^{\mathbb{N}} \rightrightarrows \mathbb{N}^{\mathbb{N}}$ *and* $g :\subseteq \mathbb{N}^{\mathbb{N}} \rightrightarrows \mathbb{N}^{\mathbb{N}}$ *be problems satisfying the reduction* $\mathsf{H}_f \leq_{\mathsf{W}} \mathsf{H}_g$. *If* f *has an overt domain, then* $f \leq_{\mathsf{W}} g$ *holds.*

However, before we prove this statement, let us first continue with our intended applications:

Theorem 3. *Let* $f, g :\subseteq \mathbb{N}^{\mathbb{N}} \rightrightarrows \mathbb{N}^{\mathbb{N}}$ *be problems with*

$$f * \mathsf{G} \leq_{\mathsf{W}} g * \mathsf{G}.$$

If f *has an overt domain, then*

$$f \leq_{\mathsf{W}} g$$

holds.

There is an interesting similarity to the Kreisel–Lacombe–Shoenfield–Tseitin theorem (see e.g. [9, Section 3.6]), as noted by Vasco Brattka. Similar to how we eliminate an appearance of G on the right hand side here, the mentioned theorem removes φ (i.e. the left inverse of G) on the right hand side of an equality there. Moreover, $(\mathsf{pre} \circ \mathsf{dom})(f)$ has to be computably enumerable here, while a dense subset of $\mathsf{dom}(f \circ \varphi)$ has to be computably enumerable there.

He also noted the striking resemblance to [6, Theorem 2.4], where $f \leq_{\mathsf{W}} g * \mathsf{C}_{\mathbb{N}}$ implies $f \leq_{\mathsf{W}} g$ for so-called *closed fractals* f. Recall that G was even equivalent to $\mathsf{C}_{\mathbb{N}}$ with respect to the *continuous* Weihrauch degrees, which makes the resemblance particularly striking. However, Theorem 3 does *neither* transfer to $\mathsf{C}_{\mathbb{N}}$ *nor* the continuous degrees as any such result would entail the contradiction $\mathsf{id} \equiv_{\mathsf{W}} \mathsf{C}_{\mathbb{N}}$. Before we return to the proof of Theorem 3, let us note the obvious question regarding the domains it requires:

Question 1. Does Theorem 3 also hold *without* the restriction to overt domains?

Proof (Theorem 3). Assume that $f * \mathsf{G} \leq_{\mathsf{W}} g * \mathsf{G}$ holds. If f has an overt domain, then by Lemma 6, there is $f' :\subseteq 2^{\mathbb{N}} \rightrightarrows \mathbb{N}^{\mathbb{N}}$ with an overt domain satisfying $f \equiv_{\mathsf{W}} f'$. By Lemma 4, we have $\mathsf{H}_{f'} \equiv_{\mathsf{W}} f' * \mathsf{G} \equiv_{\mathsf{W}} f * \mathsf{G}$ and $\mathsf{H}_g \equiv_{\mathsf{W}} g * \mathsf{G}$. We apply Proposition 1 and conclude $f \equiv_{\mathsf{W}} f' \leq_{\mathsf{W}} g$.

With this, our main result follows without any effort:

Proof (Theorem 2). The reduction $\mathsf{G}^{[n]} \leq_{\mathsf{W}} \mathsf{G}^{[n+1]}$ can be derived from the well-known results [3, Lemma 4.2 and Lemma 4.3]. In order to derive the strict reduction, assume that $n \in \mathbb{N}$ is minimal with $\mathsf{G}^{[n+1]} \leq_{\mathsf{W}} \mathsf{G}^{[n]}$. From Theorem 1, we know that n must be positive.[4] By Lemma 5, the domain of (some problem of degree) $\mathsf{G}^{[n]}$ is overt. Hence, Theorem 3 yields $\mathsf{G}^{[n]} \leq_{\mathsf{W}} \mathsf{G}^{[n-1]}$, which contradicts the minimality of n.

[4] This can also be seen directly using a simple continuity argument.

In order to derive the last missing piece, i.e. Proposition 1, let us fix two problems $f :\subseteq 2^{\mathbb{N}} \rightrightarrows \mathbb{N}^{\mathbb{N}}$ and $g :\subseteq \mathbb{N}^{\mathbb{N}} \rightrightarrows \mathbb{N}^{\mathbb{N}}$ such that f has a non-empty and overt domain. (The claim is trivial if f has an empty domain.) Similar to the previous section, we also fix two computable functions $h :\subseteq \mathbb{N}^{\mathbb{N}} \to \mathbb{N}^{\mathbb{N}}$ and $k :\subseteq \mathbb{N}^{\mathbb{N}} \to \mathbb{N}^{\mathbb{N}}$. Let $e \in \mathbb{N}$ be some code of $\pi_0 \circ k$. We investigate the following statement:

$$\mathsf{H}_f \leq_{\mathrm{W}} \mathsf{H}_g \quad \text{is realized by } h \text{ (for solutions) and } k \text{ (for instances)} \qquad (\dagger)$$

Since we are now considering H_f instead of H, our notion of being nice changes:

Definition 7. *A finite sequence $\sigma \in \mathbb{N}^*$ is nice if and only if $(\zeta \circ \pi_n)(\sigma)$ is prefix of an element in $\mathsf{dom}(f)$, for every $n > 0$.*

In the previous section, any computable sequence of increasingly good (i.e. longer and longer) comparable nice approximations led to an element in $\mathsf{dom}(\mathsf{H})$. In general, this property is no longer satisfied: Depending on the domain of f, a sequence in $(\mathsf{pre} \circ \mathsf{dom})(f)$ as mentioned does not necessarily yield an inhabitant of $\mathsf{dom}(f)$ in its limit. (Think of a domain consisting of precisely those sequences in $2^{\mathbb{N}}$ with only finitely many 1s. Then, a sequence of prefixes might yield an object with infinitely many 1s.) We will handle this problem manually later using choice on Baire space.

We also update our splitting property:

Definition 8. *A code $e \in \mathbb{N}$ satisfies the splitting property if and only if for any nice $\sigma \in \mathbb{N}^*$, there are $\tau_0, \tau_1 \in \mathbb{N}^*$ satisfying the following:*

- *τ_0 and τ_1 are nice extensions of σ,*
- *$\pi_0(\tau_0) = \pi_0(\tau_1)$,*
- *$(\zeta \circ \pi_n)(\tau_0) = (\zeta \circ \pi_n)(\tau_1) = (\zeta \circ \pi_n)(\sigma)$ for all $n \in \mathbb{N}$ with $0 < n \leq |\sigma|$,*
- *$\varphi_e^{\tau_0}$ and $\varphi_e^{\tau_1}$ are incomparable.*

The reason for our newly added condition is precisely the property that sequences in $(\mathsf{pre} \circ \mathsf{dom})(\mathsf{H}_f)$ as mentioned do not necessarily converge to an inhabitant of $\mathsf{dom}(\mathsf{H}_f)$. As soon as an initial segment $(\zeta \circ \pi_n)(\sigma)$ for $n > 0$ has been visited and initially defined (in the sense that it has positive length), it should never be touched again by our effective search. Instead, in a second pass after having constructed our tree as before, we will manually extend these prefixes in a way that is guaranteed to produce elements in $\mathsf{dom}(\mathsf{H}_f)$ in their limit.

Lemma 7. *If e violates the splitting property, then $(\dagger)$ implies $f \leq_{\mathrm{W}} g$.*

Proof. If the splitting property fails, then there is a nice sequence $\sigma \in \mathbb{N}^*$ such that $\varphi_e^{\tau_0}$ and $\varphi_e^{\tau_1}$ are comparable for any two nice extensions $\tau_0, \tau_1 \in \mathbb{N}^*$ of σ if $\pi_0(\tau_0) = \pi_0(\tau_1)$ and $(\zeta \circ \pi_n)(\tau_0) = (\zeta \circ \pi_n)(\tau_1) = (\zeta \circ \pi_n)(\sigma)$ are satisfied for all $n \in \mathbb{N}$ with $0 < n \leq |\sigma|$.

Let $(x_n)_{n \leq |\sigma|} \subseteq \mathbb{N}^{\mathbb{N}}$ be a collection $(|\sigma| + 1)$-many sequences with the following properties: First, x_0 is a computable extension of $\pi_0(\sigma)$ such that the smallest code of x_0 lies strictly above $|\sigma|$. Secondly, we define $x_n := \pi_n(\sigma) * \mathbf{0}$ for any other $n \in \mathbb{N}$ with $0 < n \leq |\sigma|$ (where, as before, $\mathbf{0}$ denotes the constant zero sequence). For any such n, this entails $\zeta(x_n) = (\zeta \circ \pi_n)(\sigma)$.

We define $j\colon \mathsf{dom}(f) \to \mathsf{dom}(\mathsf{H}_f)$ to be the computable function that maps any $z \in \mathsf{dom}(f)$ to the unique sequence $j(z)$ satisfying the following: First, the equality $(\pi_n \circ j)(z) = x_n$ holds for all $n \leq |\sigma|$. Secondly, we set $(\pi_n \circ j)(z)$ to be the canonical[5] sequence with $(\zeta \circ \pi_n \circ j)(z) = z$ for all other $n > |\sigma|$. Notice that by definition of x_0 and j, the property $(\zeta \circ \pi_{e+1} \circ j)(z) = z \in \mathsf{dom}(f)$ is guaranteed for any $z \in \mathsf{dom}(f)$ and code e of x_0. Hence, the range of j lies within $\mathsf{dom}(\mathsf{H}_f)$ and, in particular, it lies in the domain of $\pi_0 \circ k$.

Consider two $z, w \in \mathsf{dom}(f)$. We claim that $\varphi_e^{\tau_0}$ and $\varphi_e^{\tau_1}$ are comparable for any two prefixes τ_0 and τ_1 of $j(z)$ and $j(w)$, respectively, with $|\tau_0| = |\tau_1|$. The argument works like the matching part in the proof of Lemma 2. Regarding the new condition, $(\zeta \circ \pi_n)(\tau_0) = (\zeta \circ \pi_n)(\tau_1) = (\zeta \circ \pi_n)(\sigma)$ holds by construction of $(x_n)_{n \leq |\sigma|}$ for all positive $n \leq |\sigma|$ if $|\tau_0|, |\tau_1| \geq |\sigma|$. Hence, by our assumption, $\varphi_e^{\tau_0}$ and $\varphi_e^{\tau_1}$ are comparable. Recall that e was a code of $\pi_0 \circ k$. Hence, $(\pi_0 \circ k \circ j)(z) = (\pi_0 \circ k \circ j)(w)$ holds. We conclude that $\pi_0 \circ k \circ j$ is constant on the set $\mathsf{dom}(f)$. Since it must be computable, let $n \in \mathbb{N}$ be such that this function $\pi_0 \circ k \circ j$ constantly maps to φ_n on $\mathsf{dom}(f)$. Now, we describe how to extract witnesses for $f \leq_{\mathsf{W}} g$ from the gathered information.

Let $z \in \mathsf{dom}(f)$ be an arbitrary instance. From this, we construct an instance $(k \circ j)(z)$ of H_g. We know that $(\pi_0 \circ k \circ j)(z)$ has code n. Hence, the result of $(\zeta \circ \pi_{n+1} \circ k \circ j)(z)$ is an instance of g. Let w be a solution for this. Then, $\langle n, w \rangle$ solves the instance $(k \circ j)(z)$ of H_g. Hence, $h(\langle j(z), \langle n, w \rangle \rangle)$ is a solution for $j(z)$ of H_f. It is given by a pair consisting of a code $m \in \mathbb{N}$ for $(\pi_0 \circ j)(z) = x_0$ together with a solution for $(\zeta \circ \pi_{m+1} \circ j)(z)$, which is equal to z since any code of x_0 must lie above $|\sigma|$. In conclusion, the maps

$$\langle z, w \rangle \mapsto (\pi^r \circ h)(\langle j(z), \langle n, w \rangle \rangle) \text{ (for solutions)}, \; \zeta \circ \pi_{n+1} \circ k \circ j \text{ (for instances)}$$

witness the reduction $f \leq_{\mathsf{W}} g$.

Lemma 8. *If e satisfies the splitting property, then (†) cannot hold.*

Proof. We define a map $T :\subseteq 2^* \to \mathbb{N}^*$ using finite approximations $(T_n)_{n \in \mathbb{N}}$ of the same signature. For each $n \in \mathbb{N}$, the function T_n is supposed to satisfy the following invariants: $T_n(\tau)$ properly extends $T_n(\sigma)$ for any proper extension $\tau \in \mathsf{dom}(T_n)$ of $\sigma \in \mathsf{dom}(T_n)$. The range of T_n only consists of nice sequences. The range of $\pi_0 \circ T_n$ only consists of pairwise comparable sequences. The domain of T_n is a finite tree. Whenever T_n is defined on one of $\sigma * \langle 0 \rangle$ or $\sigma * \langle 1 \rangle$, for $\sigma \in \mathsf{dom}(T_n)$ with $|\sigma|$ even, then it must be defined on both. Also, $\varphi_e^{T_n(\sigma * \langle 0 \rangle)}$ and $\varphi_e^{T_n(\sigma * \langle 1 \rangle)}$ must be incomparable in this case.

Now, we begin our construction: T_0 is only defined on the empty sequence, which it maps to itself. Assume that T_n has already been defined (and that it satisfies our invariants). Let $\sigma \in 2^*$ be the n-th sequence in a computable enumeration of 2^* that visits each element infinitely often. Assume that $\sigma \in \mathsf{dom}(T_n)$

[5] By this, we mean the sequence that results from z if each member is raised by 1.

holds. Let $\rho \in \mathbb{N}^*$ be a nice extension of $T_n(\sigma)$ such that $\pi_0(\rho)$ extends each element in the range of $\pi_0 \circ T_n$ and $(\zeta \circ \pi_k \circ T_n)(\sigma) = (\zeta \circ \pi_k)(\rho)$ holds for all $k > 0$. Regarding the former, notice that the range of $\pi_0 \circ T_n$ only consists of pairwise comparable elements by our invariant.

There are two cases to distinguish: Assume that $|\sigma|$ is even. If both $\sigma * \langle 0 \rangle$ and $\sigma * \langle 1 \rangle$ are missing in $\mathsf{dom}(T_n)$, then let $\tau_0, \tau_1 \in \mathbb{N}^*$ be two nice extensions of ρ such that both $\varphi_e^{\tau_0}$ and $\varphi_e^{\tau_1}$ are incomparable, and the following are satisfied: $\pi_0(\tau_0) = \pi_0(\tau_1)$ as well as $(\zeta \circ \pi_k)(\tau_0) = (\zeta \circ \pi_k)(\tau_1) = (\zeta \circ \pi_k)(\rho)$ holds for all $k \in \mathbb{N}$ with $0 < k \leq |\rho|$. Finally, set $T_{n+1} := T_n \cup \{(\sigma * \langle 0 \rangle, \tau_0), (\sigma * \langle 1 \rangle, \tau_1)\}$. Clearly, T_{n+1} still satisfies all of our invariants. Notice that τ_0 and τ_1 can be searched for effectively as we have a computable enumeration of $(\mathsf{pre} \circ \mathsf{dom})(f)$.

Otherwise, assume $|\sigma|$ is odd. Then, there are $i, j \in \mathbb{N}$ satisfying both $i > 0$ and $2\langle i-1, j \rangle + 1 = |\sigma|$. Do the following for each $b \in 2$ with $\sigma * \langle b \rangle \notin \mathsf{dom}(T_n)$: If $(\zeta \circ \pi_i)(\rho) * \langle b \rangle$ is among the first n-many sequences of our computable enumeration of $(\mathsf{pre} \circ \mathsf{dom})(f)$, then let $\tau_b \in \mathbb{N}^*$ be an extension of ρ satisfying the equality $(\zeta \circ \pi_i)(\tau_b) = (\zeta \circ \pi_i)(\rho) * \langle b \rangle$ and $(\zeta \circ \pi_k)(\tau_b) = (\zeta \circ \pi_k)(\rho)$ for all $k \neq i$. Finally, T_{n+1} results from T_n by adding $(\sigma * \langle 0 \rangle, \tau_0)$ or $(\sigma * \langle 1 \rangle, \tau_1)$, depending on whichever has just been defined. Regarding the case, in which we add both, notice that $\pi_0(\tau_b)$ only extends $\pi_0(\rho)$ with zeros, for $b \in 2$. Hence, they are comparable and all of our invariants stay intact.

After each T_n has been defined, we set $T := \bigcup_{n \in \mathbb{N}} T_n$. Notice that the graph of T is computably enumerable since each step of our construction was effective. Also, T satisfies all of our invariants except for the property of having a finite domain, of course. However, there are some key differences to our construction in the proof of Lemma 3: It is no longer guaranteed that T is total, which also does not allow us to extend it to a total function $\bar{T}$ on $2^{\mathbb{N}}$. Moreover, it is possible that T maps to objects outside of $\mathsf{dom}(\mathsf{H}_f)$.

Our final task is to find $x \in 2^{\mathbb{N}}$ such that $T(x \restriction n)$ is defined for all $n \in \mathbb{N}$ and yields an element $y := \lim_{n \in \mathbb{N}} T(x \restriction n)$ in $\mathsf{dom}(\mathsf{H}_f)$ but for which $(\pi_0 \circ k)(y)$ is non-computable. Beginning with $\sigma := \langle \rangle \in \mathsf{dom}(T)$, assume that we have already defined a prefix $\sigma \in \mathsf{dom}(T)$ of x. Similar to the construction of T, there are two cases to consider: If $|\sigma|$ is even, let $n \in \mathbb{N}$ be such that $2n = |\sigma|$ holds. If φ_n is total, choose $b \in 2$ such that $\varphi_e^{T(\sigma * \langle b \rangle)}$ is *not* a prefix of φ_n (if interpreted as a sequence). If φ_n is not total, the choice of b does not matter. Then, continue the construction with $\sigma := \sigma * \langle b \rangle$. By definition of T it is clear that this new σ also lies in $\mathsf{dom}(T)$. Notice that in the end, this part of the construction will ensure that $\varphi_e^y = (\pi_0 \circ k)(y)$ is non-computable.

Otherwise, if $|\sigma|$ is odd, let $i, j \in \mathbb{N}$ with $i > 0$ be such that $2\langle i-1, j \rangle + 1 = |\sigma|$ holds. Consider the sequence $\tau := (\zeta \circ \pi_i \circ T)(\sigma)$. By construction of T, it is guaranteed that this sequence lies in $(\mathsf{pre} \circ \mathsf{dom})(f)$. Let $z \in \mathsf{dom}(f)$ be minimal (with respect to a well-order on $\mathsf{dom}(f)$) infinite extension of τ. Then, set $\sigma := \sigma * \langle z_{|\tau|} \rangle$, i.e. we extend σ such that $(\zeta \circ \pi_i \circ T)(\sigma) = z \restriction (|\tau| + 1)$ becomes a proper extension of τ that is still an initial segment of z. By definition of T, it is clear that this new σ also lies in $\mathsf{dom}(T)$.

While this part of the argument was highly non-computable, it is important that our *previous step*, the construction of T itself, was computably enumerable. Only because of this property is it now guaranteed that $\pi_0(y)$ is a computable sequence. This follows since $\pi_0(y)$ is comparable to each sequence in the range of $\pi_0 \circ T$ by our invariant.

Let $n > 0$ be arbitrary. We claim that $(\zeta \circ \pi_n)(y) \in \mathsf{dom}(f)$ holds: For this, let $z \in \mathsf{dom}(f)$ be the minimal infinite extension of $(\zeta \circ \pi_n \circ T)(x \upharpoonright n)$. We claim that $(\zeta \circ \pi_n)(y) = z$ holds. Let $k \geq n$ be arbitrary and assume $(\zeta \circ \pi_n \circ T)(x \upharpoonright k)$ was a prefix of z. If k is even, then $(\zeta \circ \pi_n \circ T)(x \upharpoonright k) = (\zeta \circ \pi_n \circ T)(x \upharpoonright (k+1))$ follows from the fact that $T(x \upharpoonright (k+1))$ is part of an extension of $T(x \upharpoonright k)$ in the sense of Definition 8 and that $T(x \upharpoonright k)$ has length above $k \geq n$. Otherwise, if k is odd, let $i, j \in \mathbb{N}$ with $i > 0$ be such that $2\langle i-1, j\rangle + 1 = k$ holds. If i is distinct from n, we have $(\zeta \circ \pi_n \circ T)(x \upharpoonright k) = (\zeta \circ \pi_n \circ T)(x \upharpoonright (k+1))$ by construction. If $i = n$ holds, then $z \in \mathsf{dom}(f)$ is the minimal infinite extension of $(\zeta \circ \pi_n \circ T)(x \upharpoonright k)$ (any smaller infinite extension would also have been a smaller infinite extension of $(\zeta \circ \pi_n \circ T)(x \upharpoonright n)$). Hence, $(\zeta \circ \pi_n \circ T)(x \upharpoonright (k+1))$ must be a prefix of z. In fact, by construction, it is a *longer* prefix. Notice also that this case with $i = n$ is visited infinitely often. Hence, $(\zeta \circ \pi_n)(y)$ is not just a finite prefix in $(\mathsf{pre} \circ \mathsf{dom})(f)$ but even an infinite sequence in $\mathsf{dom}(f)$.

We conclude $y \in \mathsf{dom}(\mathsf{H}_f)$ and, hence, $k(y) \in \mathsf{dom}(\mathsf{H}_g)$. However, $(\pi_0 \circ k)(y)$ is non-computable. This contradicts the definition of $\mathsf{dom}(\mathsf{H}_g)$.

Combining the last two lemmas, we complete the proof of our proposition:

Proof (Proposition 1). Assume that the reduction $\mathsf{H}_f \leq_{\mathsf{W}} \mathsf{H}_g$ holds and is realized by $h :\subseteq \mathbb{N}^{\mathbb{N}} \to \mathbb{N}^{\mathbb{N}}$ for solutions and $k :\subseteq \mathbb{N}^{\mathbb{N}} \to \mathbb{N}^{\mathbb{N}}$ for instances. Since our claim is trivial if f has an empty domain, we furthermore assume that f has a non-empty and overt domain.

If $e \in \mathbb{N}$ is a code of $\pi_0 \circ k$ and violates the splitting property, then by Lemma 7, our reduction implies $f \leq_{\mathsf{W}} g$. Otherwise, if e satisfies the splitting property, we arrive at a contradiction by Lemma 8.

Acknowledgments. I would like to thank Vasco Brattka for bringing this question to my attention and for our many discussions on the topic, as well as Christopher Sorg and the referees for their helpful suggestions. I would also like to thank the organizers of the CCA 2025 conference in Kyoto and the seminar of the Swansea theoretical computer science group for their kind invitations, which allowed me to present first results in the early stages of this project.

References

1. Brattka, V.: On the complexity of computing Gödel numbers (2023). https://doi.org/10.48550/arXiv.2302.04213, preprint: arXiv:2302.04213
2. Brattka, V., Gherardi, G.: Weihrauch degrees, omniscience principles and weak computability. J. Symb. Log. **76**(1), 143–176 (2011). https://doi.org/10.2178/jsl/1294170993

3. Brattka, V., Gherardi, G., Marcone, A.: The Bolzano-Weierstrass theorem is the jump of weak Kőnig's lemma. Ann. Pure Appl. Log. **163**(6), 623–655 (2012). https://doi.org/10.1016/j.apal.2011.10.006
4. Brattka, V., Gherardi, G., Pauly, A.: Weihrauch complexity in computable analysis. In: Handbook of Computability and Complexity in Analysis, pp. 367–417. Theory and Applications of Computability, Springer, Cham (2021). https://doi.org/10.1007/978-3-030-59234-9_11
5. Brattka, V., Pauly, A.: On the algebraic structure of Weihrauch degrees. Log. Methods Comput. Sci. **14**(4), Paper No. 4, 36 (2018). https://doi.org/10.1080/10724117.2007.11974704
6. Le Roux, S., Pauly, A.: Finite choice, convex choice and finding roots. Log. Methods Comput. Sci. **11**(4), 4:6, 30 (2015). https://doi.org/10.2168/LMCS-11(4:6)2015
7. Pauly, A.: On the topological aspects of the theory of represented spaces. Computability **5**(2), 159–180 (2016). https://doi.org/10.3233/COM-150049
8. Spector, C.: On degrees of recursive unsolvability. Ann. Math. (2) **64**, 581–592 (1956). https://doi.org/10.2307/1969604
9. Weihrauch, K.: Computability, EATCS Monographs on Theoretical Computer Science, vol. 9. Springer-Verlag, Berlin (1987). https://doi.org/10.1007/978-3-642-69965-8

Author Index

A
Ackerman, Nathanael 109
Andreev, Mikhail 124

B
Bacho, Aras 135
Bazhenov, Nikolay 150
Borsetto, Riccardo 162
Bournez, Olivier 3

C
Carl, Merlin 178
Carlucci, Lorenzo 193, 208
Celli, Giordano 193
Cruz, Julianne 224

D
de Brecht, Matthew 240

E
Ellmies, Gero 256

F
Fellin, Giulio 162, 272
Fisman, Dana 23
Fluschnik, Till 256
Freer, Cameron 109
Fujiwara, Makoto 289

G
Gjetaj, Oriola 208
Glashausser, Sho 224
Grobler, Mario 304

H
Hertling, Peter 320
Hitchcock, John M. 335

J
Janicki, Philip 350

L
Lázár, Katalin Anna 389
Li, Xiaoyuan 224
Lipin, Anton E. 358
Lutz, Neil 224, 374

M
Manea, Florin 389, 405
Martin, Spencer Park 374
Mirabi, Mostafa 109
Mitrana, Victor 48
Morawietz, Nils 304
Mustafa, Manat 150

N
Nakid Cordero, A. 421
Negri, Sara 272
Neumann, Eike 437
Niskanen, Reino 453
Normann, Dag 514

O
Orellana-Martín, David 60

P
Pauly, Arno 469
Potapov, Igor 453
Pulari, Subin 482

Q
Quinon, Paula 78

R
Ringleb, Tina 405

V. Brattka et al. (Eds.): CiE 2026, LNCS 16674, pp. 553–554, 2027.
https://doi.org/10.1007/978-3-032-31348-5

S
Sacher, Silas Cato 304
Sanders, Sam 498, 514
Schulte-Michels, Leonard 320
Schuster, Peter 272
Scott, I. 421
Seiller, Thomas 86
Sekoni, Adewale 335
Selivanova, Svetlana 135
Shafei, Hadi 335
Shen, Alexander 124
Siemer, Stefan 389, 405
Specht, Timo 389
Staiger, Ludwig 528
Sudit, Elina 23

T
Tembo, Margret 437
Topley, James 453

U
Uftring, Patrick 537
Uustalu, Tarmo 162

V
Volkov, Mikhail V. 358

W
Wan, Cheng-Syuan 162
White, Rain 374
Winkler, Maximilian 405

Y
Yun, Stanislav 150

Z
Ziegler, Martin 135
Zimerman, Oded 23

MIX
Papier aus verantwortungsvollen Quellen
Paper from responsible sources
FSC® C105338

FSC
www.fsc.org

If you have any concerns about our products,
you can contact us on
ProductSafety@springernature.com

In case Publisher is established outside the EU,
the EU authorized representative is:
Springer Nature Customer Service Center GmbH
Europaplatz 3, 69115 Heidelberg, Germany

Printed by Libri Plureos GmbH
in Hamburg, Germany